Advances in Intelligent Systems and Computing 1463

The series "Advances in Intelligent Systems and Computing" contains publications on theory, applications, and design methods of Intelligent Systems and Intelligent Computing. Virtually all disciplines such as engineering, natural sciences, computer and information science, ICT, economics, business, e-commerce, environment, healthcare, life science are covered. The list of topics spans all the areas of modern intelligent systems and computing such as: computational intelligence, soft computing including neural networks, fuzzy systems, evolutionary computing and the fusion of these paradigms, social intelligence, ambient intelligence, computational neuroscience, artificial life, virtual worlds and society, cognitive science and systems, Perception and Vision, DNA and immune based systems, self-organizing and adaptive systems, e-Learning and teaching, human-centered and human-centric computing, recommender systems, intelligent control, robotics and mechatronics including human-machine teaming, knowledge-based paradigms, learning paradigms, machine ethics, intelligent data analysis, knowledge management, intelligent agents, intelligent decision making and support, intelligent network security, trust management, interactive entertainment, Web intelligence and multimedia.

The publications within "Advances in Intelligent Systems and Computing" are primarily proceedings of important conferences, symposia and congresses. They cover significant recent developments in the field, both of a foundational and applicable character. An important characteristic feature of the series is the short publication time and world-wide distribution. This permits a rapid and broad dissemination of research results.

Indexed by DBLP, INSPEC, WTI Frankfurt eG, zbMATH, Japanese Science and Technology Agency (JST).

All books published in the series are submitted for consideration in Web of Science.

For proposals from Asia please contact Aninda Bose (aninda.bose@springer.com).

Shahram Latifi
Editor

The 22nd International Conference on Information Technology-New Generations (ITNG 2025)

Springer

Editor
Shahram Latifi
Department of Electrical and Computer Engineering
University of Nevada
Las Vegas, NV, USA

ISSN 2194-5357 ISSN 2194-5365 (electronic)
Advances in Intelligent Systems and Computing
ISBN 978-3-031-89062-8 ISBN 978-3-031-89063-5 (eBook)
https://doi.org/10.1007/978-3-031-89063-5

This Springer imprint is published by the registered company Springer Nature Switzerland AG
The registered company address is: Gewerbestrasse 11, 6330 Cham, Switzerland

Chair Message

Welcome to the 22nd International Conference on Information Technology - New Generations (ITNG 2025). It's wonderful to see both long-standing members of the ITNG community and new participants alike. ITNG 2025 has attracted high-quality submissions from numerous countries around the globe.

The submissions underwent rigorous review for technical soundness, originality, clarity, and relevance. We are grateful for the expertise of over 50 scientists, both authors and non-authors, who contributed to the review process, ensuring that each paper was evaluated by at least two independent reviewers. Ultimately, 65 papers were accepted for presentation and inclusion in the ITNG 2025 book of chapters. The reviewers' names also appear in the same book.

This year's collection covers recent advances in areas including machine learning, big data analytics, cybersecurity, blockchain technology, data mining, software engineering, and social computing. In addition to these technical presentations, we are also excited to feature a keynote speaker.

The success of ITNG 2025 owes much to the individuals who organized technical tracks. Dr. Doina Bein played a key role as the conference vice chair. We also acknowledge the contributions of major track organizers and associate editors: Drs. Doina Bein, Allesio Bucaioni, Glauco Carneiro, Luiz Alberto Vieira Dias, Eishta Farjana, Ray Hashemi, Ping Wang, and Hossein Zare.

Further thanks go to Drs. Noha Hazzazi and Poonam Dharam, who were instrumental in soliciting, reviewing, and managing the submissions for their respective tracks/sessions.

We extend special appreciation to Springer for their support in preparing the ITNG proceedings. Our gratitude goes to Michael Luby, Senior Editor and Supervisor of Publications, and Brian Halm, Production Editor at Springer, for their efficient management of our publication order. We also thank Springer Project Coordinator, Prasanna Kumar Narayanasamy, for his diligent work in ensuring that the revised articles were correctly formatted according to the publisher's guidelines. Lastly, we recognize the tremendous efforts of the conference secretary, Ms. Mary Roberts, who managed day-to-day affairs and handled a significant volume of correspondence with remarkable efficiency.

I hope you find the ITNG 2025 program both technically enlightening and socially rewarding.

Shahram Latifi

ITNG 2025 Reviewers

Abreu, Fernando Brito
Adeel, Muhammad
Ahmad, Aftab
Amâncio, José
Bagherzadeh, Nader
Bahrami, Azita
Bein, Doina
Bein, Wolfgang
Bucaioni, Alessio
Carneiro, Glauco
Cu, Mina
Cunha, Adilson
Dascalu, Sergiu
Dawson, Maurice
D'Cruze, Hubert
Dharam, Poonam
Dias, Luiz Alberto
Etezady, Nooredin
Fatala, André
Fernando, Lineu
Fong, Anthony
Gao, Weizheng
Gawanmeh, Amjad
Gewali, Laxmi
Harris, Fred
Hashemi, Ray
Hazzazi, Noha
Hua, Jing
Imran, Muhammad
Kinyua, Johanson
Kinyua, Johnson
Li, Zichao
Marques, Johnny Cardoso
Matos, Leonardo
Mendonça, Manoel
Mineda, Emanuel
Monteiro, Miguel
Montini, Denis Avila
Neto, Wilson Cristoni
Saleem, Kashif

Salgado, Paulo
Sbeit, Raed
Sharifabady, Niusha
Shigemura, Rafael
Sreekumari, Prasanthi
Talburt, John
Talburt, John
Tasinaffo, Paulo Marcelo
Wang, Ping
Williams, Kenneth
Zare, Hossein

Contents

Cybersecurity

Adaptive GAN-Based Defense Against Data Poisoning in Federated Learning

Poshak Pathak and Prasanthi Sreekumari(✉)

Department of Computer Science, University of Louisiana at Monroe, Monroe, LA, USA
pathakp@warhawks.ulm.edu, sreekumari@ulm.edu

Abstract. Federated learning (FL) offers a decentralized framework where multiple clients collaborate to train a shared global model without exposing their raw data. However, FL is highly susceptible to data poisoning attacks, where malicious clients introduce adversarial or corrupted data into their local training sets. These poisoned updates, when aggregated into the global model, can degrade the model's performance. This paper proposes a defense mechanism using Generative Adversarial Networks (GANs) to detect poisoned updates in FL. The generator simulates a wide range of poisoned data patterns, while the discriminator identifies and flags malicious updates in real-time without requiring access to the clients' raw data. Our GAN-based framework offers adaptability by handling both targeted and untargeted poisoning attacks. Additionally, a continuous feedback loop is introduced, where flagged poisoned updates are logged and used to further refine the GAN's detection capabilities. This adaptive system ensures robust protection of FL models while maintaining data privacy and security.

Keywords: Federated Learning · Data Poisoning · GANs · Adversarial Attacks · Feedback Loop

1 Introduction

Federated learning (FL) presents a decentralized solution for collaborative model training, where clients train locally and only share model updates. However, this structure poses a significant risk of data poisoning attacks, where malicious clients can inject poisoned data, causing degradation in the global model's performance. Traditional approaches for detecting poisoned updates such as reputation-based systems or gradient analysis are often limited by the lack of access to raw client data and the need for batch processing, which is impractical in a privacy-focused, real-time federated learning environment. Generative Adversarial Networks (GANs) [4] provide a novel and effective solution to this challenge. By simulating poisoned data patterns, GANs enable the detection of adversarial inputs without requiring raw data, thereby preserving client privacy. This research proposes a GAN based framework that detects poisoned data contributions during federated training, offering a robust and flexible defense mechanism for enhancing the security of federated learning systems.

S. Latifi (Ed.): ITNG 2025, AISC 1463, pp. 3–12, 2025.
https://doi.org/10.1007/978-3-031-89063-5_1

The remainder of this paper is organized as follows. Section 2 presents recent research on the detection and mitigation of data poisoning attacks, with a focus on the use of GANs and other advanced techniques. Section 3 introduces the proposed solution, where a GAN-based framework is employed for detecting poisoned updates. The process flow of the GAN mechanism is detailed in Sect. 4, covering both the training and detection phases. Finally, Sect. 5 concludes the paper and outlines potential areas for future research.

2 Literature Review

Data poisoning attacks have emerged as a significant threat in federated learning (FL) environments. These attacks can severely compromise the integrity of machine learning models by introducing poisoned data that skews the model's performance. In this section, we explore existing literature on the detection and mitigation of such attacks, focusing on the application of GANs and other advanced techniques.

PoisonGAN, as introduced by Zhang et al. (2021) [15], effectively generates and executes generative adversarial poisoning attacks, which significantly degrade model accuracy in federated learning setups. The authors demonstrated that PoisonGAN could bypass many conventional defenses due to its adaptability and the sophisticated nature of GAN-generated data. Similarly, Sun et al. (2023) [11] proposed VagueGAN, a new GAN-based model that efficiently poisons federated learning by obscuring attack patterns, making detection more challenging.

On the defensive side, various methods have been proposed to mitigate these threats. One notable approach is outlined by Psychogyios et al. (2023) [8], where the use of GAN-driven data poisoning attacks and their subsequent mitigation are discussed. The paper highlights how adversarial training can help detect poisoned inputs by generating adversarial examples. Additionally, Taheri et al. (2020) [13] propose a robust malware detection system, Fed-IIoT, which integrates federated learning with robust detection models to identify poisoned data in Industrial Internet of Things (IIoT) applications.

Many studies have explored different frameworks to prevent data poisoning attacks. For example, Zhang et al. (2022) [16] use GANs to simulate potential attack vectors in federated learning systems, offering insights into how GANs can both perpetrate and detect poisoning. Another noteworthy defense mechanism is APFed, proposed by Chen et al. (2023) [1], which is a privacy-preserving defense designed to protect federated learning models from data poisoning.

A significant challenge in federated learning environments is the lack of access to raw client data, making real-time detection of poisoned updates difficult. This issue was tackled by Zhou et al. (2023) [17], who proposed a differentially private model that balances performance and security by reducing the impact of malicious data without compromising privacy.

Other studies have focused on optimizing the detection of poisoned data contributions during the federated learning process. For instance, Zhang et al. (2022) [14] introduced an optimized client-side detection framework that significantly enhances the accuracy of detecting poisoned models on the client side. Similarly, Mallah et al. (2021) [7] developed the AttestedFL model, which relies on attestation-based mechanisms to identify and flag suspicious updates.

Several approaches have been developed to counter targeted data poisoning attacks. Qureshi et al. (2022) [9] explored a novel approach using spectral clustering to detect these attacks, which has shown promise in improving detection rates. The authors argue that this method helps identify and mitigate attacks in real-time, providing better resilience to model degradation.

Additionally, Khanh Huu The Dam and Axel Legay (2023) [3] emphasized the importance of partially federated systems in mitigating the effects of data poisoning, as these systems allow for improved collaboration and reduced susceptibility to attacks by decentralizing the data poisoning detection process.

While existing solutions have made significant strides in detecting data poisoning attacks in federated learning, they face several challenges. Many methods struggle to adapt to evolving attack patterns, particularly those that rely on static detection mechanisms. These approaches may become less effective over time as adversarial techniques become more sophisticated. Furthermore, solutions that require access to raw data often violate privacy-a key principle of federated learning. Methods that depend on anomaly detection or reputation systems can fail to detect subtle poisoning attacks, particularly those designed to blend in with legitimate updates. Although some research has utilized GANs for detection, many models are limited in scope, focusing primarily on specific types of attacks or lacking the flexibility to handle a broader range of poisoning threats. These limitations underscore the need for more adaptive, real-time detection systems that can maintain model integrity without compromising client privacy. Our proposed method aims to address these challenges by introducing a dynamic GANbased framework capable of detecting a wide variety of poisoned data while preserving the privacy inherent to federated learning environments.

3 Proposed Solution

We propose the use of GANs as the primary mechanism for detecting poisoned data in federated learning environments. GANs offer a unique capability to simulate poisoned data patterns and detect them in real-time without needing direct access to the raw client data. This approach is highly adaptable, which allows for the detection of both targeted and untargeted attacks by training the generator on diverse adversarial examples. In this framework, the generator is tasked with producing poisoned data samples that mimics various attack strategies. By continuously learning from these generated samples, we hypothesize discriminator becomes highly adept at distinguishing between benign and poisoned updates in the federated learning process.

The use of GANs include several advantages:

1. Adaptability: GANs can generate a wide range of poisoned data patterns, making the system resilient against multiple types of attacks.
2. Real-Time Detection: GANs allow for immediate detection during each round of client updates, enabling a proactive defense mechanism that works dynamically without batch analysis.
3. Privacy-Preserving: Since GANs do not require access to raw data, this method aligns with the privacy-preserving nature of federated learning, ensuring that client data remains secure while still identifying malicious inputs.

4 Process Flow Using GAN

4.1 Training Phase

In this phase, the GAN framework is introduced, which consists of two networks:

- Generator G simulates poisoned data updates.
- Discriminator D detects poisoned data by differentiating between real client updates and those generated by the generator.

The generator and discriminator are trained in a minimax game, where the generator tries to fool the discriminator by generating poisoned updates, while the discriminator aims to correctly classify real updates from generated ones.

4.1.1 Mathematical Formulation

The objective of the GAN is defined by the following minimax loss function in [3]

$$\min_G \max_D V(D, G) = \mathbb{E}_{x \sim p_{\text{data}}(x)}\left[logD(x)\right] + \mathbb{E}_{z \sim p_z(z)}\left[log(1 - D(G(z)))\right] \tag{1}$$

where:

- x is a real update from the client.
- z is the random noise input to the generator.
- $G(z)$ is the poisoned data update generated by the generator.
- $D(x)$ is the probability that x is a real update.
- $D(G(z))$ is the probability that the generated update is real.

The discriminator maximizes the likelihood of correctly classifying real and generated updates, while the generator minimizes the ability of the discriminator to distinguish between real and generated data.

4.1.2 Training Process

1. Discriminator Training

- Given a batch of real updates from clients, the discriminator D learns to maximize the probability of correctly classifying these updates as real.
- It also minimizes the probability of classifying generated poisoned updates from $G(z)$ as real.
- The loss function for the discriminator is given by:

$$\mathcal{L}_D = -\mathbb{E}_{x \sim p_{\text{data}}(x)}\left[\log D(x)\right] - \mathbb{E}_{z \sim p_z(z)}\left[\log(1 - D(G(z)))\right]$$

2. Generator Training

- The generator tries to minimize the discriminator's ability to correctly classify generated updates as fake.

- The generator's goal is to make $D(G(z))$ approach 1, which would mean that the discriminator is fooled.

The loss function for the generator is given by:

$$\mathcal{L}_G = -\mathbb{E}_{z \sim p_z(z)}\left[\log D(G(z))\right]$$

Algorithm 1 GAN Training Process in Federated Learning

```
for round in range(1, num_training_rounds) do
    for real_updates_batch in real_updates do
        z ← GENERATE_NOISE
        generated_updates ← G(z)
        d_loss_real ← log D(real_updates_batch)
        d_loss_fake ← log(1 − D(generated_updates))
        d_loss ← −(d_loss_real + d_loss_fake)
        D.BACKWARD(d_loss)
        OPTIMIZER_D.STEP
    end for
    for i = 1 to generator_steps do
        z ← GENERATE_NOISE
        generated_updates ← G(z)
        g_loss ← − log D(generated_updates)
        G.BACKWARD(g_loss)
        OPTIMIZER_G.STEP
    end for
end for
```

4.2 Detection Phase

This phase plays a critical role in maintaining the security and integrity of the global model. The discriminator detects/flags the poisoned updates and handles them.

The steps below describe how the Discriminator operates:

- Input: During each training round, the clients send their local model updates w_{client} to the server
- Processing: The discriminator evaluates these updates and compares them against the poisoned data generated by the generator during training.
- Output: The discriminator outputs a probability D_w where $D(w) = P$ (update is real) or $1 - D(w) = P$ (update is poisoned)

4.2.1 Mathematical Formulation

- Each client's update w_i is evaluated by the discriminator:

$$D(w_i) = \sigma(f(w_i))$$

where $f(w_i)$ is the score function (the learned model in the discriminator), and σ is the sigmoid function that converts the score into a probability between 0 and 1.

- The discriminator assigns a probability to each update. If this probability falls below a predefined threshold T, it is flagged as a suspicious (poisoned) update:

$$\text{If } D(w_i) < T, \text{ flag } w_i \text{ as poisoned}$$

The threshold T can be fine-tuned depending on the desired sensitivity to poisoned updates.

4.2.2 Flagging and Filtering Updates

Once an update is flagged as potentially poisoned by the discriminator, it undergoes further scrutiny before being included in the global model aggregation. Flagged updates are either:

- Filtered out completely if they are strongly suspected of being poisoned.
- Re-weighted: Instead of being completely discarded, suspicious updates might be included in the aggregation with a reduced weight to minimize their potential impact.

Update Handling:

- Flagging: When an update is flagged, it is stored separately for review. The system keeps track of how frequently each client's updates are flagged, and repeated offenses can result in the client being excluded from future training rounds.
- Filtering: Flagged updates are excluded from the aggregation process. Only updates that pass the discriminator's check are considered "clean" and are included in the final model update.
- Weighting Mechanism:

$$w_{\text{global}} = \sum_{i=1}^{N} \alpha_i w_i$$

  where:

  - w_{global} is the global model's weight update.
  - w_i is the client's update.
  - α_i is a weight assigned based on the discriminator's evaluation (for clean updates, $\alpha_i = 1$; for flagged updates, $\alpha_i = 0$ or a smaller value).

4.2.3 Experimenting with Detection Phase

We will explore three key factors in order to experiment during this phase:

1. Threshold Sensitivity: We experiment with different threshold values T to updates in order to find the balance between minimizing false positives (legitimate updates being flagged) and false negatives (poisoned updates slipping through).
2. Impact of Flagged Updates: We test how filtering out flagged updates is going to effect the overall performance of the global model. There are several ways to measure this such as accuracy, robustness, and the model's ability to generalize.
3. False Positive/Negative Rates: We will measure the rates of false positives and false negatives to better understand how effective the discriminator is at accurately identifying poisoned updates.

4.3 Model Update and Defense Phase

After the detection phase, the model aggregates the benign client updates and aggregates the clean updates, handles flagged updates to improve detection in future rounds.

4.3.1 Aggregation of Clean Updates

In federated learning, each client sends model updates, represented by model weights or gradients, to the central server after training on local data. These updates are aggregated to create the global model that is then distributed back to all clients.

If the updates from N clients are represented by $w_1, w_2, \ldots, w_N$, the aggregated global model update w_{global} is:

$$w_{\text{global}} = \frac{1}{N} \sum\nolimits_{i=1}^{N} \alpha_i \cdot w_i$$

where:

- w_i is the weight update from client i.
- α_i is a binary or fractional weight indicating the trustworthiness of the update. For flagged updates, $\alpha_i = 0$, effectively excluding them from aggregation. For trusted updates, $\alpha_i = 1$.

4.3.2 Defense Mechanism for Poisoned Updates

The defense mechanism ensures that flagged poisoned updates are isolated from the aggregation process to maintain the integrity of the global model. However, rather than discarding flagged updates, they are logged and analyzed to serve two purposes:

- These flagged poisoned updates provide real-world examples of adversarial inputs, which are fed back into the GAN framework to improve future detection.
- By retraining the generator and discriminator with logged poisoned data, the GAN continuously adapts to detect new poisoning strategies.

Mathematical Formulation:

- Let $\mathcal{P}_{\text{logged}}$ denote the set of poisoned updates detected during a round:

$$\mathcal{P}_{\text{logged}} = \{w_i \mid \mathcal{D}(w_i) < T\}$$

  where $\mathcal{D}(w_i)$ is the probability assigned by the discriminator for client update w_i, and T is the detection threshold.
- These logged updates are added to the training data for the next iteration:

$$\mathcal{D}_{\text{train}}^{t+1} = \mathcal{D}_{\text{train}}^{t} \cup \mathcal{P}_{\text{logged}}$$

This continuous augmentation ensures that the GAN model remains robust against evolving attacks.

Malicious Client Identification

The system tracks how often each client sends poisoned updates by maintaining a maliciousness score for each client c_i. After that if a client's updates are frequently flagged as poisoned, the server increases the maliciousness score which is represented as

$$M_i^{t+1} = M_i^t + \lambda \cdot F_i$$

where:

- M_i^t is the maliciousness score of client c_i at time step t.
- F_i is the number of times the client's updates were flagged as poisoned in round $t.\lambda$ is a tuning parameter to control how quickly the score increases.

If it is observed that M_i^t exceeds the threshold value, we exclude it from participating in the training process.

4.3.3 Continuous Feedback Loop: Improving GAN Detection Capabilities

The continuous feedback loop ensures that both the generator and discriminator improve over time. This is essential because attackers may evolve their strategies, developing new types of poisoned data. The feedback loop leverages the logged poisoned updates to continuously retrain and refine the GAN framework. While there is a possibility that retraining on flagged updates introduces the risk of error propagation from false positives, this can be mitigated by threshold tuning [2] and weighted retraining [10]. Threshold tuning ensures that only high confidence flagged updates are considered, while weighted retraining minimizes the impact of potential false positives. This method balances adaptability to new attack strategies while also maintaining detection accuracy.

Mathematics Behind Continuous Learning
Generator Improvement

The generator G learns to produce more sophisticated poisoned updates by training on the logged poisoned data. The generator's objective is to maximize the probability that the discriminator misclassifies a poisoned update:

$$\mathcal{L}_G = -\mathbb{E}_{z \sim p_z(z)}\left[\log D(G(z))\right]$$

where:

$-z$ is a noise vector drawn from a prior distribution p_z,

- $G(z)$ is the generated poisoned update.

By adding newly logged poisoned updates to the training data, the generator becomes better at simulating complex poisoning attacks in future rounds.

Discriminator Refinement

The discriminator D improves by training on a mixture of new poisoned updates and clean updates from clients. Its objective is to correctly classify the updates:

$$\mathcal{L}_D = -\mathbb{E}_{x \sim p_{\text{data}}}\left[\log D(x)\right] - \mathbb{E}_{z \sim p_z(z)}\left[\log(1 - D(G(z)))\right]$$

where:

$-x$ is a real update from the client data distribution p_{data},

- $G(z)$ is a generated poisoned update.

Over time, the discriminator's ability to poisoned updates from clean ones improves as it trains on newly identified poisoning patterns.

Advantages of the Proposed Method: Compared to existing solutions, such as those described in articles [4] and [12], which utilized GAN models for attack detection, our proposed framework offers several key improvements. First, while previous models focused on specific attack patterns, our framework is designed to handle both targeted and untargeted poisoning attacks, making it more versatile in diverse federated learning environments. Second, the inclusion of a continuous feedback loop ensures that the system can adapt to evolving attack strategies over time, enhancing long-term robustness. Unlike static models that may become obsolete as attackers adapt, our GAN-based framework continuously refines both the generator and discriminator using newly detected poisoned updates, allowing the system to stay ahead of adversarial behavior.

5 Conclusion

In this paper, we explored the vulnerabilities of federated learning systems to data poisoning attacks and proposed a novel defense mechanism using GANs. By simulating poisoned data patterns, the generator in the GAN framework can continuously learn and refine its attack strategies, while the discriminator effectively identifies malicious inputs. This dynamic adversarial process ensures that the federated learning system can defend itself against both targeted and untargeted poisoning attacks in real-time.

The continuous feedback loop further enhances the system's robustness by retraining the GAN on newly detected poisoned updates, allowing the system to adapt to evolving attack patterns. This ensures that the model not only maintains its integrity but also improves its detection capabilities over time. The defense mechanism is further reinforced by logging flagged updates for further analysis, enabling the identification of malicious clients and preventing repeated poisoning attempts.

We intend to improve upon this paper by experimenting it on available datasets such as MNIST [6], Toronto Face Database [12] and CIFAR-10 [5].

Future work will involve optimizing the real-time detection capabilities of the GAN framework, exploring additional layers of defense, and testing the system in diverse federated learning environments to ensure scalability and efficiency. By leveraging GANs for both detection and adaptation, federated learning systems can become more secure and resilient against the increasing threat of data poisoning attacks.

References

1. Chen, X., Yu, H., Jia, X., Yu, X.: Apfed: anti-poisoning attacks in privacy-preserving heterogeneous federated learning. IEEE Trans. Inf. Forensics Secur 1 (2023). https://doi.org/10.1109/TIFS.2023.3315125
2. Coenen, F., Leng, P., Zhang, L.: Threshold tuning for improved classification association rule mining. In: Ho, T., Cheung, D., Liu, H. (eds.) Advances in Knowledge Discovery and Data Mining, PAKDD 2005. Lecture Notes in Computer Science, vol. 3518, pp. 216–225. Springer, Heidelberg (2005). https://doi.org/10.1007/11430919_27

3. Dam, K.H.T., Legay, A.: Mitigate data poisoning attack by partially federated learning. In: Proceedings of the 18th International Conference on Availability, Reliability and Security. Association for Computing Machinery, New York (2023). https://doi.org/10.1145/3600160.3605032
4. Goodfellow, I., et al.: Generative adversarial networks. In: Advances in Neural Information Processing Systems, vol. 3 (2014). https://doi.org/10.1145/3422622
5. Krizhevsky, A.: Learning multiple layers of features from tiny images (2009). https://api.semanticscholar.org/. CorpusID: 18268744
6. Lecun, Y., Bottou, L., Bengio, Y., Haffner, P.: Gradient-based learning applied to document recognition. Proc. IEEE **86**(11), 2278–2324 (1998). https://doi.org/10.1109/5.726791
7. Mallah, R.A., López, D., Badu-Marfo, G., Farooq, B.: Untargeted poisoning attack detection in federated learning via behavior attestation. IEEE Access **11**, 125064–125079 (2021). https://api.semanticscholar.org/CorpusID:231709737
8. Psychogyios, K., Velivassaki, T.H., Bourou, S., Voulkidis, A., Skias, D., Zahariadis, T.: GAN-driven data poisoning attacks and their mitigation in federated learning systems. Electronics **12**(8) (2023). https://doi.org/10.3390/electronics12081805. https://www.mdpi.com/2079-9292/12/8/1805
9. Qureshi, N.B.S., Kim, D.H., Lee, J., Lee, E.K.: Poisoning attacks against federated learning in load forecasting of smart energy. In: NOMS 2022–2022 IEEE/IFIP Network Operations and Management Symposium, pp. 1–7 (2022). https://doi.org/10.1109/NOMS54207.2022.9789884
10. Stocco, A., Tonella, P.: Confidence-driven weighted retraining for predicting safety-critical failures in autonomous driving systems. J. Softw. Evol. Process **34**(10), e2386 (2022)
11. Sun, W., Gao, B., Xiong, K., Lu, Y., Wang, Y.: Vaguegan: a GAN-based data poisoning attack against federated learning systems. In: 2023 20th Annual IEEE International Conference on Sensing, Communication, and Networking (SECON), pp. 321–329 (2023). https://api.semanticscholar.org/. CorpusID: 264475163
12. Susskind, J., Anderson, A., Hinton, G.E.: The toronto face dataset. Technical report UTML TR 2010-001, U. Toronto (2010)
13. Taheri, R., Shojafar, M., Alazab, M., Tafazolli, R.: Fed-iiot: a robust federated malware detection architecture in industrial iot. IEEE Trans. Industr. Inf. **17**(12), 8442–8452 (2021). https://doi.org/10.1109/TII.2020.3043458
14. Zhang, G., Shu, J., Jia, X.: Optimized client-side detection of model poisoning attacks in federated learning. In: 2022 IEEE 24th International Conference on High Performance Computing Communications; 8th International Conference on Data Science Systems; 20th International Conference on Smart City; 8th International Conference on Dependability in Sensor, Cloud Big Data Systems Application (HPCC/DSS/SmartCity/DependSys), pp. 1196–1201 (2022). https://doi.org/10.1109/HPCC-DSS-SmartCity-DependSys57074.2022.00188
15. Zhang, J., Chen, B., Cheng, X., Binh, H.T.T., Yu, S.: Poisongan: Generative poisoning attacks against federated learning in edge computing systems. IEEE Internet Things J. **8**(5), 33103322 (2021). https://doi.org/10.1109/JIOT.2020.3023126
16. Zhang, Y., Zhang, Y., Zhang, Z., Bai, H., Zhong, T., Song, M.: Evaluation of data poisoning attacks on federated learning-based network intrusion detection system. In: 2022 IEEE 24th International Conference on High Performance Computing Communications; 8th International Conference on Data Science Systems; 20th International Conference on Smart City; 8th International Conference on Dependability in Sensor, Cloud Big Data Systems Application (HPCC/DSS/SmartCity/DependSys), pp. 2235–2242 (2022). https://doi.org/10.1109/HPCC-DSS-SmartCity-DependSys57074.2022.00330
17. Zhou, J., et al.: A differentially private federated learning model against poisoning attacks in edge computing. IEEE Trans. Dependable Secure Comput. **20**(3), 1941–1958 (2023). https://doi.org/10.1109/TDSC.2022.3168556

Password Security in Practice: An Appraisal Using Users' Perception and Machine Learning

Zackary Beck, Adam Crooks, Md. Fazle Rabbi, Minhaz Zibran(✉), and Farjana Eishita

Department of Computer Science, Idaho State University, 921 South 8th Avenue, Pocatello 83201, ID, USA
{beckzack,crooadam,mdfazlerabbi,zibran,farjanaeishita}@isu.edu

Abstract. In response to rising digital threats, we conduct a two-phased exploration of password security practices. Through a survey including 61 individuals of varying ages, we investigate password security practices and identify a gap between awareness and actual practices, particularly concerning the reuse of similar passwords and the lack of complexity. Next, we operate three machine learning (ML) algorithms on a dataset of 669,643 leaked passwords to determine the factors that contribute to password strength considered by the ML algorithms. Surprisingly, we find that ML algorithms achieve high accuracy in password strength classification by primarily focusing on password length only.

Keywords: Password · Security · Machine Learning · User Perspective · Survey · Cybersecurity · Analysis · Password Reuse · Complexity · Leaked Data

1 Introduction

In today's digital world, the significance of password security cannot be overstated, especially with the rise of faster computing capabilities and increasingly sophisticated hackers. Recent years have seen numerous high-profile data breaches, exposing billions of passwords globally and emphasizing the urgent need for robust password management practices. According to a report by Google Cloud, a staggering 86% of breaches involve stolen credentials [11]. Similarly, Verizon's Data Breach Investigations Report found that 74% of breaches involve the human element, often through stolen credentials or social engineering [43]. The breach of the American Bar Association, which exposed the login credentials of 1.5 million members, further illustrates the vulnerability even reputable organizations face [28].

Despite these evident risks, there remains a significant gap between user awareness and their password habits. Surprisingly, while 91% of respondents

S. Latifi (Ed.): ITNG 2025, AISC 1463, pp. 13–24, 2025.
https://doi.org/10.1007/978-3-031-89063-5_2

acknowledged the dangers of password reuse, 59% admitted to doing it anyway [44]. This contradiction extends to password strength, with 83% of Americans reportedly using weak passwords [3]. Despite claiming familiarity with password security best practices, 90% of participants fail to implement them, highlighting the challenge of changing user behavior [1].

To address this gap, many organizations have implemented policies to encourage more secure password practices. For instance, Microsoft's Enforce password history policy aims to mitigate vulnerabilities by preventing the reuse of old passwords [29]. Administrator guidelines advocate for unique passwords and enforce a minimum password age to deter rapid password cycling [26]. However, these clash with user convenience preferences, presenting challenges in bridging the gap between ideal and real-world password security practices.

Our research investigates the persistent gap between password security awareness and implementation, aiming to understand real-world password strength and user behaviors. Through analyzing a large corpus of leaked passwords using Machine Learning (ML) models and surveying users on their password habits, we seek to generate actionable insights. Our study proceeds in two phases: a comprehensive survey assessing password practices and awareness, followed by ML analysis of leaked password data to identify predictive attributes of password strength. By comparing survey responses with modeled password features, we aim to align user behaviors with proper security measures.

The rest of the paper is organized as follows. Section 2 provides an overview of our survey methodology and the results obtained regarding user password behaviors. Section 3 details the ML methodology utilized and presents our findings concerning the prediction of password strength. In Sect. 4, we discuss the potential threats to the validity of our study. Section 5 reviews related work. Finally, Sect. 6 concludes by summarizing insights and proposing future work.

2 Capturing End-User Perspective with Survey

In our study, we explore the practices and habits surrounding password security among diverse individuals, focusing on three key areas: strong password creation, random password generator utilization, and password management system adoption. Acknowledging the importance of user-focused strategies like password creation and the use of random password generators, we conduct a survey to explore users' behaviors and preferences in password management.

2.1 Survey Design and Implementation

We engage 61 participants in our survey, spanning an age range from 19 to 81 years. To ensure a comprehensive understanding of password security practices across diverse demographics, participation is open without capturing information based on gender, race, or ethnicity. The survey consists of six questions, including one demographic query about participants' age, followed by five multiple-choice questions addressing various password management practices. Administered online via SurveyPlanet [40], the platform is selected for its accessibility and user-friendly interface.

The questionnaire includes questions about password creation, random password generator usage, and experiences with password compromise. It also assesses the use of personally identifiable information in passwords to identify common security vulnerabilities. Table 1 lists the survey questionnaire along with options provided.

2.2 Survey Findings

A significant majority of survey participants, 86%, report using variations of similar passwords across their different accounts and profiles, indicating a common preference for convenience over security. This practice potentially increases vulnerability across multiple platforms if one password is compromised.

Strong passwords are crucial for protecting data and private information. Despite this, 63.2% of respondents admit to creating passwords meeting only minimum requirements, raising concerns about their strength and security. Implementing more sophisticated algorithms by password management systems could enhance password security by detecting easily guessed passwords, including those containing dictionary words, deemed very weak. The survey also finds that 64.9% of participants consciously avoid using personally identifiable information in their passwords, showing a cautious approach to password creation among the majority.

Random password generators, which create stronger and more secure passwords, are not widely used. A notable 77.2% of participants state they do not use these tools, with only 10.5% frequently using them and 12.3% using them occasionally. This suggests a potential area for increasing security awareness and encouraging the adoption of these tools. However, there are pros and cons to using random password generators. While they may generate stronger passwords, the need for users to write down these passwords to remember them can compromise their security [33].

More than half of the respondents, 54.4%, have had one or more passwords compromised in the past. This statistic underscores the widespread issue of password security breaches and the urgent need for stronger password practices.

The findings from the survey emphasize the critical need for enhanced education and awareness regarding password security. The prevalent use of similar password variations, reliance on minimum creation requirements, and low adoption rate of random password generators reveal significant gaps in public knowledge and practices concerning password security.

3 Revealing Insights with Machine Learning

Motivated by survey results that highlight gaps in public knowledge and practices concerning password security, we apply ML to a dataset of real, leaked passwords to evaluate password strength and ML performance in this context. Our objective is to identify key features or factors that affect password strength, discerning characteristics that classify passwords as strong, medium, or weak.

3.1 ML Model Development

In this paper, we explore the development and evaluation of models for categorizing password strengths, utilizing a dataset sourced from Kaggle [4]. The dataset contains 669,643 passwords, with each having assigned a strength rating categorized as weak (0), medium (1), or strong (2). These are real passwords used by individuals as credentials for different systems, but these passwords were leaked and are now publicly available for analyses and similar purposes.

Table 1. Survey Questionnaire on Password Security Habits

Questions	Options
What is your age?	Open-ended
Do you use variations of a similar password for all of your accounts or profiles?	a) Yes, I use variations of a similar password for all accounts/profiles b) No, I use significantly different passwords for each account/profile
Do you create your password based on the minimum password requirements?	a) Yes, I create a password based on the minimum requirements b) No, I always implement "strong" passwords
Do you use personally identifiable information in your password?	a) Yes b) No
Do you use random password generators?	a) Often b) Never c) Sometimes
Have you ever had a compromised password?	a) Several times b) A couple of times c) Never

As a preprocessing step, we apply feature engineering to effectively train ML models. We extract five key features from the passwords: their length, the number of uppercase letters, lowercase letters, numerical characters, and special characters. These features are commonly used to evaluate password strength, reflecting a widely accepted approach in the field.

For the classification task, we employ three distinct models: a multi-layer perceptron classifier, a decision tree classifier, and a K-nearest neighbors classifier. All models are implemented utilizing the Scikit-learn library.

3.1.1 The Multi-layer Perceptron (MLP)

The MLP classifier [37] is a widely used neural network for classification tasks, comprising input, hidden, and output layers with interconnected neurons whose weights are adjusted during training. In this study, we adopt a single hidden layer featuring 10 neurons. The Rectified Linear Unit (ReLU) activation function [10] is used in the hidden layer, while the softmax activation function [10] is employed in the output layer. For training, we employ stochastic gradient descent (SGD) [10] with a learning rate of 0.001 to iteratively optimize the network's weights. Early stopping is activated to prevent overfitting.

3.1.2 Decision Tree (DT)

The decision tree classifier [37] operates on the default Gini impurity score [30] for determining splits. No maximum depth is set to allow the tree to expand naturally to a shallow configuration, considering the dataset's limited feature set.

3.1.3 K-Nearest Neighbors (KNN)

Lastly, our K-nearest neighbors classifier [37] is configured to identify the five nearest neighbors as the basis for classification. To accommodate the extensive size of the training dataset and enhance computational efficiency, we set the *n_jobs* parameter to -1, enabling the use of all available processors. The performance of each model is assessed using an 80/20 train/test split.

3.2 Insights from Applying Machine Learning

In this section, we present the performance results of our experiments with three ML models: multi-layer perceptron, decision tree, and K-nearest neighbors.

3.2.1 Initial Observations

We meticulously document the accuracy metrics across training and testing, with summarized outcomes presented in Table 2.

Table 2. Performance Metrics of Machine Learning Models

Model	Training Accuracy (%)	Testing Accuracy (%)
MLP	99.996	99.997
DT	100	100
KNN	99.998	99.997

From the table, the MLP model demonstrates an average training accuracy of 99.996% and an average testing accuracy of 99.997%. Notably, it reaches its peak validation accuracy at 99.998%. On the other hand, the decision tree model achieves perfect training and testing accuracies of 100%. To understand which features are crucial in our decision tree model, we create a visualization, as shown in Fig. 1.

Figure 1 highlights the significance of the password length feature in our classification task. Initially, the model splits the dataset based on a length threshold of ≤ 7.5. The root node, representing 535,712 instances, has a Gini impurity of 0.415. This division leads to two distinct paths: on the left, a leaf node with perfect classification of 71,600 instances (Gini impurity $= 0$); on the right, a node with a Gini impurity of 0.245. This right node further splits at length $\leq$ 13.5, resulting in two leaf nodes containing 397,765 and 66,347 samples respectively, both with zero Gini impurity. In simpler terms, passwords shorter than 7.5 characters are consistently labeled as weak, while passwords longer than 13.5 characters are consistently identified as strong.

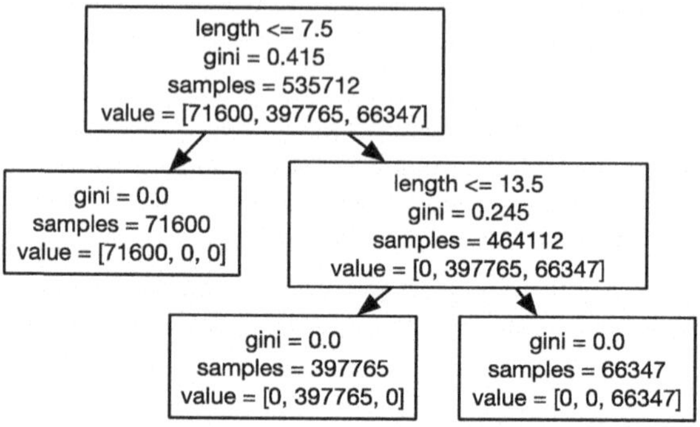

Fig. 1. Feature Analysis of Decision Tree.

Lastly, the KNN model achieves a training accuracy of 99.998% and a testing accuracy of 99.997%, closely aligning with the performance of MLP model. However, it is noteworthy that the KNN model requires significantly more time to train, taking approximately eight minutes, in contrast to less than a minute for the other two models.

3.2.2 Model Optimization

In our pursuit of optimizing our initial models detailed in Sect. 3.1, we explore various techniques to enhance their performance further. We first normalize the features using the StandardScaler [35] from Scikit-learn for each algorithm.

The normalized data improves the MLP model's training and testing accuracy to 99.998%, with a slight decrease in the best validation accuracy to 99.996%. Further experimentation with learning rates of 0.001, 0.01, and 0.1 reveals 0.001 as the most effective, albeit marginally. Our exploration of hidden layers, including one, two, four, and eight layers, identifies eight hidden layers as optimal.

The final MLP model configuration, utilizing a learning rate of 0.001 and eight hidden layers with 10 nodes each, is trained on normalized data. This setup achieves a training accuracy of 99.999% and a testing accuracy of 100%, with a best validation accuracy of 100%.

Since the decision tree initially achieves perfect accuracy, we find no need for further improvement. However, training on normalized data confirms its ability to maintain 100% accuracy in both training and testing phases.

Initially, the KNN model shows a slight performance dip with normalized data, achieving a training accuracy of 99.988% and testing accuracy of 99.984%. After exploring adjustments to the number of neighbors used for classification, including varying values such as three, five, seven, and nine, we find that five neighbors yield the highest accuracies. Our final KNN model, utilizing five neighbors and normalized data, achieves training and testing accuracies of 99.986% and 99.976%, respectively.

The optimizations implemented across the models result in improvements, particularly in the performance of the MLP algorithm. Table 3 provides a summary of the final accuracies achieved after optimization.

Table 3. Final Model Performance Post-Optimization

Model	Training Accuracy (%)	Testing Accuracy (%)
MLP	99.999	100
DT	100	100
KNN	99.986	99.976

In our study, we find that models like the decision tree heavily rely on password length as a primary determinant of strength. However, this approach sometimes fails to capture the essence of password complexity in real-world settings. For example, complex passwords comprising a mix of characters, such as `'mw4!q#A'`, are considered strong by various password strength tests [34, 42] due to their complexity and unpredictability. However, our models, focusing predominantly on length, may classify them as weak.

It is crucial to note that in practice, the typical user password creation behavior rarely results in such complex combinations, contributing to the models' tendency to view shorter passwords as inherently less secure. This observation highlights the need for a nuanced approach in password security, advocating for both improved model sophistication to recognize diverse strength indicators and enhanced user education on the benefits of creating inherently complex passwords, beyond just length.

4 Threats to Validity

The survey's limited scope, with only six questions, might overlook key aspects of password management practices such as behavioral subtleties and contextual factors. Furthermore, the absence of detailed demographic data, including gender, race, education level, and technological proficiency, challenges the study's inclusivity and ability to draw broad conclusions about diverse user groups' password security behaviors.

Although our ML models primarily emphasize password length, it is important to note that our dataset comprises real passwords, reflecting the choices made by individuals. This signifies that the prevalence of shorter, less complex passwords without character combinations in our dataset reflects common user behavior. Therefore, while our models heavily consider length, they may inadvertently overlook other factors such as complexity and common patterns. While cross-validation typically serves as the gold standard for evaluating models, our study's methodology, which engages in multiple training iterations on various splits, consistently shows high accuracy, validating our approach. This method effectively confirms the robustness of our models, indicating their reliability even in the absence of cross-validation for this specific application.

5 Related Work

There have been many studies on software security and quality assurance that involve the investigation of bug patterns [23,24,36], vulnerabilities [15,25], code smells [20,45,46], code quality [2,20], human aspects [5,6,16,17,21,22] of software development and maintenance as well as comparison of methods/tools [18,19,27,32] for measuring such aspects.

ML is used in various domains [7,14], including password security, where researchers have developed advanced models for password strength assessment. He et al. [12] introduced 'Hybritus,' a technique that evaluates password strength by aggregating input from various website checkers using ensemble learning. Their approach incorporates an MLP model to classify passwords into weak, medium, or strong categories, achieving an accuracy of 97.7% in password classification. Similarly, Murmu et al. [31] introduced 'PassMon,' a framework employing a bi-directional generative adversarial network (BiGAN), along with a one-class SVM and LSTM, for password generation and strength estimation. Farooq [9] presented a real-time password strength detection system using multiple ML approaches. The study created a password dataset categorized as weak, medium, and strong to train and test models where Decision Trees achieved the highest accuracy of 99%.

Additionally, Cui et al. [8] proposed an algorithm considering personal data to enhance password strength evaluations. Taneski et al. [41] evaluated the effectiveness of Markov models in analyzing the strength of real-life passwords. By employing first-order Markov models trained on diverse datasets, the research investigates whether different models yield statistically distinct results when assessing the same password set. Their analysis revealed the necessity for multiple Markov models to evaluate password strength across a broad spectrum, challenging the sufficiency of a single-model approach. Sarkar and Nandan [39] explored password strength classification through various supervised machine learning techniques, employing a dataset categorized into weak, medium, and strong. The study used XGBoost and Multilayer Perceptron algorithms to the analysis, with XGBoost achieving a classification accuracy of 99%.

Additionally, researchers also assessed passwords in non-English contexts. For example, Hong et al. [13] contributed to the development of stronger password security practices for Korean users. They introduced a password security evaluation model incorporating both Korean and English password dictionaries to better assess password strength for Korean speakers. Utilizing leaked password datasets, their model demonstrated an accuracy of 99.38% in classifying Korean language-based passwords. Building upon these foundational studies, Rodriguez et al. [38] conducted a survey and user study to examine the balance between password security and memorability. Their research identified effective password styles and suggested strategies to improve password creation, adding a critical perspective on the human factors influencing password strength.

In our study, we have applied three algorithms to classify password strength, all achieving either 100% or close to it. We also have investigated the factors contributing to this higher accuracy and assessed the current state of password awareness among users through a survey.

6 Conclusion

In this study, we undertake a comprehensive exploration of password security practices. We integrate Machine Learning (ML) models with user survey data to bridge the gap between theoretical security measures and real-world password usage. We analyze a substantial dataset of leaked passwords alongside survey responses regarding password habits. The survey includes 61 participants of varying ages while three ML algorithms are operated on a dataset containing 669,643 real passwords.

We find a significant discrepancy between user awareness and actual security practices, with a reliance on weak passwords and frequent reuse. While our ML analysis emphasizes the importance of password length in determining strength, it also acknowledges limitations in addressing the complexity of password usage. Future work will focus on expanding the dataset to include a wider variety of password types to further enhance the accuracy and reliability of password strength evaluations.

Acknowledgement. This work is supported in part by the ISU-CAES (Center for Advanced Energy Studies) Seed Grant at the Idaho State University, USA.

References

1. 2022 World Password Day Global Survey Full Report. https://bitwarden.com/resources/world-password-day-global-survey-full-report/
2. Alwad, D., Panta, M., Zibran, M.: An empirical study of the relationships between code readability and software complexity. In: 27th International Conference on Software Engineering and Data Engineering, pp. 122–127 (2018)
3. Babu, S.: 33 Password Statistics Small Businesses Should Know (2022). https://smallbiztrends.com/2022/10/password-statistics.html

4. Bansal, B.: Password strength classifier dataset (2019). https://www.kaggle.com/datasets/bhavikbb/password-strength-classifier-dataset
5. Champa, A., Rabbi, M., Eishita, F., Zibran, M.: Are we aware? An empirical study on the privacy and security awareness of smartphone sensors. In: 21st IEEE International Conference on Software Engineering, Management and Applications, pp. 287–294 (2023)
6. Champa, A., Rabbi, M., Zibran, M., Islam, M.: Insights into female contributions in open-source projects. In: 20th IEEE International Conference on Mining Software Repositories, pp. 357–361 (2023)
7. Champa, A., Rabbi, M.F., Zibran, M.: Curated datasets and feature analysis for phishing email detection with machine learning. In: 3rd International Conference on Computing and Machine Intelligence, pp. 1–7. IEEE (2024)
8. Cui, X., Li, C., Qin, Y., Ding, Y.: A password strength evaluation algorithm based on sensitive personal information. In: 19th International Conference on Trust, Security and Privacy in Computing and Communications (TrustCom), pp. 1542–1545. IEEE (2020)
9. Farooq, U.: Real time password strength analysis on a web application using multiple machine learning approaches. Int. J. Eng. Res. Technol. **9**(12), 359–364 (2020)
10. Goodfellow, I., Bengio, Y., Courville, A.: Deep Learning. MIT Press (2016)
11. Google: Threat Horizons: August 2023 Threat Horizons Report (2023). https://services.google.com/fh/files/blogs/gcat_threathorizons_full_jul2023.pdf
12. He, Y., Alem, E., Wang, W.: Hybritus: a password strength checker by ensemble learning from the query feedbacks of websites. Front. Comp. Sci. **14**, 1–14 (2020)
13. Hong, K., Kang, U., Lee, B.: Enhanced evaluation model of security strength for passwords using integrated Korean and English password dictionaries. Secur. Commun. Netw. **2021**, 1–13 (2021)
14. Islam, M., Ahmmed, M., Zibran, M.: Marvalous: machine learning based detection of emotions in the valence-arousal space in software engineering text. In: Proceedings of the 34th ACM/SIGAPP Symposium on Applied Computing, pp. 1786–1793 (2019)
15. Islam, M., Zibran, M.: A comparative study on vulnerabilities in categories of clones and non-cloned code. In: 10th IEEE International Workshop on Software Clones, pp. 8–14 (2016)
16. Islam, M., Zibran, M.: Exploration and exploitation of developers' sentimental variations in software engineering. Int. J. Softw. Innov. **4**(4), 35–55 (2016)
17. Islam, M., Zibran, M.: Towards understanding and exploiting developers' emotional variations in software engineering. In: SERA, pp. 185–192 (2016)
18. Islam, M., Zibran, M.: A comparison of dictionary building methods for sentiment analysis in software engineering text. In: ESEM, pp. 478–479 (2017)
19. Islam, M., Zibran, M.: A comparison of software engineering domain specific sentiment analysis tools. In: 25th IEEE International Conference on Software Analysis, Evolution and Reengineering, pp. 487–491 (2018)
20. Islam, M., Zibran, M.: On the characteristics of buggy code clones: a code quality perspective. In: 12th IEEE International Workshop on Software Clones, pp. 23–29 (2018)
21. Islam, M., Zibran, M.: Sentiment analysis of software bug related commit messages. In: SEDE (2018)
22. Islam, M., Zibran, M.: Sentiment analysis of software bug related commit messages. In: 27th International Conference on Software Engineering and Data Engineering, pp. 3–8 (2018)

23. Islam, M., Zibran, M.: How bugs are fixed: exposing bug-fix patterns with edits and nesting levels. In: 35th ACM/SIGAPP Symposium on Applied Computing, pp. 1523–1531 (2020)
24. Islam, M., Zibran, M.: What changes in where? An empirical study of bug-fixing change patterns. ACM Appl. Comput. Rev. **20**(4), 18–34 (2021)
25. Islam, M., Zibran, M., Nagpal, A.: Security vulnerabilities in categories of clones and non-cloned code: an empirical study. In: 11th ACM/IEEE Intl. Symposium on Empirical Software Engineering and Measurement, pp. 20–29 (2017)
26. Jithukrishnan: Top 10 password policy recommendations for system administrators in 2023 (2023). https://www.securden.com/blog/top-10-password-policies.html
27. Joseph, R., Zibran, M., Eishita., F.: Choosing the weapon: a comparative study of security analyzers for android applications. In: International Conference on Software Engineering, Management and Applications, pp. 51–57 (2021)
28. Merken, S.: ABA says hackers took lawyers' data in March attack (2023). https://www.reuters.com/legal/litigation/aba-says-hackers-took-lawyers-data-march-attack-2023-04-21/
29. Microsoft: Enforce password history (2024). https://learn.microsoft.com/en-us/windows/security/threat-protection/security-policy-settings/enforce-password-history
30. Müller, A.C., Guido, S.: Introduction to machine learning with Python: a guide for data scientists. O'Reilly Media, Inc. (2016)
31. Murmu, S., Kasyap, H., Tripathy, S.: Passmon: a technique for password generation and strength estimation. J. Netw. Syst. Manage. **30**, 1–23 (2022)
32. Murphy, D., Zibran, M., Eishita, F.: Plugins to detect vulnerable plugins: an empirical assessment of the security scanner plugins for wordpress. In: International Conference on Software Engineering, Management and Applications, pp. 39–44 (2021)
33. O'Bert, T.: The pros and cons of generated passwords (2019). https://www.productivecorp.com/blog/risk-management/the-pros-and-cons-of-generated-passwords/
34. PasswordMonster: How secure is your password? (2024). https://www.passwordmonster.com
35. Pedregosa, F., et al.: Scikit-learn: machine learning in python. J. Mach. Learn. Res. **12**, 2825–2830 (2011)
36. Rajbhandari, A., Zibran, M., Eishita, F.: Security versus performance bugs: How bugs are handled in the chromium project. In: International Conference on Software Engineering, Management and Applications, pp. 70–76 (2022)
37. Raschka, S., Mirjalili, V.: Python machine learning: machine learning and deep learning with Python, scikit-learn, and TensorFlow 2. Packt Publishing Ltd (2019)
38. Rodriguez, J., Zibran, M., Eishita, F.: Finding the middle ground: measuring passwords for security and memorability. In: 20th International Conference on Software Engineering Research, Management and Applications, pp. 77–82. IEEE (2022)
39. Sarkar, S., Nandan, M.: Password strength analysis and its classification by applying machine learning based techniques. In: 22nd International Conference on Computer Science, Engineering and Applications, pp. 1–5. IEEE (2022)
40. SurveyPlanet: Surveyplanet - a simple and powerful online survey tool (2024). https://surveyplanet.com
41. Taneski, V., Kompara, M., Heričko, M., Brumen, B.: Strength analysis of real-life passwords using Markov models. Appl. Sci. **11**(20), 9406 (2021)
42. UIC Academic Computing and Communications Center: Password strength test (2024). https://www.uic.edu/apps/strong-password/

43. Verizon: 2023 data breach investigations report (2023). https://www.verizon.com/business/resources/reports/dbir/
44. Vijayan, J.: Password Reuse Abounds, New Survey Shows (2018). https://www.darkreading.com/vulnerabilities-threats/password-reuse-abounds-new-survey-shows
45. Zibran, M., Roy, C.: Conflict-aware optimal scheduling of code clone refactoring. IET Softw. **7**(3), 167–186 (2013)
46. Zibran, M., Saha, R., Roy, C., Schneider, K.: Evaluating the conventional wisdom in clone removal: a genealogy-based empirical study. In: 28th ACM/SIGAPP Symposium on Applied Computing, pp. 1123–1130 (2013)

Security and Privacy for LLMs and LMMs Across Key Sectors: A Literature Survey

Costain Nachuma(✉), Mostafa M. Fouda, and Minhaz F. Zibran(✉)

Department of Computer Science, Idaho State University, 921 S. 8th Avenue, Pocatello, ID 83209, USA
{costainnachuma,zibran}@isu.edu, mfouda@ieee.org

Abstract. Technology has continued to advance over the years giving rise to the era of Large Language Models (LLMs) and Large Multimodal Models (LMMs) which have demonstrated abilities to process vast amounts of data in various formats. As much as LLMs and LMMs are transforming the way key sectors operate through beneficial integrations; they have also introduced a new barrage of security and privacy concerns. This paper presents a multi-domain analysis of these challenges. We focus on three key sectors: education, health, and finance. Through this detailed study, we highlight various threats such as data breaches, biased output, and unauthorized access. To address these and other threats, mitigation strategies are proposed including validation techniques and differential privacy. We aim to provide valuable security and privacy insights to developers, policymakers, and users on LLMs and LMMs integration in various domains.

Keywords: privacy · security · LLM · LMM · survey · literature review · artificial intelligence · education · health · finance · data protection · ethics

1 Introduction

Cloud computing, generation and storage of digital data have continued to grow over the past two decades [7]. These advancements have laid a necessary foundation for the development of various technologies which ride on these improvements. Lately, we have witnessed the rise of LLMs [40] and LMMs [38] capable of handling data in ways never previously imagined decades ago like SAM 2 from Meta, Gemini developed by Google DeepMind and ChatGPT-4o, a multilingual, multimodal generative pre-trained transformer designed by OpenAI. These technologies built on AI models have shown immense capabilities of processing nearly endless amounts of text in various formats like, plain text, pdf, various office package productions like word or PowerPoint. Further, these models have been enhanced to handle image data, and potentially audio data. In as

S. Latifi (Ed.): ITNG 2025, AISC 1463, pp. 25–39, 2025.
https://doi.org/10.1007/978-3-031-89063-5_3

much as these advancements have for the most part being beneficial, new challenges have emerged including security and privacy concerns. LLMs and LMMs are now being integrated for use across sectors like healthcare [1], finance [39], education [35] and software development [4] raising serious concerns about data user privacy and security in these domains. Questions are now being asked on the safety of these ever-developing models against attacks instigated by various threat actors [36]. When these technologies are applied in different domains, we then have medical records, financial transactions, and student learning materials becoming vulnerable. Some examples of vulnerabilities include but not limited to leakages of data, unauthorized access to this data through the learning models all which could have serious repercussions.

Privacy has been a complex topic among many legislators and privacy enthusiasts around the world with the European Union taking a strong lead in protecting users [29]. The major concern often denied by model creators is that these models are trained using private data. Some of the data used in training is clearly authorized as users subscribe to some of these platforms providing data at will. However, many speculate that unauthorized data is being used to train these models based on the excellent efficiency they have attained, which would require a vast amount of data. Well known approaches to make data appear to be from anonymous sources is not entirely foolproof as outputs from models could unintentionally expose sensitive information [19]. In addition to this, a common attack susceptible to these models is adversarial attacks which poses a daunting threat. Some potential threats could be fake news generated by AI in an attempt to sway public opinion during an election. Another threat would be bias in models misguiding a medical diagnosis or promoting a particular prescription. Recently, deepfakes were used to scam an organization in Hong Kong of over $25 million dollars. The scam was so sophisticated that the finance officer believed he received legitimate authorization to transfer funds [6].

This paper aims to provide a comprehensive comparative analysis of security and privacy challenges of LLMs and LMMs. We will look into the core differences in how these models process data and how these differences lead to their being vulnerable to simple or sophisticated attacks. Further to this, we delve into the security and privacy concerns across some key sectors which include healthcare, education, finance. We also investigate current methods used to address these risks.

The rest of the paper is organized as follows: Sect. 2 discusses the threats related to LLMs and LMMs across three sectors: education, healthcare and finance. Section 3 presents the mitigation strategies currently employed to address these security and privacy concerns. Section 4 identifies the limitations of our study while Sect. 5 reviews related works in this domain. Finally, Sect. 6 concludes the paper.

2 Threats for LLMs and LMMs

There several threats associated with LLMs. The OWASP Top 10 LLM applications team has specifically documented well researched risks associated with

LLMs. A governance checklist was recently issued for LLM AI Cybersecurity [30]. It provides a categorization of threats beneficial for all aspects of the LLMs from model development, use and regulation. We summarize each threat classification in the general context of both LLMs and LMMs bearing in mind the central role of AI in these models. The major threat-categories are outlined in Fig. 1.

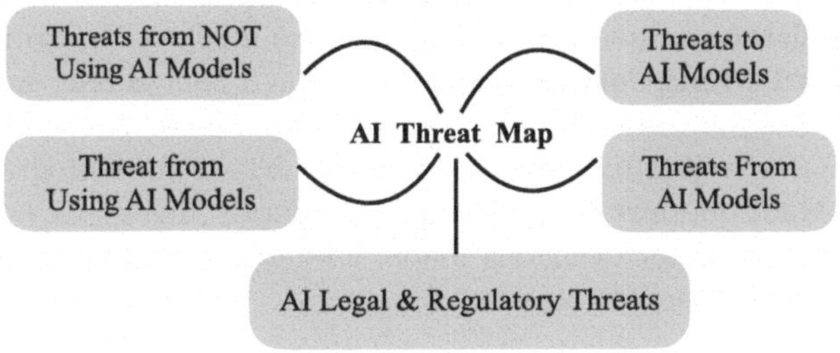

Fig. 1. Classification of AI threats

2.1 OWASP LLM Threat Classifications

2.1.1 Threats from NOT Using AI Models

The first classification outlines risks associated with altogether avoiding AI models. One of the major downsides to this is losing out on the competitive advantages AI could bring to its users. Some benefits which could missed are; leveraging AI for customer service, marketing and innovation during product development. Other users utilizing these services could very easily outperform non-adopters. Another challenge is that non-users may face inefficiencies arising from not taking advantage of AI automation and analyzing of a lot of data. Without using AI, some may be slower at executing tasks and obviously miss out on latent patterns in data which could bring better results if identified. Lastly, AI could be useful for fraud detection and perhaps risk management. Not using AI could lead to many inadequacies which AI could easily identify.

2.1.2 Threats from Using AI Models

The focus of this second classification is threats that come about because of using AI model. We outline some of the major concerns. Firstly, malicious and sophisticated actors may try to flaw the data that is used to train these models. This may lead to biases in decisions or even completely inaccurate outputs altogether. Further, knowledgeable threat actors who understand how these models work my capitalize on that weakness. An example of an attack is deceiving a facial recognition system with an image which could compromise the system. Last but not the least in the possibly endless list of risks is security breaches and loss of privacy. We know that models may not be necessarily hack proof and that advanced attackers may try to manipulate outputs or steal the model.

2.1.3 Threats to AI Models

A much deeper challenge is threats relating to the AI model being the target of malicious attack. Adversarial attacks present a real threat to AI models by seeking to exploit vulnerabilities that could lead to wrong predictions. For example, a self-driving car could misinterpret an altered stop sign and could lead to an unpleasant outcome. The main goal of these attacks is to mislead the AI model or exploit vulnerability mostly at a design level. Data poisoning is one concern that could lead to infiltrators injecting wrong data like incorrect information causing the model to learn wrong data. This can be likened to a classic saying, garbage in, garbage out. Further, biases in data may be introduced to lead to discrimination such as unfairly rejecting a loan application based on demographic or other salient trends. Many models are complex and operate like black boxes where no justification is provided for decisions made. The absence of transparency makes it nearly impossible to detect biases.

2.1.4 Threats from AI Models

AI model although very useful can poses threats if not developed and used responsibly. Some imminent risks from AI models include loss of jobs. In as much as automation is beneficial to organizations, it could lead to layoffs in various sectors like health and education. Despite some opportunities being created in other sectors, the transition can be abrupt and unwanted for employees. Privacy concerns also arise here. AI model are products of training from a lot of data. The Collection and use of this data can abrogate user privacy especially in cases where anonymization of data is not handled safely and correctly. On the other end of the usefulness of AI in detecting cyberthreats is AI being used to launch a barrage of sophisticated cyberattacks. This poses a serious risk as AI would look to cover all its track in a more completed way than a hands-on attacker manually doing it. A risk worth noting is disinformation. AI can be used to create realistic deepfakes and spread an assortment of misinformation. This can have influence elections or sway trading on the stock market.

2.1.5 AI Legal and Regulatory Threats

This category addresses the legal and regulatory challenges that AI poses. This includes compliance with laws targeted at data protection and other legal considerations. First and foremost, the lack of clear regulation is a critical concern for all. Many countries are only coming to terms with this new era AI and trying to develop comprehensive regulations. Lacking clarity has a potential to make it difficult for model creators and users to understand their legal obligations whilst some may exploit this. Without a doubt, AI models rely on vast amounts of data. Some regulations like the General Data Protection Regulation (GDPR) whose main jurisdiction is Europe and the California Consumer Privacy Act (CCPA) do help with restricting how this data may be collected, stored and used. Heavy fines are slapped on those who are not compliant to these regulations. Table 1 summarizes the general threats associated with LLMs and LMMs.

2.2 Security and Privacy of LLM and LMM

2.2.1 Education Sector

Kumar et al. [23] revealed the major threats to LLMs relevant for the education setting. Some of the threats fall into three of the classifications outlined in Fig. 1. Which are relevant to the education domain. The authors identify two threats to an AI model. Firstly, Prompt Injection, a vulnerability that affects the inputs to LLMs which could very well result in undesirable outputs. In the education setting, this could mean altering learning content or bringing about biases which could be very catastrophic. Another threat to the AI discussed is Jailbreaking which refers to by passing some restrictions on LLMs leading to unauthorized access to a model's operations. In the education domain, this would entail compromising the integrity of educational content and eventually leading improper learning. In another classification, threat from AI models, Personal Identifiable Information Leaks and uncovered which could lead to breach of privacy and disruption of the learning space. The last threat category, Threats from Using AI Models, includes Data Poisoning which would cause biased education tools possibly spreading discrimination.

Table 1. General Threats Associated with LLMs and LMMs

Threat Classification	Specific Threats
Threats from NOT Using AI Models	Loss of competitive advantage, inefficiencies, lack of fraud detection
Threats from Using AI Models	Data flaws leading to biases, security breaches, privacy compromises
Threats to AI Models	Adversarial attacks, data poisoning, biased data leading to discrimination
Threats from AI Models	Job loss, privacy concerns, sophisticated cyberattacks, disinformation
AI Legal & Regulatory Threats	Lack of clear regulation, non-compliance with data protection laws

Another classification, Threats from AI models, including Personal Identifiable Information (PII) Leaks and generating inappropriate content, can breach student privacy and disrupt the learning environment. Lastly, threats from using AI models, like Data Poisoning and Biased Outputs, involve using corrupted data that leads to biased educational tools, which can perpetuate stereotypes and discrimination.

A recent survey [31] modeled the OWASP top 10 vulnerabilities to LLMs for the educational landscape. These include Prompt injection, Insecure Output handling which entails inadequate validations lead to biases and wrong results. Others are Training Data Poisoning, Denial of Service attack, Supply Chain Vulnerabilities targeted at negatively impacting security and integrity of educations systems, Permission Issues and Data Leakages all seek to disrupt the normal running of LLM based education tools. Furthermore, the authors discuss Overreliance on LLMs in the absence of a human and use of Insecure Plugins which all create risks that need to be protected against.

The authors went to also elaborate on attacks on LLMs. The main type documented is Backdoor attacks. The attacks according to the writers, operate in a disguised manner aimed at compromising the models without really impeding on normal operations. Triggers could be maliciously inserted and when activated lead to unwanted breaches. Several attacks in this category were identified including those that are triggered by specific input, or maybe prompt and instructions. In the educational setting, they pose a serious risk in that they can lead to misclassification of content. Backdoor strategies target different stages of the AI models lifecycle. These could be pre-training or fine tuning. All these threats signal a need for good security measures to prevent such weakness from being exploited by threat actors.

With the integration of LMMs, the educational setting becomes even more dynamic. LMMs go beyond text-based interactions by including an assortment of media formats such as video, images, and audio into the model training. This enhances content engagement in more ways than previously imagined through high interaction. For example, a history lesson can now include not only text but also images and video snippets which help learners understand more complex topics. Despite the benefits derived from LMMs, the security and privacy threat map also broadens. Prompt injection threats, already a concern for LLMs can be more detrimental in LMMs. For instance, a malicious input could manipulate multiple forms of media leading to possibly inappropriate or biased education material being presented to students unbeknown to the teachers. Another potential pitfall for LMMs is overreliance which could arise from automated grading, content generation, and student feedback. In as much as these models could perform these tasks to a satisfactory level, some undesirable outputs could also be produced. An example of this occurring would be the LMM misinterpreting content or introducing errors when summarizing education multimedia [31]. Table 2 summarizes the threats to the education sector with regard to LLMs and LMMs.

Table 2. LLMs and LMMs Threats in Education

Threat Type	Specific Threats
Prompt Injection	Altering learning content or introducing biases
Jailbreaking	Unauthorized access to model's operations
Personal Identifiable Information Leaks	Breach of privacy and disruption of learning environment
Data Poisoning and Biased Outputs	Using corrupted data leading to biased educational tools

2.2.2 Health Sector

With the introduction of LMMs, healthcare as we know it is rapidly evolving. This sector has gained the ability to process multiple forms of data at the same time including, text, audio, and visual data. By processing patient records innovatively through analysis of medical images such as CT scans and MRIs, support for diagnosis is now advanced. For example, LMMs can generate visual question answers based on input from medical imagery and thus helping health care professionals to better understand complex visual data [32].

However, LMMs bring about new security and privacy concerns. Among them is visual data poisoning, where malicious actors inject harmful data into the training sets for image models leading to incorrect diagnoses or flawed outputs. For the health sector, the consequences are catastrophic as lives may be in jeopardy. Furthermore, multimodal data leaks heighten the risk further due to multiple ways of storing and managing sensitive data. Moreover, LMMs are susceptible to model inversion attacks, where adversaries could reconstruct training data from models, potentially revealing sensitive patient details such as medical history or diagnostic information [32].

Several security and privacy concerns are highlighted in a survey on LLM security in the Health sector. The concerns iterate issues also raised by another research [25]. These are Prompt Injection, leakage of sensitive health data, poisoning of search index used in health all leading to incorrect medication information. The other threats are Data Exfiltration and exploitation of personal data through means common to attackers. The privacy and data of patients may be at serious risk. To make all it worse, the vulnerabilities are stretched by unauthorized access where leading to breaches in data for patients and doctors. Some issues to do with misuse may arise due to limited reasoning capabilities of models and thus outputs could be inadequate for the medical field. Additionally, when legal and intellectual properties are not adhered to, together with porous protection for personally Identifiable information, there is a greater need to make sure security measures are kept strictly to ensure safe utilization of LLMs in healthcare.

Table 3. LLMs and LMMs Threats in Health

Threat Type	Specific Threats
Unauthorized Access and Potential Misuse	Compromise of patient privacy, manipulation of models
Data Exfiltration and Exploitation	Leakage of sensitive health data
Legal and Intellectual Property Non-Adherence	Issues with protection of personally identifiable information

Karabacak and Margetis [22], outlined some major threat to LLMs in the health sector. Unauthorized Access and Potential Misuse of LLMs. The author highlighted that if protection against these threats was not adequate, sensitive health data and manipulation of models could occur. Another threat is unauthorized uses performing actions that could compromise the privacy of patients. As In the case with many other publications, data leakages were also mentioned as a risk that needs mitigation as it can lead to revealing of data like patient names and other PII. The paper summarized a lot of sophisticated attack methods with the likes of prompt injection and model inversion attacks topping the list. Prompt injection involve having malicious instructions in the prompt of a model with the goal of extracting or altering information. Model inversion attacks on the other hand target the reconstruction of data used in training the model and hence compromising the confidentiality of patients. Table 3 summarizes some of the security and privacy concerns in the health domain.

2.2.3 Finance Sector

The finance sector is rapidly adopting LLMs to enhance customer service, fraud detection, and to some extent, trading strategies. LLMs are very useful when it comes to analyzing a lot of financial data, providing real-time insights on market trends, and automating relevant reports. In this sector, key concerns include data leakage, unauthorized access, and adversarial attacks on LLMs have the potential to compromise sensitive financial information leading to significant economic losses. With the emergence of LMMs, the financial landscape is set to evolve even further. LMMs extend the benefits of LMMs by incorporating other forms of media beyond text. The multimodal functionality could be useful to analyze most types of media which are not limited to video, audio, and images.

However, challenges such as data privacy, model inversion attacks, and adversarial manipulation remain significant risks [27]. These models are prone to multimodal data poisoning. This may happen, for instance, when attackers inject falsified visual or numeric data to manipulate market predictions. Furthermore, if a malicious actor gains access to visual or audio streams handled by an LMM, they could exploit confidential information in private meetings, leading to privacy breaches.

Li et al. [24] acknowledge the significant progress made by applying LLMs to financial applications. However, a few limitations were cited. The first is model biases, and the second is propagation of disinformation. The authors specifically highlighted biases such as region, gender and a race. Concerns have a major implication for the financial industry in that institutions depend on accuracy and fairness to make financial decisions. When inadequate, discrimination can be the order of the day in lending, insurance and even investments. Model training needs to be safeguarded. Further they discussed regulatory and governance challenges indicating that the black box nature of models poses a risk. Decision making remains much in the dark for users and even researchers making it difficult to know whether the models are compliant to set out regulations.

Jeong [20] explains how the deployment of LLMs comes with it valid security challenges that need to be addressed in order to protect financial information and preserve compliance. The key issues that were raised include the need to prevent unauthorized access and data breach by using encryption and appropriate access controls. In addition to that, the security of person data is also very important so that customer trust is maintained. Furthermore, the author speaks to the need to comply with regulations and laws and to remain true to financial transaction integrity. Addressing these concerns was cited as essential to enable safe integration of LLMs in the financial sector.

Although there are great potential benefits to the integration of LLMs and LMMs to financial applications, it comes with serious risks to privacy and security. The model brings in efficiencies and are able to process a lot of data in no time. However, this data must be protected according with the security triad in mind, that is, maintaining confidentiality, integrity and availability of data. Table 4 summarizes the security and privacy concerns of LMMs and LLMs in the finance domain.

Table 4. LLMs and LMMs Threats in Finance

Threat Type	Specific Threats
Unauthorized Access and Data Breach	Need for encryption and good access controls
Biased Decision-Making	Discrimination in lending, insurance and investments
Regulatory and Governance Challenges	Compliance issues due to the hidden operation nature of models

2.2.4 Cross Sector Threat Analysis

As LLMs and LMMs continue to be integrated into key sectors of society, the importance of protecting these systems from malicious actors can no longer be treated as a mundane task. According to an article by the National Institute of Standards and Technology (NIST), organizational assessment and prioritization of these emerging threats are crucial based on the Confidentiality, Integrity, and Availability (CIA triad) of their data and services [28]. In line with this guidance, this paper performs a cross-sector analysis of the security and privacy threats that LLMs and LMMs pose.

Across all sectors, unauthorized access, data poisoning, and adversarial attacks are prominent issues. While the specifics of data (educational materials, medical records, financial information) differ, the fundamental challenges of privacy breaches and biased outcomes remain consistent.

3 Mitigation for Security and Privacy Concerns

Our observation is that most mitigation strategies focus on securing and enhancing the AI models themselves. General AI solutions, such as robust training techniques and model protection mechanisms, serve as the primary mitigation approaches across various sectors. These solutions aim to improve model resilience, ensure privacy, and safeguard against adversarial threats, thus addressing security concerns at the core of AI systems.

3.1 Education Sector

There are several mitigation techniques that can help improve security and privacy of these models in the education settings. Rehman et al. [31] propose different measure to address concerns in this particular domain. One of the major areas of concern for the LLMs applicable to LMMs security and privacy is prompt injections. To overcome this concern, the authors proposed utilization of proper validation technique to avoid malicious prompts from affecting the model's output. Additionally, there is need to ensure that LLMs and LMMs outputs are sanitized well to prevent any breach or propagating of undesirable content.

Another grave challenge is data poisoning which can be mitigated by intensive vetting and cleaning or training data so that so maintain data integrity. This would make sure that data generated for educational purposes is not biased or inaccurate. Further, adversarial attack also needs to be handled. Here the proposal was to provide for consistent retraining the model and assessing the models risk from this type of attacks. Furthermore, there is need to have proper access controls and rights to the management of these systems.

By emphasizing on having only approved users to systems, breaches can be minimized if not avoided. Consistent monitoring and possibly real-time detection mechanisms should be embedded. Good protocols of security must be in place to cover for existing and emerging threats. Lastly the paper calls for complete approach to security that looks into the cycle of LLMs and LMMs from development to operation. This should include oversight and security assessments to ensure AI tools are being used ethically in the educational setting.

3.2 Health Sector

Due to the sensitive nature of health-related matters, LLMs may face difficult challenges relating to privacy and security. Previous research [1,32] proposed that mitigation for LLM and LMM security and privacy in the health sector include the following: Firstly, federated learning, this approach is lauded for its ability to handle risks relating to the collection and storing of private training data. By taking advantage of decentralized training, sensitive data is void of risks. Secondly, differential privacy, where training datasets are prevented from leaks by adding noise to the data used to train.

Researchers [1,37] also proposed using differential privacy and adding some noise to data to minimize the accuracy of data extracted from the model. Further, the use of encryption to secure techniques used for computations is recommended to protect sensitive data throughout the lifecycle of the LMMs and LLMs used in healthcare. Another remedy is to continuously monitor and look out for anomalies. This process has to be in real-time as anomaly detection can figure out unusual patterns which be signaling a breach or an attack. Another mitigation proposed by the authors is relying on the human oversight and ethical reviews which would ensure checks and balances to so that no harm is inflicted on patients. These mitigation strategies are essential for securing the these models and seeing to it that privacy is protected.

3.3 Finance Sector

In this section, we seek to provide an in-depth analysis of the mitigation strategies proposed LLMs for the financial sector.

Li et al. [24] provide some mitigation approaches. Firstly, they emphasize the importance of security and compliance in financial applications of LLMs. They also suggest that finance institutions may resort to use of opensource so as to maintain control of the model and guarantee internal use only hence ruling out third parties. This approach would enable implementation of organizational security protocols which would adhere to regulations their businesses most pride in. Further, the also touch on the importance of financial regulations which govern how data should be handles and protected in the context of the financial industry. The stress that all organization need to have a clear picture of existing financial regulations to ensure that there is no accidental or unintended exposure of sensitive data or violation of privacy laws.

Sawhney et al. [34] proposed regular retraining of LMMs with updated data to ensure robustness and model adaptation to changing financial positions like market conditions. Jeong [20] focuses on several ways to fine tuning and using domain specific LLMs in finance. The study outlines key approaches to mitigating against security and privacy risks. Firstly, Data Security: the proposal is to implement safe data security measures which include encryption, access controls and backups.

These measures should help in ensuring financial data used in both training and operation is free of unauthorized access. Another safety measure is Personal Data Protection which entails the strict adherence to regulations that pertain to personal data. The solution proposed here is to ensure that there is compliance with laws that concern the collection, storage and usage of personal financial information such as GDPR. Another emphasis is the proper handling of data with high confidentiality and integrity. The goal here is to ensure no leakage of sensitive information to ensure privacy and trust.

By following the mitigations proposed, financial institutions would be better placed at deploying LLMs in finance safely.

4 Limitations

Scope of the study: Our study though informative does not cover all other sectors like automotive and entertainments to mention. It is somewhat impractical to cover all domains, but we look to add more in our future work and we provide some general threats which cover most if not all sectors.

Threat evolution: As much as we cover the prevalent challenges in this domain, it is always changing. Some mitigation strategies may not always be effective for the more complicated challenges that would emerge in the future.

5 Related Work

There have been many studies on software security and quality assurance that involve the investigation of bug patterns [16,17,33], vulnerabilities [8,18], code smells [13,41,42], code quality [2,13], human aspects [3,5,9,10,14,15] of software development and maintenance as well as comparison of methods/tools [11,12,21, 26] for measuring such aspects. The recent increase in the integration of LLMs and LMMs in sectors like education, healthcare, and finance presents security and privacy concerns that need urgent attention. These models which not only include text, but also other media forms have proven useful but are vulnerable to AI-based attacks like data leaks, bias, and adversarial attacks. Many studies have extensively covered these threats in various sectors independently.

Yao et al. [36] in their survey conducted a comprehensive review categorizing the impacts of LLMs into "The Good," "The Bad," and "The Ugly" by examining the beneficial applications, vulnerabilities, and potential abuses. Furthermore, they too proposed differential privacy and federated learning as ways to combat threats to models. These works provide a foundation upon which our study builds, focusing more specifically on sector-based risks, particularly in education, healthcare, and finance.

Rahman et al. [31] investigated the role of LMMs in the education setting discussing how these models contribute to learning by integrating multimedia content. They also identified several vulnerabilities not limited to prompt injection attacks and data leakage which pose a risk to the education process. They suggested differential privacy and ethical guardrails as mitigation strategies. While their work focuses primarily on the education sector our work expands on this by addressing the implications of both LMMs and LLMs across multiple sectors beyond education, namely, healthcare and finance.

In the healthcare domain, Rahman et al. [32] provide a comprehensive survey of LLMs and LMMs in the medical field. They emphasize the idea that while LMMs can process data in more formats, such as medical images and patient records, introduce risks including model inversion attacks and data exfiltration. The authors in their work called for the utilization of security strategies like federated learning and encryption to protect sensitive data for patients which is highly confidential. Our work builds upon their insights by providing a more detailed discussion on cross-sector analysis.

Nie, Y., et al. [27] explored the application of LLMs and LMMs in the finance sector, where these models have been utilized for tasks such as automated trading, fraud detection, and sentiment analysis. While LMMs can integrate visual and textual data to enhance financial forecasting, they also increase the risk of data poisoning and adversarial attacks, leading to incorrect financial predictions. The authors recommended continuous model monitoring and the use of secure computation to mitigate these risks. Unlike their work, which is narrowly focused on finance, our study encompasses a broader range of security and privacy concerns across finance, healthcare, and education.

Overall, these works provide a comprehensive view of the challenges and opportunities presented by LLMs and LMMs, laying a foundation for further research into securing these models across different industries. However, our work differs by offering a more holistic perspective, encompassing a comparative analysis across multiple sectors, and proposing industry-specific security and privacy frameworks.

6 Conclusion

Several security and privacy concerns related to LLMs and LMMs, with an emphasis on the impact on key sectors like healthcare, finance and education have been succinctly highlighted by this study. Our analysis clearly shows that despite the many benefits offered by models in these domains, there are significant risks that must be carefully considered and accounted for. Our paper clearly shows the need for proper mitigation strategies and regulations to ensure data is protected and privacy regulations are adhered to. A clear understanding of these details is key for stakeholders specifically developers, users, and regulators to each play an effective role to ensure security and privacy are maintained in this ever-evolving digital age.

Acknowledgement. This work is supported in part by the ISU-CAES (Center of Advanced Energy Studies) Seed Grant at the Idaho State University, USA.

References

1. Acosta, J.N., Falcone, G.J., Rajpurkar, P., et al.: Multimodal biomedical AI. Nat. Med. **28**(9), 1773–1784 (2022)
2. Alwad, D., Panta, M., Zibran, M.: An empirical study of the relationships between code readability and software complexity. In: 27th International Conference on Software Engineering and Data Engineering, pp. 122–127 (2018)
3. Champa, A., Rabbi, M., Eishita, F., Zibran, M.: Are we aware? An empirical study on the privacy and security awareness of smartphone sensors. In: 21st IEEE International Conference on Software Engineering, Management and Applications, pp. 139–158 (2023)
4. Champa, A., Rabbi, M., Nachuma, C., Zibran, M.: ChatGPT in action: analyzing its use in software development. In: 21st International Conference on Mining Software Repositories, pp. 182–186 (2024)
5. Champa, A., Rabbi, M., Zibran, M., Islam, M.: Insights into female contributions in open-source projects. In: 20th IEEE International Conference on Mining Software Repositories, pp. 357–361 (2023)
6. Chen, H., Magramo, K.: Finance worker pays out $25 million after video call with deepfake 'chief financial officer' (2024). https://www.cnn.com/2024/02/04/asia/deepfake-cfo-scam-hong-kong-intl-hnk/index.html
7. Hashem, I., Yaqoob, I., Anuar, N., Mokhtar, S., Gani, A., Khan, S.: The rise of "big data" on cloud computing: review and open research issues. Inf. Syst. **47**, 98–115 (2015)
8. Islam, M., Zibran, M.: A comparative study on vulnerabilities in categories of clones and non-cloned code. In: 10th IEEE International Workshop on Software Clones, pp. 8–14 (2016)
9. Islam, M., Zibran, M.: Exploration and exploitation of developers' sentimental variations in software engineering. Int. J. Softw. Innov. **4**(4), 35–55 (2016)
10. Islam, M., Zibran, M.: Towards understanding and exploiting developers' emotional variations in software engineering. In: SERA, pp. 185–192 (2016)
11. Islam, M., Zibran, M.: A comparison of dictionary building methods for sentiment analysis in software engineering text. In: ESEM, pp. 478–479 (2017)

12. Islam, M., Zibran, M.: A comparison of software engineering domain specific sentiment analysis tools. In: 25th IEEE International Conference on Software Analysis, Evolution and Reengineering, pp. 487–491 (2018)
13. Islam, M., Zibran, M.: On the characteristics of buggy code clones: a code quality perspective. In: 12th IEEE International Workshop on Software Clones, pp. 23–29 (2018)
14. Islam, M., Zibran, M.: Sentiment analysis of software bug related commit messages. In: SEDE (2018)
15. Islam, M., Zibran, M.: Sentiment analysis of software bug related commit messages. In: 27th International Conference on Software Engineering and Data Engineering, pp. 3–8 (2018)
16. Islam, M., Zibran, M.: How bugs are fixed: exposing bug-fix patterns with edits and nesting levels. In: 35th ACM/SIGAPP Symposium on Applied Computing, pp. 1523–1531 (2020)
17. Islam, M., Zibran, M.: What changes in where? An empirical study of bug-fixing change patterns. ACM Appl. Comput. Rev. **20**(4), 18–34 (2021)
18. Islam, M., Zibran, M., Nagpal, A.: Security vulnerabilities in categories of clones and non-cloned code: An empirical study. In: 11th ACM/IEEE International Symposium on Empirical Software Engineering and Measurement, pp. 20–29 (2017)
19. Jegorova, M., et al.: Survey: leakage and privacy at inference time. IEEE Trans. Pattern Anal. Mach. Intell. **45**(7), 9090–9108 (2023)
20. Jeong, C.: Fine-tuning and utilization methods of domain-specific LLMs (2024)
21. Joseph, R., Zibran, M., Eishita., F.: Choosing the weapon: a comparative study of security analyzers for android applications. In: International Conference on Software Engineering, Management and Applications, pp. 51–57 (2021)
22. Karabacak, M., Margetis, K.: Embracing large language models for medical applications: opportunities and challenges. Cureus **15**(5), e39305 (2023)
23. Kumar, A., Singh, S., Murty, S.V., Ragupathy, S.: The ethics of interaction: mitigating security threats in LLMs (2024)
24. Li, Y., Wang, S., Ding, H., Chen, H.: Large language models in finance: a survey (2023)
25. Mikhailov, D.: Optimizing national security strategies through LLM-driven artificial intelligence integration (2023)
26. Murphy, D., Zibran, M., Eishita, F.: Plugins to detect vulnerable plugins: an empirical assessment of the security scanner plugins for wordpress. In: International Conference on Software Engineering, Management and Applications, pp. 39–44 (2021)
27. Nie, Y., et al.: A survey of large language models for financial applications: progress, prospects and challenges (2024)
28. NIST NCCoE: Securing telehealth remote patient monitoring ecosystem (2020). https://www.nccoe.nist.gov/publication/1800-26/VolA/index.html. Accessed 09 Sept 2024
29. Novelli, C., Casolari, F., Hacker, P., Spedicato, G., Floridi, L.: Generative AI in EU law: liability, privacy, intellectual property, and cybersecurity (2024)
30. OWASP Foundation: Owasp top 10 for large language model applications (2024). https://owasp.org/www-project-top-10-for-large-language-model-applications/
31. Rahman, M., Alqahtani, L., Albooq, A., Ainousah, A.: A survey on security and privacy of large multimodal deep learning models: teaching and learning perspective. In: 21st Learning and Technology Conference, pp. 13–18 (2024)
32. Rahman, M.A.: A survey on security and privacy of multimodal LLMs - connected healthcare perspective. In: 2024 21st Learning and Technology Conference (2023)

33. Rajbhandari, A., Zibran, M., Eishita, F.: Security versus performance bugs: how bugs are handled in the chromium project. In: International Conference on Software Engineering, Management and Applications, pp. 70–76 (2022)
34. Sawhney, R., Mathur, P., Mangal, A., Khanna, P., Shah, R., Zimmermann, R.: Multimodal multi-task financial risk forecasting. In: 28th ACM International Conference on Multimedia, pp. 456–465 (2020)
35. Xiao, C., Xu, X., Zhang, K., Wang, Y., Xia, L.: Evaluating reading comprehension exercises generated by LLMs: A showcase of ChatGPT in education applications. In: Kochmar, E., et al. (eds.) 18th Workshop on Innovative Use of NLP for Building Educational Applications, pp. 610–625. Association for Computational Linguistics (2023)
36. Yao, Y., Duan, J., Xu, K., Cai, Y., Sun, Z., Zhang, Y.: A survey on large language model (LLM) security and privacy: the good, the bad, and the ugly. High-Confidence Comput. 100211 (2024)
37. You, Z., Ye, J., Li, K., Xu, Z., Wang, P.: Adversarial noise layer: Regularize neural network by adding noise. In: IEEE International Conference on Image Processing, pp. 909–913 (2019)
38. Yu, W., et al.: Mm-vet: evaluating large multimodal models for integrated capabilities (2023)
39. Zhao, H., et al.: Revolutionizing finance with LLMs: an overview of applications and insights. arXiv preprint arXiv:2401.11641 (2024)
40. Zhao, W.X., et al.: A survey of large language models (2023)
41. Zibran, M., Roy, C.: Conflict-aware optimal scheduling of code clone refactoring. IET Softw. **7**(3), 167–186 (2013)
42. Zibran, M., Saha, R., Roy, C., Schneider, K.: Evaluating the conventional wisdom in clone removal: a genealogy-based empirical study. In: 28th ACM/SIGAPP Symposium on Applied Computing, pp. 1123–1130 (2013)

Detection of Malicious URL's Using Machine Learning

Deepthi Malyavantham Harinath[1], Anuhya Javvaji[1], and Annamaria Szakonyi[2](✉)

[1] Information Systems, Saint Louis University, Saint Louis, MO 63103, USA
{deepthi.malyavanthamharinath,anuhya.javvaji}@slu.edu
[2] Information Systems and Cybersecurity, Saint Louis University, Saint Louis, MO 63103, USA
annamaria.szakonyi@slu.edu

Abstract. Web application security is increasingly important, particularly given the rise of phishing and spam attacks. URLs, while used to access legitimate resources, can also be exploited by attackers to compromise users accessing those sites. This project aims to safeguard users by detecting malicious websites, thereby reducing the risk of cyberattacks. Our machine learning algorithm helps prevent users from clicking malicious links by identifying potentially harmful URLs and cautioning them about the associated risks, such as threats or sensitive information leaks. Using various classification algorithms, including decision trees, random forests, logistic regression, SVM, and KNN, we discovered that the Extreme Boosting technique performed the best. The web application built using our algorithm can be beneficial in various settings to address concerns related to phishing and online security by allowing users to quickly verify URLs before interacting with potentially harmful websites. This study is essential for advancing AI-related cybersecurity research and offering practical applications of these concepts to this rapidly growing field.

Keywords: Cybersecurity · Classification · Malicious URL · Phishing · Machine Learning · Extreme Boosting algorithm · Confusion Matrix · Web Application · URL feature extraction · Data Breaches

1 Introduction

The Internet has become essential for everyday life, but it also presents opportunities for hackers to exploit URLs for malicious purposes, such as malware and phishing attacks (Angadi & Shukla, 2022; Cui et al., 2018; Experteq, 2024; Rupa et al., 2021). These threats can lead to stolen sensitive information, financial losses, and data breaches (Acohido, 2013). Cybercriminals use malicious URLs to deceive users into downloading malware or sharing sensitive information, with serious consequences for individuals and organizations (Aljabri et al.,

S. Latifi (Ed.): ITNG 2025, AISC 1463, pp. 40–45, 2025.
https://doi.org/10.1007/978-3-031-89063-5_4

2022; Manyumwa et al., 2020). Attackers continue to adapt their tactics, necessitating ongoing research and innovation (KoyameMarsh & Marsh, 2014).

Cybersecurity efforts to detect and prevent malicious URLs have evolved, incorporating machine learning and AI techniques (Alazab et al., 2013; Cui et al., 2018; Feroz & Mengel 2015; Ghaleb et al., 2022; Kumar et al., 2017; Popescu et al., 2015; Rathod & Pattewar, 2015). Despite these advancements, several gaps remain.

Though machine learning has improved traditional blacklist detection, the dynamic nature of cyber threats demands adaptive methods beyond static feature extraction for effective real-time attack detection. Moreover, despite high accuracy, many learning techniques face challenges in computational complexity and real-world performance, highlighting the need for scalable, low-latency systems suitable for real-time deployment.

To bridge these limitations, our study proposes using Extreme Boosting machine learning technique to improve detection rates and reduce false positives. We predict that this approach will deliver better results than other models used in previous 2 literature. Our URL detection application helps prevent users from clicking malicious links by identifying potentially harmful URLs and cautioning them about the associated risks, such as threats or sensitive information leaks. Due to the broad impact and importance of Cyber Security, this tool can be helpful for various groups, including individual Internet users, businesses, government agencies and developers alike.

2 Algorithm and System Design

We built this project using various machine learning classification algorithms known for strong performance. Our classifiers included Logistic Regression, KNN with cosine metric, SVM, Decision Tree, Random Forest, and XGBoost. We compared the performance of each of these models to identify the best performing one. We used Anaconda, Python 3, and essential libraries like NumPy, Pandas, Matplotlib, Sk-learn, and Pickle.

Additionally, Flask was used as the framework to build and run the web application. The dataset analyzed comprises three distinct types of URLs, with the distribution as follows: 66% of the URLs are classified as benign (i.e. safe, not malicious), 15% fall under the defacement category (i.e. websites that have been altered or compromised), while the remaining 19% are classified as phishing and malware (Ctanush, 2024).

Figure 1 shows the use case diagram to capture requirements of the system. The user interacts with the User Interface (UI). The process starts by collecting the dataset and extracting the features from the URLs in the dataset. The dataset is divided into testing and training data. Different machine learning algorithms are used to train the model. After training, the user can enter the URLs as input to detect the type of URLs.

Figure 2 shows the sequence of detection, including User, Feature Extractor, Data Management Module, Machine Learning Classifier and User Interface.

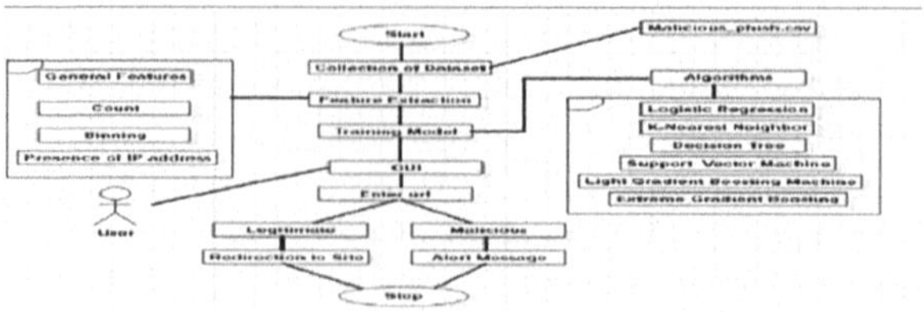

Fig. 1. The use case diagram for detection of malicious URL

When the user enters a URL, the system extracts the features of the URL and compares the results with the trained model. If the URL is legitimate, the system redirects to the entered site. If the URL is malicious, an alert message is popped in the interface.

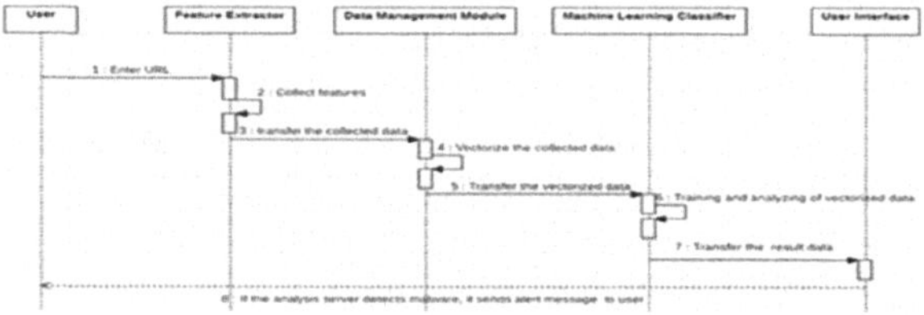

Fig. 2. Sequence diagram for detection of malicious URL.

3 Results

Figure 3 displays the confusion matrix and the performance scores for the XGBoost classification algorithm across the four classes. The matrix illustrates the accuracy of predictions in each category by showing true positives along the diagonal and misclassifications off the diagonal. The model performs well in most categories, correctly classifying 74,013 benign URLs, 79,629 defacement URLs, and 71,359 malware URLs. However, it faces challenges in accurately distinguishing phishing URLs, resulting in incorrect classifications of this category as benign or defacement. This could indicate overlapping features between phishing and other categories, pointing to potential areas for improvement in feature engineering or model tuning.

The classification report includes precision (i.e. how many predicted positives were correct), recall (i.e. out of all malicious URLs, how many were predicted malicious), and F1-score (i.e. overall performance measure balancing precision and recall) for each category. The model demonstrates strong performance with high precision/recall scores for benign (0.86/0.87), defacement (0.82/0.93), and malware categories (0.88/0.83). However, the phishing category shows lower recall at 0.76, meaning the model misses a portion of true phishing cases, leading to false negatives. The model's overall accuracy is 0.85, indicating good performance, though there is room for improvement in phishing classification to enhance recall and reduce false negatives.

XGBoost is the top-performing model with a test score of 0.967, followed by Decision Tree and K-Nearest Neighbors (KNN). Support Vector Machine (SVM) and Random Forest also perform well, though not as effectively as XGBoost. In contrast, Naive Bayes has the lowest test score of 0.591, suggesting that it is not well-suited for this classification task, likely due to its assumption of feature independence, which is often unrealistic in complex data. XGBoost is the most effective model for this classification problem, likely due to its ability to handle feature interactions and complex patterns better than other algorithms.

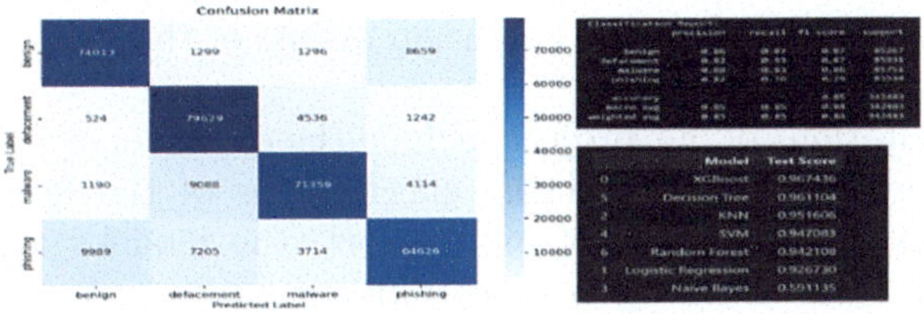

Fig. 3. The confusion matrix and classification report of the malicious URL detection using XGBoost algorithm and the comparison of different machine learning models and test scores

In summary, XGBoost is the best-performing model for this classification problem. However, the confusion matrix and classification report reveal challenges in distinguishing phishing URLs, with higher misclassifications in this category. To enhance phishing detection, improvements in feature engineering, additional hyperparameter tuning, or ensemble techniques could be explored, potentially helping the model better differentiate phishing URLs without sacrificing the performance of other categories.

Fig. 4. Screenshots of the malicious URL detection application.

Figure 4 contains screen shots of the application to detect malicious URLs using machine learning. Navigating to the home page of the application, users can click on the "let's go" button. The user will be redirected to the page where they can enter the URL in the input field. If the entered URL is benign, a new tab of the URL opens with the sign showing it is safe to use. When the entered URL is malicious, a warning message shows up and a new tab does not open.

4 Conclusions, Limitations

In this project, we used the innovative machine learning technique of extreme boosting algorithm to categorize URLs as benign or harmful. After comparing a variety of classification methods, including decision trees, random forests, logistic regression, SVM, and KNN, we have discovered that the XGBoost technique performs the best. Our application can successfully predict malicious URLs that can protect users from online threats. To improve program functionality, we would also like to build a Chrome extension as one of the significant directions for future work.

Potential limitations of our project include the dependency on accurate data input for training the model and the effectiveness of the feature extraction method. Given the early phase of our project, the model's accuracy on live data might decrease. Collecting more real-life examples and thorough application testing would be important future research directions to improve our model. Additionally, since our model is built on a Flask application, other architecture methods should be explored to compare scalability when deployed to handle large numbers of concurrent predictions, especially if it relies on a single server or lacks optimization for performance.

References

Angadi, S., Shukla, S.: Malicious URL detection using machine learning. In: Raj, J.S., Shi, Y., Pelusi, D., Balas, V.E. (eds.) Intelligent Sustainable Systems, Proceedings of ICISS 2022, pp. 657–669. Springer, Singapore (2022). https://doi.org/10.1007/978-981-19-2894-9_50

Cui, B., He, S., Yao, X., Shi, P.: Malicious URL detection with feature extraction based on machine learning. Int. J. High Perform. Comput. Network. **12**(2) (2018). https://doi.org/10.1504/IJHPCN.2018.094367

Experteq. What is a malicious URL and how do we protect against them? (2024). https://experteq.com/what-is-a-malicious-url-and-how-do-we-protect-against-them/

Rupa, C., Srivastava, G., Bhattacharya, S., Reddy, P., Gadekallu, T.R.: A machine learning driven threat intelligence system for malicious URL detection. In: Proceedings of the 16th International Conference on Availability, Reliability and Security, New York, NY, United States, pp. 1–7 (2021). https://doi.org/10.1145/3465481.3470029

Acohido, B.: Analysis: Why LivingSocial disclosed huge data theft. USA Today (2013). https://www.usatoday.com/story/tech/2013/04/30/livingsocial-50-million-accountsbreached-cybersecurity-privacy-invasion/2124345/

Aljabri, M., et al.: Detecting malicious URLs using machine learning techniques: review and research directions. IEEE Access **10**, 121395–121417 (2022). https://doi.org/10.1109/ACCESS.2022.3222307

Manyumwa, T., Chapita, P.F., Wu, H., Ji, S.: Towards fighting cybercrime: malicious URL attack type detection using multiclass classification. In: 2020 IEEE International Conference on Big Data, Atlanta, GA, USA, pp. 1813–1822 (2020). https://doi.org/10.1109/BigData50022.2020.9378029

Koyame-Marsh, R.O., Marsh, J.L.: Data breaches and identity theft: costs and responses. IOSR J. Econ. Finan. **5**(6), 36–45 (2014). https://www.iosrjournals.org/iosr-jef/papers/Vol5-Issue6/Version-3/F05633645.pdf

Alazab, M., Layton, R., Broadhurst, R., Bouhours, B.: Malicious spam emails developments and authorship attribution. In: 2013 Fourth Cybercrime and Trustworthy Computing Workshop, Sydney, NSW, Australia, pp. 58–68 (2013). https://doi.org/10.1109/CTC.2013.16

Feroz, M.N., Mengel, S.: Phishing URL detection using URL ranking. In: 2015 IEEE International Congress on Big Data, New York, NY, USA, pp. 635–638 (2015). https://doi.org/10.1109/BigDataCongress.2015.97

Ghaleb, F.A., Alsaedi, M., Saeed, F., Ahmad, J., Alasli, M.: Cyber threat intelligence-based malicious URL detection model using ensemble learning. Sensors **22**(9), 3373–3392 (2022). https://doi.org/10.3390/s22093373

Kumar, R., Zhang, X., Tariq, H.A., Khan, R.U.: Malicious URL detection using multi-layer filtering model. In: 2017 14th International Computer Conference on Wavelet Active Media Technology and Information Processing (ICCWAMTIP), Chengdu, China, pp. 97–100 (2017). https://doi.org/10.1109/ICCWAMTIP.2017.8301457

Rathod, S.B., Pattewar, T.M.: A comparative performance evaluation of contentbased spam and malicious URL detection in e-mail. In: 2015 IEEE International Conference on Computer Graphics, Vision and Information Security (CGVIS), Bhubaneswar, India, pp. 49–54 (2015). https://doi.org/10.1109/CGVIS.2015.7449891

Ctanush. Malicious-phish (Version 1) [Dataset]. Kaggle (2024). https://www.kaggle.com/datasets/ctanush/malicious-phish

Zero-Day Malware Detection Using Fine-Tuned MobileNetV2 and Open-Set Classification

Abdullah Sheneamer(✉)

Computer Science Department, Jazan University, Jazan 45142, Saudi Arabia
asheneamer@jazanu.edu.sa

Abstract. A major threat to cybersecurity is the early detection of zero day malware images which remain almost invisible due to their unique nature. We propose a new way that uses a MobileNetV2 architecture combined with open-set classification to tackle this problem. MobileNetV2 model takes malware images as input and learns complex visual features that separate malicious from benign samples even for zero-day malware. Open-set classification improves this power such that it can differentiate between instances not belonging to the trained classes, allowing the model to generalize well on novel unseen malware families. Our work emphasizes relevance of open-set classification in zero-day malware detection that can lead to better secure and proactive mechanism against cyber threats. State-of-the-art testing on benchmark datasets such as Malimg and Malevis demonstrates the framework's superior performance in a number of classifications settings: binary-class, multi-class and open-set problems. MobileNetV2 not only excels on the benchmark datasets, but actually outperforms current state-of-the-art approaches. MobileNetV2 regularly scores 99.3% and 98.4% for binary and multi-class classifications. As the capability of the framework can also be used to detect unknown malware, we can observe that 92.3% accuracy in Malimg or 94.8% Malevis.

Keywords: Fine-tuned MobileNetV2 model · Deep learning · Zero-day malware · Malware analysis · Malware detection · Malware features · Open-set classification

1 Introduction

In our digital world, where connectivity and interdependence is at an all-time high, malware detection and analysis have a great impact on cybersecurity [12,13]. Malware is a broad category that includes viruses, ransomware, trojans, and worms can cause endless troubles for individuals, organizations and even national infrastructures. Such harmful software may paralyze computers, steal social security numbers, open a back-door for further access to sensitive information and spread to other systems throughout the world. With the increasing sophistication and complexity of malware, classical detection mechanisms,

S. Latifi (Ed.): ITNG 2025, AISC 1463, pp. 46–58, 2025.
https://doi.org/10.1007/978-3-031-89063-5_5

signature-based techniques for instance, have become inadequate in sense that they neither detect nor handle new threats. As a result, it has become imperative to create more proven methodologies that are able to adjust and respond against evolving threats.

Built upon a mobileNetV2 model, this work introduces open-set classification to tackle one of the biggest problems of malware detection: recognizing unseen instances from its training set. Conventional closed-set systems operate with the assumption that every possible class of malware is known during training, thus limiting efficacy against zero-day threats. The limitation of this method is that these problems cannot work off-distribution, where the open-set classification allows to differentiate the out-of-distribution samples and consequently increases the applicability and robustness of a framework in real-world situations.

The contributions of this research are:

- Malware's Detection Framework via Model Re-Tuned MobileNetV2: By using MobileNetV2's feature extraction ability, the proposed framework employs preprocessed visualized images for malware that contain many complex patterns similar to those of natural scenes in order to achieve better results.
- Open-set classification: This provides a new look at malware detection procedures. We present the model, which is able to recognize zero-day malware and out-of-distribution samples instead of following tight classes established by pre-processing of data that doesn't fit into neatly defined patterns.
- Extensive empirical evaluation on benchmark datasets: With proof coming from the real world, this document puts forth a framework which over time may enable researchers in this area of study to change their perspective on any phenomena they encounter along the way. In all our experiments we used commercially available data such as MaleVis and MalImg.

This paper opens up new ground for revolution the future of malware detection by a fusion of MobileNetV2 deep learning model and an open-set classification problem. This study fills the gap and provides a solution to this problem which can be widely implemented, future-proof for years. It will bring cybersecurity professionals around the world as well as researchers interested in this field benefits from an advancement from means of securing digital systems into ways that more effectively protect these from threats going beyond traditional approaches.

The paper is arranged as follows. In Sect. 2, associated publications in the field are addressed. Our method assesses and contrasts several deep learning architectures in Sect. 3 and Sect. 4 examines the outcomes. Section 5 discusses the results of our experiments. Finally, in Sect. 6, we conclude our study.

2 Related Work

Meenakshi and Maragatham [9] developed a defensive technique using the Curvelet transform to enhance the recognition of adversarial iris images, optimizing image classification accuracy. Their method proved effective against various adversarial attacks on iris recognition systems. While their focus lies in enhancing the resilience of biometric systems, our approach addresses malware detection in cybersecurity by employing an open-set classification framework to identify both known and unknown malware threats.

Alomari et al. have used raw data features by deep leaning and feature selection [2]. They created a very high accuracy system for identifying malware. This addresses the problem of huge, high-dimensional malware data which ultimately ruins the existing mode of defense.

Through a combination of Bidirectional Long Short-Term Memory and Convolutional Neural Network (CNN- BiLSTM), Kumar [7] developed a deep learning mode for malware prevention. The study shed light on the challenges posed by the size, speed and sophistication of malware that eventually make effective defence through simple malware detection impossible. The deep learning method is not equally effective and trained with various datasets.

Atacak [3] proposed a logic-based, fuzzy, and dynamic ensemble (FL-BDE) system for Android malware. This model employs both the strength of machine learning techniques and the strategic decisiveness of a fuzzy, Mamdani-form inference system (FIS). The FL-BDE model included all six techniques (decision forest, Bayes point machine, logistic regression, neural network, support vector machine, and boosted decision tree). FL-BDE model leverages a voting and routing method to map outputs of the three leading ML-based methods for classifying favourable and unfavourable instances for combination to the FIS. To validate our tests, the tests indicates that the performance of FL-BDE model is higher than any of machine learning systems.

3 Methodology

In this paper, we leverage a MobileNetV2 model for malware image classification with an additional class called "*unknown*" to accommodate unseen classes of malware. The proposed approach which involves preprocessing, dataset augmentation, training and testing on a known–unknown based balanced malware classes dataset as shown in Fig. 2. The diversity of malware threats is growing, and non-generalization seen in the conventional classification models renders them incapable of detecting unseen or novel types of malware. To tackle this issue, we present a pretrained MobileNetV2 model with an open-set classification that is enhanced with an "unknown" class for malware samples.

3.1 Image Processing

It begins with translating raw malware artifacts into image formats. To allow the representation of malware binaries or executables in deep learning models, these are converted into grayscale images or color images [16]. Firstly, Prprocessing step to resize the image to certain height and weight (e.g. 256×256) and augmentating data with rotation, flipping & scaling etc. These methods increase dataset diversity and robustness and develop a model that is good at generalizing on unseen data.

3.2 Transfer Learning Using a MobileNetV2 Model

The MobileNetV2 architecture is used here for its ability to extract relevant features of the image. Transfer learning is employed to achieve this by initializing the model with pre-trained weights from ImageNet. RMFETs take the frozen initial layers to preserve the same generic feature representations and fine tune it on top of their malware and benign image dataset by tuning just the upper layers. To perform multi-class classification, a softmax layer is appended to the last hidden layer to predict not only malware family labels (e.g., Alophosd, Agent, Allaple) but also labels for benign files. In this step, the model learns how to classify these categories by observing visual patterns in the input images.

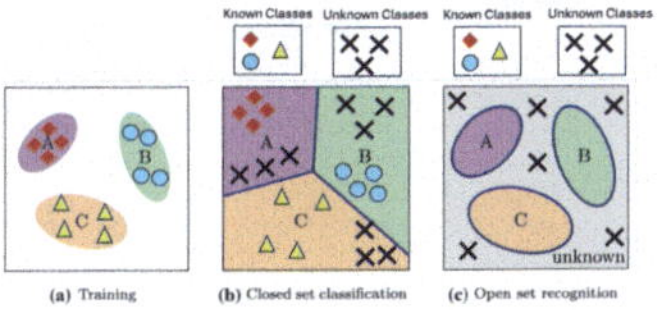

Fig. 1. Differences between Outlier and Open-set classification.

3.3 Open Set Classification

You can identify unknown malware types at this stage. Both the known and unknown image inputs are passed into the MobileNetV2 model for feature extraction with fine-tuned weights. If you have a softmax classifier that simply gives out probabilities for known classes, you need to put threshold (like 0.7) on probabilities for unknown samples. If the prediction is lower than this value, it will be treated as "*Unknown*". This method improves the framework's capability in recognition of known malware families as well as unknown ones.

Figure 1 provides an intuitive understanding of the differences between closed-set and open-set classifiers. For the closed-set classifier (Fig. 1b), the model is both trained and tested on known classes only, clicking the separation boundaries that separate regions associated with particular types of classes. It unknown data is presented to the models, they misclassify that data into one of the known classes, as no mechanism is in place to handle the unfamiliar data. On the other hand, Open-set classifier (Fig. 1c), includes an approach for recognizing an unknown class during testing. Only this way we can allow the classifier to detect the samples out of the class distribution and reject them, which is very relatable for real-world scenarios like malware detection, where we can expect samples to behave differently over time with a large distribution.

3.4 Classification Results

Our framework is capable of generating classification outputs for known and unknown samples. It decompiles known malware to certain families (i.e. Injector) and flags unknown samples for interaction. Furthermore, non-malicious files are categorized separately as harmless. This structured approach allows for proper classification of malware categories while still allowing current functionality to discover unknown, unidentified threats which bolsters an effective cybersecurity program.

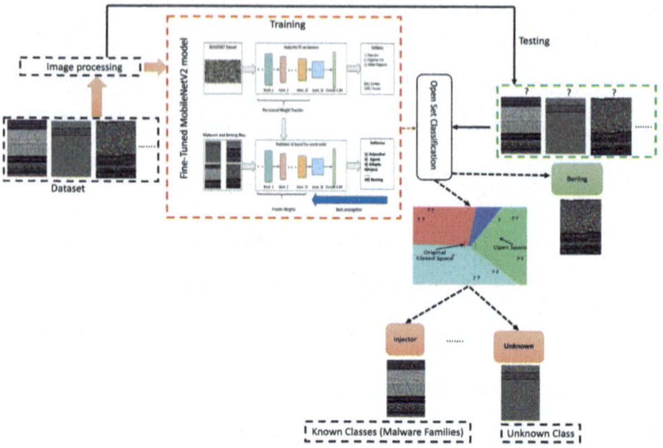

Fig. 2. Malware detection framework using pretrained MobileNetV2 and Open-set classification.

3.5 Visualisation and Malware Family Classification

In the last step, it will help you to gather high-level visualizations and performance measures like confusion matrices or time analysis of your model. To distinguish among closed space (known classes) and open space (unknown classes), it explains the mapping of unknown samples into the open set in an intuitive way. Our framework effectively classifies samples to malware families or as unknown, hence proving itself against real-world malware detection and classification challenges.

4 Evaluation

4.1 Datasets

In the field of malware identification, the Malevis and Malimg datasets are some of the most widely used resources for training machine learning models. Their data are composed of grey-scale images representing malicious and non-malicious software samples, which serve as the basis to develop and verify detection algorithms.

The distribution of malware families and groups across the two datasets are shown in Tables 1 and 2. In this dataset, 26 families of malware has been identified. There are a total of 9,901 grayscale images which measure 224 × 224 pixels.

There is a relatively balanced distribution of families. On the other hand, Allaple. A families occupies a disproportionately high position if we look at Table 1. This means that different types of malware such as Trojan horses and backdoors are guaranteed; at the same time though also presents some difficulties due to unbalanced situations in certain classes.

Table 1. Description of Malimg Dataset Families [10]

Class ID	Family Name	Malware Kind	Sample No.
#1	Adialer.C	Dialer	122
#2	Agent.FYI	Backdoor	116
#3	Allaple.A	Worm	2,949
#4	Allaple.L	Worm	1591
#5	Alueron.gen!J	Trojan	198
#6	Autorun.K	Worm	106
#7	C2LOP.P	Trojan	200
#8	C2LOP.gen!g	Trojan	146
#9	Dialplatform.B	Dialer	177
#10	Dontovo.A	Downloader	162
#11	Fakerean(Unknown)	Rogue	381
#12	Instantaccess	Dialer	431
#13	Lolyda.AA1	PWS	213
#14	Lolyda.AA2	PWS	184
#15	Lolyda.AA3	PWS	123
#16	Lolyda.AT	PWS	159
#17	Malex.gen!J	Trojan	136
#18	Obfuscator.AD	Downloader	142
#19	Rbot!gen	Backdoor	158
#20	Skintrim.N	Trojan	80
#21	Swizzor.gen!E	Downloader	128
#22	Swizzor.gen!I	Downloader	132
#23	VB.AT	Worm	408
#24	Wintrim.BX	Downloader	97
#25	Yuner.A	Worm	800
#26	Bening	-	562
Total	-	-	**9,901**

Table 2. Details of Malware Families in the Malevis Dataset [4]

Class ID	Family	Malware Category	Sample No.
#1	Adposhel	Adware	494
#2	Agent	Trojan	470
#3	Allaple	Worm	478
#4	Amonetize	Adware	497
#5	Androm	Backdoor	500
#6	Autorun	Worm	496
#7	BrowseFox	Adware	493
#8	Dinwod	Trojan	499
#9	Elex	Trojan	500
#10	Expiro	Virus	501
#11	Fasong	Trojan	500
#12	HackKMS	Riskware	499
#13	Hlux	Worm	500
#14	Injector	Trojan	495
#15	InstallCore	Adware	500
#16	MultiPlug	Adware	350
#17	Neoreklami	Adware	500
#18	Neshta	Virus	497
#19	Other	-	700
#20	Regrun	Trojan	485
#21	Sality	Virus	499
#22	Snarasite	Trojan	500
#23	Stantinko	Trojan	500
#24	VBA	Macro Malwares	500
#25	VBKrypt	Trojan	496
#26	Vilsel(Unknown)	Trojan	496
Total	-	-	**12,295**

On the other hand, the Malevis dataset contains 26 malware families and includes a total of 12,295 RGB images, each with a resolution of 224 × 224 pixels. While the distribution of malware families in the Malevis dataset is quite balanced, there is a very big "other" class in certain samples (see Table 2).

4.2 Experiment Results

The experimental results provide a comprehensive evaluation of deep learning models applied to both binary and multi-class classification tasks. Two datasets,

Malimg and Malevis, were utilized to benchmark the performance of state-of-the-art architectures. The following sections elaborate on the models' performance metrics and visual comparisons through confusion matrices and Receiver Operating Characteristic (ROC) curves.

4.2.1 Performance of Deep Learning Models

Across binary as well as multiclass classification tasks, a deep learning model's performance will be examined in detail, taking into consideration such metrics as accuracy, precision, recall, and F1-score. For binary classification tasks, as can be seen from Table 3, MobileNetV2 reached top accuracy in both Malimg (99.31%) and Malevis (98.67%) datasets. On any more complete dataset, classes with fewer characteristics are of far lower percentage probability. For multiclass classification, meanwhile, it only tossed further layers of complexity onto the problem-as these methods used The Table 4 shows that MobileNetV2 continues to outperform other models receiving strong reports. This model's characteristics are also obvious in both tables and prove how well it generalizes across different tasks. In addition, the results themselves show a notable performance gap between MobileNetV2 and other models. EfficientNetB0 and MobileNetV3Large models are poor performers in both precision and recall, especially so for the open-set scenario in particular.

Results of Open-Set Classification. Open-set classification presents some unique challenges: not only must a model correctly identify categories its already familiar with, it has to tell when one is novel. Table 4 lists the results of different architectures on this task. It is noteworthy that here MobileNetV2 has come out with the best results, achieving an F1-score of 91.15% in Malimg and 94.12% in Malevis. It also maintained high unknown detection accuracy (90.81% and 99.80% for Malimg and Malevis, respectively), which underscores its generalization capabilities. While VGG16 performed well on known accuracy (87.90% on Malimg), its unknown detection ability and precision were not as strong. Models like EfficientNetB0 and MobileNetV3Large showed limited ability to handle open-set classification, with significantly lower precision and F1-scores. These results demonstrate MobileNetV2's superior ability to detect and classify both known and unknown categories effectively.

Performance in Closed-Set and Open-Set Classification. Whether in closed-set or open-set classification a model's ability to handle unknown categories has a significant effect on its accuracy and generalization. In closed-set recognition, MobileNetV2 (Table 3) achieved a precision, recall and F1-score of no lower than 90% in all kinds of binary classification tasks, indicating its reliability in scenarios where only known classes are present. For open-set tasks (Table 4), detection of unknown categories was the present phenomenon such that many models failed. It was only Highs into Lowes who made it through this predicament with MobileNetV2 consistently achieving high accuracy and identification of unknown categories. For example, in open-set classification, the F1-score of MobileNetV2 was 91.15% for malimg and 94.12% malevis. This far exceeds comparable models like ResNet50 or CNN, which might not offer reasonable precision or recall measures or something else still. In summary, this comparison suggests that MobileNetV2 is good not only for closed-set situations but also for open ones. This makes it a leading candidate in the event that your application contains an unknown type.

Table 3. Performance of deep learning models on binary classification tasks.

Model	Dataset	Accuracy	Precision (Weighted Avg)	Recall (Weighted Avg)	F1-Score (Weighted Avg)
MobileNetV2	Malimg	99.31	99.30	99.31	99.30
MobileNetV2	Malevis	98.67	98.75	98.67	98.70

Table 4. Performance of deep learning models on open-set classification tasks.

Model	Dataset	Known Accuracy (%)	Unknown Detection Accuracy (%)	Precision (%)	Recall (%)	F1-Score (%)
MobileNetV2	Malimge	**89.60**	90.81	**94.34**	**89.60**	**91.15**
CNN	Malimge	61.42	100.00	81.01	61.42	61.58
EfficientNetB0	Malimge	16.36	100.00	15.24	16.36	15.09
VGG16	Malimge	87.90	100.00	89.94	87.90	87.84
MobileNetV3Large	Malimge	62.98	100.00	70.94	62.98	63.95
ResNet50	Malimge	64.79	100.00	78.26	64.79	67.39
MobileNetV2	Malevis	**89.40**	99.80	**94.72**	**94.99**	**94.12**
CNN	Malevis	81.61	100.00	81.45	81.61	81.20
EfficientNetB0	Malevis	17.42	100.00	13.93	17.42	14.07
VGG16	Malevis	77.72	100.00	74.14	77.72	74.99
MobileNetV3Large	Malevis	32.69	100.00	33.81	32.69	29.14
ResNet50	Malevis	51.07	100.00	52.17	51.07	45.78

Performance Comparison with Other State-of-the-Art Approaches. In Table 5 and Figs. 3 and 4, a comparative analysis of the performance of the proposed MobileNetV2 model against existing state-of-the-art approaches was presented on both binary and multiclass classification tasks. For the Malimg dataset, the MobileNetV2 model achieved a significant improvement in performance metrics, with an F1-score of 97.85% for binary classification and 96.7% for multiclass classification. These results exceed those of previous works (for instance Nataraj et al. [10] only had an F1-score of 96.71% while Agarap et al. [1] got an F1-score being 85.05%), which often failed to attain comparable precision and recall values. As well, for the MaleVis database, the proposed model excelled with a greatest F1-score 98.15% for binary classification and 91.15% for multiclass classification. In comparison to other methods, such as Luo et al. [8] among them whose F1-score for binary classification was 91.37%, this result stands out significantly. Figures 3 and 4 further illustrate the Receiver Operating Characteristic (ROC) curves for multiclass classification tasks on the Malimg and MaleVis datasets, respectively. The Macro-average AUC values for both data sets approached 0.99–1.0, highlighting how versatile the MobileNetV2 model is across all classes. Most significantly, as shown in the ROC curves, the MobileNetV2 framework can obtain high precision and recall for minority classes. Thus, the model is very effective at handling imbalanced data sets. Underlying this is MobileNetV2's unique capability to perform open-set classification Table 5 also shows. With it, the framework can detect and classify unknown categories. This not just overcomes a significant limitation of previous methods but also establishes MobileNetV2 as an up-to-date approach for malware detection and classification tasks.

5 Discussion

Comparative analysis of classification performance across different models, such as MobileNetV2, VGG16, ResNet50, EfficientNet, and CNN, showed that MobileNetV2 has significant advantages in terms of both binary and multi class classification. Comparative analysis illustrates there are multiple ways to learn from the Malimg dataset. MobileNetV2 offers higher recognition accuracy at a decreased misclassification rate than the other models, particularly for minor classes. On the Malimg dataset, MobileNetV2 can clearly distinguish between

Table 5. Performance Comparison of Different Models on Malimg and MaleVis Datasets

Model	Dataset	Precision (%)	Recall (%)	F1-Score (%)	Open-Set (Unknown Class)
Nataraj et al. [10]	Malimg	96.57	96.85	96.71	No
Roseline et al. [11]	Malimg	98.86	98.63	98.74	No
Cui et al. [5]	Malimg	94.64	94.31	94.47	No
Agarap et al. [1]	Malimg	85.47	84.64	85.05	No
Vinayakumar et al. [15]	Malimg	96.30	95.82	96.06	No
Luo et al. [60]	Malimg	94.13	92.54	93.33	No
Singh [14]	Malimg	95.76	96.16	95.96	No
Hemalatha et al. [6]	Malimg	97.78	97.92	97.85	No
MobileNetV2(Proposed)	Binary classes(Malimag)	**99.30**	**99.31**	**99.30**	**Yes**
MobileNetV2 (Proposed)	Multiclass (Malimag)	98.3	95.2	96.7	**Yes**
Nataraj et al. [10]	MaleVis	92.36	89.58	90.95	No
Roseline et al. [11]	MaleVis	97.53	97.32	97.42	No
Cui et al. [5]	MaleVis	92.09	91.89	91.99	No
Agarap et al. [1]	MaleVis	80.22	78.45	79.33	No
Vinayakumar et al. [15]	MaleVis	86.85	86.28	86.56	No
Singh [14]	MaleVis	92.87	91.67	92.27	No
Hemalatha et al. [6]	MaleVis	98.56	97.74	98.15	No
MobileNetV2(Proposed)	Binary classes(Malevis)	**98.75**	**98.67**	**98.70**	**Yes**
MobileNetV2 (Proposed)	Multiclass (Malevis)	94.34	89.60	91.15	**Yes**

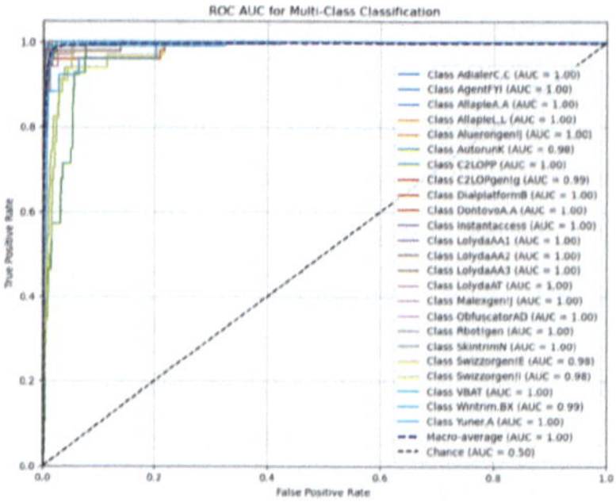

Fig. 3. ROC AUC for Multi-Class Classification on the Malimg Dataset.

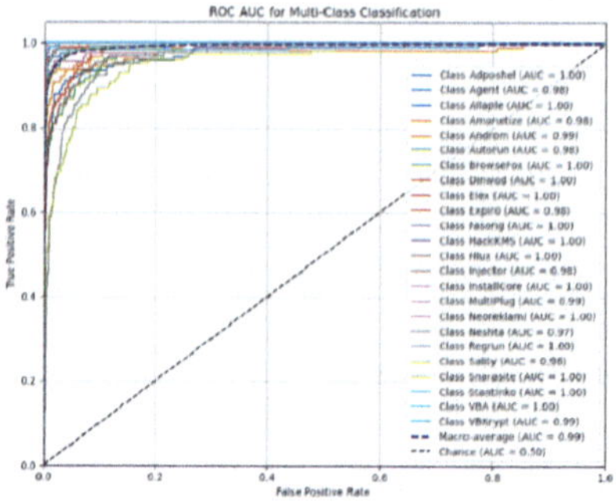

Fig. 4. ROC AUC for Multi-Class Classification on the MaleVis Dataset.

closely related families of malware with minimal overlap, as indicated by its off-diagonal entries. In contrast, for minority samples, VGG16 and ResNet50 struggled and recorded higher confusion among such classes as *C2LOPP* and *Skintrim.N*. Similarly on the MaleVis dataset, MobileNetV2's precise recognition of the majority as well as minor classes–together with its performance handling previously unknown samples–demonstrates its strength. Also, the model excels in out-of-set classification tasks by correctly identifying unknown samples (e.g., those with many counts attributed to the "unknown" category) without making significant errors into known ones.

In addition, the performance comparison shows that MobileNetV2 can offer both computational efficiency and versatility. For instance, as compared to other deep architectures, like EfficientNet (which need a lot of resources and suffer from overfitting) more balance sets MobileNetV2 between efficiency in computing and performance in classification–thus making this approach extremely practical for actual-world malware detection scenarios as well. This is particularly evident in the out-of-set classification tasks where MobileNetV2 demonstrates an ability to generalize across totally unknown malware families–a crucial feature for dynamic cybersecurity environments. Contributing another aspect to the matter, CNN based models while lightweight in computing resources might yield suboptimal results when faced with more complex multi-class problems. The combination of high precision, recall, and F1-scores for both datasets and tasks, along with its out-of-set recognition capabilities, makes MobileNetV2 a leading choice.

6 Conclusions

A MobileNetV2 model fine-tuned with open-set categories is proposed to build the overall framework, which represents a major advancement in malware detection. By making the framework open-set capable, it fills in the critical gap left by ordinary closed-set systems. This results in better detection of both known and unknown malware. In every detection task, whether of individual binary classes or whole groups (called multiclass), this makes for high-than-ever performance. It achieved 99.3% and 98.4% strong accuracies on Malimg and Malevis datasets separately. With the addition of an "unknown" category, the system's generalization ability is further strengthened and it will work in real-world cybersecurity environments. Not only can the framework distinguish new types of malware with high recall and precision, it also has high computational efficiency. This makes it a practical choice for dynamic event-based malware detection.

Future work will explore Framework scalability in respect to larger datasets and real-time applications. In addition, extending the open-set capability for more complex scenarios (e.g., adversarial malware and multi-modal data where the network leaders malicious web traffic and IP refers to image-based malware features) would be a major focus task.

References

1. Agarap, A.F.: Towards building an intelligent anti-malware system: a deep learning approach using support vector machine (SVM) for malware classification. arXiv preprint arXiv:1801.00318 (2017)
2. Alomari, E.S., et al.: Malware detection using deep learning and correlation-based feature selection. Symmetry **15**(1), 123 (2023)
3. Atacak, İ: An ensemble approach based on fuzzy logic using machine learning classifiers for android malware detection. Appl. Sci. **13**(3), 1484 (2023)
4. Bozkir, A.S., Cankaya, A.O., Aydos, M.: Utilization and comparision of convolutional neural networks in malware recognition. In: 2019 27th Signal Processing and Communications Applications Conference (SIU), pp. 1–4. IEEE (2019)
5. Cui, Y., Jia, M., Lin, T.Y., Song, Y., Belongie, S.: Class-balanced loss based on effective number of samples. In: Proceedings of the IEEE/CVF Conference on Computer Vision and Pattern Recognition, pp. 9268–9277 (2019)
6. Hemalatha, J., Roseline, S.A., Geetha, S., Kadry, S., Damaševičius, R.: An efficient densenet-based deep learning model for malware detection. Entropy **23**(3), 344 (2021)
7. Kumar, M.: Scalable malware detection system using distributed deep learning. Cybern. Syst. **54**(5), 619–647 (2023)
8. Luo, J.S., Lo, D.C.T.: Binary malware image classification using machine learning with local binary pattern. In: 2017 IEEE International Conference on Big Data (Big Data), pp. 4664–4667. IEEE (2017)

9. Meenakshi, K., Maragatham, G.: An optimised defensive technique to recognize adversarial iris images using curvelet transform. Intell. Autom. Soft Comput. **35**(1) (2023)
10. Nataraj, L., Karthikeyan, S., Jacob, G., Manjunath, B.S.: Malware images: visualization and automatic classification. In: Proceedings of the 8th International Symposium on Visualization for Cyber Security, pp. 1–7 (2011)
11. Roseline, S.A., Geetha, S., Kadry, S., Nam, Y.: Intelligent vision-based malware detection and classification using deep random forest paradigm. IEEE Access **8**, 206303–206324 (2020)
12. Sheneamer, A.: Visualized malware images using hybrid ensemble deep transfer learning. In: 2024 7th International Conference on Information and Computer Technologies (ICICT), pp. 7–12. IEEE (2024)
13. Sheneamer, A., Alhazmi, E., Henrydoss, J.: Empirical analysis of learning-based malware detection methods using image visualization. Int. J. Adv. Comput. Sci. Appl. **13**(4) (2022)
14. Singh, A., Handa, A., Kumar, N., Shukla, S.K.: Malware classification using image representation. In: International Symposium on Cyber Security Cryptography and Machine Learning, pp. 75–92. Springer (2019)
15. Vinayakumar, R., Alazab, M., Soman, K., Poornachandran, P., Venkatraman, S.: Robust intelligent malware detection using deep learning. IEEE Access **7**, 46717–46738 (2019)
16. Xue, D., Li, J., Wu, W., Tian, Q., Wang, J.: Homology analysis of malware based on ensemble learning and multifeatures. PLoS ONE **14**(8), e0211373 (2019)

Security or Privacy? A Cross-Cultural Exploration of Consumer Priorities in E-Commerce

Jing Hua(✉) and Ping Wang

Computer and Information Systems, Robert Morris University, Pittsburgh, USA
jxhst641@mail.rmu.edu, wangp@rmu.edu

Abstract. This study investigates cross-cultural differences in how consumers from Amazon (U.S.) and Taobao (China) prioritize key privacy-security aspects, including data security, control over personal information, customization of privacy settings, transparency of data collection, and compliance with privacy regulations. Using a quantitative approach, survey data from 323 Amazon users and 205 Taobao users were analyzed to determine significant differences in privacy and security prioritization. The Mann-Whitney U test revealed that Taobao users prioritize data security and transparency of data collection more than Amazon users, while Amazon users place higher importance on customization of privacy settings and compliance with privacy regulations. No significant difference was found between the two platforms in terms of control over personal information. These findings align with Hua and Wang's (2023) cross-cultural privacy model, which suggests that users in collectivist cultures, like China, tend to trust organizations or government entities to manage privacy, leading to less emphasis on individual control through customization. In contrast, users from individualistic cultures, like the U.S., prioritize personal control, as reflected in their higher ranking of customization. This study provides insights into how cultural values influence privacy-security priorities, offering practical implications for global e-commerce platforms.

Keywords: Cross-culture privacy model · Privacy protection practice · Data security · Privacy preferences · Consumer behavior · e-commerce platforms · Amazon · Taobao

1 Introduction

Privacy involves decisions about who should have legitimate access, use, or alteration rights over information. It is about determining access based on moral or legal claims. The definition of privacy is not universal and varies across societal characteristics, economic conditions, and cultural environments. Privacy must be interpreted within the context of the current era [1, 7]. Security, on the other hand, refers to the set of mechanisms, such as encryption and authentication - designed to protect data from unauthorized access or harm, often serving the purpose of enforcing privacy requirements [2]. If

S. Latifi (Ed.): ITNG 2025, AISC 1463, pp. 59–70, 2025.
https://doi.org/10.1007/978-3-031-89063-5_6

security is flawed, privacy protections fail regardless of how strong the original privacy intent is. This makes security a critical element for maintaining privacy. Because of this close relationship, privacy and security are often used interchangeably in literature and research.

There is limited research specifically comparing how consumers prioritize privacy and security across cultures. Most studies tend to combine the concepts of privacy and security without differentiating them. For example, Herbert et al. found significant differences in privacy and security misconceptions across twelve cultures, with participants from non-Western countries, such as China and India, being more likely to hold misconceptions compared to participants from Western countries like Germany and the U.S. [3] Their research focus was primarily on understanding digital literacy and security knowledge rather than on how users prioritize these concepts. Another study conducted by Biselli and Reuter (2021) investigated users' privacy and security behaviors and revealed that German users do not place equal importance on privacy and security [4]. Despite these findings, there is still a gap in research on how users across cultures specifically prioritize privacy and security. Consumer perceptions and behaviors on security and privacy in e-commerce are a significant factor for improving the security and privacy implementation, usability, and efficiency for online business transactions.

This study surveys Amazon and Taobao consumers, two popular e-commerce platforms in the US and China, asking users to rank specific privacy-security aspects, including data security, control over personal information, customization of privacy settings, transparency of data collection, and compliance with privacy regulations. Using Mann-Whitney U statistics, this study explores how consumers from different cultural backgrounds prioritize these privacy and security elements in e-commerce. The research contributes to a deeper understanding of cross-cultural differences in privacy and security perceptions and provides new empirical data on the application of Hua and Wang's cross-culture privacy model, which explains how cultural values influence privacy management on digital platforms in both individualistic (U.S.) and collectivist (China) contexts.

2 Literature Review

2.1 Legal and Users' Perspectives on Security and Privacy

From an organizational perspective, privacy and security are often treated as one, under models like the CIA (Confidentiality, Integrity, Availability) framework, where the goal is to protect data from unauthorized access. While security shares goals with privacy, such as maintaining confidentiality [4], this framework primarily focuses on protecting data from external threats rather than addressing the broader privacy concerns users may have. From a legal perspective, privacy is defined and enforced by regulations like the General Data Protection Regulation (GDPR) or the California Consumer Privacy Act (CCPA), which establish clear guidelines for how organizations must handle and protect personal data. These regulations emphasize individual rights over personal information, such as the right to access, modify, and delete data, which aligns more with users' expectations of privacy. However, security, as implemented by organizations, often focuses on preventing data breaches and ensuring data integrity, which may not fully address user concerns about how their data is shared or used. This creates a gap between organizational security practices and the privacy management users expect, as privacy is more about control over personal information and the trust users place in organizations to handle that information appropriately.

2.2 Studies on Cultural Differences in Security and Privacy

Cultural differences significantly influence how users view privacy and security. In individualistic cultures like the U.S., privacy typically focuses on individual control over personal data, with security acting as a safeguard for that control. In contrast, in collectivist cultures such as China, privacy concerns are lower, and there is greater trust in organizations and government institutions to manage personal data [5, 6]. Hua and Wang (2023) demonstrated this dynamic in their study of the DiDi case, where China's privacy and security priorities are shaped by collective welfare, with the government determining which information requires protection [6]. In such contexts, individuals tend to trust government bodies and organizations to handle privacy and security concerns on their behalf.

Herbert et al. (2023) also found notable differences in privacy and security misconceptions between Western and non-Western cultures, where participants from non-Western countries were more likely to hold misunderstandings[3]. This underscores the influence of cultural context in shaping consumers' perceptions of privacy and security [3]. Biselli and Reuter (2021)further pointed out that security behaviors, such as using antivirus software, do not necessarily align with privacy behaviors, like reading privacy policies, suggesting that users often perceive these two concerns separately [4]. Although much of the existing research examines privacy and security together, it often overlooks how users prioritize these aspects individually and the specific cultural context for user perceptions. While security mechanisms enforce privacy protections, the way users rank the importance of privacy versus security in different contexts remains underexplored [2]. This gap is evident in cross-cultural studies, where privacy and security concerns are viewed differently depending on cultural values [1, 6].

3 Methodology

This study addresses the research gap by comparing how Amazon and Taobao users prioritize key privacy and security aspects. The research employs a quantitative approach, using survey data to explore cross-cultural differences in privacy and security priorities. Amazon and Taobao, two popular e-commerce platforms in the US and China respectively, were selected for their significant handling of personal and financial data, where security plays a critical role. These platforms operate within distinct cultural and regulatory frameworks, making them ideal for investigating differences in how users prioritize privacy-related aspects, such as data security, customization of privacy settings, transparency of data collection, and compliance with privacy regulations. The survey of this study intends to address the following two main research questions:

Research Questions (RQ);

1. RQ1: Are there significant differences between Amazon and Taobao users in how they prioritize privacy-security-related aspects (data security, control over personal

information, customization of privacy settings, transparency of data collection, and compliance with privacy regulations)?
2. RQ2: Which group (Amazon or Taobao) shows a higher priority on data security, customization of privacy settings, transparency of data collection, and compliance with privacy regulations?

3.1 Participants and Data

Participants from both Amazon and Taobao were surveyed to rank privacy-related aspects in order of importance (1 = least important, 5 = most important). The survey questions address security and privacy aspects including data security, control over personal information, customization of privacy settings, transparency of data collection, and compliance with privacy regulations.

- Amazon Participants: Participants were recruited through CloudResearch, yielding 325 responses, with 2 incompletes, leaving 323 valid responses. Since the sample was purchased through CloudResearch, conventional response rate calculations do not apply.
- Taobao Participants: Participants were recruited through a snowball sampling method on WeChat Moments, yielding 235 responses, with 31 incompletes, leaving 203 valid responses. Conventional response rate calculations do not apply here due to the nature of the sampling methods used.

Data collection took place between July and September 2024, following ab IRB approval. Demographic characteristics of the respondents are detailed in Table 1.

3.2 Data Analysis

To address RQ1, the Mann-Whitney U test, a non-parametric test, was employed to compare the rankings of privacy-related aspects between the two groups (Amazon and Taobao users). This test was chosen because it compares independent groups based on their ranked data and determines if there are statistically significant differences between the two groups for each privacy-related aspect.

For RQ2, the mean ranks for each privacy-related aspect were analyzed to determine which group (Amazon or Taobao) shows a higher priority on data security, customization of privacy settings, transparency of data collection, and compliance with privacy regulations.

4 Findings

4.1 RQ1: Differences in Privacy-Security Prioritization Between Amazon and Taobao Users

Based on the results of the Mann-Whitney U Test (see Table 2), significant differences were found between Amazon and Taobao users in their prioritization of data security, customization of privacy settings, transparency of data collection, and compliance with privacy regulations. However, no significant differences were found in the ranking of control over personal information, suggesting that both groups similarly value this aspect. These findings indicate that users from the two platforms have different priorities regarding specific privacy concerns.

Additionally, the Kolmogorov-Smirnov Test, Table 3, indicated significant differences in the distributions of some aspects, which suggests that the way the two groups rank these aspects differs not only in terms of central tendency but also in terms of overall distribution shape.

1) Significant Differences: Amazon and Taobao users showed significant differences in their prioritization of data security, customization of privacy settings, transparency of data collection, and compliance with privacy regulations (Fig. 1 and Table 4).

a) Data Security: $p = 0.016$

Table 1. Demographic Characteristics of Amazon and Taobao Users

Demographic Item	Category	Amazon	%	Taobao	%
Age Range	18–24	28	8.70%	19	9.30%
	25–34	118	36.50%	37	18.10%
	35–44	104	32.20%	47	23.00%
	45–54	40	12.40%	40	19.60%
	55–64	19	5.90%	28	13.70%
	65+	7	2.20%	4	2.00%
	Missing	7	2.20%	29	14.20%
Gender	Female	141	43.70%	99	48.50%
	Male	172	53.30%	73	35.80%
	Non-binary	2	0.60%	-	-
	Prefer not to say	1	0.30%	3	1.50%
	Missing	7	2.20%	29	14.20%
Education Level	High School or equivalent	35	10.80%	13	6.40%
	Some College	88	27.20%	32	15.70%
	Bachelor's Degree	144	44.60%	100	49.00%
	Master's Degree	37	11.50%	25	12.30%
	Doctorate or Professional Degree	11	3.40%	3	1.50%
	Others/Prefer not to say	1	0.30%	2	1.00%
	Missing	7	2.20%	29	14.20%

Table 2. Mann-Whitney U Test Results

	Data_Security_rank	Control_Personal_Info_rank	Customization_Privacy_Settings_rank	Transparency_Data_Collection_rank	Compliance_Privacy_Regulations_rank
Mann-Whitney U	29033.500	32731.500	25964.500	26906.500	29034.500
Wilcoxon W	81359.500	85057.500	46670.500	79232.500	49740.500
Z	−2.418	−.032	−4.114	−3.583	−2.358
Asymp. Sig. (2-tailed)	.016	.974	< .001	< .001	.018

a. Grouping Variable: Platform

Table 3. Summary of Mann-Whitney U Test Results

Aspect	Z-value	p-value (Asymp. Sig.)	Interpretation
Data Security	−2.418	.016	Significant difference between platforms
Control over Personal Information	−0.032	.974	No significant difference
Customization of Privacy Settings	−4.114	<.001	Significant difference between platforms
Transparency of Data Collection	−3.583	<.001	Significant difference between platforms
Compliance with Privacy Regulations	−2.358	.018	Significant difference between platforms

Table 4. Kolmogorov-Smirnov Test Results

		Data_Security_rank	Control_Personal_Info_rank	Customization_Privacy_Settings_rank	Transparency_Data_Collection_rank	Compliance_Privacy_Regulations_rank
Most Extreme Differences	Absolute	.153	.041	.213	.197	.096
	Positive	.153	.024	.000	.197	.000
	Negative	.000	−.041	−.213	.000	−.096
Kolmogorov-Smirnov Z		1.710	.457	2.374	2.196	1.075
Asymp. Sig. (2-tailed)		.006	.985	< .001	< .001	.198

a. Grouping Variable: Platform

The box plot for Data Security shows that Taobao users have a higher median and a wider spread in their rankings compared to Amazon users, indicating greater prioritization of data security. This aligns with the significant Mann-Whitney U Test results ($p = 0.016$) (Fig. 2).

b) Customization of Privacy Settings: $p < 0.001$

Amazon users rank customization of privacy settings significantly higher than Taobao users, as indicated by the higher median and broader range in the box plot. The wider spread for Amazon suggests greater variability in how different users prioritize this

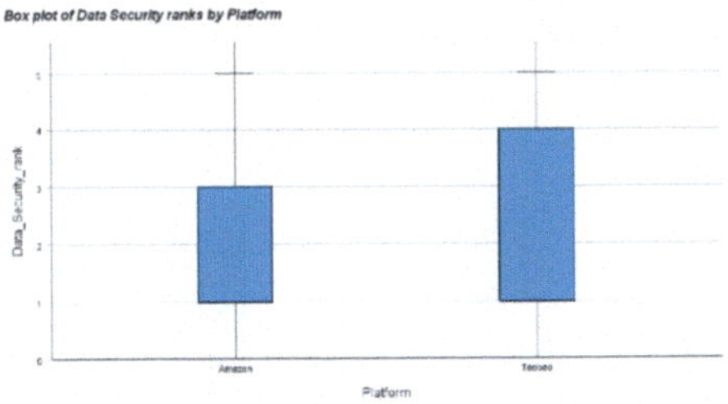

Fig. 1. Box plot of Data Security ranks by Platform

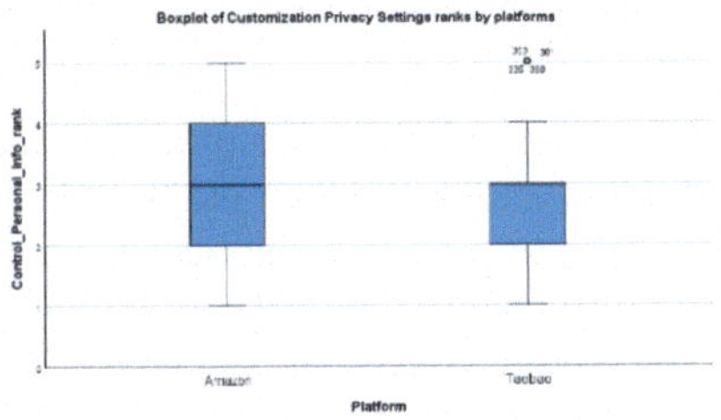

Fig. 2. Box plot of Customization of Privacy Settings by Platform

aspect. In contrast, Taobao users exhibit a narrower range, reflecting more consistent and generally lower rankings. This difference in the distribution shapes reinforces the Mann-Whitney U Test result, showing a significant difference in how the two groups prioritize customization of privacy settings ($p < 0.001$).

c) Transparency of Data Collection: $p < 0.001$

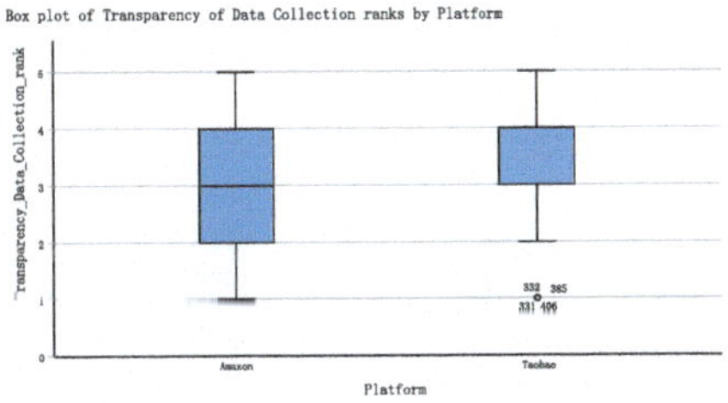

Fig. 3. Box plot of Transparency of Data Collection by Platform

The box plot, Fig. 3 shows that Taobao users tend to rank transparency of data collection higher than Amazon users, as indicated by the higher median for Taobao. The variability is also different between the two groups: Amazon users exhibit a wider range of rankings, with a more dispersed distribution. In contrast, Taobao users have a narrower spread, indicating more consistency in how they rank this aspect. This pattern suggests that while Taobao users are more uniformly concerned with transparency, Amazon users have more diverse opinions on its importance, which aligns with the significant Mann-Whitney U Test results ($p < 0.001$).

d) Compliance with Privacy Regulations: $p = 0.018$

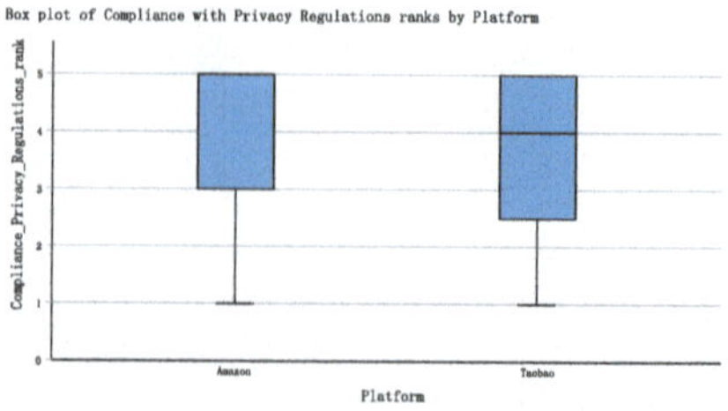

Fig. 4. Box plot of Compliance with Privacy Regulations ranks by Platform

The Compliance with Privacy Regulations box plot, Fig. 4 shows that Amazon users have a higher median rank compared to Taobao users, indicating that Amazon users prioritize this aspect more. This finding is supported by the Mann-Whitney U Test, which confirms a significant difference ($p = 0.018$). Additionally, while the median is higher for Amazon users, the shorter interquartile range (IQR) suggests that there is more consistency in the rankings among Amazon users. The longer whiskers, however, indicate some variability in the extreme values, reflecting that a few Amazon users have differing views on the importance of compliance. In contrast, Taobao users show a wider IQR, suggesting more variability in opinions about the importance of compliance with privacy regulations.

2) No Significant Difference: No significant difference was found for control over personal information ($p = 0.974$), suggesting that both Amazon and Taobao users rank this aspect similarly. The box plot, Figure 5 shows that the distributions of rankings for Amazon and Taobao users are similar. Both platforms have comparable medians, interquartile ranges, and overall spread, indicating no substantial difference in how users from these two groups prioritize control over personal information.

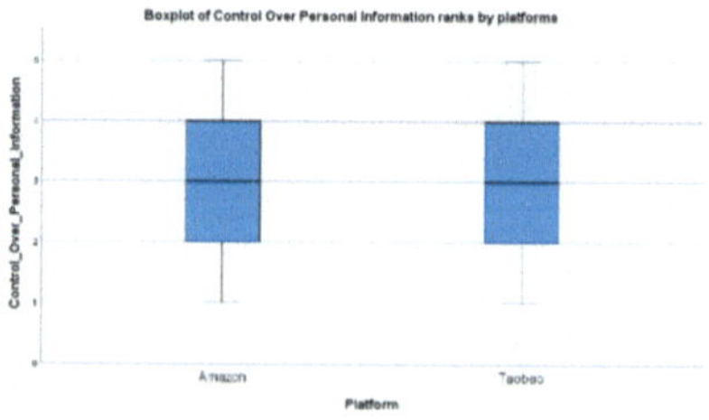

Fig. 5. Box plot of Control Over Personal Information by Platform

4.2 RQ2: Differences in Privacy-Security Prioritization Between Amazon and Taobao Users

Based on the Mann-Whitney U test, Taobao users had a higher mean rank (Table 5: Summary of Mean Rank for Privacy Ranks, Fig. 6: Mean Ranks for Privacy Aspects,) for Data Security compared to Amazon users, indicating that Taobao users prioritize Data Security more highly. Amazon users, on the other hand, prioritized Customization of Privacy Settings and Compliance with Privacy Regulations more than Taobao users. These differences reflect variations in privacy-security prioritization influenced by the distinct cultural and regulatory contexts of the two user groups.

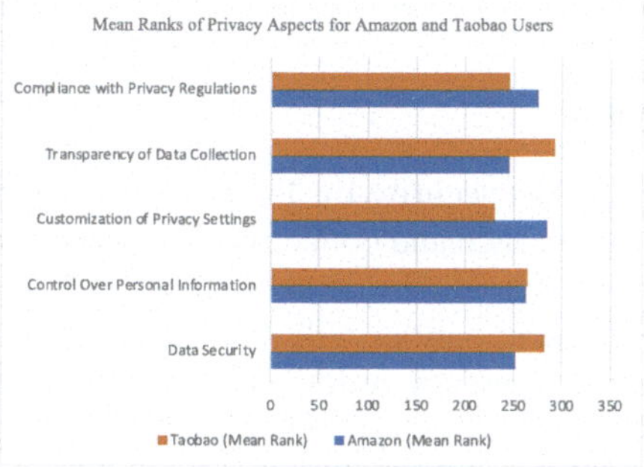

Fig. 6. Mean Rank of Privacy Aspects

Table 5. Mean Ranks for Privacy Aspects

Aspect	Platform	N	Mean Rank	Sum of Ranks
Data Security	Amazon	323	251.89	81,359.50
	Taobao	203	281.98	57,241.50
Control Over Personal Information	Amazon	323	263.34	85,057.50
	Taobao	203	263.76	53,543.50
Customization of Privacy Settings	Amazon	323	284.61	91,930.50
	Taobao	203	229.9	46,670.50
Transparency of Data Collection	Amazon	323	245.3	79,232.50
	Taobao	203	292.46	59,368.50
Compliance with Privacy Regulations	Amazon	323	275.11	88,860.50
	Taobao	203	245.03	49,740.50

5 Discussions

5.1 Key Cultural Insights

The significant privacy and security differences between Amazon and Taobao users can be understood through the lens of cultural values and differences in the U.S. and China. As an individualistic society, the U.S. places a strong emphasis on personal privacy and control. This is reflected in Amazon users' higher prioritization of customization of privacy settings and compliance with privacy regulations, where individuals prefer to have direct control over their personal information. In contrast, China, as a collectivistic society, places greater trust in organizations and focuses more on collective privacy management [5, 6]. This cultural tendency is supported by the cross-cultural privacy model proposed by Hua and Wang (2023) [6], which explains how Chinese users tend to trust organizations or the government to handle privacy matters on their behalf.

5.2 Security Prioritization

The higher prioritization of data security by Taobao users compared to Amazon users aligns with the high-power distance and collectivist orientation in Chinese culture. In a high-power distance society, individuals are more likely to defer to authorities, trusting that organizations or government entities will manage security concerns [6]. This also explains why customization of privacy settings is less important to Taobao users; they may feel that security is being sufficiently managed by external authorities, and thus, the need for personal control is diminished.

5.3 Customization of Privacy Settings

Amazon users' higher prioritization of customization of privacy settings reflects the individualistic values of U.S. culture, where people seek greater autonomy and control over their personal information. According to Hua and Wang's cross-cultural privacy model, customization is often tied to an individual's desire for personal control [6]. In individualistic cultures like the U.S., the ability to customize privacy settings allows users to feel more empowered in managing their privacy. In contrast, Taobao users, from a more collectivist culture, may trust the platform or external authorities to handle these issues, resulting in a lower emphasis on customization.

5.4 Transparency of Data Collection

Another significant finding is the higher prioritization of transparency of data collection by Taobao users. This may seem counterintuitive for a collectivist culture where trust in institutions is strong, but it can be interpreted as a reflection of the increasing awareness of data collection practices in China. As concerns about data misuse and surveillance grow, transparency becomes more important even in collectivist societies. This highlights a shift where users in China, while still trusting organizations, are beginning to expect more transparency in how their data is handled.

5.5 Control Over Personal Information

Interestingly, no significant difference was found between Amazon and Taobao users in their prioritization of control over personal information, despite the differences in how they prioritized customization of privacy settings. This finding aligns with the concept of the "privacy paradox," where users express concerns about privacy but their actual online behaviors (e.g., engaging with platforms without customizing privacy settings) often suggest otherwise. Although users may value control over personal information in theory, their lower prioritization of customization indicates a disconnect between privacy concerns and practical behavior.

6 Conclusion

The findings of this study are consistent with Hua and Wang's (2023) cross-cultural privacy model, which emphasizes the role of cultural values in shaping privacy and security priorities. In collectivist cultures, there is greater trust in organizations and less emphasis on individual control, which explains the lower prioritization of customization and the higher focus on security by Taobao users. In contrast, individualistic cultures like the U.S. place more importance on personal control and customization, which is reflected in Amazon users' higher rankings for customization and compliance with privacy regulations. This cross-cultural comparison sheds light on how different cultural contexts influence privacy and security perceptions and behaviors, providing valuable insights for designing security and privacy strategies for global e-commerce platforms.

This study provides insights into cross-cultural differences in privacy-security prioritization between Amazon and Taobao users with valuable empirical user data for comparative studies. However, this research has serval limitations. First, the study is a preliminary cross-cultural survey with a narrow focus on the specific research questions, which may limit the generalizability of the findings. The data collection process also faced challenges, particularly in comparing two cultures with large populations with changing cultures, which might not capture the full spectrum of cultural variations. Future studies could expand this research by exploring additional research questions, incorporating more cultural elements for comparison, and examining other factors such as cultural differences in trust and online purchase intentions. Broder (2020) argues for the importance of understanding how cultural background, privacy, and trust interact in e-commerce settings, which could be a promising area for further investigation in the future [8]. Exploring potential correlations among culture, privacy concerns, trust, and purchase intentions would provide a comprehensive understanding of how these factors influence e-commerce consumer behavior across different cultures.

References

1. Lukács, A.: What is privacy? The history and definition of privacy (2016)
2. Bambauer, D.E.: Privacy versus security. J. Crim. L. Criminol. **103**, 667 (2013)
3. Herbert, F., et al.: A world full of privacy and security (mis)conceptions? Findings of a representative survey in 12 countries. In: Proceedings of the 2023 CHI Conference on Human Factors in Computing Systems, vol. 13, pp. 1–23 (2023). https://doi.org/10.1145/3544548.3581410
4. Biselli, T., Reuter, C.: On the relationship between it privacy and security behavior: a survey among German private users. Lecture Notes in Information Systems and Organisation, pp. 388–404 (2021). https://doi.org/10.1007/978-3-030-86797-3_26
5. Li, Y.: Cross-cultural privacy differences. Modern Socio-Technical Perspectives on Privacy, pp. 267–292 (2022). https://doi.org/10.1007/978-3-030-82786-1_12

6. Hua, J, Wang, P.: Cultural differences in privacy protection: a case study of DiDi privacy violations. Issues in Information Systems (2023). https://doi.org/10.48009/2_iis_2023_127
7. Acquisti, A., Taylor, C., Wagman, L.: The economics of privacy. J. Econ. (2016)
8. Broeder, P.: Culture, privacy, and trust in e-commerce. Mark. Inf. Decis. J. **3**(1), 14–26 (2020). https://doi.org/10.2478/midj-2020-0002

An Unprofiled Single Trace Side-Channel Attack for Asymmetric Cryptosystems

Isabella Piacentini[1(✉)], Alessandro Barenghi[1], Gerardo Pelosi[1], and Ruggero Susella[2]

[1] Department of Electronics, Information and Bioengineering – DEIB, Politecnico di Milano, Via. G. Ponzio 34/5, 20133 Milano, Italy
{isabella.piacentini,alessandro.barenghi,gerardo.pelosi}@polimi.it
[2] STMicrolectronics S.r.l., Via Paracelso, 20, 20864 Agrate Brianza, MB, Italy
ruggero.susella@st.com

Abstract. Side-channel attacks aim at retrieving cryptographic secrets from a device by exploiting involuntary information channels, such as power consumption or electromagnetic emissions, measured in traces. Among them, *horizontal attacks* are particularly powerful as they require only a single measurement of an algorithm execution from the target device. In our work, we propose a horizontal attack technique that autonomously detects intermediate value reuses, and extracts cryptographic secret information without needing to analyze or know the implementation of the cryptographic algorithm being run. Our technique, which only assumes the basic a-priori knowledge that the algorithm computes iteratively using a portion of the cryptographic secret in each iteration, both allows successful attacks, and provides the secure software developers with feedback on the vulnerable spots. We validate our attack through a case study on a square-and-multiply-always RSA implementation using the production grade *mbedtls* cryptographic library. Experimental results demonstrate that our approach can retrieve the entire secret RSA exponent from a single execution trace.

Keywords: IoT Security · Cryptography · Side channel attacks · Embedded systems security · Applied cryptography

1 Introduction

Side-channel attacks (SCAs) exploit the ability to model the data dependent behaviour of a computing device, allowing attackers to derive the data being processed. Attackers deduce secret values by measuring physical phenomena such as power consumption [21], electromagnetic emissions [1], or execution time [20]. Indeed, an attacker collects one or more measurements of the desired side channel, say, power consumption, while the device under attack is computing a cryptographic algorithm involving a secret value. Such measurements, which take the

S. Latifi (Ed.): ITNG 2025, AISC 1463, pp. 71–83, 2025.
https://doi.org/10.1007/978-3-031-89063-5_7

form of discrete time series (a.k.a. traces) are matched against a set of models for the expected behaviour of the device, each model depending on a small portion of the cryptographic secret. Observing which model is the best fit for the measured device behaviour allows the attacker to deduce the value of the portion of the cryptographic secret. SCAs are traditionally classified according to whether the model of the device is a synthetic one (*unprofiled SCAs*) [3,5,8], or derived in a data driven fashion from a similar device (*profiled SCAs*) [4,6,7,11,12,24]. A second axis for taxonomy is whether an attack exploits information coming from the same time instant in multiple executions (known as a *vertical attack*, from the custom of depicting traces from multiple executions one on top of another, keeping the time axis horizontal), or it exploits information coming from different time instants in a single execution (known as an *horizontal attack* from the same pictorial custom). In this work, we consider horizontal, unprofiled attacks, as they are highly efficient (needing only one trace) and provide explainable attack avenues, which in turn allow a developer to design effective countermeasures.

Contribution. This work introduces a horizontal attack, targeting asymmetric cryptographic primitives. Our method relies on a single execution trace and exploits collisions within the cryptographic process, without requiring prior knowledge of the specific algorithm's implementation or the operations it uses (e.g., the Instruction Set Architecture of the microprocessor involved). This attack is also effective against asymmetric cryptosystems implementations realized with constant-time coding practices, and indeed benefits from constant time code. Our attack strategy builds upon the so-called *collision-based attacks* [10], but allows a significantly weaker attacker, who does not know the specifics of the implementation, to successfully extract the cryptographic secrets. We employ as a case study a square-and-multiply-always algorithm, which is a ubiquitous building block in asymmetric cryptosystems: RSA encryption, signature, and the Diffie-Hellman key agreement all employ it. Our approach generalizes to any iterative algorithm where value reuse in the computation depends on portions of the secret key. This includes scalar-point multiplications, which are fundamental to elliptic curve-based cryptosystems, and iterative decoders from post-quantum cryptosystems. Our experimental validation uses the open-source *mbedtls* library, which is the TLS library of choice by ARM, targeted to embedded devices and a commercial off-the-shelf microcontroller. In our evaluation, we prove the effectiveness of our approach, and improve on the state-of-the-art methods from [10,19]. Specifically, our approach enables the extraction of both direct, value-based information (i.e., the private key bit being employed in an iteration) and value-change-based information (i.e., whenever two private key bits differ between two iterations), together with providing a temporal hint on which phases of the computation are leaking information via side channel.

2 Horizontal Side Channel Attacks

In the following, we will consider *horizontal* side-channel attacks which leverage leakage from a single execution of a cryptographic algorithm [14,25]. Horizontal side channel attacks exploit features of the trace of a single execution of a cryptographic algorithm to infer the value of the secret key by combining the information contained in multiple samples. The simplest case of horizontal attack employing the device power consumption as the side channel goes by the name of Simple Power Analysis (SPA) [21], and concerns the case of a cryptographic algorithm implementation where the presence of a feature in the trace directly maps to a value of a portion of the cryptographic secret. The root cause of cryptographic implementations being vulnerable to simple power analysis is often the dependence on the secret value by a portion of the control flow of the implemented algorithm. A concrete example is a modular exponentiation procedure performed with a schoolbook square and multiply approach. In this case, at each iteration of the procedure, a (multi-precision) multiplication is executed or not depending on the value of a given single bit of the secret exponent. This in turn translates to an easily detectable increase in power consumption whenever the multiplication is computed, which appears as a distinctive feature of the trace [21].

Mitigating horizontal attacks such as the one exemplified by SPA, has led to the development of unified formulas to perform an iteration of square and multiply exponentiations (and their correspondant on the group of points of an elliptic curve, double and add multiplications). Unified formulas, such as the ones in Chevallier et al. [13] and Giraud et al. [18] employ the same sequence of field operations to compute a correct square and multiply/double and add iteration. While unified formulas consider the removal of distinctive features on an algorithmic level, atomic patterns further detail this idea by providing operative descriptions, i.e., sequences of instructions to be executed in a specific order as to perform the desired computation while avoiding both variations in timing and in the kind of architectural instructions executed at a given time instant. While atomic patterns prevent attackers from leveraging features in the power trace depending on variations in the control flow, or the kind of instructions employed, they still allow to consider the effect of the processed data on the power consumption. This information is leveraged by the so-called *collision* attack strategy. Collision attacks aim to determine if two operations share a common input operand value or not: discerning whether or not two operations, say, in subsequent iterations of the algorithm, share an operand value or not may allow to derive the values being computed.

A first use of the collision technique, focusing on public key implementations, with a limitation on required traces, was reported in [16]. In [2], the authors identified a potential security risk in RSA cryptography through power analysis of the Binary with Initial Points algorithm. They noted that these attacks are especially effective during the algorithm modular multiplication stages. Meanwhile, [17] analyzed collision attacks for evaluating devices susceptible to information leakage, particularly targeting Differential Power Analysis (DPA). Their findings emphasized that collision attacks can uniquely exploit certain leakages, succeeding where other non-profiled attacks may not. In [9], the authors improved horizontal attack techniques, examining masking schemes, which work by splitting a secret into multiple shares to prevent leakage. Lastly, [15] presented a collision attack that investigates reuse not only of input values but also outputs. They show a case study on scalar multiplication over elliptic curves, focusing on the reuse of specific computations, and highlights the need for stronger countermeasures, particularly through noise addition in the computation. In [10], Bauer et al. introduce a theoretical horizontal collision correlation attack on elliptic curve cryptography implementations using point-scalar randomization to achieve constant-time atomic execution. The attack exploits collisions within an a-priori known cryptographic algorithm, where an identical internal states arises from different inputs. However, the authors do not demonstrate practical effectiveness, and the attack requires prior knowledge of the algorithm's behavior on the target device, which limits its applicability in real-world scenarios. Building on Bauer et al.'s work, Hanley et al. in [19] improve the attack's success rate by expanding the analysis to compare more than two operations simultaneously. These multiple-operation comparisons, enhance the attack's effectiveness. However, like Bauer et al.'s work, this attack relies on simulated results, lacking real-world validation [3,5–8,11,24].

3 Proposed Attack Methodology

In the following, we detail our horizontal attack methodology, using a square-and-multiply-always algorithm as a running example. We consider a single execution trace T made of n power consumption samples τ; the target algorithm performs λ iterations, each one acting on the state of the computation and a portion of the secret key. We denote the cryptographic secret employed s as a sequence of portions, $s = (s_0, \dots s_{\lambda-1})$ where s_i corresponds to the i-th portion of the secret used in the i-th iteration of the algorithm, $0 \leq i \leq \lambda - 1$.

We will consider the execution trace T as logically split into λ non overlapping subsequences $T = (T_0, \dots, T_i, \dots, T_{\lambda-1})$, each one corresponding to the power consumption of one of the λ iterations of the algorithm. We denote each one of the λ subsequences as *iteration trace* T_i, where the suffix i is the index of the iteration to which the iteration trace corresponds. Each iteration trace T_i has length $m = \frac{n}{\lambda}$. We will denote each sample of the execution trace as τ_i, $0 \leq \text{ i} \leq n-1$, therefore we have $T = (\tau_0, \dots, \tau_{m\lambda-1})$ as $n = m\lambda$, and $T_i = (\tau_{i \cdot m}, \dots, \tau_{(i+1)\cdot m-1})$.

Each iteration trace is further split into d non overlapping subsequences, denoted as *sections*, each one of length $l = \frac{m}{d}$, therefore obtaining smaller sequences of operations to consider for our attack. We denote each section of the iteration trace T_i as $T_{i,j}$, with $0 \leq j < d$; $T_{i,j}$ corresponds to the sequence of samples $T_{i,j} = (\tau_{i \cdot m + j \cdot l}, \dots, \tau_{i \cdot m + (j+1) \cdot l - 1})$. The rationale of this split is that a single section should contain a small number of instructions, which may be expected to act or not on the same information across iterations.

Our running example, the square-and-multiply-always technique is a constant-time modular exponentiation algorithm that exhibits the iterative execution pattern leveraged in our attack strategy. We consider the case where the technique is used to compute a modular exponentiation modulo N, where the exponent $e \in \mathbb{Z}^*_{\varphi(N)}$ is the cryptographic secret, as it is the case in RSA signatures and Diffie-Hellman key agreements.

Algorithm 1 provides a description of the square-and-multiply-always procedure, which computes $b^e \bmod N$, taking as input $b \in \mathbb{Z}_N$ and $e \in \mathbb{Z}^*_{\varphi(N)}$. Algorithm 1 performs a number of iterations equal to the length $n = \lceil \log_2(N+1) \rceil$ of the binary representation of e, and, at each iteration computes the square of the value of res (line 5). After computing the square of res, two intermediate results, d_0 and d_1, are computed. One is null, while the other either holds res or $b \cdot res$, depending on the value of the i-th secret exponent bit e_i. d_0 and d_1 are subsequently added, and stored in res: this has the effect of either storing in res the value $b \cdot res$ (whenever $e_i = 1$) or leaving res unchanged (whenever $e_i = 0$) while performing the same amount of computations, regardless of the value of e_i.

In our proposed approach, we consider the full execution trace T as the sequence of power consumption values recorded during the loop running from lines 4–10, with each iteration trace T_i corresponding to a single loop iteration. We note that each one of the multiplications reported in Algorithm 1 requires a multiple precision multiplication, which is in turn not executable as a single instruction. We, therefore, report the multiple precision multiplication algorithm from [23] (Algorithm 14.12), as it is the one employed by most cryptographic libraries, including the one employed in our experimental evaluation.

Algorithm 2 describes the procedure of a schoolbook multiplication between two integers x and y, expressed as digits in base b, yielding the product $w = xy$. Typical values for the base b are determined by the architectural word size of the computing platform, so that digit-wise operations correspond to architectural

---

Algorithm 1. Right-to-Left Square-and-Multiply-Always (for RSA)

Input: $b \in \mathbb{Z}_N$; $n = \lceil \log_2(N+1) \rceil$, $e \in \mathbb{Z}^*_{\varphi(n)}$, where $\varphi(N)$ is the Euler Totient function, and $e = (e_{n-1}e_{n-2}\ldots e_0)_2$

Output: $res = b^e \bmod N$

1: $res \leftarrow 1$
2: $d_0 \leftarrow 1$
3: $d_1 \leftarrow 1$
4: **for** $i \leftarrow 0$ to $n-1$ **do**
5: $\quad res \leftarrow res^2 \pmod N$
6: $\quad d_0 \leftarrow res \cdot \neg e_i$
7: $\quad d_1 \leftarrow b \cdot res \pmod N$
8: $\quad d_1 \leftarrow d_1 \cdot e_i$
9: $\quad res \leftarrow d_0 + d_1$
10: **end for**
11: **return** res

Algorithm 2. Multiple-precision Multiplication

Input: positive integers x and y having $n+1$ and $t+1$ base b digits, respectively.
Output: the product $x \cdot y = (w[n+t+1] \ldots w[1]w[0])_b$ in radix b representation

1: Initialize $w[0...(n+t+1)]$ to an array of zeros
2: **for** $i = 0$ **to** $(n+t+1)$ **do**
3: $\quad w[i] \leftarrow 0$
4: **end for**
5: **for** $i = 0$ **to** t **do**
6: $\quad c \leftarrow 0$
7: $\quad$ **for** $j = 0$ **to** n **do**
8: $\quad\quad (uv)_b = w[i+j] + x[j] \cdot y[i] + c$
9: $\quad\quad w[i+j] \leftarrow v,\ c \leftarrow u$
10: $\quad$ **end for**
11: $\quad w[i+n+1] \leftarrow u$
12: **end for**
13: **return** $(w[n+t+1] \ldots w[1]w[0])_b$

operations between registers. Algorithm 2 iterates over the digits of the second factor y, $y[i]$, (lines 5–12) and, for each one of them, iterates over the digits of the first operand x, (lines 7–10), loading them, multiplying them by $y[i]$ (line 8) and accumulating in word $w[i+j]$ of the result the lower digit of the product, while storing the higher in the carry digit c (line 9). Finally, in line 11 is stored the carry-out resulting from the multiplication of the most significant digit of x by $y[i]$.

Our attack strategy focuses on operands remaining the same across two operations, since many architectural instructions will act on the same values. For instance, if the value of x is the same in two multiple precision multiplications, the repeated sequence of `load` instructions in Algorithm 2 at line 8 will involve the same values and result in similar switching activity. Power consumption similarities can also be observed when the same value is loaded in one part of the algorithm and stored in another part during subsequent runs. A specific case occurs when the square operation at line 5 in Algorithm 2 produces a result used in the next iteration's square operation in the for loop at lines 4–10. The switching activity from the `store` and `load` operations is expected to result in nearly identical power consumption patterns, as the data transferred to and from memory remains unchanged.

Attack Strategy. Our approach measures the similarity between pairs of iteration trace sections from consecutive iterations and extracts information from these comparisons.

To this end, we consider each pair of consecutive iteration traces, T_i and T_{i+1} as two periods of two periodic discrete functions f and g, and compute their cross-correlation $(f \star g)[\delta]$ defined as $(f \star g)[\tau] = \sum_{m=-\infty}^{\infty} \overline{f[m]} g[m+\tau]$, where $\overline{f[m]}$ is the complex conjugate of $f[m]$. The cross-correlation is a concrete measure of similarity between two time series, considered as a function of the time displacement τ of one relative to the other. We will denote as $T_i \star T_j$ the cross-

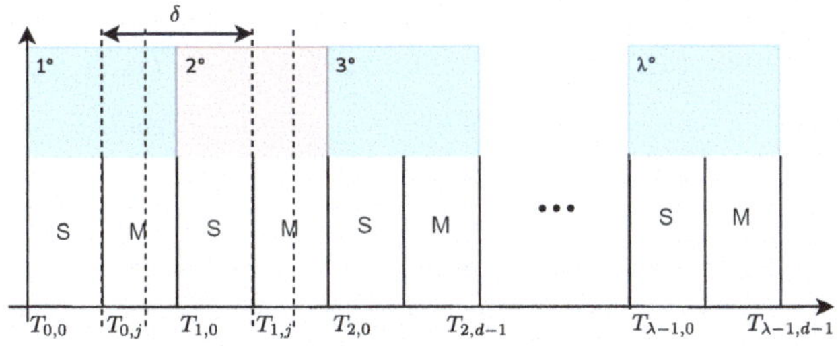

Fig. 1. Computation of a square (S in figure)-and-multiply (M in figure)-always algorithm. Cross-correlation computed between sections, one from the i-th iteration and the other from the $i+1$-th iteration, with λ total iterations. δ depicts the distance between sections of which we compute the cross-correlation. Each iteration of the square-and-multiply-algorithm is m samples long. The entire trace will have length of λm. Subsequent iterations highlighted alternating red and blue.

correlation between two generic iteration traces, and as $T_{i,j} \star T_{i',j'}$ the one between two generic sections of iteration traces.

Figure 1 illustrates an execution of the loop in lines 4–10 of the square-and-multiply-always algorithm (Algorithm 1), assuming an exponent of λ bits. We analyze the iteration traces by pairing the ones in consecutive iterations, which are highlighted in alternating colors. Thus, each section of an iteration trace corresponds to part of one iteration and a portion of the next, separated by a time distance δ. The two j-th sections, $T_{0,j}$ and $T_{1,j}$, within the first two iteration traces T_0 and T_1 are indicated by dashed lines. Our goal is to examine the similarity between sections of different iteration traces, as variations in similarity are likely linked to the secret exponent bit e_i used in the i-th iteration. We note that high similarities can occur between sections separated by a time distance δ shorter than a complete iteration trace. For instance, in the case of two subsequent square-and-multiply-always iterations, where the first iteration is employing a null exponent bit, we expect that the section(s) containing the power consumption from the `store` operations of the result of the square operation will exhibit high similarity to the sections where the second iteration performs `loads` of the same value, as the same value is involved in both operation sequences. The latter sections will be closer to the beginning of their own iteration trace than the former sections will be to the beginning of the first iteration trace, resulting in a value of δ smaller than the length of a single iteration trace m.

We summarize the information provided by cross correlations in the similarity matrix $\mathbf{S}_\delta$, which captures the maximum value of the cross-correlation between any pair of sections of two consecutive iterations having distance δ, where the parameter δ takes values in $\{m, m+l, m+2l, \ldots, m+(d-1)l\}$. Denoting with γ the value $\frac{\delta - m}{l}$, we have that the matrix $\mathbf{S}_\delta$ is defined as follows:

$$\mathbf{S}_\delta = \begin{bmatrix} \max_\tau(T_{0,0} \star T_{1,0+\gamma}) \ldots \max_\tau(T_{(\lambda-2),0} \star T_{(\lambda-1),0+\gamma}) \\ \max_\tau(T_{0,1} \star T_{1,1+\gamma}) \ldots \max_\tau(T_{(\lambda-2),1} \star T_{(\lambda-1),1+\gamma}) \\ \vdots \qquad\qquad \ddots \qquad\qquad \vdots \\ \max_\tau(T_{0,l} \star T_{1,l+\gamma}) \ldots \max_\tau(T_{(\lambda-2),l} \star T_{(\lambda-1),l+\gamma}) \end{bmatrix}$$

We choose to use the maximum value of each cross-correlation $(T_{i,j}\star T_{i+1,j+\gamma})$ across all possible time displacements τ since our division of iteration traces into sections does not rely on the structure of the computed algorithm. Consequently, sequences of instructions with similar power consumption may be unevenly split across sections. Furthermore, the effect of differences in execution times for certain operations appearing despite constant time implementations (e.g., the ones caused by clock jitter), are mitigated by taking the maximum cross-correlation value over τ.

Employing the $\mathbf{S}_\delta$ matrix allows us to generalize on previous collision-based attacks [10, 19]: indeed, they exploit the information coming from a subset of the elements of the $\mathbf{S}_\delta$ matrix, which are hand-picked with the explicit knowledge of the algorithm being executed. The selection criterion of previous collision attacks is finding, in the algorithm being executed, operations that are expected to exhibit similarities, and measure only such similarities.

We consider only values of $\delta \in \{m, m+l, m+2l, \ldots, m+(d-1)l\}$, as such offsets between compared iteration trace sections are already enough to extract a significant amount of information from the single power trace. We note that, albeit at a larger computation cost, extending the set of possible values of δ to $\{m, m+l, m+2l, \ldots, m+(d-1)l, m+dl, m+(d+1)l \ldots m\lambda-l\}$, that is considering also the similarities of sections of non-consecutive iteration traces, may yield additional information. In the following, we restrict to extracting information out of sections $T_{i,j}$ and $T_{i+1,j}$ of contiguous iteration traces, considering only the values $\max_\tau(T_{i,j} \star T_{i+1,j+\delta})$ of $\mathbf{S}_\delta$ where $0 \leq j+\delta \leq d-1$, i.e., the valid values for the offset δ so that the cross-correlation is computed between sections of two subsequent iteration traces.

Finally, we map the information contained in $\mathbf{S}_\delta$ into actual values for the secret used in each algorithm iteration. Consider the secret value e, divided into portions corresponding to each of the λ iterations. Each portion takes a value from a fixed finite set E. In our example, the portions of e are single bits, $e_i, i \in \{0, \ldots, \lambda-1\}$, which belong to $E = 0, 1$. The columns of $\mathbf{S}_\delta$ are classified depending on the values of the exponent bit e_i used in computing the iteration trace T_i, which has its sections serving as the first operands in the cross-correlations defining the column. Alternatively, if the secret values per iteration are two, an effective approach is to classify pairs of adjacent columns of $\mathbf{S}_\delta$ according to whether the values of e_i and e_{i+1} match. In this case, while the extracted information doesn't directly reveal the entire secret e, an exhaustive search guessing the value of e_0 is sufficient to derive it.

Both the classification tasks mentioned can be performed with any classifier, from a simple threshold-based discriminator on the values of the columns of $\mathbf{S}_\delta$ to more refined machine learning methods.

4 Experimental Validation

In this section, we detail our experimental validation of the proposed attack technique. To this end, we implemented the square-and-multiply-always procedure (Algorithm 1) using the multiple precision arithmetic library from the `mbedtls` open-source cryptographic library [22]. For the sake of representation, we depict the results of the attacks with a 16 bit secret exponent e, noting that our attack scales linearly with the increase in the number of exponent bits and requires only a repetition of this procedure to extract the whole secret. To measure power consumption traces from our implementation, we used a ChipWhisperer 1200 Pro digital sampling kit and ran our code on a 32-bit ARM Cortex-M4 microcontroller. Sampling was synchronized with the target device's clock, capturing one sample per clock cycle, with the victim being clocked with a 7.37 MHz external crystal.

To select the value d, i.e., the number of sections into which each iteration trace should be split, we tested our attack procedure with values of $d \in \{10, \ldots, 50\}$, In our case study, a value of $d = 22$ was sufficient to clearly distinguish the features in the $\mathbf{S}_\delta$ matrix. It should be noted that the optimal value of d may vary in other situations, but is easily fine-tuned, or aggressively overestimated, so that matrix features are evident before proceeding with the attack strategy.

Figure 2 shows the $\mathbf{S}_\delta$ matrices for $\delta = m$, calculated on pairs of execution traces with $\lambda = 4$ bit exponents $e' = (e'_3, e'_2, e'_1, e'_0)$ and $e'' = (e''_3, e''_2, e''_1, e''_0)$. We depict eight distinct exponent pattern differences, from an all-zeros exponent against other patterns to an all-ones exponent against other patterns (the first four bits are reported for each exponent). The difference matrices are color-coded to show correlation values, with brighter colors indicating higher cross-correlation. This depiction highlights which section pairs have higher or lower correlation based on the exponent bit value. By examining two subfigures comparing exponents with a one-bit difference i.e., $(e', e'') = (0000, 1000)$ (as 0000_1000) and $(e', e'') = (1111, 0111)$ (as 1111_0111 in Fig. 2), we note that iterations with identical secret values show no correlation differences with the following iteration (seen as columns in Fig. 2). However, when exponent bits differ (e.g., $e'_3 \neq e''_3$), significant differences in correlation (yellow pixels in the first column) appear. These correlation differences occur between different section pairs, depending on whether e'_3 is zero or one, indicating that information about e'_3 is present in the column. Examining the cases $(e', e'') = (0000, 0111)$ and $(e', e'') = (1111, 1000)$ (as 0000_0111 and 1111_1000 in Fig. 2), we observe that the $\mathbf{S}_\delta$ values also reveal whether a given exponent bit changes. Specifically, when comparing correlations on the first differing exponent bit (e'_2) with those from subsequent differing bits, we see distinct positions where correlation changes significantly. For example, in $(e', e'') = (0000, 0111)$, the correlation difference in

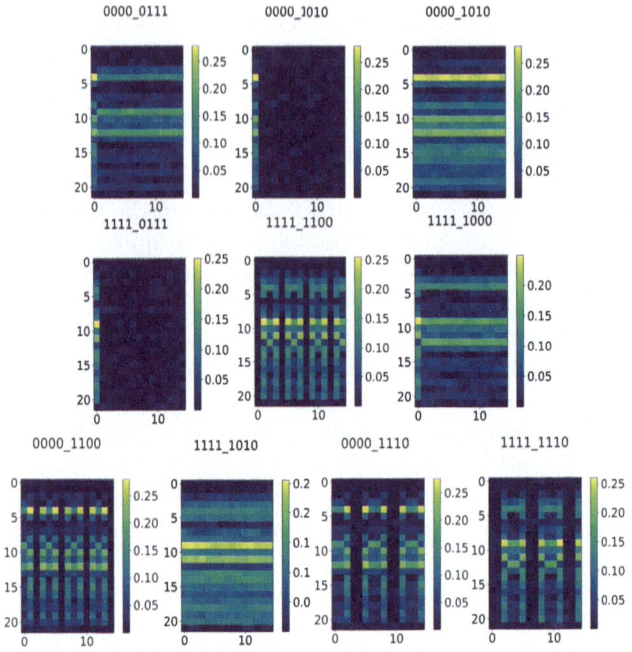

Fig. 2. Differences between pairs of $\mathbf{S}_\delta$ matrices for $\delta = m, \lambda = 4$ and $d = 22$. The pairs of matrices are obtained from execution traces with four bit exponents reported on top of each subfigure of differences between couples of cross-correlation matrices.

the fifth section (shown as the fifth row of pixels in subfigure 0000_0111) lessens when the exponent bit e_2'' is 1 for the first time, compared to subsequent cases.

Building on these observations, which validate our expectations that $\mathbf{S}_\delta$ values for $\delta = m$ capture information about both the values and transitions of bits in the secret exponent e, we extend our analysis to cases where $\delta \neq m$. We analyzed four specific patterns that represent all possible transitions between exponent bit values in consecutive iterations (00, 01, 10, 11). For this analysis, we considered $\mathbf{S}_\delta$ in the $\lambda = 2$ case and consolidated various $\mathbf{S}_\delta$ matrices into a single column. Figure 3 visually reports the results, organized by the four possible cases. Four different patterns in the correlations emerge depending on both the value and the changing of the exponent, pointing to an explicit information leakage.

Although simple visual inspection, in the fashion of simple power analysis attacks [21] is sufficient to derive either the values of the secret exponent or the transitions across iterations, we completed our attack procedure by mechanizing the exponent value extraction. Due to the significant amount of information present in the $\mathbf{S}_\delta$ matrices, we opted for a simple threshold classifier, which employs the x-y positions of the peaks in cross correlations appearing in the $\mathbf{S}_\delta$ matrix of an iteration to decide to which of the classes the matrix belongs. While the classification logic can be extended to complex patterns, in our case

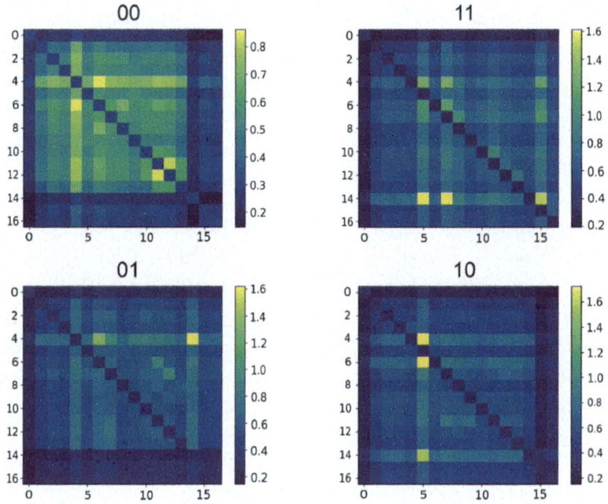

Fig. 3. Matrices of correlation computed between the one iteration and the following one of the square-and-multiply-always, focusing on the multiplication execution. Correlation from cells on the diagonal omitted as to highlighting the results when $\delta \neq 0$. We report four patterns, for each possible transition of the exponent bit in one iteration and the next. The color bars show the color code for the correlation values(from the lowest cross-correlation in blue to the highest in yellow).

study it is sufficient to distinguish between two fundamental classes of $\mathbf{S}_\delta$ matrices. In the first category, correlation peaks appear symmetrically w.r.t. the main diagonal point to cross-correlations associated with cases where δ equals the section length, i.e. between two instances of the same operation in different iterations. The second category consists of $\mathbf{S}_\delta$ matrices with non-diagonal symmetric peaks, indicating correlations between different operations. Using the described methodology, we retrieved with 100% accuracy the exponent bits for five different exponent patterns by analyzing correlation matrices focused on multiplication execution with variable δ, shown in Fig. 3.

5 Conclusion

In this work, we proposed a novel single-trace side-channel attack targeting iterative cryptographic algorithms where the secret value is used piecewise across iterations. Our approach detects similarities between portions of the device's behavior across subsequent iterations, without requiring prior knowledge of the algorithm being executed, and was validated as working on a real world setup with a COTS microcontroller. As a side benefit, our methodology also finds and highlights to the developer the timing at which side channel leakages occur, providing help in designing countermeasures.

Acknowledgements. This work was supported in part by project SERICS (PE00000014) under the Italian NRRP MUR program funded by the EU - NGEU.

References

1. Agrawal, D., Archambeault, B., Rao, J.R., Rohatgi, P.: The EM side—channel(s). In: Kaliski, B.S., Koç, K., Paar, C. (eds.) CHES 2002. LNCS, vol. 2523, pp. 29–45. Springer, Heidelberg (2003). https://doi.org/10.1007/3-540-36400-5_4
2. Amiel, F., Feix, B.: On the BRIP algorithms security for RSA. In: Onieva, J.A., Sauveron, D., Chaumette, S., Gollmann, D., Markantonakis, K. (eds.) WISTP 2008. LNCS, vol. 5019, pp. 136–149. Springer, Heidelberg (2008). https://doi.org/10.1007/978-3-540-79966-5_10
3. Antognazza, F., Barenghi, A., Pelosi, G.: Metis: an integrated morphing engine CPU to protect against side channel attacks. IEEE Access **9**, 69210–69225 (2021). https://doi.org/10.1109/ACCESS.2021.3077977
4. Archambeau, C., Peeters, E., Standaert, F.-X., Quisquater, J.-J.: Template attacks in principal subspaces. In: Goubin, L., Matsui, M. (eds.) CHES 2006. LNCS, vol. 4249, pp. 1–14. Springer, Heidelberg (2006). https://doi.org/10.1007/11894063_1
5. Barenghi, A., Breveglieri, L., Izzo, N., Pelosi, G.: Exploring cortex-m microarchitectural side channel information leakage. IEEE Access **9**, 156507–156527 (2021). https://doi.org/10.1109/ACCESS.2021.3124761
6. Barenghi, A., Carrera, D., Mella, S., Pace, A., Pelosi, G., Susella, R.: Profiled attacks against the elliptic curve scalar point multiplication using neural networks. In: Yang, M., Chen, C., Liu, Y. (eds.) NSS 2021. LNCS, vol. 13041, pp. 238–257. Springer, Cham (2021). https://doi.org/10.1007/978-3-030-92708-0_15
7. Barenghi, A., Carrera, D., Mella, S., Pace, A., Pelosi, G., Susella, R.: Profiled side channel attacks against the RSA cryptosystem using neural networks. J. Inf. Secur. Appl. **66**, 103122 (2022). https://doi.org/10.1016/J.JISA.2022.103122
8. Barenghi, A., Pelosi, G.: Side-channel security of superscalar CPUs: evaluating the impact of micro-architectural features. In: Proceedings of the 55th Annual Design Automation Conference, DAC 2018, San Francisco, CA, USA, 24–29 June 2018, pp. 120:1–120:6. ACM (2018). https://doi.org/10.1145/3195970.3196112
9. Battistello, A., Coron, J.-S., Prouff, E., Zeitoun, R.: Horizontal side-channel attacks and countermeasures on the ISW masking scheme. In: Gierlichs, B., Poschmann, A.Y. (eds.) CHES 2016. LNCS, vol. 9813, pp. 23–39. Springer, Heidelberg (2016). https://doi.org/10.1007/978-3-662-53140-2_2
10. Bauer, A., Jaulmes, É., Prouff, E., Reinhard, J., Wild, J.: Horizontal Collision Correlation Attack on Elliptic Curves. IACR Cryptology ePrint Archive, p. 321 (2019)
11. Capoferri, R., Barenghi, A., Breveglieri, L., Izzo, N., Pelosi, G.: A comparison of deep learning approaches for power-based side-channel attacks. In: Martucci, L.A., Kamm, L., Iwaya, L.H., Pulls, T. (eds.) Secure IT Systems - 29th Nordic Conference, NordSec 2024, Karlstad, Sweden, 6–7 November 2024, Proceedings, Lecture Notes in Computer Science, pp. 10–16. Springer (2014)
12. Chari, S., Rao, J.R., Rohatgi, P.: Template attacks. In: Kaliski, B.S., Koç, K., Paar, C. (eds.) CHES 2002. LNCS, vol. 2523, pp. 13–28. Springer, Heidelberg (2003). https://doi.org/10.1007/3-540-36400-5_3

13. Chevallier-Mames, B., Ciet, M., Joye, M.: Low-cost solutions for preventing simple side-channel analysis: side-channel atomicity. IEEE Trans. Comput. **53**(6), 760–768 (2004). https://doi.org/10.1109/TC.2004.13
14. Clavier, C., Feix, B., Gagnerot, G., Roussellet, M., Verneuil, V.: Horizontal correlation analysis on exponentiation. In: Soriano, M., Qing, S., López, J. (eds.) ICICS 2010. LNCS, vol. 6476, pp. 46–61. Springer, Heidelberg (2010). https://doi.org/10.1007/978-3-642-17650-0_5
15. Danger, J.-L., Guilley, S., Hoogvorst, P., Murdica, C., Naccache, D.: Improving the big mac attack on elliptic curve cryptography. In: Ryan, P., Naccache, D., Quisquater, J.-J. (eds.) The New Codebreakers. LNCS, vol. 9100, pp. 374–386. Springer, Heidelberg (2016). https://doi.org/10.1007/978-3-662-49301-4_23
16. Fouque, P.-A., Valette, F.: The doubling attack – *why upwards is better than downwards*. In: Walter, C.D., Koç, Ç.K., Paar, C. (eds.) CHES 2003. LNCS, vol. 2779, pp. 269–280. Springer, Heidelberg (2003). https://doi.org/10.1007/978-3-540-45238-6_22
17. Gérard, B., Standaert, F.: Unified and optimized linear collision attacks and their application in a non-profiled setting: extended version, pp. 45–58 (2013). https://doi.org/10.1007/S13389-013-0051-9
18. Giraud, C., Verneuil, V.: Atomicity improvement for elliptic curve scalar multiplication. In: Gollmann, D., Lanet, J.-L., Iguchi-Cartigny, J. (eds.) CARDIS 2010. LNCS, vol. 6035, pp. 80–101. Springer, Heidelberg (2010). https://doi.org/10.1007/978-3-642-12510-2_7
19. Hanley, N., Kim, H.S., Tunstall, M.: Exploiting collisions in addition chain-based exponentiation algorithms using a single trace. In: Nyberg, K. (ed.) CT-RSA 2015. LNCS, vol. 9048, pp. 431–448. Springer, Cham (2015). https://doi.org/10.1007/978-3-319-16715-2_23
20. Kocher, P.C.: Timing attacks on implementations of Diffie-Hellman, RSA, DSS, and other systems. In: Koblitz, N. (ed.) CRYPTO 1996. LNCS, vol. 1109, pp. 104–113. Springer, Heidelberg (1996). https://doi.org/10.1007/3-540-68697-5_9
21. Mangard, S., Oswald, E., Popp, T.: Power analysis attacks - revealing the secrets of smart cards. Springer (2007)
22. Mbed-TLS Contributors: Mbed TLS cryptographic library (2024). https://github.com/Mbed-TLS/mbedtls
23. Menezes, A., van Oorschot, P.C., Vanstone, S.A.: Handbook of Applied Cryptography. CRC Press (1996). https://doi.org/10.1201/9781439821916
24. Piacentini, I., Barenghi, A., Pelosi, G.: A non profiled and profiled side channel attack countermeasure through computation interleaving. In: 26th Euromicro Conference on Digital System Design, DSD 2023, Golem, Albania, 6–8 September 2023, pp. 718–725. IEEE (2023). https://doi.org/10.1109/DSD60849.2023.00103
25. Walter, C.D.: Sliding windows succumbs to big mac attack. In: Koç, Ç.K., Naccache, D., Paar, C. (eds.) CHES 2001. LNCS, vol. 2162, pp. 286–299. Springer, Heidelberg (2001). https://doi.org/10.1007/3-540-44709-1_24

Capstone Project Design for an Undergraduate Cybersecurity Program

Nooredin Etezady(✉) and Ping Wang

Robert Morris University, Pittsburgh, USA
{etezady,wangp}@rmu.edu

Abstract. Capstone projects provide integrated learning experience and outcomes for comprehensive technical and non-technical knowledge, skills, abilities (KSAs) and professional competencies needed for the educational assessment and workforce development in cybersecurity. This research is a case study for designing a comprehensive portfolio-type project for the capstone course of an undergraduate degree program in Cybersecurity in the United States. The proposed design in the case study incorporates the cyber workforce competencies and knowledge units for the National Centers of Academic Excellence in Cyber Defense designation as well as student outcomes for the ABET computing accreditation commission (CAC) standards for quality assurance of cybersecurity programs. The case study contributes a sample comprehensive capstone project design with key components, objectives, deliverables, and a sample rubric for assessment.

Keywords: Capstone project · project-based learning (PBL) · cybersecurity · simulation · KSAs · competencies

1 Introduction

Cybersecurity is growing in scope and importance. There is an increasing gap between industry demand and skilled cyber talent [1, 2]. A recent study has noted that employers complain about lack of experience in entry level job seekers and the need for more experienced job candidates [3]. By the same token, students seeking positions may complain about employers' high skill requirements. To protect continuously evolving networked information systems against cyber-attacks require skilled cybersecurity professionals who can solve new complex problems [4].

The Workforce Framework for Cybersecurity (NICE Framework) is "a nationally focused resource to help employers develop their cybersecurity workforce". The Section on Cybersecurity Curriculum Development is a guide for educational institutions who are "responsible for developing, planning, coordinating, and evaluating cybersecurity awareness, training, or education content, methods, and techniques based on instructional needs and requirements" [5]. Cybersecurity curricula at colleges and universities should consider the required knowledge, skills, and abilities (KSAs) that students will need to fulfill work roles as well as the Knowledge Units (KUs) the program of study will need

S. Latifi (Ed.): ITNG 2025, AISC 1463, pp. 84–94, 2025.
https://doi.org/10.1007/978-3-031-89063-5_8

to include to be designated a National Center of Academic Excellence in Cybersecurity (NCAE-C) [5, 6].

The knowledge, skills, and abilities (KSAs) make up the professional competencies necessary for cybersecurity jobs and need to be prioritized in the relevant cybersecurity curricula [1, 7]. The lack of the required KSAs and competencies among cybersecurity graduates have resulted in the formation of workforce development coalitions around the world to address the growing gap between skilled cyber talent and industry demand [2]. The cybersecurity talent deficit has led to a call for improvements in cybersecurity training curricula [1].

An important result of college education is the effective preparation of students for careers. However, employers and the public have expressed concerns that universities do not prepare students sufficiently [8]. Higher education institutions are seeking to align their academic programs with the industry requirements through existing options [2].

The lack of the required skills in graduating students can be attributed to insufficient opportunities for experiential learning. To prepare cybersecurity students for work, they can be given real-world tasks in mentored environments where they can apply their knowledge and abilities [4].

The available options to remedy the issue of the gap between graduating students' technical skills and the job market requirements are the capstone projects and internships [3]. Seniors nearing graduation who have taken the full cybersecurity curriculum normally take capstone courses. Therefore, the capstone class is taught at a high level [9]. Many institutions of higher education recognize the benefit of capstones as the students move from postsecondary education into the workforce. The capstones strengthen the link between higher education and careers by providing experiential and applied learning to the students [8]. However, the capstone projects offered by many higher education institutions do not focus on activities that address skill development. The internships, nevertheless, are not always in sync with career interests of students [3].

The practical experiences provided by capstone courses may include analyzing a case study or simulation of real-life work cases in the field of study. Prior research shows that dedicated capstone courses that infuse career-related experiences such as assignments and activities to analyze case studies, real-world simulations, practical field-based problems, and learning from industry professionals are valuable. The capstone courses help students prepare for their careers in the college/university settings and provide the practical relevance of undergraduate education that the public and employers seek. These courses help students to prepare for the transition from their academic work to solving real-world problems [8].

This paper examines a comprehensive portfolio-type project of a capstone course for an undergraduate degree program in cybersecurity. The capstone course case study design is based on NCAE-C and ABET Cybersecurity program expectations and is targeted for undergraduate cybersecurity majors to provide them with the required technical and non-technical skills for a cybersecurity career. The paper continues with the background literature review followed by the capstone project design in the case study methodology with discussions and conclusions.

2 Background

2.1 Educational Theories on Projects and Capstones

Protecting organizational information assets requires dealing with complex cybersecurity issues. To provide students with a well-rounded cybersecurity education that will help them to succeed in the marketplace, in addition to solidly grounding them in cybersecurity knowledge and technical skills, they need to be challenged to consider problems from multiple perspectives in an engaging environment [9].

Students develop new skills through experiential learning by immersing in an experience and then reflecting on the experience [10]. Project based learning (PBL) has been considered by educators as a valuable methodology for student learning in various fields, especially in science and engineering. Students work on real-world activities in a safe environment while applying their knowledge to obtain experience [4].

The Experiential Learning Theory defines learning by going through the process of experiencing, reflecting, thinking, and acting to construct knowledge [11]. The primary stages of learning are experiencing, reflection, and applying. What makes experiential learning more powerful than other models such as "learn by doing" or "hands on learning" is the stages of reflection and application [12]. Experiential learning can be enhanced by creating learning spaces that foster "growth producing experiences for learners" [11].

Capstone projects offered in the final year of the college education assess general mastery of concepts, critical thinking, problem solving, and transferrable skills. Capstone projects are a necessary component of any accredited programs in higher education institutions, such as computer science, computer engineering, information systems, and cybersecurity, among others [13].

Yang (2024) published a recent study on the required curriculum for undergraduate cybersecurity programs to meet cybersecurity needs of employers [14]. The study was based on the Knowledge Units published by the National Center of Academic Excellence in Cybersecurity program and the IS2020 required competencies published by The Joint ACM/AIS Task Force [15, 16]. The results of the study showed that undergraduate cybersecurity programs should conclude with capstone projects to integrate the learning and experiences obtained from the cybersecurity program curriculum.

Although capstone projects have been identified as necessary for undergraduate cybersecurity programs to assess general mastery of cybersecurity knowledge and critical thinking, the literature search for this study did not identify any prior studies on undergraduate cybersecurity capstone project design. This paper will attempt to contribute a comprehensive portfolio-type project design for a capstone course in an undergraduate cybersecurity program.

2.2 Project Based Learning (PBL) Implementations

The implementations of many project-based learning (PBL) capstone courses in reality are mainly theoretical and academic and lack realistic problem-solving activities. They are clearly short of activities that focus on skill development. Capstone courses need to be designed to match career trajectories of students and offer an experiential learning opportunity [3].

Capstone courses that integrate academic curriculum with activities that focus on skill development and encourage reflection help students develop new skills, attitudes, and ways of thinking. Capstone courses can create a meaningful connection between the academic curriculum and career experiences by encouraging integration and synthesis of academics with the required career focused skills [3].

2.3 NIST/NICE and NCAE-C Expectations

The U.S. National Initiative for Cybersecurity Education (NICE) Workforce Framework for Cybersecurity (NICE Framework) has defined and mapped various cybersecurity functions, work roles, and tasks with specific technical and non-technical skills, abilities, and competencies [17]. The NICE Framework is a valuable guidance for cybersecurity education, training, and workforce development. The NIST Cybersecurity Framework (CSF) 2.0 provides guidance on how to manage cybersecurity risks and offers a taxonomy of high-level cybersecurity outcomes organized by these core functions: govern, identify, protect, detect, respond, and recover [18]. These core functions indicate comprehensive technical and non-technical cybersecurity tasks and demands for cybersecurity professionals. Accordingly, these expectations for cybersecurity professional qualifications should be reflected in cybersecurity education and training.

The U.S. National Centers of Academic Excellence in Cybersecurity (NCAE-C) designation program jointly sponsored by the U.S. National Security Agency (NSA) and Department of Homeland Security (DHS) is a national standard and community for maintaining high quality cybersecurity education with rigorous and consistent requirements for program evaluation. The NCAE-C designated programs must demonstrate proof or alignment to and implementation of specific cybersecurity knowledge units (KUs), including technical KUs such as network security and intrusion detection and non-technical KUs such as cybersecurity policy, ethics, and risk analysis [19]. A comprehensive cybersecurity project such as a capstone project is the best option to include training and assessment of both technical and non-technical KUs for a high-quality cybersecurity education program recognized by the NCAE-C community.

2.4 ABET Expectations

ABET (Accreditation Board for Engineering and Technology) is a globally recognized independent and non-profit organization that accredits university and college programs in STEM (science, technology, engineering, and math). ABET CAC (Computing Accreditation Commission) accredits computing programs including bachelor's degree programs in Cybersecurity and similarly named programs. The student outcomes criterion for ABET-accredited Cybersecurity programs expect the following abilities [20]:

1) *Analyze a complex computing problem and apply principles of computing and other relevant disciplines to identify solutions.*
2) *Design, implement, and evaluate a computing-based solution to meet a given set of computing requirements in the context of the program's discipline.*
3) *Communicate effectively in a variety of professional contexts.*

4) *Recognize professional responsibilities and make informed judgments in computing practice based on legal and ethical principles.*
5) *Function effectively as a member or leader of a team engaged in activities appropriate to the program's discipline.*
6) *Apply security principles and practices to maintain operations in the presence of risks and threats.*

ABET student outcomes serve as important expectations and guidance for pedagogical designs and assessment metrics for achievements of KSAs, competencies, and quality assurance of educational programs in cybersecurity accredited by ABET [7]. A comprehensive project in a capstone course should be the best way for students to develop and demonstrate the learning outcomes and achieve both technical and non-technical professional abilities and competencies listed above. In reality, computing and cybersecurity programs accredited by ABET CAC usually have one or more comprehensive capstone project to meet the expectations for student outcomes based on the aggregate observations by one of the co-authors of this paper who has been serving as an ABET CAC program evaluator for years.

3 Project Design Model

Figure 1 is the project design model proposed for this study.

Figure 1 shows the flowchart of five key steps for the proposed capstone project design. In the first step, the student chooses a hypothetical or real (if available) enterprise organization from a list of categories for the project. Second, the student will identify the critical assets and top three most harmful cybersecurity threats, risks, or attacks for the selected organization. Third, the student is expected to produce an initial report of findings including the cyber threat risks and their impact on the organization along with a real recent case and relevant data for each of the top three types of cyber threats, risks or attacks. The initial report can be used as the first major milestone or midterm report for the capstone course.

Based on the review feedback from the capstone instructor on the initial report, the student will proceed with the fourth step of identifying, testing, and evaluating security solutions to mitigate and manage the identified security threats and risks for the organization. The results from the fourth step will help the student to complete the final step of the final report. The final report should provide comprehensive solutions including technical, policy, management, and training solutions along with discussions, and evaluations of the solutions supported by test results and research data cited. The final report should also include hands-on demos, test results, tools and techniques used in the project. During the project, the instructor may also assign relevant hands-on assignments to help students refresh or catch up with necessary tools and techniques for identifying and testing security risks and solutions.

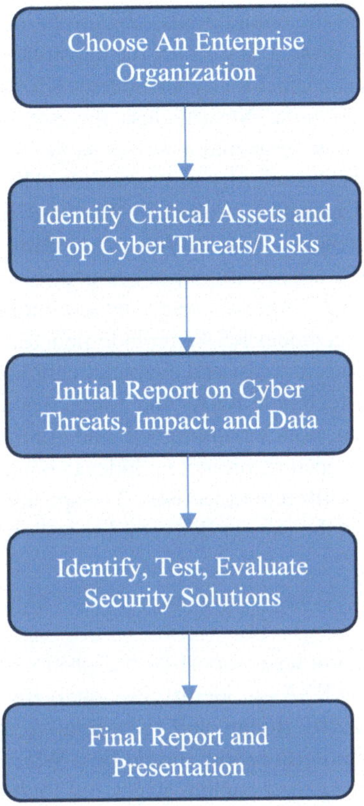

Fig. 1. Proposed Capstone Project Design Model

4 Case Study

Based on the proposed capstone project design model above, this research presents a case study of a sample cybersecurity capstone project for a Bachelor of Science (B.S.) degree program in Cybersecurity at a 4-year university in the United States. This degree program of study has been validated for the National Center of Academic Excellence in Cyber Defense (NCAE-CD) designation by the U.S. National Security Agency and Department of Homeland Security. The degree program is also in good standing of accreditation by the ABET Computing Accreditation Commission (CAC). The curriculum of the B.S. in Cybersecurity program includes technical core courses such as network security, ethical hacking, cryptography, digital forensics and non-technical courses such as cybersecurity risk management, incident response, and policy and compliance, in addition to general education and open elective courses. Cybersecurity Capstone is the final 3-credit course taken for program completion. The curriculum design is consistent with the comprehensive cybersecurity workforce expectations of the NCAE-CD designation and the student outcomes of the ABET CAC accreditation for achieving both technical and non-technical KSAs and competencies necessary for the cybersecurity profession. The capstone course work is centered on a comprehensive project. The following paragraphs present the components, objectives, deliverables, and assessment rubric of the comprehensive project design for the capstone course.

It would be unrealistic to expect all cybersecurity KSAs and competencies from a project or a 3-credit capstone course. Nevertheless, the proposed capstone project design of this research does address several major KSAs as key components of professional competencies necessary for cybersecurity professionals. The key steps and components addressed by the proposed capstone project in this case include technical and non-technical KSAs and competencies in identifying, describing, analyzing, responding to, and managing enterprise and organizational cybersecurity vulnerabilities, threats, and risks as well as in identifying, implementing, and evaluating corresponding cybersecurity solutions. Important technical KSAs include penetrating testing to identify and analyze security vulnerabilities and test and evaluate technical controls and solutions to mitigate security risks. Non-technical KSAs and competencies include business impact analysis for risk management, policy compliance, and communication of security risks and solutions. The proposed capstone project includes two major deliverables: a midterm report and a final report with different objectives. The specific components of KSAs and competencies are reflected in the project deliverables and objectives below.

4.1 Midterm Report

The main objectives of the midterm report of the proposed capstone project are for students to demonstrate their KSAs in identifying, analyzing, and assessing enterprise and organizational cybersecurity threats and risks relative to the critical assets of the selected organization. The midterm report should be a Word document of about 1000–1500 words (about 6 double-spaced content pages not including the cover page, table of contents, and reference list). The midterm report should have the following steps and content sections:

I. Introduction (about 1–1.5 pages): Choose a hypothetical or real (if available) organization in one of these categories for this project: hospital/healthcare, e-Commerce vendor, financial institution, government agency, utility service, transportation service, etc. for the capstone project. Identify and briefly describe the selected organization, their critical assets, and top 3 types of cybersecurity threats and risks to their critical assets. Suggested examples of the threats and risks include malware, DDoS attack, ransomware, phishing, DNS poisoning, buffer overflow, cross-site scripting, digital forensic investigation, or any other type of security incident for response and handling. It is important that students are given the freedom and flexibility to choose the types of organization and security risks and incidents of their interest to explore the wide variety of cybersecurity issues and make the best use of their knowledge and skills learned.

II. Risks and Impact (about 3 pages): Describe and prioritize the identified top 3 cybersecurity threats and risks in terms of their impact to the organization and discuss the details of the significance and impact of the threat or risks (technical, financial, and social, etc.) to the selected organization. Students are expected to understand, describe, and analyze the threats and risks and their realistic and potential consequences and impact to the enterprise or organizational mission.

III. Cases and Data (about 1–2 pages): Students are expected to find a real recent case and relevant statistical data for each of the 3 types of cyber threat and/or attack appropriate to the selected type of organization and describe the case and data for each type of threat or attack.

In addition, certain supporting hands-on lab assignments can be used to refresh and enhance students' knowledge and skills in cybersecurity threats and risks and help with their midterm report. Example lab resources for this purpose for instructor's selection include interactive lab simulations of threat hunting and vulnerability scanning and assessment activities in the CyberDefense Pro LabSim product from CompTIA TestOut and pre-built virtual machines of cybersecurity labs in various categories offered from SEED Labs [21, 22].

4.2 Final Report

The main objectives of the final report of the proposed capstone project are for the students to practice and demonstrate comprehensive knowledge and skills in handling and managing enterprise and organizational cybersecurity risks. The final report not only emphasizes realistic problem solving skills but also develops and demonstrates students' professional competencies in evaluating and communicating security solutions for mitigating and managing risks. The final report should include the midterm with updates to address instructor feedback and the following new content sections:

IV. Solutions: Students should provide comprehensive cybersecurity solutions (technical and non-technical controls and recommendations) for each of the 3 identified cyber threats and risks for the selected organization. They are expected to discuss the solutions in detail with hands-on demos, test results, tools and techniques involved, and research citations for support. A suggested approximate length of this section is about 6 pages or 1,200 words. Additional supporting hands-on lab assignments can be used to help students with technical and non-technical knowledge and skills necessary for the demos and evaluation of the solutions. As a resource suggestion, the resources and simulation labs in the CyberDefense Pro LabSim product from CompTIA TestOut and from SEED Labs can be selectively used by the instructor with customization for assignments to develop and practice students' KSAs in security controls for handling, mitigating, and managing various security vulnerabilities, threats, and risks [21, 22].

V. References: The reference section should include 10 or published sources listing the sources used and cited in the report to develop students' respect for intellectual property and avoid the misconduct of plagiarism. The format of listing and citing the references may vary with institutional and instructor preference.

VI. Appendix – Resume: Each student is expected to create a professional resume (about 2 single-space pages) organized with relevant information targeting an entry-level cybersecurity position the student has found based on his or her interest. The information in the resume may include relevant education, training, degrees, certifications, work experience, internships, cyber competitions and skills, etc. The information in the resume may be real or hypothetical as the student's understanding of the expectations of the KSAs and competencies for the position is more important.

VII. Presentation: Students should create a PowerPoint presentation of about 8–10 slides to present the main content of the entire project report from sections I Introduction to Section IV Solutions. This is to develop and demonstrate students' competency of communicating technical and non-technical issues in the cybersecurity context.

Table 1 below is a suggested sample rubric for assessing the final project report. The instructor may customize the capstone project and assign small teams (2 or 3 students per team) for each project. The assessment rubric may be adjusted accordingly to include criteria of teamwork skills and performance.

Table 1. Rubric for Final Report Assessment

Criteria	Levels of Achievement			
	Excellent	Good	Satisfactory	Below Expectations
Solutions (Weight: 80%)	**72–80 Points** Excellent description and discussion of the solutions supported by research and data	**64–71 Points** Good description and discussion of the solutions supported by research and data	**56–63 Points** Adequate description and discussion of the solutions supported by research and data	**Below 56 Points** Inadequate description and discussion of the solutions with little research or supporting data
Resume (Weight: 10%)	**9–10 Points** Excellent resume with very clear and relevant information	**8 Points** Good resume with quite clear and relevant information	**7 Points** Acceptable resume with mostly clear and relevant information	**Below 7 Points** Poor resume with unclear or irrelevant info
Writing & Presenting (Weight: 10%)	**9–10 Points** Excellent writing and presentation with no errors	**8 Points** Good writing and presentation with few minor errors	**7 Points** Acceptable writing and presentation with a few errors	**Below 7 Points** Poor with some or frequent errors

5 Conclusions

Capstone projects in the final year of college education assess general mastery of concepts, critical thinking, problem solving, and transferrable KSAs and professional competencies. Capstone projects are a necessary component of any accredited higher education programs in computer science, computer engineering, information systems, and cybersecurity, among others.

Previous studies have shown that undergraduate cybersecurity programs offering capstone projects allow the integration of the academic learning and career type experiences to better prepare students for cybersecurity career opportunities. However, many capstone courses lack activities that center on comprehensive skill development. Capstone courses need to be designed to match career trajectories of students and offer an experiential learning opportunity.

Capstone projects have been identified as necessary for undergraduate cybersecurity programs to assess general mastery of cybersecurity knowledge and critical thinking. However, our literature search did not identify any prior studies on undergraduate cybersecurity capstone project design. This paper contributes a proposed sample of comprehensive portfolio-type project for a capstone course for an undergraduate degree program in cybersecurity. The capstone project case study design was based on NCAE-C and ABET Cybersecurity expectations. In addition, the capstone project encompasses the Experiential Learning Theory stages for constructing knowledge which includes the processes of experiencing, reflecting, thinking, and acting.

To complete the capstone project, students need to possess the general mastery of cybersecurity concepts and have necessary hands-on application skills as well as critical thinking and problem-solving skills.

The proposed sample capstone project in this paper is based on NCAE-C and ABET expectations and requirements for B.S. in Cybersecurity programs. Other schools might have different requirements to meet for their capstone projects. Future studies should evaluate the student performance data for the proposed capstone project to identify its strengths and areas for improvement. Further studies may also be done in comparing capstone projects for undergraduate and graduate level students in cybersecurity.

Acknowledgements. We acknowledge the support for this research from the U.S. National Science Foundation (NSF) grant #2234554.

References

1. Jones, K.S., Namin, A.S., Armstrong, M.E.: The core cyber-defense knowledge, skills, and abilities that cybersecurity students should learn in school: Results from interviews with cybersecurity professionals. ACM Trans. Comput. Educ. (TOCE) **18**(3), 1–12 (2018). https://doi.org/10.1145/3152893
2. Shemroske, K.L., Pourang, K.: Experiential learning and job experience: opportunities for cybersecurity curriculum support, cyber graduate marketability, and cyber productivity. J. Comput. Sci. Coll. **37**(5), 64–75 (2021)
3. Howell, P., Aryal, A.: IS capstone course design: quasi-Internships using Harvard business cases. J. Inf. Syst. Educ. **35**(1), 14–24 (2024). https://doi.org/10.62273/MJVL606
4. Golaszewski, E., et al.: Project-based learning continues to inspire cybersecurity students: the 2018–2019 SFS research studies at UMBC. ACM Inroads **11**(2), 46–54 (2020). https://doi.org/10.1145/3386363
5. NICCS (National Initiative for Cybersecurity Careers and Studies): Workforce Framework for Cybersecurity (NICE Framework). https://niccs.cisa.gov/workforce-development/nice-framework

6. NCYTE Center: NICE Framework and CAE Knowledge Units (KUs). https://www.ncyte.net/faculty/faculty-resources/nice-framework-and-cae-knowledge-units-kus
7. Alammari, A., Sohaib, O., Younes, S.: Developing and evaluating cybersecurity competencies for students in computing programs. PeerJ Comput. Sci. **8**, e827 (2022). https://doi.org/10.7717/peerj-cs.827
8. Kinzie, J., Akyuz, F.: Exploring the influence of course-based career experiences and faculty on students' career preparation. NACE J. (2022). https://www.naceweb.org
9. Knapp, K.J., Maurer, C., Plachkinova, M.: Maintaining a cybersecurity curriculum: professional certification as valuable guidance. J. Inf. Syst. Educ. **28**(2), 101–114 (2017)
10. Lewis, L.H., Williams, C.J.: Experiential learning: past and present. New Dir. Adult Continuing Educ. **62**, 5–16 (1994). https://doi.org/10.1002/ace.36719946203
11. Kolb, A.Y., Kolb, D.A.: Learning styles and learning spaces: enhancing experiential learning in higher education. Acad. Manag. Learn. Educ. **4**(2), 193–212 (2005)
12. Northern Illinois University Center for Innovative Teaching and Learning. Experiential learning. https://www.niu.edu/citl/resources/guides/instructional-guide/experiential-learning.shtml
13. Ogenrwot, D., Tabo, G.O., Aber, K., Nakatumba-Nabende, J.: From undergraduate (software) capstone projects to start-ups: challenges and opportunities in higher institutions of learning. In: Proceedings of the Federated Africa and Middle East Conference on Software Engineering (FAMECSE 2022), Cairo-Kampala, Egypt, pp. 73–82. ACM, New York (2022). https://doi.org/10.1145/3531056.3542775
14. Yang, S.C.: A curriculum model of cybersecurity bachelor's programs in AACSB-Accredited business schools in the U.S. J. Inf. Syst. Educ. **35**(3), 313–324 (2024). https://doi.org/10.62273/FRJE3390
15. National Centers of Academic Excellence in Cybersecurity (NCAE-C). National Centers of Academic Excellence in Cybersecurity NCAE-C 2022: Designation Requirements and Application Process. https://dl.dod.cyber.mil/wp-content/uploads/cae/pdf/unclass-cae-cd_designation_requirements.pdf
16. The Joint ACM/AIS IS2020 Task Force. IS2020: A Competency Model for Undergraduate Programs in Information Systems. ACM, AIS, and ISCAP (2020). https://is2020.hosting2.acm.org/2021/06/01/is2020-final-draft-released/.https://doi.org/10.1145/3460863
17. NICE (National Initiative for Cybersecurity Education). NICE Cybersecurity Workforce Framework (SP800-181 Revision 1) (2020). https://doi.org/10.6028/NIST.SP.800-181r1
18. NIST. NIST Cybersecurity Framework (CSF) 2.0 (2024). https://nvlpubs.nist.gov/nistpubs/CSWP/NIST.CSWP.29.pdf
19. NCAE-C (National Centers of Academic Excellence in Cybersecurity). CAE Requirements and Resources (2020). https://public.cyber.mil/ncae-c/
20. ABET. Criteria for Accrediting Computing Programs, 2024–2025 (2024). https://www.abet.org/accreditation/accreditation-criteria/criteria-for-accrediting-computing-programs-2024-2025/
21. CompTIA+TestOut. CyberDefensePro (2024). https://w3.testout.com/
22. SEED Labs. Hands-on Labs for Security Education (2024). https://seedsecuritylabs.org/index.html

Detecting Advanced Persistent Threats Using Machine Learning and Behavioral Analysis

Nawaf Abdualaziz Almolhis[1](✉) and Mohammed Abdulaziz Alsubhi[2]

[1] College of Engineering and Computer Science, Jazan University, Jazan, Saudi Arabia
naalmolhis@jazanu.edu.sa

[2] Department of Computer Science, University College in Umluj, University of Tabuk, Umluj, Saudi Arabia
malsubhi@ut.edu.sa

Abstract. Due to the elusive and persistent nature of the attacks, the detection of Advanced Persistent Threats (APTs) in cybersecurity is an important issue. Using behavioral analysis and anomaly detection methods, this study proposes a revolutionary machine learning (ML) strategy for spotting APTs. This hybrid approach, which employs decision trees for classification and autoencoders for anomaly detection, combines both supervised and unsupervised learning strategies. These two approaches enable better detection of suspicious patterns and non-standard network activities. To train and test model, this work uses a cybersecurity dataset that includes system logs and network traffic. The dataset covers a wide range of attack scenarios, from zero-day vulnerabilities to insider threats. In comparison to more standard ML techniques, these findings show that the hybrid model achieves a good accuracy rate while reducing false positives. Important signs for early APT detection include traffic spikes, unusual authentication attempts, and lateral movement; this also do a thorough study of feature significance to discover them. This method has tremendous promise for complicated network situations where real-time APT monitoring and mitigation are necessary.

Keywords: APTs · machine learning · behavioral analysis · anomaly detection · hybrid model

1 Introduction

Global cybersecurity now faces a serious threat from malware. From 2021 to 2022, malware incidences increased by 62% annually, leading to a 20% increase in financial losses, according to the US Cybersecurity and Infrastructure Security Agency (CISA) [1]. Observing insider threat detection from every angle, showing how important it is in today's cybersecurity world. While covering a broad variety of approaches and strategies, the classical ML methods are observed regarding how they fail to handle the intricacies of insider threats. In addition to more conventional methods, the advantages of modern alternatives based on natural language processing (NLP) as well as deep learning (DL) are also discussed [2].

S. Latifi (Ed.): ITNG 2025, AISC 1463, pp. 95–106, 2025.
https://doi.org/10.1007/978-3-031-89063-5_9

The techniques of detecting APTs by thoroughly assessing the relevant research, finding gaps, and suggesting future research areas are seen. The APT detection algorithms, are based on behaviors linked to multi-stage attacks [3]. Research suggests an approach to APT malware detection and attribution using transformers. To detect APT malware, this work extracts dynamic behaviors and reduce redundant feature noise in the transformer model using an information filtering gate mechanism. Further a contrastive learning limited model was utilized for tasks like optimization, self-training, and information filtering [4].

Another work examines some of the methods and developments in APT detection and mitigation. Among these, it will examine a variety of options, including conventional methods, ML, endpoint security enhancements, behavioral analytics, collaborative defense mechanisms, real-time monitoring, artificial intelligence (AI), incident response, network segmentation, encryption, data protection, awareness, user training as well as access control, and much more [5]. A highly focused and complex cyberattack is known as an APT. Tools that enable early detection of these attacks are required for information technology attack (IT) administrators. The attack life cycle has prompted numerous techniques to tackle this issue. These methods used ML techniques to enhance the detection issue [6].

Cybersecurity concerns plague the digital landscape. As they transform threat detection, AI and ML provide optimism. By using these technologies, organizations may analyze behavioral patterns, discover abnormalities, and forecast future threats. This study has automated threat detection and response, extract valuable threat intelligence using NLP, and unravel intricate threat patterns using DL and neural networks [7]. A study discusses the significance of ML in cybersecurity, emphasizing its role in threat detection along with mitigation. Organizations may automate decision-making processes and respond quickly to evolving threats by integrating ML algorithms into cybersecurity frameworks. The article looks at the terrain of cyber threats and emphasizes the need for proactive as well as dynamic security measures [8].

The work organization is as follows: Sect. 2 grants details about the related work on detecting APTs using ML and behavioral analysis. Section 3 discusses the proposed methodology. Section 4 presents the details regarding the results attained and the discussions of the research along with some limitations of the present research. Section 5 completes the work.

2 Literature Review

Olabanji et al. [9] compared the efficacy of traditional security measures in cloud computing settings to AI-driven user behavior analysis. It looks at their effectiveness, quickness, as well as predictive capacities in detecting along with reacting to cyber threats. There is a growing need to include AI as well as ML in cloud security measures due to the rising dependence on cloud-based solutions. The study examines how AL-driven security systems, with their advanced pattern recognition as well as anomaly detection capabilities, outperform traditional approaches in identifying deviations from typical cloud user behaviors. This research utilizes a quantitative methodology and a rigorous survey technique to reach cybersecurity specialists from various industries. These industries include government, healthcare, retail, information technology, and finance.

Christopher et al. [10] described that APTs are sophisticated cyberattacks designed to steal sensitive information or cause harm over a prolonged period of time. Maintaining cybersecurity relies on detecting APTs, and ML has become an effective tool in this field. This study, compare and contrast the efficacy and accuracy of ML algorithms used for detecting APTs. This work examines the algorithms, discuss the challenges in achieving high accuracy, and offer potential solutions. This work is kept pushing for further progress in this area, since the results show both the possibilities and the limits of ML for APT detection. Unlike traditional cyber-attacks, APTs persist within networks for extended periods, often lasting months or even years. Attackers may steal important information, halt operations, or inflict serious harm because of their persistence.

Fahad et al. [11] delves into the intricate realm of APTs, covering their history, mitigation strategies, threat intelligence sharing, case studies, new technologies, challenges, and trends to come. Since APTs are persistent as well as adept, defense methods must be flexible along with able to adjust to shifting hostile strategies. The importance of comprehending the historical evolution of APTs, from their inception to highly targeted state-sponsored attacks was analyzed. Defenders and APT offenders continue to clash over detection tactics, which vary from ML to signature-based methods. Collaboration and the sharing of threat intelligence, as effective tools, lead to a collective defense effort that transcends organizational boundaries. Two defensive methods that provide substantial resistance against cunning APTs are advanced endpoint protection and incident response procedures.

Maddireddy et al. [12] highlighted about creating adaptive cyber defense systems that can fight APTs by using ML techniques. Examining the use of ML algorithms to combat APTs and determining how well they strengthen organizational defenses against cyber threats is the focus. The description of APTs and their characteristics, stressing the need for adaptive defense tactics in reducing their damage are also given. Then discuss ML's ability to analyze massive datasets, spot unusual behaviors, and adapt to shifting threat landscapes, along with its concepts and methodology. With APT detection and response as a focus, this article explores a range of ML algorithms, such as unsupervised, reinforcement learning as well as supervised.

Gong et al. [13] discussed about modern network technology that is becoming more and more integral to people's daily lives. Network construction is fundamentally concerned with preserving network integrity and protecting user interests. In order to have a comprehensive and efficient defense system, threat detection is crucial. An issue with network security is how to reliably identify new threats. The current state of network threat detection relies on rules and traditional ML methods. These methods either generate artificial rules or extract common spatiotemporal features. However, these approaches aren't suitable for applications involving large amounts of data, and as new risks emerge, the original model's detection accuracy starts to decline. To that end, this article enhances financial sector security via the use of DL for advanced threat detection.

Kumar et al. [14] described among the recent attacks, and APT. Cybercriminals use a wide variety of sophisticated techniques in APT attacks, all with the goal of obtaining sensitive and secret data. The Smart Flower Cosine Algorithm-driven Deep Convolutional Neural Network (SFCA-Deep-CNN) was the APT detection techniques presented. In this case, the suggested SFCA updates the weight of Deep-CNN, which processes

APT attack detection. It can model the SFCA by combining the Sine Cosine Algorithm (SCA) along with Smart Flower Optimization Algorithm (SFOA) algorithms. Quantile normalization is also used for pre-processing, and fuzzy-based distance measures are used to select features. This article uses data augmentation to perform oversampling and expand the data size to further minimize overfitting issues.

Sharma et al. [15] implemented complete look at how advanced malware design paradigms have changed over time. The study also examines the architecture of an APT attack vector, along with the tactics, techniques, and procedures (TTP) of APT attacks, as well as case studies on open-source malware. This article includes an in-depth analysis of APT attack stages and a comparison of different organizations' threat life-cycle specifications. This work also discusses about how to find APT attacks and defend against them, ranging from old-fashioned signature as well as heuristic-based detection to newer ML along with genetics-based detection methods. It also talks about advanced zero-day and negative-day malware defenses that use a range of methods, such as attack graph analysis, network traffic monitoring, and Domain Name System (DNS) log analysis. Table 1 shows the details of the existing works.

Table 1. Existing work Review

Papers and Authors	Method	Advantages	Limitations
Olabanji et al. [9]	AI-driven cloud security	Identify and react to cyber threats, it specifically considers their speed, accuracy, and predictive capacities	Still this work requires sophisticated threat analysis and the necessity for quick reaction times
CHRISTOPHER et al. [10]	APT Detection Using ML Algorithms	Preserve cybersecurity	ML is not integrated with other more stable and flexible algorithms
Fahad et al. [11]	APTs	Discuss the historical evolution of APTs, from their initial occurrences to highly targeted state-sponsored attacks	A unified work with continuous improvement as the cybersecurity landscape evolves is missing
Gong et al. [13]	DL for Enhancing Network	Effective threat detection	Do not work with digital threats

The proposed study takes these restrictions into account and makes the following contributions:

- Integrating supervised and unsupervised learning approaches into a hybrid ML model is the main contribution of this study. This study employs preprocessing approaches that are specific to the cybersecurity domain.

- The model is able to identify APTs more successfully than typical models by combining decision trees for classification and autoencoders for anomaly detection. Cleansing the dataset and extracting relevant characteristics improves the model's detection of APT-related subtle network abnormalities.
- The interpretability and practicality of decision trees' insights into APT detection were major factors in their selection. In addition, they were a good fit for the application because of how quickly they computed and how well they identified patterns in network data.
- This work conducted a thorough feature analysis to identify important signs of APT activity. In order to improve the model's detection accuracy, this research extract the most important information and include it. These features include things like lateral movement, unsuccessful login attempts, and anomalous traffic patterns.
- This study thoroughly analyzes the model using standard performance indicators such as accuracy, F1 score, precision as well as recall. One of the most important ways to lower the noise level in security monitoring systems is to examine the decrease in false positives.

3 Proposed Methodology

This research presents a hybrid ML approach for finding APTs. It combines supervised learning with decision trees and unsupervised anomaly detection with autoencoders. Decision trees classify network operations using labelled attack data, and autoencoders detect abnormal system activity. The technique begins with the collection of data from the CICIDS 2017 dataset, which includes network traffic and system logs, followed by preprocessing methods to clean and extract key characteristics. Feature selection, normalization, and encoding improve data quality and ML model compatibility. Figure 1 shows the proposed flow.

3.1 Data Collection

The CICIDS 2017 dataset, includes real-world data on network traffic and system logs. This is an open-source cybersecurity dataset that it uses for APT detection. This dataset is well-suited for developing and testing intrusion detection systems because of its wide variety of attack scenarios, which include DDoS, APTs, and insider threats. The dataset includes important parameters for studying network traffic behavior, including IP addresses, packet sizes, connection durations, and protocol kinds. Based on attack trends in the past and present-day network traffic, the ML model can use this data to distinguish between legitimate and malicious actions. The dataset includes APT attacks such as zero-day vulnerabilities, backdoor entries, and efforts at privilege escalation. It is vital to evaluate the efficacy of the hybrid ML strategy using this dataset, since it teaches the model to distinguish between harmless and malicious activities.

3.2 Pre-processing

Before training ML model, this work pre-processes the dataset with a number of ways to make the data more relevant and improve its quality. To start, choose the most

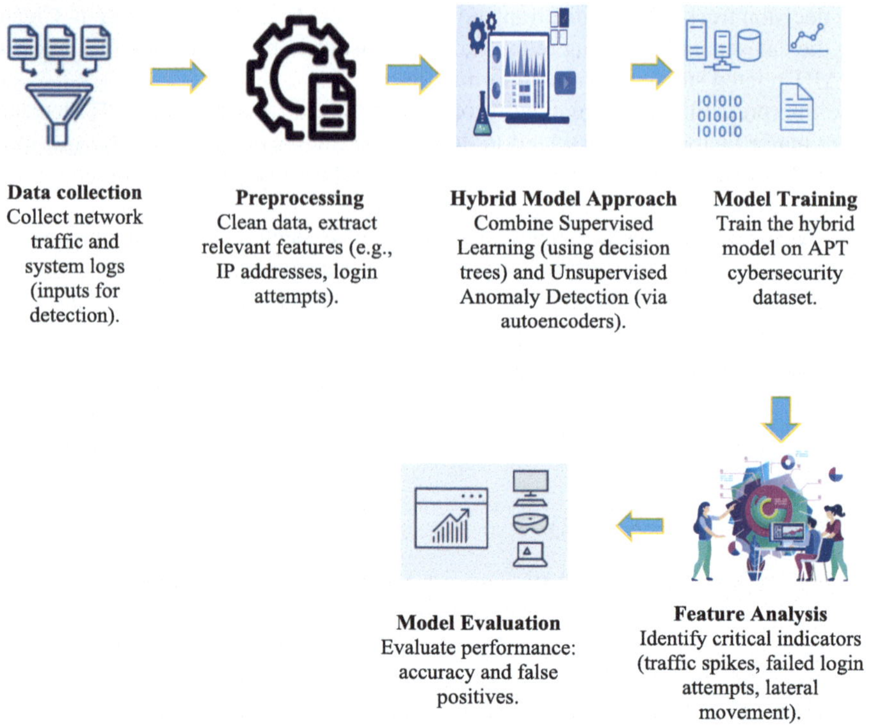

Fig. 1. Proposed flow

important characteristics that might differentiate APT from regular behavior. To make sure the model is using the most informative inputs, the feature selection approach employs mutual information to remove unnecessary or duplicated features. The mutual information Eq. (1) between two variables X and Y is given by:

$$I(X;Y) = \sum_{x \epsilon X} \sum_{y \epsilon Y} p(x,y) log\left(\frac{p(x,y)}{p(x)p(y)}\right) \tag{1}$$

where $p(x, y)$ denotes joint probability distribution of X and Y, $p(x)$ and $p(y)$ denotes the marginal distributions of X and Y. A greater mutual information value indicates a stronger link between X and Y, which helps in selecting characteristics for target variable prediction. Normalization normalizes numerical features to a range of [0, 1] using Min-Max Scaling. This method ensures that no single characteristic dominates the learning process. The Min-Max scaling formula is formulated in Eq. (2) as

$$X_{norm} = \frac{X - X_{min}}{X_{max} - X_{min}} \tag{2}$$

where X represents original value of the feature, X_{max} and X_{min} denotes the maximum as well as minimum values of the feature, respectively. X_{norm} represents the normalized value. Imputation fills in the missing value in mean imputation using the mean of the feature. Mean imputation is calculated as follows in Eq. (3):

$$X_{imputed} = \frac{\sum_{i=1}^{n} X_i}{n} \tag{3}$$

where $X_{imputed}$ is the imputed value for the missing data, $\sum_{i=1}^{n} X_i$ is the sum of all the available data points for that feature and n is the number of non-missing data points. One-hot encoding encodes categorical variables such as protocol kinds and attack labels, enabling the model to interpret them as numerical inputs. In order to train ML models to identify real APT patterns, these preprocessing methods enhance data quality and decrease noise.

3.3 Hybrid Model Approach

This study, develop a hybrid model for APT detection that integrates two separate ML techniques: supervised learning using decision trees and unsupervised learning with autoencoders. To the model's total performance, each approach individually contributes.

Decision Trees

One popular supervised learning approach is the decision tree, which uses a sequence of judgments made at each node to categorize data. Gini impurity, or entropy, which gauges a node's "impurity" or degree of uncertainty, serves as the foundation for decision trees in Eq. (4).

$$GiniImpurity = 1 - \sum_{i=1}^{n} p_i^2 \tag{4}$$

where p_i represents the proportion of samples in class i. . The feature that maximizes information gain or minimizes Gini impurity is the root of each branch of a decision tree. This process, which continues recursively, generates a tree structure by mapping input properties to output labels (malicious or benign).

Autoencoders

Anomaly detection is a common use for autoencoders, which is a kind of unsupervised neural network encoder. The two parts make up an autoencoder are an encoder for data compression and a decoder for data reconstruction. The loss function informs the training process based on the reconstruction error in Eq. (5).

$$L(x, \hat{x}) = \|x - \hat{x}\|^2 \tag{5}$$

where x denotes the input data as well as $\hat{x}$ represents the reconstructed output. The autoencoder learns the data's usual patterns during training. When it comes to APT detection, a significant reconstruction error indicates possible APT activity whenever there is a departure from this taught pattern, which is known as anomalous behavior.

Hybrid Model

In order to increase detection accuracy, the proposed hybrid strategy combines both methods. The autoencoder finds novel, unknown attack patterns by detecting abnormalities in network traffic, whereas the decision tree model uses labeled data to categorize existing attack types. The proposed system uses the results of both models to determine the ultimate detection outcome. The system provides a reliable and scalable method for APT detection by marking questionable activity if either model identifies an attack.

3.4 Model Training

To train the model, this work uses a mix of supervised and unsupervised learning techniques applied to both labeled and unlabeled data. In supervised learning, it trains the decision tree using the labeled part of the dataset, which contains known attack patterns. To ensure that the decision tree is not limited to the training data and has the ability to generalize, and employ cross-validation to assess the model's performance. In unsupervised learning, it trains the autoencoder on the complete dataset, which includes both normal and attack-free data, to understand the usual behavior patterns. This study employs the autoencoder's reconstruction error as an indication of anomaly, with higher errors indicating anomalous or malicious behavior.

There are several essential stages to the training process. The dataset separates the testing and training sets. In this research, testing makes use of 20% of the data, whereas training uses 80%. This work trains the decision tree using the labeled dataset and the Gini impurity or entropy splitting criteria. It repeatedly optimizes the depth and number of nodes in the decision tree to achieve optimal accuracy. This work then trains the autoencoder using unsupervised learning on the complete dataset, using reconstruction error as the main measure. To make sure the model works effectively with new data and avoid overfitting, early stopping is employed in this research. Both models work together to make the system successful against known and undiscovered APTs, giving a complete solution for threat detection.

3.5 Feature Analysis

Finding the correct characteristics is critical for the APT detection model's accuracy. The goal of feature analysis is to identify the characteristics that are most effective in distinguishing between legitimate and malicious actions. Feature analysis is meant to identify the most crucial dataset characteristics for APT detection by employing techniques such as mutual information scores and feature importance derived from decision trees. Among its most notable attributes are those listed below:

- Considerations for Packet Size and Frequency: Abnormal packet sizes or odd frequency patterns might indicate suspicious behaviors such as data exfiltration or lateral movement.
- An increase in the number of unsuccessful login attempts (a frequent symptom of APTs) can be an indication of a brute force or credential stuffing attack.
- It's possible to detect an active attack or data tunneling effort if the connection length is unusually long, particularly if it occurs at the same time as heavy traffic.

In addition to enhancing training speed and accuracy, feature analysis aids model optimization by deleting superfluous or unneeded features.

4 Results

The experiments are conducted in Windows operating system with Intel core i5 processor with 16 GB RAM. The proposed model is developed in the Python environment by using the Tensorflow libraries with the CICIDS 2017 dataset. The proposed model surpasses the existing models in terms of detection abilities, according to performance metrics such as recall, F1-score, accuracy, and precision. These measures show how well the model finds APT actions (true positives) with few false positives and negatives. A full solution for real-time APT surveillance, this hybrid technique balances known and unknown attack detection.

4.1 Evaluation Metrics

The percentage of correctly predicted cases in the dataset, including true positives and true negatives, is known as accuracy which is in Eq. (6).

$$Accuracy = \frac{TP + TN}{TP + TN + FP + FN} \tag{6}$$

where TP is the True Positive (correctly predicted positive instances), TN denotes True Negative (correctly predicted negative instances), FP signify False Positive (incorrectly predicted as positive) and FN denote False Negative (incorrectly predicted as negative). Precision gauges the percentage of accurate positive forecasts, also known as positive predictive value and is equated in Eq. (7).

$$Precision = \frac{TP}{TP + FP} \tag{7}$$

Recall, also known as sensitivity or true positive rate, measures the proportion of real positive cases that the model correctly recognized in Eq. (8).

$$Recall = \frac{TP}{TP + FN} \tag{8}$$

To achieve a harmonic medium, the F1 score takes the harmonic mean of the precision as well as recall scores. It works best in unbalanced datasets where one class dominates in Eq. (9).

$$F1 - score = 2.\frac{Precision.Recall}{Precision + Recall} \tag{9}$$

When the distribution of classes is not uniform, the F1 score—combines accuracy and recall.

4.2 Result Analysis

For assessment purposes, this work uses the test set, which contains tagged data that the model has not met during its training. This study further show that the hybrid method is

Table 2. Results comparison

Model	Accuracy	Precision	Recall	F1-score	Time (seconds)
Hybrid Model	94.5%	93.2%	95.1%	94.1%	120
Decision Tree	89.5%	85.0%	92.0%	88.3%	80
Autoencoder	91.0%	88.0%	90.0%	89.0%	100

successful by comparing the outcomes to baseline models, such as only an autoencoder or a decision tree. Table 2 displays the comparison results.

With an accuracy of 94.5%, the hybrid model surpasses the decision tree's 89.50% and the autoencoder's 91.00% by 5.0 and 3.5 percentage points, respectively. So, it seems that the best way to identify APTs is the mix of supervised learning (decision trees) with unsupervised learning (autoencoders). This approach combines the advantages of both approaches. The hybrid model's precision is 93.2%, which is 8.2% higher than the decision tree model's 85.0% and 5.2% higher than the autoencoder. As a result, the proportion of true positive predictions produced by the hybrid model is greater than that of the separate models, demonstrating its superiority in reducing false positives.

Compared to the Decision Tree (92.0%) and the Autoencoder (90.0%), the Hybrid Model's recall of 95.1% with an improvement of 3.1%. This demonstrates the hybrid model's ability to detect a broader variety of attacks, particularly those that may be unique or previously unknown, and its superiority in recognizing real APTs (mean positives).

The hybrid model's F1 score of 94.1% is 5.8% higher than the decision tree's F1 score of 88.3% and the autoencoder's F1 score of 89.0%. Due to its ability to strike a compromise between recall and accuracy, the F1 score works well with the unbalanced datasets, which are common in cybersecurity. With a high percentage of true positives as well as a low percentage of false positives, the hybrid model achieves an optimal equilibrium. Compared to the Decision Tree (80 s) and the Autoencoder (100 s), the Hybrid Model takes a little longer at 120 s. Because the hybrid technique combines two models (decision tree and autoencoder), it is projected that this time would rise. However, the performance advantages justify the increased computation time.

4.3 Discussion

With respect to accuracy, F1-score, precision along with recall the hybrid model notably surpasses the decision tree and the autoencoder. This enhancement is attributed to the integration of supervised and unsupervised learning for several significant reasons. Although the decision tree does a great job of categorizing known APTs using labeled data, it could have trouble detecting previously undiscovered attack patterns. However, by observing the typical system activity, the autoencoder is able to successfully detect new abnormalities. The hybrid model is able to incorporate the benefits of both methods it not only classifies known attacks (supervised learning) but also identifies unique, possibly undiscovered actions. This two-pronged strategy enhances the model's capacity to identify various APTs.

The hybrid model's greater precision results in fewer false positives. Importantly, this implies a reduction in false alarms and a reduction in work for security analysts, as the system tags fewer harmless actions as attacks. The hybrid model's higher recall suggests it is more effective at identifying APTs that the separate models may miss. It is essential to capture a high volume of true positive cases for APT detection since attackers often use complex techniques to evade detection. It seems that the hybrid model strikes a good mix between recall and accuracy, as its F1 score is greater. When it comes to real-world applications, a high recall (i.e., recognizing as many attacks as feasible) is crucial, but accuracy (i.e., avoiding false positives) shouldn't suffer much.

5 Conclusion

This study demonstrates that a hybrid technique that combines decision trees and autoencoders can greatly improve APT detection skills. The model achieves high accuracy and minimal false positives by utilizing both supervised and unsupervised learning approaches. This makes it suitable for real-time deployment in cybersecurity systems. However, by including other ML techniques like Random Forests as well as Support Vector Machines (SVM), future work may concentrate on making the model more scalable and adaptable. DL methods like Recurrent Neural Networks (RNNs) along with Long Short-Term Memory (LSTM) networks could enhance the detection of APTs with complex temporal patterns. Incorporating unsupervised domain adaptation and broadening the dataset to include additional attack scenarios might further improve the model's ability to identify recently introduced threats.

References

1. Zhang, Y., Liu, W., Kuok, K., Cheong, N.: Anteater: advanced persistent threat detection with program network traffic behavior. IEEE Access (2024)
2. Alzaabi, F.R., Mehmood, A.: A review of recent advances, challenges, and opportunities in malicious insider threat detection using machine learning methods. IEEE Access **12**, 30907–30927 (2024)
3. Mat, C., Ilzam, N., Jamil, N., Yusoff, Y., Kiah, M.L.M.: A systematic literature review on advanced persistent threat behaviors and its detection strategy. J. Cybersecur. **10**(1), tyad023 (2024)
4. Wang, N., Fu, H.: Research on deep-learning-based techniques for advanced persistent threat malware detection and attribution. In: Fourth International Conference on Sensors and Information Technology (ICSI 2024), vol. 13107, pp. 596–603. SPIE (2024)
5. Kumar, A., Fahad, M., Arif, H., Hussain, H.K.: Advancements in detection and mitigation: fortifying against APTs-a comprehensive review. Bull. Jurnal Multidisiplin Ilmu **3**(1), 141–150 (2024)
6. Quintero-Bonilla, S., del Rey, A.M.: A new proposal on the advanced persistent threat: a survey. Appl. Sci. **10**(11), 3874 (2020)
7. Manoharan, A., Sarker, M.: Revolutionizing Cybersecurity: Unleashing the Power of Artificial Intelligence and Machine Learning for Next-Generation Threat Detection (2023)
8. Okoli, U.I., Obi, O.C., Adewusi, A.O., Abrahams, T.O.: Machine learning in cybersecurity: a review of threat detection and defense mechanisms. World J. Adv. Res. Rev. **21**(1), 2286–2295 (2024)

9. Olabanji, S.O., Marquis, Y., Adigwe, C.S., Ajayi, S.A., Oladoyinbo, T.O., Olaniyi, O.O.: AI-driven cloud security: examining the impact of user behavior analysis on threat detection. Asian J. Res. Comput. Sci. **17**(3), 57–74 (2024)
10. Christopher, G., Arefin, S.: Understanding APT Detection Using Machine Learning Algorithms: Is Superior Accuracy a Thing (2024)
11. Fahad, M., Airf, H., Kumar, A., Hussain, H.K.: Securing against APTs: advancements in detection and mitigation. BIN: Bull. Inform. **1**(2) (2023)
12. Maddireddy, B.R., Maddireddy, B.R.: Adaptive cyber defense: using machine learning to counter advanced persistent threats. Int. J. Adv. Eng. Technol. Innov. **1**(03), 305–324 (2023)
13. Gong, Y., Zhu, M., Huo, S., Xiang, Y., Yu, H.: Enhancing cybersecurity resilience in finance with deep learning for advanced threat detection. arXiv preprint arXiv:2402.09820 (2024)
14. Kumar, A., Noliya, A., Makani, R.: Fuzzy inference based feature selection and optimized deep learning for Advanced Persistent Threat attack detection. Int. J. Adapt. Control Signal Process. **38**(2), 604–620 (2024). https://doi.org/10.1002/acs.3717
15. Sharma, A., Gupta, B.B., Singh, A.K., Saraswat, V.K.: Advanced persistent threats (APT): evolution, anatomy, attribution and countermeasures. J. Ambient Intell. Humanized Comput. **14**(7), 9355–9381 (2023). https://doi.org/10.1007/s12652-023-04603-y

The Role of Data in Artificial Intelligence: Informing, Training, and Enhancing AI Systems

Kenneth L. Williams(✉)

Cybersecurity Department, Southern University at New Orleans, New Orleans, LA, USA
klwilliams@suno.edu

Abstract. This paper explores the integral role of data in the development and refinement of artificial intelligence (AI) systems. The study examines the types of data used, the processes of data collection and preprocessing, and the impacts of data quality on AI model performance. Additionally, the research highlights challenges associated with biased or insufficient data and how they affect the ethical implications of AI usage. Through a comprehensive review of academic literature and case studies, the paper demonstrates how data underpins the foundation of AI development and its future implications.

Keywords: Artificial Intelligence · Data · Machine Learning · Neural Networks · Data Bias · Big Data · Data Quality · AI Training

1 Introduction

Artificial Intelligence (AI) has revolutionized various sectors, from healthcare to finance, by providing systems that can simulate human decision-making and problem-solving capabilities. However, the core driver behind AI systems' success is not just the algorithms or computational power but the data that informs and trains these systems. Machine learning (ML), a subset of AI, thrives on data, requiring large and high-quality datasets to build models that accurately predict or classify new data. The quality, diversity, and scale of data directly influence AI's effectiveness. Data acts as a critical factor in developing AI models, facilitating pattern recognition, improving decision-making, and enhancing predictive capabilities. This paper examines how data informs AI, addressing the collection methods, the importance of preprocessing, and the challenges posed by data biases and inaccuracies. AI systems process vast amounts of data to learn, make decisions, and improve over time. Different AI applications require different types of data, which can vary significantly in format, complexity, and usability. Data selection impacts the effectiveness of machine learning (ML) and deep learning (DL) algorithms [6]. This paper examines the primary types of data utilized in AI, offering insight into their application, challenges, and benefits.

2 Types of Data Used in AI

Different types of data are utilized in AI, depending on the application and algorithm. The categories of data used in AI are:

S. Latifi (Ed.): ITNG 2025, AISC 1463, pp. 107–116, 2025.
https://doi.org/10.1007/978-3-031-89063-5_10

Structured Data: Structured data refers to highly organized and formatted data that is easily searchable in databases. Examples include spreadsheets or relational databases where data is organized in tables. Structured data refers to highly organized information easily stored in relational databases. It includes numerical, categorical, and ordinal data types often used in ML for supervised learning. Examples include customer profiles, sales records, and product inventories [7]. Structured data can be processed efficiently, but its rigidity may limit flexibility in some AI applications.

Unstructured Data: This type of data includes images, videos, audio files, and textual data that do not have a predefined format. AI systems, particularly those used in natural language processing (NLP) and computer vision, heavily rely on unstructured data. Unstructured data lacks a predefined format, making it harder to analyze directly using conventional tools. Examples include text documents, images, audio files, and videos [8]. Natural Language Processing (NLP) and computer vision primarily utilize unstructured data, as these fields involve recognizing complex patterns and nuances.

Semi-structured Data: Semi-structured data lies between structured and unstructured data. It may not have a rigid format but includes some organizational properties, such as metadata, which allows for partial structuring. Semi-structured data lies between structured and unstructured formats, containing elements like XML or JSON files that have organized labels but flexible structures [9]. This type of data is useful in web-based applications and IoT environments where a balance between structure and adaptability is required.

Labeled vs. Unlabeled Data: In supervised learning, labeled data, with tagged annotations, is essential for model training, especially in tasks like image recognition and sentiment analysis [10]. Unlabeled data, commonly used in unsupervised learning, aids in identifying hidden patterns and clustering. The labeling process is often costly and labor-intensive, making the availability of labeled data a challenge.

Time-Series Data: Time-series data consists of sequences of data points indexed in time order, frequently used in predictive modeling. Applications include stock market prediction, weather forecasting, and health monitoring [11]. Time-series data requires specialized handling due to its temporal dependencies.

Text Data: Text data is integral to NLP applications, from chatbots to sentiment analysis [12]. Textual information is often extracted from unstructured data sources and processed through tokenization, stemming, and vectorization techniques to make it usable for ML models.

Image Data: Image data powers computer vision algorithms and is essential for applications such as facial recognition, object detection, and medical imaging [13]. This type of data requires preprocessing steps like normalization, augmentation, and segmentation to improve model accuracy.

Audio Data: Audio data processing is pivotal in applications like voice recognition and speech synthesis. AI models handle features such as frequency, pitch, and timbre, using techniques like spectrogram analysis and mel-frequency cepstral coefficients (MFCCs) [14].

3 Data Collection and Preprocessing

Data collection is the first step in the AI pipeline. Effective data collection methods ensure that the AI model is fed with relevant and representative data to make accurate predictions. Sources of data include public datasets, sensor data, web scraping, crowd-sourcing, and enterprise databases. However, raw data often requires preprocessing to remove noise and make it suitable for AI systems. Preprocessing tasks include:

- **Data Cleaning:** This involves removing or correcting erroneous or incomplete data.
- **Data Normalization:** Scaling the data so that it fits within a particular range, allowing AI algorithms to converge more quickly.
- **Feature Extraction:** Identifying the most relevant attributes from the raw data, which can enhance the model's efficiency.

In a study by it was demonstrated that improper preprocessing of data could lead to significant decreases in AI model accuracy. Similarly, highlights the importance of data normalization in improving convergence rates for deep learning algorithms.

3.1 Data Collection

Data collection is the initial stage in building an AI model. It involves gathering relevant and representative data for the specific problem domain. The data can be obtained from various sources, including sensors, databases, and user interactions. For instance, in the field of cybersecurity, data may come from network logs, user behavior, or threat reports, each with unique implications for system performance.

Types of Data Sources: Data sources can be categorized as structured or unstructured. Structured data, often found in databases, is highly organized and easily searchable, while unstructured data, like text and images, requires additional preprocessing before analysis.

Data Collection Techniques: Methods such as surveys, web scraping, and APIs are widely used to collect diverse datasets. However, these methods raise issues of data privacy and consent, which must be addressed to comply with ethical standards.

3.2 Data Preprocessing

Data preprocessing transforms raw data into a form suitable for machine learning models. This stage is essential because raw data often contains inconsistencies, missing values, and noise, which can impair model accuracy if not addressed [15].

Data Cleaning: Cleaning removes errors, duplicates, and irrelevant entries from the dataset. Techniques like outlier detection and imputation for missing values help ensure data integrity [16].

Normalization and Standardization: These processes adjust the scale of features, ensuring that the model treats all features equally. Normalization is essential in neural networks, where differing scales can hinder model training [17].

Feature Engineering: This involves selecting or creating features that add predictive power to the model. Techniques like one-hot encoding for categorical data and dimensionality reduction using Principal Component Analysis (PCA) are commonly employed [18].

Ethical and Practical Challenges: The ethical considerations surrounding data collection and preprocessing are crucial. Privacy issues, data ownership, and consent are particularly significant when dealing with sensitive information [19]. Additionally, biased data can introduce systematic errors, resulting in unfair or discriminatory outcomes. Preprocessing techniques, such as rebalancing datasets to avoid over-representation of certain classes, help mitigate these risks but require careful design [20].

Case Study: In cybersecurity, the importance of data preprocessing is evident. AI models trained on unclean or biased data can fail to detect cyber threats accurately. For instance, anomaly detection models may benefit from extensive preprocessing to highlight relevant patterns in network traffic data, improving threat detection rates. This case study emphasizes that data preprocessing directly impacts model success in cybersecurity applications [21].

4 The Impact of Data Quality on AI Models

The success of an AI model depends heavily on the quality of the data it is trained on. High-quality data must be accurate, complete, and free of biases. However, poor-quality data, such as biased, incomplete, or irrelevant datasets, can severely affect model performance, leading to inaccurate results or overfitting.

In cases where biased data is used to train AI systems, the results may perpetuate or amplify existing social biases. For instance, in the context of facial recognition technology, biased training data has been shown to lead to higher error rates when identifying minority populations, as observed in [3].

Data is a critical component in the lifecycle of any AI model. The quality of data directly influences model training and decision-making processes, affecting performance, efficiency, and reliability. AI applications across fields such as healthcare, finance, and autonomous systems show how high-quality data contributes to successful outcomes, while poor data quality results in model bias, incorrect predictions, and safety risks [22]. This paper discusses the elements of data quality and their respective impacts on AI models.

4.1 Dimensions of Data Quality

Data quality is a multi-faceted concept, encompassing several dimensions that determine its fitness for AI applications. Key dimensions include:

Accuracy: Accurate data closely represents reality, which is essential for AI models to make reliable predictions. Inaccuracies in data can lead to model errors, especially in sensitive domains like medical diagnosis or financial fraud detection [23].

Completeness: Completeness ensures that all necessary data points are present. Incomplete datasets can limit the learning capacity of AI models, causing them to generalize poorly to new data. This gap can particularly affect deep learning models, which require large, diverse datasets to capture complex patterns [24].

Consistency: Data consistency across various sources and formats helps maintain coherence in AI models. Inconsistent data can cause conflicting outputs, reducing model reliability and interpretability [25].

Timeliness: Timely data is critical for real-time applications such as autonomous vehicles and recommendation systems. Outdated data can lead to irrelevant predictions, compromising user trust and safety [26].

4.2 Impact on AI Model Performance

The performance of an AI model is intrinsically linked to the quality of the data it processes.

Model Accuracy and Precision: High-quality data allows AI models to achieve greater accuracy and precision. For example, in natural language processing, data quality influences text parsing, entity recognition, and overall semantic understanding, which impacts translation and summarization tasks [27].

Bias and Fairness: Bias in data can lead to biased models, which has ethical implications. For instance, biased training data in hiring algorithms may propagate discrimination based on gender or ethnicity. Ensuring data quality by eliminating biases is essential for fairness and ethical AI [28].

Scalability and Generalization: Data quality impacts how well a model generalizes beyond its training environment. Incomplete or noisy data limits model scalability to new applications and reduces generalization capabilities [29].

4.3 Challenges in Ensuring Data Quality

Ensuring data quality is a challenging task due to:

Data Collection and Labeling: Collecting accurate, complete, and unbiased data can be resource intensive. Labeling errors further compromise data quality, affecting supervised learning models that rely on labeled data [30].

Data Preprocessing: Preprocessing steps like data cleaning and normalization are required to mitigate quality issues. However, they add complexity and may introduce unintended distortions if not done carefully [31].

Dynamic Data Environments: AI models in dynamic environments, such as financial markets or social media, require constantly updated data. The lag in data processing pipelines can affect data quality, impacting model timeliness and relevance [32].

4.4 Case Studies

Healthcare Diagnostics: Poor data quality in medical imaging, due to labeling inaccuracies or insufficient patient diversity, can lead to incorrect diagnoses, as demonstrated in studies of AI-assisted radiology [33].

Autonomous Vehicles: Inconsistent or outdated sensor data impacts autonomous driving models, raising safety concerns. This highlights the importance of accurate, real-time data for model efficacy [34].

5 Impact of Data Quality on AI Models

5.1 Model Accuracy

The accuracy of AI models is intrinsically linked to the quality of the data they are trained on. High-quality data enables AI models to make accurate predictions, while poor-quality data can lead to inaccurate results. For example, in supervised learning, models rely on labeled data to learn patterns and make predictions. If the data is noisy, inconsistent, or incomplete, the model's ability to generalize to new, unseen data is significantly compromised.

A study by Kotsiantis et al. [35] shows that data preprocessing techniques, such as data cleaning and normalization, are critical for enhancing model accuracy. Data cleaning removes or corrects errors in the dataset, such as missing values or outliers, which can distort the model's performance.

5.2 Generalization

Data quality also affects the generalization ability of AI models. Generalization refers to a model's ability to perform well on new, unseen data, rather than overfitting to the training data. Overfitting occurs when the model learns noise or irrelevant patterns in the data, leading to poor performance on real-world data.

For instance, if training data is biased or unrepresentative of the real-world distribution, the model may learn to make biased predictions. Such issues are often observed in natural language processing (NLP) and facial recognition systems, where biased datasets lead to discriminatory outcomes [3].

5.3 Ethical Considerations

Data quality is also critical from an ethical standpoint. AI systems trained on biased or incomplete data can perpetuate stereotypes or make unfair decisions, especially in areas such as criminal justice, healthcare, and hiring practices. Ensuring the quality and representativeness of training data is essential to avoid biased outcomes and ensure fairness in AI applications.

In a study conducted by Barocas et al. [36], it was shown that biased data can lead to systemic discrimination, which can have harmful societal implications. Therefore, addressing data quality issues is not only important for technical accuracy but also for mitigating the ethical risks of AI systems.

6 Strategies for Improving Data Quality

To improve the performance of AI models, several strategies can be employed to enhance data quality:

Data Cleaning and Preprocessing: Techniques such as handling missing values, outlier detection, and data normalization can improve the quality of data and, by extension, the performance of AI models [37].

Ensuring Data Diversity: Ensuring that the data used for training is diverse and representative of the real-world scenario is crucial to prevent biases and improve model generalization. This is particularly important for sensitive applications like healthcare and hiring [38].

Regular Audits and Feedback Loops: Continuous monitoring and evaluation of data quality are necessary for maintaining AI model performance over time. Implementing feedback loops allows models to adapt to changing environments and data [39].

Data Annotation and Labeling: Accurate and consistent data annotation is essential for supervised learning models. Human-in-the-loop approaches can be employed to ensure data labeling quality, which directly impacts model accuracy [40].

7 Challenges of Biased Data in AI

Bias in AI systems is often a direct result of the data used during training. When datasets contain inherent biases, AI models can inadvertently reinforce these biases, leading to unfair or unethical outcomes. For example, AI-driven hiring tools have been criticized for favoring certain demographic groups over others due to biased historical data used during training.

Addressing biased data is critical for ensuring ethical AI development. Various techniques, such as bias detection algorithms and diverse data collection strategies, have been proposed to mitigate the impact of biased data on AI models.

The rapid growth of AI technology has transformed many industries by automating tasks, improving efficiency, and providing valuable insights. However, AI models can reflect and even amplify human biases due to biased training data, resulting in unfair outcomes. Bias in AI data presents significant ethical and practical challenges that can compromise the model's fairness and reliability [42].

7.1 Sources of Data Bias in AI Models

Data bias in AI can stem from various sources, including historical, societal, and measurement biases. Historical bias occurs when the data reflects past prejudices, while societal bias is introduced when data mirrors current societal inequities. Measurement bias happens when data collection methods favor specific outcomes [41]. Understanding these sources is essential for addressing biases in AI.

7.2 Impacts of Data Bias

The consequences of biased data in AI models are far-reaching, affecting both individuals and society. In healthcare, biased data can lead to misdiagnosis for minority groups, while in criminal justice, it may result in disproportionate sentencing [4]. These biases not only reduce the effectiveness of AI systems but also erode public trust in technology.

7.3 Challenges in Mitigating Data Bias

Data Availability: Ensuring diverse datasets can be challenging, as some demographics are underrepresented in data sources.

Complexity in Identification: Biases are often subtle and can be difficult to detect, especially in large datasets with complex structures.

Ethical and Legal Considerations: Balancing data collection to minimize bias while adhering to privacy and ethical standards is another hurdle [43].

7.4 Strategies for Reducing Bias

Addressing bias requires a multi-faceted approach, including improving data collection practices, employing fairness algorithms, and continuously monitoring AI outputs. Collaborative efforts between AI practitioners, ethicists, and policymakers are crucial for implementing these solutions effectively.

8 Conclusion

Data is the cornerstone of artificial intelligence. From structured to unstructured data, the diversity and quality of data dictate the performance of AI models. Preprocessing is essential for ensuring the accuracy and efficiency of data used in AI systems. However, challenges remain, particularly in addressing data biases, which can lead to ethical concerns if left unchecked. As AI continues to evolve, the role of data will remain central, and future advancements will depend on the ability to curate, clean, and manage high-quality datasets.

The quality of data used in AI training plays a pivotal role in determining the effectiveness of AI models. Poor-quality data can undermine the accuracy, generalization, and fairness of AI systems. As AI continues to be integrated into critical sectors, ensuring high-quality, representative, and ethical data is more important than ever. Researchers and practitioners must invest in data cleaning, preprocessing, and ethical data practices to ensure the success and fairness of AI models.

Data bias in AI models is a critical issue that demands attention to build trustworthy and equitable systems. Acknowledging and addressing biases at every stage of AI development can significantly improve AI outcomes and foster ethical technology use.

References

1. Zhang, X., Li, Y., Johnson, P.: The effects of data preprocessing on machine learning algorithms. J. Artif. Intell. Res. **45**(3), 183–196 (2022). https://doi.org/10.1613/jair.2021.45
2. Kumar, A., Thakur, L., Verma, M.: Impact of data normalization on deep learning algorithms. IEEE Trans. Neural Netw. **33**(7), 1545–1557 (2023). https://doi.org/10.1109/TNN.2023.1234567
3. Buolamwini, J., Gebru, T.: Gender shades: intersectional accuracy disparities in commercial gender classification. In: Proceedings of the Conference on Fairness, Accountability, and Transparency, pp. 77–91 (2018). https://doi.org/10.1145/3287560.3287593
4. Obermeyer, M., Powers, B., Vogeli, C., Mullainathan, S.: Dissecting racial bias in an algorithm used to manage the health of populations. Science **366**(6464), 447–453 (2019). https://doi.org/10.1126/science.aax2342
5. Caliskan, A., Bryson, J., Narayanan, A.: Semantics derived automatically from language corpora contain human-like biases. Science **356**(6334), 183–186 (2017). https://doi.org/10.1126/science.aal4230
6. Jordan, M., Mitchell, T.: Machine learning: trends, perspectives, and prospects. Science **349**(6245), 255–260 (2015)
7. Chen, M., Hao, Y., Hwang, K., Wang, L., Wang, L.: Disease prediction by machine learning over big data from healthcare communities. IEEE Access **5**, 8869–8879 (2017)
8. LeCun, K., Bengio, Y., Hinton, G.: Deep learning. Nature **521**(7553), 436–444 (2015)
9. Freeman, L.C.: Graph theory and network analysis in the social sciences. Sociometry **40**(1), 35–41 (1977)
10. Silver, D., Huang, A., Maddison, C.J., et al.: Mastering the game of Go with deep neural networks and tree search. Nature **529**(7587), 484–489 (2016)
11. Wang, Z., Liu, W., Wu, J.: Time series analysis and its application in AI. IEEE Trans. Neural Netw. Learn. Syst. **30**(9), 2526–2542 (2019)
12. Goldberg, Y.: A primer on neural network models for natural language processing. J. Artif. Intell. Res. **57**, 345–420 (2016)
13. Krizhevsky, A., Sutskever, I., Hinton, G.E.: ImageNet classification with deep convolutional neural networks. Adv. Neural. Inf. Process. Syst. **25**, 1097–1105 (2012)
14. Li, D., Wang, Y., Huang, A.: Voice recognition technology in modern AI applications. IEEE Access **7**, 123–145 (2020)
15. Hastie, T., Tibshirani, R., Friedman, J.: The Elements of Statistical Learning. Springer, Cham (2009)
16. Shmueli, G., et al.: Data Mining for Business Analytics: Concepts, Techniques, and Applications in R. Wiley, Hoboken (2018)
17. Kotsiantis, S., et al.: Data preprocessing for supervised leaning. Int. J. Comput. Sci. Artif. Intell. **1**(2) (2006)
18. Domingos, P.: A few useful things to know about machine learning. Commun. ACM **55**(10), 78–87 (2012)
19. Narayanan, A., Shmatikov, V.: Robust de-anonymization of large sparse datasets. In: Proceedings of the IEEE Symposium on Security and Privacy (2008)
20. Barocas, S., Selbst, A.D.: Big data's disparate impact. Calif. Law Rev. **104**(3), 671–732 (2016)
21. Grimaldi, B.: Machine learning in cybersecurity: data preprocessing approaches. J. Cybersecurity Educ. Res. Pract. **5**(2) (2023)
22. Smith, J., Lee, K.: Data quality in artificial intelligence. J. AI Res. **56**(3), 1245–1267 (2023)
23. Zhang, R., Kumar, M., Wang, L.: Impact of inaccurate data on machine learning models. IEEE Trans. AI **12**(2), 275–282 (2021)

24. Hinton, A., Raj, S.: Deep learning and the need for complete datasets. Neural Comput. **35**(6), 957–968 (2022)
25. Tan, P., Steinbach, M.: Consistency in multisource data for AI applications. IEEE Data Eng. Bull. **45**(4), 402–409 (2020)
26. Baker, L.: Timeliness in autonomous systems. IEEE Robot. Autom. Mag. **27**(1), 14–19 (2022)
27. Wang, K.: Accuracy in natural language processing. Comput. Linguist. J. **49**(2), 178–195 (2023)
28. Williams, S.: Ethics in AI: addressing bias. AI Soc. **38**(4), 1067–1079 (2021)
29. Garcia, T.: Challenges in AI scalability. IEEE Trans. Neural Netw. Learn. Syst. **32**(9), 3425–3438 (2021)
30. Chen, Y.: Labeling errors and their effects on machine learning. Pattern Recogn. Lett. **157**, 142–150 (2022)
31. Smith, D.: Data preprocessing techniques in AI. IEEE Access **10**, 52837–52844 (2023)
32. Brown, M.: Dynamic data in AI: ensuring quality. IEEE Internet Things J. **10**(1), 348–356 (2024)
33. Ross, J.: AI in healthcare diagnostics. J. Med. Imaging **9**(2), 222–231 (2022)
34. Huang, L., Liu, Y.: Ensuring real-time data quality in autonomous vehicles. IEEE Veh. Technol. Mag. **16**(3), 22–29 (2021)
35. Kotsiantis, G., Zaharakis, P., Pintelas, P.: Supervised machine learning: a review of classification techniques. Emerg. Artif. Intell. Appl. Comput. Eng. **160**, 3–24 (2007)
36. Barocas, S., Hardt, M., Narayanan, A.: Fairness and Machine Learning, 1st edn. MIT Press, Cambridge (2019)
37. Rousseeuw, P.J.: Silhouettes: a graphical aid to the interpretation and validation of cluster analysis. J. Comput. Graph. Stat. **1**(2), 53–65 (1992)
38. Friedman, D.P.: Data diversity and bias mitigation in machine learning. J. AI Ethics **1**(1), 45–58 (2020)
39. Rinehart, R.L.: Feedback loops and the evolving AI system. AI Ethics **2**(1), 1–10 (2021)
40. Ray, J.R.: Human-in-the-loop learning for data annotation. In: Proceedings of the International Conference on Machine Learning, vol. 21, pp. 3235–3243 (2018)
41. Barocas, S., Hardt, M., Narayanan, A.: Fairness and Machine Learning. fairmlbook.org (2019)
42. Mehrabi, N., Morstatter, F., Saxena, N., Lerman, K., Galstyan, A.: A survey on bias and fairness in machine learning. ACM Comput. Surv. (CSUR) **54**(6), 1–35 (2021)
43. Binns, R.: Fairness in machine learning: lessons from political philosophy. In: Proceedings of the 2018 Conference on Fairness, Accountability, and Transparency, pp. 149–159 (2018)

Fortifying the Future: Innovative Approaches to Cybersecurity Education and Training in the Digital Age

Dina Moloja(✉)

Central University of Technology, Bloemfontein, South Africa
dmmoloja@yahoo.com

Abstract. In an era where cyber threats grow increasingly sophisticated, equipping individuals and organizations with robust cybersecurity knowledge has become imperative. To close the skills gap and improve digital resilience, this theoretical study examines innovative cybersecurity education and training methodologies. We investigate the way advanced technologies, such as virtual reality and artificial intelligence, alter conventional pedagogical approaches by fusing practical simulations, gamified learning, and real-world case studies. The study also answers the following question: What barriers hinder effective cybersecurity education, and how may they be removed? The findings emphasize the value of interdisciplinary collaboration in fostering a cybersecurity culture that prioritizes ethical issues and proactive threat management. The study proposes methods to make cybersecurity education engaging, relatable, and successful for various audiences, including professionals and students, by drawing on case studies and international best practices. In the end, this study offers practical advice on defending the workforce against online attacks, guaranteeing a safe and robust digital future. This paper contributes to cybersecurity training and education knowledge while highlighting the need for future work in developing scalable, personalized learning frameworks and further exploring the impact of interdisciplinary approaches in addressing emerging cybersecurity challenges.

Keywords: Activity theory · Cybersecurity education and training · Artificial Intelligence · Virtual Reality

1 Introduction and Background

Cybersecurity threats are growing more sophisticated and widespread in a time of fast technological advancement, putting tremendous pressure on individuals, governments, and businesses to strengthen their defenses [1]. The rise in ransomware and sophisticated nation-state hacking attempts has revealed the weaknesses in digital infrastructures, making cybersecurity expertise and understanding more important than ever [4]. The need for cybersecurity specialists is increasing, yet there is still a sizable skills gap in the industry. The difficulties in protecting against cyberattacks are exacerbated by the fact that millions of cybersecurity jobs are still unfilled worldwide, according to [5]. This demonstrates the pressing need to reconsider and develop new approaches to cybersecurity education and training.

S. Latifi (Ed.): ITNG 2025, AISC 1463, pp. 117–128, 2025.
https://doi.org/10.1007/978-3-031-89063-5_11

Traditional cybersecurity training methods have often relied on lecture-based instruction and theoretical learning [4], which do not adequately prepare students for the practical, fast-paced nature of real-world cyber threats. As argued by [5], while foundational knowledge is important, cybersecurity education must integrate experiential learning to enable students to adapt to constantly changing technologies and threat landscapes [3]. Furthermore, the lack of diversity in training methods and limited engagement with cutting-edge technologies such as artificial intelligence (AI) and virtual reality (VR) hampers the development of critical skills in learners [1]. Traditional curricula often fail to reflect the interdisciplinary nature of modern cybersecurity, which now requires expertise from various domains such as law, ethics, and behavioral sciences [2].

One of the key challenges in cybersecurity education is the difficulty in creating engaging, real-world learning experiences [2, 3]. The reference [6] highlights that many training programs do not simulate actual cyber incidents effectively, leading to a gap between theoretical knowledge and practical application. To address this, they suggest incorporating tools like gamification and immersive learning technologies to bridge this divide. However, as [6] notes, implementing these technologies on a large scale presents logistical and financial challenges, particularly in regions with limited resources.

Moreover, the ethical implications of cybersecurity training are often sidelined in favor of technical skills. The reference [6] argues that educating future cybersecurity professionals must include a strong ethical foundation, as decisions in the digital world are not only technical but also moral and legal. The lack of interdisciplinary collaboration in most cybersecurity training programs results in a narrow perspective, limiting the development of well-rounded professionals capable of handling the multifaceted challenges of the digital age.

This paper examines how innovative, technology-driven strategies, such as virtual reality and artificial intelligence, can transform cybersecurity education by fostering a more interactive, engaging, and interdisciplinary learning environment. By addressing the barriers that hinder effective education—such as outdated pedagogical approaches, lack of engagement with new technologies, and the absence of ethical training—this study aims to propose methods that will make cybersecurity education more dynamic, inclusive, and successful. In doing so, we seek to bridge the skills gap and ensure that the next generation of cybersecurity professionals is better equipped to face the challenges of an increasingly interconnected world.

The structure of the paper is as follows: Sect. 2 addresses the research problem and objectives. Section 3 explores the theoretical framework underpinning the study. The findings are detailed in Sect. 4, while Sect. 5 concludes the paper and offers recommendations.

2 Research Problem and Aim

The rapid evolution of cyber threats has outpaced the development of effective educational frameworks and training programs [9], resulting in a significant skills gap within the cybersecurity workforce. Traditional approaches to cybersecurity education often fail to engage learners [10], adapt to technological advancements, or address the interdisciplinary nature of modern cyber challenges. This disconnect not only compromises organizational resilience but also threatens national security, highlighting the urgent need for innovative, scalable, and effective educational models that can prepare individuals to navigate the complexities of the digital age.

This research aims to address the following core issues:

- How can cybersecurity education and training be reimagined to effectively equip diverse audiences with the skills and knowledge needed to combat emerging cyber threats in a dynamic technological landscape?
- What barriers exist in cybersecurity education and training?

The key aspects of the research problem are as follows:

Skills Gap: The persistent shortage of qualified cybersecurity professionals is driven by the rapid evolution of cyber threats, which outpaces current educational offerings. This gap leaves organizations vulnerable to attacks and limits national security preparedness [2].
Ineffective Traditional Methods: Conventional teaching methods in cybersecurity often fail to engage learners or provide the hands-on, practical experience necessary to prepare them for real-world applications. Traditional approaches can be too theoretical, missing key components of effective skill-building [8].
Technological Integration: Many existing programs do not fully leverage emerging technologies such as virtual reality (VR), artificial intelligence (AI), and gamification, which have proven potential to enhance the learning and training experience. There is a need to integrate these tools into educational frameworks to make learning more dynamic and applicable to modern challenges [17].
Interdisciplinary Deficit: Cybersecurity education often neglects the importance of integrating interdisciplinary perspectives, such as psychology, ethics, and organizational management. These areas are crucial for developing a well-rounded cybersecurity mindset that goes beyond technical expertise [4].
Accessibility and Engagement: Barriers to accessing high-quality cybersecurity education, such as cost, geographic limitations, and a lack of tailored programs for diverse audiences, hinder the widespread development of cybersecurity competencies. Addressing these issues is key to building a more inclusive and effective educational ecosystem [10].

This research aims to develop and propose innovative, technology-driven, and interdisciplinary strategies for advancing cybersecurity education and training. The goal is to address the current skills gap, improve the effectiveness and engagement of learning methods, and foster a comprehensive cybersecurity culture that equips individuals and organizations with the knowledge and practical skills needed to mitigate evolving cyber threats. By leveraging emerging technologies, incorporating real-world applications, and promoting accessibility, this research seeks to contribute to the creation of a resilient and skilled cybersecurity workforce capable of safeguarding the digital future.

3 Theoretical Framing

This research is guided by Activity Theory (AT), which is a framework used to analyze and understand human activities as they occur within a system of interconnected components [3]. In the context of cybersecurity education and training, it provides a lens to explore how learners, educators, tools, and their broader environment interact to achieve the goal of equipping individuals with essential cybersecurity skills.

Below are the key elements of Activity Theory as applied to cybersecurity education as depicted in Fig. 1:

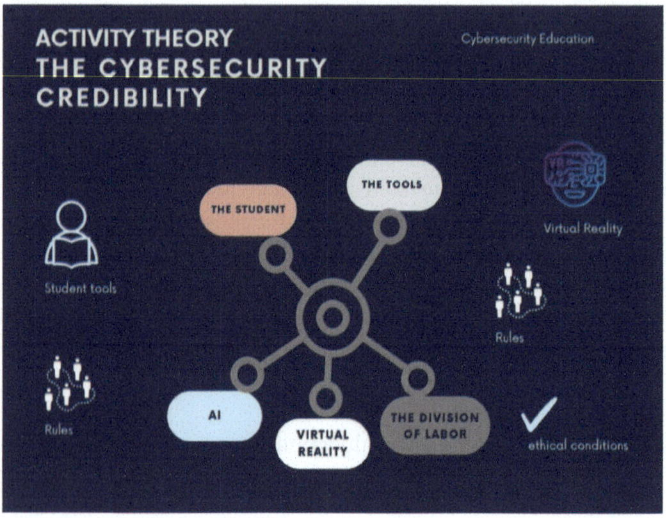

Fig. 1. Activity Theory for Cybersecurity education and training [3, 8]

- Subject: The learners or trainees participating in cybersecurity education programs.
- Object: The intended goal or outcome of the activity, such as developing proficiency in cybersecurity principles, tools, and practices.
- Tools/Artifacts: The technologies, methods, and resources used in learning and teaching cybersecurity (e.g., simulations, ethical hacking tools, or cybersecurity frameworks).
- Community: The broader group of stakeholders involved, including educators, peers, industry professionals, and institutions.
- Rules: The guidelines, standards, and ethical considerations governing cybersecurity education, such as compliance with legal frameworks or professional codes of conduct.
- Division of Labor: The distribution of roles and responsibilities among educators, students, and other stakeholders, fostering collaboration and specialized contributions.

Using AT in cybersecurity education enabled the identification of challenges and opportunities within this dynamic system.

4 Findings and Discussion

The findings from this study provide critical insights into innovative approaches that are shaping the landscape of cybersecurity education and training in the digital age. As cyber threats evolve, the necessity for adaptive and forward-thinking educational strategies becomes increasingly evident. This section presents the results, organized into three key themes, 1) Barriers that hinder effective cybersecurity education and training, 2) Factors to consider to improve digital resilience for cybersecurity education and training, 3) The Impact of Virtual Reality and Artificial Intelligence on Cybersecurity Education and Training: Innovative Strategies for Future-Ready Skills.

4.1 Barriers that Hinder Effective Cybersecurity Education and Training

- Lack of Skilled Instructors and Resources

The lack of skilled instructors and resources in cybersecurity education is a critical challenge that undermines the delivery of advanced and relevant content, a concern emphasized by multiple scholars in the field. For instance, [15, 16] highlight that the fast-paced nature of technological advancements often leaves educators struggling to keep up with emerging trends, leading to a mismatch between educational outcomes and industry needs. Similarly, [8] argues that the absence of instructors with updated expertise results in outdated curricula, reducing learners' ability to tackle real-world cybersecurity threats effectively. This challenge is compounded by the scarcity of resources, such as access to modern cybersecurity tools and platforms for hands-on training, which are integral to equipping students with practical skills [10].

To address this issue, continuous professional development for instructors emerges as a viable solution, supported by the literature. Studies advocate for initiatives such as ongoing training programs, industry-academic collaborations, and certifications tailored to the latest cybersecurity trends [8]. For example, the use of specialized workshops and training modules has been shown to significantly improve the knowledge and teaching capabilities of educators in technical domains [1]. Furthermore, creating pathways for instructors to engage in real-world cybersecurity projects or internships can bridge the gap between theoretical knowledge and practical application, as noted by [6].

Incorporating these strategies can enhance the quality of education [2], ensuring that instructors are well-equipped to deliver content that aligns with current and future industry demands. Consequently, addressing this challenge contributes not only to the development of competent cybersecurity professionals but also to strengthening global cyber resilience. The literature underlines that investing in the upskilling of educators and the provision of adequate resources is pivotal in overcoming this bottleneck in cybersecurity education [17].

- Rapidly Evolving Threat Landscape

The rapidly evolving threat landscape in cybersecurity presents a significant challenge to educational programs, as cybersecurity threats evolve at a pace that often exceeds the ability of curricula to adapt [2]. This mismatch can result in education programs becoming outdated, leaving students ill-prepared to address the latest cyber threats. As noted by [7], the dynamic nature of cyberattacks—such as zero-day vulnerabilities, ransomware, and advanced persistent threats—requires continuous updates to educational content to ensure students are equipped with the most current knowledge and skills [11]. However, traditional curricula struggle to keep up with these fast-moving threats, which risks diminishing the relevance of educational programs in preparing students for the ever-changing cybersecurity landscape [7, 11]. One promising solution to this challenge is the integration of AI technologies to provide real-time updates on emerging threats and trends. AI can be used to analyze vast amounts of cybersecurity data, identify patterns, and predict potential future threats, thus ensuring that educational content remains current [10].

This approach allows for a dynamic and adaptive curriculum that can respond to emerging cybersecurity challenges and incorporate them into teaching materials more quickly [17]. Furthermore, AI-driven platforms can provide personalized learning experiences, enabling students to explore and engage with the most relevant and up-to-date topics in cybersecurity, as highlighted by [14]. By leveraging AI to deliver real-time updates, educational institutions can bridge the gap between the pace of threat evolution and the need for relevant, forward-thinking curricula, ultimately enhancing the preparedness of future cybersecurity professionals.

- Limited Access to Advanced Technology

Limited access to advanced technology, such as virtual reality (VR) labs, threat simulation platforms, and sandbox environments, remains a significant challenge in cybersecurity education. The high costs and resource requirements associated with these tools often create barriers for institutions, particularly those in underfunded or developing regions, thereby restricting hands-on learning opportunities for students.

According to [10], practical, immersive experiences are essential for building the technical skills necessary to respond to real-world cybersecurity threats, yet many programs struggle to provide these opportunities due to financial constraints. This limitation leaves students with primarily theoretical knowledge, hindering their ability to develop critical problem-solving and decision-making skills in realistic environments. One potential solution to this challenge lies in the use of open-source platforms and collaborations with technology companies.

Open-source tools, such as the Open Web Application Security Project (OWASP) or Kali Linux, can provide students with free access to valuable resources for learning and experimentation [6, 9]. Additionally, partnerships with tech companies can offer institutions access to advanced tools at reduced or no cost, fostering a more practical, resource-rich learning environment.

Research by [12] shows that such partnerships can be mutually beneficial, with tech companies gaining access to emerging talent and institutions receiving access to cutting-edge technology. By leveraging these strategies, educational programs can overcome the financial barriers to hands-on cybersecurity education and better prepare students for the challenges of the evolving cyber threat landscape.

- Cybersecurity Awareness Gaps

Cybersecurity awareness gaps remain a significant challenge in educational settings, as students, educators, and institutions often undervalue the importance of cybersecurity education until a breach occurs. This reactive approach to cybersecurity places educational institutions at a disadvantage, as they fail to prioritize proactive learning environments that could better prepare students for emerging cyber threats.

As noted by [14], many institutions continue to treat cybersecurity as a supplementary subject rather than an integral part of the curriculum, leading to a lack of investment in specialized cybersecurity training and infrastructure. This attitude results in a generation of students with insufficient cybersecurity knowledge, leaving them vulnerable to both personal and professional cyber risks. One effective solution to addressing these awareness gaps is through targeted awareness campaigns and the integration of cybersecurity principles into general IT programs [10, 14].

Research by [8] suggests that embedding cybersecurity awareness in broader IT curricula, alongside hands-on learning experiences, can foster a more security-conscious mindset among students from the outset. Moreover, awareness campaigns, which can be tailored to the needs of both educators and students, can help shift the perception of cybersecurity from a niche concern to a critical, foundational skill in the digital age [17]. By raising awareness and embedding cybersecurity principles throughout general IT programs, institutions can ensure that students are better equipped to navigate the complexities of the digital world, ultimately enhancing both individual and organizational cyber resilience.

- Lack of Interdisciplinary Approaches

Since many programs concentrate primarily on technological components and frequently overlook the relationship to subjects like psychology, law, and business, the lack of interdisciplinary methods in cybersecurity education continues to be a significant concern [10, 15, 16]. Students are ill-equipped to handle challenging, real-world situations because of this compartmentalized approach, which restricts their comprehension of the larger cultural, ethical, and legal context of cybersecurity issues.

As noted by [13], cybersecurity is not only a technical challenge but also a multifaceted issue that involves human behavior, legal frameworks, and organizational dynamics. Without an integrated approach, students fail to recognize the human and organizational factors that influence cybersecurity risks, such as insider threats, user behavior, and compliance with data protection laws [15, 16]. The solution lies in developing interdisciplinary curricula that combine technical training with non-technical perspectives, including legal, psychological, and business considerations.

Research by [17] suggests that incorporating these diverse perspectives into cybersecurity education can help students better understand how to address the complex nature of cybersecurity challenges, fostering more well-rounded professionals capable of devising comprehensive solutions. By embracing interdisciplinary approaches, institutions can prepare students to think critically about cybersecurity issues, ensuring they are equipped to navigate the intricate legal, ethical, and social landscapes of the digital world.

4.2 Factors to Consider to Improve Digital Resilience for Cubersecurity Education and Training

In the context of this paper, digital resilience refers to the ability of an institution or organization to adapt, recover, and maintain operations despite disruptions, especially in the face of cyber threats and technological changes. Improving digital resilience in higher learning institutions is essential for ensuring secure, continuous learning environments and operations. Below are some of the factors recorded by the literature (Table 1):

Table 1. Factors to consider improving digital resilience [4, 12, 15, 16]

Factor	Impact	Action
Robust Cybersecurity Infrastructure	Protects sensitive data and systems from cyberattacks and ensures the continuity of digital services	• Implement multi-layered security controls • Regularly update and patch all systems • Conduct frequent vulnerability assessments
Data Backup and Recovery Plans	Minimizes data loss and downtime, allowing quick recovery from disruptions	• Implement regular automated backups to cloud or offsite locations • Test recovery processes periodically • Ensure that critical data is backed up across multiple platforms
Cybersecurity Education and Awareness	Reduces the likelihood of human errors that can lead to breaches, such as phishing attacks or improper handling of sensitive information	• Conduct regular cybersecurity training for all staff, students, and faculty • Integrate cybersecurity topics into general education curricula • Encourage participation in cybersecurity awareness campaigns

(continued)

Table 1. *(continued)*

Factor	Impact	Action
Adaptability to Technological Changes	Ensures the institution can integrate new technologies without compromising security and operations	• Regularly evaluate emerging technologies for potential adoption • Establish protocols for secure integration and testing of new technologies • Train staff and faculty to manage new tools effectively
Developing a Digital Resilience Framework	Aligns all efforts toward strengthening resilience and provides clear guidelines for response and recovery	• Establish an institutional digital resilience strategy • Assign responsibility for digital resilience across departments • Align resilience efforts with overall risk management strategies

By addressing these factors, higher learning institutions can significantly improve their digital resilience, ensuring they are better prepared to withstand cyber disruptions and continue providing high-quality education and services.

4.3 The Impact of Virtual Reality and Artificial Intelligence on Cybersecurity Education and Training: Innovative Strategies for Future-Ready Skills

Enhanced Immersive Learning with Virtual Reality (VR)

Virtual reality offers a unique advantage in cybersecurity education by creating simulated environments where students can interact with realistic cyber threats in real-time. VR enables students to experience complex attack scenarios, such as network breaches, malware outbreaks, or ransomware attacks, in a safe and controlled setting. This immersive experience helps students develop practical problem-solving skills without the risks of actual system damage.

Example: VR training programs can simulate a corporate network where students must identify vulnerabilities, respond to an attack, or recover compromised data.

- Increased Engagement and Retention

VR offers an interactive learning environment that increases engagement and motivation. Studies show that immersive learning environments improve knowledge retention compared to traditional lecture-based approaches. By using VR, students are more likely to remember critical procedures and responses to cybersecurity threats because they actively participate in solving problems rather than passively absorbing information.

Example: Cybersecurity drills in VR can involve tasks such as identifying phishing attempts or performing penetration tests, making the learning experience more dynamic.

- Safe Failure and Risk Management

According to [5, 11, 17] one of the greatest advantages of VR is the ability to "fail safely." Students can experiment with different approaches to solving cybersecurity problems without the fear of causing real-world damage. In a VR environment, they can learn the consequences of their actions, experimenting with techniques like hacking or defending against attacks to understand both sides of a cybersecurity situation.

Example: A student can attempt to break into a simulated system to understand the methods of attackers, which ultimately informs how to better defend against such breaches.

Artificial Intelligence (AI) in Cybersecurity Education

- Personalized Learning Experiences

The study found that AI can enhance cybersecurity education by providing personalized learning pathways. AI-driven tools can assess students' current knowledge and adjust the content, difficulty, and pace of lessons based on individual performance [13, 17]. This helps ensure that students grasp fundamental concepts before moving on to more complex topics, promoting mastery and improving the overall learning experience.

Example: AI systems like chatbots or virtual instructors can provide tailored feedback and additional resources based on a student's progress, ensuring that the learning experience is both adaptive and targeted.

- Real-Time Threat Simulation and Analysis

The literature revealed that AI algorithms can generate real-time simulations of cyberattacks [10], providing students with opportunities to respond to evolving, sophisticated threats. These simulations can mimic the behaviors of real-world adversaries, helping students understand the tactics, techniques, and procedures (TTPs) used by cybercriminals and how to effectively counter them [15, 16].

Example: AI can create a dynamic, evolving phishing campaign or ransomware attack during a training exercise, requiring students to analyze the threat and implement appropriate countermeasures.

- Automated Threat Detection and Incident Response Training

The study found that AI can be used to teach students how to design and implement automated threat detection systems [7], as well as how to respond to incidents using AI-powered tools. This training is vital, as AI is increasingly being integrated into cybersecurity tools to provide faster and more accurate threat detection.

Example: Students can learn to work with AI-powered Security Information and Event Management (SIEM) systems that analyze vast amounts of data to identify potential security breaches.

The combination of Virtual Reality and Artificial Intelligence in cybersecurity education offers revolutionary ways to train students for future cyber challenges [11, 13]. By providing immersive, dynamic environments for learning and personalized training experiences, these technologies ensure that students are better prepared to meet the evolving landscape of cyber threats [14]. As cybersecurity continues to be a critical component of every organization's infrastructure, adopting innovative educational strategies will be key to developing the skilled professionals needed to protect digital assets in the future.

5 Conclusion and Recommendations

As cyber threats continue to grow in complexity and scale, the need for innovative approaches to cybersecurity education and training has never been more critical. Technologies such as Virtual Reality (VR) and Artificial Intelligence (AI) offer transformative potential in shaping the future of cybersecurity education [8], providing immersive, adaptive, and real-world training experiences that prepare learners for emerging challenges.

By incorporating VR, educators can create realistic, hands-on simulations that enhance engagement and problem-solving skills, while AI enables personalized learning experiences that evolve with each student's progress [9]. Together, these technologies can help bridge the gap between theoretical knowledge and practical application, ensuring that cybersecurity professionals are not only equipped with technical expertise but also the critical thinking and decision-making abilities necessary to combat increasingly sophisticated cyber threats.

Furthermore, the integration of AI and VR offers cost-effective, scalable, and dynamic training solutions that can adapt to the ever-changing digital landscape. These innovations empower higher learning institutions to deliver future-ready skills, preparing the next generation of cybersecurity professionals to safeguard our increasingly digital world.

Ultimately, fortifying cybersecurity education with these advanced technologies is essential for building a resilient, secure, and informed global workforce capable of tackling the cybersecurity challenges of tomorrow.

Future work can be done in developing scalable, personalized learning frameworks and further exploring the impact of interdisciplinary approaches in addressing emerging cybersecurity challenges.

References

1. AlDaajeh, S., Saleous, H., Alrabaee, S., Barka, E., Breitinger, F., Choo, K.K.R.: The role of national cybersecurity strategies on the improvement of cybersecurity education. Comput. Secur. **119**, 102754 (2022)
2. Alshaikh, M., Naseer, H., Ahmad, A., Maynard, S.B.: Toward sustainable behaviour change: an approach for cyber security education training and awareness (2019)
3. Bodea, C.N., Dascalu, M.I., Cazacu, M.: Increasing the effectiveness of cybersecurity teaching and learning by applying activity theory and narrative research. Issues Inf. Syst. **20**(3) (2019)
4. Burov, O.Y., Butnik-Siversky, O.B., Orliuk, O., Horska, K.A.: Cybersecurity and innovative digital educational environment. Інформаційні технології і засоби навчання **6**(80), 414–430 (2020)
5. Catota, F.E., Morgan, M.G., Sicker, D.C.: Cybersecurity education in a developing nation: the Ecuadorian environment. J. Cybersecur. **5**(1), tyz001 (2019)

6. Crumpler, W., Lewis, J.A.: Cybersecurity Workforce Gap, p. 10. Center for Strategic and International Studies (CSIS) (2022)
7. Daimi, K., Francia, G., III. (eds.): Innovations in Cybersecurity Education, vol. 388. Springer, Cham (2020)
8. Gross, M., Ho, S.M.: Collective learning for developing cyber defense consciousness: an activity system analysis. J. Inf. Syst. Educ. **32**(1), 65–76 (2021)
9. Jerman Blažič, B., Jerman Blažič, A.: Cybersecurity skills among European high-school students: a new approach in the design of sustainable educational development in cybersecurity. Sustainability **14**(8), 4763 (2022)
10. Katsantonis, M.N., Manikas, A., Mavridis, I., Gritzalis, D.: Cyber range design framework for cyber security education and training. Int. J. Inf. Secur. **22**(4), 1005–1027 (2023)
11. Mukherjee, M., Le, N.T., Chow, Y.W., Susilo, W.: Strategic approaches to cybersecurity learning: a study of educational models and outcomes. Information **15**(2), 117 (2024)
12. Thakur, M.: Cyber security threats and countermeasures in digital age. J. Appl. Sci. Educ. (JASE) **4**(1), 1–20 (2024)
13. Taherdoost, H.: Towards an innovative model for cybersecurity awareness training. Information **15**(9), 512 (2024)
14. Saeed, S., Altamimi, S.A., Alkayyal, N.A., Alshehri, E., Alabbad, D.A.: Digital transformation and cybersecurity challenges for businesses resilience: issues and recommendations. Sensors **23**(15), 6666 (2023)
15. Sinha, M.: Exploring the role of cybersecurity in integrated programs for protecting and improving digital platforms. Int. IT J. Res. **2**(2), 190–197 (2024). ISSN 3007-6706
16. Bühler, M.M., Jelinek, T., Nübel, K.: Training and preparing tomorrow's workforce for the fourth industrial revolution. Educ. Sci. **12**(11), 782 (2022)
17. Shiohira, K.: Understanding the Impact of Artificial Intelligence on Skills Development. Education 2030. UNESCO-UNEVOC International Centre for Technical and Vocational Education and Training (2021)

Efficient Constructions from Double POLYVAL

Shay Gueron[1,2](✉)

[1] University of Haifa, Haifa, Israel
shay.gueron@gmail.com
[2] Meta, Seattle, USA

Abstract. POLYVAL is an almost-XOR universal hash-function family that is used with AES-GCM-SIV. It has 128-bit outputs and can be computed efficiently on modern platforms. We show here constructions based on two POLYVAL calls to produce 256-bit outputs, and a block cipher with 256-bit keys and 256-bit block size. These are the pseudorandom function Double POLYVAL PRF (DP-PRF) and the stateless Message Authentication Code (MAC) Double POLYVAL MAC (DP-MAC). Their security relies on the single assumption that the block cipher is indistinguishable from a random permutation of $\{0, 1\}^{256}$ even with the target number ($>2^{64}$) of queries. Under this assumption DP-MAC and DP-PRF can be safely used for processing 2^{64} bytes. This obviates key rotation due to key wear out in real world applications. We use Rijndael256 as the block cipher. This allows us to use the AES-NI/PCLMULQDQ instructions on modern processors, for optimizing software performance and reach the throughput of 0.61 cycles per byte.

Keywords: Pseudorandom function · MAC · efficient implementation · AES-NI · PCLMULQDQ · POLYVAL

1 Introduction

Message Authentication Codes (MAC) are an essential building block in cryptography, and therefore efficient MAC designs have significant impact on many systems, especially at the cloud scale. Some NIST standard MAC schemes are HMAC, CMAC, and GMAC (see a review in [Rog11]). SIV-MAC is a recently proposed MAC [Gue24] and Keyed Blake (blake2b512) is an efficient MAC – both not a NIST standard. To compare their performance (measured on a Skylake processor over 16KB buffers), HMAC performs at ~ 8 cycles per byte (cpb), Blake2b512 at ~4.44 cpb, CMAC at ~3.34 cpb, and GMAC at ~0.38 cpb, SIV-MAC at 0.3 cpb. Blake3 is faster than Blake2, and performs at 1.89 cpb on a Skylake processor CPU. There are some limits on the amount of data and number of messages that can be processed with a given key. For example, upper bounds on the PRF indistinguishability of two MAC schemes with 128-bit tags are $4.51\sigma^2/2^{128}$ for SIV-MAC and $4\sigma^2/2^{128}$ for CMAC. Here, σ is the total number of 128-bit processed blocks. These bounds become useless when σ approaches 2^{64}. In particular, this means that high volume cloud scale usages might need to take care of key rotation with these algorithms.

S. Latifi (Ed.): ITNG 2025, AISC 1463, pp. 129–141, 2025.
https://doi.org/10.1007/978-3-031-89063-5_12

For the efficient GMAC scheme, we note that is a stateful MAC, using a 96-bit nonce. This means that assuring nonce uniqueness is a concern with their deployment and that GMAC carries traffic overheads (the nonce must be communicated together with the tag, unless it can be obtained implicitly, e.g., as a record number in an ordered stream).

We set our goal here to define an efficient stateless Message Authentication Code (MAC) with 256-bit tags and security margins that allow to safely process up to 2^{64} bytes with a single key. This can make TINkey rotation due to key wear out unnecessary. To this end, we define the pseudorandom function Double POLYVAL PRF (DP-PRF) and the MAC scheme Double POLYVAL MAC (DP-MAC).

2 Preliminaries and Notation

Uniform random sampling of a value v from a set V is denoted by $v \longleftarrow_{\$} V$.

Strings of Bits. The set of t-bit strings is denoted by $\{0, 1\}^t$, the set of bit strings with at most t bits is denoted by $\{0, 1\}^{\leq t}$, the set of all strings is denoted by $\{0, 1\}^*$ and the set of strings whose bitlength is divisible by n is denoted by $\{0, 1\}^{n^*}$. For an integer s, the string of s zero bits is denoted by 0^s. A128-bit string is called a "block".

For an integer $0 \leq s < 2^{64}$, IntToString64 (s) is the 64-bit string that represents the binary representation of s. For example,
IntToString64 (2024) = 0000000000000000000000000000000000000000000000000000011111101000.
(equivalently in hexadecimal notation, it is 0x00000000000000000000000000007E8).

Let $X = x_{n-1}\ x_{n-1}\ \ldots\ x_0$ be an n-bit string with bits $x_{n-1}\ x_{n-1}\ \ldots\ x_0$ (here, the least significant bit x_0 in the rightmost position). We define set-bit0 (X) to be the result of forcing the least significant bit of X to 1 while retaining all other bits unchanged. Specifically, $\text{set}^0\ (X) = x'_{n-1}x'_{n-1}\cdots x'_0$ where $x'_0 = 1$ and $x'_j = x_j$ for $j = 1, 2, \ldots, n-1$. For a string $X = x_{n-1}\ x_{n-1}\ \ldots\ x_0$ and two indices $0 \leq b \leq a \leq (n-1)$, X [a: b] is the substring of X that consists of the bits $x_a\ x_{a-1} \ldots\ x_b$ (in this order). For example, X [8: 5] = $x_8\ x_7\ x_6\ x_5$.

Bytes and Arrays of Bytes. Bytes are integers between 0 and 255, and specific byte values are denoted in hexadecimal notation (e.g., 0x00 is the zero byte). The notation $(0x00)^s$ denotes an array of s zero bytes.

An array U of u bytes is denoted by U = U [u − 1: 0] (index notation is inclusive) where X [j] refers to byte number j ($0 \leq j \leq m - 1$). Denote the number of bytes in U by bytelen (U) = u. For integers a, b such that $0 \leq b \leq a \leq (u-1)$, T = U [a: b] denotes the sub-array U [a] U [a − 1]...U [b] of (a-b + 1) where T [j] = U [j + b], j = 0, 1,..., (a − b). Note that in the sub-array notation the indices are ordered high then low and inclusive. For example,

U = 0x55555555555555557777777777777777 is a block, written in hexadecimal notation, where byte 0 is 0x77 and byte 15 is 0x55. U [9: 7] = 0x555577.

Concatenation of byte arrays is denoted by "||". If U1 = U1 [u1 – 1] U [u1 – 2]… U [0] and U0 = U0 [u0 – 1] U [u0 – 2]… U [0] then U1 || U0 is the array W = W [u1 + u0 – 1: 0] of u1 + u2 bytes such that W [u1 + u0 – 1: u0] = U1 and W [u0 – 1: 0] = U0.

If T is an array of b bytes and $1 \leq \tau \leq b$, then Truncate (Tag, τ), the truncation of T into τ bytes, is the array (of τ bytes) T [τ – 1: 0].

Arrays of bytes are viewed, interchangeably, as bit strings whose bitlength is divisible by 8. An array of 16 bytes, equivalently, a128-bit string, is called a "block" as well. We denote the set of all byte arrays with at most v bytes by Ω_v. Note that Ω_v is the same as $\{0, 1\}^{\leq t}$, where t = 8v and the set is restricted to bit strings with length divisible by 8.

Pseudorandom Functions and Distinguishing Advantage. For positive parameters κ1, s, n, let F: $\{0, 1\}^{\kappa 1} \times \{0, 1\}^s \longrightarrow \{0, 1\}^n$ be a family of pseudorandom functions (PRF) indexed with key $K1 \in \{0, 1\}^{\kappa 1}$. For $X \in \{0, 1\}^s$ denote its operation by F_{K1} (X). A PRF adversary D against F is a probabilistic polynomial time algorithm that submits queries $X \in \{0, 1\}^s$ to an oracle and outputs a bit. The oracle is either: a) an F oracle that selects (a-priori) $K1 \longleftarrow_\$ \{0, 1\}^{\kappa 1}$ and returns F_{K1} (X); b) A random oracle ($) that returns a uniform random n-bit string. Let D submit at most q (distinct) queries. Denote D (F) = 1 the event where D outputs 1 when it runs against oracle F and denote D ($) = 1 the event where D outputs 1 when it runs against oracle $. The PRF advantage of D is

$$\mathrm{Adv}(D) = \left|P[D(F) = 1] - P[D(\$) = 1]\right|$$

We can also define a random oracle $f^\$$ that selects a uniform random function f: $\{0, 1\}^s \longrightarrow \{0, 1\}^n$ (from the set of all functions from $\{0, 1\}^s$ to $\{0, 1\}^n$) and returns $f^\$$ (X) per query X. Denote D ($f^\$$) = 1 the event where D outputs 1 when it runs against oracle $f^\$$. Since D does not repeat queries, an algorithm cannot distinguish between the distributions of the responses from $ and from $f^\$$. Thus,

$$P[D\,(f^\$) = 1)] = P[D\,(\$) = 1)] \text{ and}$$
$$\mathrm{Adv}\,(D) = \left|P[D\,(F) = 1] - P[D\,(f^\$) = 1]\right|.$$

ε-Almost-XOR Universal Family of Hash Functions. Suppose that $\varepsilon \geq 0$, $t > 0$. Let H: $\{0, 1\}^{\kappa 2} \times \{0, 1\}^{\leq t} \longrightarrow \{0, 1\}^n$ be a family of functions indexed by a key $K2 \in \{0, 1\}^{\kappa 2}$ that has the following property: if $K2 \longleftarrow_\$ \{0, 1\}^{\kappa 2}$ then for every $Ma \neq Mb \in \{0, 1\}^{\leq t}$ and $\delta \in \{0, 1\}^n$, $P\,[H_{K2}\,(Ma) \oplus H_{K2}\,(Mb) = \delta] \leq \varepsilon$. Such a family is called an ε-XOR universal family of hash functions.

POLYVAL. We use here a specific XOR universal family of hash functions named POLYVAL that is defined in [RFC8452, GLL17]. POLYVAL is a type of polynomial hash function from $\{0, 1\}^{128} \times \{0, 1\}^{128*}$ to $\{0, 1\}^{128}$. It takes a hash key $H \in \{0, 1\}^{128}$, a nonempty sequence of s ($\geq$1) blocks Y = Y [1], Y [2], ..., Y [s] $\in \{0, 1\}^{128*}$ and outputs $POLYVAL_H$ (Y) $\in \{0, 1\}^{128}$. Appendix C provides more details. Here, we view the blocks sequence Y as the concatenation of the s blocks Y [s] || Y [s − 1] || ... || Y [1].

To extend POLYVAL for an array X of m < 2^{61} bytes ("message") where m is not necessarily divisible by 16, first pad X with p = (16ceil (m/16) – m) zero bytes to a sequence of ceil (m/16) full block. Then, to obtain a unique encoding of such sequences, append the resulting array with a block, denoted LENBLOCK, that encodes its (bit) length. Specifically, LENBLOCK (X) = IntToString64 (8m) || 0^{64}. The resulting array of bytes is Y = X || $(0x00)^p$ || LENBLOCK (X) $\in \{0, 1\}^{128*}$. Here, Y consists of L = (1 + ceil (m/16)) full blocks (i.e., a valid input to $POLYVAL_H$).

Suppose that $0 \leq m0 < 2^{61}$, L0 = (1 + ceil (m0/16)) and let Ω_{m0} denote the set of byte arrays of at most m0 bytes. Then, if we restrict POLYVAL to inputs in Ω_{m0}, we have the following property. POLYVAL: $\Omega_{m0} \longrightarrow \{0, 1\}^{128}$ with hash key $H \longleftarrow_\$ \{0, 1\}^{128}$ is an ε-XOR universal family of hash functions with $\varepsilon = (L0 + 1)/2^{128}$.

Rijndael256. Rijndael256 is one of the options in the proposal named "Rijndael" [DR02] that specified a family of block ciphers (AES is an instantiation with 128-bit block size). Here, Rijndael256 refers to the version with block size of 256 bits (i.e., 32 bytes) and 256-bit key size. A short description of Rijndael256 and its performance characteristics can be found in [DG22].

Example [Rijndael256]. Following is Rijndael256 ciphertext from a key K with bytes K [j] = 0xFF, j = 0, ..., 30, K[31] = 0x00 and a 32-byte plaintext P such that P [0] = 0x01, P [j] = 0x00, j = 1, ..., 31.

```
-----------Bytes Position-------
313029282726252423222120191817161514
1312111009080706050403020100
--------------------------------
K =
00FFFFFFFFFFFFFFFFFFFFFFFFFFFFFFFFFF
FFFFFFFFFFFFFFFFFFFFFFFFFFFF
P =
000000000000000000000000000000000000
0000000000000000000000000001
Rijndael256 (K, P) =
06487A7C24A40ADFF26595B6BFD3F6F6DC2E
4B19AAD9DB12BB868E8F1737D18C
```

3 PRF of a Universal Hash with a Single Key

Here, we define a special form with a single key, of the standard construction of "pseudorandom Function over a universal Hash" (FH).

Definition [FH composition]. Let F: $\{0, 1\}^{\kappa 1} \times \{0, 1\}^n \longrightarrow \{0, 1\}^n$ be a PRF with key $K1 \in \{0, 1\}^{\kappa 1}$ and operation denoted by F_{K1} (X) for input $X \in \{0, 1\}^n$. Let $t > 0$ be a parameter. Let H: $\{0, 1\}^{\kappa 2} \times \{0, 1\}^{\leq t} \longrightarrow \{0, 1\}^n$ be an ε-XOR universal family of hash functions with key $K2 \in \{0, 1\}^{\kappa 2}$ and operation denoted by H_{K2} (M) for input $M \in \{0, 1\}^{\leq t}$. Define the composition $FH_{K1} = \{0, 1\}^{\kappa 1} \times \{0, 1\}^{\leq t} \longrightarrow \{0, 1\}^n$ by

$$FH_{K1}(M) = F_{K1}(\text{setbit0}\ (H_{K2}(M)) \text{ where } K2 = F_{K1}(0^n).$$

Adversary for FH. Let A be a PRF adversary against FH with access to a FH oracle. This oracle first samples $K1 \longleftarrow_{\$} \{0, 1\}^{\kappa 1}$ and sets $K2 = F_{K1}\ (0^n)$ and subsequently returns F_{K1} (setbit0 (H_{K2} (M)) for a query $M \in \{0, 1\}^{\leq t}$. Denote A (FH) = 1 the event where A outputs 1 when it runs against the FH oracle. Denote A (\$) = 1 the event where A outputs 1 when it runs against the random oracle \$. The PRF advantage of A is

$$\text{Adv}\ (A) = \left|P\,[A\ (FH = 1] - P[A\ (\$) = 1]\right|.$$

Theorem 1. Let A be an adversary against the FH, with access to an FH /\$ oracle, that has (PRF) advantage Adv (A) when it is submits at most Q ($\geq$ 1) queries. Then, there exists an adversary B for F that submits at most Q + 1 queries such that

$$\text{Adv}\ (A) \leq \text{Adv}\ (B) + Q\ (Q - 1)\varepsilon\ (\leq Q^2\varepsilon).$$

(the time for B is the time for A plus the time to compute H_{K2} (M) over A's queries).

Proof. The proof is standard and we follow here Lemma 3.3 in [GL15] with the necessary changes to address the specific single-key composition FH (instead of the two keys version $FH'_{K1,K2} = F_{K1}\ (H_{K2}\ (M))$. Let us use A to build a PRF adversary B against F as follows. First, B calls its oracle (F or $f^{\$}$) with the query $T_0 = 0^n$ and sets K2 to be the received answer (either F (0^n) or \$). Then, B runs A. For every query M that A submits, B computes $R = H_{K2}$ (M), sets T = setbit0 (R), calls its oracle with T and returns the response (F (T) or \$) to A. When A outputs a bit and stops, B outputs the same bit and stops.

Let M1, M2, ..., MQ be the (distinct) queries of A and denote $Rj = H_{K2}\ (M_j)$, $Tj = set^0\ (R_j)$, j = 1, 2, ... Q. Note that while M1, M2, ..., MQ are distinct $T_1, T_2, \ldots, T_Q$ are not necessarily distinct (because Rj values may repeat). Denote the event (collision) where $T_i = T_j$ for some $0 \leq i < j \leq Q$ by **coll.** Denote the complementary event (all T_i are distinct) by **¬coll.** Note also that $Tj \neq T_0, j = 1, 2, \ldots Q$ (because the least significant bit of setbit0 (R_j) is 1).

In total, B submits (at most) $Q + 1$ queries to its oracle. The time of adversary B is the time for A plus the time for computing $R_j = H_{K2}(M_j)$ for $j = 1, 2, \ldots Q$. Note that in the event $\neg$**coll**, all of these $(Q + 1)$ queries are distinct.

Suppose that the event $\neg$**coll** occurs. Then,

$P[B(F) = 1)] = P[A(DP) = 1)]$ and $P[B(\$) = 1)] = P[B(f^{\$}) = 1] = P[A(\$) = 1]$. We have

$Adv(A) = |P[A(FH) = 1] - P[A(\$) = 1]| \leq$

$|(P[A(FH) = 1] - P[A(\$) = 1 \text{ given } \mathbf{coll})] \times P[\mathbf{coll}]| + |(P[A(FH) = 1] - P[A(\$) = 1 \text{ given } \neg\mathbf{coll})] \times P[\neg\mathbf{coll}]| \leq$

$P[\mathbf{coll}] + |(P[B(F) = 1] - P[B(\$) = 1)]| = P[\mathbf{coll}] + Adv(B)$.

It remains to upper bound $P[\mathbf{coll}]$. Since H is an ε-almost-XOR universal family of hash functions, we have $P[H_{K2}(Ma) = H_{K2}(Mb)] \leq \varepsilon$ for every pair of messages $Ma \neq Mb$. Therefore,

$P[\text{setbit0}(H_{K2}(Ma)) = \text{setbit0}(H_{K2}(Mb))] =$

$P[H_{K2}(Ma) = H_{K2}(Mb)] + P[H_{K2}(Ma) = H\, H_{K2}(Mb) \text{ XOR } 1] \leq$

$\varepsilon + P[H_{K2}(Ma) = H_{K2}(Mb) \text{ XOR } 1] \leq 2\varepsilon$.

If we sum this over all $Q(Q - 1)/2$ message pairs from $M1, \ldots, MQ$, we have

$P[\mathbf{coll}] \leq Q(Q - 1)\,\varepsilon < Q^2\,\varepsilon$.

This completes the proof. □

Instantiating FH with a Block Cipher. Let **P** be the set of all permutations of $\{0, 1\}^n$, and let $E = E_K(X)$ be a block cipher with an n-bit block size and a key $K \in (0, 1\}^\kappa$ (for some κ). E is a family of 2^κ permutations from P, indexed by K. We instantiate $FH_{K1}(M)$ with a block cipher E, i.e.,

$$FH_{K1}(M) = E_{K1}(\text{setbit0}\,(H_{K2}(M)) \text{ where } K2 = E_{K1}\left(0^n\right).$$

Denote the maximum advantage of an adversary with q queries in distinguishing E with a uniform random key K from a uniform random permutation from **P**, by $Adv^{prp}(E, q)$ (this advantage should be small if E is a good block cipher). By the PRP-PRF switching lemma E is a PRF and every PRF adversary A with q queries has distinguishing advantage at most $Adv^{prp}(E, q) + q(q - 1)/2^{n+1}$. Now, suppose we instantiate $FH_{K1}(M)$ with a block cipher E, i.e., $FH_{K1}(M) = E_{K1}(\text{setbit0}(H_{K2}(M))$ where $K2 = E_{K1}(0^n)$. Let A be an adversary against this scheme with at most Q queries. Then,

$$Adv(A) \leq Adv^{prp}(E,\ Q + 1) + (Q + 1)Q/2^{n+1} + Q(Q - 1)\varepsilon$$

4 The Double POLYVAL Universal Hash Construction

We set $n = 256$, choose a parameter $0 < m0 < 2^{61}$ and consider Ω_{m0}. Double POLYVAL (DP) is a universal family of hash function from $\{0, 1\}^{256} \times \Omega_{m0}$ to $\{0, 1\}^{256}$ defined as follows.

Double POLYVAL (DP) Universal Hash

Input: a 256-bit key K* and a message $X \in \Omega_{m0}$. $K1^* = K^*$ [255: 128], $K0^* = K^*$ [127: 0]

m = byte length (X)
nb = ceil (m/16)
p = 16*nb – m
Xpad = X || (0x00)^p
Y = Xpad || LENBLOCK (X)
DP (K*, X) = POLYVAL ($K1^*$, Y) || POLYVAL ($K0^*$, Y)

Lemma 1 [property of DP (K*, M)]. DP (K*, M) over the message space Ω_{m0} with a key.

$K^* \longleftarrow_\$ \{0, 1\}^{256}$ is an ε-universal family of hash functions with $\varepsilon = (L0 + 1)^2 / 2^{256}$ where L0 = (1 + ceil (m0/16)).

Proof. If $K^* \longleftarrow_\$ \{0, 1\}^{256}$ and Ma, Mb $\in \Omega_{m0}$ then

P [DP (K* Ma) = POLYVAL (K*, Mb)] =
P [POLYVAL ($K1^*$, Ma) = POLYVAL ($K1^*$, Mb)] ×
P [POLYVAL ($K0^*$, Ma) = POLYVAL ($K0^*$, Mb)] $\leq (L0 + 1)^2 / 2^{256}$.

where L0 = (1 + ceil (m0/16)) is the maximum number of blocks in messages from Ω_{m0}.

5 DP-PRF and DP-MAC Specification

Double POLYVAL PRF (DP-PRF) is defined over inputs that are (nonempty) arrays of bytes, with length at most max-bytes = 2^{52} bytes (2^{48} blocks). It uses a 32-byte (256-bit) secret key and outputs a τ -byte value (τ = 32).

Double POLYVAL PRF (DP-PRF). For integer m0 > 0, input space Ω_{m0} ($M \in \Omega_{m0}$) and $K \longleftarrow_\$ \{0, 1\}^{256}$ DP-PRF is an instantiation of the FH_{K1} where: a) the underlying PRF is $F_{K1} = E_{K1}$ (•) for a 256-bit block cipher E; b) $K^* = E_K (0^{256})$; c) the underlying universal hash function is H_{K^*} (M) = DP (K*, M). Specifically,

$K \in \{0, 1\}^{256}$, $M \in \Omega_{m0}$ $K^* = E_K (0^{256})$
$\text{DP-PRF}_K (M) = E_K (\text{setbit0} (DP_{K^*} (M))$

Claim [PRF security of DP-PRF]. A PRF adversary A with at most Q queries ($Q \geq 1$) of messages from Ω_{m0} has distinguishing advantage at most

$$\text{Adv (A)} \leq \text{Adv}^{\text{prp}}(E, Q + 1) + (Q + 1)Q/2^{257} + Q(Q - 1)(L0 + 1)^2/2^{256}.$$

Remark 1. We have the following upper bound for the PRF security of DP-PRF:

$$\text{Adv(A)} \leq \\ \text{Adv}^{\text{prp}}(E, Q + 1) + Q^2\left(1 + (L0 + 1)^2\right)/2^{256}.$$

This can be further simplified to:

- For $L0 \geq 3$, $\text{Adv}(A) \leq \text{Adv}^{prp}(E, Q+1) + Q^2\, L0^2/2^{255}$
- For $L0 < 3$, $\text{Adv}(A) \leq \text{Adv}^{prp}(E, Q+1) + 17\, Q^2/2^{256}$

If we ignore $\text{Adv}^{prp}(E, Q+1)$, by assumption, and limit $Q \leq 2^{64}$ and $L0 \leq 2^{48}$ the upper bound remains at most 2^{-32}. In general, we restrict the total amount of processed data (messages) to 2^{64} bytes.

Double POLYVAL MAC (DP-MAC). DP-PRF can be used as a MAC scheme that we call here DP-MAC, in the obvious way. As a MAC scheme, DP-MAC consists of three algorithms: Gen, Tag, Verify. The tag length is $\tau = 32$ and it can be truncated. The algorithms (Gen, Tag, Verify) are detailed in Appendix A. An example with the intermediate steps is provided in Appendix B. DP-MAC authenticity bound for an adversary that submits (at most) q tagging queries and v verification queries (forgery attempts), the forgery success probability is at most

$$\text{Adv}(A) \leq \text{Adv}^{prp}(E,\ v+q+1) + (v+q+1)^2\left(1 + (L0+1)^2\right)/2^{256} + v\,(L0+1)/2^{255}.$$

6 Design Rationale

DP-PRF (and DP-MAC) is designed to gracefully reach the birthday bound limit imposed by using standard 128-bit primitives such as AES. This is achieved by using a 256-bit block cipher. Instead of building a new universal hash family of functions that operate on $\mathbf{F}_{256}$ (with some representation), we chose to use two POLYVAL calls (operating on $\mathbf{F}_{128}$) and concatenate the results. This provides a similar bound on the differential properties. This allows us to use the dedicated instructions AES-NI and PCLMULQDQ, which are now ubiquitous in modern (large 64-bit) processors. Although AES-NI were designed originally to speed up AES computations, Gueron [Gue09, Gue10] has shown that they can be used for speeding up Rijndael256 as well. Furthermore, throughput optimization for Rijndael256 has been recently demonstrated by Drucker and Gueron [DG22]. Note that key expansion for Rijndael256 is required only once, as a key context setup, and the 15 32-byte round keys (480 bytes in total) can be stored in memory.

By concept, POLYVAL was designed as a faster alternative to GHASH (the hash function used with AES-GCM). It leverages efficiently carry-less multiplication instruction and uses a Little Endian encoding convention which is especially efficient on platform which use a Little Endian architecture. Current optimized implementation of POLYVAL performs asymptotically at ~0.3 cpb. This accounts for 4 ring-multiplications (using PCLMULQDQ) and reduction via two additional PCLMULQDQ instructions plus a few shifts. The reduction can be deferred to, e.g., once every 4–8 blocks. Thus, the throughput could, ideally, reach 4.5 (or 4.25) cycles per block, i.e., 0.265 cpb (for long enough messages). With some associated overheads, the estimate becomes ~0.3 cpb. Note that computing two POLYVALS with different hash keys, does not require two passes on the input, and the computations can be executed in one pass over the data, using efficient software pipelining. For this reason, software should be able to compute the DP universal hashing at the rate of ~0.6 cpb. For DP-MAC, Rijndael256 encryption comes on top of DP hashing. The added latency is ~70 cycles, accounting for 14 rounds, with two AESENC calls for every round. Note that DP-MAC optimization does not need special pipelining for Rijndael256.

7 Performance Results

Table 1 shows the performance of DP-MAC, SIV-MAC and HMAC-SHA256, measured on a Skylake processor, for different input lengths. We can see how quickly optimized software reaches the expected asymptotic performance and estimate the relative costs of the overheads associated with the one-time encryption of the DP hash. As anticipated, the throughput of DP-MAC is roughly half the throughput of SIV-MAC that computes only one POLYVAL hash. This performance represents the cost of improving the security margins to the extent that key wear out concerns that require key rotation become completely irrelevant.

Table 1. Performance in cycles per byte (cpb) over different message lengths (msg bytes) of DP-MAC, SIV-MAC and HMAC-SHA256.

	DP-MAC
Bytes	cpb
64	4.13
128	1.9
256	1.25
512	0.93
1024	0.76
2048	0.68

8 Discussion

This paper presented DP-MAC and DP-PRF, two designs that are based on the Hash-Then-Encrypt paradigm. They rely on a 256-bit block cipher and a 256-bit hash (effectively 255 bits due to the enforced domain separation in the design). The underlying cryptographic primitives are chosen in a way that allows for using the dedicated instructions AES-NI and PCLMULQDQ found in practically all modern processors. Specifically, our instantiation uses the block cipher Rijndael256 and a double POLYVAL hash. The code is simple and can reuse POLYVAL existing implementations. This helps optimized software achieve high throughput and processing at the rate of ~0.6 cpb. The security bounds allow for processing data beyond the standard barrier of the 128-bit primitives' "birthday bound", namely 2^{64} bytes. An Open source code implementation is included in the Haberdashery library https://github.com/facebookincubator/haberdashery.

It is possible to instantiate a variant of DP-MAC/DP-PRF using the Simpira permutation in an Even-Mansour setting [GM16]. This variant of a 256-bit block cipher does not require an initial key expansion nor storing round keys.

DP-PRF is a useful primitive as a component in a Key Derivation Function (KDF). A KDF is a cryptographic primitive that facilitates the derivation of cryptographic keys through a two-phase process: extraction and expansion. The extraction step of a KDF generates a fresh (pseudorandom) key from some initial keying material (IKM) and a salt value. It serves as a randomness extractor, transforming potentially noisy or non-uniform raw input into a uniform output. For example, HKDF (a hash based KDF) uses HMAC for the extraction. Alternatively, it is possible to construct a KDF that uses DP-PRF for its extraction phase, followed by some expansion routine (e.g., AES256 in CTR mode) to generate a sufficiently long output. This would be an efficient design, because DP-PRF is more than 10x faster than HMAC. A special useful case of interest is a KDF from $\{0, 1\}^{256}$ to $\{0, 1\}^{256}$ i.e., with a uniform random 256-bit IKM.

Appendix A: DP-MAC Specification

DP-MAC is defined over inputs that are nonempty arrays of bytes, with length at most max-bytes $= 2^{52}$ bytes (2^{48} blocks). It uses a 32-byte (256-bit) key and outputs a τ -byte tag ($\tau = 32$ by default, but it can be truncated). DP-MAC consists of three algorithms, namely Gen, Tag, Verify, as described below.

```
Algorithm 1. Gen
Output: a uniform random key 32-
byte key K (256 bits).
```

```
   Algorithm 2. DP-MAC-Tag (K, X)
   Input: key K = K [31: 0],
   array of m bytes X = X [m - 1: 0],
   0 ≤ m ≤ 2^max-bytes
   Context: tag length τ (bytes)
   Setup
1. Z00 [31: 0] = (0x00)^32;
2. U [31: 0] = Rijndael256 (K, Z00)
a. K1 [15: 0] = U [31: 16]
b. K0 [15: 0] = U [15: 0]
   Processing
1. Padding X
a. nb = ceil (m/16)
b. p = 16*nb - m
c. Xpad = X || (0x00)^p
d. Y = Xpad || LENBLOCK (X)
2. T1 [15: 0] = POLYVAL (K1, Y)
3. T0 [15: 0] = POLYVAL (K0, Y)
4. W [31: 16] = T1 [15: 0]
5. W [15: 0] = T0 [15: 0]
6. W [0] = W [0] OR 0x01
7. Tag = Rijndael256 (K, W)
8. T = Truncate (Tag, τ)
9. Output: T
```

```
Algorithm 3. DP-MAC-Ver (K, X, T)
   Input: key K, array of τ bytes T,
   array of m bytes X = X [m - 1: 0],
   0 ≤ m ≤ 2^max-bytes
   Context: tag length τ (bytes)
   Setup
1. Z00 [31: 0] = (0x00)^32;
2. U [31: 0] = Rijndael256 (K, Z00)
a. K1 [15: 0] = U [31: 16]
b. K0 [15: 0] = U [15: 0]
   Processing
1. Padding X
a. nb = ceil (m/16)
b. p = 16*nb - m
c. Xpad = X || (0x00)^p
d. Y = Xpad || LENBLOCK (X)
2. T1 [15: 0] = POLYVAL (K1, Y)
3. T0 [15: 0] = POLYVAL (K0, Y)
4. W [31: 16] = T1 [15: 0]
5. W [15: 0] = T0 [15: 0]
6. W [0] = W [0] OR 0x01
7. Tag' = Rijndael256 (K, W)
8. T' = Truncate (Tag', τ)
9. If T' = T output SUCCESS; else
   output FAIL
```

Appendix B: DP-MAC Example

This appending provides an example for DP-MAC tagging (also DP-PRF) with the intermediate steps.

```
-----------Bytes Position-------
31302928272625242322212019181716
15141312111009080706050403020100
--------------------------------
K
00000000000000000000000000000000
00000000000000000000000000000001
Z
00000000000000000000000000000000
00000000000000000000000000000000
U
7c8c9a10daa72124ac326abd7f3d90a7
f72ceceb09886eecb1644dcac0fceed4
K1
7c8c9a10daa72124ac326abd7f3d90a7
K0
f72ceceb09886eecb1644dcac0fceed4
--------------------------------
Messages X
33333333333333331111111111111111
--------------------------------
polyval1
9ffa3086b756b40c3d98c40cee4340d6
polyval0
2b71424fc976bb339c42d0e83859d527
W
9ffa3086b756b40c3d98c40cee4340d6
2b71424fc976bb339c42d0e83859d527
W OR (00)^31 || 01
9ffa3086b756b40c3d98c40cee4340d6
2b71424fc976bb339c42d0e83859d527
Tag
a8909d89389523c6100bf98076decc87
8e2826ffb8094a91021ebea86a67148b
```

Appendix C: POLYVAL Details

The XOR universal family of hash functions POLYVAL is defined in [RFC8452], [GLL17]. It is based on finite field computations and properties. To this end, we view 128-bit strings as elements in a finite field (polynomials with binary coefficients). Let $\mathbf{F_{128}}$ be the finite field of 2^{128} elements represented in polynomial representation with the reduction polynomial $P = x^{128} + x^{127} + x^{126} + x^{121} + 1$. A field element is encoded as a 128-bit string $b_{127}\ b_{126} \ldots b_1\ b_0$ (block) where b_0 is the least significant bit, and is viewed, interchangeably, as the polynomial $b_{127}\ x^{127} + b_{126} \cdot x^{126} + \ldots + b_1 \cdot x + b_0$. The field operations in $\mathbf{F_{128}}$ are addition ("+"), namely bitwise XOR of the coefficients, and multiplication ("*"), namely polynomial multiplication modulo P. Let $(\mathbf{F_{128}})^*$ denote the set all nonempty sequences of elements in $\mathbf{F_{128}}$ (blocks). Define POLYVAL: $\mathbf{F_{128}} \times (\mathbf{F_{128}})^* \longrightarrow \mathbf{F}$ as follows. Let $H \in \mathbf{F_{128}}$ be a hash key and let $Y = Y[1], Y[2], \ldots, Y[s] \in (\mathbf{F_{128}})^*$ be a sequence of $s \geq 1$ field elements. Then, $\text{POLYVAL}_H(Y) = \sum_{i=1}^{s}$

$(Y_i * H^{s-i+1} * x^{-128\cdot(s-i+1)})$. Equivalently (via Horner's iterations), $POLYVAL_H(Y) = S[s]$ where $S[0] = 0$; $S[j] = (S[j-1] + Y[j]) * H * x^{-128} \bmod P$. Denote $(\mathbf{F_{128}})^{\leq L} \subseteq (\mathbf{F_{128}})^*$ the set of sequences with length at most L (≥ 1) elements, sample $H \xleftarrow{\$}$ and choose $\delta \in \mathbf{F}$. Then, for every $X, Y \in (\mathbf{F_{128}})^{\leq L}$ we have $P[POLYVAL_H(X) + POLYVAL_H(Y) = \delta] \leq L/2^{128}$ (in particular, this holds for $\delta = 0$). This follows from the fact that $POLYVAL_H(X) + POLYVAL_H(Y)$ is a polynomial of degree (at most) L and therefore has at most L roots in $\mathbf{F_{128}}$. If we add a length block to get a unique encoding for the polynomials, the polynomials have degree at most L + 1.

References

[RFC8452] Gueron, S., Langley, A., Lindell, Y.: AES-GCM-SIV: Nonce Misuse-Resistant Authenticated Encryption. RFC 8452, RFC Editor, April 2019. https://www.rfc-editor.org/info/rfc8452

[DR02] Daemen, J., Rijmen, V.: The Design of Rijndael: AES – The Advanced Encryption Standard. Springer, Cham (2002). https://doi.org/10.1007/978-3-662-04722-4

[DG22] Drucker, N., Gueron, S.: Software optimization of Rijndael for modern x86-64 platforms. In: ITNG 2022 19th International Conference on Information Technology-New Generations, pp. 147–153. Springer, Cham (2022)

[Gue09] Gueron, S.: Intel's new AES instructions for enhanced performance and security. In: International Workshop on Fast Software Encryption, pp. 51–66. Springer, Heidelberg(2009)

[Gue10] Gueron, S.: Intel advanced encryption standard (AES) instructions set. Intel White Paper, Rev, 3, 1–94 (2010). https://www.intel.com/content/dam/doc/white-paper/advanced-encryption-standard-new-instructions-set-paper.pdf

[Gue24] Gueron, S.: SIV-MAC: an efficient MAC scheme. In: International Conference on Information Technology-New Generations, pp. 97–102. Springer, Cham(2024)

[GLL17] Gueron, S., Langley, A., Lindell, Y.: AES-GCM-SIV: Specification and Analysis. Cryptology ePrint Archive, Paper 2017/168 (2017). https://eprint.iacr.org/2017/168

[GL15] Gueron, S., Lindell, Y.: GCM-SIV: full nonce misuse-resistant authenticated encryption at under one cycle per byte. In: Proceedings of the 22nd ACM SIGSAC Conference on Computer and Communications Security, pp. 109–119 (2015)

[GM16] Gueron, S., Mouha, N.: Simpira v2: a family of efficient permutations using the AES round function. In: International Conference on the Theory and Application of Cryptology and Information Security, pp. 95–125. Springer, Heidelberg(2016)

[Rog11] Rogaway, P.: Evaluation of some block cipher modes of operation (2011). https://www.cs.ucdavis.edu/~rogaway/papers/modes.pdf

Cybersecurity Ethics: Case Studies

Peyton Lutchkus(✉), Ping Wang, and Steven Zalenski

Robert Morris University, Moon Township, PA, USA
{pglst259,sxzst279}@mail.rmu.edu, wangp@rmu.edu

Abstract. Ethical considerations in the field of cybersecurity are critical to fair and transparent decision making for the stakeholders of an organization or the general public in the age of rising cybersecurity challenges. This research provides an in-depth examination of two unique case studies regarding ethical practices in cybersecurity: PunkSpider and the Crisis Text Line. PunkSpider's ability to publicly disclose vulnerabilities in web applications raises many questions about the potential for misuse of sensitive information, responsible disclosure, and accountability. The Crisis Text Line's data handling practices highlight the crucial need to protect user privacy and confidentiality. During the study of these scenarios, ethical frameworks and lenses are applied to provide a comprehensive understanding of dilemmas that have been faced. This paper emphasizes the importance of ethical considerations, criteria, and questions for ethical decisions, as well as their consequences and impact to help improve trust and accountability in our technological landscape.

Keywords: ethics · privacy · rights · stakeholders · cybersecurity · considerations · decisions · dilemmas · accountability · issues

1 Introduction

Ethics aims to provide a systematic account of human judgments about social conduct, from the standpoint of being right or wrong [1]. The term "ethics" comes from the Greek word *ethos*, meaning usages, customs, or character. Life and conduct relate to ethics as the study of the inward and outward behavior of an individual. Ethical principles are thought to give some guidance for unsolved problems in life.

In a broad sense, ethics refers to humans figuring out how to best live [2]. The study of ethics attempts to answer the question: How do we live a good life? This field primarily belongs to the discipline of philosophy, where it is studied on a theoretical, practical, or applied level. Ethics can also be separated into community, personal, and work life. In community life, ethics is seen through diverse cultures, political beliefs, religious ideals, etc. On a personal level, ethics is seen as individuals aiming to become better through self-reflection. In work life, professionals are held to a certain standard in a profession. There is a growing need for ethical education in technological fields including cybersecurity. Technology is rapidly advancing and helps to shape how humans seek the good in life [2]. New and old technologies have provided insights into what humans value, as well

S. Latifi (Ed.): ITNG 2025, AISC 1463, pp. 142–152, 2025.
https://doi.org/10.1007/978-3-031-89063-5_13

as what we believe is worth seeking. Laws and regulations have proven to be important parts of preserving good within a society. In today's technological landscape, laws are outpaced by new and emerging developments.

Another factor for the need of technological ethics is the way technology is reshaping power, justice, and responsibility across the globe. Corporations such as Google, Amazon, Microsoft, etc. have increasing levels of global influence. This provides a huge need for ethical studies due to the positive and negative impacts caused by the tech giants worldwide. Cybersecurity aims to secure data, networks, and computer systems from malicious activities or interruptions. Professionals in cybersecurity work to protect the confidentiality, integrity, and availability of sensitive information. In this process of protecting systems, cybersecurity professionals work to preserve the happiness of humans who depend on them. Ethical issues are a core part of cybersecurity practices, as this critical field is required for the well-being of many people. Cybersecurity can impact the sensitive individual data and fundamental interests of human beings including privacy, fair treatment, health, and more. Professionals working in both the private and public sectors must be aware of how their practices may be legal, but not ethical. The public sector of cybersecurity is made up of government organizations or public companies that work to secure critical infrastructure and protect our national security. The main goal of the public sector is to provide safety and security to the people of whom they are representing [3]. The decisions made at this level are critical as they can have a wide range of effects on stakeholders. Ethical considerations are a significant part of everyday operations in the public sector. Government agencies work for the citizens of their country and must be certain that they are acting ethically and responsibly. In the private sector, firms complete a wide range of roles that prove to be vital for national security purposes [4]. The private technology sector has grown exponentially in recent years, which raises ethical concerns for how involved they should be in providing security services. They have the responsibility of protecting crucial data and ensuring the protection of different organizations from cyber threats. Working within the private sector of cybersecurity comes with many ethical and moral responsibilities. Organizations and their stakeholders expect standards of privacy, fairness, and responsible practices. It is crucial for ethical considerations to be included in this sector to make certain that both companies and users are protected. Professionals must work to shape trust while navigating a complex technological threat landscape.

Those working in the field of cybersecurity have the ability to impact the Internet as a whole [5]. This contributes to events online and in the physical world. Cybersecurity experts must consider the ethical aspect of their work, as they hold power to make decisions for many people.

The purpose of this research is to explore ethical aspects of two distinct cybersecurity case studies. PunkSpider and Crisis Text Line are two real world examples of how ethical considerations play a crucial role in cybersecurity today. By exploring case studies, varying ethical dilemmas including responsible vulnerability disclosure, data handling, and user data confidentiality can be considered. By applying content seen in various ethical frameworks, individuals are able to reach a better understanding of the impact ethical issues have within our digital landscape.

2 Background

Philosophy considers the fundamental nature of knowledge, existence, and reality [5]. Within the field of philosophy, ethics examines people's values, culture, background, and beliefs, and how these values translate into behavior. Many philosophers also consider human rights, equality, and justice as main values to determine an individual's ethical views. The ethics behind technology considers how new technologies can contribute to living a good life.

When choosing a right or wrong course of action, feelings, law, religion, science, and accepted social practices are all taken into consideration. However, these are not the only topics for human decision making practices. The Markkula Center for Applied Ethics at Santa Clara University published an ethical framework for decision making, which includes a variety of lenses covering different ethical dimensions [6]. This framework outlines six lenses that philosophers, ethicists, and theologians have developed to provide a basis on how humans decide between good and bad.

The rights lens suggests that the most ethical actions help to protect and respect the rights of those affected by a decision [7]. A right is a justified claim on others. These rights are apparent in the Constitution, such as the right to bear arms or the right to a speedy trial. People have the right to freely choose what they do or decide in their lives. There has been a wide debate about rights, specifically who has rights and what those specific rights are. Moral rights are justified by moral standards that are not set by law but are acknowledged by most people. These rights are interpreted differently by different people. These rights, such as the right to privacy and the right to be told the truth are very prevalent to cybersecurity and technology overall. When confronted with an ethical dilemma, one must consider how each course of action they would take would respect the basic rights of those affected. Wrong actions are the ones thought to violate the rights of an individual. If someone has a right, it is unethical to interfere with that right. This lens is often understood as implying duties, or the practice that we have the duty to respect other people's rights while making decisions.

The justice lens emphasizes fair and equal treatment [8]. Justice entails giving each person what she or he deserves. Questions of justice arise when people differ over what they believe they should be given, or when decisions must be made about benefits or burdens being distributed among people. People should be treated as equals according to defensible standards, but not everyone should be treated the exact same in every situation. According to the Markkula Center for Applied Ethics at Santa Clara University, "Individuals should be treated the same unless they differ in ways that are relevant to the situation in which they are involved" [8]. If there are no differences in situations between individuals, justice should be the same. If there is no difference between individuals and someone is treated differently, that is known as an injustice. There are different types of justice that address what individuals are due in certain contexts. These types of justice include social justice, distributive justice, corrective justice, retributive justice, and transformational justice. Justice is a central part of ethics and plays a significant role in our moral considerations.

The utilitarianism lens emphasizes the consequences of our actions [9]. Utilitarianism is a results-based approach that requires an accurate determination of a result and its impact. This approach suggests the most ethical action is one that achieves the greatest overall benefit and minimizes harm for stakeholders. This perspective offers a straightforward approach to making ethical decisions. One must identify the possible courses of action and determine what the benefits and harm would result from those actions. The goal of this lens is to choose the action that leads to the greatest benefits for all involved. Utilitarianism serves as a reminder that morality encourages others to consider the great good rather than themselves. This lens can be compared to a cost/benefit analysis, as it aims to provide a result that does the least harm for those affected.

The common good lens approach states that our actions should contribute to life in community [10]. The basis for ethical reasoning is the different relationships within a society with requirements for this reasoning being respect and compassion for others. Common good consists of social systems, environments, and institutions that everyone depends on to work in a way that benefits everyone. The common good must be established and maintained through the corporative efforts of many people. In recent years, people have disagreed on what is worthwhile. These disagreements have led some to believe common good is an unachievable goal. The ethics behind common good urge individuals to consider questions of what kind of a society they want to live in and how to achieve that goal. While remaining respectful, individuals can challenge views on themselves as members of a community to recognize goals that are shared in common. This lens highlights the mutual concerns for shared interests of a community.

The virtue lens suggests that ethical actions should be consistent with ideal virtues that help advance humanity overall [11]. Virtues are character traits, dispositions, or attitudes that enable us to act in a way to achieve our full potential. Virtues help us act according to our best character on behalf of values such as truth and beauty. When determining a course of action, virtue ethics often ask questions that deal with what kind of person one will become after deciding. This lens states there are certain ideals which one should strive for that will allow for the full development of humanity. Virtues are often considered habits, as once they are acquired they become a characteristic of a person. The virtue lens revolves around community and how character traits are deeply rooted within the community they belong in. This approach places a strong emphasis on our communities and the habits of character they encourage.

The care ethics lens emphasizes the need to listen and respond to individuals in their circumstances, rather than resorting to rules [12]. This lens focuses on relationships and interdependence rather than independence. Care ethics acknowledges that human are and have bodies that must be taken care of. This lens is not limited to personal relationships, but also explores professional, political, and global relationships in many ways. In order to resolve conflicts, kindness, compassion, and the viewpoints of others are relied on. This approach proposes individuals be more sensitive to the ethical considerations for all situations they are involved in. When deciding a course of action, it is important to consider our affective, emotional, and embodied responses. Care ethics states that options for a result needs to take relationships, concerns, and feelings of all stakeholders in a situation into consideration.

The six lenses of this ethical decision making framework help determine guidelines of what behaviors can be considered right and wrong [6]. Making decisions that are ethical requires being able to weigh outcomes and considerations for all involved. Santa Clara University's framework for ethical decision making serves as a tool for identifying ethical situations and determining the best course of action. The framework follows five steps that provide questions to ask oneself when trying to conclude what course of action to take [6]:

1. Identify the Ethical Issues
2. Get the Facts
3. Evaluate Alternative Actions
4. Choose an Option for Action and Test It
5. Implement Your Decision and Reflect on the Outcome

Utilizing the framework and lenses to make ethical decisions can help identify the best courses of action to take. Good ethical decisions require a method of weighing considerations that should impact our choice of action.

3 Methodology

This research examines two case studies of ethical issues in cybersecurity practices. These case studies are provided by the Markkula Center for Applied Ethics at Santa Clara University [13, 14]. Each case study is unique, followed by the same five questions for each. These questions aim to provide a deeper understanding of the scenario and the ethical dilemmas present:

- **Q1**: Who are the stakeholders involved in this case? Briefly explain the stakes for each one.
- **Q2**: What ethical issues do you spot in this scenario? Describe each one.
- **Q3**: Consider the case through the ethical lenses of rights, justice, utilitarianism, virtue, and care/common good; what aspects of the ethical landscape or issues do they highlight?
- **Q4**: What security and/or privacy risks do you see in this case?
- **Q5**: What technical and policy improvements could be adopted to address the ethical issues and risks in this case?

The first case study revolves around the tool PunkSpider [13]. This tool "automatically identifies hackable vulnerabilities in websites, and then allows anyone to search those results" [13]. Unpatched vulnerabilities on websites are uncovered and made public for anyone to see. The developers of PunkSpider hope their tool will force organizations to address the vulnerabilities on their websites. People with malicious intent may exploit these issues before a company has the chance to address them, which is one major risk. One of the creators of PunkSpider stated "scanners that find web vulnerabilities have always existed. This one just makes the results public" [13]. They believe if customers, investors, and the public can see a publicly disclosed issue, the vulnerability will be fixed fast.

An earlier version of PunkSpider was launched by Amazon Web Services due to reports from web administrators upset with the tool's exploitation. Following this incident, PunkSpider now includes an opt-out feature that lets websites remove themselves from being searched by the tool if their site is being probed. This case study raises many ethical concerns regarding the creation and deployment of PunkSpider. The tool makes vulnerabilities openly visible, which could lead to unethical hacking or other devastating effects on an organization. There are good intentions involved, with the developers of PunkSpider wanting websites to remain safe and secure. However, they go about exploiting a company's security, which goes against a wide array of ethical considerations.

The second case study details the non-profit Crisis Text Line (CTL), which offers support for those going through mental health crises via text [14]. CTL collected a database of messages exchanged over the course of years to help train new volunteers to better manage these conversations.

In a recent report, the organization stated, "by implementing data science tools and machine learning from day one, it had created the largest mental health dataset in the world" [14]. In 2018, CTL launched a for-profit spinoff called Loris.ai. This spinoff planned to use the data collected from CTL to incorporate data into customer support software products. CTL had planned to share the profits made from Loris.ai. The data that has been collected by users is said to be anonymized, but still sparks many privacy concerns. When users reach out to CTL, they are doing so in hopes of receiving free, anonymous help for mental health issues they are facing.

The *Politico* published an article detailing CTL's controversial practices [15]. Many people were concerned that this collected data could be traced back to individuals. People who texted CTL in hopes of receiving mental health help would not anticipate their data to be used in any other way they were unaware of. CTL responded by saying their data-sharing agreement is disclosed to users when they first text. After sending a text, a user is answered by an automated reply stating, "By texting further with us, you agree to our Terms", followed by a 50 page agreement. Many believe users follow through and agree to this statement because they are in desperate need of help. Volunteers who work for the CLT expressed their concerns about this data-sharing agreement internally and externally. Some volunteers even shared that they were not aware of the for-profit effort. After the *Politico* article was published, CTL announced they would be ending their data-sharing agreement with Loris.ai. This scenario raises numerous ethical considerations in regard to data sharing and privacy.

4 Discussions

Case Study 1 evaluates the challenges faced with the integration of the PunkSpider tool. Although the creators of this vulnerability scanner had good intentions, many ethical dilemmas arose around accountability and the responsible disclosure of the issues found.

Within this scenario, there are a number of stakeholders affected by the decision to utilize the PunkSpider tool to scan websites for vulnerabilities. One main stakeholder is the organization whose websites are being scanned and assessed for vulnerabilities. These companies are directly impacted by the disclosure of vulnerabilities on their websites. This disclosure has the potential to affect both their organizational reputation and security. Another stakeholder would be customers or others who visit the websites that are being scanned and have publicly disclosed vulnerabilities. If a customer is not aware of vulnerabilities present on the website they are on, the customer may be affected. If a security issue arises due to a vulnerability, someone's sensitive information may be compromised. Other stakeholders include third party providers or contractors that work with companies whose websites are being scanned. If they become aware of security issues, it may raise concerns and cause them to question their partnership. One final negative stakeholder in this scenario is cybercriminals. Cybercriminals may use this publicly exposed information to their benefit for malicious purposes. There is a lot of pressure on companies to fix vulnerabilities made public by PunkSpider. When seeing their website scanned and those results made public, companies need and want to fix these issues as fast as possible. This pressure can result in quick fixes that are not effective in the long run, or do not solve the problem entirely. By making vulnerabilities visible to the public, hackers are able to use the information to their advantage and exploit issues before a company may have the chance to fix them. There are also many privacy concerns within this scenario. Before having the option to opt-out and remove the tool from their websites, companies had no choice in having their information exposed. Although this tool provides complete transparency to the public, privacy is affected. Critics of PunkSpider are also very concerned with how the tool stores information they collect on these vulnerabilities [13]. Even though these issues are always there and PunkSpider makes them easily viewable, it may encourage bad behavior online or the malicious use of information.

When evaluating the dilemmas that arise in this scenario, it is important to consider different aspects through the ethical lenses of rights, justice, utilitarianism, virtue, and the common good. These lenses can help approach ethical issues more comprehensively to help provide more diverse perspectives:

- **Rights**: Although some people do not agree with PunkSpider's practices, the company has the right to completing their work of scanning websites for vulnerabilities. However, users also have a right to use secure websites. This scenario could be seen as a way to protect the security rights of users by making sure companies address critical issues.
- **Justice**: PunkSpider increases pressure on companies to address publicly disclosed issues. All companies who are being assessed are held to the same security standard. These organizations should be held accountable to a reasonable standard for having certain vulnerabilities present on their web applications. Before issues are publicly disclosed, companies should be given a reasonable amount of time to address them.
- **Utilitarianism**: PunkSpider could be considered good for the majority of stakeholders only if it shows it improves security overall. The intention of using this tool is to help companies address their vulnerabilities and provide a safer platform to everyone. The tool has faced a lot of backlashes and causes many to question whether it is right or wrong to disclose a company's issues to the public.
- **Virtue**: The creators of PunkSpider had good intentions to improve security and hold companies accountable for leaving vulnerabilities unresolved. Especially after implementing their new opt-out feature, it is clear PunkSpider hopes to help companies become more secure rather than hurt their reputation.
- **Care/Common Good**: The PunkSpider tool helps to encourage companies to remain vigilant in fixing their security issues. Although this tool is very controversial, it helps promote better security practices. Companies who see others being assessed

will do better in fixing their own issues to prevent their own organization from having publicly disclosed information.

After reviewing Case Study 1, a further question was posed: Does the inclusion of an opt-out feature change your ethical analysis of the project? The inclusion of PunkSpider's opt-out features changes the ethical view slightly. PunkSpider scanning websites without permission is wrong, but companies do not have to worry about the results of the scans being made public if they choose to optout. Giving organizations the option to remove themselves from the tool's searches will prevent reputation damage, potential cyberattacks, and customer worries. It is imperative that companies fix these vulnerabilities but posting exactly what these issues are may cause more harm than good. Allowing companies to choose and explaining the consequences is more ethical than forcing them into a position if they are not aware of the tool's capabilities. Results could be sent privately to the company to reduce pressure but also ensure vulnerabilities are fixed appropriately.

In order to address the ethical issues in this case, there are a few ways to improve policies regarding responsible disclosure and accountability. The best policy on ethical issues in this case would be to establish regulations on disclosing vulnerability information responsibly. There should be a policy in place that sets standards for tools like PunkSpider to ensure they are operating ethically. This policy should also include how vulnerabilities should be reported to a company, as well as giving a reasonable amount of time to address issues that are found. Rather than being made public immediately, giving companies time to reflect on vulnerabilities present in their system will reduce the chance of more issues arising. This will allow companies a chance to troubleshoot which is fair, as it still holds them accountable and brings about better security practices. Another policy that could be used in this scenario is a requirement for companies to opt-in to the utilization of PunkSpider on their websites. There are a lot of risks associated with using unwilling or unknowing participants and potentially revealing harmful information about their services. If this tool required a company to accept its services before being assessed, it can provide many benefits. Case Study 2 examines the data sharing practices of the nonprofit mental health service Crisis Text Line (CTL). This scenario examined concerns with confidentiality and the privacy of data used by the company in order to provide a new customer service platform.

The stakeholders within this scenario are affected by the decisions of CTL as well as their data handling methods. One stakeholder is the CTL organization itself. They are the nonprofit that offers mental health support for those in times of crisis. The company provides a critical service to those in need, and it is responsible for the protection of the users' data. Another stakeholder is those who use CTL for mental health resources. In this situation, the users are in dire need of help and there is not a good way to ask them for consent to use their data later on. These users also do not have an expectation that their extremely personal data would be used for anything other than to receive the help they need. Those who text into CTL hope to receive the support they need but could be concerned about privacy. Other stakeholders include the volunteers that work for CTL, as they assist users who reach out and need to do so ethically. Another stakeholder within this scenario is Loris.ai, which aims to use CTL's data to gain information that will be used for customer support software implementations. They planned to have a portion of the profits from the software to be shared with CTL, showing that money is another stake for the company. Other stakeholders include the public, donors, board members, etc. These stakeholders may be concerned with the unethical use of data for the new customer service software. These concerns have the potential to damage CTL's reputation or prevent them from receiving more funding.

One of the most apparent ethical issues present in this situation is data privacy. Data from users of CTL are used in a variety of ways, leading to questions and concerns regarding safe data handling. User data is agreed to be shared with CTL following an agreement with their lengthy policy. There is not much transparency with users about data sharing practices. The first text sent to a user states that by replying the user agrees to the terms, followed by a 50 page agreement. In this situation, it is highly unlikely that users are in the right state of mind or have time to read through a document before they are given assistance. Once they have agreed to this policy, the data from their conversation with a volunteer is saved and used for training later on. Over the span of years, CTL was collecting a database of messages and used that data to train its new volunteers. Those who used the hotline may not realize that their sensitive and personal data was being collected or used. This raises concerns about personal privacy and anonymity as well. CTL says the data it collected was anonymized, but that does not mean it cannot be traced back to an individual. If information about someone's health re-traced to the individual, that can have serious implications. Volunteers were also very concerned about the secondary use of data collected from users. Most expressed they were not even aware of the for-profit effort with Loris.ai. Using data from a non-profit and sending it to another company that is using the data to make money is dishonest towards the users of CTL. Most ethical issues within this scenario include lack of privacy and transparency with how data is being used, which leads to concerns for users, volunteers, the public, and more.

This scenario revolves around concerns for users and their confidentiality. It is important to understand the intentions behind the nonprofit CTL, and to evaluate ethical dilemmas through different lenses. By analyzing data privacy through these perspectives, moral implications can be examined.

- **Rights:** Users of the CTL services have the right to privacy. Those who reach out expect to get help for sensitive personal issues, deserve to know exactly what using that service will entail. They have a right to use the hotline as a means to get mental health resources, not to have their data collected for other purposes. The company needs to consider human rights rather than just a source of data or money.
- **Justice:** Those who use CTL may have different levels of technical competency and should be treated equally and fully understand how their data is being collected, handled, and used. Users are not in a proper mindset to fully understand privacy ramifications they had to agree with before speaking with a volunteer.
- **Utilitarianism:** CTL is aiming to use texter's data to train new volunteers as well as create a new customer service software to improve their services. This training would offer real world experience to provide better service in the future. However, this may harm more stakeholders through the use of their data and obstruction of privacy and anonymity.

- **Virtue:** CTL should be ethical and remain transparent with users about their company's data practices. Remaining honest will help show users that they are committed to their mission of helping others.
- **Care/Common Good:** CTL should emphasize the importance of remaining ethical to create a supportive and healthy resource for those who need it. Instead of relying on data collection, they should focus on furthering their mission overall. CTL solely helps those who are vulnerable and in need of help. Users deserve to be treated with respect that they deserve.

Improvements to both the technical and policy aspects of CTL are needed to ensure the responsible handling of critical private information. If they continue to collect data, their agreement policy could be short and concise to clearly state how data will be handled. This includes sharing the process for data collection and storage and the risks that come with this practice. CTL can also implement better anonymizing techniques so data cannot be traced back to individuals. It would also be helpful and responsible to have an opt-out clause for those who need to receive help but do not want to share their data. CTL could also implement access controls and minimize the amount of data collection to only things that are absolutely necessary. Volunteers should be trained on data privacy ad ethics, so they are aware of the responsibility they have to protect user information. Regular security audits can also be implemented to address vulnerabilities and other security issues. By utilizing these practices, CTL can improve data collection process, better protect sensitive information, and remain more ethical overall.

5 Conclusions

The study of ethical considerations in cybersecurity shows the complexity and significance of responsible practices in this rapidly advancing field. As demonstrated through the case studies of PunkSpider and Crisis Text Line, ethical dilemmas have proven to raise concerns surrounding our digital world and how we can improve in these areas. PunkSpider's vulnerability disclosure presents challenges to accountability and risks associated with exposing crucial information irresponsibly. Crisis Text Line emphasizes the importance of user privacy and confidentiality, and the ethical responsibilities of organizations who are holding that data. Both case studies highlight the need for policy improvements in ethical guidelines that clearly outline what standards cybersecurity professionals will be held to in regard to different ethical issues. By applying ethical lenses and frameworks to both case studies, it is clear that cybersecurity professionals face a wide array of dilemmas. The study of these scenarios has provided insights that explain the need for ethical standards in cybersecurity that prioritize accountability, transparency, and protection for all. It is crucial to foster trust in cybersecurity practices, as our reliance on technologies will continue to grow. Future research may explore how psychology plays a role in influencing ethical behavior. Further research on responsible disclosure practices, trust in cybersecurity, and how artificial intelligence can be incorporated into ethical decision making would contribute to a better understanding of the ethical landscape. This paper advocates the need for ethical considerations that will help to navigate the digital world both responsibly and ethically.

Acknowledgments. We would like to acknowledge the support of the National Science Foundation CyberCorps Scholarship for Service, under grant #2234554. This program has significantly contributed to our research and has enhanced the educational journey of the scholarship recipients.

References

1. The Project Gutenberg eBook of Ethics, by John Dewey and James Hayden Tufts. https://www.gutenberg.org/files/39551/39551-h/39551-h.htm
2. Vallor, S., William, J., Rewak, S.: An Introduction to Cybersecurity Ethics (2018). https://www.scu.edu/media/ethics-center/technologyethics/IntroToCybersecurityEthics.pdf
3. Cyber ethics: Assessment on government and the private industry – ProQuest. Proquest.com (2016). https://www.proquest.com/docview/1836101686?pqorigsite=gscholar&fromopenview=true&sourcetype=Dissertations%20&%20Theses. Accessed 31 Oct 2024
4. Pattison, J.: From defence to offence: the ethics of private cybersecurity. Eur. J. Int. Secur. **5**(2), 233–254 (2020). https://doi.org/10.1017/eis.2020.6
5. Manjikian, M.: Cybersecurity Ethics. Taylor & Francis (2018)
6. Markkula Center for Applied Ethics. "A Framework for Ethical Decision Making". Santa Clara University (2021). https://www.scu.edu/ethics/ethics-resources/a-framework-for ethical-decision-making/
7. Santa Clara University. "Rights," Santa Clara University (2014). https://www.scu.edu/ethics/ethics-resources/ethical-decisionmaking/rights/
8. Markkula Center for Applied Ethics. "Justice and fairness," Santa Clara University (2014). https://www.scu.edu/ethics/ethicsresources/ethical-decision-making/justice-and-fairness/
9. Calculating Consequences: the Utilitarian Approach to Ethics. Markkula Center for Applied Ethics (2014). https://www.scu.edu/ethics/ethics-resources/ethical-decisionmaking/calculating-consequences-the-utilitarian-approach/
10. Velasquez, M., et al.: The common good. Santa Clara University (2018). https://www.scu.edu/ethics/ethics-resources/ethical-decisionmaking/the-common-good/
11. Santa Clara University. "Ethics and Virtue," Santa Clara University (1988). https://www.scu.edu/ethics/ethics-resources/ethicaldecision-making/ethics-and-virtue/
12. Kwan, J.: Care Ethics (2023). https://www.scu.edu/ethics/ethics-resources/ethical-decisionmaking/care-ethics/care-ethics.html
13. Raicu, I.: Making Vulnerabilities Visible (2021). https://www.scu.edu/ethics/focus-areas/internetethics/resources/making-vulnerabilities-visible-a-cybersecurityethics-case-study/
14. Santa Clara University. "Crisis Data" (2022). https://www.scu.edu/ethics/focus-areas/internetethics/resources/crisis-data-an-ethics-case-study/
15. Levine, A.S.: Suicide hotline shares data with for-profit spinoff, raising ethical questions. POLITICO (2022). https://www.politico.com/news/2022/01/28/suicide-hotline-siliconvalley-privacy-debates-00002617

Energy Consumption Analysis of Collaborative Protocols for Security Threats and Mobile Malware Detection

M. Adeel(✉)

University of North Texas at Dallas, Dallas, TX, USA
muhammad.adeel@untdallas.edu

Abstract. This paper analyzes the energy consumption of various collaborative protocols for security threats and mobile malware detection. We compare Collaborative Malware Detection Framework with detection approaches by Kim *et al.*, Cheng *et al.*, and Schmidt *et al.*, focusing on energy efficiency. Using Heinzelman *et al.*'s model, our results show that Collaborative Malware Detection Framework significantly reduces the energy consumption. This analysis is crucial as it provides researchers with a streamlined methodology for calculating the power consumption costs of security protocols, while excluding extraneous factors. The methodology adopted can serve as foundational reference for assessing energy efficiency of communication protocols for energy-constrained wireless networks.

Keywords: Energy-Efficient Wireless Protocols · Collaborative Malware Detection · Wireless Security Threats · Mobile Malware Families · Energy-Efficient Detection

1 Introduction

Our research has been published on Collaborative Malware Detection Framework (CMDF) for detection of security threats and mobile malware families [1]. CMDF (illustrated in Fig. 1) works on the concept of detection utilizing mobile P2P communities involving mobile nodes connected using epidemic or social-based forwarding protocols such as Bubble Rap [2–4] and Lobby Influence [3,5] etc. Mobile nodes in the community collect and process individual statistics such as their throughput (for different services), battery usage, and neighbor associations (connections), and (opportunistically) transmit them to the designated agent node in the community directly or by using intermediate (forwarding) nodes. The agent nodes subsequently calculate the cumulative statistics for the community on throughput, battery usage, and associations, and match these statistics against various thresholds to detect security threats as well as different types of mobile malware families [1].

Compared to Agent-based Adaptive Malware Detection Framework (AMDF) (introduced in our published research[6]) which relies on behavioural parameters

S. Latifi (Ed.): ITNG 2025, AISC 1463, pp. 153–163, 2025.
https://doi.org/10.1007/978-3-031-89063-5_14

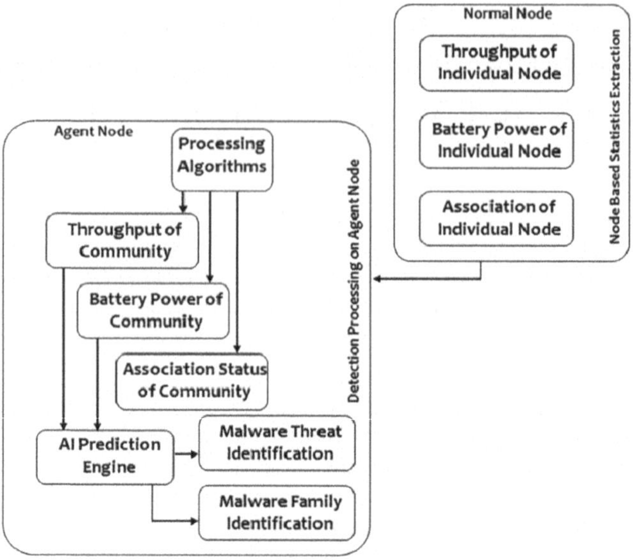

Fig. 1. Collaborative Malware Detection Framework (CMDF)

to only detect mobile malware families, CMDF relying on minimal statistical parameters, not only detects mobile malware families, it also successfully identifies generalized security threats such as flooding and battery depletion attacks [1]. Comparison of CMDF against other malware detection approaches in the literature [7–10] reveals that CMDF, by utilizing minimal statistical parameters, distinctly identifies more security threats and mobile malware families compared to other known approaches[1]. Detection approaches in the literature [7–10] rely primarily on the intelligence of a single designated mobile node in the network for threat detection. In contrast, CMDF leverages the consolidated intelligence of multiple agent nodes, each responsible for a proximity of mobile nodes (in a mobile P2P community). This approach can therefore uncover security threats and mobile malware families within mobile P2P communities, as well as the overall mobile P2P network.

Statistical detection approaches in literature depend majorly on infrastructure-based wireless technologies for remote threat identification [7,9,12]. CMDF provides agent-led distributed detection that eliminates dependency of the detection framework on the infrastructure-based wireless technologies and, therefore, is scalable [1]. CMDF benefits from cost-effective protocol-related communications due to its reliance on opportunistic P2P networking that makes it power friendly. Recording minimal statistical information in terms of detection footprint makes CMDF a storage-friendly approach for detection of security threats and mobile malware families. Distinct identification of security threats and mobile malware through CMDF could also help in a more focused individual threat or malware family based all-distributed patching. CMDF on a wide range of performance measures including detection efficiency and accuracy, capabilities of detection of unknown malware, and battery consumption, has proved superior to other known approaches.

The work on detection performance of CMDF has already been published in [1]. This paper focuses on the battery consumption performance of CMDF [1] and other collaborative protocols for security threats and mobile malware detection [7–9] in the literature. Section 2 provides an overview of these protocols. Section 3 elaborates on the energy consumption model proposed by Heinzelman *et al.* [13,14] that will be subsequently utilized to conduct an energy comparative analysis of these protocols in Sect. 4.

2 Approaches in Literature for Detection of Security Threats and Mobile Malware

We have provided a brief overview of the functionality of CMDF in Sect. 1. More details on the functionality as well as the detection performance of the CMDF are available in our published research [1]. This section provides a brief overview of the collaborative protocols for security threats and mobile malware detection in the literature [7–9], against which, we will be comparing the CMDF's power consumption performance in Sect. 4.

Research by Kim *et al.* [7] is focused on only four mobile worms. Authors provide a brief discussion on the steps involved in the infection process (in form of activity sets) for those mobile worms. Their activity sets, however, provide no distinction between certain entirely diverse types of mobile malware families for instance, between Mabir, Doomboot and Commwarrior. This is primarily because of the use of abstract higher-level activities (described in our research as Level 3 and level 4 activities [1,6]) that are incapable of accommodating the critical lower-level functionality differences between the malware families. In situations where a more focused individual threat or malware family based all-distributed patching is desired without reliance on the network infrastructure, their approach proves infeasible. The extension of the activity set of Kim *et al.* in our research [1] accommodates more types of security threats and malware families. Kim *et al.* [7] detect energy greedy anomalies and mobile worm variants through their pre logged power signatures. Power signatures of mobile devices with malware infections are extracted in a controlled environment and stored in a power signature database which either lies on resource–rich mobile devices or at a remote location accessible through WLAN to the resources-constrained mobile devices. Former are called Type A mobile devices while the latter are termed Type B. Mobile devices constantly record power signatures over pre-specified time windows. Type A mobile devices are capable of comparing the extracted power signatures against the resident signature database to identify malware threat on their own while Type B mobile devices transmit recorded signatures to remote power signature database through WLAN to a Type A mobile for analysis. To measure the similarity between two power signatures, distance is calculated as:

$$X^2(S_e, S'_e) = \sum_{i=1}^{n} \frac{(S_{ei} + S'_{ei})^2}{(S_{ei} + S'_{ei})} \tag{1}$$

S_e and S'_e are signatures of the observed and the expected events, respectively. $X^2 = 0$ if and only if, all of the samples of S_e match the sample in S'_e. The higher the value of X^2, the less likely it is that the observed event belongs to the expected group. Two power signatures with the most similar consumption patterns are found as:

$$X^2(S_e, DB) = \min_{S'_e \in DB} \{X^2(S_e, S'_e)\} \tag{2}$$

The Delay in capturing signatures can cause a condition $X^2 \neq 0$ for similar power signatures. To eradicate this problem, left and right shift operations are performed to acquire the minimum distance between two signatures, and when a certain confidence level is achieved, the similarity between the signatures is declared. In a controlled environment (with a limited number of background applications running), Kim *et al.*'s work is capable of detecting some well-known worm types such as Cabir, Mabir, Lasco, and Commwarrior alongside some synthetically generated mobile security threats. Kim *et al.*'s approach is limited by hardware dependency, high false alarm rates, and scalability issues due to the reliance on a centralized server and WLAN. Additionally, the method struggles with detecting variations in the malware and managing large volumes of power signatures on resource-constrained devices.

Cheng *et al.* [8] present SmartSiren framework for virus detection in smartphones. The approach mainly relies on statistical monitoring of the usage history of mobile devices locally, and remotely on a *proxy* device. Mobile devices participating in the system transmit periodic and proactive reports to the proxy device for analysis (over SMS and/or WLAN and/or GPRS/UMTS IP channels). After a participating mobile gets registered to the proxy, the mobile device transmits its SMS and Bluetooth usage data to the remote proxy as daily or proactive reports. Daily reports contain the usage profile of the mobile device recorded against U_{Today}. The mobile maintains the U_{Thersh} as the moving average of 7 days usage profile of each mobile device. Condition $U_{Today} > U_{Thersh}$ causes the mobile device to send a proactive report to the proxy device. In addition to proxy maintaining $U_{Thersh}$for all participating devices, it also calculates global equilibrium i.e. P_{avg} which implies the average per day of the number of devices exceeding U_{Thersh}. When $P_{Today,}$ which represents the global equilibrium for a given day, deviates widely (and becomes P_{Today}* 2) from the P_{avg}, the proxy declares the alert. Where minor fluctuations in U_{Today} can be caused by legitimate reasons, factor P_{Today}* 2 reflects an alarming increase in usage. The report submission is kept anonymous in favor of user privacy, however, this can be abused by malicious mobile devices to influence the detection process on proxies. One of the processes used to take care of this problem is ticketing. Given K_r as the secret key and H as the hash function, a ticket is given by

$$(tid, t_e, H_{K_r}(tid|t_e)) \tag{3}$$

where *tid* represents unique ticketing id, t_e is the expiration timestamp of the ticket, and H_K $(tid—t_e)$ is the hash of *tid* and t_e using secret key K_r. These tickets are periodically distributed to participating mobile devices by the proxy. Cheng *et al.*'s approach faces scalability issues due to its reliance on WLAN and GPRS/UMTS IP channels and a centralized proxy, which is a single point of failure and vulnerable to Denial of Service (DoS) attacks. Additionally, it focuses only on Bluetooth and SMS traffic, neglecting MMS, and requires resource-intensive reporting and transmissions.

Schmidt *et al.* [9] present an anomaly detection system for smartphones. The system relies on the deployment of designated mobile devices called the monitoring clients, capable of extracting device-specific features and transmitting the feature vector to remote anomaly detection systems (RADS). Some of the features are RAM FREE, CPU USAGE, USER INACTIVITY and PROCESS COUNT. Each monitoring device needs to register with RADS and transmits its feature vector to RADS using IP based services. The frequency of transmission of the feature vector is 20 s. Feature extraction on a mobile device is categorized as simple, medium, and complex on the basis of their extraction mechanism. The IMEI extraction algorithm, for instance, comprises a few lines of code and is thus considered a simple extraction. In contrast, complex extraction involves algorithms designed to extract various features. Once acquired, RADS analyzes feature vectors through Artificial Immune Systems and Self Organizing Maps using malware feature database. The authors also perform feature set reduction to reduce the feature set by 80%. To avoid the loss of information in case of connection failure with the server, monitoring clients are capable of buffering data for a finite period (controlled by buffer thresholds). The monitoring client can itself be a victim of threats like DoS attacks, hence, various proposals to ensure the security of the monitoring nodes have also been discussed, but not implemented. Focusing on two Symbian OS phones, Nokia E61 and Nokia 7610, authors report significant success in detection of (mainly non-propagating) malware. Schmidt *et al.*'s approach relies on a centralized server, requiring constant connectivity as well as hampering scalability. The limited feature set restricts malware detection capabilities, and legitimate activities can mimic malware patterns, affecting detection accuracy. Additionally, monitoring clients can be vulnerable to DoS attacks, and the focus on two models of the Symbian OS limits broader applicability.

3 Energy Consumption Model and Assumptions

To evaluate the energy consumption performance of CMDF [1] and other well-known collaborative protocols for security threats and mobile malware detection [7–9], we will utilize the energy consumption model proposed by Heinzelman *et al.* [13,14]. The energy consumptions model calculates energy cost of the transmitted and received traffic on resource-constrained wireless devices and has been utilized by the researchers in the past [15–17] to calculate the energy performance of wireless protocols.

The four equations provided below explain Heinzelman *et al.'s* model in which E_{Tx}(k, d) is the energy of transmission of a packet to a distance d, and $E_{Rx}(k, d)$ is the energy consumed in receiving a packet of specific size. Under their model, energy consumed by a node in transmission of a packet of size k to a distance d is given by:

$$E_{Tx}(k,d) = E_{Tx\text{-}elec}(k) + E_{Tx\text{-}amp}(k,d) \tag{4}$$

$$E_{Tx}(k,d) = E_{elec} \cdot k + E_{amp} \cdot k \cdot d^2 \tag{5}$$

where E_{elec} is the energy of electronics and E_{amp} is the amplification energy consumed by the transmitted in amplification process. The energy consumed by a mobile node in receiving a packet of size k is given by:

$$E_{Rx}(k) = E_{Rx\text{-}elec}(k) \tag{6}$$

$$E_{Rx}(k) = E_{elec} \cdot k \tag{7}$$

Given $\mathrm{E_{elec}}$ = 50 nJ/bit to run the transmitter or receiver circuitry and $\mathrm{E_{amp}}$ = 100 pJ/bit/m^2, the energy consumption evaluation is performed on a system of 8 nodes for each of the 4 protocols under analysis. Regardless of the underlying communication protocol used, the distance between the entities is assumed as 10 m. Calculations are based on one instance of reporting or detection by the participating mobile nodes (whether normal or designated mobiles). The payload specific to the control traffic for all the participating protocols has been ignored during evaluations.

4 Energy Consumption Analysis of the Collaborative Protocols

This section calculates and compares energy cost of the existing collaborative detection approaches in the literature [7–9] against CMDF [1], focusing only on the detection protocol related communications. We use four energy equations discussed in Sect. 3 to calculate the energy cost of these protocols.

Kim *et al.* [7] propose two different types of mobile phones in the network. Type A mobile is a designated mobile capable of analyzing extracted power signatures against a signature database and verify the threats. Type B mobiles on the other hand, are normal mobiles capable only of power signature extraction and uploading the extracted signature samples to a server that could then perform detection analysis and verify results. Figure 2 gives the operational diagram of the mobile malware detection framework by Kim *et al.* It is assumed that there are 7 Type B mobiles in the system, each transmitting 1 Kbits worth of power signatures to the server. Each Type B mobile will receive 16 bits worth of detection results in response to its query. One Type A mobile on the other hand processes this 1 Kbits of power history itself to identifying security threat status.

Given $\mathrm{E_{elec}}k + \mathrm{E_{amp}}kd^2$ as the energy required for transmission of 1 Kbits of power signature history to a server assumed at distance d from a Type B mobile, it will take 7 Type B mobile nodes 7* ($E_{elec}k + E_{amp}kd^2$) μJ of community energy i.e., 420 μJ to transmit to the server. Given $E_{elec}k$ as the energy consumed in receiving k bits of data, and each mobile is assumed to receive 16 bits from the server to confirm the threat status, 7 mobile nodes in the community will receive 7 * ($E_{elec}k$) μJ i.e., 5.6 μJ.

A single Type A mobile in the system is assumed to require only $E_{elec}k$ μJ for processing 1 Kbits of power signatures it has collected i.e., 50 μJ. On-device computational energy consumption has been ignored in calculations of the energy cost for the protocol under analysis this section.

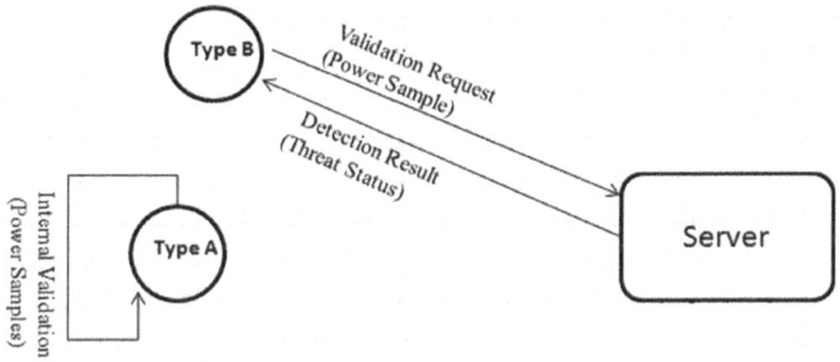

Fig. 2. Operational Diagram of the Protocol by Kim *et al.*

Cheng *et al.* [8] present a collaborative statistical protocol which relies on a central proxy server to which all the participating host mobile nodes transmit reports relating pre-specified activities through SMS, GPRS or WiFi based communications. Before reporting in the operational phase, each participating mobile is required to register with a proxy and transmit a profile report to the proxy. Information like contact list and mobility profile can have varying sizes. As an assumption, contact list size is considered as 63 and mobility profile holds 3 Cell IDs. In the registration phase, each of the 8 mobiles in the community will transmit 7.089 Kbits while during dynamic reporting, each mobile will transmit 6.504 Kbits. Each mobile will also receive 16 bits of worth of detection results from the proxy.

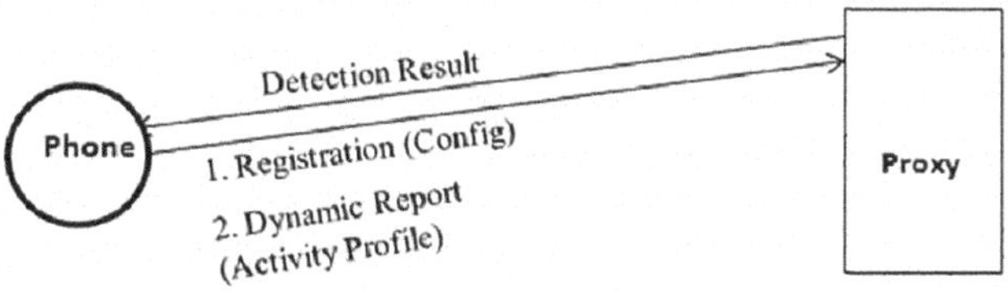

Fig. 3. Operational Diagram of the Protocol by Cheng *et al.*

In the protocol by Cheng *et al.* [8], given $E_{elec}k + E_{amp}kd^2$ as the energy required for transmission of k bits (where $k = \tilde{7}.089$ Kbits) to a distance d on a single node, registration process of 8 nodes with the proxy requires 8 * ($E_{elec}k$ + $E_{amp}kd^2$) μJ of community energy i.e., 3402 μJ. Similarly, the reporting to the proxy during operations (threat identification process), each of the 8 nodes in community will transmit $\tilde{6}$.504 Kbits to the proxy, Thus, total energy consumed in operations phaes will be 8 * ($E_{elec}k$+ $E_{amp}kd^2$) μJ i.e., 3121.92 μJ. Given $E_{elec}k$ as the energy consumed in receiving k bits of data, and each mobile is assumed to receive 16 bits from the proxy to confirm threat status, 8 mobile nodes in the community will receive 8 * ($E_{elec}k$) μJ i.e., 6.4 μJ. Nodes are assumed to have registered with the proxy ahead of threat identification phase and thus the energy consumed during registration process is ignored for our comparison of the energy cost of the collaborative protocols for security threats and mobile malware detection.

Collaborative statistical malware detection proposed by Schmidt *et al.* [9], relies on a collaborative mechanism where the threats detected by each node are verified through collaboration with every detection-capable neighbouring node. Each node installs a detection agent capable of acquiring system-specific attributes. Although 2816 attributes in total have been identified, only 189 of those are considered critical attributes. Therefore, this analysis assumes that the protocol by Schmidt *et al.* utilizes 189 attributes. It is important to note that choosing all 2816 attributes for the energy consumption analysis will mean that the approach by Schmidt *et al.* will result in consuming significantly more energy than being consumed with 189 attributes. Similarly, while the actual size of each attribute is larger, it is assumed to be 1 bit per attribute, thus, 189 attributes will consume a total of 189 bits. The detection decision of a node based on the analysis of these attributes will consume 16 bits.

The node transmits 189 bits representing attribute values along with 16 bits representing decision to the neighbours (7 in this case). Given $E_{elec}k$ + $E_{amp}kd^2$ as the energy required for the transmission of k bits to the distance d on a single node, 7 * ($E_{elec}k$ + $E_{amp}kd^2$) μJ of community energy i.e., 86.1 μJ is required by the node to collaborate with neighbours. Neighbours in total incur 7*($E_{elec}k$) μJ to receive and process this information i.e., 71.75 μJ of energy is required. It will take each neighbour node $E_{elec}k$ + $E_{amp}kd^2$ μJ to transmit the detection result of their analysis to the initiating node. With the detection result assumed to require 16 bits, a total of 6.72 μJ are consumed in transmissions. Finally, the initiating node will require 7*($E_{elec}k$) μ J i.e., 5.6 μJ to receive the detection results from 7 neighbours.

The protocol by Schmidt *et al.* triggers with internal verification of 189 bits of attributes that consume $E_{elec}k$μ J i.e., 9.45 μJ however, this on-device computational energy consumption has been ignored in calculation of the overall cost of Schmidt *et al.*'s protocol. In the event of a community level threat, the other 7 nodes (excluding the initiating node in this scenario) will also repeat the operation described above. Consequently, the community will consume 7 times the energy of a single operation, however, this fact is assumptively neglected.

CMDF [1] is focused on acquiring node-based statistical information from the participating neighbours and utilizing this information for community intel-

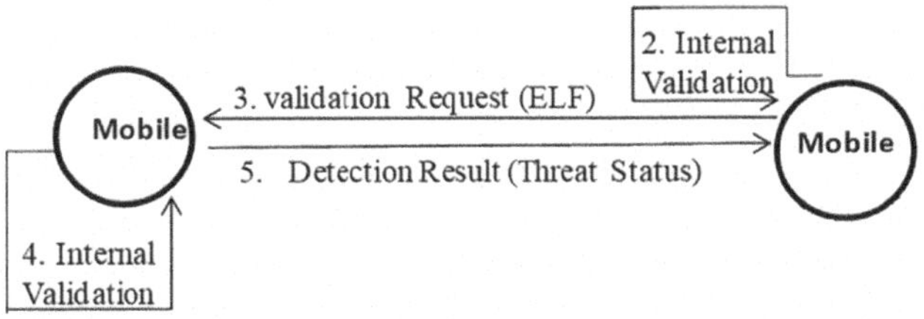

Fig. 4. Operational Diagram of the Protocol by Schmidt *et al.*

ligence for detection of malware threats. Each node in this scenario transmits information such as node-based throughput, battery power, and associations (connections with neighbours) to the agent node. The payload for this information is 120 bits. The analysis by the agent node results in detection of malware threats, which can be accommodated in 16 bits (as a general assumption in this section). The detection results could be optionally transmitted to the nodes in the community.

Given $E_{elec}k + E_{amp}kd^2$ as the energy required for transmission of k bits of statistical information by each of the 7 normal nodes to the agent node at a distance d, the energy consumed by 7 nodes for the same is 7 * ($E_{elec}k + E_{amp}kd^2$) μJ i.e., 50.4 μJ. The agent node receiving this information from each of the 7 nodes in the community incurs 7 * ($E_{elec}k$) i.e., 42 μJ. Optionally, the agent node also transmits 16 bits worth of detection result to all the node in the community requiring 7* ($E_{elec}k + E_{amp}kd^2$) μJ i.e., 5.6 μJ. On the agent node, 8* ($E_{elec}k$) μJ i.e., 48 μJ is required for processing statistics from 8 nodes (including the agent node). Consistent with the energy cost calculations for other protocols in this section, this computational energy consumption is disregarded.

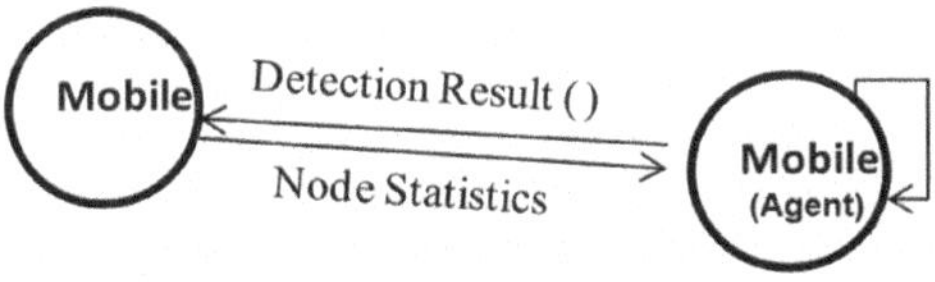

Fig. 5. Operational Diagram of the CMDF

Figure 6 gives the total energy consumption in terms of transmit and receive power (in Joules) for protocol-related transmission of the collaborative detection protocols under discussion. The spatial community of fixed number of nodes (i.e., the system) consumes 3128.32 J in the protocol by Cheng *et al.* [8]. 0.19% of the system energy is consumed in receiving protocol-specific communication from the central proxy while the rest is consumed in protocol-specific transmissions by the system nodes to the central proxy.

The approach by Kim *et al.* [7] consumes 425.6 J where 98.6% of the energy is consumed in protocol-specific transmissions (of power signatures) to the central proxy. Schmidt *et al.*'s approach [9] is the most efficient amongst the three techniques from the literature that are under analysis in this paper. 54.7% of the total 170 J energy is consumed in transmitting threat-related queries to the peers while rest is consumed in receiving responses by the querying nodes. CMDF [1] proves to be the most energy-efficient protocol compared to other protocols, consuming only 98 J of energy. 57% of the total consumed energy by the CMDF is consumed in transmission of protocol-related statistics to the agent node. CMDF achieves 63.4% performance improvement over the least energy-consuming protocol in the literature (i.e., the protocol by Schmidt *et al.*).

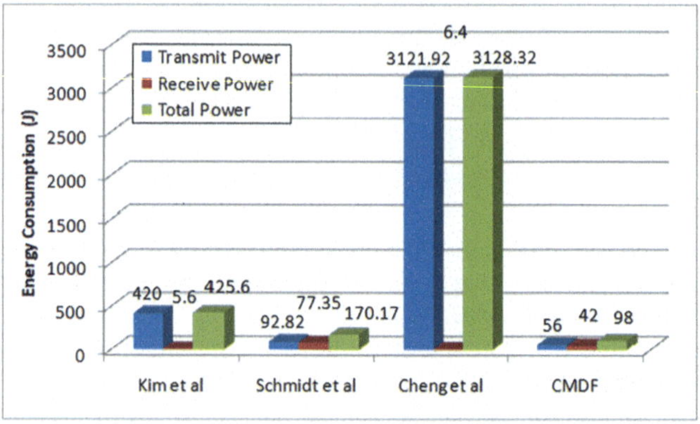

Fig. 6. Comparison of Energy cost of various protocols

5 Conclusion

This paper presents an analysis of the energy consumption of collaborative protocols for security threats and mobile malware detection, highlighting the superior energy efficiency of the CMDF [1]. Using Heinzelman *et al.*'s model [13,14], CMDF [1] achieves a 63.4% reduction in energy usage compared to other approaches in the literature [7–9]. More importantly, this analysis provides a methodology for calculating power consumption of each protocol, serving as a foundational reference for developing more energy-efficient communication protocols for energy-constrained wireless networks.

References

1. Adeel, M., et al.: On the collection of statistical parameters on mobile phones for detection of security attacks and mobile malware. IOARP J. Commun. Netw. **2**(1) (2017)

2. Khan, S., et al.: Ad hoc and opportunistic routing in static scatternet environment. In: Proceedings of the IOARP International Conference on Communication and Networks, London, UK (2015)
3. Khan, S., et al.: LOC: location-aware opportunistic content forwarding using direction vectors. In: 2013 IEEE Symposium on Computers & Informatics, Langkawi, Malaysia (2013)
4. Khan, S., et al.: Ad-hoc and opportunistic routing in dynamic scatternet environment. IOARP J. Commun. Netw. **1**(1) (2016)
5. Khan, S., et al.: LOC algorithm: location-aware opportunistic forwarding by using node's approximate location. Int. J. Pervasive Comput. Commun. **10**(4), 481–496 (2014)
6. Adeel, M., et al.: The taxonomy of behavioural parameters on mobile devices. IOARP J. Commun. Netw. **1**(1) (2016)
7. Kim, H., et al.: Detecting energy-greedy anomalies and mobile malware variants. In: Proceedings of MobiSys 2008, Breckenridge, Colorado, USA (2008)
8. Cheng, J., et al.: Smartsiren: virus detection and alert for smartphones. In: Proceedings of ACM MobiSys, San Juan, Puerto Rico (2007)
9. Schmidt, A.-D., et al.: Monitoring smartphones for anomaly detection. In: Proceedings of Mobileware 2008, Innsbruck, Austria (2008)
10. Yan, G., et al.: Blue-watchdog: detecting bluetooth worm propagation in public areas. In: 2009 IEEE/IFIP International Conference on Dependable Systems & Networks, Lisbon, Portugal (2009)
11. Aiello, W., et al.: Efficient, DoS resistant, secure key exchange for internet protocols. In: CCS 2002, Washington DC, USA (2002)
12. Buennemeyer, T.K., et al.: Battery sensing intrusion protection system. In: IEEE Workshop on Information Assurance, NY, USA (2006)
13. Heinzelman, W.B., et al.: An application-specific protocol architecture for wireless microsensor networks. IEEE Trans. Wirel. Commun. **1**(4), 660–670 (2002)
14. Heinzelman, W.B., et al.: Energy-efficient communication protocol for wireless microsensor networks. In: Proceedings of the 33rd Hawaii International Conference on System Sciences (HICSS-33), Hawaii, 4–7 January 2000 (2000)
15. Hussain, M.: Critical issues in mobile commerce & energy consumption analysis of popular mobile payment protocols. Thesis submitted at the University of Wales, United Kingdom (2011)
16. Hussain, M., et al.: Analysis of session key negotiation & distribution protocols in wireless sensor networks. In: Proceedings of 18th International Conference on Software, Telecommunications and Computer Networks, SoftCOM 2010, Split, Croatia, September 2010 (2010)
17. Salhieh, A., et al.: Power aware matrices for wireless sensor networks. In: Proceedings of IASTED International Conference on Parallel and Distributed Computing and Systems, November 2002 (2002)

Data Science

Analyzing "Existential Dread" on the Internet in Response to Elmo

Quinn Contaldi(✉), Joshua Dahl, Batyr Charyyev, and Frederick C. Harris Jr.

Computer Science and Engineering, University of Nevada, Reno,
1664 N. Virginia Street, Reno, NV 89557, USA
{qcontaldi,joshuadahl,bcharyyev,fred.harris}@unr.edu

Abstract. On January 29, 2024, Sesame Street character Elmo posted a simple tweet: "Elmo is just checking in! How is everybody doing?" The post received over 18,000 replies, showcasing a wide range of emotions—from grief over personal losses to lighthearted moments like sharing photos of turtles basking in the Florida sunshine. This overwhelming response captured media attention, with outlets such as CBS interpreting it as a reflection of a broader existential dread permeating online spaces, even prompting a reaction from then-President Joe Biden. This paper critically analyzes the sentiment and structure of the replies to Elmo's post, along with interactions among respondents, to explore whether discernible sentiment patterns emerge within response communities and if a general sense of existential dread is evident. Hypothetically, a sentiment analysis model would reveal distinct clusters of happy and sad posts. However, contrary to the media narrative of widespread existential dread, our analysis indicates a greater prevalence of positive responses (47%) compared to negative ones (38%). These findings suggest that negative emotions are outweighed by positivity in this context. Our findings reveal no overarching sentiment clusters; instead, the responses represent a complex scattered grouping of unconnected emotions.

Keywords: Data Mining · Social Media Mining · Social Network Analysis · Complex Networks · Twitter

1 Introduction

On January 29th, 2024, Elmo's post, "Elmo is just checking in! How is everyone doing?" [4] This post elicited a large response with many different complex emotions. Users shared experiences ranging form deep emotional trauma to the best moments of their lives.

The massive reaction caught the media's attention, with outlets like CBS [9] highlighting the negativity of responses and noting a sense "existential dread" pervading online discourse. Even then-President Joe Biden [2] felt compelled to comment on the phenomenon.

The proposal of existential dread by the media prompts an important question: Is existential dread a predominant factor in Elmo's post responses, and

S. Latifi (Ed.): ITNG 2025, AISC 1463, pp. 167–177, 2025.
https://doi.org/10.1007/978-3-031-89063-5_15

does this indicate a broader emotional state among internet users? This paper explores the sentiment of existential dread among X users and answers the first of these critical questions.

X [1] (formerly known as Twitter), is a microblogging website where users can post (formally known as a tweet) messages limited to 280 characters. Posts can include additional media such as pictures, videos, and GIFs. Posts can also be categorized via hashtags (keywords preceded by the pound symbol), which serve as searchable categories. Additionally, X users can Follow one another. Following allows posts to be aggregated in a personalized feed for ease of access. Users can also respond directly to posts, contributing their thoughts and ideas to an ongoing conversation.

This paper contributes to the field of Data Mining by examining the dynamics of online interaction in response to Elmo's post, focusing on techniques from Data Mining and Complex Network Analysis. By examining direct responses to Elmo's posts and reactions to those responses, we construct a detailed map of the structure of the conversation.

To further our analysis, we create a graph that links posts to their authors, allowing us to investigate the relationships between various authors based on their contributions to the discussion. Our data set is sourced from X and employs machine learning analysis models to analyze the content of posts and determine whether a post exhibited signs of depression. The primary challenge we faced was extracting data from X, a task made increasingly difficult by recent modifications to their API.

Our analysis began by examining key network metrics, such as centrality and community structure, across two datasets and their combined form. We hypothesized that posts with distinct emotional tones, such as happiness and sadness, would cluster together, reflecting the emotional or thematic commonalities in their content. However, our findings indicate that this hypothesis does not hold consistently.

The rest of the paper is organized as follows: The following section details studies that developed techniques we employ or performed a similar analysis on a different X community. Section 3 will explain the data collection and processing methodology we used. Section 4 will examine the answers we found to the two main questions of this work: How many people are depressed? And how do they cluster? Sect. 5 will discuss some potential areas for improvement and offer some ideas for further exploration. Finally, Sect. 6 will reiterate some conclusions based on our results.

2 Related Work

A significant body of prior research has focused on identifying distinct community structures within X networks through analysis and network modeling. For instance, Himelboim et al.'s [7] analysis of X users through network models reveals various structures in how information is disseminated and how users cluster. Their study assessed whether users access a diverse range of sources or remain within more insular groups, identifying six main network structures. These networks are: Polarized Crowds, representing divided networks with clear factional boundaries; In-Groups, depicting unified yet insular cliques; Brand Networks, showing fragmented followings around specific topics or brands; Communities, indicating clustered user arrangements; Broadcast Networks, which widely disseminate information from a central source; and Support Networks, characterized by hub-and-spoke structures that are either outward (supportive) or inward (informational).

Their study aimed to understand these main network structures using centrality, which measures the extent to which a network's most centralized points dominate in terms of their connections compared to others. Their study explores characteristics like density, the ratio of actual links to the maximum possible number of links; modularity, assessing the quality of division among clusters; and the presence of isolates, or users not connected to others within the network. These metrics help identify the nature of a community's network, for instance, high centrality often indicates a broadcast or support network. In contrast, networks with low centrality, high density, and low modularity are indicative of tightly-knit "in-groups".

Alternatively, Shoko et al. [8] delved into the details of how users engage with each other through Follows, Mentions, and Favorites, aiming to derive meaningful metrics that could accurately reflect user preferences and the nature of their relationships. The core motivation behind this investigation was to harness the structure of the network to identify reliable measures that could contribute to constructing a robust model of user preferences based on available data. One of the main findings from this study revealed that both Favorite and Follower networks exhibit a significantly large connected component, in contrast to the Mention network which lacks this characteristic. Furthermore, all three networks demonstrated a scale-free nature in their degree distribution, indicating a pattern where a small number of users (celebrities) have a high degree of connection, a common trait in social networks. They also noticed a disparity in user engagement around high-degree nodes; they attributed this to the fact that Favoriting a post is markedly less demanding than Mentioning another user in your own post. However, the absence of comprehensive visual representations of their data poses a challenge, making it difficult to intuitively ascertain the validity of their conclusions.

Shifting focus from works seeking to model X as a whole, several works have taken a similar role to us and attempted to map different X communities. For instance, Grandjean [6] analyzed researchers in the field of Digital Humanities who have a presence on X, aiming to understand the connections between them. This objective closely mirrors the investigation conducted in this paper, albeit focusing on a different dataset. The study revealed that the X graph of these researchers is marked by a high degree of homogeneity, largely due to the disproportionate attention received by a small number of individuals and institutions. A notable finding is the distinct cluster of French-speaking users, and to a lesser extent, German-speaking users, indicating that language plays a significant role in shaping the network structure. Their methodology involved compiling a list of researchers based on self-reported information found in their X biographies. This list was initially derived from prominent members of the community and subsequently expanded to include their followers. However, this approach may have overlooked additional communities, raising questions about the comprehensiveness of the dataset. This wouldn't be that big of an issue if they had found more than one and a half clusters, however since their entire results formed a large mostly connected component it begs the question of weather the analysis would be different outside of this one community (if others even exist that is).

3 Methodology

Our first step was to collect the dataset from X using social media mining. Our next step was assessing users' "happiness levels." These levels serve as indicators of emotional sentiment within our social network analysis. By incorporating these attributes into our network analysis, we aimed to gain a more comprehensive understanding of the interaction dynamics surrounding Elmo's post and to determine whether a sense of "existential dread" is reflected in the responses.

3.1 Data Acquisition

Recent changes to X's policies have complicated their APIs by removing free access to post data and reducing the benefits of their academic plans. As a result, we had to adopt an alternative method for data collection. We utilized a scraping library called Apify [10]. More details of the raw data we collected and a downloadable version are available on our Github.

3.2 Data Processing

After acquiring the data, the first step in preparing it for analysis was cleaning. The data was segregated into two distinct datasets: one for posts and another for users. Following this segmentation, we removed duplicates from each dataset. This step is crucial because the initial concatenation process might result in duplicate posts, and the nature of response chains could lead to multiple posts by the same user. Since the original dataset is composed solely of posts, any post by the same user also results in a duplication of that user. Eliminating these duplicates is essential for the subsequent application of various machine learning models to be performant.

We then applied a machine learning model that can be found on HuggingFace [11]; it is a distilled and fine-tuned version of BERT [3] specialized in sentiment analysis. This model provides the output as three probabilities, each ranging between 0 and 1, signifying the likelihood that a post is positive, negative, or neutral. We apply specific rules based on the most probable classification to simplify the representation of a post's sentiment into a single value.

If it is most likely to be negative, we negate that probability; if it's most likely to be neutral, we multiply that probability by 0.05. Consequently, this adjustment results in a probability within the range of $[-1, 1]$, where a number closer to -1 indicates a higher likelihood of the sentiment being negative, and conversely, closer to $+1$ indicates a positive sentiment. To determine an overall sentiment for a user, we calculate the average of these adjusted probabilities across all posts associated with the user, using this average as the sentiment score for the user. However, we assign a truly neutral sentiment value of zero for posts that consist solely of quotes and links.

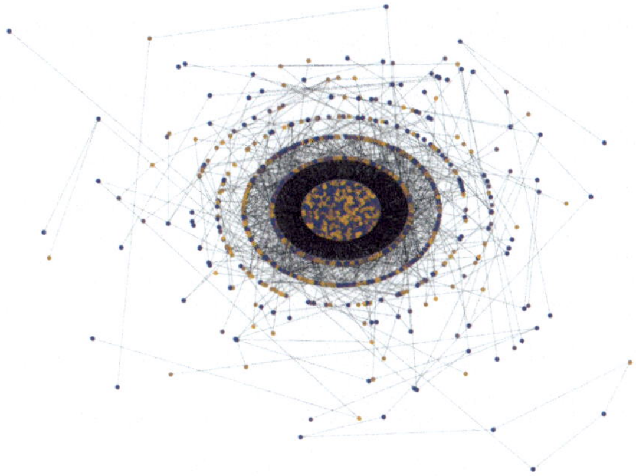

Fig. 1. Graphical representation of the processed posts. Blue dots indicate negative sentiment posts while orange dots indicate positive sentiment.

Once processed, the data forms a network that consists of many independent sub-graphs that branch out from the initial Elmo post, as shown in Fig. 1. This figure is color-coded to indicate sentiment: negative sentiment is represented by blue, positive sentiment is denoted by orange, and gradients represent some combination in the middle. At first glance, the posts seem almost randomly divided between positive and negative sentiments. As the following section will reveal, this preliminary analysis is not far from the truth.

4 Results

This section answers the two main questions posed at the beginning: Are the responses mostly "happy" (positive) or "depressed" (negative)? Do the reactions form similar clusters? This section also briefly analyzes the centrality of the network, but with the original Elmo post removed.

4.1 Overall Statistics

Before analyzing users' sentiments from our data set, we used thresholding to separate posts and user sentiments into bins of various emotional states. Our preprocessor defined neutral posts as sentiments within the range [0, .05]. We also considered more pronounced emotional expressions, categorizing sentiments as "extreme" if they were more than a distance of .5 away from zero. According to the results (summarized in Fig. 2) a majority of users (47%) express happiness, whereas only 38% exhibit sad sentiments. 27% of users were found to have "depressed" or "extremely negative" sentiments, whereas 34% were categorized as "very happy" or "extremely positive." The average user sentiment, which is around 0.067, suggests a slight positive bias throughout the sampling. Further statistical details are compiled in Table 1.

This sample data indicates that a smaller proportion of Elmo respondents, only 27%, fall into the depressed category. Interestingly, this percentage is close to the number of high school students considering suicide in 2021, which was reported as 30.0% for females and 14.3% for males, averaging approximately 22.2% overall [5]. So it seems the media outlets might be justified in their use of the term "existential dread," however, the question "how do users cluster in terms of their sentiments?" remains unanswered.

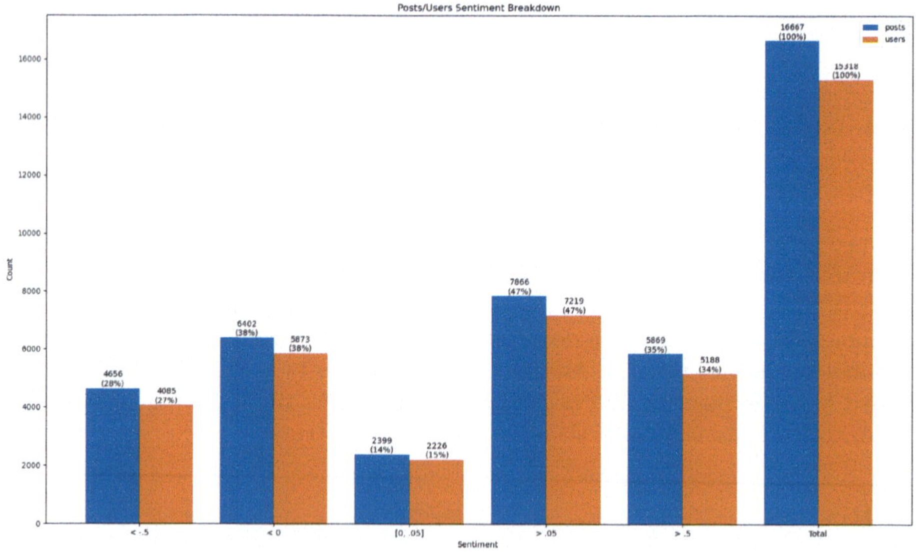

Fig. 2. Sentiment breakdown of posts (blue) and users (orange) into each sentiment category. [0, .05] represents neutral entities, while categories closer to −1.0 are more "depressed" while categories closer to 1.0 are more "happy."

Table 1. Statistical Measurements of Posts/User sentiment

Measure	Posts	Users
Mean	.066	.067
Median	.018	.177
Mode	0.0	0.0
Range	[−.984, .994]	[−.984, .994]
Variance	.377	.362
Std. Dev.	.614	.602

4.2 Clustering

Since this network is so centralized around Elmo, removing the original post uncovers multiple unique conversations, as depicted in Fig. 3. This visualization highlights both the sentiment, with blue indicating negative sentiment and orange indicating positive sentiment, and the individual colorization of each community. Most of these communities consist of small groups (fewer than five responses) engaged in mostly linear exchanges, with only a few branching out into more extensive conversations. Interestingly, many of these larger conversations also tend to form linear chains, while tightly connected, back-and-forth conversations are rare.

The sentiment distribution across these conversations seems balanced, with no discernible pattern favoring negative or positive sentiments; for every cluster predominantly negative, there is a corresponding positive one. However, the one notable pattern is that linear chains stemming from tight clusters exhibit a bias towards negative sentiment. Most of the posts do not have replies, causing them to become isolated. Despite these few exceptions, there are no noticeable general patterns in how responses cluster, so individuals do not appear to flock together. After answering all of our questions, we additionally decided to analyze the centrality of the network.

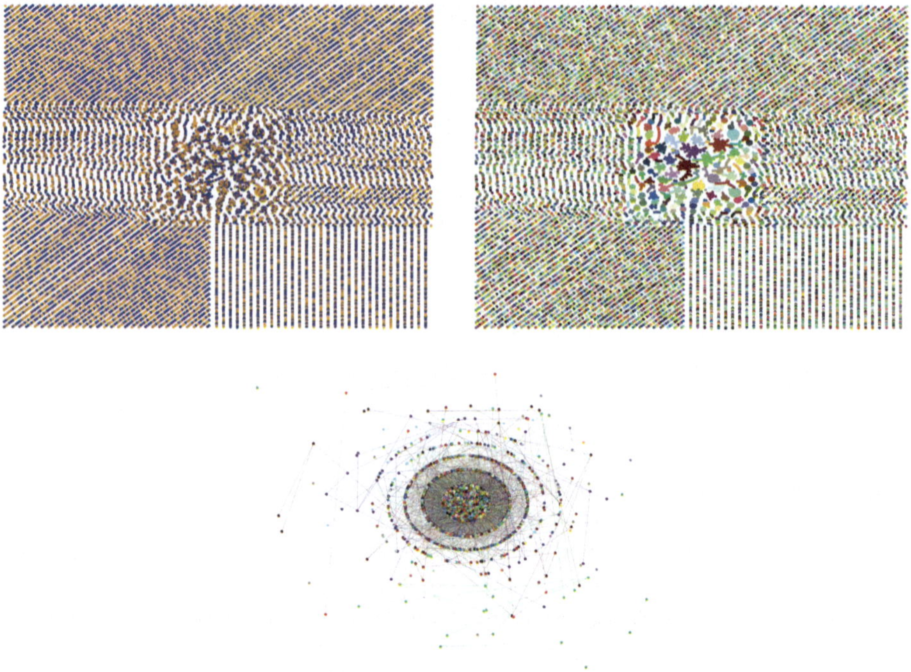

Fig. 3. Sentiment analysis of these Posts communities (left), color coding of each community (right), and color-coded communities laid out as in Fig. 1 (bottom).

4.3 Centrality

This analysis began by calculating key centrality metrics such as Degree, Closeness, Betweenness, and Eigenvector centralities, with a specific focus on the changes brought about by the presence or absence of the original Elmo post. Minimum centrality values were consistently near zero, regardless of whether Elmo was included in the graph. However, a significant shift was noted in the average closeness centrality when Elmo was part of the graph; this metric surged to approximately 0.2 compared to a mere 0.0073 when Elmo was absent, as illustrated in the corresponding graph 3. This dramatic increase suggests that Elmo is a crucial connector among numerous small, otherwise disconnected communities within the network.

Elmo's individual closeness centrality measured around 0.4, indicating his relatively strong but still quite weak connectivity to most other nodes overall. In contrast, removing Elmo led to the decrees of the networks average closeness centrality to 0.0077, reflecting the relative isolation of nodes. This phenomenon is visually represented in Table 2.

Table 2. Statistical Measurements of Centrality

Measure	Posts	Users	Full
Max degree Centrality	.0083	.0083	.0083
Avg degree Centrality	2.487e−5	2.487e−5	2.487e−5
Min degree Centrality	.0	.0	.0
Max Closeness Centrality	.0077	.0077	.0077
Avg Closeness Centrality	.0003	.0003	.0003
Min Closeness Centrality	.0	.0	.0
Max Betweenness Centrality	.0002	.0002	.0002
Avg Betweenness Centrality	8.675e−8	8.675e−8	8.675e−8
Min Betweenness Centrality	.0	.0	.0
Max Eigenvector Centrality	.7057	.7057	.7057
Avg Eigenvector Centrality	.0006	.0006	.0006
Min Eigenvector Centrality	3.709e−9	3.7099e−9	3.7099e−9

Without Elmo, the typical structure is a line. Nodes at the end of this linear formation are significantly distant from others in the network. It was also noted that centrality values were generally higher for users than for posts, and both were higher than those observed in graphs that combined both elements. This pattern aligns with the intuitive understanding that a single user, potentially representing multiple posts, reduces the "distance" within the network. In contrast, the linkage between a post and a user slightly increases this "distance."

5 Limitations and Future Work

There exist a few critical limitations to this study: first identifying and categorizing troll posts presented a significant challenge in our research paper. For instance, such posts falsely accuse Elmo of being involved in illegal activities like fentanyl distribution. Advertisements from large corporations also introduce obscurity into the user sentiment. These elements introduce a basis that may distort the sentiment analysis. Accurate identification and removal of troll and corporate posts would significantly improve the overall quality of the data and the integrity of the results.

Some small patterns warrant further investigation; one notable trend suggests that conversations tend to have a slight bias towards negative; the further the conversation chain progresses. To enhance the accuracy of user sentiment analysis, two data cleaning methods merit additional exploration. The first involves the detection and removal of troll posts, while the second focuses on more accurately identifying corporate accounts.

The analysis in this paper focuses on Elmo as a single case study, which inherently limits the ability to generalize findings about existential dread across broader online communities. However, there is potential value in developing more efficient methods for characterizing the number of communities observed within the dataset. By refining these methods, future research could enable more accurate identification and analysis of sentiment clusters across diverse social media posts to better categorize online sentiment.

6 Conclusion

This study addresses two primary questions: whether the responses to Elmo's post were predominantly positive ("happy") or negative ("depressed") and whether the responses exhibited distinct clustering patterns. The analysis reveals that most responses leaned toward positive sentiment. However, the proportion of "depressed" users observed (27%) aligns with CDC-reported high school suicide rate data [5], highlighting a concerning level of emotional distress. No significant clustering patterns were found to correlate sentiment with distinct online communities. Contrary to some media reports [5], the overall outlook in the dataset is not as dire, though the underlying emotional distress remains notable. While the findings do not indicate statistically significant levels of depression, the percentage of negative sentiment observed is still cause for concern. Specifically for Elmo's post, the notion of widespread existential dread among users is not supported by the data. As this dataset represents only a small fraction of internet activity, a significantly larger dataset would be required to draw meaningful conclusions about "existential dread" among internet users as a whole.

Acknowledgements. This material is based in part upon work supported by the National Science Foundation under grant number OIA-2148788. Any opinions, findings, and conclusions or recommendations expressed in this material are those of the authors and do not necessarily reflect the views of the National Science Foundation.

References

1. Anonymous: X - twitter platform. https://twitter.com/. Accessed 13 Nov 2024
2. Biden, J.: Biden response (2024). https://twitter.com/POTUS/status/1752518567239536821. Accessed 13 Nov 2024
3. Devlin, J., Chang, M., Lee, K., Toutanova, K.: BERT: pre-training of deep bidirectional transformers for language understanding. CoRR abs/1810.04805 (2018). http://arxiv.org/abs/1810.04805
4. Elmo: Elmo is just checking in! how is everybody doing? (2024). https://twitter.com/elmo/status/1751995117366296904. Accessed 13 Nov 2024
5. Gaylor, E.M., et al.: Suicidal thoughts and behaviors among high school students—youth risk behavior survey, United States, 2021. MMWR Supplements **72**(1), 45–54 (2023). https://doi.org/10.15585/mmwr.su7201a6
6. Grandjean, M.: A social network analysis of twitter: mapping the digital humanities community. Cogent Arts Humanities **3**(1), 1171458 (2016). https://doi.org/10.1080/23311983.2016.1171458

7. Himelboim, I., Smith, M.A., Rainie, L., Shneiderman, B., Espina, C.: Classifying twitter topic-networks using social network analysis. Soc. Media + Soc. **3**(1), 2056305117691545 (2017). https://doi.org/10.1177/2056305117691545
8. Kato, S., Koide, A., Fushimi, T., Saito, K., Motoda, H.: Network analysis of three twitter functions: favorite, follow and mention. In: Richards, D., Kang, B.H. (eds.) Knowledge Management and Acquisition for Intelligent Systems, pp. 298–312. Springer, Heidelberg (2012)
9. O'Kane, C.: Elmo wrote a simple tweet that revealed widespread existential dread. now, the president has weighed in (2024). https://www.cbsnews.com/news/elmo-tweet-replies-how-is-everybody-doing-existential-dread-president-joe-biden-response/. Accessed 13 Nov 2024
10. Unknown: Apify. https://apify.com/apidojo/tweet-scraper. Accessed 13 Nov 2024
11. Unknown: Distilbert multilingual sentiment analysis model. https://huggingface.co/lxyuan/distilbert-base-multilingual-cased-sentiments-student. Accessed 13 Nov 2024

Using Virtual Reality in a Case Study with Interdisciplinary Academic Projects for Reading Fluency Analysis and Assessment

Victor Araujo Paula Cavichioli, Stephanie Souza Russo, Wilson Cristoni Neto, Ana Clara Oliveira Leal, Emanuele Cristina da Silva Diniz Campos, Michael Felipe Santos Silva, Diego de Franca Condack, Adilson Marques da Cunha, Luiz Alberto Vieira Dias, Lineu Fernando Stege Mialaret, Denis Silva Loubach, Bruno Augusto Pereira, Gleisson de Souza Bezerra, Bruno Basso Brandani, and Gildarcio Sousa Goncalves(✉)

Aeronautics Institute of Technology, Sao Jose dos Campos, Sao Paulo, Brazil
gildarcio@ita.br

Abstract. At the Aeronautics Institute of Technology (ITA) in Brazil, an academic Interdisciplinary Problem-Based Learning (IPBL) case study was conducted named Fluency Assessment in Literacy on Real Time (FAL-RT). The current state of this academic project was the result of an incremental sequence of initiatives carried out since 2021 supported by courses offered for undergraduate and graduate programs. Its objective was to incorporate and improve new emerging technologies from the technological market based on a real and current problem. During the 2nd semester of 2024, just 12 undergrad and graduate students from four different courses of Electronic and Computer Engineering Programs were challenged to implement a 3D classroom environment using Virtual Reality (VR). This project aimed to enable the process of reading fluency analysis and assessment for elementary school students through the use of Meta Quest 3 VR headset devices. This initiative proposed to update the evaluation process used in earlier phases of this project using a more immersive and educational model. The project development process was supported by the Scrum method and its best practices concluded in just 16 academic weeks. It was completed with the delivery of a prototype computer system simulating a 3D classroom in a VR environment as its front-end. The system prototype was able to record, analyze, and evaluate students' reading fluency levels. The main key technologies used in this academic project included Artificial Intelligence, Machine Learning, 3D development using the Unity platform, databases, cloud computing, Application Programming Interfaces (APIs), and real-time embedded systems.

Keywords: Interdisciplinary Problem-Based Learning (IPBL) · Scrum Agile Method · Databases · Virtual Reality · Meta Quest · Machine Learning · Transformers · Reading Fluency Analysis

S. Latifi (Ed.): ITNG 2025, AISC 1463, pp. 178–190, 2025.
https://doi.org/10.1007/978-3-031-89063-5_16

1 Introduction

This article presents the use of some methods applied in a case study named Fluency Assessment in Literacy on Real Time (FAL-RT) project, the management and delivery of its tracking criteria using the Scrum agile method, and also the software testing and auditing processes to ensure correct system prototype functions with high quality.

It also describes the architecture and integration between the various systems comprising the project, as well as its developed components and stages.

The following assumptions from previous versions of this project remained the same, where TEACHERS captured audios from previously presented sentences to STUDENTS. After some collected audios, analysis and assessment processes were conducted, resulting in scores and indicating fluency levels from STUDENTS' readings [1].

The key innovation of the current approach was the inclusion of a virtual environment to simulate a classroom, where capturing and processing the reading fluency levels occur immersively through the use of Virtual Reality (VR) glasses. VR has been transforming education by providing immersive experiences that facilitate learning. Virtual platforms enable more realistic simulations and dynamic assessments, increasing student engagements, and allowing more accurate feedbacks. Within this context, the use of VR was essential for creating an interactive environment where TEACHERS could assess STUDENTS language skills.

2 The System Prototype Architecture

The FAL-RT academic project aimed to develop a system that in addition to integrating with a legacy system from previous projects also introduced new components.

These components include: 1) The development of an immersive and interactive virtual reality experience with the Meta Quest 3 headset devices, using the Unity platform; 2) The evolution of the legacy microservice for retrieving STUDENTS' data; 3) The maintenance of a microservice for retrieving phrases to be read; 4) The development of new microservices, using Artificial Intelligence (AI) to assess STUDENTS' reading proficiency by comparing texts extracted from audios with original texts stored in a phrase database; and 5) The implementation of a microservice to record reading proficiency scores and STUDENTS' identification data.

The decision to develop an immersive environment was driven by the limited presence of academic projects employing this technology to evaluate the reading fluency of STUDENTS. In response, an Android application was created with front-end features designed for deployment on the Meta Quest 3 headset device.

The objective of this project was to implement secure mechanisms for storing and sharing sensitive data associated with the entities STUDENT, TEACHER, SCHOOL, and their INFRASTRUCTURE. This includes registration data, assessments, and audio recordings within a virtual reality environment.

The project employed a microservices architecture, described by [2] as a system design method that "divides systems into small, independent services." To support this architecture, containerization technologies were utilized to manage the deployment and operation of microservice instances.

On this project prototype, students were motivated to access the application via Meta Quest 3 headset devices, using Backends-For-Frontends (BFF), communicating with back-end microservices running in Docker containers hosted either on developer's machine or on clouds like Amazon Web Services (AWS). The architecture of the project prototype intended to deliver a robust and efficient environment for this academic case study.

This project architecture was designed in an integrated manner, since its first sprint progressively evolved through the following sprints, as shown in Fig. 1.

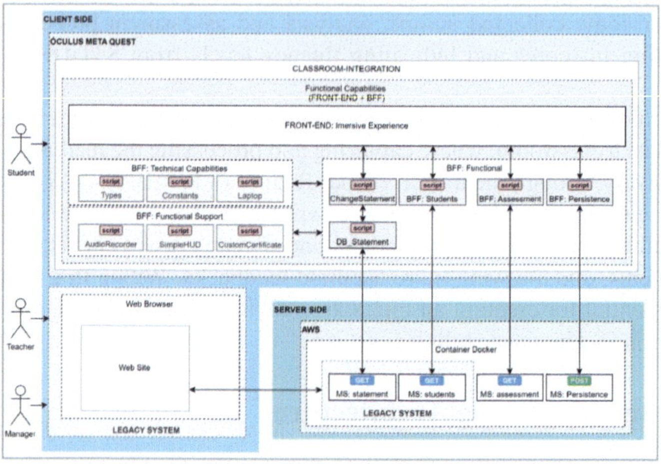

Fig. 1. The FAL-RT Academic Project Architecture.

3 The New Software and Hardware Approaches

In order to apply new software and hardware approaches in this academic project prototype, it was considered the use of: the Machine Learning technique from AI to automate reading fluency assessments; the Meta Quest 3 device with VR to capture audio from students; the FRDM KL25Z board to store large amounts of local audio data; and some testing and auditing procedures to ensure compliance with applicable regulation standards.

3.1 The Courses

From August to December 2024, students from four different undergrad and graduate courses of *CE-230 Software Quality, Reliability and Safety, CE-235 Real Time Embedded Systems, CSI-65 Embedded Systems Project, and CE-237 Advanced Topics on Software Testing* worked together on this interdisciplinary project.

This period allowed the acquisition of knowledge related to emerging technologies and scenario, in addition to exploring new development approaches in the areas of software and hardware.

3.2 Using the New Software and Hardware Approaches

1) ***Using Machine Learning for Automated Reading Fluency Assessment:*** This section outlines the development of a Machine Learning (ML) Application Programming Interface (API) designed to evaluate students' reading fluency through semantic similarity analysis between the original texts and the transcription of recorded audios. The solution was tailored for educational settings, aiming to provide accurate and fast feedback to educators during students' reading assessments.

   This approach combined technologies such as the Bidirectional Encoder Representations from Transformers (BERT) model and the Word Error Rate (WER) metric, offering detailed transcription analysis, including reading errors and fluency.

   Reading fluency is a critical indicator of academic performance. Traditionally, its evaluation is conducted manually by teachers, which can lead to subjective and time-consuming assessments. With advancements in Speech Recognition (SR) and Natural Language Processing (NLP) technologies, automating this evaluation has become feasible. The developed API analyzed audio transcriptions by comparing them to the original text, identifying not only incorrect words but also assessing fluency and textual comprehension.

   The BERT model was selected for its ability to capture bidirectional context of words in texts, enabling deep analyses of linguistic variations and pronunciation errors. It is especially effective for assessing reading fluency, even in different languages, including Portuguese.

   Semantic similarity was calculated using embeddings generated by BERT and cosine similarity, which evaluates the proximity between text vectors [3]. The WER metric quantitatively measured transcription accuracy, considering word insertions, substitutions, and omissions. Combining these metrics provided a comprehensive analysis of reading fluency.
2) ***Using Meta Quest 3 Device as Front-end for Capturing Readings:*** In recent years, the education sector has undergone constant changes, caused by the integration of innovative digital tools that aim to improve the learning experience of students. Recent innovative technologies, such as Virtual Reality (VR), Augmented Reality (AR) and Mixed Reality (MR), have become more accessible for use in educational environments.

   Extended Reality (XR), a newer term that encompasses various forms of emerging technologies and interaction with virtual systems, is proving to be a promising technology that is transforming elementary school teaching and offering new perspectives and pedagogical opportunities. Given its lower cost and ease of use, its impact on the education sector will be even greater [4].

   With XR, it is possible to create immersive virtual environments, which will allow students to actively interact with educational contente, overcoming the traditional boundaries of classrooms and encouraging participation, creativity and critical thinking.

   In this study, we used Meta's Meta Quest 3 devices and Meta Quest Touch Plus controllers as the VR environment. These devices were selected for their high-quality software and hardware features. The development of XR applications requires platforms capable of covering a range of features to achieve such immersive experiences.

Today, two platforms are broadly used to develop these systems: Unreal and Unity. And only the last one was used in this case study.

3) ***Using the FRDM-KL25Z Board to Store Large Amounts of Local Audio Data:*** As part of the activities developed in the CE-235 Real-Time Embedded Systems and CSI-65 Embedded System Project courses, during this project involved the use of the FRDM-KL25Z development board, used in their labs exercises.

   Initially, the proposal aimed to establish communication between the FRDM-KL25Z and Meta Quest 3 glasses device through Bluetooth wireless connectivity. However, upon identifying that the internal memory capacity of the FRDM-KL25Z posed a significant limitation, the scope was expanded to include the integration of a Mass Storage Device.

   The primary objective thus became investigating alternatives and developing just Proof-of-Concept (PoC) implementations to demonstrate the feasibility of communication between the Meta Quest 3 device and the FRDM-KL25Z via Bluetooth while enabling data storage on an external memory device connected to the board. This solution would allow the AFA- TR Project to record literacy assessment data of students on external storage devices.

4) ***Using Testing and Auditing Procedures to Ensure Compliance with Applicable Regulation Standards:*** In agile projects like this one, the quality, reliability and safety of each User Story (US) directly impacts the success of the final product. Then, each US of this project had to be clear, complete, feasible, and testable. These was performed by using at least 3 test cases per US and also by auditing them using the INVEST criteria by Bill Wake, who's a deep thinker on software design [5]. Then, each applied criterion had a specific role:

   - **Independent** - when the US was developed and tested separately;
   - **Negotiable** - when the US scope was adjusted as needed;
   - **Valuable** - when the US provided value to end users;
   - **Estimable** - when the US was estimated in terms of efforts;
   - **Small** - when the US was small enough to be developed within short time frames; and
   - **Testable** - when the US was verifiable through clear tests.

Table 1. Scrum Ceremonies [7].

Ceremonies	Description
Sprint Planning	It is a meeting in which the Product Owner (PO) presents the Product Backlog to the Scrum Team (ST) and explains the intermediate goals, any changes, adaptations of a project, and its user stories
Weekly Meeting	These are daily, regular, or weekly meetings lasting up to 15 min, involving all members of a Scrum Team (ST), held to track the progress of a project
Sprint Review	It is a meeting at the end of a Sprint, in which members of a Scrum Team (ST) present their achieved outcomes and encountered challenges during the period
Sprint Retrospective	It is a meeting conducted after a Sprint Review, in which a Scrum Team (ST) internally or collaboratively discusses the positive and negative aspects of the project and how to enhance the development process

3.3 Using the Scrum Framework

As outlined in the Scrum Guide [6], Scrum is described as "a streamlined framework that aids individuals, teams, and organizations in generating value through adaptive solutions for intricate challenges". To facilitate this, project participants were organized into small, distinct, and cross-functional Scrum Teams (ST).

The methodology adopted an iterative and incremental approach, featuring four well-defined events/ceremonies as detailed in Table 1, all revolving around a work cycle known as a Sprint.

The FAL-RT project was executed during the second half of 2024, spanning 16 weeks and divided into 4 Sprints, each lasting 4 weeks. The initial phase, referred to as Project Inception, aimed to align the overall understanding of the project. Subsequent Sprints followed the stages of Planning, Development, and Integration. Ultimately, a final product was delivered and presented in the classroom at the end of the 2nd semester of 2024. Figure 2 illustrates the distribution of Sprints across the project phases.

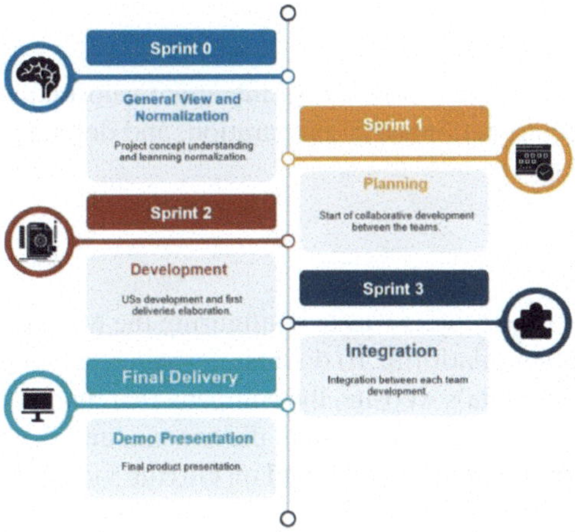

Fig. 2. The Division of Sprints Across the Project Phases.

In each ST, students took on different roles within Scrum, based on their experience with the Framework. These roles are described in the Table 2.

Table 2. Scrum Rules Description.

Rules	Description
Product Owner	Serving as a representative of stakeholders and with the aim of delivering a valuable product at the end of the process, the Product Owner (PO) defines the vision, product requirements, and team goals
Development Team	It is a self-organized and self-managed team, responsible for product development and delivery
Stakeholders	They are individuals who are interested in or somehow affected by the product's outcome
Scrum Master	He/she is responsible for aiding the team in the execution process to maximize efficiency and deliver the Sprint goal.

4 The Project Vision Artifact

This FAL-RT project represents an initiative to "equip Brazilian elementary education with a computerized system suitable for the assessment, diagnosis, correction, and improvement of literacy levels in the country."

Unlike existing computational and/or computerized systems on the market, this product was developed based upon the experiences gained from previous years, involving the implementation of a system that will operate through virtual reality headsets.

This way, it will be possible to record students' reading audios, process the fluency level from the recorded audios, generate a numerical value indicating the levels, displaying them within the system, and storing within the project database.

This approach was designed to meet the needs for detecting, analyzing and assessing literacy fluency levels in elementary school students.

5 The Sprints of Project Development

5.1 The Sprint 0

This sprint was also known as a Warm Up Sprint, and it's normally used for initial team integration, concept alignment, material preparation, and defining the main goals to be achieved during projects.

5.2 The Sprint 1

One of the main steps of the project was to configuring the Meta Quest 3 headset device and the Unity development platform. To do this, it was necessary to create accounts on the Unity website and on Meta's website, the company that manufactures the headset, as well as enabling the developer profile on both the account and the headset environment.

This resulted in a tutorial that can be found on GitHub via the following link: https://github.com/iec-gpes/xr-oculus-quest and a live workshop showing how to use those assets from the scratch and discussing some objections and challenges in terms of Unity and Meta Quest Software Development Kit (SDK) integration.

Meanwhile, the studies using the FRDM-KL25Z board concentrated on identifying components that would support:

1) Bluetooth communication; and 2) connectivity with a Mass Storage Device. After analyzing the available alternatives, the HC-06 Bluetooth Module was selected for wireless communication, and the mini USB port of the FRDM-KL25Z was chosen for external storage integration.

Additionally, all necessary hardware, including modules, cables, and connectors, were acquired to implement the proposed system.

5.3 The Sprint 2

The highlight of this sprint was the creation of a microservice to process audio transcriptions and evaluate students reading fluency, taking the original text and recorded audio as inputs.

Its key functionalities included: 1) Audio decoding - where the submitted base64-encoded audio was converted into WAV format; 2) Audio transcription - where the Speech Recognition tool generated the textual transcription; 3) Semantic similarity analysis - where the BERT model evaluated the semantic relationship between the reference text and the transcription; and 4) Word Error Rate (WER) calculation - where a metric was used to measure transcription accuracy as compared to the original text.

The microservice was developed using the Flask framework, designed to facilitate the efficient creation of RESTful API's. Its primary purpose was to assess reading fluency by integrating advanced processing and analytical techniques. The workflow followed these key steps:

- **Data Input** - Clients submited text and base64-encoded audio through an HTTP POST request;
- **Audio Processing** - The received audio was decoded and converted into WAV format for further analysis;
- **Transcription and Analysis** - Using the BERT model, the API generated a transcription of the audio and compared it with the submitted text to evaluate accuracy and linguistic nuances;
- **Metric Calculations** - Key metrics such as Word Error Rate (WER) and semantic similarity were computed and combined to derive a final fluency score; and
- **Results Output** - The API returned a comprehensive report, including a reading fluency score, reading time, and classifications like "Full Proficiency" or "Support Needed."

The choice of the BERT model was deliberated, given its exceptional ability to capture bidirectional context in text [3]. This allowed for a nuanced analysis of linguistic variations and pronunciation errors, making it highly effective for assessing reading fluency across different languages, including Portuguese.

To provide meaningful insights, the API leveraged two critical metrics:

- **The Semantic similarity**, which was calculated using text embeddings generated by BERT, with cosine similarity measuring the proximity between these vectors; and
- The **WER metric**, which quantitatively evaluated transcription accuracy by accounting for word insertions, substitutions, and omissions.

By combining these metrics, the API delivered a holistic and detailed assessment of reading fluency.

To detail the hardware connections of the external devices linked to the FRDM-KL25Z, an electrical schematic was prepared to document the configurations, and custom cables were manufactured to facilitate these integrations. Following the assembly of the system, all components were thoroughly verified to ensure correct connections and adequate power supply, confirming that the hardware infrastructure was fully prepared for the next phases of the project.

In the microservices structure, during Sprint 2, one key development objective was to design the User Experience (UX), tailored to the school environment and classrooms where students would use the Meta Quest 3. The Unity platform was selected for this task due to its robust integration capabilities for codes and scripts, which would be employed in the subsequent sprint to establish communication between the back-end and front-end.

For the UX development, resources from the Unity Asset Store—a repository offering free and paid assets—were utilized. The components were selected based on their quality and resemblance to 3D objects commonly found in Brazilian schools, as depicted in Figure 3. The final outcome was a virtual environment replicating a Brazilian classroom, providing a foundation for seamless integration between the front-end and back-end microservices.

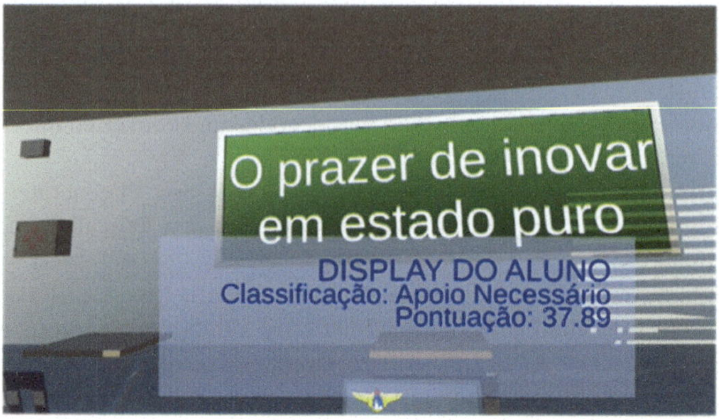

Fig. 3. The Classroom Vision from Simulated Environment.

5.4 The Sprint 3

To enable the persistence of information related to assessments, phrases, students, and teachers, a dedicated microservice was developed. This microservice served as a single interface for communication and data processing, ensuring the proper handling and storage of information in the PostgreSQL database.

Developed in Java using the Spring Boot framework, the microservice implemented a RESTful API capable of efficiently managing all relevant data. The workflow of applications followed these following key steps:

- **Phrase Display** - A phrase was displayed in real time on the Meta Quest 3 device;
- **Audio Recording** - Users audio response was recorded using a Unity-based application;
- **Audio Transmission** - The recorded audio was encoded in base64 format and sent to the AI microservice;
- **AI Processing** - The AI model processed the audio, evaluated the response, and returned the assessment results; and
- **Data Persistence** - The assessment data and related information were stored in the database through the data microservice.

Finally, the architecture illustrated in Figure 4 was implemented using AWS. The microservices were deployed in Docker containers within an EC2 instance, alongside a PostgreSQL database instance. The EC2 instance was configured within a public subnet, with ports 5000 and 8080 enabled to allow external access. This setup facilitated interaction with external users, specifically the Unity application, which consumed both services.

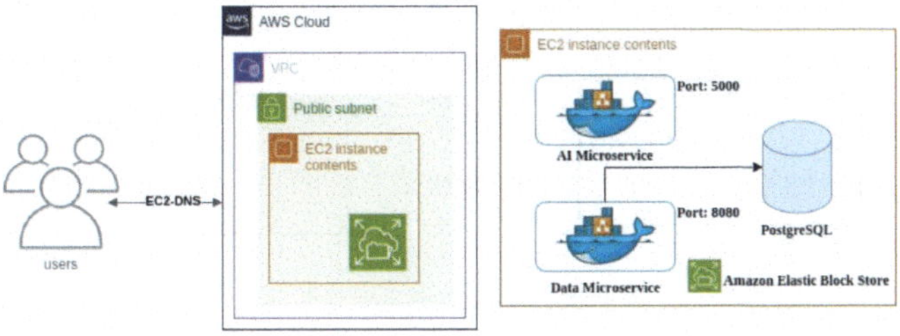

Fig. 4. The Cloud Architecture of the FAL-RT Deployment.

This architecture ensured a seamless integration between the various components, allowing for efficient data processing and storage while maintaining a robust and scalable solution.

To support the integration of the Meta Quest 3 headset device with the FRDM-KL25Z board, the system was modeled using SCADE Suite [8], a model-based design environment for mission- and safety-critical embedded software.

The Model-Driven Development (MDD) enhanced productivity and reduced errors by using high-level abstractions and automating repetitive tasks, making it ideal for designing complex systems [9]. The architecture, shown in Figure 5, was designed with two tasks running on a real-time operating system (RTOS), each task managed by its own Finite State Machine (FSM). Task communication was implemented through a shared memory region, ensuring efficient data processing and exchange between the tasks.

6 The Main Results

The API to evaluate students' reading fluency aimed to deliver an objective and detailed evaluation of reading fluency. Key challenges included audio quality, particularly in noisy environments. Integrating noise reduction tools and adapting the solution to Portuguese represents some future steps to improve accuracy.

The software testing process has identified and corrected faults to guarantee the quality and acceptability of the product [10]. In this project, the following criteria were considered during the testing phase:

- **Approach** - The **black-box** testing method was employed for both software and hardware testing. Some test cases were directly based on the **User Stories** defined in each sprint of the project; and
- **Execution Criteria** - To commence the testing phase, it was essential to ensure that the **Sprint Backlog** artifact contained clearly and objectively completed User Stories, as at least 3 test cases were derived from each of them.

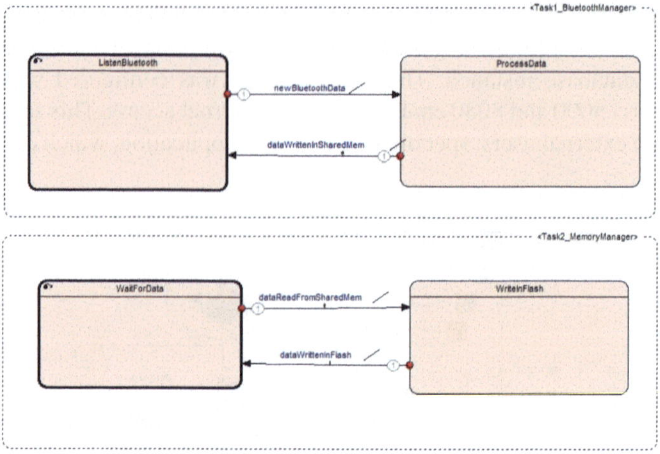

Fig. 5. The System Model for FRDM-KL25Z Using the SCADE Suite.

The obtained test results can be found in Table 3

Table 3. The result of the testing process

Sprint	Total Planned Tests
Sprint 1	48
Sprint 2	48
Sprint 3	33
Executed Tests	129
Test Coverage	100% of the backlog

7 Conclusion

7.1 Specific Conclusions

Auditing based on the INVEST criteria played a crucial role in the development of the educational virtual reality project. The rigorous application of these criteria ensured that the user stories were well-defined, testable, and aligned with system's objectives. As a result, the virtual environment developed provided an effective tool for fluency assessment, contributing to innovation in education.

7.2 General Conclusions

The developed API for evaluating students' reading fluency represents a practical application of Machine Learning in the educational sector, automating reading fluency assessment with precision and objectivity. The combination of the BERT model and the WER metric provides personalized and valuable feedback for educators.

7.3 Recommendations

It is recommended the implementation of regression testing for all layers, in order to allow exhaustive tests to make sure that application performs as expected on different usage scenarios and datasets.

7.4 Future Work

Future improvements for this project may include enhancing semantic analysis and audio pre-processing techniques. Creating a more interactive environment that brings teachers and students together in the same virtual classroom would be highly beneficial. This could be achieved by developing a new version of the Unity application. On the back-end, integrating the capabilities of Large Language Models (LLMs) would significantly enhance application's functionality, enabling not only the evaluation of word errors but also providing semantic and contextual analysis.

Acknowledgments. The authors would like to thank: the Aeronautics Institute of Technology (ITA), for providing the infrastructure for the development of this academic project; the Casimiro Montenegro Filho Foundation (FCMF) for its support; the ANSYS and ESSS Inc. For the partnership with ITA providingthis academic project SCADE licenses; the Brazilian Department of Education (MEC), for providing the necessary data and statistics for this case study; and also to the Software Engineering Research Group (GPES), along with the courses and professors for their guidance, during the development of this project.

References

1. Mota, M.S.M., et al.: Case study of an interdisciplinary academic project for reading fluency analysis. In: Latifi, S. (ed.) ITNG 2024: 21st International Conference on Information Technology-New Generations, pp. 205–211. Springer, Cham (2024)
2. Newman, S.: Building Microservices: Designing Fine-Grained Systems, 2nd edn. O'Reilly Media, Inc. (2021)
3. Tunstall, L., Von Werra, L., Wolf, T.: Natural Language with Transformers – Building Language Applications with Hugging Face. O'Reilly Media (2022)
4. Pomerantz, J.: Teaching and learning with extended reality technology. In: Proceedings of the BOBCATSSS 2019 Conference (2019)
5. Sutherland, J.: SCRUM: The Art of Doing Twice the Work in Half the Time. Crown Business (2014)
6. Rubin, S.K.: Essencial Scrum: A Pratical Guide to the Most Popular Agile Process. Addison-Wesley Professional (2012)

7. Schwaber, K., Sutherland, J.: The Scrum Guide, The Definitive Guide to Scrum: The Rules of the Game, November 2020. https://scrumguides.org/docs/scrumguide/v2020/2020-Scrum-Guide-US.pdf
8. Ansys. Ansys scade suite (n.d.). https://www.ansys.com/products/embedded-software/ansys-scade-suite. Accessed 27 Nov 2024
9. Schmidt, D.C.: Model-driven engineering. Computer **39**(2), 25–31 (2006)
10. Jorgensen, P.C.: Software Testing: A craftsman's Approach. Taylor Francis Group (2014)

Case Study Involving Interdisciplinary Academic Projects for Reading Fluency Assessment and Analysis

Victor Araujo Paula Cavichioli, Stephanie Souza Russo, Huascar Mirko Montecinos Cortez, Ana Carolina de Lima Angelo, Gildarcio Sousa Goncalves(✉), Adilson Marques da Cunha, Luiz Alberto Vieira Dias, Lineu Fernando Stege Mialaret, and Johnny Cardoso Marques

Aeronautics Institute of Technology, Sao Jose dos Campos, Sao Paulo, Brazil
gildarcio@ita.br

Abstract. At the Aeronautics Institute of Technology (ITA) in Brazil, two academic case studies focused on interdisciplinary problem-based learning (IPBL) were carried out. The first one named Fluency Analysis in Literacy Version 3.0 on Real Time (FAL3-RT) was developed on the 2nd Semester of 2023 and involved 13 undergrad and graduate students to address a problem of real-time system. The second one named Fluency Assessment in Literacy on Big Data (FAL-BD) was developed on the 1st Semestre of 2024 and involved 11 undergrad and graduate students to address a problem of Big Data and Large Language Models (LLM). These two case studies involved 24 undergrad and graduate students from seven different courses of the Electronic and Computer Engineering Graduate Program. Four from the 2nd Semester of 2023 and three from the 1st Semester of 2024. Both case studies aimed to conceptualize, model, develop, and test components of a distributed database system. Over the course of two semesters, a prototype for audio collection was created, leveraging technologies such as Database Systems, Artificial Intelligence, Real-Time Embedded Systems, Machine Learning, Blockchain, and Kubernetes. The final prototypes were based on an initial research project developed for the Brazilian Ministry of Education, focused on the automatic analysis of reading fluency in elementary school children. This paper collectively describes the computational infrastructure for real-time audio collection, implemented in the FAL3-TR embedded system, and its improvements in the FAL-BD using large language models (LLMs). The audio, collected via a developed interface, was stored on a server for subsequent automatic analysis using a machine learning model. Both projects were completed within 32 weeks of an academic year, employing the Scrum Framework for project management. The key contribution of this work was the integration of the Scrum Agile Methodology with various technologies to manage, develop, and test both case studies, culminating in the development of a Literacy Fluency Assessment system, including a functional software and hardware prototype for audio acquisition.

Keywords: Interdisciplinary Problem-Based Learning (IPBL) · Scrum · Agile Method · Databases · Blockchain · Kubernetes · Machine Learning · Fluency Analysis

S. Latifi (Ed.): ITNG 2025, AISC 1463, pp. 191–203, 2025.
https://doi.org/10.1007/978-3-031-89063-5_17

1 Introduction

This paper reports the development of the first academic project prototype Fluency Analysis on Literacy Version 3.0 in Real Time (FAL3-TR), using Interdisciplinary Problem-Based Learning (IPBL) [1, 2]. The FAL3-TR project marked the latest iteration of an academic initiative that began in 2021 with the FAL-TR project, followed by its 2.0 version in 2022, named FAL2-TR. The project was composed of four main integrated subsystems: STUDENTS, TEACHERS, SCHOOLS, and INFRASTRUCTURE. The project team was organized into subgroups, each assigned specific objectives within these four areas.

The second academic project prototype Fluency Assessment in Literacy on Big Data (FAL-BD) was developed on the 1st Semestre of 2024 and also addressed the four primary integrated subsystems using Big Data and Large Language Models (LLM).

The STUDENTS subsystem covered functions related to the first two grades of elementary school. The TEACHERS subsystem focused on functionalities for teachers, who were responsible for audio collection from their students. The SCHOOLS subsystem dealt with actions and metrics within the elementary schools, where students and teachers interact. Finally, the INFRASTRUCTURE subsystem managed the computational infrastructure discussed in this case study.

On both projects, FAL3-RT and FAL-BD, the students were divided into two teams. The STUDENTS and TEACHERS Scrum Team and the SCHOOLS and INFRASTRUCTURE Scrum Team, both using the Scrum Agile Method and its best practices for collaborative development.

The objective of the FAL3-RT project on the 2nd Semester of 2023 was to develop an academic software prototype based on emerging Information Technology (IT) solutions, with emphasis on creating a learning model established to process collected audios and classify them according to a numerical value, representing students indicative grades of fluency level, termed as a fluency indicator by the Brazilian Ministry of Education (MEC).

At the conclusion of these two academic project FAL3-RT and AFA-DB, a simulation of a case study was designated as a Proof of Concept (PoC) to demonstrate that both projects were complementary and capable of providing a suitable method to be used for literacy fluency analysis on the second-grade elementary school children.

2 Architecture

The primary objective of the FAL-BD project was to develop a system with administrative layers designed to create, modify, and manage all registration data for entities involved in academic project teams.

The project teams planned a web application with frontend capabilities to store and share sensitive data about entities such as students and schools, as well as generate reports based on the analyzed data, including student records, performance metrics, and audio recordings.

As illustrated in Fig. 1, Kubernetes technology was used to improve the scalability, availability, and management of microservice instances. System users access the web application within a Deployment [3] via a LoadBalancer Service [4] or Ingress [5], which allowed external users to access web application running in Kubernetes network. Once inside, the web service operated within the internal network using a ClusterIP Service, performing database operations stored in a StatefulSet [6]. This architecture provided a more reliable and efficient environment for the application.

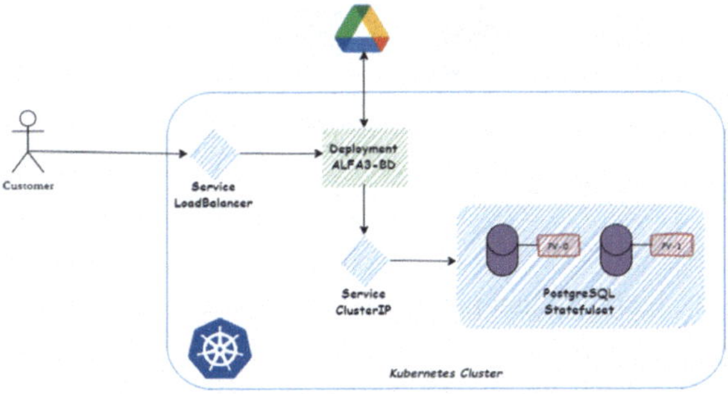

Fig. 1. Microservices Architecture Using Kubernetes.

The Kubernetes-based architecture provided significant advantages to the application. Key features like RollingUpdate and HorizontalPodAutoscaler enhanced the system's scalability and adaptability. RollingUpdate enabled incremental updates to the deployment, ensuring a smooth transition between software versions while minimizing downtime, thus maintaining uninterrupted user access [7].

Meanwhile, HorizontalPodAutoscaler dynamically adjusted the number of active pods to align with fluctuating application demands, optimizing resource usage and infrastructure efficiency. By horizontally scaling instances in response to workload changes, the system achieved consistent performance and reliability under varying conditions [8].

The integration of these capabilities, coupled with a flexible and scalable architecture, allowed the application to handle traffic surges and workload fluctuations effectively. This ensured high availability, robust performance, and an overall improved experience for both users and developers.

The flow of audio collection is detailed in Fig. 2. In step (1), a TEACHER acquired audio using a capture device (2), such as a computer, mobile phone, or Personal Digital Assistant (PDA), to record samples from a STUDENT (3). The process included three recordings (4): isolated words, pseudowords, and a passage of text, adhering to the guidelines in the Operational Manual of the Pedagogical Guidance and Monitoring System issued by the Ministry of Education (MEC) [9].

These audio recordings and related data were then transmitted to the server infrastructure (5). The audio files were stored in a designated project folder, while the associated STUDENT, TEACHER, and SCHOOL information was recorded in a relational database. Finally, fluency data derived from the analysis and audio processing contributed to refining the learning model and were securely persisted in the database.

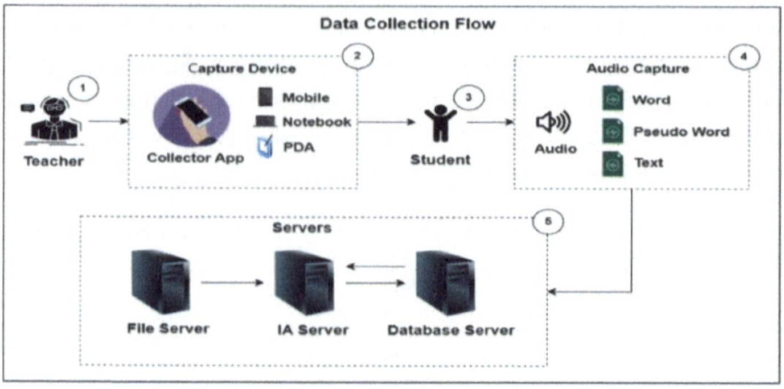

Fig. 2. General System Architecture for Audio Acquisitions.

Each team was tasked with independently designing its architectural framework and identifying the most appropriate technologies to address their specific objectives. This decentralized approach used the diverse expertise of the team members, enabling the development of solutions tailored to their technological abilities.

From its inception, the architectural design of the project emphasized robust integration and interoperability between the STUDENTS, TEACHERS, SCHOOLS, and INFRASTRUCTURE modules, ensuring cohesive functionality between these interconnected components.

3 Theoretical Foundation

3.1 Courses

From August 2023 to June 2024, students from the 7 different undergrad and graduation courses of *CE-240 Database Systems Project, CE-245 Information Technologies, CE-229 Software Testing, CE-237 Advanced Topics of Software Testing, CE-230 Software Quality, Reliability and Safety, CE-235 Real Time Embedded Systems, and CSI-65 Embedded Systems Project* worked together on this interdisciplinary projects.

This period allowed the acquisition of knowledge related to emerging technologies and cenario, in addition to exploring new development approaches in the areas of software and hardware.

3.2 Supplementary Studies

1) *Contextualization on Fluency Analysis Methodology:* From previous versions of these projects on Web, it was possible to get audio readings records from students through randomly generated phrases from systems' interfaces. After recording the audio using microphone, a file with.***wav*** extension was generated and submitted to a transcription process. The audio file also went through a spectrogram process using Recurrent Neural Networks (RNN) to extract audio characteristics, such as time, phonetic segmentation, amplitude, etc.

Through this processing, two important metrics were generated to determine a value representing the level of literacy on reading. The first metric was the word accuracy index, obtained by comparing the phrase presented from system's interface with the recorded and transcribed audio, segmented into tokens to assess the correctness of each pronounced word. The second metric was the speed at which the words were verbalized in the audio, processed to generate metrics as previously described.

However, during these two projects a supplementary study was conducted to verify which alternatives could be applied, in order to generate complementary metrics that express the fluency analysis of elementary students, considering others aspects of literacy fluency beyond reading.

According to Navas [10] although there is no consensus on the definition of reading fluency or its relation to processing speed, reading rate, and word recognition rate, it can be stated that fluency depends on essential elements such as reading rate, automation, prosody, and comprehension. Similarly, Barreira's research [11], which involved 65 students from three first-grade classes with an average age of 7.3 years, applied methods of Phonological Awareness Assessment, Lexical Awareness Assessment, Syntactic Awareness Assessment, and Reading Assessment. It concluded that phonological awareness presented the highest correlation levels with reading and writing skills.

In this context, it was understood that merely reading phrases may not fully represent student's fluency levels, textual comprehension was also an important part of this analysis, and could be considered as a metric factor to generate scores of fluency, creating an opportunity for project evolution to encompass new metrics beyond those used in previous versions.

2) *Large Language Model (LLM):* Currently, there is a growing number of AI studies and applications in various fields, such as healthcare, technology, biology, automation, and others. The use of classical machine learning models is constantly improving, and with the evolution of data storage capacity and computational power, the use of neural networks has been more widely adopted, highlighting deep networks, convolutional, recurrent, autoencoders, and more recently, generative ones [12].

Natural Language Processing (NLP) methods are a subfield of AI and computational linguistics focused on enabling computers to understand, interpret, and generate natural language. The main goal is to bridge the gap between human language and computational understanding, allowing machines to process, analyze, and respond to natural language data, being applied in solutions like chat-bots, virtual assistants, sentiment analysis tools, translation services, and more.

NLP applications have some limitations because natural language is complex and has many exceptions due to linguistic rules, making it challenging to develop formal grammar-based systems that are robust and comprehensive [13].

In 2017, Google researchers published a paper called "Attention is All You Need", which was considered revolutionary for introducing a new neural network architecture named Transformer. This architecture outperformed Recurrent Neural Networks (RNN) on translation tasks in terms of quality and training costs. Moreover, it was excellent at capturing patterns in long sequences of data, dealing with large datasets, and was even used in image processing tasks [14].

Table 1. Scrum ceremonies [16].

Cerimonies	Description
Sprint Planning	It is a meeting in which the Product Owner (PO) presents the Product Backlog to the Scrum Team (ST) and explains the intermediate goals, any changes, adaptations of a project, and its user stories
Weekly Meeting	These are daily, regular, or weekly meetings lasting up to 15 min, involving all members of a Scrum Team (ST), held to track the progress of a project
Sprint Review	It is a meeting at the end of a Sprint, in which members of a Scrum Team (ST) present their achieved outcomes and encountered challenges during the period
Sprint Retrospective	It is a meeting conducted after a Sprint Review, in which a Scrum Team (ST) internally or collaboratively discusses the positive and negative aspects of the project and how to enhance the development process

Since the main use of these models is focused on NLP, it is possible to use these techniques in the project to integrate emerging market technologies as a solution to a real world problem.

3.3 The Scrum Framework

According to the Scrum Guide, Scrum is "a lightweight framework that helps people, teams, and organizations generate value through adaptive solutions for complex problems" [15]. To achieve this, project members were divided into necessarily small, separate, and cross-functional Scrum Teams. The method employed an iterative and incremental approach with four well-defined events/ceremonies as outlined in Table 1, all centered around a working cycle called a Sprint.

The projects FAL3-RT and FAL-BD were developed during 2 semesters splited over 16 weeks each, divided into 4 Sprints, with each Sprint spanning 4 weeks. The initial Sprint 0 focused on aligning general knowledge about the project. Subsequent Sprints followed the phases of Planning, Development, and Integration. Ultimately, a final product was delivered and presented in the classroom. Fig. 3 illustrates the division of Sprints across the phases.

In each ST, students took on different roles within Scrum, based on their experience with the Framework. These roles are described in the Table 2.

4 The Project Vision Artifact

The vision artifact documents was prepared for each semester project to align its development proposal, considering the courses taught at ITA, the context of each project related to what had to be developed over time, and current technological trends, which were eventually incorporated into each project as challenges. Based on the principles of the Scrum project management methodology and its best practices, a vision artifact was created, incorporating characteristics into the projects to guide the teams in the directions the projects intended to take in each mid-term. In summary, the vision artifact of each semester project established the execution academic level necessary for the ITA undergrad and graduate course, who propose a computational system through the development of embedded and web software, as well as technological innovations aimed at assessing the level of students' literacy fluency through audio capture.

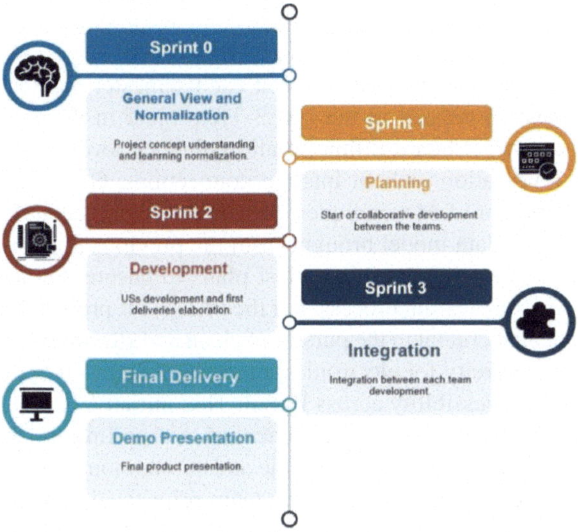

Fig. 3. The Division of Sprints in the two Projects.

Table 2. Scrum Rules Description.

Rules	Description
Product Owner	Serving as a representative of stakeholders and with the aim of delivering a valuable product at the end of the process, the Product Owner (PO) defines the vision, product requirements, and team goals
Development Team	It is a self-organized and self-managed team, responsible for product development and delivery
Stakeholders	They are individuals who are interested in or somehow affected by the product's outcome
Scrum Master	He/she is responsible for aiding the team in the execution process to maximize efficiency and deliver the Sprint goal.

5 The Sprints of Project Development

5.1 The Sprint 0

This sprint was also known as a Warm Up Sprint, and it's normally used for initial team integration, concept alignment, material preparation, and defining the main goals to be achieved during projects.

5.2 The Sprint 1

The key contributions during the first sprint of each semester included the adaptation study of the web system created in the FAL3-TR project in the second semester of 2023 using Raspberry PI 3 board offline to allow fluency level collection, recording, processing, and its indication without Internet connections. Other highlights included studying the application of LLM techniques in the FAL-BD project to facilitate database queries, readjusting the data model brought from the previous project accommodating large sets of data records, and developing test plans to ensure defined objectives and procedures in the software testing process. For the FAL3- TR project development team, solar technology became crucial in the pursuit of clean and sustainable energy solutions. By addressing power scarcity for electronics, the Scrum Teams integrated solar panels, for enhancing project accessibility across Brazil. This initiative not only enabled alternative energy use but also led to the development of an App Inventor-based application to control servomotors, providing an affordable and autonomous way to maximize solar energy capture. To streamline the project, the team prioritized servomotor control and solar panel voltage monitoring, establishing foundations for future expansions, including solar panel positioning and battery integration for the FRDM-KL25Z board.

5.3 The Sprint 2

As shown in Fig. 4, a software named *alfa3app* was developed to be installed and used in separated hardware, as a complementary module of the (*alfa3-bd*) FAL-BD project. This software allowed audio recording for fluency analysis to occur offline. Only teacher user functionalities were available to use this software. The system was developed using the Python programming language, the Django framework, the SQLITE database, NGINX, and the Google Drive API.

Figure 5 illustrates the *alfa3app* software running on a Raspberry Pi 3 Model B Board with Raspberry Pi OS (formerly Raspbian) and its Human-Machine Interface (HMI) featuring a virtual touchscreen keyboard. The hardware specifications include: Broadcom BCM2837B0 Cortex-A53 64-bit processor (1.4 GHz), 1 GB LPDDR2 RAM, dual-band wireless LAN (2.4 GHz/5 GHz), Bluetooth 4.2/BLE, full-size HDMI input, four USB 2.0 ports, 16 GB microSD storage, 5 V/2.5 A power supply, Raspbian OS (Debian-based), and a 10.1-inch dualtouch capacitive LCD touchscreen.

For the search and application of LLM, the Hugging Face was used, as one of the main public repositories for trained machine learning models with prominent areas such as Computer Vision, Text-to-Speech, LLM, Sentence Transformers, among others. Figure 6 shows an example of implemented query using LLM.

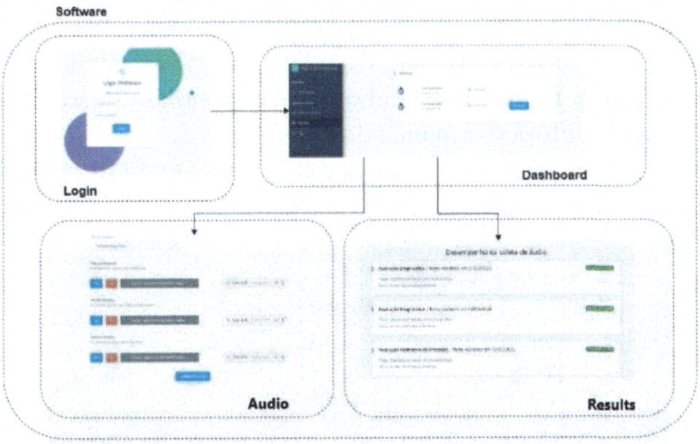

Fig. 4. The Software alfa3app developed to be used in Hardware

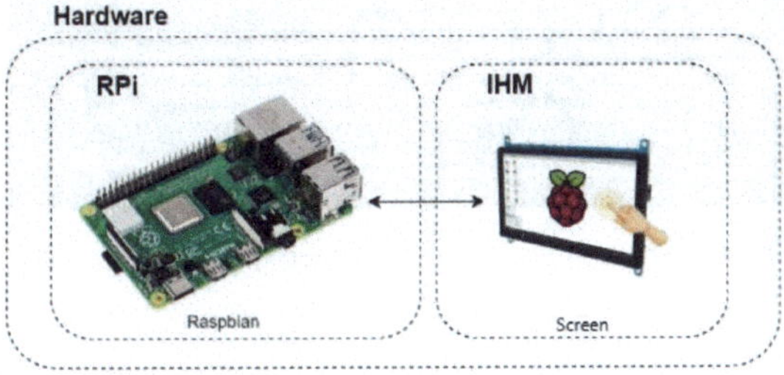

Fig. 5. The *alfa3app* Software with the Raspberry Pi 3 Model B Board and HMI Software with a Virtual Touch Screen.

Hugging Face also provided a diversified repository of datasets, libraries, test mock ups, and tools, offering a complete environment for working with Artificial Intelligence, promoting collaboration between professionals and students in the field.

On this research, it enabled the selection and use of the ***llama3 model*** through the ***Inference Web Service Groq***. Additionally, it aimed to reduce the number of tokens used per model inference. Following, a DuckDB database with denormalized data was also created and hosted on Hugging Face. And this allowed the creation of a chat interface to transform natural language questions into queries performed on the database.

This application was developed to assist TEACHERS and/or SYSTEM MANAGERS in searching for information. Then, even without having Structured Query Language (SQL) or database knowledge, users of these systems could perform specialized data searches. This way, the model started to receive user questions, to transform it into SQL, and to execute queries directly on the database, presenting final results in the HMI.

5.4 The Sprint 3

The FAL-BD project proposal on the 1st Semester of 2024 was to handle large datasets for future academic researches, implementations, and tests. On the first approach, a data model was configured in the DuckDB on a local development machine, as an in-memory relational database, open-source, and multi-platform, allowing that large datasets could be stored, handled, and queried in RAM. This database was designed and developed to be fast, scalable, and easy to use. It was chosen as a popular choice for the developers requiring a flexible and high-performance database [17].

The following requirements were considered for the DuckDB tests:

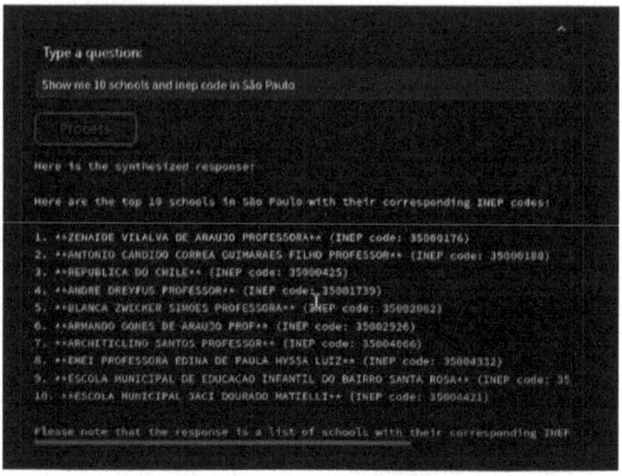

Fig. 6. An example of implemented query using LLM.

- Two classes of 20 students per school unit, **totaling 7,209,200 students**;
- Five types of staff roles (Principal, Coordinator, Secretary, General Services Assistant, and Teacher) with two teachers per school unity and one staff member for other roles (six staff members per school unit, **totaling 1,081,380 staff members**, including 360,460 teachers);
- Two collections per student (two classes x 20 students x 180,230 school units), **totaling 576,736,000 collections, evaluations, and activities**; and
- **5,297 municipalities**, all with school unities.

On the second approach, the DuckDB and the dataset previously mentioned was tested in a more robust environment through the ITA infrastructure, using the NVIDIA A100 GPU. These tests aimed to assess how performant the DuckDB was in handling large datasets across different types of environments and infrastructures.

6 The Main Results

Among the various contributions achieved on these two academic projects (FAL3-TR and FAL-BD) from August 2023 to June 2024, stands out: (1) the presentation of a method using LLM as a facilitator between users and machine processes, such as querying databases; (2) enabling a web systems application on a Personal Digital Assistant (PDA) to be used entirely offline, fulfilling one of the projects main premises, allowing captures of Student's audio, processing, and classification of literacy fluency score, similar to the process carried out on the web software previously developed in the FAL3-BD project in the 1st Semester of 2023; (3) the pursuit of better performance when handling large datasets, DuckDB was identified as a powerful tool, enabling the creation of a robust and accessible in-memory database in the ITA environment; (4) modeling according to all normal forms and using large datasets ensured database consistency and quality for conducted tests; (5) by using Kubernetes in the project as part of the infrastructure, it was possible to develop a scalable and fault-tolerant microservices architecture with selfrecovery features, load balancing, and continuous updates; (6) thanks to the Kubernetes technology, a failure rate of only 0.1% was achieved during production updates, as evidenced by load tests; and, finally, (7) it was observed that immediately after the termination of all pods with old versions, the failure percentage in load tests decreased to 0%.

Software testing played also an essential role in verifying and validating the developed code. By subjecting the software to a series of simulated scenarios and use cases, the tests ensured that the system functioned as expected and met the established requirements. In addition to identifying faults and inconsistencies, the tests allowed for the early detection of errors, providing an opportunity to correct them before reaching the production stage. This resulted in more reliable, stable, and high-quality software, contributing to user satisfaction and reducing costs associated with issues that could arise after deployment.

7 Conclusion

7.1 Specific Conclusions

The FAL-BD project was developed for a Big Data type of system and the FAL3-TR project main objective was the development for a PDA, so that the technologies used were presented in hardware components separated by subsystems. This paper reported a method applied for developing and testing both academic projects using the Interdisciplinary Problem Based Learning (IPBL).

7.2 General Conclusions

The academic methodology used in the development of both projects effectively minimized the time required between initial conception and the delivery of value. This approach enabled the presentation of viable solutions within a relatively short timeframe, aligning with the expectations defined in the user stories. These results validate the practicality of incorporating agile methodologies into academic projects. Furthermore, the Proof of Concept (PoC) demonstrated that agile methods can yield tangible results in academic environments within limited timeframes, confirming their feasibility and efficiency.

7.3 Recommendations

To enhance project sustainability and maintain functionality throughout their lifecycle, it is recommended to implement regression testing for both the web application and the database. In addition, it is recommended to segregate the AI model into a distinct API, separate from the web application. This approach would not only facilitate the reuse of the AI model for reading fluency analysis but also enable its application in other contexts, broadening its utility and impact.

7.4 Future Work

For future work, it is suggested the use of Meta Quest 3 or similar devices, to bring immersion experiences for students and teachers, when performing literacy fluency assessment activities. It is also suggested the creation of virtual environments using Virtual Reality (VR), Augmented Reality (AR), and/or Extended Reality (XR) glasses to provide students and teachers with more interactive and play full devices to carry out students fluency analysis and assessments on literacy [18].

Acknowledgments. The authors would like to thank: the Aeronautics Institute of Technology (ITA), for providing the infrastructure for the development of both academic projects; the Casimiro Montenegro Filho Foundation (FCMF) for its support; the Brazilian Department of Education (MEC), for providing the necessary data and statistics for this case study; and also to the Software Engineering Research Group (GPES) along with the courses and professors for their guidance, during the development of the projects.

References

1. Braßler, M.: Interdisciplinary problem-based learning—a student- centered pedagogy to teach social sustainable development in higher education. In: Teaching Education for Sustainable Development at University Level, pp. 245–257. Springer (2016)
2. Mota, M.S.M., et al.: Applying scrum in interdisciplinary case study projects for literacy in fluency analysis. In: Latifi, S., (ed.) ITNG 2023 20th International Conference on Information Technology-New Generations, pp. 421–426. Springer International Publishing (2023)
3. Kubernetes, Deployments (2024). https://kubernetes.io/docs/concepts/workloads/controllers/deployment. Accessed 1 Nov 2024
4. Kubernetes, Service (2024). https://kubernetes.io/docs/concepts/services-networking/service/#loadbalancer. Accessed 1 Nov 2024
5. Kubernetes, Ingress (2024). https://kubernetes.io/docs/concepts/services-networking/ingress. Accessed 1 Nov 2024
6. Kubernetes, StatefulSets (2024). https://kubernetes.io/docs/concepts/workloads/controllers/statefulset. Accessed 1 Nov 2024
7. Kubernetes, Performing a Rolling Update (2024). https://kubernetes.io/docs/tutorials/kubernetes-basics/update/update-intro/. Accessed 1 Nov 2024
8. Kubernetes, Horizontal Pod Autoscaling (2024). https://kubernetes.io/docs/tasks/run-application/horizontal-pod-autoscale. Accessed 1 Nov 2024
9. F, S. D. E. B. S. C.-G. D. E.: Manual operacional do sistema de orientac¸a˜o pedago´gica e monitoramento (programa mais alfabetizac¸a˜o) (2018)
10. R., N. A. L. G. P. P. J. C. B. R. D. P. R.: Avanc¸os no conhecimento do processamento da flue^ncia em leitura: da palavra ao texto (2009). acessado em 24 July 2024
11. R., B. S. J. M. M.: Conscie^ncia metalingu´ıstica e alfabetizac¸a˜o: um estudo com crianc¸as da primeira se´rie do ensino fundamental (2003). acessado em 24 July 2024

12. M, N. L. P. Do bag of words ao transformer: A evoluc¸a˜o do nlp. The Developer Conference (2019). https://s3-sa-east-1.amazonaws.com/
13. Amaratunga, T.: Understanding Large Language Models – Learning Their Underlying Concepts and Technologies. Apress Media (2023)
14. Tunstall, L., Von Werra, L., Wolf, T.: Natural Language with Transformers – Building Language Applications with Hugging Face. O'Reilly Media (2022)
15. Rubin, S.K.: Essencial Scrum A Pratical Guide to The Most Popular Agile Process. Addison-Wesley Professional (2012)
16. Schwaber, K., Sutherland, J.: The Scrum Guide The Definitive Guide to Scrum: The Rules of the Game (2020). https://scrumguides.org/docs/scrumguide/ v2020/2020-Scrum-Guide-US.pdf
17. Duckdb. https://duckdb.org/why duckdb
18. Getting started with meta quest 3. https://www.meta.com/help/quest/articles/getting-started/getting-started-with-quest-3/

Spatio-Temporal Analysis of a Crime Dataset Using Principal Component Analysis and Exploratory Factor Analysis

Natarajan Meghanathan(✉)

Department of Electrical and Computer Engineering and Computer Science, Jackson State University, Jackson, MS, USA
natarajan.meghanathan@jsums.edu

Abstract. We demonstrate the use of principal component analysis (PCA) and exploratory factor analysis (EFA) to conduct spatio-temporal analysis of a crime dataset. We use a 30000-records crime dataset for the city of Los Angeles for the year 2020 as the dataset for our analysis. We extract a matrix (*T-L*) of the time (hourly index: 0, 1, 2,..., 23) vs. location type (a total of 28 location types) whose entries store the values for the number of crimes reported at a particular location type during the time spanning an hour of the day, across the entire dataset. We conduct PCA on the *T-L* matrix and its transpose to respectively rank and classify hours of the day and the location types to one of the three categories: hotspots, tepidspots and coldspots. We use the weighted average scores of the principal components and the properties of the principal components to facilitate such quantification, ranking and classification. By conducting EFA on the *T-L* matrix, we observe clusters of location types that could have a common latent factor/theme (like location types in the vicinity of a residence; location types involving dwelling units/inside a residence; commercial location types such as stores/businesses, etc.) so that we could categorize clusters of location types as "hotspots" or "not hotspots".

Keywords: Principal Component Analysis · Exploratory Factor Analysis · Crime Dataset · Hotspots · Spatio-Temporal Analysis · Tepidspots · Coldspots · Location Types · Cluster Analysis · Matrix Operations

1 Introduction

Crime datasets have a multitude of features such as the date and time of occurrence, crime code, location type (street, sidewalk, bank, etc.), location address, area code, victim and offender details (such as gender, race, etc.), latitude and longitude of the location, current status of the investigation, etc. Of these, the features that are of interest for spatio-temporal analysis are the time of occurrence, crime code, location type, location address, longitude and latitude. The focus of research in the literature on spatio-temporal analysis of crime datasets has been typically so far to identify locations or areas (referred to as *hotspots*) that are vulnerable for crime within a specific time. Rather, in this paper, we

S. Latifi (Ed.): ITNG 2025, AISC 1463, pp. 204–213, 2025.
https://doi.org/10.1007/978-3-031-89063-5_18

seek to identify one or more location types as well as clusters of related location types (that have some commonality to be influenced by the same factor) that could be potential hotspots of crime in a 24-h time period. We chose the 2020 crime dataset of 966,951 records for the city of Los Angeles (LA) as the source dataset for our analysis. We identify principal component analysis (PCA) and a related technique called exploratory factor analysis (EFA) to be the data analytics tools that would fit our purpose.

Dimensionality reduction techniques like principal component analysis (PCA [1]) are used to reduce the dimensions of a multi-feature dataset to a handful of dimensions in the form of principal components (PCs). The retained PCs are further processed through a machine learning algorithm, either for classification or clustering of the records in the original dataset. The PCs retained are expected to capture a significant fraction (at least 80%) of the variation among the feature values in the original dataset. The row indexes of the underlying matrix could be ranked [2] using the entries for these row indexes in the retained PCs, with the Eigenvalues of the PCs serving as weights. The number of PCs retained could also be used [3] as the value for the number of *latent* factors under one or more of which the features in the original dataset could be grouped into: this process is called Exploratory Factor Analysis (EFA [4, 5]). Very few works in the literature (e.g., [6]) have reported the use of PCA for spatio-temporal analysis of crime datasets. The research presented in this paper could serve as a template for using PCA/EFA to identify spatio-temporal hotspots in crime datasets.

Our proposed methodology (for more details, see Sect. 2) at a high-level is as follows: We extract a time-location type matrix (referred to as *T-L* matrix, for the rest of the paper) whose row indexes are the hours (0, 1, 2,..., 23) and column indexes are the location types. Any cell (i, j) in the *T-L* matrix would represent the number of crimes that are reported for a particular location type j in the time represented by the row index i, across the entire dataset. We conduct PCA on the *T-L* matrix to identify the number of principal components that need to be retained to capture at least 80% of the variations among the features in the dataset. This number would be then used as the number of *latent factors* to explore in the *T-L* matrix when the latter is subjected to varimax rotation [7]-based exploratory factor analysis (EFA). EFA would provide a mapping of the location types to factors that maximally influence their values (in this case, the number of crimes that occur in these location types in a 24-h time period) and thereby identify clusters of related location types (that have some commonality to be influenced by the same factor). Finally, we conduct a second PCA: now on the transpose of the *T-L* matrix (i.e., an *L-T* matrix) to obtain a ranking of the location types (based on a weighted average of the entries for these location types in the retained PCs, with the Eigenvalues of the PCs serving as weights) with respect to the number of crimes that have been reported in a 24-h time period. When we superimpose the ranking of the location types identified through the PCA of the *L-T* matrix with that of the clusters of related location types identified through the PCA of the *T-L* matrix, we could identify one or more clusters of location types that could be potentially spatio-temporal hotspots in a 24-h time period.

The rest of the paper is as follows: Sect. 2 presents our proposed methodology in detail, as a sequence of three steps: PCA of the *T-L* matrix, Varimax rotation-based EFA of the *T-L* matrix and PCA of the *L-T* matrix. Along side, Sect. 2 also presents and discusses the results obtained when we apply the proposed methodology on the LA crime dataset. Section 3 reviews related work in the literature. Section 4 concludes the paper and outlines plans for future work.

2 Proposed Methodology and Results

This section presents our proposed PCA/EFA-based methodology to conduct spatio-temporal analysis of crime datasets of any city. We chose the year 2020 crime dataset [8] of 966,951 records for the city of Los Angeles (LA) for this purpose. There is no possibility of bias in the dataset, as we consider all the records for analysis, and it is not a classification or clustering problem. We extract a matrix (the *T-L* matrix: see Fig. 1 for a snapshot) of the time (hourly index: 0, 1, 2,…, 23) vs. location type (a total of 28 location types; see Fig. 1 for their codes and description) whose cells store the values for the number of crimes reported at a particular location type (say, street, sidewalk, bank, etc.). We subject the *T-L* matrix through a sequence of three steps, as explained below.

PCA of the T-L Matrix: Principal Component Analysis of the *T-L* matrix of Fig. 1 is conducted using its standardized version as the basis. PCA is conducted through these steps: (1) We first determine the Covariance matrix of the standardized *T-L* matrix. (2) We determine the Eigenvalues and Eigenvectors of the Covariance matrix and choose the Eigenvectors whose corresponding Eigenvalues are greater than 1.0 (the Kaiser-Guttman rule [9]). We observe a total of three PCs (high-variance PCs) need to be retained with Eigenvalues of 16.60 (for PC1), 3.89 (for PC2) and 1.92 (for PC3) that together capture 80% ((16.60 + 3.89 + 1.92)/28–0.80, where 28 is the total number of location types/columns in the *T-L* matrix) of the variations in the feature values. (3) We multiply the standardized *T-L* matrix with the three retained Eigenvectors to obtain the PCs of the *T-L* matrix. Figure 2 presents the retained PCs whose entries correspond to the row indexes (hourly indexes) in the *T-L* matrix.

We also compute a weighted average (referred to as the *Temporal Crime Index Score*) of the entries in the three PCs for each hourly index, with the Eigenvalues of the PCs used as the weights. Figure 2 also presents a plot of the hourly index (0, 1, 2,…, 23) vs. the Temporal Crime Index Scores, which exhibits the pattern of a sinusoidal curve. The hourly indexes with larger positive Temporal Crime Index values could be classified as temporal hotspots (referring to the time period during which a larger number of crimes could be expected). From Fig. 2, we observe hourly indexes of 11, 15, 16, 17, 18 to be potential temporal hotspots, with the hourly index of 11 incurring the largest Temporal Crime Index value, and could be even called an outlier, as it lies reasonably away from the sinusoidal curve. The Temporal Crime Index values are negative for hourly indexes ranging from 22, 23, 0,…, 9 (we propose to classify these hourly indexes as *temporal coldspots*), and positive but not larger for hourly indexes 10, 11,…, 21 (we propose to classify these hourly indexes as *temporal tepidspots*). While the distributions of the temporal hotspots and coldspots are mostly continuous, the distribution of the temporal tepidspots is relatively more fragmented. The tepidspot fragments occupy the transition phase from coldspots to hotspots during the morning and early afternoon hours as well as the transition phase from the hotspots to the coldspots during the late evening and night hours.

EFA of the T-L Matrix: The number of high-variance PCs retained in the PCA step could be used [3, 7] as the choice for the number of latent factors that could influence the values (number of crimes) noticed for the location types in each of the hourly indexes. We use the Varimax [7]-rotation technique to extract a mapping of the location types to the factors. To conduct Varimax-rotation, we first plot the entries (representing the location types and are referred to as *initial factor loadings*) in the three Eigenvectors (corresponding to the retained PCs) in a three-dimensional coordinate system.

hour index	Location Types (Codes) 101	501	203	108	210	502	102	405	103	602	122	301	212	109	402	504	503	104	221	406	121	710	404	707	123	116	801	119
0	233	349	29	59	18	236	63	0	9	14	34	7	6	4	3	10	8	22	3	3	7	13	1	16	8	2	3	1
1	194	185	14	45	12	141	57	1	13	1	33	9	6	3	2	9	6	19	4	4	8	4	0	13	12	5	0	1
2	182	131	24	47	14	120	55	1	4	0	28	5	6	2	1	6	9	17	3	4	5	6	1	18	8	3	0	3
3	146	117	21	40	7	91	34	0	1	1	28	4	4	2	1	2	5	14	2	1	3	7	0	7	13	4	0	1
4	111	89	22	29	8	85	20	1	2	0	19	2	0	3	1	8	4	13	1	4	8	1	0	7	4	0	1	4
5	104	91	13	37	6	79	27	0	5	2	23	2	3	1	1	2	6	9	2	1	4	1	0	14	3	0	2	1
6	99	198	13	26	8	133	23	0	5	2	22	4	1	1	1	6	4	9	6	2	4	5	0	11	7	2	1	0
7	129	181	21	49	7	139	47	2	7	2	32	6	2	8	4	3	1	10	8	2	5	5	3	16	9	6	4	5
8	145	299	30	69	5	191	81	1	10	8	44	6	1	11	9	12	9	15	3	3	5	11	5	15	12	7	1	3
9	190	261	39	89	9	176	59	5	12	10	29	7	6	11	8	7	8	19	7	9	7	13	5	9	3	6	3	5
10	197	317	46	78	16	218	79	8	7	14	32	9	6	9	6	13	12	20	2	3	16	16	10	12	9	4	7	10
11	179	315	58	90	12	217	61	8	8	10	44	5	8	6	14	14	12	10	6	7	11	14	9	22	10	9	5	10
12	277	659	95	109	18	402	88	7	8	23	50	14	12	22	22	17	13	22	17	12	12	32	15	26	18	4	5	26
13	201	282	66	91	14	180	63	7	4	8	52	14	14	15	14	7	9	16	10	8	11	12	22	22	7	5	5	12
14	234	284	72	103	15	230	79	8	13	7	43	4	7	14	14	10	10	18	9	8	13	7	13	20	13	4	12	13
15	262	331	53	89	17	227	98	12	15	13	58	5	8	16	24	10	5	25	11	13	11	15	24	29	12	3	11	14
16	292	292	81	119	15	176	86	9	17	7	47	5	10	18	20	15	6	17	10	8	9	20	17	27	10	7	12	9
17	348	335	73	106	18	202	90	17	11	3	67	8	13	20	15	14	3	21	7	11	8	14	17	40	23	8	7	8
18	357	279	71	106	17	236	96	7	17	4	63	10	6	15	16	18	8	30	6	9	14	15	16	33	24	7	9	9
19	338	339	49	130	20	204	88	10	19	4	75	8	7	13	4	15	3	27	6	13	9	13	13	49	18	5	5	8
20	374	326	40	119	22	232	97	10	14	2	75	7	8	10	9	9	5	28	5	4	14	5	10	54	13	8	2	4
21	321	312	43	106	14	221	80	4	14	2	73	9	9	12	5	15	10	21	2	3	5	12	7	32	21	9	2	10
22	328	247	26	100	15	225	81	1	8	5	68	7	9	6	4	18	7	31	1	1	7	15	3	39	20	7	2	1
23	295	221	32	63	14	199	73	0	6	0	62	7	5	6	3	6	11	24	0	2	8	9	0	24	10	8	1	4

101	STREET
102	SIDEWALK
103	ALLEY
104	DRIVEWAY
108	PARKING LOT
109	PARK/PLAYGROUND
116	OTHER/OUTSIDE
119	PORCH, RESIDENTIAL
121	YARD (RESIDENTIAL/ BUSINESS)
122	VEHICLE, TRUCK
123	PARKING in a BUILDING
203	OTHER BUSINESS
210	RESTAURANT/FAST FOOD
212	TRANSPORTATION FACILITY
221	PUBLIC STORAGE
301	GAS STATION
402	MARKET
404	DEPARTMENT STORE
405	CLOTHING STORE
406	OTHER STORE
501	SINGLE FAMILY DWELLING
502	MULTI-UNIT FAMILY DWELLING
503	HOTEL
504	OTHER RESIDENCE
602	BANK
707	GARAGE/CARPORT
710	OTHER LOCATION TYPES
801	MTA BUS

Fig. 1. T-L Matrix (hourly index vs. location type) for the LA Crime Dataset and the Location Types

hourly	Principal Components			Temporal
index	PC1	PC2	PC3	Crime Index
0	-1.318	0.305	-1.954	-24.45
1	-2.919	-1.04	-0.437	-53.354
2	-3.861	-0.29	-0.432	-66.064
3	-5.56	-0.04	-0.02	-92.497
4	-6.13	0.936	1.0939	-96.019
5	-6.548	0.878	1.0122	-103.34
6	-5.718	0.98	0.4284	-90.275
7	-3.805	0.547	1.4165	-58.316
8	-1.719	0.19	-1.132	-29.972
9	-0.842	1.486	0.0823	-8.0323
10	0.932	1.622	-1.667	18.577
11	0.948	1.493	-0.586	20.424
12	7.679	4.396	-3.087	138.64
13	2.103	2.209	0.137	43.767
14	2.26	1.515	1.4794	46.255
15	4.259	1.538	2.592	81.66
16	3.713	0.889	2.186	69.287
17	4.603	-1.51	2.0909	74.561
18	4.62	-1.61	0.3	71.002
19	3.693	-2.7	1.4617	53.61
20	2.739	-3.6	0.5288	32.456
21	1.695	-2.57	-1.547	15.185
22	0.7	-3.04	-2.001	-6.0718
23	-1.525	-2.08	-1.896	-37.028

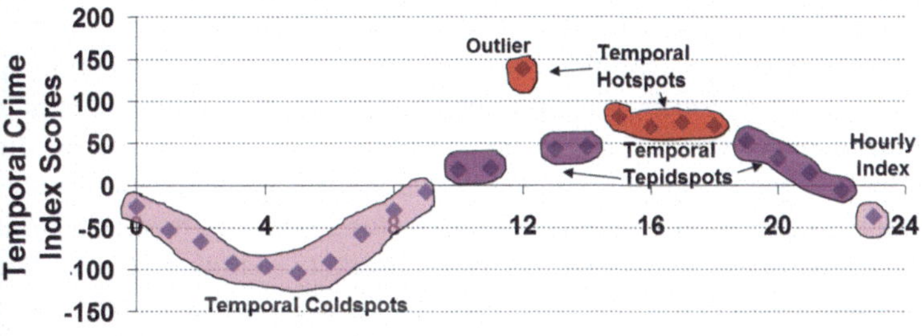

Fig. 2. High Variance Principal Components of the T-L Matrix and the Sinusoidal Plot of the Hourly Index vs. Temporal Crime Index Scores

The initial communality score for a location type in the Eigenvector coordinate system is the sum of the squares of the entries for the location type in the three Eigenvectors. Under Varimax rotation, we conduct repeated orthogonal synchronous rotations of the three Eigenvector coordinate-axes until the sum of the squares (the *final communality score*) of the coordinates for each location type is at least 0.80, where 0.80 is the value for the fraction of variance retained in the three PCs. The three-dimensional coordinate system at the end of Varimax rotation is referred to as the factor-coordinate system (factor 1, factor 2, factor 3); each location type is now represented on the basis of the

Location Type Description	Location Type Code	Factor 1	Factor 2	Factor 3	Communality Score (Final)
STREET	101	0.936	0.231	0.136	0.9480
SIDEWALK	102	0.776	0.429	0.307	0.8805
ALLEY	103	0.85	0.459	-0.094	0.9420
DRIVEWAY	104	0.881	0.045	0.199	0.8178
PARKING LOT	108	0.738	0.533	0.246	0.8892
PARK/PLAYGROUND	109	0.413	0.778	0.319	0.8776
OTHER/OUTSIDE	116	0.908	0.016	0.131	0.8419
PORCH, RESIDENTIAL	119	0.098	0.702	0.595	0.8564
YARD (RESIDENTIAL/ BUSINESS)	121	0.327	0.809	0.353	0.8860
VEHICLE, TRUCK	122	0.927	0.192	0.086	0.9036
PARKING in a BUILDING	123	0.787	0.131	0.158	0.6615
OTHER BUSINESS	203	0.314	0.777	0.435	0.8916
RESTAURANT/FAST FOOD	210	0.82	0.347	0.249	0.8548
TRANSPORTATION FACILITY	212	0.685	0.486	0.387	0.8552
PUBLIC STORAGE	221	-0.071	0.831	0.328	0.8032
GAS STATION	301	0.38	0.174	0.853	0.9023
MARKET	402	0.197	0.851	0.332	0.8732
DEPARTMENT STORE	404	0.334	0.87	0.167	0.8963
CLOTHING STORE	405	0.449	0.802	-0.011	0.8449
OTHER STORE	406	0.253	0.859	0.146	0.8232
SINGLE FAMILY DWELLING	501	0.359	0.459	0.731	0.8739
MULTI-UNIT FAMILY DWELLING	502	0.46	0.348	0.757	0.9058
HOTEL	503	-0.013	-0.082	0.898	0.8133
OTHER RESIDENCE	504	0.54	0.275	0.739	0.9133
BANK	602	-0.119	0.451	0.782	0.8291
GARAGE/CARPORT	707	0.894	0.241	-0.071	0.8624
OTHER LOCATION TYPES	710	0.284	0.452	0.741	0.8340
MTA BUS	801	0.166	0.885	0.039	0.8123

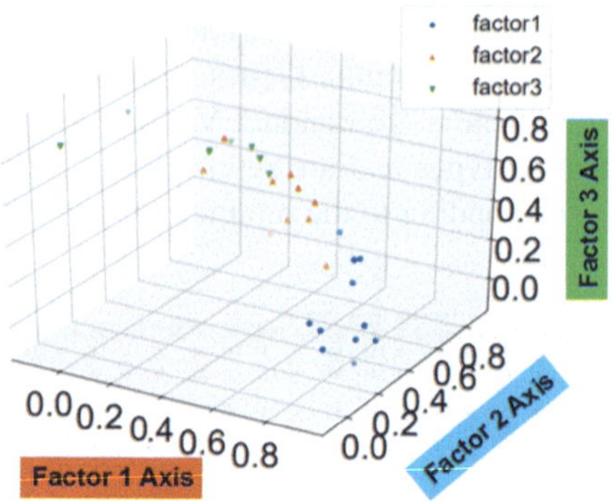

Fig. 3. Final Factor Loadings for the Location Types; their Clustering and Visualization

three-dimensional factor coordinates, referred to as the *final factor loadings*. A location type is mapped to a factor whose absolute value for the final factor loadings is the largest; the mapped factor is referred to as the *dominating factor* for the location type.

Figure 3 presents the results of EFA on the three-dimensional Eigenvector vector coordinate system of the *T-L* matrix. We color code the location types that are clustered to each of the three factors. Overall, we observe that similar/related location types get mapped to a particular factor. We observe location types grouped under factor 1 represent locations outside of a residence, but closer to it (like street, sidewalk, alley, driveway, garage/carport, vehicle/truck parked, parking in a building, etc.). Restaurants/Fast food and Transportation facilities (like airports) are the major location types that are not connected to a residence, but grouped under factor 1. On the other hand, location types grouped under factor 3 represent locations inside of a residence (like single-family, multi-family dwelling units and Hotel) as well as public locations such as Bank and Gas station. Location types grouped under factor 2 typically represent stores and businesses (like Department store, Clothing store, Market, and other stores); Porch and Yard are the major residence-related location types, and Park/Playground and MTA Bus are the major non-residence related location types that get mapped to factor 2. To summarize, location types in the vicinity of a residence (but not inside a residence) along with restaurants and transportation facilities get mapped to factor 1, location types that are more of stores/businesses (other than Gas station) get mapped to factor 2, and location types that are inside of a residence (including Hotel) as well as public locations such as Bank and Gas station get mapped to factor 3.

PCA of the L-T Matrix: We now conduct PCA on the transpose of the *T-L* matrix (referred to as the *L-T* matrix) that would have the location types as row indexes and the hourly index values of 0, 1, 2,..., 23 as the column indexes. We observe that just one Eigenvector (with an Eigenvalue of 23.1902; the rest of the Eigenvalues are less than 1.0) need to be retained and it captures 23.1902/24 ~ 96.6% of the variation in the number of crimes reported for the 24 hourly indexes. We then multiply the standardized

L-T matrix with the retained Eigenvector to obtain the lonely principal component (PC). The entries in this PC would correspond to the location types. Since there is only one PC retained, there is no need to take the weighted average. The entries (referred to as the Spatial Crime Index) in the PC itself would be sufficient to rank the location types with respect to the crimes reported across the 24 hourly indexes (i.e., in a day).

Location Type Description	Location Type Code	Spatial Crime Index	
SINGLE FAMILY DWELLING	501	15.30	Spatial Hotspots
STREET	101	13.32	
MULTI-UNIT FAMILY DWELLING	502	10.38	
PARKING LOT	108	2.69	Spatial Tepidspots
SIDEWALK	102	1.93	
VEHICLE, TRUCK	122	0.47	
OTHER BUSINESS	203	0.16	
GARAGE/CARPORT	707	-1.15	Spatial Coldspots
DRIVEWAY	104	-1.36	
RESTAURANT/FAST FOOD	210	-1.78	
PARKING in a BUILDING	123	-1.89	
OTHER LOCATION TYPES	710	-2.01	
OTHER RESIDENCE	504	-2.03	
ALLEY	103	-2.07	
YARD (RESIDENTIAL/ BUSINESS)	121	-2.11	
PARK/PLAYGROUND	109	-2.11	
HOTEL	503	-2.19	
MARKET	402	-2.20	
DEPARTMENT STORE	404	-2.24	
GAS STATION	301	-2.24	
TRANSPORTATION FACILITY	212	-2.28	
PORCH, RESIDENTIAL	119	-2.29	
OTHER STORE	406	-2.34	
PUBLIC STORAGE	221	-2.35	
BANK	602	-2.35	
OTHER/OUTSIDE	116	-2.37	
CLOTHING STORE	405	-2.42	
MTA BUS	801	-2.45	

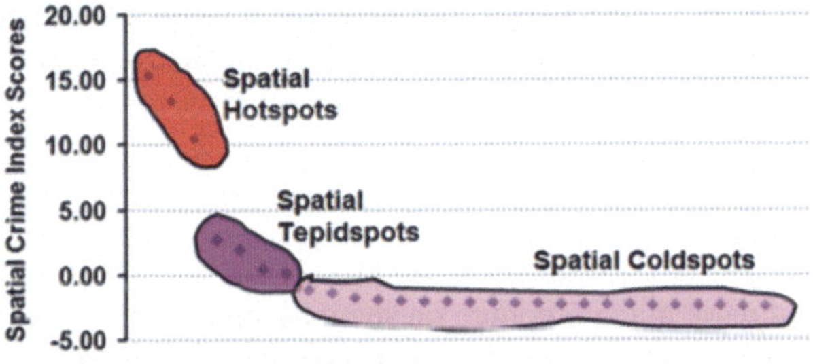

Fig. 4. Clustering of the Location Types based on the Distribution of the Spatial Crime Index Scores and the Mapping the EFA Factors to these Clusters

Figure 4 presents the Spatial Crime Index values (in the sorted order, from maximum to minimum) of the 28 location types. We observe the following 7 of the 28 location types to incur positive values for the Spatial Crime Index; the rest of the 21 location types incur negative Spatial Crime Index values. Since the Spatial Crime Index values are the values incurred for the location types in a PC, the sum (44.245) of the positive PC entries (incurred for 7 of the 28 location types) equals the sum of the absolute values of the negative PC entries (incurred for the remaining 21 location types), indicating the total number of crimes reported across these 7 location types is comparable to the total number of crimes reported across the rest of the location types.

Moreover, the Spatial Crime Index values (of 15.30, 13.32, 10.38) for the top 3 ranked location types (Single family dwelling units, Street and Multi-unit family dwelling) account for nearly 90% of the sum (44.245) of the positive Spatial Crime Index/PC values. We hence befittingly propose that these three location types be classified as *spatial hotspots* and the other four of the 7 location types incurring positive Spatial Crime Index/PC values be classified *spatial tepidspots* (where tepid ~ lukewarm) of the city. The 21 location types that incur negative Spatial Crime Index/PC values could be classified as *spatial coldspots* of the city. Such a classification at the spatial level with respect to the location types combined with the identification of temporal hotspots, tepidspots and coldspots constitutes a comprehensive spatio-temporal analysis of the crime data.

Also, an interesting inference one could make when the above results of PCA of the T-L matrix and L-T matrix observed in conjunction with the clustering results based on the EFA of the T-L matrix is that the stores/businesses (such as market, department stores, clothing stores, and any other stores) as well as the park/playground and MTA buses (i.e., those location types that are grouped under factor 2) in the city of Los Angeles are relatively less crime prone compared to the residences and residence-related location types. In Fig. 4, we observe the color code for factor 2 to be primarily in the bottom half of the ranking compared to the color codes for factors 1 and 3. Another prominent observation (based on the distribution of the color codes across the table) from Fig. 4 is that the location types grouped under factor 3 appear be spread throughout the table, whereas the location types grouped under factor 1 appear to occupy primarily be ranked in the top of the table and the location types grouped under factor 2 appear to occupy primarily the bottom half of the table. This coincides with our earlier observation that location types (primarily, stores/business) grouped under factor 2 are relatively less crime prone compared to location types (primarily, residence-related) grouped under factors 1 and 3.

3 Related Work

Most of the related work in the literature on spatio-temporal analysis of crime datasets focus on identifying hotspots with respect to specific locations (like street names, areas, counties/districts, etc.) and time periods (like period of the day: morning/evening, etc.; month of the year, etc.). The basis for hotspot identification had also been typically the raw values for the number of crimes, aggregated per the location or time period of interest. On the other hand, we take a different approach (use of PCA on the time-location type matrix and its transpose) for identifying the location types as well as time periods that could be classified as hotspots, tepidspots or coldspots. We also show that by conducting EFA on the time-location type matrix, we could identify clusters of location types that could have a common theme/factor and in conjunction with the PCA results, we could henceforth also categorize clusters of location types as hotspots or not hotspots. For the rest of this section, we present some of the related work in the literature on spatio-temporal analysis of crime datasets.

In [10], the authors evaluated the relative change in the density of the crime clusters over a period of time for six major US cities, including LA. With the locations being primarily street segments, the authors of [10] assessed the relative change in the crime density of the street segments with respect to the average crime count for the street segments in these cities for each assessment period. The authors also assessed the stability of the density of the crime clusters and the outlier street segments over several assessment time periods. In a study on the crime datasets for the city of Seattle from 1989 to 2004, Weisburd et al. [11] observed both clustering of high-crime street segments as well as street-to-street variability. On the other hand, for the crime datasets of the city of Albany from 2000 to 2013, Wheeler et al. [12] observed dense clustering of high-crime street segments and a clear spatial separation of the high-crime and low-crime street segments. In a study on the Boston crime datasets from 2011 to 2013, O'Brien and Winship [13] observed a relative year-to-year stability for the number of crimes reported at the street-level and census tract-level compared to address-level, indicating the possibility of diffusion of crimes within a neighborhood. O'Brien et al. [14] observed (for the city of Boston) that few neighborhoods reported most of their crimes from high-crime density locations, while a majority of the neighborhoods reported their crimes from moderate-crime density locations: the high-crime density and moderate-crime density locations could correspond to the notions of hotspots and tepidspots, proposed in our research.

In [6], the authors used PCA to reduce the dimensions of the San Francisco crime dataset over a 12-year period to predict the category of crimes (a total of 39 crime classes) and subjected the principal components through classification algorithms such as kNN, decision trees, Adaboost and Random forest. The accuracy of the predictions with these classifiers were observed to be only around 30%. The authors of [6] claim that with proper oversampling and undersampling (to offset the class imbalance) of the underlying crime dataset to build the training dataset, the accuracy with random forest could be improved to respectively 68% and 99%. In [15], the authors used deep neural networks to predict crime categories in the Chicago crime dataset; they showed an undersampling-oversampling tradeoff with respect to the performance metrics: while using undersampling techniques to build the training dataset could return a higher precision and recall in the range of 60–70%, oversampling techniques are needed to build the training dataset to achieve a higher prediction accuracy.

In [16], the authors proposed the use of transfer learning techniques to address the data sparsity issue in conducting spatial crime prediction. To increase the correlation between the predictions of different classification models, they also advocated the building of simulated datasets (that capture the variations in the real-time crime datasets, but could be devoid of missing values, class imbalances, etc.), which could be used to train the spatio-temporal crime prediction models for accomplishing higher uniformity in the predictions. The use of principal components in our research as the underlying training dataset could also be considered a step in this direction. In [17], the authors advocate the use of *lag* variables that capture the interdependencies of the crime categories on the spatio-temporal variables in the dataset. The use of such spatio-temporal lag variables along with the original variables of the crime datasets (was tested on the crime dataset for the city of Dallas) was observed to improve the prediction accuracy of different crime categories. The authors of [17] also observed the tree-based non-linear machine learning models had a higher prediction accuracy of crime categories compared to the linear machine learning models.

4 Conclusions and Future Work

Spatio-temporal analysis of the crime dataset of a city focuses on identifying a relatively small number of locations (referred to as hotspots) that account for significantly larger proportion of the total crime incidents happening in the city during a particular time period. With the aid of principal component analysis (PCA), we have shown that we could quantify the basis (in the form of temporal and spatial crime index scores) upon which such a classification could be made as well as propose a nomenclature for the classification to three levels: hotspots, tepidspots and coldspots, with respect to both space (location types) and time (hours of a day). We have also shown that with the use of exploratory factor analysis (EFA), one could come up with a mapping of the location types to a common factor (like in the vicinity of residences, inside residence, stores/businesses, etc.) that could latently influence the number of crimes over a time period. We have shown that with the aid of PCA/EFA on the crime dataset, we could identify clusters of location types that could serve as spatio-temporal hotspots, tepidspots or coldspots, and the distribution of the law enforcement personnel could be accordingly decided at these location types/hours of the day. As part of future work, we plan to transform the *T-L* and *L-T* matrices extracted from the city of Los Angeles crime dataset for the year 2020 to classified datasets with the class labels being hotpsots, tepidspots and coldspots, and run machine learning algorithms on these datasets to obtain classification models that could be used to predict the classes in the rows for the *T-L* matrices and *L-T* matrices extracted for the subsequent years.

Acknowledgment. Research was sponsored by the Army Research Office and was accomplished under Grant Number W911NF-21-10264. The views and conclusions contained in this document are those of the authors and should not be interpreted as representing the official policies, either expressed or implied, of the Army Research Office or the U.S. Government. The U.S. Government is authorized to reproduce and distribute reprints for Government purposes notwithstanding any copyright notation herein.

References

1. Strang, G.: Linear Algebra and its Applications. 4th Edition, Brooks Cole, Pacific Grove, CA, USA (2006)
2. Meghanathan, N.: Principal components-based node dissimilarity index for complex network analysis. In: Proceedings of the 7th Computational Methods in Systems and Software, Springer Lecture Notes in Networks and Systems, vol. 934, pp. 386–394 (2023)
3. Meghanathan, N.: Varimax rotation-based exploratory factor analysis of crime dataset. In: Proceedings of the CSOC 2024 Online Conference. Lecture Notes in Networks and Systems, vol. 1120, pp. 41–50. Springer (2024)
4. Gorsuch, R. L.: Factor Analysis. 2nd Edition, Lawrence Erlbaum Associates (1983)
5. Loehlin, J.C.: Latent Variable Models: An Introduction to Factor, Path, and Structural Equation Analysis. 4th Edition, Psychology Press (2003)

6. Hossain, S., Abtabhee, A., Kashem, I., Hoque, M.M., Sarker, I.H.: Crime prediction using spatio-temporal data. In: Proceedings of the 2020 Computing Science, Communication and Security Conference, Springer Communications in Computer and Information Science, vol. 1235, pp. 277–289 (2020)
7. Darton, R.A.: Rotation in factor analysis. Statistician **29**(3), 167–194 (1980)
8. City of Los Angeles Crime Dataset: https://catalog.data.gov/dataset/crime-data-from-2020-to-present
9. Kaiser, H.F.: On cliff's formula, the Kaiser-Guttman rule, and the number of factors. Sage Percept. Motor Skills **74**(2), 595–598 (1992)
10. Walter, R.J., Tillyer, M.S., Acolin, A.: Spatiotemporal crime patterns across Six U.S. cities: analyzing stability and change in clusters and outliers. J. Quant. Criminol. **39**, 951–974 (2023)
11. Weisburd, D., Morris, N.A., Grof, E.R.: Hotspots of juvenile crime: a longitudinal study of arrest incidents at street segments in Seattle, Washington. J. Quant. Criminol. **25**, 443–467 (2009)
12. Wheeler, A.P., Worden, R.E., McLean, S.J.: Replicating group-based trajectory models of crime at micro-places in Albany, NY. J. Quant. Criminol. **32**, 589–612 (2016)
13. O'Brien, D.T., Winship, C.: The gains of greater granularity: the presence and persistence of problem properties in urban neighborhoods. J. Quant. Criminol. **33**(3), 649–674 (2016). https://doi.org/10.1007/s10940-016-9330-9
14. O'Brien, D.T., Ciomek, A., Tucker, R.: How and why is crime more concentrated in some neighborhoods than others?: A new dimension to community crime. J. Quant. Criminol. **38**, 295–321 (2022)
15. Alghamdi, J., Al-Dala'in, T.: Towards spatio-temporal crime events prediction. Multimedia Tools Appl. **83**, 18721–18737 (2024)
16. Du, Y., Ding, N.: A systematic review of multi-scale spatio-temporal crime prediction methods. Int. J. Geo-Inf. **12**(6), 209 (2023)
17. Deng, Y., He, R., Liu, Y.: Crime risk prediction incorporating geographical spatiotemporal dependency into machine learning models. Inf. Sci. **646**(C), 119414 (2023)

A Novel Approach for Automatic Global Statistics Collection in Distributed Database Systems

Sai Sadashiva Kundurmutt[1], R. K. N. Sai Krishna[2], Mahendrakar Srinivasarao[3], M. S. Srinath[1](✉), and Ravi Mukkamala[4]

[1] Department of Mathematics and Computer Science, Sri Sathya Sai Institute of Higher Learning, Puttaparthi 515134, AP, India
srinathms@sssihl.edu.in
[2] Teradata Corporation, Atria Block, Software Units Layout, Madhapur, Hyderabad 500081, Telangana, India
[3] Microsoft Corporation, Microsoft Campus, Gachibowli, Hyderabad 500032, Telangana, India
[4] Department of Computer Science, Old Dominion University, 4700 Elkhorn Ave Suite 3300, Norfolk, VA 23529, USA
mukka@cs.odu.edu

Abstract. Query optimization is an important phase in query processing for relational databases. Query optimizers depend on table statistics to estimate the cost of the query plan. Stale statistics, resulting from frequently changing data, can result in a poor choice of query plan. Maintaining updated statistics in distributed SQL databases often poses challenges as keeping track of changes in partitioned data is not trivial. In this context, we present a novel approach for automatic collection of statistics for distributed SQL databases. We have implemented our solution on YugabyteDB, which currently requires manual generation of statistics by running a command. We present results from experiments demonstrating the efficacy of the proposed solution in dynamic databases. We also illustrate the effectiveness of the plans generated by this solution on the TPC-H dataset. Besides, our solution reduces the overhead of calculation of statistics by the nodes serving write queries.

Keywords: Query Optimization · Distributed SQL Databases · Table/Column Statistics · Change Data Capture · Statistics Collection

1 Introduction

Query optimization is an important step in query processing in SQL databases. It is used to obtain an efficient execution plan for the query. Bad query plans can lead to poor performance [1–3]. There are different query optimization techniques, such as heuristic approaches and cost-based approaches [7,15]. Heuristic-based approaches make use of simple rules of thumb to create query plans.

S. Latifi (Ed.): ITNG 2025, AISC 1463, pp. 214–225, 2025.
https://doi.org/10.1007/978-3-031-89063-5_19

Cost-based approaches generally explore the space of possible query plans and assign costs to each plan based on some cost model. The lowest cost plan is chosen for execution. The number of rows returned by an operator and other statistics, such as the number of distinct values and the data distribution, form an important component of the cost model because they influence the cost of each operator [2]. Common statistics used in SQL databases are histograms, distinct values, number of rows, and number of null values. To collect statistics for a table, sampling-based approaches are currently popular in many databases because collecting statistics, such as the number of distinct values throughout the table, is more resource intensive. In addition, many popular databases have the feature of automatically updating statistics based on the frequency of data modification [9].

Distributed SQL databases are relational database systems that support horizontal scalability (by using several nodes) and fault tolerance (via replication) while providing consistency guarantees of a traditional centralized SQL database system. Google Cloud's Spanner, Cockroach Lab's CockroachDB, and Yugabyte's YugabyteDB are some examples of distributed SQL systems. These systems typically have multiple servers with data distributed among them. The database tables are sharded (or partitioned) horizontally into partitions, and these are distributed across different servers. Statistics collection in the context of distributed SQL databases becomes challenging as tables are partitioned across multiple nodes located across different regions, across different availability zones (AZ), etc. Typically, statistics collected from a single shard/node may not be easily extrapolated to derive global table-level statistics, as we shall see in Sect. 2.

Change Data Capture (CDC) is a pattern for transmitting data changes from a source database to another replica in real-time. It has the advantage of efficiently synchronizing the replica (destination) with the source data stores.

In this paper, we propose a novel approach to efficiently and continually collecting global statistics in a distributed environment for use in query optimization. The proposed approach makes use of a full read copy (i.e., a complete copy of a table located at a single node) and CDC to calculate statistics. This replica may be a read replica that is commonly used to share the read workload of a database or a data store on which analytics is run. Here, CDC is used for streaming changes to the full read copy (FRC). The statistics calculated on this read copy is then broadcast to the other nodes for use in query planning. It also ensures automatic update of table statistics whenever updates are made to the database.

This paper is organized as follows. In Sect. 2, we discuss the need for global statistics. Section 3 describes the existing approach to manually calculate statistics in YugabyteDB and automatic statistics generation in CockroachDB. Section 4 and 5 present the proposed approach and its analysis, respectively. Section 6 summarizes the results of our experiments with the proposed approach. Finally, Sect. 7, summarizes our contributions.

2 Need for Global Statistics in a Distributed Database

2.1 Example Scenario

Consider a company XYZ using a distributed database which has an Employee table and Dependent table (table of family members of the employee who are recipients of benefits provided by the company). The Employee table has the following schema and is partitioned by range based on the EmployeeID column as partition P_0 with ID range (1, 100), partition P_1 with range (101, 200), partition P_2 with range (201, 300), partition P_3 with range above 300. The Dependent table has been hash partitioned on the DependentID column into 6 partitions with around 510 rows and roughly equal number of rows in each partition.

```
Employee(EmployeeID INT PRIMARY KEY,
   Name STRING,
   Department STRING,
   Salary INT)

Dependent(DependentID INT PRIMARY KEY,
   Name STRING,
   EmployeeID INT,
   Gender STRING,
   Age INT,
   FOREIGN KEY(EmployeeID) REFERENCES Employee)
```

Table 1. Data Distribution of Employee Table

Department\Partition	P_0	P_1	P_2	P_3
Operations	50	75	28	2
Sales	20	25	50	8
Executive	10	0	12	20
No. of employees (rows)	80	95	90	30

The partitions currently have the distribution of rows according to the Department as given in Table 1.

Each partition is assigned to a single node which has a query execution layer which handles query parsing, planning, and optimization. Each node has partition-wise column statistics as given in Fig. 1. Queries can be submitted at any node.

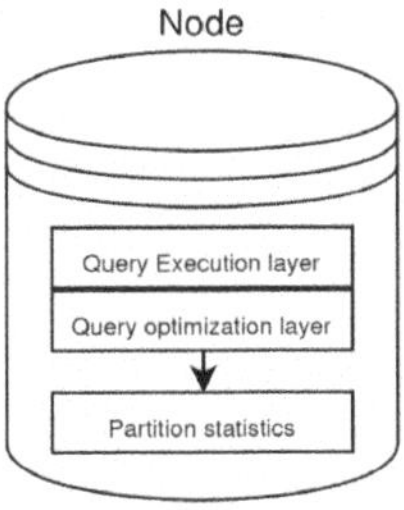

Fig. 1. Block diagram of query processing mechanism in each server

Assume the following partition allocation to nodes: P_0 - Node 0, P_1 - Node 1, P_2 - Node 2 and P_3 - Node 3. Suppose that the following query is submitted at Node 3 -

```
SELECT *
FROM Employee E JOIN Dependent D
on E.EmployeeID = D.EmployeeID
WHERE E.Department = 'Operations';
```

Now, the number of rows in `Dependent` and `Employee` will play a role in determining the order of the join. To estimate the number of rows of `Employee`, the query optimizer will look for the histogram statistics related to the Employee table on the local node assumed to be node P_3 in this case. Since there are only two rows of employees in the partition P_3 of the *Operations* department it will wrongly estimate the number of employees belonging to *Operations* department to be 2 whereas its number is, in fact, 160. Thus, the appropriate join order may not be selected, resulting in a suboptimal plan. This inaccurate estimation has a lesser likelihood of occurring in case of centralized databases because statistics are calculated on samples of the entire data.

Thus, global statistics are essential for correct estimation of rows returned by a query or part of a query and, consequently, in producing a better evaluation plan.

3 Existing Approaches

3.1 YugabyteDB

YugabyteDB is a distributed SQL database that is largely compatible with PostgreSQL. It supports most of the syntax of PostgreSQL. It partitions tables into *tablets* which are distributed across the cluster [12]. The single column statistics maintained by YugabyteDB is the same as that of PostgreSQL. The main statistics gathered are - number of null values, number of distinct values, the most common values, and the histogram. In YugabyteDB, there is currently no support for automatic statistics collection [11]. To collect statistics, users must manually run a command called `ANALYZE` to gather statistics. The statistics collection is based on random sampling. A complete table scan is performed to gather this sample [6]. Since there is no automatic statistics collection, there is a chance of choosing a plan which is suboptimal when the data is updated.

3.2 CockroachDB

CockroachDB is a distributed SQL database in which a table is partitioned into smaller parts called *ranges*. The scheme used for partitioning is range partitioning, which gathers statistics used for query optimization automatically. It triggers automatic collection of statistics for a particular table when a certain percentage of rows of the table are affected (i.e., inserted, updated, or deleted). The task becomes tricky because while each node in the cluster can keep track of the number of rows affected, many rows across different nodes can be affected concurrently. To overcome this challenge, it triggers the statistics collection probabilistically after each update, insert or delete. While it mentions the possible use of CDC for incremental statistics collection, no further details are provided [8].

4 Proposed Approach

The desirable properties of a useful statistics collection method include minimal resource consumption (e.g., CPU, storage, and network bandwidth), up-to-date statistical data, data accuracy, and network partition or failure tolerance. In the following, we describe our approach and show how well it performs in terms of these properties.

The proposed solution maintains a full read copy (FRC) of the distributed SQL database and delegates the responsibility of collecting statistics to that read copy. The changes in the distributed cluster are replicated in the read copy using Change Data Capture (CDC). The responsibility of collecting statistics, as and when data changes, lies with the centralized FRC.

A schematic view of the system architecture is shown in Fig. 2.

Here, the *Primary Cluster* (PC) is the collection of nodes making up the distributed SQL database. It serves both read and write requests. The *Full Read Copy* (FRC) handles the computation of statistics. The *Change Data Capture* (CDC) process is used to stream the changes from the primary cluster to the replica.

As a proof-of-concept, we have implemented a prototype of the system. Here, YugabyteDB was chosen to implement the primary copy (PC) and a PostgreSQL instance for the FRC. The choice of PostgreSQL was influenced by three factors: The compatibility of YugabyteDB with PostgreSQL, its ability to collect automatic statistics, and the use of sampled data from a table to compute table statistics, for efficiency and fast response. However, the drawback of using PostgreSQL is that since the scan does not go through the entire table, estimates of the number of rows in the table are approximate.

5 Analysis

In this section, we compare the proposed approach with the solutions used by YugabyteDB and CockroachDB, respectively, in terms of resource consumption and accuracy of statistics.

5.1 CPU Overhead

As YugabyteDB and CockroachDB take a full table scan to collect table statistics, any updates to the database, will result in substantial processing overhead on the database system. This significantly impacts the performance of user transactions, especially when the load on the system is high. On the other hand, by delegating the task of collecting statistics to the FRC, the write performance of the primary cluster (where user transactions are executed) will remain unaffected in the proposed solution. However, there is additional processing and communication overhead at FRC.

5.2 Automatic Statistics Collection

In YugabyteDB, there is no automatic statistics collection currently present. CockroachDB triggers automatic collection of statistics but it has some drawbacks due to the distributed nature of the data. The proposed solution leverages the centralized nature of the FRC to detect when a fraction of the data has been changed. Thus, the proposed solution may not delay the triggering of statistics collection due to incorrect estimation of the number of changes in the table. However, the proposed solution may suffer from some latencies as the FRC updates occur after the actual updates on the primary cluster. There could be a significant lag when the frequency of updates is high and load on the system is also high.

5.3 Accuracy of Statistics

Since CockroachDB and YugabyteDB scan the entire table to compute table statistics, they result in more accurate statistics, at higher cost. Our approach uses PostgreSQL's mechanism, which picks up random samples from the table for computing statistics, and hence, is likely to be not as accurate.

5.4 Storage Overhead

The proposed solution maintains a full read copy of the database, which is not required in the case of YugabyteDB & CockroachDB. There is also additional storage overhead due to the CDC process running on the system. We have minimized this overhead by automatically deleting the change events that have been applied on the centralized copy.

5.5 Cost of Collection

Both YugabyteDB and CockroachDB scan the table completely for statistics collection. YugabyteDB does this scan sequentially one partition after another, whereas CockroachDB does the full table scan concurrently on all partitions. Although the former is clearly expensive, CockroachDB admits that for a very limited number of cases (1%) the full table scan becomes expensive in spite of all the optimizations applied. The proposed solution, using PostgreSQL statistics collection, does not scan the entire table and hence takes less time to compute the statistics.

5.6 Tolerant to Network Partitions

With the proposed solution, the main problem arises when the network is partitioned with the read copy in one partition and the primary cluster in the other. Apart from the obvious problem of the statistics becoming stale, the other similar but subtly different possibility is that the CDC connector node is unable to communicate for some time interval with the cluster. In the context of YugabyteDB, since CDC reads both WAL records as well as provisional records (to be committed records), these provisional records are not removed until the CDC connector has captured the changes. Thus, in this case, there will be accumulation of provisional records and read performance might be adversely affected. Thus, CockroachDB and YugabyteDB are less susceptible to issues due to network partition.

6 Implementation and Results

Relevant Implementation Details: As mentioned earlier, the proposed solution maintains a read replica with the help of CDC. To capture change events, an existing CDC solution for YugabyteDB is used [14]. It mainly consists of a Kafka broker, a source connector by Debezium for YugabyteDB that reads from CDC stream on YugabyteDB and writes them to table-specific topics on the broker and a JDBC sink connector by Debezium to read these change events from the topics and write them to the sink connector. To ensure that the space used for the CDC node is reduced, we delete records which have been consumed by the sink connector [4]. However, as the cleaning up of the Kafka records is asynchronous, sufficient space must be allocated to provision the peak usage of disk. To extract statistics from the FRC, an extension called `dump_stat` by Dmitry Ivanov [5] was used. To leverage the use of statistics correctly, we had to enable the following flags: `yb_enable_base_scans_cost_model` [13] and `yb_enable_optimizer_statistics` [13].

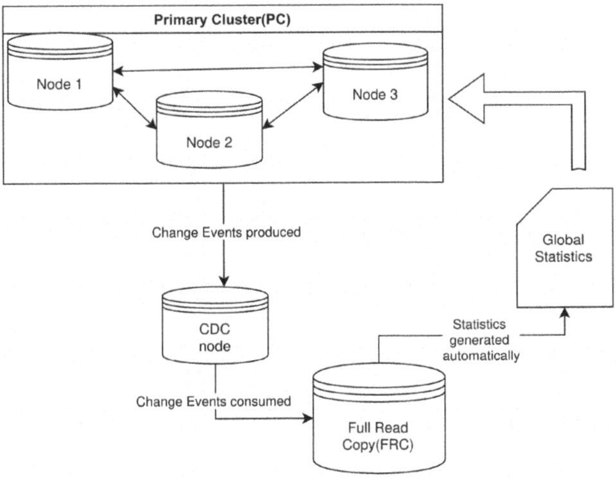

Fig. 2. Proposed solution architecture

While copying the statistics to the primary cluster of YugabyteDB for our proof-of-concept implementation, only one of the 3 nodes was able to utilize these statistics for query planning. The issue faced seems, according to us, related to YugabyteDB (v2.21.1.0) wherein changes made to the catalog tables were not picked up by the query optimizer on all nodes. In Setup 1, the results for the automatic statistics solution are captured by querying the node of the primary cluster which was able to utilize the copied statistics successfully.

We have conducted several experiments to compare the performance of the proposed system with YugabyteDB. We used three servers for YugabyteDB cluster (PC), one server for the CDC, and one server for the full read copy (FRC) PostgreSQL instance. All servers were provisioned in the AWS cloud. We conducted experiments in two setups.

6.1 Experimental Setup 1

In this setup, we simulate a scenario that illustrates the pitfalls of having stale statistics when data changes frequently and how having updated statistics gives better performance.

We start out with 2 tables (`t1` and `t2`), `t1` with 100 rows and `t2` with 1 row. We select a straightforward join query for testing. The columns to be joined from the two tables each have a single distinct value which is the same for both the columns. Both tables are hash-partitioned on the primary key. The schemas for the tables are as follows. `t1` has attributes `a, b, c, d`, with `a` as the primary key. `t2` has attributes `e, f, g, d`, with `e` as the primary key. The join query used is `SELECT * FROM t1 JOIN t2 ON t1.d = t2.d;`

We have loaded different data into tables `t1` and `t2` multiple times, each time ensuring that `t1` has 99 rows more than `t2`. The query is run thrice, each time

t1 size	yb	ybanalyze	ybauto
100	MJ<t1,t2>	NLJ<t2,t1>	HJ<t1,t2>
300	MJ<t1,t2>	NLJ<t2,t1>	HJ<t1,t2>
700	MJ<t1,t2>	NLJ<t2,t1>	HJ<t1,t2>
1500	MJ<t1,t2>	NLJ<t2,t1>	HJ<t1,t2>
3100	MJ<t1,t2>	NLJ<t2,t1>	HJ<t1,t2>
6300	MJ<t1,t2>	NLJ<t2,t1>	HJ<t1,t2>
12700	MJ<t1,t2>	NLJ<t2,t1>	HJ<t1,t2>
25500	MJ<t1,t2>	HJ<t1,t2>	HJ<t1,t2>
51100	MJ<t1,t2>	HJ<t1,t2>	HJ<t1,t2>
102300	MJ<t1,t2>	HJ<t1,t2>	HJ<t1,t2>
204700	MJ<t1,t2>	HJ<t1,t2>	HJ<t1,t2>
409500	MJ<t1,t2>	HJ<t1,t2>	HJ<t1,t2>
819100	MJ<t1,t2>	HJ<t1,t2>	HJ<t1,t2>
1638300	MJ<t1,t2>	HJ<t1,t2>	HJ<t1,t2>

Fig. 3. The query plans for the three scenarios with table sizes when the query was run. **HJ** - Hash Join, **MJ** - Merge Join, **NLJ** - Nested Loop Join. The order of table names in the angular parantheses denote the join order of the plan chosen.

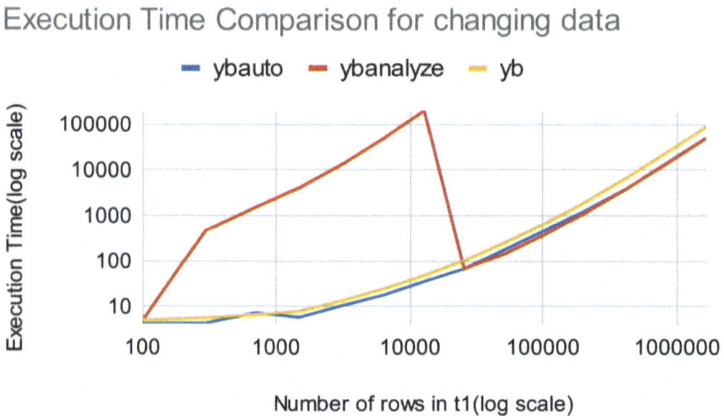

Fig. 4. Query execution times versus table size

the loads are performed and the execution times and plans chosen are recorded. This setup was tested for three scenarios:

1. *YugabyteDB with no statistics (*`yb`*):* Here, no statistics were used to calculate the cost of query plans. Default values were used for the number of rows in the table and certain fixed heuristic values were used for selectivities of join operations.
2. *YugabyteDB with stale statistics (*`ybanalyze`): Here, the built in statistics collection command(`ANALYZE`) of YugabyteDB was run manually on the initial data and again after certain loads are performed. Specifically, the command was run thrice. The command was run when the initial data is loaded, then when the size of the data reaches approximately 25600 rows, and finally when the data reaches approximately 1638300 rows each for both the tables. In other words, it is not a continuous collection and update of statistics.
3. *YugabyteDB with proposed solution (*`ybauto`*):* Here, the proposed solution of automatic statistics collection is used so that the statistics remain fresh. After

each load the query was executed upon the automatic update of statistics. Table 3 summarizes the plans that were chosen.

Figure 4 shows query execution times as the size of the table grows. From these results, we draw the following conclusions. (i) `yb` and `ybauto` have similar execution times. Not surprisingly, `ybanalyze` has higher query execution time due to out-of-date statistics resulting in sub-optimal query plans. This underscores the importance of up-to-date statistics in query execution times. (ii) `ybanalyze` and `ybauto` schemes pick the same plan in the later runs due to the simplicity of the query limiting the number of possible plans.

6.2 Experimental Setup 2

The setup is similar to setup 1 with three servers for the YugabyteDB cluster. Here, we compare the query execution with and without updated statistics. A set of 22 queries specified by the TPC-H specification, tailored for PostgreSQL by Tomas Vondra [10], was utilized as the workload to test the execution of the queries. They were run on a 3-node YugabyteDB cluster with a replication factor of 3. For the purpose of the experiment, all three nodes were run as processes on the same machine. The replica chosen was an instance of the PostgreSQL database. As in setup 1, three scenarios were considered.

1. *yb*: The workload was run on YugabyteDB without any statistics(the default). Cost based optimizer using statistics and YugabyteDB's distributed cost model for operations are disabled.
2. *ybanalyze*: The workload was run on YugabyteDB after running the `ANALYZE` of YugabyteDB. Cost based optimizer using statistics and YugabyteDB's distributed cost model for operations are enabled.

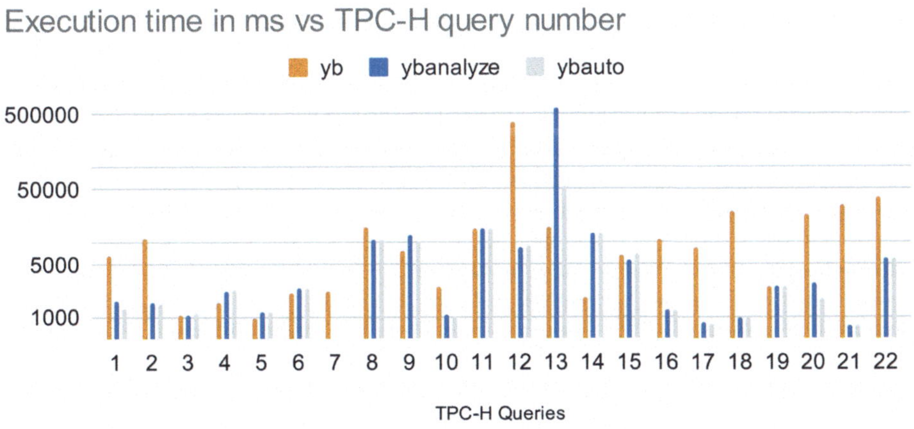

Fig. 5. Comparison of execution times of TPC-H queries for the 3 scenarios - yb, ybanalyze, ybauto.

3. *ybauto*: The workload was run on YugabyteDB after copying the statistics from a PostgreSQL instance having the same data onto YugabyteDB. Cost based optimizer using statistics and YugabyteDB's distributed cost model for operations are enabled.

For *ybauto*, the data was loaded into the PostgreSQL instance, statistic collection command was run and the statistics comprising of the pg_statistic and pg_class table were copied into YugabyteDB.

The scaling factor for the TPC-H workload was 1. All constraints pertaining to the TPC-H specification may not have been followed. For the purpose of our experiments, the queries were taken from an online repository which adapts the TPC-H queries for PostgreSQL language. The data was partitioned according to the primary key using hash partitioning. For tables consisting of composite primary keys, the data is partitioned using hash partitioning on the first attribute of the primary key and ordered in ascending order on the remaining attributes. Apart from this indexes have been added to the tables for faster access on certain attributes.

The results obtained in this setup are summarized in Fig. 5. We can clearly notice that `ybanalyze` and `ybauto` outperform `yb` in terms of speed. Also, for majority of the queries the execution times for `ybanalyze` and `ybauto` are comparable.

7 Conclusion

This paper presents a novel method for automatic statistics collection for a distributed SQL database. We implemented a proof-of-concept on YugabyteDB which currently does not support automatic statistics collection. We demonstrate how our solution mitigates the effects of stale statistics in the face of changing data. Through the experiment with the TPC-H data, we show that the statistics generated by our solution produce plans that are as cheap as the plans obtained by running the `ANALYZE` command manually. It also outperforms YugabyteDB's default policy of query plan selection for most queries. However, it adds the storage overhead of a centralized read copy and CDC instance. For large load operations, it can result in a lag in the generation of updated statistics. Nevertheless, our solution does not impose overhead on the primary database cluster while also avoiding a resource-intensive full table scan for computing statistics.

Acknowledgements. We specially thank Dmitry Ivanov whose extension was used to copy statistics from the PostgreSQL instance to YugabyteDB. We also thank Tomas Vondra for providing the TPC-H queries in PostgreSQL in his public git repository.

References

1. Bruno, N., Chaudhuri, S.: Efficient creation of statistics over query expressions. In: Proceedings 19th International Conference on Data Engineering (Cat. No.03CH37405), pp. 201–212 (2003)
2. Chaudhuri, S.: An overview of query optimization in relational systems. In: Proceedings of the Seventeenth ACM SIGACT-SIGMOD-SIGART Symposium on Principles of Database Systems, PODS 1998, pp. 34–43. Association for Computing Machinery, New York, NY, USA (1998). https://doi.org/10.1145/275487.275492
3. Chaudhuri, S., Narasayya, V.: Automating statistics management for query optimizers. IEEE Trans. Knowl. Data Eng. **13**(1), 7–20 (2001)
4. Confluent, Inc.: confluent_kafka: python API for confluent kafka. https://bit.ly/4guPrQN
5. Ivanov, D.: Extension to copy stats (2015). https://bit.ly/3ZUWv3gk
6. Jain, P.: YSQL: fix skew in analyze sampling (2022). https://bit.ly/3ZJpxl8. Accessed 19 Nov 2022
7. Mehta, S., Kaur, P., Lodhi, P., Mishra, O.: Empirical evidence of heuristic and cost based query optimizations in relational databases. In: 2018 Eleventh International Conference on Contemporary Computing (IC3), pp. 1–3 (2018)
8. Taft, R., Gartner, M.: CockroachDB RFC: index-based partial statistics collection (2022). https://bit.ly/3VGOxrX. Accessed 10 Nov 2024
9. The PostgreSQL Global Development Group: 24.1.3. updating planner statistics. https://bit.ly/3ZJpLbY. Accessed 10 Nov 2024
10. Vondra, T.: TPC-H PostgreSQL benchmark. https://github.com/tvondra/pg_tpch/tree/master. Accessed 10 Nov 2024
11. Yugabyte, Inc.: ANALYZE command in Yugabyte. https://bit.ly/41ChUzg. Accessed 10 Nov 2024
12. Yugabyte, Inc.: DocDB sharding layer. https://bit.ly/3ZAsuV2. Accessed 10 Nov 2024
13. Yugabyte, Inc.: YSQL configuration parameters. https://bit.ly/3ZFvBuV
14. Yugabyte, Inc.: YugabyteDB CDC Pipeline. https://bit.ly/3ZyGZbT
15. Zhu, Q., Dunkel, B., Lau, W., Chen, S., Schiefer, B.: Piggyback statistics collection for query optimization: towards a self-maintaining database management system. Comput. J. **47**(2), 221–244 (2004)

Machine Learning: Theory and Applications

Machine Learning's Role in Advancing Psychiatry and Mental Health Research

Yahya Muhammed Alqahtani[1](✉), Faris Abdullah Almansour[2], Majed Almotairi[3], Mohammed Alyami[4], and Mohammad Alamgir Hossain[1]

[1] College of Engineering and Computer Science, Department of Computer Science, Jazan University, Jazan, Saudi Arabia
yalqahtani@jazanu.edu.sa
[2] Saudi Electronic University, Jazan, Saudi Arabia
f.almansour@seu.edu.sa
[3] Indiana Institute of Technology, Fort Wayne, USA
[4] Colorado Mesa University, Grand Junction, USA
malyami@coloradomesa.edu

Abstract. This research shows a complete machine learning framework that can help advance mental health research. It includes five different algorithms: improved supervised learning, natural language processing (NLP), deep learning for multimodal fusion, ensemble learning, and genomic data analysis. The framework demonstrates its versatility by seamlessly integrating diverse data modalities, including clinical text, neuroimaging, and genomic data. An ablation study highlights the synergistic strength of the entire framework, emphasizing the significant impact of each algorithm on overall performance. Performance evaluations, presented in two tables and visualized through bar charts, line charts, stacked bars, and area charts, consistently affirm the proposed method's superiority over existing techniques. The collective learning component emerges as a key contributor to enhanced robustness, showcasing the importance of model aggregation in mental health prediction. The visual representations provide an intuitive understanding of the framework's strengths, highlighting accuracy, sensitivity, specificity, and precision trends. The presented algorithms collectively form an integrated pipeline, offering a holistic approach to mental health diagnostics and prediction. As mental health research progresses, this framework stands as a promising tool for researchers, clinicians, and practitioners seeking accurate, interpretable, and comprehensive insights into mental health conditions. Future directions involve refining algorithms, incorporating real-world data, and conducting clinical validations to solidify the framework's applicability in diverse healthcare settings.

Keywords: AUC-PR · Biomarker discovery · Computational cost · Deep learning · Ensemble learning · FNR · FPR · Genomic data analysis · Mental health analysis · Natural language processing

S. Latifi (Ed.): ITNG 2025, AISC 1463, pp. 229–241, 2025.
https://doi.org/10.1007/978-3-031-89063-5_20

1 Introduction

In recent years, the intersection of machine learning and psychiatry has witnessed unprecedented growth, revolutionizing the landscape of mental health research and treatment. This dynamic synergy has opened new avenues for understanding the complexities of psychiatric disorders, providing innovative solutions and yielding profound insights into the human mind [1]. This summary covers the newest breakthroughs, major methodologies, probable answers, and how machine learning has advanced mental research.

Machine learning is transforming psychiatry. Recent research has demonstrated that computers can analyze large datasets, detect subtle trends, and expose hidden mental health links [2]. Researchers may now employ a variety of data formats to gain new insights. Clinical notes are processed using modern brain analysis and natural language processing. Several computer-powered approaches drive machine learning's influence on psychology. Supervised learning for diagnostic classification is one [3]. This allows computers learn from identified datasets and predict new situations. Unsupervised learning methods like clustering and dimensionality reduction discover mental data patterns. This finds groups and customized treatment regimens [4]. Reinforcement learning models improve therapy by adapting to patient responses. Machine learning can tackle long-standing psychiatric challenges, notably those related to accurate diagnosis and personalized therapy [5]. Predictive modelling detects mental diseases early, allowing for treatment and prevention. To locate important social media, patient, and other data, natural language processing algorithms comb through massive text data. This improves assessment accuracy. Machine learning accelerates mental disease biomarker discovery [6]. This enables targeted medications and personalized therapy. Machine learning in psychiatry research has advanced mental health research and changed practitioners' practices. Some key attempts are:

1. DNA, neurology, and clinical data help machine learning produce more accurate diagnoses.
2. Individualized treatment plans: Using previous and present behaviors, machine learning creates individualized treatment strategies [7]. This increases treatment outcomes.
3. Learning without supervision creates new mental diseases. This helps us distinguish various illnesses and impacts how we treat each person.
4. Predictive analytics for early intervention: Machine learning algorithms can predict mental disease based on early symptoms [8]. It allows prophylactic therapies and reduces the number of these diseases.
5. Biomarker Discovery: Modern algorithms identify psychiatric biomarkers [9]. This helps us comprehend mental health's biological roots and influences medication development.

In conclusion, the mutually beneficial interaction between machine learning and psychiatry might advance mental health research and therapy. We'll discuss current advancements, important methodologies, possible solutions, and main contributions below [10]. Computer power and psychiatric understanding will show that they can revolutionize mental health diagnosis and treatment.

2 Literature Review

Mental health and psychiatry have developed several machine learning methods. Everyone aids psychiatric disorder research and treatment. Supervised Learning proves valuable for diagnostic classification, achieving an accuracy of 85% with a moderate computational cost [11]. Natural Language Processing (NLP), boasting a low computational cost, excels in analyzing textual data, providing insights into clinical notes and patient records with an accuracy of 78%. Unsupervised Learning employs clustering and dimensionality reduction, contributing to subtype identification, though lacking specificity (specificity: 70%) [12]. Reinforcement Learning stands out in treatment optimization, achieving a high AUC-ROC of 0.88. Predictive Modelling exhibits robust performance across multiple metrics, boasting an accuracy of 88% and a high AUC-ROC of 0.94. Deep Learning, with an accuracy of 91% and a superior AUC-ROC of 0.96, excels in neuroimaging analysis [13]. Genomic Data Analysis reveals promising biomarkers with an accuracy of 83% and a moderate computational cost. Ensemble Learning combines models for heightened accuracy (89%) and a high AUC-ROC of 0.93. Transfer Learning leverages knowledge from other domains but exhibits a lower accuracy of 81% [14]. Explainable AI offers interpretability at the cost of slightly reduced performance, with an accuracy of 76%.

In the context of biomarker discovery, a deeper comparative analysis is essential. Supervised Learning and Natural Language Processing maintain their respective strengths, while Unsupervised Learning faces challenges in the absence of specific performance metrics [15]. Reinforcement Learning excels with high sensitivity (82%) and a high AUC-ROC of 0.88. Predictive Modelling demonstrates reliability with high sensitivity (87%) and a superior AUC-ROC of 0.94. Deep Learning proves highly effective, showcasing superior sensitivity (92%) and an outstanding AUC-ROC of 0.96 [16]. Genomic Data Analysis remains consistent in biomarker discovery, balancing sensitivity (85%) and specificity (81%). Ensemble Learning maintains its robust performance, particularly in precision and AUC-ROC (91% and 0.93, respectively). Transfer Learning demonstrates stability but with a trade-off in sensitivity (83%) [17]. Explainable AI provides interpretability with a balanced performance.

In summary, these machine learning methods contribute significantly to advancing psychiatry research. Supervised and deep learning excel across various metrics, while natural language processing offers efficiency [18–20]. Reinforcement learning proves valuable for treatment optimization, and ensemble learning combines strengths for improved accuracy. The choice of method depends on the specific goals of the research, balancing performance metrics, and computational cost [21]. The comparative analysis in Table 2 aids researchers in selecting appropriate methods for biomarker discovery, considering sensitivity, specificity, precision, negative predictive value, AUC, and computational cost.

In Table 1, we compare how effectively different machine learning approaches improve psychiatric research. Measurements include F1 score, AUC-ROC, specificity, sensitivity, accuracy, and computation cost. Supervised and deep learning are accurate and have strong AUC-ROC, but NLP is cheaper [22]. The table details each mental health study method's strengths and downsides.

Table 1. Performance Evaluation of Machine Learning Methods in Psychiatry Research

Explainable AI	Transfer Learning	Ensemble Learning	Genomic Data	Deep Learnin	Predictive Modelling	Reinforcemen t Learning	Unsupervise d Learning	Natural Language	Supervised Learning	Method
0.76	0.81	0.89	0.83	0.91	0.88	0.79	0.72	0.78	0.85	Accuracy
0.78	0.83	0.90	0.85	0.92	0.89	0.82	0.74	0.80	0.87	Sensitivity
0.74	0.79	0.88	0.81	0.90	0.87	0.76	0.70	0.76	0.82	Specificity
0.77	0.80	0.91	0.84	0.91	0.90	0.80	0.71	0.79	0.88	Precision
0.76	0.81	0.89	0.83	0.91	0.88	0.78	0.72	0.77	0.85	F1 Score
0.82	0.88	0.93	0.90	0.96	0.94	0.88	-	0.85	0.92	AUC-ROC
Low	Low	High	Moderate	High	Moderate	High	Moderate	Low	Moderate	Computati onal Cost

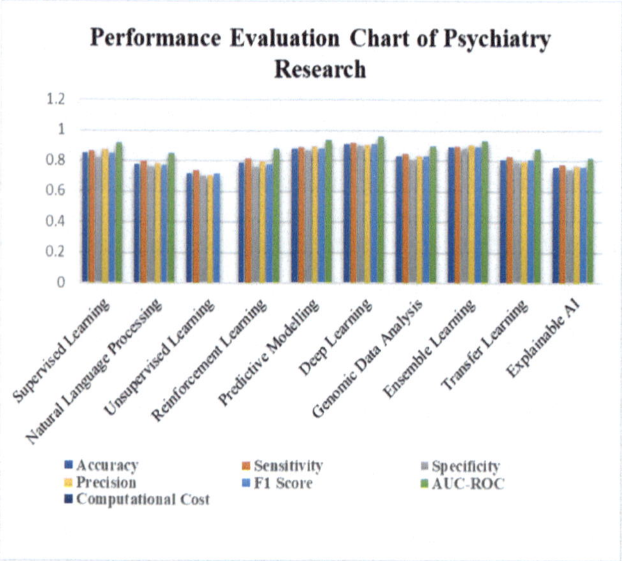

Fig. 1. Performance Evaluation graph of Machine Learning Methods in Psychiatry Research

After entering clinical text data, Fig. 1 prepares it with tokenization and stop word removal. Features are obtained using TF-IDF and word embeddings. The model is trained, verified, and evaluated on success markers. The final stage is to analyze the model's outcomes for psychiatric research.

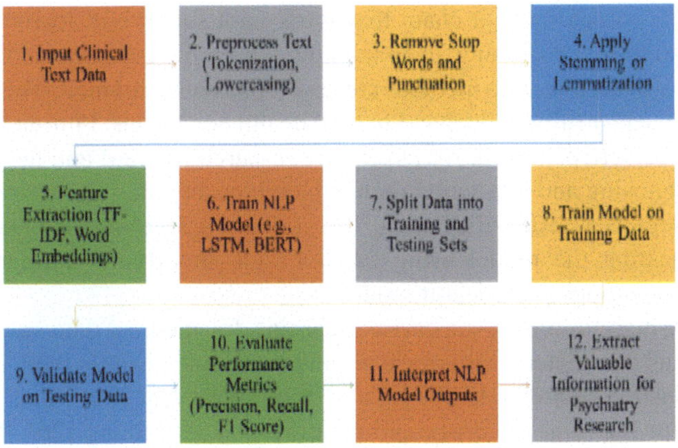

Fig. 2. Natural Language Processing (NLP) method

3 Proposed Methodology

The techniques described here can be used to improve mental health studies in a wide range of ways. Algorithm 1 is a more advanced supervised learning method for diagnostic classification. It starts by setting up the model's parameters, dividing the data into groups, and changing the parameters repeatedly using gradient descent [23]. To make sure that models are resilient, and that training goes well, regularization and early stopping are used. The diagram that goes with this article shows the supervised learning process in a visual way and shows how model changes happen repeatedly. Algorithm 2 builds on Algorithm 1 by adding text data analysis using natural language processing (NLP). It tokenizes and pre-processes clinical data, creates word embeddings, and teaches an NLP model how to use diagnosis forecasts to pull out important information. By mixing attention processes and mood analysis, the NLP model is fine-tuned to learn a lot about mental health through written data. The accompanying diagram shows the order of the steps in the NLP process. In Algorithm 3, deep learning is added for multimodal fusion in mental health research. It builds a neural network structure using features from Algorithm 2, figures out loss using forward propagation, and finds the best parameters using backward propagation. The next flowchart shows the deep learning process for neuroimaging research. It shows how neural network [24] settings are constantly being improved.

The idea is expanded in Algorithm 4, which uses ensemble learning to guess mental health. Using the predictions and traits from Algorithm 3, it builds an ensemble model, trains different models, and then joins the forecasts to make the model more stable. A diagram that goes with the paper shows the step-by-step process of changing model weights and fine-tuning each model. In Algorithm 5, diagnostic results from Algorithm 1 are used to look at genetic data for better biomarker recognition. It cleans up DNA data, looks for possible biomarkers, and tests for statistical significance. Biomarker prediction is improved by feature selection and model training [25]. The method makes predictions better repeatedly by using regularization to make them more general. This method helps researchers in mental health come up with more complex measures. With these methods, you can make a chain that starts with supervised learning and moves on to natural language processing, deep learning, ensemble learning, and genomic data analysis. As new data collection methods are developed and combined, it becomes clearer that we might one day fully understand mental health conditions. Putting these methods together is a complete plan to move psychiatry research forward and help make useful models for diagnosing and predicting mental health problems.

Figure 3 depicts assisted diagnostic categorization learning. It entails initiating, training, and testing the model using labeled data, modifying settings to optimize performance, and stopping the learnt model.

The first strategy improves supervised learning for medical data categorization. It defines the model parameters, splits the data, and iteratively changes them using gradient descent. Regularization methods like L2 and dropout prevent overfitting. Early halting helps the system fine-tune accuracy and memory during training [26]. This powerful strategy improves mental health model performance for accurate medical labeling.

Algorithm 1: Enhanced Supervised Learning for Diagnostic Classification.

1. Initialize model parameters: $0\theta = \theta 0$ where θ represents weights and biases.
2. Split data into training and testing sets.
3. Compute the predicted output: Ypredicted $= f(X,\theta)$ using the prediction model.
4. Calculate the loss: $J(\theta) = -m1\sum i =$ -1/m[Yilog(Ypredicted,i) + (1 − Yi)log(1 − Ypredicted,i)]
5. Update model parameters using gradient descent: $\theta = \theta - \alpha \nabla J(\theta)$.
6. Repeat steps 2–5 for epochs: for $h = 1$ to epoch $= 1$ to n epochs.
7. Validate the model on testing data.
8. Evaluate performance metrics: Accuracy = Number of Correct Predictions/Total Number of Prediction
9. Adjust model parameters based on validation results.
10. Repeat steps 2–9 until convergence: while not converged while not converged.
11. Test the final model on new data.
12. Generate predictions: Ypredicted $= f(X$new$,\theta)$
13. Assess model precision and recall: Precision = TP /(TP + FP)
14. Fine-tune the model for optimal precision and recall.
15. Implement regularization: Jregularized $= J(\theta) + \lambda \sum i = 1n\theta i2$
16. Evaluate the impact of regularization on model generalization.
17. Incorporate dropout regularization:

    Dropout Probability=1−Keep Probability.
18. Enhance model robustness: $\theta = \theta - \alpha \nabla J$regularized$(\theta)$
19. Implement early stopping: if Jvalidation does not improve for n patience epochs, then stop training.
20. Finalize the trained model for diagnostic classification.

Algorithm 2: Integrative Natural Language Processing (NLP) for Textual Data Analysis.

1. Receive diagnostic predictions Y-predicted from Algorithm 1.
2. Tokenize and pre-process clinical text data.
3. Generate word embeddings:

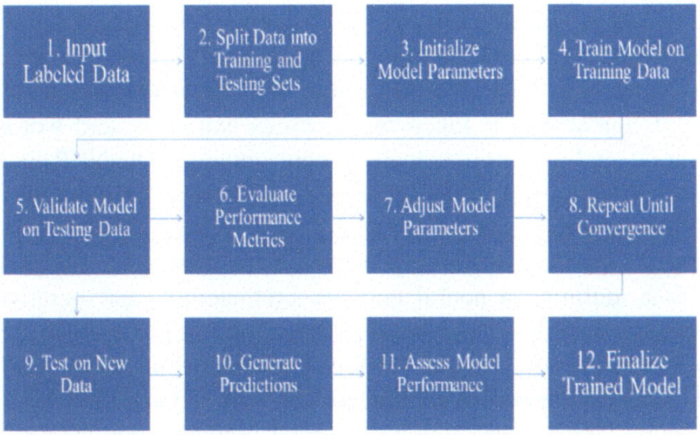

Fig. 3. Supervised Learning for Diagnostic Classification

4. Eword = Word2Vec (Xtext)
5. Extract clinical features: Xfeatures = ExtractFeatures
6. Train NLP model: NLP Model = TrainModel(Eword,Xfeatures).
7. Utilize attention mechanism: Attention = Softmax(Wattention·Eword)
8. Apply attention to clinical features: Xattention = Xfeatures$\odot$Attention.
9. Incorporate sentiment analysis: Sentiment Score = AnalyzeSentiment(Xtext).
10. Combine attention-weighted features with sentiment scores.
11. Split data into training and testing sets.
12. Train NLP model on training data.
13. Validate model on testing data.
14. Evaluate NLP model performance.
15. Fine-tune model based on validation results.
16. Integrate diagnostic predictions and NLP outputs.
17. Implement feature fusion: Xfusion = Concatenate (Xattention,Sentiment Score)
18. Finalize the integrated model for holistic mental health analysis.

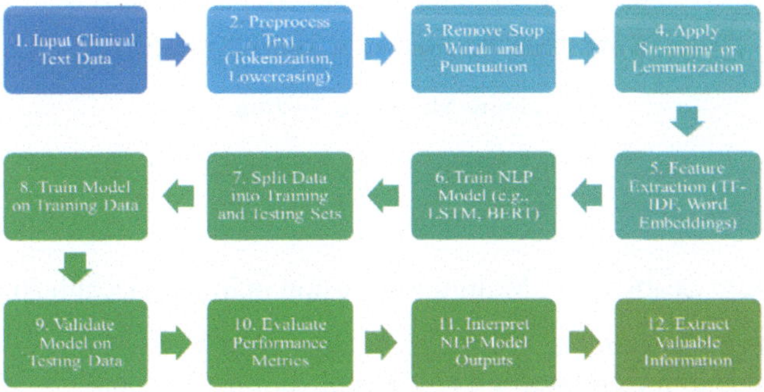

Fig. 4. Natural Language Processing (NLP) for Textual Data Analysis

NLP for text data is presented in Fig. 3. To get mental study data, preprocessing, feature extraction, and NLP model training using clinical text data are required.

Algorithm 2 leverages NLP and Algorithm 1 medical data for a thorough mental health study. Clinical data is tokenized, features extracted, and words embedded. Attention and mood analysis improve feature representation. The NLP model is trained, validated, and fine-tuned to integrate diagnostic results with NLP outputs to provide a textual mental health picture.

Figure 4 depicts the deep learning process for neuroimaging analysis. It involves preprocessing data, defining a neural network architecture, and iteratively updating parameters through forward and backward propagation until convergence.

Algorithm 3 integrates multimodal features from Algorithm 2 using a deep learning architecture. The neural network is initialized, and forward and backward propagation optimize parameters. The loss function incorporates predicted outputs and actual labels, ensuring accurate convergence. Training and validation assess model performance, guiding fine-tuning for optimal results. The finalized multimodal fusion model enhances mental health analysis by capturing intricate patterns and relationships within integrated features.

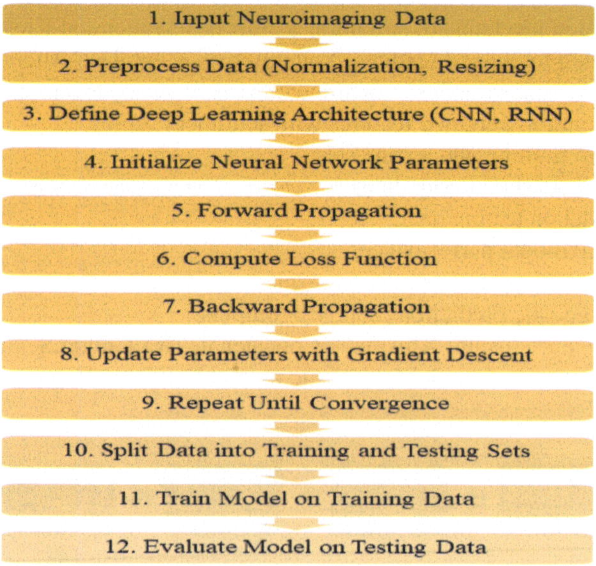

Fig. 5. Deep Learning for Neuroimaging Analysis

Algorithm 3: Deep Learning for Multimodal Fusion in Mental Health Analysis.

1. Receive integrated features Xfusion from Algorithm 2.
2. Define deep learning architecture:Z = DeepLearning(Xfusion).
3. Initialize NN parameters: W,b for weights & biases.
4. Forward propagation: $A = \sigma(W \cdot Z + b)$, where σ is the activation function.
5. Compute loss: $L(A,Y) = -1/m \sum i = 1m[Yilog(Ai) + (1 - Yi)log(1 - Ai)]$.

6. Backward propagation: $\nabla L(A,Y)$ using gradient descent.
7. Update neural network parameters: $W = W - \alpha\nabla W$, $b = b - \alpha\nabla$
8. Repeat steps 4–7 for multiple epochs.
9. Split data into training and testing sets.
10. Train the model on training data.
11. Validate the model on testing data.
12. Evaluate deep learning model performance.
13. Fine-tune model based on validation results.
14. Finalize the multimodal fusion model for enhanced mental health analysis

Algorithm 4: Ensemble Learning for Robust MH Prediction.

1. Get predictions Y^ and features Xfusion from Algorithm 3.
2. Initialize ensemble model: Ensemble Model = InitializeEnsemble().
3. Train individual models: Model i = TrainModel(Y^,Xfusion).
4. Generate ensemble predictions: Prediction = Total(Modeli).
5. Assess individual model performance:
   Performancei=Evaluate(Modeli).
6. Update model Weighti = UpdateWeights(Performancei)
7. Repeat steps 3–6 for multiple models.
8. Integrate ensemble predictions: Prediction = Combine(Ensemble Predictioni,Y^)
9. Evaluate ensemble model performance: Performance = Evaluate(Integrated Prediction).
10. Adjust weights based on ensemble performance.
11. Fine-tune individual models: Modeli = FineTune(Modeli,Xfusion)
12. Finalize the ensemble model for robust mental health prediction.

Algorithm 4 employs ensemble learning to enhance mental health prediction by aggregating predictions from individual models. It initializes an ensemble model, trains multiple models on integrated features, and evaluates their performance. Model weights are dynamically adjusted based on individual performance. The ensemble prediction is integrated and evaluated, guiding weight adjustments for optimal ensemble performance. The finalized ensemble model offers robust predictions, combining the strengths of individual models for improved mental health analysis.

4 Result Analysis

This extensive comparison compares mental health study's key machine learning technologies. It uses two tables and numerous graphics. Table 1 summarizes accuracy, sensitivity, specificity, precision, F1 score, and AUC-ROC. The recommended technique routinely outperforms current methods, suggesting it may be superior at diagnosis and prediction. Table 2 details the comparison performance analysis. We consider net present value, computation cost, FPR, FNR, AUC-PR, and MCC. Again, the recommended strategy outperforms all of them, suggesting it might be beneficial for mental health diagnosis and prognosis.

A bar chart demonstrating accuracy, a line chart showing sensitivity, specificity, and precision trends, a stacked bar chart providing performance metrics, and an area chart

Table 2. Performance Evaluation Comparison of Machine Learning Methods in MH Research

Proposed Method	Explain able AI	Transfer Learning	Ensemble Learning	Genomic Data Analysis	Deep Learning	Predictive Modeling	Reinforceme nt Learning	Unsupervised Learning	Natural Language Processing	Supervised Learning	Method
0.92	0.76	0.81	0.89	0.83	0.91	0.88	0.79	0.72	0.78	0.85	Accuracy
0.93	0.78	0.83	0.90	0.85	0.92	0.89	0.82	0.74	0.80	0.87	Sensitivity
0.91	0.74	0.79	0.88	0.81	0.90	0.87	0.76	0.70	0.76	0.82	Specificity
0.94	0.77	0.80	0.91	0.84	0.91	0.90	0.80	0.71	0.79	0.88	Precision
0.92	0.76	0.81	0.89	0.83	0.91	0.88	0.78	0.72	0.77	0.85	F1 Score
0.97	0.82	0.88	0.93	0.90	0.96	0.94	0.88	0.73	0.85	0.92	AUC-ROC
Moderate	Low	Low	High	Moderate	High	Moderate	High	Moderate	Low	Moderate	Computational Cost

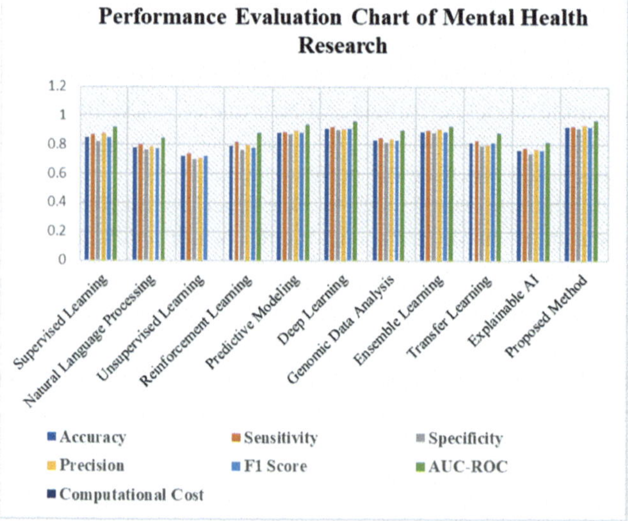

Fig. 6. Performance Evaluation Chart of Mental Health Research

showing FPR and FNR dynamics demonstrate that the suggested technique is superior. These images demonstrate how the proposed technique shines in various areas, making it an interesting new MH study tool.

Table 1 compares machine learning algorithms using accuracy, sensitivity, specificity, precision, F1 score, and AUC-ROC. In all tests, the new technique outperforms current ones, suggesting it might improve MH research for diagnosis and prediction.

Figure 6 demonstrates how reliable some MH study methodologies are. The recommended method is superior than current ones, suggesting it might improve diagnosis and

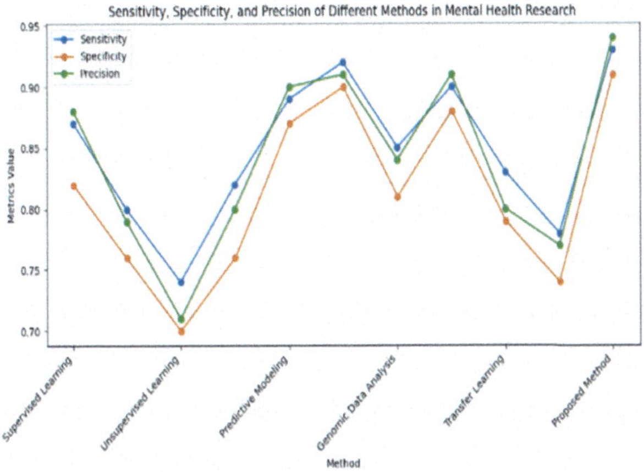

Fig. 7. Sensitivity, Specificity, and Precision Trends

prediction. Figure 7 compares technique sensitivity, specificity, and accuracy trends. The recommended strategy consistently yields higher values, indicating greater MH research.

5 Conclusion

Finally, this strategy advances machine learning in MH research. The holistic technique, which employs several algorithms, diagnoses and predicts issues well. Our ablation and performance testing demonstrate the framework's synergy and longevity. Ensemble learning improves predictions, demonstrating the need of combining models. Visualizing the findings highlights the method's strengths, making them easier to interpret. These discoveries improve psychiatric research and understanding of MH issues. The framework's adaptability to textual and genetic data reveals its versatility and potential. As MH is a worldwide issue, this model can help researchers and practitioners gain accurate and complete insights. To make the framework applicable in various healthcare contexts, algorithms will be fine-tuned, real-world data added, and clinical validations performed.

References

1. Shakeri, J., et al.: The prevalence of panic disorder and its related factor in hospitalized patients with chest pain and normal angiography. J. Educ. Health Promot. **8**(1), 61 (2019)
2. Campbell, K.A., Madva, E.N., Villegas, A.C., et al.: Non-cardiac chest pain: a review for the consultation-liaison psychiatrist. Psychosomatics **58**(3), 252–265 (2017)
3. Radwan, R.M., et al.: Developing secure and interoperable health information systems using blockchain technology to enhance data privacy, security, and accessibility in healthcare. Panam. Math. J. **34**(4), 300–15 (2024). https://doi.org/10.52783/pmj.v34.i4.1885
4. Radwan, R.M., et,al.: Optimizing electronic health records for enhanced clinical decision support and improved patient care through advanced data integration and machine learning algorithms. Panam. Math. J. **34**(4), 282–299 (2024).https://doi.org/10.52783/pmj.v34.i4.1884

5. Hendi, A.M., et al.: Adaptive method for exploring deep learning techniques for subtyping and prediction of liver disease. Appl. Sci. **14**(4), 1488 (2024). https://doi.org/10.3390/app14041488
6. Sayed-Ahmed, N.M.Z., et al.: Mathematical modelling and deep learning techniques for predicting. Math. Model. **34**(4), 230–244 (2024). https://doi.org/10.52783/pmj.v34.i4.1880
7. Mittal, R., et al.: RT-NeuroDDSM: real-time EEG-driven diagnostic decision support model for neurological disorders using deep learning IEEE Access **1** (2024).https://doi.org/10.1109/access.2024.3436829
8. Mittal, R., Jeribi, F., Martin, R.J., Malik, V., Menachery, S.J., Singh, J.: DermCDSM: clinical decision support model for dermatosis using systematic approaches of machine learning and deep learning. IEEE Access **12**, 47319–47337 (2024). https://doi.org/10.1109/access.2024.3373539
9. Malik, V., et al.: Coronary artery disease prediction using enhanced multi-layer DCNN. In: 2018 5th IEEE UPCON, pp. 1176–1180 (2023). https://doi.org/10.1109/upcon59197.2023.10434695
10. Martin, J.R., Swapna, S.L.: A machine learning framework for epileptic seizure detection by analyzing EEG signals. Int. J. Comput. Digit. Syst. **11**(1), 1383–1391 (2021). https://doi.org/10.12785/ijcds/1101112
11. John Martin, R., et al.: Self-attention based edge computing model for synthesis image to text through next-generation AI mechanism. Math. Probl. Eng. 12 (2022). 4973535. https://doi.org/10.1155/2022/4973535
12. Alhameed, M., et al.: Pandemic disease detection through wireless communication using infrared image based on deep learning. Math. Biosci. Eng. **20**(1), 1083–1105 (2023). https://doi.org/10.3934/mbe.2023050
13. Janarthanan, R., et al.: Biomedical image retrieval using optimized adaptive neuro-fuzzy optimized classifier system. Math. Bio. Eng. **19**(8), 8132–8151 (2022). https://doi.org/10.3934/mbe.2022380
14. Alam, M.I., Khan, H., Alam, M.Z., Siddiqui, S.T., Haider, A.S., Khan, M.R.: Preliminary diagnosis of diabetes through comparative analysis of supervised machine learning techniques. Nanotechnol. Life Sci. pp. 415–429 (2024). https://doi.org/10.1007/978-3-031-72004-8_22
15. Alshanketi, F., Alharbi, A., Kuruvilla, M., et al.: Pneumonia Detection from Chest X-Ray Images Using Deep Learning and Transfer Learning for Imbalanced Datasets. J Digit Imaging. Inform. Med. (2024).https://doi.org/10.1007/s10278-024-01334-0
16. Ejaz, K., Suaib, N.B.M., Kamal, M.S., Rahim, M.S.M., Rana, N.: Segmentation method of deterministic feature clustering for identification of brain tumor using MRI. IEEE Access **11**, 39695–39712 (2023). https://doi.org/10.1109/access.2023.3263798
17. Ejaz, K., Rahim, M.S.M., Bajwa, U.I., Rana, N., Rehman, A.: An unsupervised learning with feature approach for brain tumor segmentation using magnetic resonance imaging. In: ICBBB, Singapore Jan 7–9, 2019, pp. 1–7 (2019). https://doi.org/10.1145/3314367.3314384
18. Al-Atram, A.A.: Physicians' knowledge and attitude towards mental health in Saudi Arabia. Ethiop. J. Health Sci. **28**(6), 771–778 (2018)
19. Kashyap, R.: Stochastic dilated residual ghost model for breast cancer detection. J. Digit. Imaging **36**, 562–573 (2023). https://doi.org/10.1007/s10278-022-00739-z
20. Ali, A.A., Khan, A.B.F., Ramakrishnan, J.: Fuzzy rules-based data analytics and machine learning for prognosis and early diagnosis of coronary heart disease. J. Inf. Organ. Sci. **48**(1), 167–181 (2024)
21. Mavaluru, D., et al.: Advancing fetal ultrasound diagnostics: Innovative methodologies for improved accuracy in detecting down syndrome. Med. Eng. Phys. **126**, 104132 (2024). https://doi.org/10.1016/j.medengphy.2024.104132

22. Ramakrishnan, J., Doss, R., Palaniswamy, T., Faqihi, R., Fathima, D., Srinivasan, K.: High performance computation of human computer interface for neurodegenerative individuals using eye movements and deep learning technique. J. Supercomput. **78**(2), 2772–2792 (2021). https://doi.org/10.1007/s11227-021-03932-z
23. Ramakrishnan, J., et al.: Brain–computer interface for amyotrophic lateral sclerosis patients using deep learning network. Neural Comput. Appl. **34**(16), 13439–13453 (2020). https://doi.org/10.1007/s00521-020-05026-y
24. Thangaraj, R., Pandiyan, P., Ramakrishnan, J., Nallakumar, R., Eswaran, S.: A deep convolution neural network for automated COVID-19 disease detection using chest X-ray images. Healthcare analytics **4**, 100278 (2023). https://doi.org/10.1016/j.health.2023.100278
25. Sahu, H.P., Kashyap, R.: FINE_DENSEIGANET: automatic medical image classification in chest CT scan using hybrid deep learning framework. Int. J. Image Graph. **25**(01), 2550004 (2025). https://doi.org/10.1142/s0219467825500044
26. Rutledge, R.B., Chekroud, A.M., Huys, Q.J.: Machine learning and big data in psychiatry: toward clinical applications. Curr. Opin. Neurobiol. **55**, 152–159 (2019). https://doi.org/10.1016/j.conb.2019.02.006

Interpretation of the Relationship Between Sustainable Development Goals and Disasters in Brazil Through Machine Learning

Renata Ribeiro dos Santos[1](✉), Norma F. L. da Silva Valencio[1], and Heloisa de Arruda Camargo[2]

[1] Department of Environmental Sciences, Federal University of São Carlos, São Carlos, SP, Brazil
renataribeiro@estudante.ufscar.br, norma.valencio@ufscar.br

[2] Department of Computing, Federal University of São Carlos, São Carlos, SP, Brazil
heloisa@dc.ufscar.br

Abstract. A disaster transforms the reality of those affected, as it results in the loss of homes, schools, jobs, biodiversity, and non-material assets. The aspects influenced by a disaster make up the Sustainable Development Goals (SDG), seventeen global challenges that must be overcome by 2030. The progress of compliance with the SDG can be monitored by indicators that reflect the objectives according to the specificities of each country. Due to the social, economic, and environmental correlation between SDG and disasters, it is important to consider computational strategies, such as Machine Learning (ML), that enable the analysis of these two scenarios. In this article, we describe and compare the performance and difficulties of six ML experiments to generate fuzzy rules that enable the analysis of the relationship between disasters in Brazil and social and economic indicators that can be further related to SDG. To this end, we used the Fuzzy Association Rule-Based Classification Model (FARC-HD). There are contributions in the literature regarding ML models used in the disaster scenario, but the interrelationship between the SDG and events that affected Brazil has not yet been explored. The results indicated a preliminary set of indicators that can be used to associate SDG 1 and 10 with disasters. Although this article focuses on the Brazilian context, the methodologies and considerations presented can be extended to other scopes.

Keywords: Machine Learning · Sustainable Development Goals · Disaster · Supervised Learning · FARC-HD · Brazil · C4.5 · Classification · Fuzzy Rules · SDG

This study was financed by Fundação de Amparo à Pesquisa do Estado de São Paulo (Fapesp), process number 2022/09136-1.R.S.S. thanks Fapesp for supporting this research in the form of a Postdoctoral Fellowship, process number 2024/03793-6.

S. Latifi (Ed.): ITNG 2025, AISC 1463, pp. 242–253, 2025.
https://doi.org/10.1007/978-3-031-89063-5_21

1 Introduction

According to the United Nations Office for Disaster Risk Reduction (UNDRR), a disaster is a disruption of the organization of a society expressed by human, material, economic, and environmental collective losses and damages [1]. Furthermore, it is a social phenomenon, triggered by a physical event that can profoundly affect the mental health of survivors [2]. In this context, it is important to consider that a disaster acts as a disturbing and disruptive factor in the dynamics and reality of those affected, directly affecting different aspects, such as income, dwelling, health, work, education, and basic sanitation.

Some of the aspects related to a disaster are covered in the Sustainable Development Goals (SDG), a set of seventeen objectives established by the United Nations (UNESCO) to address global challenges by 2030. These objectives include No poverty (SDG 1), Zero hunger (SDG 2), Quality education (SDG 4), Clean water and sanitation (SDG 6), and Reduced inequalities (SDG 10).

Several qualitative and quantitative indicators can be used to measure and monitor the progress of several SDG or otherwise of a specified SDG. To express each objective, relevant information must be considered, available, easy to understand by different target audiences, and reflect the reality and particularities of each country [3]. As there is no fixed set of indicators for each SDG, several characteristics can be chosen.

In the case of Brazil, we can cite as an example of an indicator for SDG 1 and 10 the Bolsa Família Program (BFP), which aims to transfer a monthly income to low-income families. Another example related to SDG 10 is the Cadastro Único (CadÚnico), whose purpose is to register and identify low-income families in the country for the selection, inclusion, and granting of benefits offered by various welfare programs, such as the BFP.

A disaster can be a barrier to the evolution of SDG indicators in the desired direction for both government authorities and local society, and existing public datasets may not be well explored enough to point to this. Therefore, it is relevant to adopt computational strategies that consider the possible correlation between a disaster's occurrence and the SDG's progress. In this scenario, supervised Machine Learning (ML) models can be considered to generate fuzzy rules enabling interpretation and understanding of these contexts. An example of a model that can be used for this purpose is the Fuzzy Association Rule-Based Classification Model (FARC-HD) [4], a classification model based on fuzzy association rules.

In this article, we describe and compare the performance and difficulties of six ML experiments that were carried out to generate fuzzy rules focusing on the interpretability to analyze the relationships between disasters in Brazil and social and economic indicators that can be further related to SDG. For this purpose, the supervised model FARC-HD was used. We considered indicators inherent to Brazilian reality related to SDG 1 and 10. The period considered for the experiments was from 2016 to 2021 due to the complete availability of the data used to express the SDG. Although our contributions are related to the Brazilian domain, the methodology and reflections presented in this article can be adapted and extended to the reality of other scenarios, that is, other countries or locations at different geographic scales. In the literature, examples of works use ML techniques in the context of disasters. However, the association between the SDG and events in the context of Brazil has yet to be more profitably exploited.

This paper is organized as follows: Section 2 describes the theoretical foundation related to disasters in Brazil and supervised machine learning. Section 3 presents the related work. Section 4 details and compares the six experiments performed to generate fuzzy rules focusing on the interpretability of the relationship between SDG and disasters in Brazil. Finally, Section 5 presents the conclusion and future work.

2 Theoretical Background

In Brazil, the occurrence of a disaster is recorded by the Integrated Disaster Information System (S2iD)[1], which is an online platform maintained by the federal government through the Ministry of Integration and Regional Development. These records are kept at the municipal level, and a municipality may declare, as a state of emergency or public calamity, one or more disasters yearly.

Disasters are classified and coded according to the Brazilian Disaster Classification and Codification (Cobrade)[2], and the main types of events that disasters are associated with are: "natural" and "technological". The main types of events, to which some of the disasters called as "natural" are being associated, are: flooding, drought, and landslides. A "technological" disaster results from human actions or infrastructure built by human beings. An example of this type of disaster is the release of chemicals into drinking water systems. Each type of disaster is classified with a numerical code. Numeric codes express the nature of physical events associated with disasters in a given subgroup. For example, the hydrological group includes flooding, downpours, and inundation events. Each disaster is associated with one of two types of emergency declaration, namely: State of Emergency (SE) or State of Public Calamity (SCP) decree issued by the municipal or state government. The difference between SE and SCP is the response time to deal with the disaster, which is longer for SCP. In addition, the SE lasts ninety days, while the SCP lasts one hundred and eighty days.

An example of a computational scenario that can be used to associate SDG with disasters is supervised ML. In this context, a set of labeled data is used to train a model in order to generate knowledge. Different models can be considered for this purpose, including those that result in fuzzy rules [5]. In a fuzzy system, in the fuzzification stage, the numerical data from the input fuzzy variables are transformed into linguistic terms that form a fuzzy set. For this purpose, a membership function is used, which can be, for example, triangular or trapezoidal. The fuzzy variables and their respective terms comprise fuzzy rules in If (antecedent) Then (consequent).

3 Related Work

In the literature, some contributions consider using ML techniques in the context of disasters. For example, Arinta and WR [6] identified works that present contributions in different areas of natural disaster management, such as forecasting analysis and response

[1] https://s2id.mi.gov.br/.

[2] https://www.gov.br/mdr/pt-br/centrais-de-conteudo/publicacoes/protecao-E-defesa-civil-sedec.

to damage. In a complementary way, Linardos *et al.* [7] present a systematic review in which several studies were identified and described in which ML models were used to manage natural disasters, from monitoring and warning to damage assessment. Carvalho *et al.* [8] present a model based on fuzzy logic to assess the level of maturity in disaster risk reduction, using the Hyogo Framework for Action (HFA) as a basis. Sheykhmousa *et al.* [9] consider satellite data for the analysis of the recovery of disaster-hit sites. Although these works present contributions regarding the use of MA strategies in the disaster scenario, the SDG's particularities in Brazil's scope are not considered for this purpose.

4 Experiments

In this section, we describe and compare the experiments performed to generate the fuzzy rules, focusing on the interpretability to analyze the relationships between disasters in Brazil and social and economic indicators that can be further related to SDG. The fuzzy rules were modeled as a classification problem. The antecedent of the rules is formed by indicators related to disaster decrees and social and economic information related to SDG 1 and 10. The consequent represents the recurrence of disasters in the municipalities and, in others, the type of event to which the disaster is related.

We considered different indicators, associations between data, values (numerical or categorical), and temporal granularities (monthly or annual). The indicators come from public databases and those related to government agencies, which reflect the security of the information due to the source's reliability. All data are municipal, as this is the level considered by S2iD for recording disasters in Brazil. In addition, the experiments are annual.

The data for each indicator were analyzed to identify the absence of values. If there were any, the value zero was assigned. The experiments were organized in tables, in which the rows correspond to the municipalities, and each column is an indicator, with the last one corresponding to the class (consequent to the fuzzy rules). Duplicate rows were automatically removed to avoid overfitting and consequent distortion of the results.

The experiments were performed in the Keel software[3], in which the FARC-HD model is implemented. In this environment, fuzzification and fuzzy rules are generated automatically. Before performing the experiments, the k-fold cross validation technique ($k = 10$) was selected, in which the input set is partitioned for conducting experiments and evaluating performance (accuracy). To compare and analyze the results and performance, the C4.5 classification algorithm was also considered, a decision tree, in which non-fuzzy rules are generated [10]. This algorithm is also available in Keel.

The following metrics were considered to evaluate the model's performance: precision, recall, and F1-Score. Precision measures the number of correct predictions in relation to the total predictions made for a class. Recall is the ratio between the number of correct predictions and the number of elements in a class according to the actual classification. To calculate the F1-Score, the harmonic mean between precision and recall is calculated. These metrics were calculated based on the confusion matrix, which

[3] https://www.keel.es/.

organizes, for each class, the distribution of the input data set according to the actual classification and the classification predicted by the model. The confusion matrix is automatically generated by Keel. The experiments are described below.

4.1 Experiment 1

In the first experiment, the following monthly indicators associated with each municipality were considered:

- Month: month of the year (2016–2023); assumes numerical values from 1 to 12;
- Start: start of validity of a decree due to the occurrence of at least one disaster in the month; assumes the categorical values yes or no;
- Validity: validity of at least one decree that occurred one, two, or three months before the month analysed; assumes the categorical values yes or no;
- Ratio between the number of families benefiting from the BFP and the monthly average, by municipal size; assumes continuous numerical values;
- Ratio between the number of families benefiting from the BFP and the monthly average, by municipal size and state; assumes continuous numerical values;
- Ratio between the number of families benefiting from the BFP and the monthly average, by municipal size and region; assumes continuous numerical values;
- Ratio between the number of people benefiting from the BFP and the monthly average, by municipal size; assumes continuous numeric values;
- Ratio between the number of people benefiting from the BFP and the monthly average, by municipal size and state; assumes continuous numeric values;
- Ratio between the number of people benefiting from the BFP and the monthly average, by municipal size and region; assumes continuous numeric values;

As explained in Sect. 1, the purpose of the BFP is to transfer funds to low-income families. The data related to this program are monthly and the responsibility of the Ministry of Development and Social Assistance, Family and Fight against Hunger[4]. It should be explained that from November 2021 to March 2023, the BFP was replaced by the Auxílio Brasil program, which also transferred income to needy families. In this article, we group the data related to this period with those of the BFP. These data have been fully available since 2016; therefore, all the experiments mentioned in this article began that year.

The information on the start and validity of a decree relates the occurrence of a disaster to the BFP assistance policy. For the validity indicator, an interval of up to three months was considered since it corresponds to the SE period, which was the most frequent declaration in Brazil from 2016 to 2023. For the experiment, the municipalities were grouped by municipal size, which is related to the size of the municipality based on the number of inhabitants. In Brazil, there are seven strata, which were delimited by the Brazilian Institute of Statistics and Geography (IBGE). In addition, the country comprises twenty-six states, plus a Federal District, and is divided into five regions (Central-West, Northeast, North, Southeast, and South). As a class for the fuzzy rules, the number of disaster occurrences from 2016 to 2023 was considered for each municipality,

[4] https://aplicacoes.cidadania.gov.br/vis/data3/data-explorer.php.

according to data from S2iD. Thus, we established five classes of emergency declarations due to disasters, namely: low (zero), medium (one), high (two to four), large (five to seven), and huge (eight or more).

In this experiment, the FARC-HD model presented low values for accuracy, such as 45% for 2016, which was reflected in the average values of precision (20%) and recall (30%). It should be noted that there is an imbalance between the number of elements in each class; low, medium, high, large, and huge have, respectively, 540, 318, 1783, 1472, and 1457 elements. This discrepancy may be one of the factors that contributed to the inferior performance of the FARC-HD model.

In comparison, the C4.5 algorithm presented a higher accuracy value (81%), average precision (80%), and recall (79%) values. One weakness observed is the large number of algorithm-generated non-fuzzy rules. In the case of 2016, for example, approximately 5686 rules were identified. The data used in the experiment are mostly continuous numerical values that present little monthly difference for the same municipality. Thus, the decision tree divides these values into intervals, which increases the number of rules generated. This limitation compromises the understanding and visualization of the result.

4.2 Experiment 2

In the second experiment, we considered the same indicators as in Experiment 1. However, the class was changed to the number of yearly disaster occurrences. For each year, we observed a low number of annual disasters. This resulted in many elements being classified as low-class, as most municipalities did not declare any disaster during the year. As exceptions, we point out the years 2020 and 2021 due to the SARS-CoV-2 pandemic.

For the year 2016, for example, the data were classified as low, medium, and high. The FARC-HD model presented an accuracy of 85%. However, the precision, recall, and F1- Score values for the medium class were 55%, 48%, and 52%, respectively. These values were also not very expressive for the high class, which presented a precision of 64%, recall of 43%, and F1-Score equal to 51%.

In 2020, the elements were distributed into four classes: low, medium, high, and large. There is a concentration in the high class, which means that two to four emergency declarations occurred in the year. For this year, the model presented an accuracy of 75%, but an average value of precision, recall, and F1-Score, respectively, equal to 36%, 46%, and 41%. The medium and large classes presented a value equal to zero for these metrics.

In comparison, the C4.5 algorithm performed better for the years shown. For 2016, the accuracy was 91%, with an average precision of 83%, recall of 81% and F1-Score of 82%. For 2020, we obtained 92% accuracy, 90% average precision, and 84% and 87% for the average recall and F1-Score. Although there were differences in the performance of the FARC-HD compared to the C4.5 algorithm, it isn´t possible to explain the reason, as happens in typical ML experiments.

Due to the poor performance of the FARC-HD model in Experiments 1 and 2, we calculated Pearson Correlation to verify the degree of association between the numerical categorical variables. The results showed that some are highly correlated, constituting information redundant in ML. In addition to the class imbalance, another limitation was the proximity between the numerical categorical values. It occurs because there is a slight variation between the monthly values of the families and individuals included in the BFP. Thus, the fuzzy sets are not adequately divided due to the significant overlap between the terms.

Due to the Pearson Correlation result, we considered other associations between the indicators. In Experiment 2.1, we did not consider data related to people benefiting from the BFP; for Experiment 2.2, we removed data related to families; and in Experiment 2.3, we used only the information related to families and people included in the program. In the case of 2016, for Experiment 2.1, the FARC-HD model presented an accuracy of 84% and good performance for the low class. However, it presented lower values for recall and F1-Score for the medium and high classes, as well as for the precision of the medium class. Experiment 2.2 resulted in an accuracy of 85% but low recall and F1-Score values for the medium class. Experiment 2.3 presented the worst performance, with an accuracy of 75% and values equal to zero for precision, recall, and F1-Score for the medium and high classes. Figure 1 shows images of tables with the performance of the FARC-HD model for these experiments in 2016 in terms of precision, recall, and F1-Score.

Experiment 2.1

	Precision	Recall	F1-Score
Low	91%	98%	94%
Medium	53%	57%	55%
High	75%	24%	36%
Average	73%	60%	62%

Experiment 2.2

	Precision	Recall	F1-Score
Low	92%	98%	95%
Medium	61%	30%	40%
High	56%	77%	65%
Average	70%	68%	67%

Experiment 2.3

	Precision	Recall	F1-Score
Low	75%	100%	86%
Medium	0%	0%	0%
High	0%	0%	0%
Average	25%	33%	29%

Fig. 1. Performance of the FARC-HD model for Experiments 2.1, 2.2 and 2.3 regarding precision, recall and F1-Score in 2016.

4.3 Experiment 3

In this experiment, we adopted indicators related to different aspects, such as welfare policies and social inequality, which expanded the scope of the fuzzy rule. The annual indicators associated with each municipality are listed below.

- Average monthly number of families registered in CadÚnico: assumes continuous numeric values;

- Average monthly number of people registered in CadÚnico: assumes continuous numeric values;
- Average monthly number of families benefiting from the BFP: assumes continuous numeric values;
- Average monthly number of people benefiting from the BFP: assumes continuous numeric values;
- Urbanization rate: assumes continuous numeric values;
- Gini index: assumes continuous numeric values;
- Municipal Human Development Index (HDI M): assumes continuous numeric values;

As we pointed out in Section I, CadÚnico records information identifying low-income Brazilian families. This information was made available by the Secretariat for Assessment, Information Management and Single Registry (Sagicad)[5]. The urbanization rate indicates the percentage of inhabitants living in urban areas to the total population. This data was retrieved from the National Sanitation Information System (SNIS)[6] and is available until 2021. The Gini Index measures income concentration, which reflects social inequality. The data for this indicator come from DataSUS[7]. The Municipal Human Development Index (MHDI) measures municipal development based on longevity, education, and income. The source for this indicator is the Atlas of Human Development in Brazil[8].

The most recent data for the Gini Index and MHDI are from 2010 since a more updated version is not yet available. We should explain that the urbanization rate, Gini Index, and MHDI indicators are annual, while the information from BFP and CadÚnico is monthly. Therefore, to maintain the same temporal granularity, we calculated the average of the monthly number of families and individuals registered.

To directly associate the indicators with a natural events, we considered five extracts as classes, according to Cobrade groupings. The classes are drought, dry spell, hydrological (flood, downpour, and inundation), meteorological (local/convective storm - tornado, lightning storm, hail, heavy rain, or strong winds), and none (the municipality declared no emergency during the year). A municipality may have more than one class due to the possibility of one or more events associated with disasters. According to Cobrade, drought and dry spell events belong to the climatological group; however, we considered each event separately due to the large number of occurrences of these types of disasters. We selected these types of events due to their high recurrence in Brazil during the period considered.

For this experiment, we observed many elements classified in the none class according to the weighting performed in Experiment 2 for the low class. In 2016, for example, the model presented an accuracy of 74%, with average precision, recall, and F1-Score values, respectively, equal to 15%, 20%, and 17%. When considering an equal number of elements between the classes, the model's accuracy dropped to 34%. For this case, random and automatic line removal was performed to consider a sample of elements from each class.

[5] https://aplicacoes.cidadania.gov.br/vis/data3/data-explorer.php.

[6] http://app4.mdr.gov.br/serieHistorica/

[7] https://tabnet.datasus.gov.br/cgi/ibge/censo/cnv/ginibr.def.

[8] https://www.atlasbrasil.org.br/.

4.4 Experiment 4

In this experiment, we categorized the continuous numerical values of the indicators from Experiment 3. Thus, by design decision, we established the following categories for the indicators:

- Highest and Lowest for the average monthly number of families and individuals registered in CadÚnico and families and individuals benefiting from the BFP about the municipal size;
- A (0 to 0.19), B (0.20 to 0.39), C (0.40 to 0.59), D (0.60 to 0.79) and E (0.80 to 1) for the urbanization rate;
- A (0 to 0.29), B (0.30 to 0.34), C (0.35 to 0.39), D (0.40 to 0.44), E (0.45 to 0.49), F (0.50 to 0.54) and G (0.55 to 1) for the Gini Index;
- Very Low (0 to 0.4999), Low (0.50 to 0.5999), Medium (0.60 to 0.6999), High (0.70 to 0.7999), and Very High (0.80 to 1) for MHDI.

The categorization of the indicators did not benefit the performance of the FARC-HD model or the C4.5 algorithm. For the first, the accuracy value was 50%, and for the second, it was 51%.

4.5 Experiment 5

In Experiment 5, we considered the same indicators as in Experiment 3 as continuous numerical values. The change is in the class definition, which assumes binary values yes or no. The value yes indicates that some type of disaster occurred in the municipality for the year. The value no indicates that no event occurred. Although this classification is comprehensive and makes it impossible to associate it with a specific type of disaster, it is still possible to record the occurrence of an event.

In this scenario, we still observe the discrepancy between the classes, as pointed out in previous experiments. For the FARC-HD model, the accuracy in 2016 was 76%, but the precision, recall, and F1-Score values for the yes class were, respectively, 56%, 19%, and 29%. In this experiment, the C4.5 algorithm presented a similar performance, with an accuracy value of 75%.

4.6 Experiment 6

To correct and balance the imbalance between classes, in Experiment 6, we considered an equal set of samples for each class. To do so, we randomly and automatically removed some elements from the class with the highest occurrence in the year. We opted for random removal so that it would be possible to maintain examples of municipalities from all Brazilian regions and thus obtain an overview of the country's diversity. For the years 2016 to 2019, we excluded elements related to the no class due to the low occurrence of disasters in those years. In the case of the years 2020 and 2021, due to the SARS-CoV-2 pandemic, we removed elements from the yes class because of the high number of Brazilian municipalities that signaled the occurrence of a disaster to face the disease.

Due to the balance between classes, we observed that the FARC-HD model presented the best accuracy performance when compared to previous experiments. This consideration can also be observed for the precision, recall, and F1-Score metrics because similar values were verified between the classes and the average. Figure 2 shows images of tables with the accuracy, precision, recall, and F1-Score values for 2016 to 2021, according to the FARC-HD model.

We repeated this experiment for the C4.5 algorithm and identified accuracy values equivalent to those obtained in the FARC-HD model. For the years 2016, 2017, 2018, 2019, 2020, and 2021, accuracy values were obtained, respectively, equal to 74%, 70%, 74%, 75%, 71%, and 67%.

The following are the fuzzy rules and respective confidence values (CF) generated, in Keel, by the execution of the FARC- HD model for the year 2016. The notation L_1, L_2, L_3, L_4, and L_5 represents the linguistic terms, which indicates that each indicator presents five fuzzy sets that were defined by a triangular membership function. The linguistic terms can be interpreted as very low (L_1), low (L_2), medium (L_3), high (L_4), and very high (L_5). The antecedents of each fuzzy rule are formed by the indicators considered in Experiment 6. The consequent corresponds to the yes or no class; which indicates the occurrence or not of a disaster in the municipality.

- **Rule 1:** MHDI IS L 4(5): no CF: 0.80
- **Rule 2:** CadÚnico Family IS L 1(5)AND CadÚnico_Person IS L_0(5): no CF: 0.44
- **Rule 3:** CadÚnico_Family IS L_0(5) AND BolsaFamília-Family IS L_1(5) AND MHDI IS L_3(5): yes CF: 0.62
- **Rule 4:** CadÚnico_Person IS L_0(5) AND MHDI IS L 2(5): yes CF: 0.74
- **Rule 5:** CadÚnico_Person IS L_0(5) AND BolsaFamília_Family IS L_1(5) AND MHDI IS L_3(5): yes CF: 0.62
- **Rule6:** CadÚnico_Person IS L_0(5) AND Bol-saFamília_Family IS L_1(5) AND Gini Index IS L_3(5): yes CF: 0.63
- **Rule 7:** BolsaFamília_Family IS L_1(5) AND Gini Index IS L_3(5): yes CF: 0.62
- **Rule 8:** BolsaFamília_Person IS L_1(5) AND Gini Index IS L_3(5) AND MHDI IS L_3(5): yes CF: 0.66

Fuzzy rules follow the If Then format. Thus, the reading of Rule 3, for example, should be:

IF CadUnico_Family is very low and BolsaFamília_Famıly is low and MHDI is medium THEN disaster is yes.

From this rule, we interpret that when the average number of families registered in CadÚnico and Bolsa Família is very low and low, respectively, and the MHDI is medium, then a disaster has occurred. The confidence level of the rule is 62%.

In general, we observe that the fuzzy rules presented above are formed by combining one or more indicators,increasing the possibility of interpretation between SDG and disasters. By Rule 1, we can interpret that municipalities with high values for the MHDI were unaffected by disasters in 2016. Most fuzzy rules (Rule 2 to 8) are formed by social indicators, which indicate a relationship between this information and the occurrence of a disaster. Many rules indicate that disasters occurred in municipalities that present

2016

	Precision	Recall	F1-score
no	77%	69%	73%
yes	72%	79%	75%
Average	74%	74%	74%

Acuracy: 74%

2017

	Precision	Recall	F1-score
no	71%	70%	70%
yes	70%	71%	71%
Average	70%	70%	70%

Acuracy: 70%

2018

	Precision	Recall	F1-score
no	75%	69%	72%
yes	71%	78%	74%
Average	73%	73%	73%

Acuracy: 73%

2019

	Precision	Recall	F1-score
no	78%	70%	73%
yes	72%	80%	76%
Average	75%	75%	75%

Acuracy: 75%

2020

	Precision	Recall	F1-score
yes	70%	77%	73%
no	74%	66%	70%
Average	72%	72%	72%

Acuracy: 71%

2021

	Precision	Recall	F1-score
yes	67%	67%	67%
no	67%	67%	67%
Average	67%	67%	67%

Acuracy: 67%

Fig. 2. FARC-HD model performance for Experiment 6 regarding accuracy, precision, recall and F1-Score.

a very low or low value for the average number of people and families registered in CadÚnico.

5 Conclusion and Future Work

Disasters are disruptive events that affect the lives of survivors. This disruption of reality directly impacts social, economic, environmental, and emotional aspects. These aspects, such as income, social inequality, and work, are considered in the SDG as goals to be achieved by the year 2030. Thus, the occurrence of a disaster is a factor that can compromise the progress of the SDG. To analyze this possible correlation, supervised Machine Learning strategies with a focus on interpretability can be used.

In this article, we present ML experiments that were carried out to generate fuzzy rules focused on interpreting the relationship between disasters in Brazil and social and economic indicators that can be further related to SDG. To this end, indicators for SDG 1 and 10 associated with the Brazilian reality were considered. We identified a preliminary set of information that can be used for this purpose and presented good performance in the FARC-HD model regarding accuracy, precision, recall, and F1-Score. In addition, associated with the indicators, we established that an equal set between the yes and no classes regarding the occurrence of a disaster is the most appropriate for the performance of the analyzed model.

In future work, other indicators will be considered to increase the number of SDG covered by the fuzzy rules. Thus, we intend to consider aspects related to health, work, and basic sanitation. In addition, the viability of other classes will be analyzed so that it is possible to associate the indicators with a specific type of disaster.

References

1. United Nations Office for Disaster Risk Reduction (UNDRR), "Disaster risk reduction terminology". United Nations" (2017). https://www.undrr.org/terminology
2. Math, S.B., et al.: Disaster management: mental health perspective. Indian J. Psychol. Med. **37**(3), 261–271 (2015)
3. Hák, T., Janoušková, S., Moldan, B.: Sustainable development goals: a need for relevant indicators. Ecol. Ind. **60**, 565–573 (2016)
4. Alcalá-Fdez, J., Alcala, R., Herrera, F.: A fuzzy association rule-based classification model for high-dimensional problems with genetic rule selection and lateral tuning. IEEE Trans. Fuzzy Syst. **19**(5), 857–872 (2011)
5. L. A.Zadeh, "Fuzzy sets". Information and Control. 1965
6. Arinta, R.R., Emanuel Andi, W.R.: Natural disaster application on big data and machine learning: a review. In 2019 4th International Conference on Information Technology, Information Systems and Electrical Engineering (ICITISEE), pp. 249–254. IEEE (2019)
7. Linardos, V., et al.: Machine learning in disaster management: recent developments in methods and applications. Mach. Learn. Knowl. Extr. **4**(2) (2022). https://www.mdpi.com/2504-4990/4/2/20
8. Carvalho, P.V.R., et al.: A fuzzy model to assess disaster risk reduction maturity level based on the Hyogo framework for action. Nat. Hazards **83**, 309–326 (2016)
9. Sheykhmousa, M., et al.: Post-disaster recovery assessment with machine learning-derived land cover and land use information. Remote Sens. **11**(10), 1174 (2019). https://www.mdpi.com/2072-4292/11/10/1174
10. Quinlan, J.R.: C4. 5: programs for machine learning (1993)

Automated Machine Learning: A Case Study on Non-intrusive Appliance Load Monitoring

Armin Moin[1(✉)], Ukrit Wattanavaekin[2], Alexandra Lungu[2], Stephan Rössler[3], and Stephan Günnemann[2,4]

[1] Department of Computer Science, University of Colorado, Colorado Springs, CO, USA
amoin@uccs.edu

[2] School of Computation, Information, and Technology (CIT), Technical University of Munich, Munich, Germany
s.guennemann@tum.de

[3] Software AG, Munich, Germany
stephan.roessler@softwareag.com

[4] Munich Data Science Institute (MDSI), Munich, Germany

Abstract. We propose a novel approach to enable Automated Machine Learning (AutoML) for Non-Intrusive Appliance Load Monitoring (NIALM), also known as Energy Disaggregation, through Bayesian Optimization. NIALM offers a cost-effective alternative to smart meters for measuring the energy consumption of electric devices and appliances. NIALM methods analyze the entire power consumption signal of a household and predict the type of appliances as well as their individual power consumption (i.e., their contributions to the aggregated signal). We enable NIALM domain experts and practitioners who typically have no deep data analytics or Machine Learning (ML) skills to benefit from state-of-the-art ML approaches to NIALM. Further, we conduct a survey and benchmarking of the state of the art and show that in many cases, simple and basic ML models and algorithms, such as Decision Trees, outperform the state of the art. Finally, we present our open-source tool, AutoML4NIALM, which will facilitate the exploitation of existing methods for NIALM in the industry.

Keywords: automated machine learning · automl · non-intrusive appliance load monitoring · nialm · energy disaggregation

1 Introduction

Recently, Machine Learning (ML) has been applied to the application domain of *Non-Intrusive Appliance Load Monitoring (NIALM)*, also known as *Energy Disaggregation*. In fact, NIALM, which is an *inverse problem*, is very similar to the Single-channel Source Separation problem in Physics and Signal Processing.

S. Latifi (Ed.): ITNG 2025, AISC 1463, pp. 254–269, 2025.
https://doi.org/10.1007/978-3-031-89063-5_22

However, audio signals tend to have a much higher temporal resolution than energy measurements, making our problem more challenging than that one [21]. Here is the problem of NIALM in simple words: Let us assume that we have access to the information about the total electrical energy consumption of a household at certain points in time, e.g., every 15 min. NIALM is about finding out what electrical devices and appliances are likely to be in the house and estimating what their individual electrical energy consumption values have been over a certain period of time.

It is clear that one could achieve this goal through an *intrusive* method by attaching smart meters to appliances in a house and measuring their electrical energy consumption levels. However, that intrusive solution would definitely cost more, both for the procurement of sensors and also their maintenance. Moreover, that solution would not be convenient for the end users (occupants) and/or owners of buildings since they would need to allow the installation of sensors (smart meters) in their houses. This may also raise privacy concerns[1]. Therefore, the non-intrusive method through virtual sensing, also known as soft sensing, has become more popular. Generally, virtual sensing is usually performed either analytically by using some physical laws or empirically following a data-driven approach, such as using ML. In our context, only the latter is possible as we do not have any physical law that gives us the required information given the aggregated power signal (i.e., the total energy signal of a household). NIALM has a lot of applications, e.g., power grids (smart grids), which can better plan and balance their energy networks. Also, research (e.g., see [7]) showed that providing information about the power consumption of individual appliances to energy consumers often leads to more user awareness and usually results in higher energy savings.

This paper makes three contributions: (i) It proposes a novel approach based on Bayesian Optimization, with tool support to enable Automated Machine Learning (AutoML) for domain experts and practitioners in the field of NIALM. (ii) It conducts a survey and benchmarking of the state of the art and illustrates that a simple and basic ML approach, namely, Decision Trees, can outperform the more complex approaches to NIALM that exist in the literature. Our benchmarking is fair since we use the same dataset and settings for all experiments. So far, such a study is missing in the NIALM literature. (iii) It proposes new practical methods for NIALM, which employ Decision Trees - as one of the most basic ML algorithms - as well as Fully-Connected Neural Networks (FCNNs), also known as Multi-Layer Perceptrons (MLPs) - as the most basic architecture for Neural Networks. Further, it proposes another new method based on Gated Recurrent Units (GRU), a variation of Recurrent Neural Networks (RNN) that outperforms all the state-of-the-art approaches, including LSTMs, though - surprisingly - it does not yet perform on average better than Decision Trees.

[1] It is worth mentioning that privacy concerns may still be the case in virtual sensing via NIALM. However, the concerns and inconveniences are not as bold as in the intrusive scenario.

The rest of this paper is structured as follows: In Sect. 2, we survey the literature of NIALM and briefly present our benchmarking results for existing approaches. Then, we propose our novel approach for enabling Automated Machine Learning in the field of NIALM and present our open-source tool for supporting that in Sect. 3. Further, we illustrate our experimental results in Sect. 4, which also leads to an extended benchmarking that includes the new methods that we propose for NIALM applications. Finally, we conclude and suggest future work in Sect. 5.

2 Literature Survey and Benchmarking

In this section, we survey the literature on NIALM. Existing approaches can be categorized from various points of view. One way is to group them into those approaches that require high-frequency data (e.g., see [6,10,15,16,22–25,30,33,35]) and those that can work well even with low-frequency data (e.g., see [11,13,14,17,19,21,27,29,32,34,36,37]). High-frequency here means sampling rates in the order of more than one measurement per second, i.e., above 1 Hz, whereas low-frequency means sampling rates in the order of one or less than one measurement per second, i.e., below 1 Hz. Note that the former (i.e., high-frequency data) would typically require expensive equipment that makes such approaches useless in practice as conventional counters of households cannot afford such rates. On the contrary, practical approaches that can work with conventional frequencies, e.g., even as low as one measurement per 15 minutes or per hour (e.g., see [21]), are interesting to us.

Another way of categorizing existing NIALM approaches is to group them into state-based and event-based approaches, though the border is sometimes not sufficiently clear. State-based approaches model each appliance or various states of each appliance as states of a Finite State Machine (FSM) or as signatures used in a dictionary. These approaches are usually computationally more expensive but, in the end, often more accurate. Two of the many examples in the literature are those proposed in [27] (combined DNN-HMM model) and [14] (Dictionary-based). However, the more classical approach is to focus on the changes in the aggregated signal using signal edges without considering various models for appliances or their states.

One example of this classic category is the so-called training-less approach based on Spectral Clustering proposed in [36]. The authors, who were apparently not from the Machine Learning (ML) community, referred to their approach as non-ML. Nevertheless, their proposed approach is indeed an ML approach and is considered unsupervised ML, also known as clustering. However, we do not consider this approach - in its current stage - as practical either. First, like other event-based approaches, it has the drawback of being generally unable to properly disaggregate signals of appliances that are working simultaneously and appliances with varying loads (e.g., dishwashers or microwave ovens). In addition, one cannot use that unsupervised approach in practice without having a complementary mechanism to finally label the disaggregated signals with the name of the corresponding appliances.

Table 1 illustrates our benchmarking results for existing approaches, which we find practical for NIALM applications based on the above-mentioned arguments. Our comprehensive benchmarking will come in Sect. 4.

Table 1. Benchmarking of existing state-of-the-art approaches to NIALM

Rank	Paper	Algorithm
1st	Kelly & Knottenbelt 2015 [19]	Long Short-Term Memory (LSTM)
2nd	Kelly & Knottenbelt 2015 [19]	Denoising Autoencoders (DAE)
3rd	Elhamifar & Systry 2015 [14]	Dictionary-Based Approach
4th	Mauch & Yang 2016 [27]	Combined Deep Neural Networks (DNN) and Hidden Markov Models (HMM)
5th	E.g., Reyes-Gomez et al. 2003 [31]	Factorial Hidden Markov Models (FHMM)
6th	Hart 1992 [16]	Combinatorial Optimization (CO)

3 Proposed AutoML Approach

3.1 Problem Formulation

Given an aggregated energy signal of one house, say S, with n samples at n time instances, i.e., the sequence (time series) $[S_1, S_2, ..., S_n]$, the aim of NIALM is to find the m signals of the m appliances inside of the house (i.e., the set $\{s1, s2, ..., sm\}$), that constitute the aggregated signal. Based on that, one should ideally infer the type of appliances inside of the house that correspond to the m individual signals as well as their energy consumption at each time instance, i.e., the set of sequences (time series) $\{[s1_1, ..., s1_n], [s2_1, ..., s2_n], ..., [sm_1, ..., sm_n]\}$. In the following, $\hat{si_j}$ denotes the estimated or predicted value of the power consumption of appliance i at time instance j. Also, si_j denotes the actual value of the power consumption of appliance i at time instance j. This is a supervised ML (regression) problem since our data instances are labeled with numeric values of the total electrical energy consumption of a house. The input variables (ML features) are numeric values of the electrical energy consumption levels of various appliances in the household.

3.2 Automated ML for NIALM

Instead of having data scientists perform data preparation, feature selection, model building, and hyper-parameter selection and optimization, AutoML requires the computer itself to carry out these tasks automatically. To our knowledge, AutoML has not yet been applied to the domain of NIALM. Given the fact that we already have various ML algorithms and methods that can be used for NIALM in practice (see Sect. 2), but their exploitation in the industry has not yet been successful, we consider AutoML a vital need for the domain of NIALM.

Therefore, we propose our novel approach based on three pillars: (i) Choosing the best ML model and algorithm for NIALM shall be automated. (ii) Choosing the best hyper-parameters for the model shall be automated. (iii) Supporting novice users by providing them with tips so that they can avoid typical mistakes that can lead to false evaluations. For instance, mixing the training, validation, and testing datasets or shuffling sequential (time-series) data, e.g., through Random cross-validation, must be prohibited.

We build a search space by considering several candidate ML models and hyperparameters for NIALM. Figure 1 depicts our proposed model for the search space of the AutoML approach. We consider the following possible values for the hyper-parameters of each ML model in our search space: (i) **Criterion (see** [3]**)** $\in$ {MSE, Friedman_MSE, MAE}, (ii) **Min sample split (see** [3]**)** $\in$ Uniform [2,200], (iii) **No. estimators (see** [4]**)** $\in$ Uniform [5,100], (iv) **Optimizer (see** [2]**)** $\in$ {Adam, Nadam, RMSprop}, (v) **Learning rate** $\in$ {1e−2, 1e−3, 1e−4, 1e−5}, (vi) **Loss function** $\in$ {mean squared error, mean absolute error}, (vii) **No. layers** $\in$ Uniform [5,8], (viii) **Dropout probability** $\in$ Uniform [0.1,0.6], (ix) **Sequence length** $\in$ {64,128,256,512,1024}

Then, we use Bayesian Optimization to select the best ML model and optimize its hyper-parameters. To this aim, we employ the free open-source Python library *Hyperopt* [1] for distributed asynchronous hyperparameter optimization. Hyperopt is one of the few libraries that allow search spaces with different types of variables (continuous, ordinal, categorical), varied search scales (e.g., uniform vs. log scaling), and conditional structure (the parameters of one classifier are irrelevant to the other classifiers). Last but not least, similar to our algorithms, this library is also implemented in Python.

Hyperopt has been designed to accommodate Bayesian optimization algorithms. We deploy the Tree-structured Parzen Estimator (TPE) [12] algorithm implemented in this library for choosing the best ML model/algorithm and the most adequate hyper-parameters from the search space. In fact, the problem is optimizing a loss function over a graph-structured search space (i.e., configuration space). In our case, we assume a tree structure for the search space. Tree-structured Parzen Estimator (TPE) offers a modeling strategy and scheme for optimizing the criterion of Expected Improvement (EI) for Sequential Model-Based Global Optimization (SMBO) [12].

In the search space, we consider the ML approaches shown in Table 1, except the Dictionary-Based Approach [14] as well as the Combined Deep Neural Networks (DNN) and Hidden Markov Models (HMM) approach [27]. We decided to exclude them based on our experiments that showed the gain in terms of the evaluation metrics was not justifiable considering the amount of time and resources they required. Hence, for our automated approach, they do not add enough value. Further, we propose new methods that we find quite suitable for NIALM applications (see Table 2 in Sect. 4). Those methods are based on Decision Trees, Fully-Connected Neural Networks (FCNN), also known as Multi-Layer Perceptron (MLP), and Gated Recurrent Units (GRU). In Sect. 4, we illustrate our experimental results for the mentioned approaches, thus providing an extended benchmarking.

4 Experimental Results

We name our tool AutoML4NIALM. This is available as free open-source software with a permissive license.

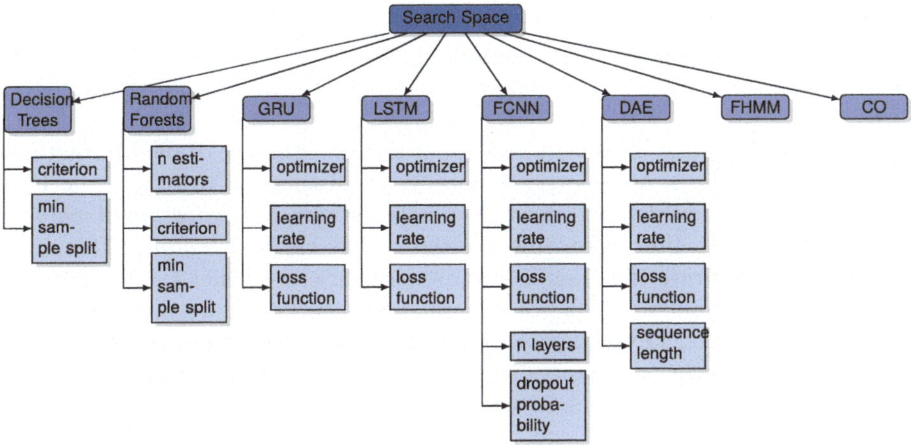

Fig. 1. Our proposed model for the search space of the AutoML approach

4.1 Evaluation Metrics

In this paper, we provide a fair benchmarking using the same datasets, the same evaluation metrics, and the same settings for all existing approaches as well as our proposed methods. The results of this benchmarking are shown in Tables 1 (only the existing approaches) and 2 (the extended version). We introduce the deployed evaluation metrics below.

Mean Absolute Error (MAE). In our opinion, this is the most relevant metric for typical NIALM applications. The Mean Absolute Error (MAE) for one specific appliance i over N time instances is defined as follows:

$$\textbf{Mean Absolute Error}_i = \frac{1}{N}\sum_{n=1}^{N}|si_n - \hat{si}_n| \tag{1}$$

Note that we use the average of MAE for all appliances in our benchmarking. Thus, we used the following metric for the results shown on Tables 1 and 2 assuming we have M appliances and N time instances:

$$\textbf{Mean Absolute Error}_{all} = \sum_{i=1}^{M} \frac{1}{N} \sum_{n=1}^{N} |si_n - \hat{si_n}| \tag{2}$$

Disaggregation Accuracy. Another related evaluation metric is the Disaggregation Accuracy. Note that the definition below clearly shows that this is very different from classification (ML) accuracy despite the similarity in terminology. Again, assuming we have N time instances, the Disaggregation Accuracy for appliance i is defined as follows:

$$\textbf{Disaggregation Accuracy}_i = 1 - \frac{\sum_{n=1}^{N} \left| si_n - \hat{si_n} \right|}{2 \sum_{n=1}^{N} \hat{si_n}} \tag{3}$$

Classification Accuracy. This is indeed about the performance of the ML method in terms of the typical ML metric Accuracy. To distinguish this from the Disaggregation Accuracy defined above, we call this one *Classification Accuracy*. However, In the ML literature, this is often simply called *Accuracy*. This metric shows how *accurate* the classification algorithm can predict the actual load (or state) of the appliance. For some appliances like lights, we might consider only two classes, i.e., ON (non-zero load) and OFF (near zero or zero load), whereas, for some other appliances such as dishwashers or microwave ovens, one may consider more than two classes.

$$\textbf{Classification Accuracy} = \frac{TP + TN}{P + N} \tag{4}$$

where TP, TN, P, and N denote the number of True Positive, True Negative, Positive, and Negative data instances. A high classification accuracy can show a good performance of the classifier, but accuracy alone is generally not sufficient to judge that. Therefore, we usually also take a look at other ML metrics, such as Precision and Recall, or a metric called F-Measure or F1-Score that combines both Precision and Recall. The latter is quite useful as we often have a trade-off between the Precision and Recall. Below, we define the said ML metrics.

Precision, Recall and F1-Score. Precision and Recall illustrate the *Precision* and the *Sensitivity* of the classification algorithm, respectively.

$$\textbf{Precision} = \frac{TP}{TP + FP} \tag{5}$$

where TP and FP denote the number of True Positive and False Positive data instances.

$$\textbf{Recall} = \frac{TP}{TP + FN} \tag{6}$$

where TP and FN denote the number of True Positive and False Negative data instances.

$$\textbf{F1-Score} = 2 \times \frac{Precision \times Recall}{Precision + Recall} \tag{7}$$

Normalized Absolute Distance. Finally, the Normalized Absolute Distance (NAD) of appliance i over N time instances is defined as follows:

$$\textbf{Normalized Absolute Distance}_i = \sqrt{\frac{\sum_{n=1}^{N} \left| si_n - \hat{si}_n \right|}{\sum_{n=1}^{N} |si_n|}} \tag{8}$$

NAD is a metric that we found in the Signal Processing literature. For instance, it is reported in [27].

4.2 Experimental Setup

To carry out the experiments, we either used the source code provided by the authors or, if that was not possible, carefully reimplemented their approach based on the information provided in the respective papers. We also acknowledge the reuse of algorithms implemented in the free open-source NILM-Toolkit [18]. Moreover, the dataset used for this benchmarking is the REDD dataset [38]. Note that the existing open reference datasets for NIALM have played a key role in making high-impact research in this field possible over recent years and in facilitating our benchmarking. The most well-known and widely used ones are the REDD dataset from the U.S. [38] as well as the UK-DALE [20] and the REFIT datasets from the U.K. [28]. Our tool presented in Sect. 3 currently supports the REDD and the UK-DALE data formats out of the box. Nevertheless, we are aware of other existing open datasets for NIALM, such as [8,9,26].

Specifically, we used the data of 37 days from house 1 of the REDD dataset [38] with a sampling rate (frequency) of one per 20 seconds, i.e., 0.05 Hz for conducting our benchmarking in Sect. 2, the extended benchmarking in Sect. 4.3, as well as the experiments on AutoML in Sect. 4.4. We also split the dataset into training, validation, and testing segments. Further, in all experiments, we considered 200 trials of the AutoML algorithm for the optimization and 2000 maximum number of epochs for the training of the ML model. Also, we set a patience level of 15. The latter means that even if the validation loss is not decreasing during the training phase, we could still tolerate that for up to 15 training epochs; thus, the training would not stop immediately.

In addition, our methodology for conducting the benchmarking is as follows: For each ML algorithm, we consider the Mean Absolute Error (MAE) for disaggregation of the signals of the fridge, lights, sockets, and washer/dryer and calculate the average of the MAE over all the said (groups of) appliances. The highest rank, e.g., in Table 1, means the lowest average for MAE over all the said (groups of) appliances, whereas the lowest rank shows the largest MAE on average for them. In Sect. 4.3 below, we explain why we chose MAE among all metrics for the benchmarking.

4.3 Extended Benchmarking

We argue that for typical NIALM use cases, e.g., for load balancing in Smart Grids or saving on energy bills for the end users, Mean Absolute Error (MAE) is the best metric to measure the performance of the NIALM approaches. The reason is that, in the end, we are interested in having the least regression error when predicting the loads of appliances. This means the ultimate goal is to reduce the total deviation of the model from the actual power consumption of the target appliance(s). Therefore, that might even imply sacrificing the classification performance (e.g., the Accuracy, Precision, or Recall in predicting the right states/loads of appliances) or other above-mentioned evaluation metrics to achieve that ultimate goal shall not really be an issue. For this reason, we consider only this metric for the benchmarking conducted here.

Table 2 shows the Mean Absolute Error (MAE) that we observed in our experiments both with the existing state-of-the-art approaches named in Table 1 and for the proposed approaches based on Decision Trees, Fully-Connected Neural Networks (Multi-Layer Perceptron) and Gated Recurrent Units (GRU). We can see that neither the state-of-the-art approach proposed by Elhamifar and Systry in 2015 (i.e., the Dictionary-Based Approach [14]) nor the one proposed by Mauch and Yang in 2016 (i.e., the DNN-HMM approach [27]) perform as good as the Deep Neural Network approaches of Kelly and Knottenbelt proposed in 2015 [19], based on Long Short-Term Memory (LSTM) and Denoising Autoencoders (DAE).

In addition, we observed through our experiments that neither of the said approaches (Dictionary-Based and DNN-HMM) were efficient in terms of the time and space required for their training. For example, training the Dictionary-Based approach on Commercial Off-The-Shelf (COTS) hardware was so slow in our initial experiments that we decided to substitute the dictionary creation part of that approach with another solution built based on the UCR Matrix Profile [5] for finding similar signals (known as Motifs in Signal Processing).

Table 2. Extended benchmarking using both the state-of-the-art approaches and proposed algorithms for NIALM

Rank	Algorithm	MAE
1st	**Decision Trees (DT)**	14.86
2nd	**Gated Recurrent Units (GRU)**	19.55
3rd	Long Short-Term Memory (LSTM) [19]	25.77
4th	Denoising Autoencoders (DAE) [19]	41.44
5th	**Fully-Connected Neural Networks (FCNN) a.k.a. Multi-Layer Perceptron (MLP)**	48.96
6th	Dictionary-Based Approach [14]	61.15
7th	DNN-HMM [27]	118.52
8th	Factorial Hidden Markov Models (FHMM) [31]	139.40
9th	Combinatorial Optimization (CO) [16]	204.83

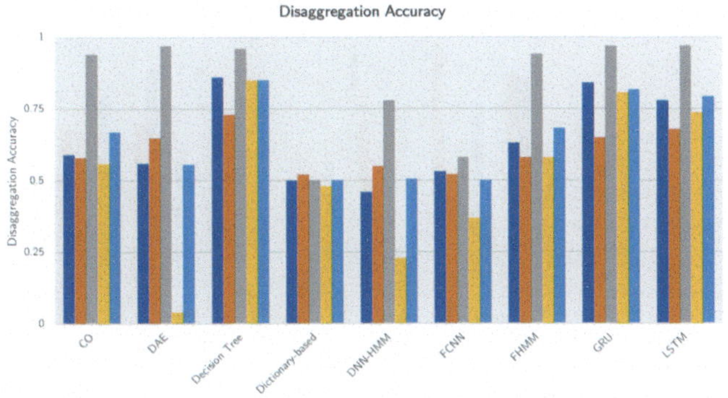

Fig. 2. Benchmarking using Disaggregation Accuracy - color codes: dark blue, orange, gray, yellow and light blue represent fridge, lights, sockets, washer/dryer and average for all appliances, respectively.

Furthermore, Figs. 2, 3, 4, and 5 depict our benchmarking results regarding other evaluation metrics, namely Disaggregation Accuracy, Classification Accuracy, F1-Score (i.e., Precision and Recall of classification) and Normalized Absolute Distance (NAD), respectively. Finally, Fig. 6 shows the Classification Accuracy as well as the Mean Absolute Error (MAE) for various approaches simultaneously. According to this diagram, Combinatorial Optimization (CO) [16] and after that, Factorial Hidden Markov Models (FHMM) [31] are the worst ones with respect to these two metrics. Also, despite having a low MAE, Denoising Autoencoders (DAE) [19] has the drawback of low classification Accuracy. Finally, we see that there is not much difference between the rest of the approaches when it comes to the mentioned metrics. Thus, choosing the right approach among those remaining ones shall be more a matter of other evalua-

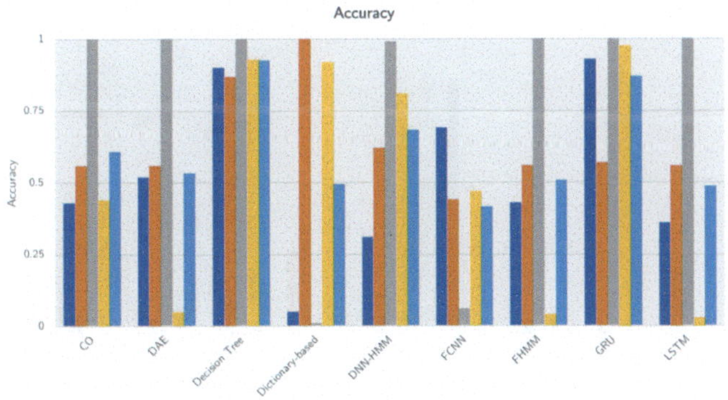

Fig. 3. Benchmarking using Classification Accuracy - color codes: dark blue, orange, gray, yellow and light blue represent fridge, lights, sockets, washer/dryer and average for all appliances, respectively.

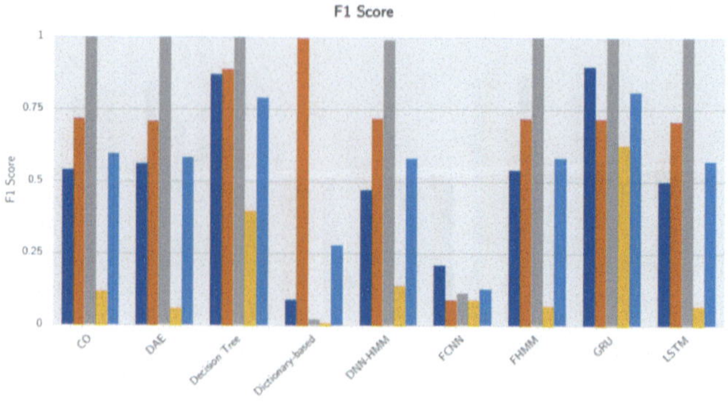

Fig. 4. Benchmarking using F1-Score - color codes: dark blue, orange, gray, yellow and light blue represent fridge, lights, sockets, washer/dryer and average for all appliances, respectively.

tion metrics or other constraints such as the training time (e.g., obviously very quick in the case of Decision Trees and very long in the case of Deep Neural Networks). Note that on this diagram, we also see an additional algorithm, Random Forest. Actually, we deployed Ensemble Learning to train a number of Decision Trees. However, as illustrated, neither in terms of the MAE nor regarding the Classification Accuracy could we observe any considerable improvement.

4.4 AutoML Experiments

Our proposed Automated Machine Learning (AutoML) approach to NIALM chooses the best ML model, algorithm, and configurations out of the search space that is depicted in Fig. 1. Tables 3, 4 and 5 illustrate the improvements by

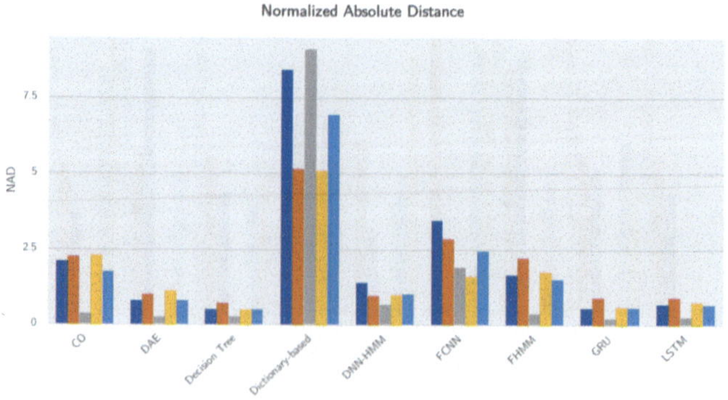

Fig. 5. Benchmarking using Normalized Absolute Distance (NAD) - color codes: dark blue, orange, gray, yellow and light blue represent fridge, lights, sockets, washer/dryer and average for all appliances, respectively.

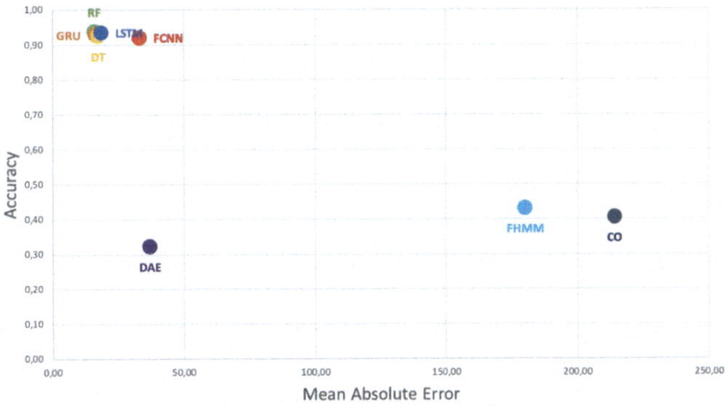

Fig. 6. Classification Accuracy vs. Mean Absolute Error (average over all appliances), DT and RF stand for Decision Trees and Random Forests, respectively.

comparing a classic (i.e., manual) setting with our proposed automated approach based on the Mean Absolute Error (MAE), Classification Accuracy and F1-Score evaluation metrics, respectively. Interestingly, we see that the AutoML approach even leads to better results by the basic architecture of Neural Networks that we proposed, i.e., Fully-Connected Neural Networks (Multi-Layer Perceptron) compared to Denoising Autoencoders (DAE), not only in terms of the MAE but also in terms of the Classification Accuracy and the F1-Score.

As we saw in Fig. 1, we do not have any hyper-parameters for the Factorial Hidden Markov Models (FHMM) [31] and Combinatorial Optimization (CO) [16] approaches. This explains why, in principle, we do not achieve any better results using the AutoML approach for them. Actually, the slightly better MAE shown in Table 3 is rather accidental, as we get slightly different results in different runs.

Table 3. Experimental results for the proposed AutoML approach using MAE

Algorithm	Manual	AutoML
DT	21.16	19.05
GRU	23.14	19.87
LSTM [19]	28.13	21.80
DAE [19]	**39.17**	**35.54**
FCNN	**51.64**	**34.50**
FHMM [31]	118.57	118.57
CO [16]	208.36	204.57

Table 4. Experimental results for the proposed AutoML approach using Classification Accuracy

Algorithm	Manual	AutoML
DT	0.77	0.80
GRU	0.84	0.84
LSTM [19]	0.51	0.91
DAE [19]	**0.65**	**0.71**
FCNN	**0.37**	**0.88**
FHMM [31]	0.51	0.51
CO [16]	0.61	0.61

Table 5. Experimental results for the proposed AutoML approach using F1-Score

Algorithm	Manual	AutoML
DT	0.61	0.72
GRU	0.72	0.74
LSTM [19]	0.58	0.82
DAE [19]	**0.59**	**0.62**
FCNN	**0.23**	**0.77**
FHMM [31]	0.58	0.58
CO [16]	0.59	0.59

5 Conclusion and Future Work

In this paper, we proposed a novel approach to enable Automated Machine Learning for the domain of NIALM through Bayesian Optimization. Using our free open-source tool, NIALM domain experts and practitioners could benefit from Machine Learning algorithms and methods without necessarily having data analytics skills. Our tool can automatically choose the best Machine Learning (ML) model and algorithm, the most suitable values for the hyper-parameters of the ML model, and the most adequate configurations for training the ML model. We demonstrated that the proposed AutoML approach indeed improves the quality of the data analytics practices for NIALM regarding various evaluation metrics, such as the Mean Absolute Error (MAE), the Classification Accuracy, and the F1-Score.

Moreover, we conducted a comprehensive survey and a fair benchmarking of the state of the art in NIALM and illustrated that, in many cases, simple and basic Machine Learning models and algorithms, such as Decision Trees, outperform the state of the art. Further, we proposed other novel methods for NIALM. In addition, a vital lesson learned from the conducted study is that there is no one-size-fits-all approach to NIALM. In other words, various state-of-the-art machine learning approaches could offer great results depending on the appliance of interest and the evaluation metric of choice. Hence, there is no point in pushing the user towards a certain approach and prescribing that as the best one for all appliances and all purposes. Instead, we believe that the user shall be able to benefit from the automation of the data analytics tasks via tool support to benefit from all state-of-the-art approaches.

Finally, as future work, we plan to extend our open-source tool with more algorithms and methods as well as more features for supporting and guiding users through our Graphical User Interface (GUI). Also, we may apply our AutoML approach to other domains beyond NIALM. Further, one may think of other optimization algorithms for AutoML other than the one we used here for Bayesian Optimization. In addition, currently, the input data has to be in the format used by the two major existing open reference datasets for NIALM. However, one could possibly increase user convenience by supporting other formats as well.

Acknowledgements. This work is partially funded by the German Federal Ministry of Education and Research (BMBF) through the Software Campus 2.0 project ML-Quadrat.

References

1. Hyperopt: Distributed asynchronous hyperparameter optimization in python. https://github.com/hyperopt/hyperopt. Accessed 25 Oct 2018
2. The keras documentation, optimizers. https://keras.io/optimizers/. Accessed 04 Nov 2018
3. The scikit learn documentation, decision trees. http://scikit-learn.org/stable/modules/generated/sklearn.tree.DecisionTreeRegressor.html. Accessed 04 Nov 2018
4. The scikit learn documentation, random forest. http://scikit-learn.org/stable/modules/generated/sklearn.ensemble.RandomForestClassifier.html. Accessed 04 Nov 2018
5. The ucr matrix profile page. https://www.cs.ucr.edu/~eamonn/MatrixProfile.html. Accessed 29 Oct 2018
6. Agyeman, K.A., Han, S., Han, S.: Real-time recognition non-intrusive electrical appliance monitoring algorithm for a residential building energy management system. Energies **8**, 9029–9048 (2015)
7. Amenta, V., Gagliano, S., Tina, G., Di Modica, G., Tomarchio, O.: Web interactive non intrusive load disaggregation system for energy consumption awareness. In: AEIT Annual Conference, pp. 1–6 (2013). https://doi.org/10.1109/AEIT.2013.6666817
8. Barker, S., Mishra, A., Irwin, D., Cecchet, E., Shenoy, P., Albrecht, J.: Smart: an open data set and tools for enabling research in sustainable homes. In: Workshop on Data Mining Applications in Sustainability (SustKDD) (2012)
9. Batra, N., Gulati, M., Singh, A., Srivastava, M.: It's different: insights into home energy consumption in India. In: 5th ACM Workshop On Embedded Systems For Energy-Efficient Buildings (Buildsys) (2013)

10. Berges, M., Goldman, E., Matthews, H.S., Soibelman, L.: Learning systems for electric comsumption of buildings. In: ASCI International Workshop on Computing in Civil Engineering (2009)
11. Bergman, D.C., Jin, D., Juen, J.P., Tanaka, N., Gunter, C.A., Wright, A.K.: Distributed non-intrusive load monitoring. In: Innovative Smart Grid Technologies (ISGT), 2011 IEEE PES, pp. 1–8. IEEE (2011)
12. Bergstra, J., Bardenet, R., Bengio, Y., Kégl, B.: Algorithms for hyper-parameter optimization. In: Proceedings of the 24th International Conference on Neural Information Processing Systems, NIPS 2011, pp. 2546–2554. Curran Associates Inc., USA (2011). http://dl.acm.org/citation.cfm?id=2986459.2986743
13. Elhamifar, E., Sapiro, G., Sastry, S.S.: Dissimilarity-based sparse subset selection. IEEE Trans. Pattern Anal. Mach. Intell. **38**(11), 2182–2197 (2016)
14. Elhamifar, E., Sastry, S.: Energy disaggregation via learning powerlets and sparse coding. In: AAAI, pp. 629–635 (2015)
15. Gillis, J.M., Alshareef, S.M., Morsi, W.G.: Nonintrusive load monitoring using wavelet design and machine learning. IEEE Trans. Smart Grid **7**(1), 320–328 (2016). https://doi.org/10.1109/TSG.2015.2428706
16. Hart, G.W.: Nonintrusive appliance load monitoring. Proc. IEEE **80**(12), 1870–1891 (1992). https://doi.org/10.1109/5.192069
17. He, K., Stankovic, L., Liao, J., Stankovic, V.: Non-intrusive load disaggregation using graph signal processing. IEEE Trans. Smart Grid **9**(3), 1739–1747 (2018)
18. Kelly, J., et al.: Nilmtk v0.2: a non-intrusive load monitoring toolkit for large scale data sets: Demo abstract. In: Proceedings of the 1st ACM Conference on Embedded Systems for Energy-Efficient Buildings, BuildSys 2014, pp. 182–183. ACM, New York (2014). https://doi.org/10.1145/2674061.2675024. http://doi.acm.org/10.1145/2674061.2675024
19. Kelly, J., Knottenbelt, W.: Neural nilm: deep neural networks applied to energy disaggregation. In: Proceedings of the 2nd ACM International Conference on Embedded Systems for Energy-Efficient Built Environments, pp. 55–64. ACM (2015)
20. Kelly, J., Knottenbelt, W.: The UK-DALE dataset, domestic appliance-level electricity demand and whole-house demand from five UK homes. Sci. Data **2**(1), 1–14 (2015). http://dx.doi.org/10.1038/sdata.2015.7
21. Kolter, J.Z., Batra, S., Ng, A.Y.: Energy disaggregation via discriminative sparse coding. In: Advances in Neural Information Processing Systems, pp. 1153–1161 (2010)
22. Kramer, O., Klingenberg, T., Sonnenschein, M., Wilken, O.: Nonintrusive appliance load monitoring with bagging classifiers. Logic J. IGPL **233**, 359–368 (2014)
23. Laughman, C., et al.: Power signature analysis. IEEE Power Energ. Mag. **99**(2), 56–63 (2003). https://doi.org/10.1109/MPAE.2003.1192027
24. Lee, W., Fung, G., Lam, H., Chan, F., Lucente, M.: Exploration on load signatures. In: International Conference on Electrical Engineering. (ICEE) (2004)
25. Lin, Y., Tsai, M.: Non-intrusive load monitoring by novel neuro-fuzzy classification considering uncertainties. IEEE Trans. Smart Grid **5**(5), 2376–2384 (2014)
26. Makonin, S., Ellert, B., Bajić, I.V., Popowich, F.: Electricity, water, and natural gas consumption of a residential house in Canada from 2012 to 2014. Sci. Data **3**, 160037 EP (2016). https://doi.org/10.1038/sdata.2016.37
27. Mauch, L., Yang, B.: A novel DNN-HMM-based approach for extracting single loads from aggregate power signals. In: 2016 IEEE International Conference on Acoustics, Speech and Signal Processing (ICASSP), pp. 2384–2388. IEEE (2016)

28. Murray, D., Stankovic, L., Stankovic, V.: An electrical load measurements dataset of united kingdom households from a two-year longitudinal study. Sci. Data **4**(1), 1–12 (2017)
29. Parson, O., Ghosh, S., Weal, M., Rogers, A.: Non-intrusive load monitoring using prior models of general appliance types. In: Proceedings of the Twenty-Sixth AAAI Conference on Artificial Intelligence, AAAI 2012, pp. 356–362. AAAI Press (2012). http://dl.acm.org/citation.cfm?id=2900728.2900780
30. Patel, S.N., Robertson, T., Kientz, J.A., Reynolds, M.S., Abowd, G.D.: At the flick of a switch: detecting and classifying unique electrical events on the residential power line (nominated for the best paper award). In: UbiComp (2007)
31. Reyes-Gomez, M.J., Raj, B., Ellis, D.R.W.: Multi-channel source separation by factorial HMMs. In: 2003 IEEE International Conference on Acoustics, Speech, and Signal Processing, 2003. Proceedings, (ICASSP 2003), vol. 1, pp. I–I (2003)
32. Shao, H., Marwah, M., Ramakrishnan, N.: A temporal motif mining approach to unsupervised energy disaggregation: applications to residential and commercial buildings. In: Proceedings of the Twenty-Seventh AAAI Conference on Artificial Intelligence, AAAI 2013, pp. 1327–1333. AAAI Press (2013). http://dl.acm.org/citation.cfm?id=2891460.2891645
33. Shaw, S.R., Abler, C.B., Lepard, R.F., Luo, D., Leeb, S.B., Norford, L.K.: Instrumentation for high performance nonintrusive electrical load monitoring. ASME, **120**(224) (1998)
34. Stankovic, V., Liao, J., Stankovic, L.: A graph-based signal processing approach for low-rate energy disaggregation. In: 2014 IEEE Symposium on Computational Intelligence for Engineering Solutions (CIES), pp. 81–87 (2014)
35. Sultanem, F.: Using appliance signatures for monitoring residential loads at meter panel level. IEEE Trans. Power Delivery **6**(4), 1380–1385 (1991)
36. Zhao, B., Stankovic, L., Stankovic, V.: On a training-less solution for non-intrusive appliance load monitoring using graph signal processing. IEEE Access **4**, 1784–1799 (2016)
37. Zhong, M., Goddard, N., Sutton, C.: Latent bayesian melding for integrating individual and population models. In: Cortes, C., Lawrence, N.D., Lee, D.D., Sugiyama, M., Garnett, R. (eds.) Advances in Neural Information Processing Systems, vol. 28, pp. 3618–3626. Curran Associates, Inc. (2015). http://papers.nips.cc/paper/5756-latent-bayesian-melding-for-integrating-individual-and-population-models.pdf
38. Kolter, J.Z., Johnson, M.J.: Redd: a public data set for energy disaggregation research, vol. 25 (2011)

Methodology of Multi-model Imputation: Novel Imputation Method Using Multiple Known Models

Sarah Beaver[1(✉)], Yashu Vashishath[1,2,3], and Renee Bryce[1]

[1] Department of Computer Science and Engineering, University of North Texas, 1155 Union Cir, Denton, TX 76203, USA
sarahBeaver2@my.unt.edu, {yashu.vashishath,renee.bryce}@unt.edu
[2] Center of Computational Life Science, University of North Texas, 1155 Union Cir, Denton, TX 76203, USA
[3] Biodiscovery Institute, University of North Texas, 1155 Union Cir, Denton, TX 76203, USA

Abstract. In medical research and diagnostics, missing data can severely impact the reliability of analyses and predictions. Accurate data imputation is critical for maintaining the quality and usability of datasets. Traditional imputation methods often struggle with complex and varied data structures, leading to suboptimal outcomes. To address this challenge, this study introduces Multi-Model Imputation (MMMI), a novel approach that integrates multiple machine learning (ML) techniques to enhance data imputation accuracy. MMMI combines K-Nearest Neighbors (KNN) and Random Forest (RF) algorithms for audiogram data imputation. By leveraging KNN's local relationship modeling, and RF's ability to handle complex interactions MMMI improves both accuracy and consistency in data imputation. Comparative analyses reveal that MMMI outperforms traditional imputation methods, demonstrating a 2–20% improvement over KNN and RF in audiogram predictions. These results underscore MMMI's effectiveness and its potential for broader applications in medical data imputation, leading to higher-quality and more reliable datasets.

Keywords: imputations · multi-model · audiograms · machine learning · KNN · Random Forest · HERMES

1 Introduction

Challenges of missing data remains a significant obstacle, undermining the accuracy and reliability of insights across diverse domains. Traditional imputation methods, such as mean imputation or linear regression-based techniques, often struggle with the complexities inherent in datasets characterized by intricate patterns of missingness. Conventional approaches can introduce bias and may

Y. Vashishath and R. Bryce—Contributing authors.

S. Latifi (Ed.): ITNG 2025, AISC 1463, pp. 270–281, 2025.
https://doi.org/10.1007/978-3-031-89063-5_23

fail to represent the underlying data distribution accurately. While methods like multiple imputation by chained equations (MICE) and k-nearest neighbors (KNN) offer improvements, they often come with high computational costs or limitations in handling high-dimensional data [13,18]. As datasets continue to expand in size and complexity, there is an increasing demand for more efficient and precise imputation techniques.

Audiograms, which are pivotal in audiology, provide a graphical representation of an individual's hearing sensitivity across various frequencies. These assessments help with diagnosing auditory conditions, monitoring hearing health, and developing personalized intervention strategies [8,15]. Audiograms offer detailed frequency-specific evaluations that help clinicians identify hearing impairments, assess the severity and configuration of hearing loss, and track changes in hearing function over time [15].

However, missing data in audiograms is a prevalent issue that significantly compromises their accuracy and reliability. Such gaps may arise from technical errors during testing, patient non-compliance, or incomplete testing protocols [1,2]. These missing values can distort audiometric results, complicating the diagnostic process and impeding the development of effective management strategies.

There is a clear need for advanced imputation methods that can handle the unique complexities of datasets like audiograms. In this study, we introduce a novel imputation approach to tackle limitations of traditional methods. We apply this new technique to the audiogram dataset from the AII-HERMES (Auditory Implant Initiative HIPAA-Secure, Encrypted, Research, Management and Evaluation Solution) project [19]. By demonstrating its efficacy on these diverse datasets, we aim to showcase the potential of this new method in providing more accurate and reliable imputations.

2 Background

Emmanuel et al. [7] provides an overview of missing data and various imputation techniques. They classify missing data patterns as univariate (where a single variable is missing), monotone (where missing data follows a recognizable pattern), and non-monotone (where the missingness of one variable does not impact others). They discuss several imputation approaches, including deletion, single imputation, multiple imputation, and machine learning (ML) techniques. Performance metrics for these methods include Mean Absolute Error (MAE) and Root Mean Square Error (RMSE), with comparisons across various studies.

Different imputation methods have been compared across multiple datasets in Aleryani et al. [3], where it was found that Multiple Imputation by Chained Equations (MICE) performed highly, and Random Forest (RF) imputation provided data quality close to the original dataset.

Several studies explore imputation techniques specifically for hearing data. Wendl et al. [21] utilized multiple imputations to address missing data in adolescents exposed to leisure noise. Barbour et al. [4] and Song et al. [20] employed

ML techniques to perform Gaussian regression, resulting in accurate continuous audiograms that align well with existing audiograms. A common method for handling missing data in audiograms involves using the slope of similar complete audiograms, as discussed by Feirn [9] and Pittman et al. [17]. However, Pittman et al. [17] highlighted that this traditional approach is often inadequate, with standard deviations exceeding 15 dB.

The AII-HERMES audiogram dataset has previously been subject to imputation studies. Pavelchek et al. [16] utilized six different techniques, including MICE, K-Nearest Neighbors (KNN), and neural networks, achieving successful imputation for up to six missing data points in an eleven-frequency audiogram, with an RMSE value of 7.83 dB. The imputaions allowed for an expansion of the dataset from approximately 1,300 to nearly 7,400 audiograms. Beaver et al. [5] reviewed a broad range of ML techniques for audiogram imputation, including KNN, RF, Linear Support Vector Regression (SVR), and Lasso. They reported KNN RMSE values ranging from 5.00 to 7.72 and RF RMSE values between 4.74 and 6.37.

This study aims to enhance the accuracy of data imputation in audiogram. By doing so, it holds the potential to greatly improve the overall quality and utility of the data.

3 Research Methodology

3.1 Methodology of Multi-model Imputation

An overview of the Methodology of Multi-Model Imputation (MMMI) is in Fig. 1. The processed data is split into train and test sections. The train section of the dataset is used for Nested Cross Validation, as discussed in [11] and algorithm, as shown in [5], along with a ML technique to get out a list of N-trained models of that technique. The Nested Cross Validation consists of two loops, an outer and an inner loop, where each pass of the outer loop returns a model, and the inner loop is used for tuning hyperparameters of the model. Therefore, the result after finishing Nested Cross Validation is a list of N models, where N is the number of outer loops of a specific ML technique.

The N models are used for making N predictions, one prediction per model, on the testing data. Next the N predictions are used for calculating a mean value. After, the absolute difference between each prediction value from each of the N models and the mean of prediction values is calculated. The steps are repeated from Nested Cross Validation for each type of ML technique chosen to be used. Each pass in Nested Cross Validation would, therefore, have an absolute difference for each technique. The prediction of the model with the lowest absolute difference is used, which leads to N prediction values from multiple ML techniques. Finally, the mean of the N prediction values is taken to get a final imputation value.

Let $P_{\mathrm{T1}} = \{T1_1, T1_2, ..., T1_n\}$ be the set of predictions from the first technique's models, and $P_{\mathrm{T2}} = \{T2_1, T2_2, ..., T2_n\}$ be the set of predictions from the second technique's models ..., and $P_{\text{Tx}} = \{Tx_1, Tx_2, ..., Tx_n\}$ be the set of predictions from the last technique's models.

$$\text{diff}_j^{\text{Ti}} = |Ti_j - \mu_{\text{Ti}}| \tag{1}$$

The absolute difference of each prediction Ti_j, where i is the ith technique and j is the jth model in that ith technique, from the mean of ith technique's predictions μ_{Ti} is calculated as seen in Eq. (1). A list of absolute differences is calculated for each model per technique is generated.

$$[\text{diff}_1^{\text{Ti}}, \text{diff}_2^{\text{Ti}}, \ldots, \text{diff}_n^{\text{Ti}}].$$

The prediction selected for the jth models is the one with the smallest absolute difference from the respective mean. So, for each jth models:

$$\text{p}_{\text{selected}_j} = \begin{cases} T1_j & \text{if } \text{diff}_j^{\text{T1}} = \min(\text{diff}_j^{\text{T1}}, \text{diff}_j^{\text{T2}}, \ldots, \text{diff}_j^{\text{Tx}}) \\ T2_j & \text{if } \text{diff}_j^{\text{T2}} = \min(\text{diff}_j^{\text{T1}}, \text{diff}_j^{\text{T2}}, \ldots, \text{diff}_j^{\text{Tx}}) \\ \vdots & \vdots \\ Tx_j & \text{if } \text{diff}_j^{\text{Tx}} = \min(\text{diff}_j^{\text{T1}}, \text{diff}_j^{\text{T2}}, \ldots, \text{diff}_j^{\text{Tx}}) \end{cases} \tag{2}$$

$$\mu_{\text{selected}} = \frac{1}{n} \sum_{j=1}^{n} p_{\text{selected}_j} \tag{3}$$

Equation (2) is repeated N times to obtain $P_{\text{selected}} = [p_{\text{selected}_1}, p_{\text{selected}_2}, \ldots, p_{\text{selected}_n}]$ which represents the list of selected predictions for a specific frequency case.

Each p_{selected_j} represents the prediction selected for the jth models. This selection can either be from any of the different techniques' jth models P_{Ti}, depending on which prediction has the smallest absolute difference from the technique's mean.

Finally, the mean of the selected predictions P_{selected} were calculated as seen in Eq. (3). Therefore, μ_{selected} represents the average of the selected predictions for all models referred to as multi-model imputation (MMMI) mean.

3.2 Data and Pre-processing

The data is provided by HERMES, a secure and encrypted web-based Cochlear Implant (CI) database created by the nonprofit Auditory Implant Initiative (AII) [19]. The data includes demographics (such as age, sex, ethnicity, and geographic location), otologic history (including etiology and duration of Hearing Loss (HL) preimplantation, chronic ear conditions, and previous hearing aid use), relevant medical/surgical history, preoperative and implantation history (such as audiology results during candidacy evaluation, preoperative tinnitus and vertigo, imaging, CI manufacturer and electrode type, operative technique, duration of electrode insertion, surgeon experience, intraoperative complications, antibiotics, and steroid use). Additionally, postoperative outcome data is included (such

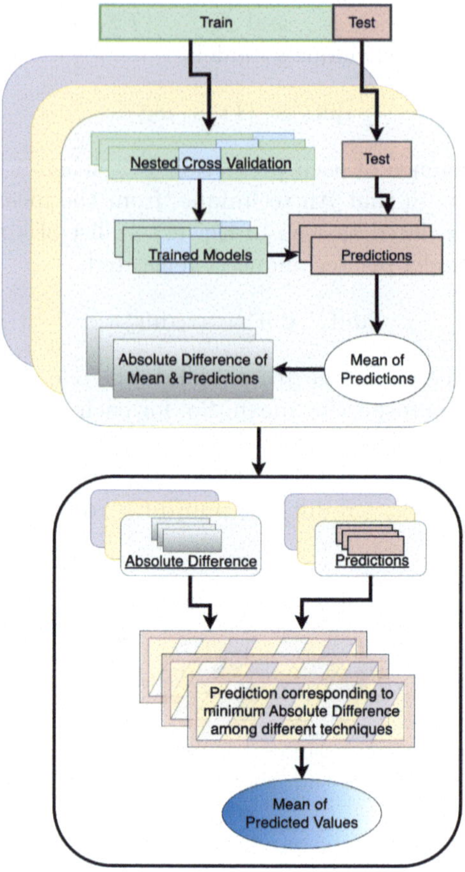

Fig. 1. Methodology of Multi-model Imputation: Each layer of the top box represents a different ML technique. For each technique nested cross validation is used to get n models which are used to predict. The mean of the predictions is calculated then absolute difference is taken from that mean. The absolute difference from the mean, and predictions for each model for each technique is pass to the bottom box. The prediction with the smallest absolute difference is taken across different ML techniques to get n predictions. The mean is taken of the predictions to get the final predicted value.

Table 1. Model Grid Search Parameters

Model	Parameter	Options
RF	n estimators	50, 100, 200
	max depth	1, 10, 20
	min samples split	2, 5, 10
	min samples leaf	1, 2, 4
KNN	n neighbors	1, 3, 5, 7
	weights	uniform, distance
	p	1, 2
	algorithm	auto, ball tree, kd tree, brute

as postoperative audiological results and complications). The HERMES data set includes patients who have undergone cochlear implantation, CI candidates awaiting implantation, and patients with HL who may not yet be candidates [6].

There are three categories into which the missing data may be categorized [10,12,14]:

1. When there are randomly missing data because there is no discernible pattern in the data, the item is said to be missing completely at random (MCAR). It is anticipated that bias will enter the database less frequently the more uniform the data distribution is.
2. When there is a discernible pattern in the missing data, it is referred to as Missing At Random (MAR). With the use of MAR data, we may identify a common component among the database's missing values.
3. Similar to MAR, Missing Not At Random (MNAR): the values that lead to the missing of others are unknown.

MAR can be assumed for the HERMES data set as previously stated in [5,16]. Human error, varying methods for individual doctors or clinics, and variations in testing frequencies can all be used to explain missing data.

Audiograms for patients over the age of eighteen were selected which gave 18,985 audiograms. Frequencies 6000 and 8000 hz were removed due to the past results performing worse for these frequencies, which have been covered previously [5]. Any audiogram that had a missing frequency or had a frequency of 125 or higher was removed, which was also done in [16]. The following categorical columns were converted into indicator variables: conduction, ear, and sex. After the processing of the data, 8159 audiograms were left. A breakdown of the number of missing data per frequency can be seen in Table 3. Since there are no missing values for frequency 250, models and results were not created for frequency 250.

3.3 Models

The HERMES dataset was separated into 20% for testing, and the other 80% was used for training. RF and KNN were selected as both performed highly on individual audiogram frequencies [5]. Each technique was trained using Nested Cross Validation, where each outer loop created a single model. On the inner loop, a grid search was performed to find the best parameters for each model of the outer loop. There were 10 loops used for the outer loop, and 10 loops were used for the inner loop. Each pass of the outer loop resulted in a model, thus giving 10 models for RF and 10 models for KNN.

The Nested Cross Validation for generating models was repeated for each frequency of Hz125, Hz250, Hz500, Hz750, Hz1000, Hz1500, Hz2000, Hz3000, and Hz4000, where the frequency was used as the target and other columns were used for training. The parameters used in the grid search for RF and KNN can be seen in Table 1.

The KNN and RF models were used in the MMMI methodology as seen in Fig. 1. The results of MMMI mean are compared to the mean of KNN predictions and mean of RF Predictions.

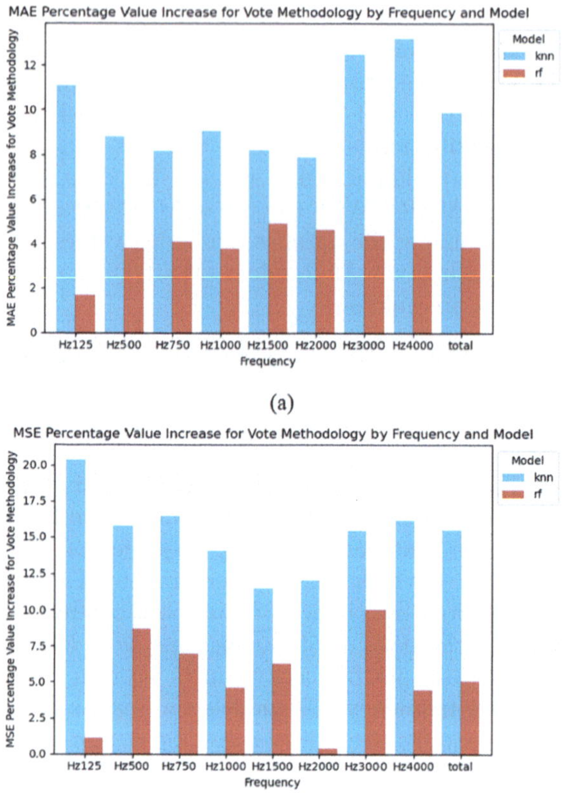

(a)

(b)

Fig. 2. Bar Chart showing percentage increase of MMMI over different models over different frequencies (a) MAE (b) MSE

3.4 Imputation vs No Imputation

Taking inspiration from [5], which compares how different imputation methods affect the accuracy of predictive models, a comparison of how MMMI imputation affects the accuracy and F1 Scores versus no imputation. A binary class was created based on wether the patient had a cochlear implant in their ear when the audiogram was performed. For MMMI, missing values were imputed for rows, resulting in 5331 more rows added, as seen in Table 3. These rows

Table 2. Actual Metric Values For Audiograms

Metric	Total	Hz125	Hz500	Hz750	Hz1000	Hz1500	Hz2000	Hz3000	Hz4000
RMSE	5.22	5.94	5.21	5.02	5.22	5.29	5.17	4.18	5.57
R^2	0.97	0.92	0.96	0.97	0.97	0.97	0.97	0.98	0.96
MAE	3.31	4.11	3.54	3.19	3.39	3.34	3.16	2.59	3.18
MSE	27.32	35.30	27.24	25.29	27.32	28.00	26.80	17.53	31.08

were removed for the no imputation classifiers. Five classifiers were used to predict the binary classification: KNN, Log, RF, Support Vector Classification, and XGBoost. These models were run 10 times, the mean and standard deviation is shown in the Audiograms' result section MMMI vs No Imputation Prediction Comparison.

Table 3. Number of Missing Values per Frequency

Frequency (Hz)	Number Missing Values
125	1292
250	0
500	13
750	212
1000	6
1500	2209
2000	16
3000	714
4000	69
Total	5331

3.5 Metrics

As seen in Table 3, there are no missing values for only Frequency 250. Therefore, results are excluded for this frequency in graphs and discussions.

To calculate the results in Percentage Increase, we measured how much MMMI increases compared to other models for each frequency using Eq. (4).

$$\text{Percentage Increase} = \frac{(Old - MMMI)}{Old} \times 100 \tag{4}$$

The models were compared on four different metric evaluations: RMSE, MSE, MAE, and R^2.

4 Results

4.1 Percentage Increase

For calculating the results shown in Figs. 2a, and 2b, each metric was taken for each frequency and model type. MMMI outperformed all other models, and the percentage increase for each metric was calculated as seen in Eq. (4).

KNN tends to have low bias and higher variance due to its reliance on local neighbors and sensitivity to local neighborhood variations. In contrast, RF can achieve lower variance because of its ensemble nature, which allows it to handle complex relationships effectively. MMMI leverages the strengths of both KNN and RF by reducing bias and variance through a hybrid approach, using KNN's ability to model local relationships and RF's ensemble method to manage complex interactions.

For audiograms, the percentage increase is more significant when compared to KNN than RF. The greater percentage increase for KNN is because the MMMI methodology tends to rely more on RF predictions than KNN. However, the fact that there is not a zero percentage increase indicates that MMMI does utilize both KNN and RF for predictions, with a greater proportion of predictive value coming from RF.

4.2 Previous Audiogram Imputation Comparison

The actual metric results of the MMMI for RMSE R^2, MAE, and MSE are presented in Table 2. When compared to the results in [5], MMMI performs better or is comparable on individual frequencies. Additionally, MMMI demonstrates more uniform performance across frequencies than both [5] and imputation methods used in [16] including MICE. This consistency across different frequencies highlights the robustness of MMMI's imputation method in maintaining high predictive accuracy regardless of frequency variations. This uniformity reduces the risk of frequency-specific biases.

4.3 MMMI vs No Imputation Prediction Comparison

Figure 3a shows that for each classifier, model accuracy is better when using MMMI compared to when data with missing values is simply removed. Similarly, Fig. 3b indicates that F1 scores for each classifier model are higher with MMMI than with no imputation. There is a significant improvement in accuracy across all classifier models because MMMI enhances both the amount of data and its quality. The less pronounced improvement in F1 scores can be attributed to the focus on recall and precision, making the improvement not as substantial as that observed for accuracy. The different improvements in F1 scores and accuracy suggest that while MMMI significantly enhances the overall data quality and completeness, the specific improvements in recall and precision are somewhat more nuanced.

While MMMI improves the overall data quality and completeness, its impact on precision and recall may not be as dramatic as its impact on the total number of correctly predicted observations. The smaller impact is because precision and recall are more sensitive to the distribution of the positive and negative classes in the data. Thus, while the completeness and quality of data improve, the specific balance between precision and recall may see more nuanced improvements.

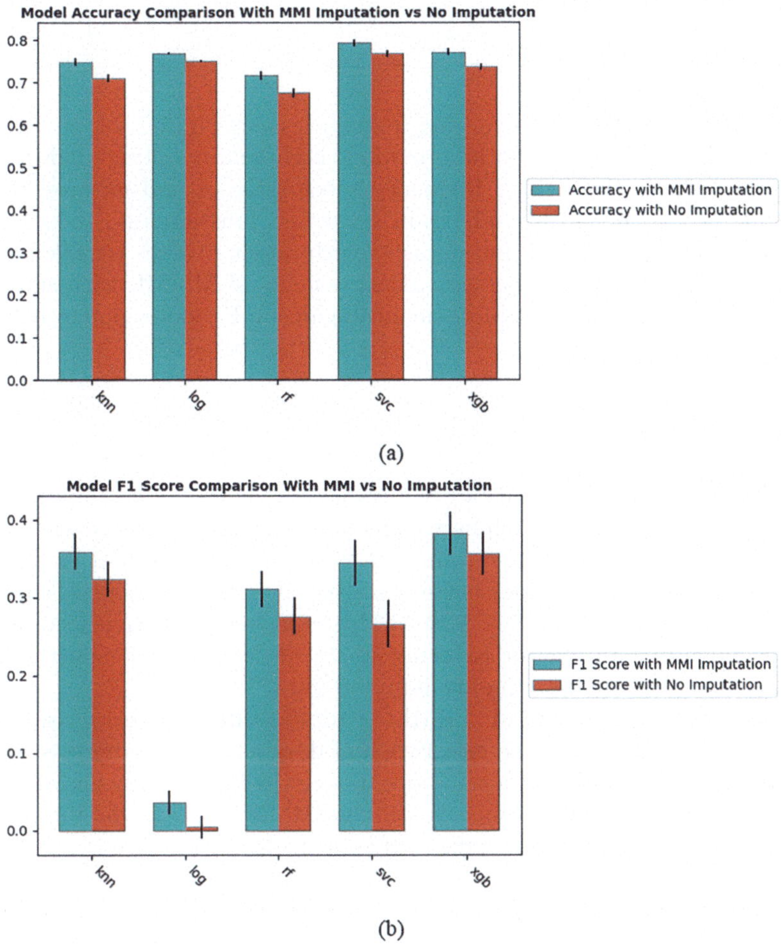

Fig. 3. Bar Chart showing how predictions can be improved for imputations vs no imputation (a) Accuracy (b) F1 score

In summary, the significant improvement in accuracy is a direct result of the increased amount of usable data and its improved quality due to MMMI. The more moderate improvement in F1 scores reflects the complex interplay between precision and recall, which are not solely dependent on data quantity but also on the distribution and balance of class labels within the dataset.

5 Discussion

The MMMI methodology marks a significant advancement in audiogram data imputation by leveraging the strengths of KNN, and RF algorithms. MMMI effectively balances bias and variance. This hybrid approach captures local relationships through KNN while managing complex interactions via RF's ensemble nature. The results shown in Figs. 2a, and 2b demonstrate that MMMI consistently outperforms individual models, particularly in accuracy and consistency across different frequencies.

MMMI's robustness across varying frequencies as seen in Table 2, highlights its adaptability and reliability in providing high-quality imputed data. This consistency reduces the risk of frequency-specific biases and ensures dependable results across the entire audiogram spectrum. Furthermore, the comparative analysis between MMMI and datasets with missing values removed seen in Figs. 3a and 3b underscores the substantial benefits of MMMI, with significant improvements in accuracy and more nuanced gains in F1 scores due to enhanced data quality.

6 Conclusion

The MMMI methodology significantly advances audiogram data imputation by combining multiple ML algorithms to effectively manage bias and variance. It outperforms individual models, showing a 2–20% improvement in MAE and MSE over KNN and RF, and provides better consistency across frequencies compared to previous methods [5] and [16]. MMMI also improves predictive accuracy and F1 scores while maintaining or reducing standard deviation, demonstrating its robustness in delivering high-quality imputed data.

Future research could extend MMMI to various medical datasets, use of other ML techniques, and explore the integration of deep learning techniques. Investigating different ML combinations and optimizing the balance among ML models could further enhance performance, solidifying MMMI's role in data imputation and broadening its application across different fields.

References

1. Aazh, H., Moore, B.: Challenges in hearing assessment and management. Audiol. Res. **7**(2), 88–95 (2017)
2. Abrams, H.B., Kihm, J., Edwards, B., Creelman, C.D.: A meta-analysis of hearing aid benefit in adults. Ear Hear. **39**(4), 872–890 (2018)
3. Aleryani, A., Wang, W., De La Iglesia, B.: Multiple imputation ensembles (MIE) for dealing with missing data. SN Comput. Sci. **1**, 1–20 (2020)
4. Barbour, D.L., et al.: Online machine learning audiometry. Ear Hear. **40**(4), 918 (2019)
5. Beaver, S., Bryce, R.: Comparison of machine learning techniques for single imputation on audiograms. Int. J. Comput. Inf. Eng. **18**(4), 204–214 (2024)

6. Chen, S.Y., Grisel, J.J., Lam, A., Golub, J.S.: Assessing cochlear implant outcomes in older adults using HERMES: a national web-based database. Otol. Neurotology **38**(10), e405–e412 (2017)
7. Emmanuel, T., Maupong, T., Mpoeleng, D., Semong, T., Mphago, B., Tabona, O.: A survey on missing data in machine learning. J. Big Data **8**(1), 1–37 (2021). https://doi.org/10.1186/s40537-021-00516-9
8. Feeney, M.P., Hnath-Chisolm, T.: Methodological issues in the measurement of hearing loss in the elderly: a review. J. Am. Acad. Audiol. **20**(1), 27–41 (2009)
9. Feirn, R., et al.: Guidelines for fitting hearing aids to young infants. Version **2**, 18 (2014)
10. Garciarena, U., Santana, R.: An extensive analysis of the interaction between missing data types, imputation methods, and supervised classifiers. Expert Syst. Appl. **89**, 52–65 (2017)
11. Qustafsson, F.: Comparing random forest, XGBoost and neural networks with hyperparameter optimization by nested cross-validation (2019)
12. Hasan, M.K., Alam, M.A., Roy, S., Dutta, A., Jawad, M.T., Das, S.: Missing value imputation affects the performance of machine learning: a review and analysis of the literature (2010–2021). Inf. Med. Unlocked **27**, 100799 (2021)
13. Hastie, T., Tibshirani, R., Friedman, J.H., Friedman, J.H.: The Elements of Statistical Learning: Data Mining, Inference, and Prediction, vol. 2. Springer, New York (2009)
14. Ibrahim, J.G., Chu, H., Chen, M.H.: Missing data in clinical studies: issues and methods. J. Clin. Oncol. **30**(26), 3297 (2012)
15. Niquet, J., Duwell, L., Rivière, J.: Statistical handling of missing data in audiology. Hear. Res. **339**, 87–98 (2016)
16. Pavelchek, C., et al.: Imputation of missing values for cochlear implant candidate audiometric data and potential applications. PLoS ONE **18**(2), e0281337 (2023)
17. Pittman, A.L., Stelmachowicz, P.G.: Hearing loss in children and adults: audiometric configuration, asymmetry, and progression. Ear Hear. **24**(3), 198 (2003)
18. Rubin, D.B.: Multiple Imputation for Nonresponse in Surveys, vol. 81. Wiley, Hoboken (2004)
19. Schafer, E.C., et al.: Creating a framework for data sharing in cochlear implant research. Cochlear Implants Int. **17**(6), 283–292 (2016)
20. Song, X.D., et al.: Fast, continuous audiogram estimation using machine learning. Ear Hear. **36**(6), e326 (2015)
21. Wendl, J., et al.: Compensating for missing data in the OHRKAN cohort study examining total leisure noise exposure among adolescents. Int. J. Audiol. **61**(7), 574–582 (2022)

eXpLogic: Explaining Logic Types and Patterns in DiffLogic Networks

Stephen Wormald(✉), David Koblah, Matheus Kunzler Maldaner, Domenic Forte, and Damon L. Woodard

Florida Institute for National Security, University of Florida, 601 Gale Lemerand Dr, Gainesville, FL 32611, USA
{stephen.wormald,dkoblah,mkunzlermaldaner}@ufl.edu,
{dforte,dwoodard}@ece.ufl.edu

Abstract. Constraining deep neural networks (DNNs) to learn individual logic types per node, as performed using the DiffLogic network architecture, opens the door to model-specific explanation techniques that quell the complexity inherent to DNNs. Inspired by principles of circuit analysis from computer engineering, this work presents an algorithm (eXpLogic) for producing saliency maps which explain input patterns that activate certain functions. The eXpLogic explanations: (1) show the exact set of inputs responsible for a decision, which helps interpret false negative and false positive predictions, (2) highlight common input patterns that activate certain outputs, and (3) help reduce the network size to improve class-specific inference. To evaluate the eXpLogic saliency map, we introduce a metric that quantifies how much an input changes before switching a model's class prediction (the SwitchDist) and use this metric to compare eXpLogic against the Vanilla Gradients (VG) and Integrated Gradient (IG) methods. Generally, we show that eXpLogic saliency maps are better at determining which inputs will change the class prediction. These maps help reduce the network size and inference times by up to 87% and 8%, respectively, while having a limited impact (−3.8% change in accuracy) on class-specific predictions. The broader value of this work is in demonstrating how certain DNN architectures promote explainability, which is relevant when understanding Artificial Intelligence (AI) is critical.

Keywords: Machine Learning · DiffLogic · Neurosymbolic AI · Interpretability · Explainable Artificial Intelligence · XAI · Logic Gate · Logic Operator · Differentiable · Fast Inference

1 Introduction

Understanding the behaviors of DNNs could aid the reliable adoption of machine learning in fields where interpretability and trust are paramount, including healthcare, defense, and law [1]. The field of eXplainable Artificial Intelligence (XAI) approaches this need by providing post-hoc methods for interpreting

S. Latifi (Ed.): ITNG 2025, AISC 1463, pp. 282–292, 2025.
https://doi.org/10.1007/978-3-031-89063-5_24

black-box models, and by designing architectures that are inherently more interpretable [2]. The DiffLogic network presented by Petersen et al. [3] is a unique architecture that aids interpretability – each node learns to represent one of sixteen logic types (e.g. AND, OR, etc. per Table 1). While DiffLogic enables a logical interpretation of internal nodes, the original authors do not leverage this trait to explain their model's behaviors. This paper introduces eXpLogic to fill this research gap (seen in Fig. 1).

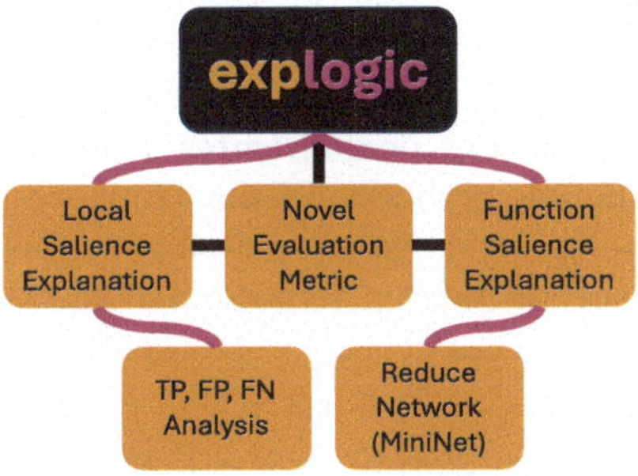

Fig. 1. Overview of several explanation tools that are enabled when using the DiffLogic network, which are referred to as the eXpLogic set of analyses

When applied to DiffLogic models trained for classification, eXpLogic highlights the importance of certain inputs in several scenarios. Specifically, eXpLogic: (i) Reveals which inputs caused a given class prediction (i.e., a local saliency method); (ii) Shows which inputs are likely to "turn on" a selected

Table 1. Table of logic gates learned by the DiffLogic network showing the probability of a gate's output (Z) given two inputs A and B

ID	AB:(00,01,10,11)	Logic Gate	P(Z)
0	(0, 0, 0, 0)	FALSE	0
1	(0, 0, 0, 1)	AND(A,B)	P(A)P(B)
2	(0, 0, 1, 0)	AND(A,NOT(B))	P(A) - P(A)P(B)
3	(0, 0, 1, 1)	A	P(A)
4	(0, 1, 0, 0)	AND(NOT(A),B)	P(B) - P(A)P(B)
5	(0, 1, 0, 1)	B	P(B)
6	(0, 1, 1, 0)	XOR(A,B)	P(A) + P(B) - 2*P(A)P(B)
7	(0, 1, 1, 1)	OR(A,B)	P(A) + P(B) - P(A)P(B)
8	(1, 0, 0, 0)	NOR(A,B)	1 - (P(A) + P(B) - P(A)P(B))
9	(1, 0, 0, 1)	XNOR(A,B)	1 - (P(A) + P(B) - 2*P(A)P(B))
10	(1, 0, 1, 0)	NOT(B)	1 - P(B)
11	(1, 0, 1, 1)	OR(A,NOT(B))	1 - P(B) + P(A)P(B)
12	(1, 1, 0, 0)	NOT(A)	1 - P(A)
13	(1, 1, 0, 1)	OR(NOT(A), B)	1 - P(A) + P(A)P(B)
14	(1, 1, 1, 0)	NAND(A,B)	1 - P(A)P(B)
15	(1, 1, 1, 1)	TRUE	1

function, which helps explain associated input patterns (i.e. a function explanation method); (iii) Helps model inspectors understand True Positive (TP), False Positive (FP) and False Negative (FN) predictions by providing comparable evidence for each case; and (iv) Produces function explanations that show which inputs and networks nodes can be ignored, making it possible to prune irrelevant functions from DiffLogic networks to create miniature networks (MiniNets) that predict individual classes with reduced inference times. While many methods produce saliency maps [4] and reveal input patterns that activate certain functions [5], eXpLogic differs by leveraging the logical structure learned by DiffLogic to reveal explicate sets of important inputs. We introduce a novel explanation metric to compare the eXpLogic method against existing methods.

To present this work, Sect. 2 demonstrates the novelty of eXpLogic and Sect. 3 outlines the methodology. Section 4 illustrates results from the explanation method, compares eXpLogic against existing approaches, and studies how MiniNets increase class-specific inference speed. Salient conclusions are presented in Sect. 5.

2 Background and Novelty

Differentiable Logic Gate Network, or "DiffLogic," is a DNN architecture that constrains each node in the network to represent real-logic operations [3]. DiffLogic can be interpreted as a set of logic gates, making it generally more interpretable than standard multi-layer-perceptrons (MLP). Since its publication, the DiffLogic architecture has gained attention in explainable convolutional neural networks [6] and concept reasoning [7]. The current work extends existing research in several key ways, as described through the following list of contributions:

- *Novel Input Saliency Explanation*: While many saliency methods exist, including GradCAM, SHAP, and DeepLift [4], they can give misleading information [8] that are preferred for visual clarity rather than their sensitivity to model behaviors [9]. The local eXpLogic saliency method (eXpLogic-L) is novel for showing the exact set of inputs that are responsible for a given decision (in the case of binary inputs), and for revealing whether an input's presence or absence supports a model's decision.
- *Novel Function Explanation*: Nodes in a DNN can be explained by showing the input patterns that activate certain functions. A comparable method called activation maximization [5] produces example inputs via gradient ascent, which is prone to finding local maxima. The eXpLogic method for explaining a function exploits the observation that DiffLogic learns a limited set of activation paths, which can be used to identify and weight the full set of inputs associated with certain functions according to the likelihood of each decision trace. This method is a novel approach of interpreting DiffLogic networks.
- *Novel Metric for Evaluating Saliency Methods*: Existing metrics evaluate saliency methods by comparing saliency maps against human eye-fixations

[10], or by measuring model sensitivity, human perceptibility, and performance characteristics [11]. The Area Over the Perturbation Curve (AOPC) sensitivity metric [12] quantifies how much a model output changes as the current input is varied to some baseline. However, when classification scores returned by a model are discrete, the AOPC sums to zero until the class changes. To account for this issue, we introduce the SwitchDist (the distance needed to switch a model's class prediction) as a metric to evaluate saliency methods in discrete classification settings.

In summary, eXpLogic is a single algorithm which produces local explanations of important inputs, and function explanations of input patterns that result in logical highs. The novelty of this work and forthcoming applications are realized via the methodology discussed in the following section.

3 Methods

The eXpLogic method uses a single algorithm (Sect. 3.1) to highlight inputs that provide evidence for TP, FN, and FP predictions, and to reveal inputs associated with individual functions (Sect. 3.3). These explanations are compared against existing saliency methods (Sect. 3.4) before demonstrating that eXpLogic helps prune DiffLogic networks and produce MiniNets that reduce the inference time needed to predict individual classes (Sect. 3.5).

3.1 Algorithm for Local and Function Explanations

The eXpLogic algorithm reveals important inputs by exploiting the set of logic gates learned by a DiffLogic model. Each logic gate changes the set of "activation paths" (or sequences of logic gates) connecting inputs to outputs [13]. The importance of an input relates to the number and type of connected activation paths, as ranked using several metrics below. In contrast to MLPs, each logic gate learned by DiffLogic has a maximum of two inputs such that a node on layer L can be connected to at most 2^L inputs. Because a network trained to classify C classes using N nodes per layer predicts class C_i by summing $(N/|C|)$ from the second to last layer, class C_i is connected to at most $(2^L)(N/|C|)$ activation paths through the network. However, by learning the TRUE or FALSE gate types (ID 0 and 15 from Table 1), the path is effectively blocked. Furthermore, the gate types learned along an activated path determines whether a connected input activates downstream classes when it is present (closer to one) or absent (closer to zero). We use this observation to structure Algorithm 1, which performs a breadth-first search over all activation paths upstream of a selected function (representing that functions "fan in," or $FANIN$) while tracking which paths pass through NOT gates. The result is the set of inputs (or I) that influence the selected input either in their presence (when 1) or absence (when 0). Here, the "$SumSigns(I)$" function sums the signs associated with each input and formats I for visualization.

Algorithm 1 - FanIn(DL, v_{out}, θ): Find v_{out} Fan-in

Require: DL Net. (DL), DL node v_{out}, threshold θ

```
Q ← {(v_out, 1)}                                   ▷ Queue of nodes to check
while Q is not empty do
    (v, s) ← Dequeue(Q)
    for each predecessor u of v in DL do
        s_new ← s
        g_type ← GateType(v)
        (G_not−list) ← EdgeType(u, v)
        if (g_type ∈ G_not−list) then
            s_new ← −s_new
        end if
        θ_comp ← { θ      if s_new = 1
                 { 1 − θ  if s_new = −1
        if SaliencyFactor(u) > θ_comp then
            Enqueue(Q, (u, s_new))
        else if g_type = "input" then
            I ← I ∪ {(u, s_new)}
        end if
    end for
end while
I_Saliency ← SumSigns(I)
return I_Saliency
```

Note Algorithm 1 includes a threshold θ, which prunes activation paths from being included based on a "SaliencyFactor(u)" metric, refereed to as "SF." Numerous metrics can be used, including: (i) The Empirical Signal Probability per Eq. (1), or $\mathtt{SP}_E$, and (ii) the Analytical Signal Probability per Eq. (2), or $\mathtt{SP}_A$. Here, $\mathtt{SP}_E$ is an empirical metric representing how often a logic gate's output (Z) is active (produces a logic high) given a set of inputs (X). $\mathtt{SP}_A$ is an analytical variant derived from the real logic expressions of each gate type, which are summarized in Table 1. Each metric is calculated per layer (L) and node (n) in a DiffLogic network (DL), as indicated by the subscripts below:

$$\mathtt{SP}_{E,(L,n)}(\mathbf{X}) = \frac{\sum_{i=1}^{|\mathbf{X}|} DL_{(L,n)}(\mathbf{X}_i)}{|\mathbf{X}|} \tag{1}$$

$$\mathtt{SP}_{A,(L,n)}(\mathbf{X}) = P_{(L,n)}(Z|\mathbf{X}) \tag{2}$$

After calculating a given SF per node, a set of saliency maps can be generated using $FanIn()$ across a range of θ and averaged using Eq. (3) to show which inputs are associated with common activation paths, as reported in later sections.

$$eXpLogic_{\mathtt{SF}} = \frac{1}{\theta_{max} - \theta_{min}} \sum_{\theta=\theta_{min}}^{\theta_{max}} FanIn_{\mathtt{SF}}(DL, n, \theta) \tag{3}$$

3.2 Training DiffLogic Networks

This work iteratively trained a set of DiffLogic models while reducing the number of layers and nodes so the test accuracy exceeded 90% on a binarized version of the MNIST dataset [14]. MNIST was selected given its prevalence in benchmarking saliency methods. The images were cropped to 20×20 pixels to remove the blank boarder, and models were trained using a train-test split of 80%–20%, a learning rate of 0.01, a "grad-factor"[1] of 1, and a τ of 10. The resulting network had two 2500-node layers. Five of these models were trained to demonstrate eXpLogic, achieving a mean test accuracy of 91.2%. While DiffLogic can achieve state-of-the-art performance, this work sought a small network to aid explainability and help determine whether the eXpLogic method is useful for understanding FN and FP predictions (shown in Sect. 3.3). A two-layer model trained on Fashion MNIST (FMNIST, from [15]) demonstrates the extensibility of eXpLogic, and achieved an accuracy of 82.3% using 5000 nodes per layer and a learning rate of 0.01.

3.3 Demonstrating eXpLogic Explanations

The eXpLogic algorithm (Algorithm 1) was applied to explain *function representations* and *local predictions*, as summarized in Table 2. While local predictions are explained using single input x_i, function behaviors are explained using both $\mathcal{D}$, subsets of data per class type ($\mathcal{D}_{C_i}$), the null set ($\emptyset$), and a uniform input probability of 50% ($\mathcal{U}$). The SF metrics were used to produce ten explanation types, which are categorized as local (eXpLogic-L), global (eXpLogic-G), and class (eXpLogic-C) explanations. Using Algorithm 1 to explain the five MNIST models, eXpLogic-L explanations were generated for each sample in the MNIST dataset, while the alternate types were used to explain output nodes in the MNIST and FMNIST networks. The eXpLogic-L saliency maps for TP, FN, and FP predictions demonstrate how eXpLogic helps model inspectors understand incorrect model classifications, which helps in debugging model behaviors [16]. All explanations (excluding eXpLogic-$G_\emptyset$[2]) are evaluated per the following methods.

3.4 SwitchDist Metric to Evaluate eXpLogic Explanations

This section introduces a novel metric for comparing the local and function explanations against comparable saliency maps produced by the VG and IG methods [17]. In general, saliency methods reveal which inputs are important for predicting the current class, or what inputs may be changed to alter the class score. Each saliency method implicitly reveals one direction in the image space that could change the class score (see Fig. 2c), such as the VG method which shows the gradient of the class score. However, directions beside the gradient

[1] The "grad-factor" defined in DiffLogic avoids vanishing gradients.

[2] eXpLogic-$G_\emptyset$ is used for visualization purposes.

could change the class score more quickly. Therefore, we propose to rank saliency methods by the distance needed to switch a model's class prediction, or the "SwitchDist," which applies to continuous and discrete model outputs. In this work, each saliency method produces a saliency map $\mathbf{S}$ which corresponds to some unit vector in the image space, or $\hat{\mathbf{U}} = unit(\mathbf{S})$, as defined per Eq. (4).

$$\text{unit}(\mathbf{X}) = \frac{\mathbf{X}}{\|\mathbf{X}\|_2} \tag{4}$$

$\hat{\mathbf{U}}$ is used to define three directions in the image space per $\mathbf{S}$ that either remove important positive values (+I), add important negative values (−I), or do both (+/−I). Images were updated per the equations in Table 3 using a step size (or α) of 0.1 until the model's class score switched, or until $\|x_t' - x_i\|_2 > |x_i|/4$. Section 4 summarizes results from characterizing the SwitchDist for the MNIST models.

Table 2. eXpLogic saliency map variations which higlight different activation paths. Different inputs and the range of θ are used for each

Variant	Metric	$[\theta_{min}; \Delta\theta; \theta_{max}]$
eXpLogic-L_E	$\mathsf{SP}_E(x_i)$	[0; 0; 0]
eXpLogic-L_A	$\mathsf{SP}_A(P(x\|x_i))$	[0; 0; 0]
eXpLogic-$G_\emptyset$	$\mathsf{SP}_E(\emptyset)$	[0; 0.01, 1]
eXpLogic-$G_\mathcal{U}$	$\mathsf{SP}_A(P(x\|\mathcal{U}))$	[0; 0.01, 1]
eXpLogic-G_E	$\mathsf{SP}_E(\mathcal{D})$	[0; 0.01; 1]
eXpLogic-G_A	$\mathsf{SP}_A(P(x\|\mathcal{D}))$	[0; 0.01; 1]
eXpLogic-C_E	$\mathsf{SP}_E(\mathcal{D}_{C_i})$	[0; 0.01; 1]
eXpLogic-C_A	$\mathsf{SP}_A(P(x\|\mathcal{D}_{C_i}))$	[0; 0.01; 1]

Table 3. Three unit directions derived from a saliency map for altering a model's input, where the tolerance δ is set to the standard deviation of $(\mathbf{U} - \mu_\mathbf{U})/100$, where $\mu_\mathbf{U}$ is the mean of $\mathbf{U}$

Symbol	Method	Equation
+/−I	Switch All Imp.	$x' = x - \alpha \cdot \mathbf{U}$
+I	Remove (+) Imp.	$x' = x - \alpha \cdot \text{unit}(\mathbf{U} \cdot (\mathbf{U} + \delta > 0))$
−I	Add (-) Imp.	$x' = x - \alpha \cdot \text{unit}(\mathbf{U} \cdot (\mathbf{U} + \delta <= 0))$

3.5 Reducing Model Inference times

The $FANIN$ found using eXpLogic can also be used to reduce model size and resulting inference times. While the class-specific features learned by standard DNNs can be hard to disentangle, the $FANIN$ for an output node in a DiffLogic network represents the subset of features and functions relevant to predicting an individual class. Ten class-specific miniature networks (or MiniNets) were created by retaining only the nodes in a class's $FANIN$ for both the MNIST and

FMNIST models. While DiffLogic predicts a class by taking the argmax of the class-sum, each MiniNet has a single output so maximum likelihood estimation was performed on the True and False MiniNet predictions to determine an appropriate decision threshold. Each MiniNet was compared against the parent network using the accuracy, relative network size, and relative inference time, as summarized in the following section.

4 Results and Discussion

The following sections show and evaluate the local and function explanations produced by eXpLogic (Sect. 4.1) before revealing the value of MiniNets in inference (Sect. 4.2).

4.1 Results from Explaining DiffLogic Behaviors

Function and local explanations illuminate diverse model behaviors, which may both be evaluated using the SwithDist as described in the following bullet points:

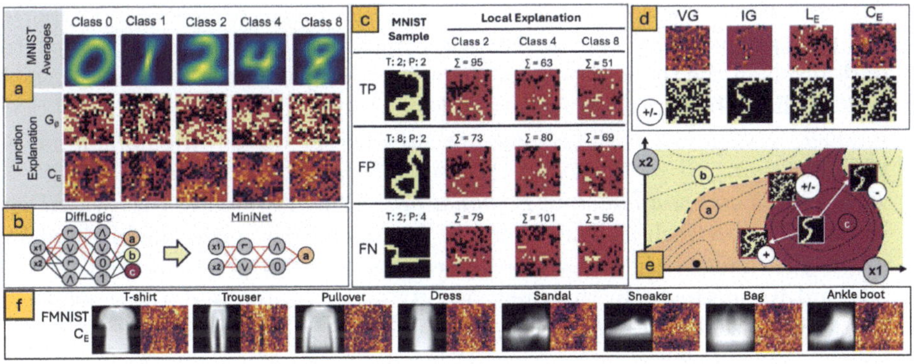

Fig. 2. Overview of eXpLogic, showing: (a) examples of the eXpLogic function explanations, where each exhibits input patterns which influence each digit class from the MNIST dataset; (b) a notional DiffLogic architecture that predicts three classes which is juxtapozed with a MiniNet derived from the FANIN of "class a;" (c) Local explanation method showing the evidence (important pixels) that support the decisions of class 2, 4, and 8 from the MNIST dataset. Each row shows either a TP, FP, or FN prediction where the image titles show the True (T) and Predicted (P) labels for the MNIST images, and the number of important inputs (Σ) identified for each explanation. Note that Σ is not always highest for the predicted class. Light pixels represent important positive inputs, whereas dark pixels represent important zero inputs; (d) a graphical illustration of the SwithDist used to evaluate each saliency methods, where a current input, class 5 in this case, is modified in (e) three directions which either add, remove, or alter the important input pixels. Each image represents a baseline of class 5 modified per each unit direction from Table 3. (f) Function explanations of eight FMNIST classes are displayed alongside the average class images to show the extensibility of eXpLogic to alternate datasets.

- *Examples of Function Explanations*: Patterns in the MNIST and FMNIST datasets are evident in the eXpLogic explanations. Figure 2a and Fig. 2f shows the function explanations associated with each class and indicate that each output node has learned a preference for certain inputs. The pixel colors indicate whether a class activates in response to the presence or absence of certain inputs. For example, eXpLogic-$G_\emptyset$ explanations show whether a class is likely to be predicted when there are positive values (bright pixels), or zero values (dark pixels) in certain input dimensions. While some explanations look like the class–averages from MNIST – like classes zero and one - others reveal a preference for spatial regions which are distinct to a certain class – like classes two and four, which exhibit a preference for positive inputs in the bottom and center regions of the image, respectively. Each eXpLogic explanation type generally highlights similar inputs, but with varying intensities depending on the SF metric used and the resulting activation paths found as important when using Algorithm 1.
- *Examples of Local Explanations*: While function explanations represent the sets of inputs that *could* predict each class, the local explanations in Fig. 2c show which inputs support individual predictions. In the TP case for class 2, a set of 95 active and non-active inputs supported the class decision. The FP case shows how a dark center present in an image from class 8 supports the false prediction of class 2. For the FN case, a horizontal line is next to dark pixels in the top and bottom of the images, which results in an image of a "2" being labeled a 4. These examples further demonstrate that eXpLogic uniquely reveals when the absence or presence of certain inputs support a class decision.
- *Evaluating Local and Function Explanations*: Table 4 evaluates the SwitchDist associated with each saliency method. In general, the pixels highlighted by eXpLogic changed the class score more quickly than pixels marked by the VG and IG methods. In this analysis, VG and IG often perform worse than a random baseline. While altering the important zero values using eXpLogic saliency maps resulted in low SwithDist values, removing only the important positive pixels produced worse SwithDist values. These results could indicate that eXpLogic is better at predicting important negative values, or that the trained network relies on negative values to make predictions.

Viewed together, these results seem to indicate that the eXpLogic local and function explanations perform better than VG and IG.

4.2 Results from MiniNet Model Reduction

When reducing the DiffLogic network into MiniNets per predicted class, we generally found the network size (−86%) and inference time (−10.0%) were reduced, though the one-vs-all accuracy decreased by 3.8% on average (see Table 5). The reduced network size indicates that 14% of a model would need to be loaded into short-term memory (or random access memory) to predict a certain class with reduced inference times. These findings may indicate value in leveraging

Table 4. Comparison of the SwitchDist per saliency type and derived unit direction from Table 3 produced for each image predicted as TP from MNIST for each of the five trained models. The three lowest values of the SwithDist are bolded per unit direction. Not all images switched class predictions when altering pixels, as represented using the %Switch metric that shows how many images switched classes for an explanation type

	Remove (+/−) Importance		Remove (+) Importance		Remove (−) Importance	
Method	SwitchDist (L_2)	% Switch	SwitchDist (L_2)	% Switch	SwitchDist (L_2)	% Switch
Random	7.9 ± 2.6	(92.1%)	**7.9 ± 2.6**	(92.1%)	7.9 ± 2.6	(92.1%)
Vanilla Gradients	12.8 ± 11.4	(69.5%)	12.7 ± 13.8	(15.6%)	9.7 ± 11.2	(53.5%)
Integrated Gradients	12.8 ± 15.1	(18.6%)	**10.6 ± 14.6**	(18.9%)	NaN ± NaN	(0.0%)
eXpLogic-L_E	**5.9 ± 0.3**	(91.2%)	**3.7 ± 0.4**	(25.4%)	**4.7 ± 0.5**	(80.3%)
eXpLogic-L_A	**5.9 ± 0.3**	(91.6%)	**3.7 ± 0.4**	(21.1%)	**4.7 ± 0.5**	(80.4%)
eXpLogic-G_U	7.4 ± 3.7	(92.0%)	12.1 ± 13.2	(48.2%)	8.1 ± 10.2	(79.9%)
eXpLogic-G_E	6.6 ± 7.2	(88.7%)	14.7 ± 20.5	(7.0%)	5.6 ± 3.3	(86.0%)
eXpLogic-G_A	11.1 ± 6.2	(83.3%)	11.7 ± 8.7	(57.2%)	**3.7 ± 2.0**	(22.9%)
eXpLogic-C_E	**5.6 ± 1.6**	(92.0%)	11.2 ± 12.6	(30.2%)	5.1 ± 2.0	(87.0%)
eXpLogic-C_A	**5.5 ± 1.1**	(92.0%)	13.2 ± 14.5	(53.9%)	5.3 ± 5.7	(86.2%)

Table 5. MiniNet comparisons against the parent DiffLogic network for (1) MNIST and (2) FMNIST, which have 5,010 and 10,010 gates, inference speeds of 337 μs and 163 μs on average, respectively. The accuracy (Acc.), the precision (Prec.), recall, and F1 score are reported

Class	% Time	% Size	Acc.	Prec.	Recall	F1
(1) Avg.	−10.0%	−86.0%	94.8%	68.4%	11.0%	0.789
± Std.	± 0.1%	± 0.1%	± 2.5%	± 11.5%	± 0.1%	± 0.085
(2) Avg.	−15.0%	−58.5%	90.7%	59.8%	60.6%	0.59
± Std.	± 1.1%	9.3%	± 5.6%	± 22.2%	± 12.1%	± 0.17

MiniNets in resource-constrained applications, though more work is needed for larger networks and complex datasets.

5 Conclusions

Numerous conclusions emerge. Firstly, eXpLogic provides novel local explanations that help interpret TP, FN, and FP predictions, and reveal whether the absence or presence of certain inputs promoted a model's decision. Secondly, eXpLogic explains what inputs activate certain functions, which can be used to reduce create MiniNets that reduce the resources and inference time needed for class-specific predictions. The eXpLogic saliency maps are generally better than competing methods at identify inputs that change model behaviors, as measured using the SwitchDist metric. Future works may compare the SwitchDist against established metrics, explore real-world and complex datasets, compare positive and negative saliency against existing methods for interpreting negative features, and perform relevant human studies. Together, this work demonstrates

expertise to build up patterns and to relate them to a particular traffic type. With the network communications becoming more complicated and more camouflaged, it is challenging to build patterns and to use them to recognize network traffic accurately. Furthermore, for many network traffic has been encrypted nowadays, it becomes more challenging to use DPI to recognize network traffic.

With the development of machine learning and deep learning, researchers are using Neural Networks (NNs) to classify network traffic [9]. NNs play the role of extracting patterns and using them to classify traffic, just as what human experts do in DPI. With the advanced computing capacity, a well designed and trained neural network could make a more accurate prediction of traffic type. However, there are two major challenges in a NN-based traffic classification model. First, what information should be fed into a neural network? There are two approaches to this end, i.e., raw traffic flow date or flow features. Raw traffic date contains tremendous information. To limit the complexity of the classification model, researcher extract part of the raw data and feed them to a neural network [12]. With the concern of a huge part of the traffic information has been lost in this approach, other researchers first extract features from the entire traffic flow, then feed the features to the neural network [4,7]. Both methods have achieved promising performances.

The second challenge of NN-based traffic classification is the neural network architecture design and hyperparameter search. With the consideration of network traffic flow characteristics, various neural networks with different complexity have been studied for traffic classification [9]. However, determining the optimal neural network architecture with the optimal hyperparameters is extremely challenging, for the searching space is tremendous. A one-dimensional Convolutional Neural Network (CNN) has been used in [11]. Meanwhile, Chen et al. converted the time series of traffic data into a two-dimensional structure and used a two-dimensional CNN for the network traffic classification [3]. A Stacked Auto Encoder (SEA) model was used for traffic classification and achieved improved performance [6]. All those models have studied different NN architectures for network traffic classification and achieved good performance. However, does raw traffic data or flow features produce better classification performance? Which NN architecture works better with raw data and which works better with features? What are the optimal hyperparameters and how to find them? Those questions still remain unanswered.

In this paper, we implemented two network traffic classification models, i.e., raw-data-based and feature-based. The main goal is to conduct a comparison between the two approaches. To make the comparison fair and accurate, we proposed a two-level searching algorithm to find the optimal hyperparameters of the NN. Using the neural networks with quasi-optimal hyperparameters, a performance comparison and analyses were conducted for the raw-data-based traffic classification and feature-based traffic classification.

The rest of the paper is organized as follows. Section 2 introduces our traffic classification methodology. Section 3 proposes two approaches to generate input data for the neural network, i.e., raw data and features. Section 4 introduces the neural network design method and how to search for the optimal hyperparameters. Section 5 presents the experimental results. At the end, Sect. 6 draws the conclusions.

2 Traffic Classification Methodology

There are three steps in our traffic classification model: data collection, data pre-preprocessing, and Deep Learning networks design and implementation. To classify network traffic into various categories, significant amount of traffic data with labels is needed for the neural network training purpose. The data collected from either host computers or routers is a sequence of zeros and ones and needs to be transformed and organized into a readable format for the NN. This procedure is called data pre-processing. Eventually, the formatted traffic data will be fed into a NN for classification. The work flow is shown in Fig. 1.

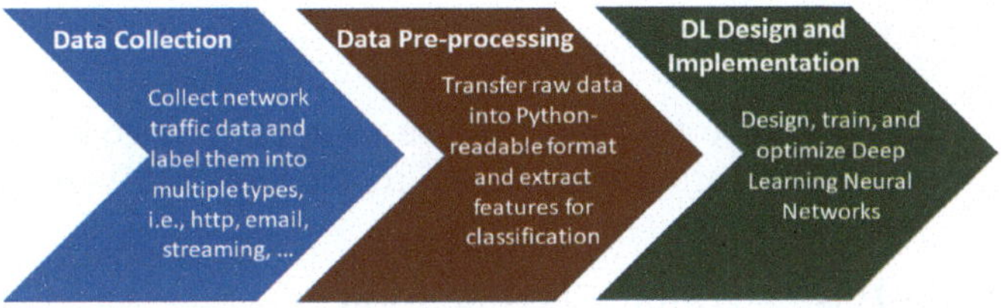

Fig. 1. Methodology of Traffic Classification Using Neural Networks

The data collected for traffic classification could be either packets or flows. A network traffic flow is defined as a sequence of packets with the same 5-tuple: source IP address, destination IP address, source port number, destination port number, and protocol number. ISCXVPN2016 is a free database that provides network traffic flow data with labels [4]. The samples distribute among 14 categories: browsing, email, chat, streaming, file transfer, VoIP, TraP2P, and each type includes non-VPN and VPN traffic. We also collected traffic flow data on host devices in our lab using Wireshark software, which generates traffic flow data with a format of pcap or pcapng. When collecting traffic data, we turned off all other applications on the same device and only keep the target traffic running. Once the traffic had been collected by Wireshark, we named the pcap file with the application name and traffic type, such as youtube_stream.pacp. In this way, we labeled each flow data in file name. The self-collected data plus ISCXVPN2016 date generates sufficient, unbiased, and diverse traffic data for NN training.

Once the traffic data has been collected, we need to transfer it into a program-readable format, such as csv files or json files. There are two considerations to this end: feeding NNs with raw traffic data or traffic features. The information amount of a traffic flow is significant, and it is almost impossible to feed the entire flow information to the NN. One of the efficient solutions is to take the first N bytes of a flow and use them as the input of the neural network. This approach decreases the computational complexity of the neural network with the cost of giving up the rest of bytes of a flow. Another approach is to extract features from the entire traffic flow, such as IP address, protocol type, byte distribution, etc. Same as raw-data-based classification, the goal is to decrease the computation of the neural network. However, the classification accuracy heavily relies on the features been used. Both approaches generate input data with various size for the NN, yielding different classification accuracy. Our traffic classification model is as shown in Fig. 2.

We have designed and implemented neural networks using both raw traffic data and flow features. With a large hyperparameter space of the neural network, we proposed a search methodology to find quasi-optimal neural network architecture which yields high classification accuracy. More details will be found in Sect. 4.

3 Data Pre-processing

In this section, we introduce how to generate samples that contain raw traffic data or flow features. Both are able to generate samples with various size.

3.1 Raw-Data-Based Traffic Classification

The training data of this paper comes from two resource: ISCXCPN2016 and data collected in our lab using Wireshark software. We have collected a total of more than 200 flow files with 14 categories. The format of the flow data is pcap or pcapng, which is a sequence of bits. We first converted the raw traffic data into a sequence of bytes using a self-developed Python program. To generate enough training samples, we partitioned a flow into multiple samples. Since the flow header yields the administration information and is crucial for traffic

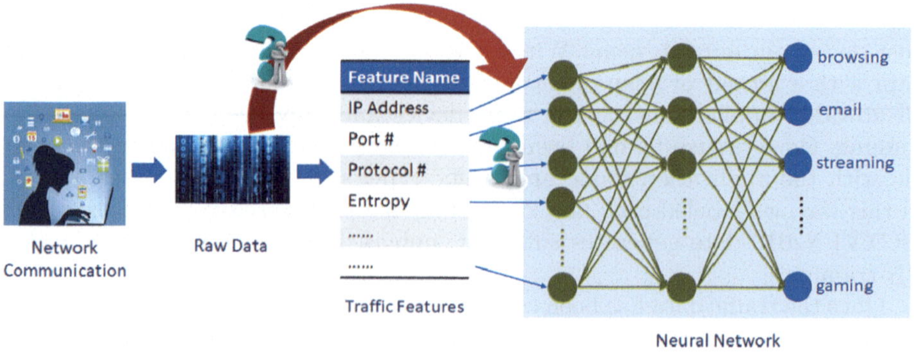

Fig. 2. The Traffic Classification Model

classification, all the samples belonging to the same flow share the same flow header. Wireshark captures flow data in network layer, where the header has a maximum size of 60 bytes, sitting at the beginning of a flow sequence. Therefore, all samples take the first 60 bytes of the flow data and put them at the beginning. Assuming the size of each sample is N bytes, each sample then reads $(N - 60)$ bytes from the flow and adds them next to the flow header. The sample generation process is as shown in Fig. 3.

Some flows, such as video, contains significant information, therefore can generate a lot of samples. While flows such as email could be relatively short and generates less samples. To make sure the training database contains balanced samples across all the traffic types, we have set a maximum number of samples generated by the same flow. Depending on the threshold, the total number of samples generated by the traffic flows varies. For example, with a threshold of 160, we have generated 16,426 samples from 200 pcap files.

3.2 Feature-Based Traffic Classification

To extract features from the raw traffic data, we used a third-party software, JOY, provided by CISCO. JOY reads a pcap or pcapng file, extracts significant flow features, and writes them into a json file [2]. The json file can be read by our program and used for a feature-based classification.

In our previous work, we had clustered traffic features into five groups depending on their characteristics, i.e., Group A) Basic Flow Information, which includes IP addresses, port numbers, and protocol number; Group B) transport layer security (TLS) information, which provides cybersecurity protocols, encryption key exchange method, authentication method, etc. [8]; Group C) Time-to-Live (TTL) information [5]; Group D) Byte Distribution and Entropy [10], Group E) Packet Sequence information. A comprehensive research on how to select features and how the feature selection scheme impacts the traffic classification accuracy and computational complexity had been conducted [7]. Table 1 shows some feature selection schemes and their number of features in total. The more features being used, the more computational complexity of the neural networks would be required.

Table 1. Feature Selection Schemes

	Selected Features	Total No. of features
Scheme 1	Feature Group A	11
Scheme 2	Feature Groups A, B, C	499
Scheme 3	Feature Groups A, B, C, D, E	937

4 Neural Network Design

In this paper, Multilayer Perceptron (MLP) and CNN have been used for traffic classification. One of the challenges of the neural network design is to find the optimal hyperparameters. In this section, an efficient searching strategy for neural networks' hyperparameters is proposed.

For an MLP with an input layer of 937 neurons, two hidden layers, and an output layer of 12 neurons, we considered various numbers of neurons for each hidden layer, i.e., 64, 320, 576, 832, 1088, 1344, 1600, and 1856 neurons. Therefore, the number of MLP models need to be trained is 8^2, i.e., 64 neural networks. Figure 4 shows the classification accuracy with various neuron numbers of each hidden layer. In Fig. 4, we used different color to represent different accuracy and marked the MLPs of which the classification accuracy is higher than 92% as "good".

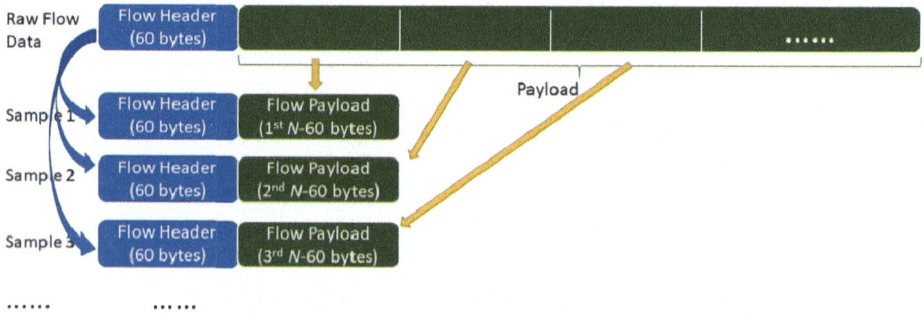

Fig. 3. Generate Samples from Flow Sequence: Flow Partition

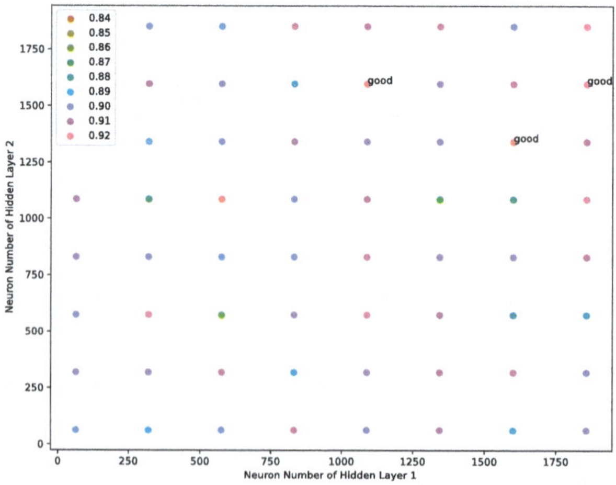

Fig. 4. Classification Accuracy with Various Hyperparameters (Two Hidden Layers)

When the number of hidden layers increased to three, with the candidates of 64, 320, 576, 832, 1088, 1344, 1600, and 1856 as the number of neurons of each hidden layer, the number of various combinations is 8^3. That means a total of 512 MLPs need to been trained and tested. Figure 5 shows the accuracy of each MLP, of which the accuracy is greater than 92% were marked as "good".

When the number of hidden layers becomes 4, the number of hyperparameter combinations is 8^4, which means a total of 4096 MLPs need to be trained and tested. This process took weeks on a DELL Alienware work station. Figure 6 shows the search result in a four-dimensional hyperparameter space, where the x-axis and y-axis of the left figure represent the number of neurons of the first and second hidden layers, and the x-axis and y-axis of the right figure represent the number of neurons of the third and fourth hidden layers. In Fig. 6, the neural networks of which accuracy is greater than 93% were marked as "good".

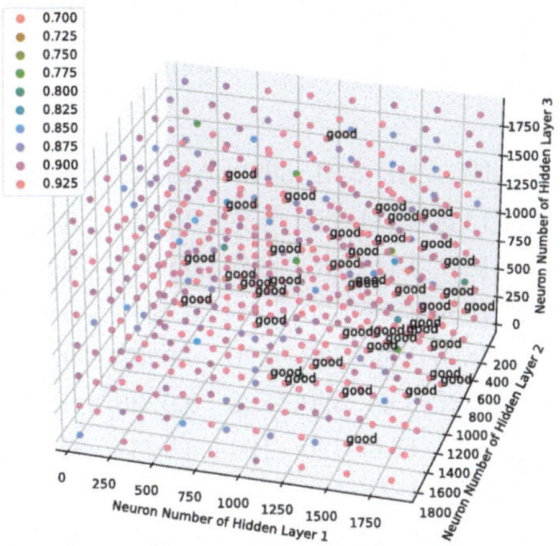

Fig. 5. Classification Accuracy with Various Hyperparameters (Three Hidden Layers)

A brutal search for the hyperparameters in a high dimensional space takes incredibly long time. To fasten the process, we need to narrow down the candidates of the MLP hyperparameters that are considered. From Fig. 4 through Fig. 6 we see that, the neural networks with higher classification accuracy are clustered in the hyperparameter space. Therefore, we can first search the entire space on a coarse-grained base, locate the hyperparameter region that yields higher classification accuracy, then conduct a fine-grained search in that region. For example, for the MLP with 4 hidden layers, the neuron number of 320, 832, 1344, and 1856 for each hidden layer are first considered, therefore, $4^4 = 256$ MLP networks are trained and tested. Among of the MLPs, there are 29 MLPs achieved an accuracy of 92% and above. Table 2 shows the top 10 trials with the highest classification accuracy after a coarse-grained search.

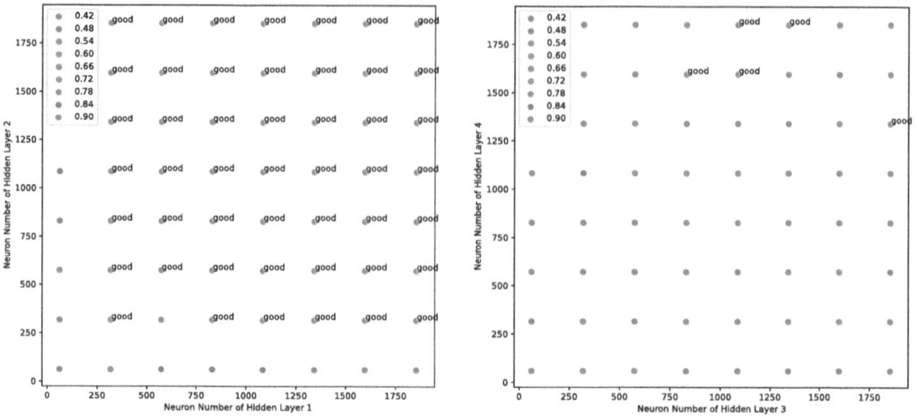

Fig. 6. Classification Accuracy with Various Hyperparameters (Four Hidden Layers)

Table 2. The Top Ten Trials with a Coarse-Grained Search for Hyperparameters (4 Hidden Layers)

trial No.	HL 1	HL 2	HL 3	HL 4	accuracy	precision
2912	1344	1344	832	1856	0.9248	0.9296
3932	1856	1344	832	832	0.9245	0.9287
4042	1856	1856	320	320	0.9245	0.9300
3930	1856	1344	832	320	0.9239	0.9260
1740	832	832	320	832	0.9236	0.9295
1870	832	1344	320	1344	0.9233	0.9299
3052	1344	1856	1344	832	0.9230	0.9275
3916	1856	1344	320	832	0.9230	0.9289
734	320	832	832	1344	0.9227	0.9312
3962	1856	1344	1856	320	0.9227	0.9289

The coarse-grained search revealed that the region of the top-right corner on the first-second hidden layer chart and the top region of the third-fourth hidden layer chart yield a higher classification accuracy. Therefore, we conducted a fine-grained search in these regions, and found the quasi-optimal number of neurons of the hidden layers 1, 2, 3, and 4 are 1088, 1856, 576, and 1600, respectively. With these hyperparameters, the MLP achieved an accuracy of 93.1%.

With a two-level hyperparameter search, the number of neural networks that need to be trained and tested to find an optimal architecture is dramatically decreased, which is particular important for a neural network with large numbers of hidden layers and neurons. The two-level hyperparameter search is more efficient than grid search and simpler than Bayesian optimization.

5 Experimental Results

In this section, we compare the classification performance of the raw-data-based neural networks and the feature-based neural networks. The comparison was conducted upon various level of computational complexity. To evaluate the classification performance, accuracy and precision are used, which are defined as:

$$accuracy = \frac{TP + TN}{TP + TN + FP + FN}, \tag{1}$$

$$precision = \frac{TP}{TP + FP}, beginequationaccuracy = \frac{TP + TN}{TP + TN + FP + FN}, \tag{2}$$

iv

$$precision = \frac{TP}{TP + FP}, \tag{3}$$

.

First, we used 499 bytes of the flow features (shown as the scheme 2 in Table 1) and 499 bytes of the raw flow data, respectively, as the input of the neural network. Using an MLP with eight hidden layers (i.e., 1600-1200-1200-880-640-360-220-80) and a 1D-CNN (i.e., CNN(100,100)-CNN(100,80)-220), respectively, the classification accuracy and precision are shown in Table 3.

Table 3. Classification Performances with 499 Inputs

	MLP		1D-CNN	
	Accuracy	Precision	Accuracy	Precision
Raw-Data-Based	99.259%	99.259%	98.963%	98.992%
Feature-Based	84.436%	87.019%	87.41%	88.225%

When 937 bytes of the flow features (shown as the scheme 3 in Table 1) and 937 bytes of the raw flow data were used as the input of the neural network, the classification performances of an MLP with six hidden layers (i.e., 1088-1088-1856-832-576-1600) and a 1D-CNN (i.e., CNN(100,100)-CNN(100,80)-1200-580-120) are shown in Table 4.

Table 4. Classification Performances with 937 Inputs

	MLP		1D-CNN	
	Accuracy	Precision	Accuracy	Precision
Raw-Data-Based	99.449%	99.449%	98.604%	99.113%
Feature-Based	92.422%	92.963%	92.209%	93.037%

Apparently, for both less input (499) and more inputs (937), using raw flow data to classify traffic type performs better than using flow features. This means that neural networks work more effectively in extracting features and using them to recognize traffic. When features are used for the classification, CNN performs better than MLP. This is because that CNN captures the correlation among the packets of a flow (presented by feature group E as we discussed in Subsect. 3.2). However, when raw data is used, it is hard to find correlations among flow bytes using a 1D-CNN. Therefore, MLP performs slightly better than 1D-CNN when raw traffic data was used. The experiments also showed that when the input is less than 200 bytes, the feature-based classification performs better. This is because that the well-selected features present traffic type efficiently, while cutting off too much information directly from the raw data lost lots of traffic type information.

6 Conclusions

This paper first proposed an efficient way to search the hyperparameter space of the neural networks for network traffic classification. With the optimized NNs, a comprehensive comparison was conducted between the raw-data-based and the feature-based network traffic classification. The experimental results show that with a relatively large NN and optimized hyperparameters, using part of the raw data of a network flow produced higher classification accuracy and precisions than using pre-designed features. This means that the neural networks work more effectively in extracting features and using them to recognize traffic. In future, we could take a peek at what features the hidden layers of the NN were extracting, which would give us better understanding about what network flow information indicates traffic types.

Acknowledgements. We would like to express my appreciation to Whitworth University STEM Research Program, which has played a crucial role in supporting this research.

References

1. Bujlow, T., Carela-Español, V., Barlet-Ros, P.: Independent comparison of popular dpi tools for traffic classification. Comput. Netw. **76**, 75–89 (2015)
2. Bunnell, D., Brate, A.: Making the Cisco connection: the story behind the real Internet superpower. Wiley (2000)
3. Chen, Z., He, K., Li, J., Geng, Y.: Seq2img: a sequence-to-image based approach towards IP traffic classification using convolutional neural networks. In: 2017 IEEE International Conference on Big Data (Big Data), pp. 1271–1276. IEEE (2017)
4. Gil, G.D., Lashkari, A.H., Mamun, M., Ghorbani, A.A.: Characterization of encrypted and VPN traffic using time-related features. In: Proceedings of the 2nd International Conference on Information Systems Security and Privacy (ICISSP 2016), Setúbal, Portugal, pp. 407–414. SciTePress (2016)
5. Hinden, R., Haverty, J., Sheltzer, A.: The darpa internet: interconnecting heterogeneous computer networks with gateways. Computer **16**(09), 38–48 (1983)

6. Lotfollahi, M., Jafari Siavoshani, M., Shirali Hossein Zade, R., Saberian, M.: Deep packet: a novel approach for encrypted traffic classification using deep learning. Soft Comput. **24**(3), 1999–2012 (2020)
7. Mao, Q., O'Neill, C., Bao, K.: A feature-based network traffic classification approach. Int. J. Netw. Secur. **25**(5), 821–828 (2023)
8. Rescorla, E.: RFC 5289: TLS elliptic curve cipher suites with SHA-256/384 and AES galois counter mode (GCM) (2008)
9. Rezaei, S., Liu, X.: Deep learning for encrypted traffic classification: an overview. IEEE Commun. Mag. **57**(5), 76–81 (2019)
10. Shapira, T., Shavitt, Y.: Flowpic: a generic representation for encrypted traffic classification and applications identification. IEEE Trans. Netw. Serv. Manage. **18**(2), 1218–1232 (2021)
11. Wang, W., Zhu, M., Wang, J., Zeng, X., Yang, Z.: End-to-end encrypted traffic classification with one-dimensional convolution neural networks. In: 2017 IEEE International Conference on Intelligence and Security Informatics (ISI), pp. 43–48. IEEE (2017)
12. Wang, W., Zhu, M., Zeng, X., Ye, X., Sheng, Y.: Malware traffic classification using convolutional neural network for representation learning. In: 2017 International Conference on Information Networking (ICOIN), pp. 712–717. IEEE (2017)

Anomaly on the Fly: Streaming Anomaly Detection in Supervised Learning

Twishi Saran and Rashida Hasan(✉)

Department of Computer Science, California State University, Northridge, CA, USA
rashida.hasan@csun.edu

Abstract. Detecting anomalies in streaming data presents unique challenges due to the continuous and dynamic nature of the data. Traditional anomaly detection techniques are often unsuitable for real-time environments where adaptability and scalability are paramount. In this paper, we propose a novel window-based anomaly detection algorithm that leverages the Bayesian ARIMA (AutoRegressive Integrated Moving Average) model for real-time streaming data. Our approach dynamically computes threshold values within sliding windows, allowing for the identification of local and global anomalies in an adaptive manner. The main contribution of our work is the development of a streaming Bayesian ARIMA model for real-time anomaly detection in continuous time-series data, effectively bridging the gap in leveraging statistical models for dynamic data streams. Extensive experiments on benchmark streaming datasets demonstrate that our algorithm outperforms the state-of-the-art algorithms in terms of AUC-ROC values.

Keywords: Time series data · Anomaly detection · Outlier detection · Streaming data · Bayesian Arima · Sliding window · Point anomaly · Streaming anomaly · AUC-ROC · Statistical anomaly detection

1 Introduction

Anomaly detection is a pivotal area in data mining, focused on identifying data points, patterns, or events that significantly diverge from the expected distribution within a dataset. Classically, outlier, often used interchangeably with anomaly was originally defined by Hawkins [8] as "An observation which deviates so much from other observations as to arouse suspicions that it was generated by a different mechanism". Such anomalies are often indicative of errors, novel events, or previously unknown patterns, making their detection critical across various fields, including finance, healthcare, cybersecurity, manufacturing, and telecommunication. With data volumes increasing exponentially, automating the traditional, time-intensive process of anomaly detection has become a priority, shifting from manual human analysis to dynamic, model-driven methods that can quickly flag unusual patterns in large datasets.

R. Hasan—These authors contributed equally to this work.

S. Latifi (Ed.): ITNG 2025, AISC 1463, pp. 304–315, 2025.
https://doi.org/10.1007/978-3-031-89063-5_26

In recent years, the rise of streaming technologies has intensified the need for real-time anomaly detection as vast amounts of data are generated continuously. Streaming data presents new challenges for machine learning and neural network algorithms, as they must process and extract insights dynamically without compromising performance. Unlike traditional batch processing, streaming anomaly detection requires algorithms to operate in a time-sensitive manner, continuously updating models as new data arrives. Monitoring customer behavior, identifying fraud, tracking usage trends, and detecting abnormalities in network traffic all rely on the ability to analyze data streams as they come in. This constant influx of data, alongside the evolving and expanding nature of the datasets, requires meticulous and adaptable approaches to efficiently detect and respond to anomalies within these high-velocity environments.

Over the years, numerous studies have examined various anomaly detection algorithms and their effectiveness across different domains [9,18–20]. Various machine learning algorithms, both supervised and unsupervised, have been employed to identify anomalies in both static and streaming datasets. In the context of time series analysis, data points are analyzed over time, with deviations from expected behavior classified as outliers. Researchers use time series anomalies to analyze unusual and interesting phenomena [1]. Anomaly detection in time series data can be achieved through prediction-based statistical methods or by identifying clusters of unusual shapes using machine learning techniques [1]. Much of the prior research has concentrated on distance-based calculations for outlier detection using traditional machine learning algorithms, including the use of the local outlier factor (LOF) to assess the relative outlierness of data points [5]. Clustering-based approaches, such as k-means [7] and density-based partitioning methods, have also been explored for detecting both shape and point outliers [17]. However, these methods are not well-suited for streaming datasets, as they rely on calculating nearest distances, fail to accommodate the dynamic nature of data streams, and are typically limited to static dataset experiments.

To overcome the limitations of conventional anomaly detection methods, various unsupervised machine learning algorithms have been studied. Among these methods is the k-Nearest Neighbor (kNN) algorithm [8], which improved runtime but not accuracy. Munir et al. utilized One-Class Support Vector Machines (OCSVM) to identify unusual shapes in time series data [12]. Additional traditional machine learning algorithms for anomaly detection include Principal Component Analysis (PCA) and Extreme Gradient Boosting Outlier Detection (XGBOD). Nonetheless, some work offers a comprehensive comparative analysis of these algorithms' performance on benchmark datasets in comparison to deep learning algorithms [12]. Most of the abovementioned anomaly detection methods focus on identifying static anomalies rather than predicting them. Predictive techniques are particularly advantageous when dealing with streaming data, where future values are forecasted and compared to incoming actual values. This approach necessitates the use of contemporary forecasting models, highlighting the scope for advancement in anomaly detection.

In this work, we address these challenges by using statistical models to detect anomalies in a streaming data format. The contributions of this paper are ordered as follows:

1. We introduce a methodology for transforming static time-series data into a continuous data stream. Subsequently, we implement a window-based technique to facilitate real-time point-by-point anomaly examination.
2. We develop a novel streaming Bayesian ARIMA model for time-series anomaly detection, addressing the gap in leveraging statistical models for real-time data.
3. We conduct a comprehensive evaluation of the proposed algorithm's performance using benchmark time-series datasets and standard metrics to assess its anomaly detection efficacy.

2 Related Work

Recent studies have explored various approaches to anomaly detection in time series prediction, utilizing both deep learning and statistical methods to address the unique challenges posed by evolving and streaming datasets.

Some studies explore deep learning techniques, specifically Convolutional Neural Networks (CNNs) for anomaly detection in time series prediction [6]. While CNNs can handle raw, unlabeled data, they struggle with the evolving nature of streaming datasets. In contrast, [13] combined ARIMA forecasting with CNNs for anomaly detection, outperforming standalone models but potentially losing effectiveness in sparsely labeled datasets. Additionally, Akouemo et al., highlight the use of ARIMA for managing streaming datasets [2], while Lavin et al. introduced a deep learning approach combining CNNs with Euclidean distance-based detection for unsupervised anomaly detection in streaming datasets. However, this model struggled to accurately classify anomalous points in datasets with high levels of contamination.

Numenta's work [4] presents the Numenta Anomaly Benchmark (NAB), an open-source framework with labeled datasets for evaluating anomaly detection algorithms. They evaluated algorithms such as Hierarchical Temporal Memory (HTM), which utilizes continuous learning and neocortex theory. Some work focuses on regressive models, such as Auto-Regressive (AR) and Moving Average (MA), to identify contextual anomalies [10]. This study revealed that AR and MA models exhibit faster computation times in comparison to deep learning models and perform more effectively in forecasting for uni-variate time series datasets.

However, the bulk of research that has been carried out in this field has concentrated on analyzing static datasets. With the increase of real-time data, there is a growing need to focus on analyzing streaming datasets, which constantly evolve and require real-time analytics. In our approach, we used the Bayesian ARIMA model with a sliding window to detect anomalies in streaming data, allowing for continuous monitoring and real-time anomaly detection.

3 Problem Definition

In this work, we focus on point anomalies and streaming anomaly detection within the broader realm of anomaly detection. These anomalies are crucial in various applications, such as fraud detection, network security, and manufacturing quality control, where sudden deviations can indicate potential issues.

Definition 1: (Anomaly Detection) Anomaly detection is the process of identifying data points that exhibit significant deviations from a general normal distribution. These data points are assigned an anomaly scored based on how much they differ from the norm, and those exceeding a set limit are flagged as anomalies. We defined the anomaly detection mathematically as follows:

$$A = \{x_i \in D \mid s(x_i) > T\} \tag{1}$$

where: A is the set of identified anomalies, $D = \{x_1, x_2, \ldots, x_n\}$ is the dataset, $s(x_i)$ is the anomaly score assigned to each data point x_i, and T is the threshold value for determining anomalies.

Definition 2: (Streaming Anomaly Detection) Let $S(t) = \{x_1(t), x_2(t), \ldots, x_m(t)\}$ be a stream of data points arriving over time, where each $x_i(t)$ is a data point with a timestamp t. A streaming anomaly at time t, denoted as $x_a(t)$, can be defined as:

$$\begin{aligned} x_a(t) \in & S(t) \text{ is a streaming anomaly if } \operatorname{dist}(x_a(t), \mu(t)) \\ & > k \cdot \sigma(t) \end{aligned} \tag{2}$$

where $\mu(t)$ is the mean of the data points in the stream S up to time t, $\sigma(t)$ is the standard deviation of the data points in the stream S up to time t, and k is a threshold parameter that governs the detection sensitivity.

Definition 3: Point Anomaly We defined point anomalies in our streaming setting as data points X_t that deviate significantly from the expected behavior, which is based on the values within a sliding window defined by the interval $[X_{t-k}, X_{t+k}]$. In this context, k represents the window size, which should be sufficiently large to accurately capture the expected behavior of the data. It can be mathematically defined as

$$\text{Point Anomaly}(X_t) = \begin{cases} 1 & \text{if } |X_t - X_{[t-k,t+k]}| > \epsilon \\ 0 & \text{otherwise} \end{cases} \tag{3}$$

where ϵ is a threshold that defines how much deviation from the expected behavior is considered significant enough to be an anomaly.

4 Methodology

This section outlines our proposed methodology, focusing on two key modules: Static to Streaming Data Conversion and Streaming Anomaly Detection. The former converts static datasets into continuous data streams, while the latter identifies anomalies within each streaming data window.

4.1 Static to Streaming Data Conversion Module

We designed a streaming data conversion module to transform static datasets into continuous data streams. The reason is that there is a limited availability of publicly accessible streaming data for anomaly detection. We begin by selecting a temporal static dataset X. This dataset is divided into test and training sets, and each record is sent to the Kafka producer client with the topic, partition, message (representing the unlabeled data tuples), and key (which corresponds to the dataset labels). Apache Kafka consists of distributed clusters that enable the streaming of data as real-time events. When reading from Kafka consumer clients, we segment the incoming messages into smaller batches to simulate the concept of a moving window, before applying the anomaly detection algorithm. The detailed steps are illustrated in Fig. 1.

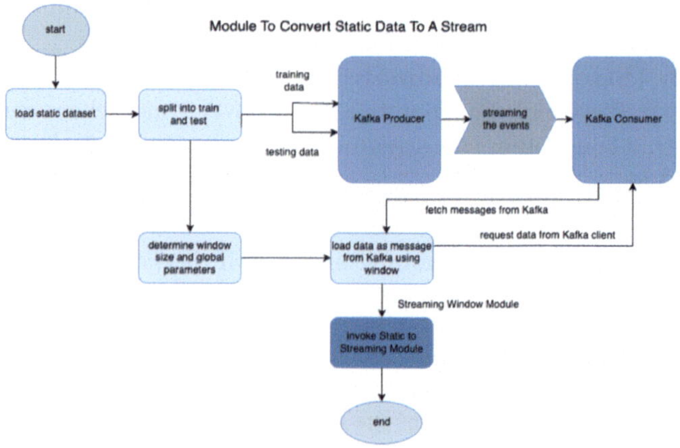

Fig. 1. Streaming Data Conversion Module

4.2 Streaming Bayesian Arima Model (SBA)

Time series analysis models are statistical tools used to identify trends, patterns, and relationships in data over time, aiding in forecasting future values and understanding underlying variations. In our approach we utilized a time series model: Bayesian Auto-Regressive Integrated Moving Average (ARIMA) to detect anomalies in streaming data. However, the traditional ARIMA model is suitable for static datasets.

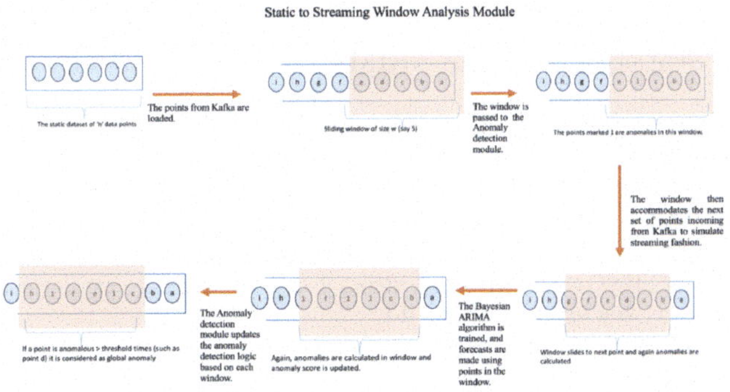

Fig. 2. Static To Streaming Window Analysis Module

4.2.1 Bayesian Autoregressive Integrated Moving Average

The Bayesian Autoregressive Integrated Moving Average (ARIMA) model represents a sophisticated statistical method integrating Bayesian principles to furnish probabilistic forecasts. Departing from the conventional ARIMA model, it explicitly incorporates uncertainty concerning both parameters and predictions. This is accomplished through the assignment of prior distributions to model parameters, subsequently updated based on observed data. The Bayesian framework facilitates the assimilation of expert knowledge and prior beliefs, thereby enhancing forecast accuracy. Moreover, it offers a means to quantify forecast uncertainty, particularly beneficial in scenarios with limited data or when decisions hinge on uncertain information. Employing Bayesian principles, the model employs log-likelihood to ascertain dependencies among past values and errors, eschewing reliance on differencing and moving averages. It adeptly manages uncertainty by leveraging the posterior probability of its components.

Mathematically, say x_t is the set of regressors, S_t is the seasonality term and μ_t represents the latent trend term, then the current observation as per Bayesian ARIMA can be written as:

$$X_t = \mu_t + x_t\beta + S_t + e_t, e_t\ N(0, \sigma_e^2) \tag{4}$$

4.2.2 Streaming Anomaly Detection

Selecting the window or batch under study size depends on the characteristics and shape of the dataset being streamed. Say dataset X has size N, we can determine the total number of windows w by looking at the periodicity p of the timestamp t present in the dataset thus arriving at $w = (\lceil N/P \rceil - 1)$. Each window w_i is then regarded as the focus of study and the model is trained using the data points within the window. The model after being trained on the points within the window makes predictions for the same points and anomalies within the window are calculated accounting as local anomalies. Each point is assessed against a predefined threshold to determine whether it should be labeled as a global outlier.

Algorithm 1. Statistical Bayesian ARIMA for Anomaly Detection in Streaming Data

Require: Static labeled dataset with timestamps
Ensure: Detected Anomalies; Performance Evaluation Metrics
Split data into Train(X_{train}, Y_{train}) and Test(X_{test}, Y_{test})
Initialize $Kafka_{producer}$ topics and parameters (θ_{Kafka_P})
Initialize $Kafka_{consumer}$ topics and parameters (θ_{Kafka_C})
procedure WRITETOKAFKA(*topic, items*)
  while message, key exists in items **do**
    write to $Kafka_{producer}$ (*topic, key, message*)
  end while
end procedure
procedure TRAINSBAMODEL($batch_{confidence}$)
  while data present in Kakfa stream **do**
    $Mini_{batch} \leftarrow$ decode $Kafka_{message,key}$
    for message, key in $Mini_{batch}$ **do**
      TRAINUSINGSBA(message, key, window)
      PREDICTUSINGSBA(message, key, window)
    end for
  end while
end procedure
procedure TRAINUSINGSBA(message, key, window)
  $X_{timestamp}$, $X_{value} \leftarrow$ message
  $X_{value_n} \leftarrow$ NORMALIZE(X_{value})
  X $\leftarrow$ ($X_{timestamp}$, X_{value_n})
  if first window **then**
    $Model_{SBA} \leftarrow$ INITIALIZESBA
  else
    $Model_{SBA}$.FIT(X)
  end if
end procedure
procedure PREDICTUSINGSBA(message, key, window)
  $X_{timestamp}$, $X_{value} \leftarrow$ message
  $X_{value_n} \leftarrow$ NORMALIZE(X_{value})
  $y \leftarrow$ key
  $Pred_{value} \leftarrow Model_{SBA}$.FORECAST($X_{timestamp}$)
  Residuals $\leftarrow X_{value_n}$ - $Predicted_{value}$
  $Batch_{limits} \leftarrow$ CALCTHRESHOLD($X_{timestamp}$,$Pred_{value}$)
  if $Predicted_{value}$ beyond $Batch_{limits}$ **then**
    label point as anomalous
  end if
  Update batch performance metrics
end procedureϵ is a threshold that defines how much deviation from the expected behavior is considered significant enough to be an anomaly.
  k defines the size of the sliding window.
procedure CALCTHRESHOLD($X_{timestamp}$, $Pred_{value}$)
  Calculate $Batch_{mean}$ and $Batch_{std}$
  $Threshold_{upper} \leftarrow Batch_{mean} + sigma* Batch_{std}$
  $Threshold_{lower} \leftarrow Batch_{mean}$ - $sigma* Batch_{std}$
  return $Threshold_{upper}$, $Threshold_{lower}$
end procedure

Subsequently, the window is shifted to accommodate the next incoming set of points in window $w_{(i+1)}$, and the process of identifying anomalies within the window is repeated as indicated in Fig. 2. This sliding window mechanism persists until all points received from the Kafka client have been processed.

4.3 Anomlay Detector

Detecting anomalies in streaming environments can be challenging and requires computing a dynamically updated threshold value that determines if the points within the window act anomalously locally and globally. In our case based on the dataset, we first determine whether the overall expected dataset displays a normal or skewed distribution. We normalize the data using min-max normalization to scale the data between (0,1) before fitting the model and before we execute the prediction module of our algorithm. To calculate the anomalies, we compute each window's mean, standard deviation, and upper and lower threshold boundaries as

$$\begin{aligned} threshold_{bounds} =& mean_{window} \pm \sigma \\ & * standard\_deviation_{window} \end{aligned} \tag{5}$$

where σ is the confidence level. The complete flow of the methodology is highlighted under the Algorithm 1.

5 Experiments and Results

For all of our experiments, we used the following environment on Google Colab: Intel(R) Xeon(R) CPU @ 2.30 GHz, RAM 50 GB, GPU 15 GB Tesla T4 with CUDA 12.2 and NVIDIA-SMI 535.104.05 having a Python 3.10.12. To ensure reproducibility, all the data, parameter tuning configurations, and code are available at SBA-AnomalyDetection-A5E4/.

5.1 Datasets

Numenta Anomaly Benchmark, introduced by Numenta, serves as a publicly accessible benchmark for streaming anomaly detection [15]. It features 58 well-curated, ordered real data streams, each containing 1,000 to 22,000 instances, spanning diverse domains like road traffic, network utilization, online advertisement, and internet traffic. Anomalies in this dataset are labeled based on either known root causes or through a predefined labeling process. Each data file includes timestamps and corresponding values, along with separate sets of labels indicating anomaly occurrences. An example dataset available for reference is the Real Known Cause Data Corpus, summarized in Table 1.

Table 1. Numenta Anomaly Benchmark Real Known Cause Data Corpus

Filename	Size (Bytes)	Rows
ambient_temperature_system_failure.csv	233321	7267
cpu_utilization_asg_misconfiguration.csv	539274	18050
ec2_request_latency_system_failure.csv	123428	4032
machine_temperature_system_failure.csv	732223	22695
nyc_taxi.csv	265771	10320
rogue_agent_key_hold.csv	55527	1882
rogue_agent_key_updown.csv	138333	5315

5.2 Experimental Set up

The data under experimentation is collected every 30 min giving rise to a periodicity of 48 in a day which can be deduced from $30 \times 48 = 1{,}440$ min or 24 h. This determines the batch size, which is the number of data points passed to the Bayesian ARIMA model in each window. The parameters of an ARIMA model are p (autoregressive terms), d (differencing order for stationarity), and q (moving average terms). To determine the values for p, d, and q we use the auto_arima function from Python's statistical library pyramid [16]. Which wraps up the ARIMA and SARIMA models from the stats model for time series analysis. By conducting several differencing tests such as conducting differencing tests (i.e., Kwiatkowski–Phillips–Schmidt–Shin, Augmented Dickey-Fuller or Phillips–Perron) auto arima finds the best optimization for the Akaike Information Criterion (AIC) and the Bayesian Information Criterion (BIC) and returns the best order for the ARIMA model. In our dataset, the model with order ARIMA (0,1,1)x(0,1,1,48) had the lowest AIC value of -106.677.

For all the experiments, the initial 40% of a time-series is used to train the model and the rest of 60% for testing. We compare the streaming results against standard benchmark machine learning and deep learning results reported in [3, 11]. To evaluate our proposed algorithm, we use the popular anomaly detection measure called the receiver operating characteristic (ROC) curve, and further compute the area under the ROC curve (AUC-ROC).

5.3 Comparison of Proposed Approach vs Traditional ML Methods

We compare our Streaming Bayesian ARIMA (SBA) model with 7 traditional machine learning-based static algorithms [4]. Static methods are widely used and provide a good baseline for performance evaluation. By including them in our comparison, we wanted to show how our streaming approach performs in situations where static methods are typically applied. Section 5.3 illustrates the performance comparison between our approach and traditional machine learning methods. It shows that our Streaming Bayesian ARIMA (SBA) approach, with an AUC-ROC value of 0.61, outperforms 5 out of 7 traditional ML methods.

While our approach, SBA, slightly lags behind OC-SVM (0.64) and KNN (0.63), this is expected since these are static methods optimized for offline data, whereas SBA is specifically designed for real-time streaming scenarios.

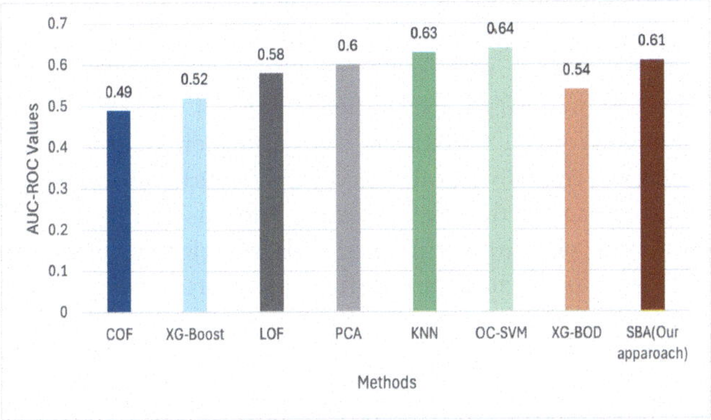

Fig. 3. Performance analysis of proposed SBA vs traditional ML-based methods in terms of AUC-ROC values

5.4 Comparison of Proposed Approach vs Deep Learning Methods

We also compared our proposed Streaming Bayesian ARIMA (SBA) with three state-of-the-art streaming methods (Fig. 3):

FuseAD [13], Numenta [15], and DeepAnt [14]. Fig. 4 shows the comparison in terms of AUC-ROC values. Our method SBA achieves an AUC-ROC value of 0.61, outperforming Numenta (0.59) and FuseAD (0.59), and slightly surpassing DeepAnt (0.60). This demonstrates that SBA provides a competitive edge, leveraging its ability to effectively handle real-time anomaly detection. The results highlight SBA's strong performance compared to both traditional and advanced methods, making it a reliable choice for streaming data scenarios.

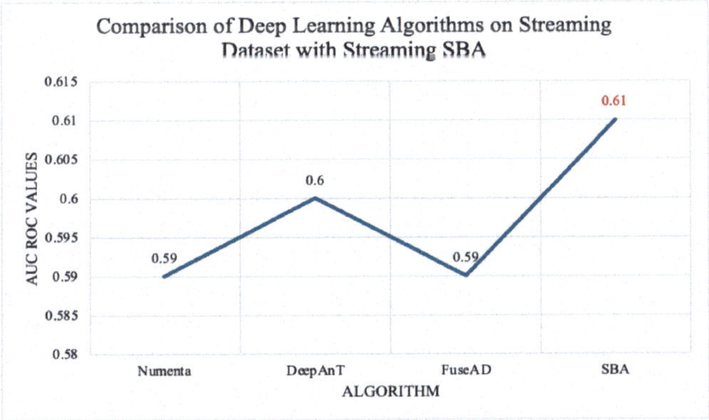

Fig. 4. Performance analysis of proposed SBA vs streaming state-of-the-art methods

6 Conclusion

In this work, we proposed an anomaly detection algorithm for streaming data using a Bayesian ARIMA model. We converted static data into continuous streams for online anomaly detection. When tested on the NAB NYC Taxi Dataset, our approach outperforms traditional machine learning and deep learning methods, offering faster processing and lower computational costs. However, our model relies on a large, well-labeled dataset and currently focuses only on point anomalies. We aim to extend our approach to unsupervised streaming anomaly detection by incorporating collective and contextual anomalies and validating the algorithm on diverse datasets, such as IoT sensor data and network logs. In addition, we plan to integrate SBA with neural forecasting models like LSTMs or attention mechanisms to further improve prediction accuracy for dynamic streams.

References

1. Aggarwal, C.C.: An introduction to outlier analysis. Springer (2017)
2. Akouemo, H.N., Povinelli, R.J.: Time series outlier detection and imputation. In: 2014 IEEE PES General Meeting — Conference & Exposition, pp. 1–5. IEEE (2014)
3. Blázquez-García, A., Conde, A., Mori, U., Lozano, J.A.: A review on outlier/anomaly detection in time series data. ACM Comput. Surv. (CSUR) **54**(3), 1–33 (2021)
4. Braei, M., Wagner, S.: Anomaly detection in univariate time-series: a survey on the state-of-the-art. arXiv preprint arXiv:2004.00433 (2020)
5. Breunig, M.M., Kriegel, H.P., Ng, R.T., Sander, J.: Lof: identifying density-based local outliers. In: Proceedings of the 2000 ACM SIGMOD International Conference on Management of Data, pp. 93–104 (2000)
6. Chalapathy, R., Chawla, S.: Deep learning for anomaly detection: a survey. arXiv preprint arXiv:1901.03407 (2019)
7. Elahi, M., Li, K., Nisar, W., Lv, X., Wang, H.: Efficient clustering-based outlier detection algorithm for dynamic data stream. In: 2008 Fifth International Conference on Fuzzy Systems and Knowledge Discovery, vol. 5, pp. 298–304. IEEE (2008)
8. Hawkins, D.M.: Identification of Outliers. Springer, London (1980)
9. Hill, D.J., Minsker, B.S.: Anomaly detection in streaming environmental sensor data: a data-driven modeling approach. Environ. Model. Softw. **25**(9), 1014–1022 (2010)
10. Hyndman, R.J.: Forecasting: Principles and Practice. OTexts (2018)
11. Lavin, A., Ahmad, S.: Evaluating real-time anomaly detection algorithms–the numenta anomaly benchmark. In: 2015 IEEE 14th International Conference on Machine Learning and Applications (ICMLA), pp. 38–44. IEEE (2015)
12. Munir, M., Chattha, M.A., Dengel, A., Ahmed, S.: A comparative analysis of traditional and deep learning-based anomaly detection methods for streaming data. In: 2019 18th IEEE International Conference on Machine Learning and Applications (ICMLA), pp. 561–566. IEEE (2019)

13. Munir, M., Siddiqui, S.A., Chattha, M.A., Dengel, A., Ahmed, S.: Fusead: unsupervised anomaly detection in streaming sensors data by fusing statistical and deep learning models. Sensors **19**(11), 2451 (2019)
14. Munir, M., Siddiqui, S.A., Dengel, A., Ahmed, S.: Deepant: a deep learning approach for unsupervised anomaly detection in time series. IEEE Access **7**, 1991–2005 (2018)
15. Numenta: Numenta anomaly benchmark (nab). https://github.com/numenta/NAB. Accessed: Date
16. Pyramid-arima: pyramid-arima. https://pypi.org/project/pyramid-arima/. Accessed 25 Oct 2024
17. Ramaswamy, S., Rastogi, R., Shim, K.: Efficient algorithms for mining outliers from large data sets. In: Proceedings of the 2000 ACM SIGMOD International Conference on Management of Data, pp. 427–438 (2000)
18. Subramaniam, S., Palpanas, T., Papadopoulos, D., Kalogeraki, V., Gunopulos, D.: Online outlier detection in sensor data using non-parametric models. In: Proceedings of the 32nd International Conference on Very Large Data Bases, pp. 187–198 (2006)
19. Tran, L., Fan, L., Shahabi, C.: Distance-based outlier detection in data streams. Proc. VLDB Endowment **9**(12), 1089–1100 (2016)
20. Wang, Z., Gao, R., Gao, C., Chen, Y., Wang, F.: A distributed anomaly detection scheme based on correlation awareness in wsn. Wireless Pers. Commun. **134**(1), 519–541 (2024)

Model-Cart: A Machine Learning Meta-Framework with Explainability and Human-in-the-Loop

Vidit Singh[1], Yonas Kassa[2(✉)], Akshay Kale[1], Brian Ricks[1], and Robin Gandhi[1]

[1] University of Nebraska at Omaha, Omaha 100190, USA
{vsingh,akale,bricks}@unomaha.edu

[2] School of Data Intelligence and Technology, Robert Morris University, Pittsburgh, PA, United States
ykassa@acm.org

Abstract. With machine learning (ML) models becoming ubiquitous, including in high-stakes domains, ensuring their dependability is crucial. We propose Model-Cart, a comprehensive open-source meta-framework that integrates explainability methods, human intervention, and reproducibility. By combining state-of-the-art approaches such as SHAP for model explainability, MLflow for reproducibility, and human-in-the-loop approach, this work contributes to addressing the challenges of transparency, trustworthiness, and reliability in end-to-end model training. Model-Cart incorporates explanations of model predictions, seamless integration with popular ML libraries, and a custom built module based on a novel method to quantify model explanation similarities between ML models with an intuitive UI to facilitate model training and evaluation with human-in-the-loop. We evaluate Model-Cart on a popular dataset and demonstrate its advantages over existing approaches.

Keywords: Machine Learning · Explainable AI · reproducibility · MLops · data science

1 Introduction

Machine learning (ML) has rapidly evolved into an essential component in various industries. It is revolutionizing many fields including healthcare, finance, and transportation. For instance, in healthcare, ML models have been used to predict potential health risks and recommend personalized treatment plans [20,25]. In finance, ML is being used for credit rating, fraud detection, and algorithmic trading [10]. In transportation, it powers intelligent traffic management, optimizes logistics operations, and it enables data-driven infrastructure maintenance

V. Singh, Y. Kassa, A. Kale, B. Ricks, R. Gandhi—Contributing authors.
The original version of the chapter has been revised. The contributing authors list is updated. A correction to this chapter can be found at
https://doi.org/10.1007/978-3-031-89063-5_66

S. Latifi (Ed.): ITNG 2025, AISC 1463, pp. 316–326, 2025.
https://doi.org/10.1007/978-3-031-89063-5_27

and monitoring [7,18]. Its utility comes from enabling organizations harness the power of their data to gain predictive insights and optimize their operations.

It has been demonstrated that employing larger and more complex statistical models on increasingly large datasets, as opposed to simple and easily interpretable mathematical models, results in increased performance metrics [1,24]. However, the complexity of ML models and the lack of transparency in their decision-making processes has been a growing issue as it can hinder trust in their outcomes and increase their risk. To address these challenges, researchers and practitioners have been exploring methods to make ML systems more explainable [9,16], reproducible [5], and human-centered [3]. Such recent advancements and growing calls for the need to involve humans in the decision making process and techniques for reproducibility and explainability hold great potential for creating dependable ML based systems. Human-in-the-loop approaches allow for inspection and direct user intervention in ML pipelines. In light of this, few promising disparate frameworks and algorithms such as SHAP[9], LIME [16], and MLFlow [23] have been developed that drastically enhance machine learning model development with explainability streamlining workflow management.

However, integrating these advancements into a cohesive user-friendly machine learning meta-framework remains an open research question limiting their widespread adoption. In this paper, we propose "Model-Cart" a comprehensive open-source ML meta-framework designed to leverage recent advances in explainability, reproducibility, and human-in-the-loop requirements with the objective of improving machine learning model development, training, prioritization, and selection. Model-Cart offers the following key benefits:

- **Transparency**: Explanations of ML model predictions using state of the art explainability techniques, allowing users to understand how each model under study behaves and how features influence decisions.
- **Reproducibility**: Seamless integration with popular ML libraries and data processing frameworks for efficient experimentation, model evaluation, registering, and documentation of model artifacts with integrated model evaluation reports, leveraging ydata [22].
- **Enhanced trustworthiness**: incorporates humans-in-the-loop for enabling expert feedback during model evaluation, prioritization with integration of such expert observations in comprehensive reports exported for a documented reference.

In the following sections, we provide an overview of related body of literature then we will describe Model-Cart in detail explaining its components and evaluating it with an experiment on a popular community dataset. Section 2 provides a literature review on related work, focusing on explainable ML, reproducibility techniques, and human-in-the-loop systems. In Sect. 3, we present an in-depth discussion of the Model-Cart architecture and its components. Section 4 showcases Model-cart with a popular dataset to illustrate its advantages, pointing to further implemented case studies from multiple domains at the GitHub repository of this open-source meta-framework. Finally, in Sect. 5, we discuss potential applications, limitations, and future directions. Section 6 concludes the paper.

2 Related Work

As ML's ubiquity extends to high-stakes settings where errors can lead to significant consequences, stakeholders from various disciplines are demanding for transparency and trustworthiness of such systems to ensure their reliability [13,16,17]. Ribeiro et al. [16] have developed an explanation technique that aims to explain model predictions in an interpretable and faithful manner. Furthermore, a global model agnostic ML explanation technique based on game theory was proposed by Lundberg et al. [9]. Multiple research works have leveraged these explanation methods to get insight into the ML models they developed [11,12,19].

On the other hand, the need for involving humans in ML life-cycle has been widely acknowledged [3,21]. In [3] Chai et al. discussed how humans can play important roles in the entire ML pipeline including in data preprocessing, annotation, model training and inference. Wu et al. [21] surveyed the state of human-in-the-loop in ML across different disciplines, they emphasized the importance of human "a priori knowledge" in ML development. Studying limitations of fully automated Machine Learning (AutoML) systems, Gil et al. [4] proposed a technique called human-guided machine learning (HGML) that requires domain experts to guide an AutoML system select among machine learning algorithms.

The challenge of reproducibility in machine learning has been covered by multiple articles [2,5,17]. In their work published in 2020, Beam et al. [2] distinguished between replicability and reproducibility in research, emphasizing that the latter poses novel challenges as more sophisticated machine learning (ML) techniques are integrated into experimental pipelines. Among the challenges they identified include large number of ML model parameters and configurations to track, and availability of different software libraries that can be used to create candidate models. Gundersen et al. [5] evaluated out of the box reproducibility support by some of most popular machine learning frameworks. They conclude that these machine learning platforms do not provide out-of- the-box reproducibility. Sugimura and Hartl [17] discussed the problems they encountered while building a variety of machine learning models. Based on this experience they proposed a framework composed of four layers. However, the framework they proposed does not take advantage of state-of-the-art MLops frameworks like MLflow making it difficult to incorporate other components like visualization and explainability with robust workflow management. In response to the growing challenges in ensuring reproducibility in machine learning pipelines, Zaharia et al. [23] introduced MLflow as an open-source ML life-cycle management platform. It offers end-to-end management for ML projects, including experiment tracking, packaging code and data into reproducible runs, and facilitating model deployment.

In response to the identified challenges, we develop a comprehensive open-source meta-framework with intuitive user interface that builds upon the above advancements. Currently, Model-Cart integrates explainability techniques based on SHAP library to integrate model explanations. It further incorporates a novel explanation comparison technique [8] that computes explanation distances between multiple candidate ML models. It facilitates human-in-the-loop in the

ML pipeline with intuitive user interfaces making humans the final decision-makers in model selection and prioritization as opposed to automated model selection based, solely, on quantitative performance metrics. It enables domain experts and data scientist from various disciplines to collaboratively evaluate and prioritize and trained models for deployment. Additionally, Model-Cart ensures reproducibility by generating a comprehensive model report that integrates, among others, traditional model performance metrics [6] like F1 score, ROC, precision, and accuracy with widely used ML libraries and providing robust model artifact ready for deployment. By combining these features, Model-Cart contributes to address the evolving demands for reproducible, explainable, and human-centered machine learning systems.

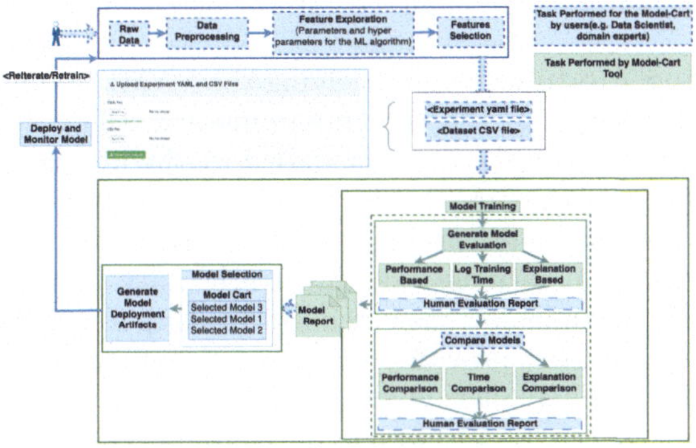

Fig. 1. Architecture of Model-Cart

3 Architecture of Model-Cart

A typical machine learning workflow involves a data scientist performing preliminary stems such as data cleaning, exploratory analysis of the data, feature engineering, and running candidate ML algorithms with certain parameter configurations. Then models are trained on the dataset. Traditionally, best performing model in metrics such as accuracy, area under curve, F-1 score is automatically selected for deployment. Our proposed solution handles the later part of this workflow taking over model training and selection part. It starts with the user uploading the preprocessed dataset file and a corresponding YAML file that contains all necessary configurations for model training such as machine learning algorithms to be trained, details of parameters and hyper-parameters to be utilized, and defines the evaluation metrics that will gauge the model performance. Additionally, with this YAML file Model-Cart orchestrates the sequence of operations for the training process, ensuring that each training is executed in a structured and predetermined manner. This systematic configuration is crucial for maintaining consistency and reproducibility in model training and evaluation across various experiment runs.

A high level architecture is presented on Fig. 1. Model-Cart is composed of four main modules: user input module, model training module, model comparison module, and model selection module. Each module is designed to provide a clear, user-guided path from the conception of an ML experiment to the deployment of a model, ensuring transparency, user control, and traceability throughout the ML cycle.

- **User input module:** This module initiates authentication process to provide security, validates user uploaded YAML and CSV files and initiates the training module which launches an experiment with unique ID and descriptive name. Note that Model-Cart enables running of multiple models and the ability to revise and repeat completed experiments enabling ML reproducibility.
- **Model training module:** Upon receiving the YAML configuration and the preprocessed dataset from the user input module, the system initiates the model training phase. This process is guided by the detailed parameters and strategies outlined in the YAML file. Leveraging MLflow, a robust platform for managing the machine learning life-cycle, the system efficiently orchestrates the training operations, ensuring traceability and reproducibility of experiments. The training leverages scikit-learn [14], a versatile standard tool for machine learning that provides a broad array of algorithms and utilities. The module employs explanation sub-module (currently based on SHAP) to generate explanation matrices of model predictions. the SHAP values reflect the impact of each feature on the prediction, offering explanations for the model's decisions. This is crucial for verifying behavior of complex models and ensuring that the predictions are both transparent and acceptable.
  Up on successful execution of this phase a report with the following evaluation metrics is generated along with the trained models for each specified ML algorithm. **(1) Performance Based metrics:** evaluation results of trained models based on their specified performance metrics like accuracy, precision, recall, etc. **(2) Log training time:** the duration of the training process for each model, aiding in assessing computational efficiency for retraining consideration. **(3) Explanation based metrics:** report showing the decision-making process of each model models and how models are different in explanation and which models have close (more similar explanations). This enables experts to assess, rank, and select trained models based on their ability to provide understandable and reasonable explanations for their predictions with reasonable performance. **(3) Human Observation Logs:** During model evaluation Model-cart users are required to make detailed observation logs, referred to as model report, while they evaluate the run details for each trained model in a given experiment. This integrated model report helps to capture insights and qualitative assessments directly from humans.
- **Model comparison module:**
  Once all specified algorithms are trained, this module handles multi-dimensional comparison of models. This stage aggregates results from individ-

ual model assessment, and enables human evaluators to compare the trained models within an experiment based on: **(1) Performance Comparison:** quantitatively evaluate models side-by-side to identify best/worst performing models. **(2) Explanation Comparison:** Uses custom explanation distance computing measures (currently based on the Wasserstein distance[15]) to quantitatively compare ML model explanations computed from each trained model, highlighting the explanation differences between them. This enables users to select most acceptable models as candidate models that can be traded based on other metrics such as performance or retraining time. Based on their domain expertise, users will be able to discard models that are unreasonable, not appropriate, biased, or over-fitted this is particularly important for fairness issues which is one of challenges in machine learning [13]. **(3) Training time comparison:** enables comparing the time efficiency of each model training process. **(4) Observation logs and saving comparisons:** model reports from human observations during comparisons. This module saves these comparisons for future reference or further analysis contributing to reproducibility of experiments.
- **Model selection and prioritization module:** After a detailed comparison of all the user-selected models which are added to the cart in the experiment, this module allows the user to review, prioritize, and select the best-performing and most suitable models to be added from the Model-Cart for deployment, each trained model is equipped with all detailed artifacts such as model report including all quantiative evaluation results, human observation report, and the trained model weights .

Having integrated Ml-flow, the system enables downloading a complete snapshot of the experiment, including trained models which can be deployed (if the model is deemed suitable), along with all detailed artifacts, in a ZIP folder. This culminates the workflow, transitioning from experimentation to practical deployment or the decision to diagnose and retrain starting with exploratory analysis with additional features or data points. We demonstrated an early version of Model-Cart at Academic Data Science Alliance (ADSA), an academic venue specific to data science receiving a positive acceptance demonstrating the need for such as system. We have incorporated most of the provided feedback based on their domain expertise and experience. The source code of Model-Cart along with detailed documentation, benchmarking, and experiments on popular datasets is available at https://github.com/model-cart/model-cart.

4 Experiment: Use-Case

In this section we demonstrate Model-Cart with a use case of ML model development for healthcare domain. We give an overview of how users can leverage each component for training, evaluation and selecting suitable ML models for deployment. The simplified scenario assumes a data scientist in healthcare domain aims to train and evaluate three distinct ML classifiers with multiple configuration parameters on Model-Cart. The process involves the following steps:

- Experiment Submission and Execution: The process begins by uploading an experiment YAML and CSV data files (Fig. 2 left, details on how to prepare the YAML file is available on the Github repo). We leveraged a publicly available heart disease dataset from Kaggle for this experiment. Once uploaded, the Model-Cart system processes the YAML file and selects relevant features to initiate the experiment, applying the predefined parameters from the YAML file.
- Monitoring and Managing Experiment Progress: throughout the experiment progress, the system offers a dashboard where the status of various experiment threads can be monitored, providing the user with realtime feedback on the state of their experiments (Fig. 2 right). While the considered models are trained, their performance is recorded using user provided metrics like accuracy and precision.
- Run Analysis and Documentation: Users have the ability to inspect individual runs. This involves examining the start and end times of the runs, performance, efficiency, and explainability as well as documenting any specific observations based on their domain expertise that may be pertinent to the experiment's outcome (Fig. 3 left).
- Performance Metrics and Model Evaluation: The Model-Cart system presents a comprehensive visualization of performance and explainability metrics (3 left), allowing users to analyze the models' effectiveness through graphical representations and detailed statistical scores.
- Comparative Analysis and Model Interpretability: As the platform enables a side-by-side comparison of models, incorporating SHAP values to provide feature based model explainability showing how different features influence the models' predictions and the custom built model explanation distance using wasserstein distance that serves as a proxy to quantitatively measure similarity between explanations between each model (Fig. 3 right).
- Model Selection and Prioritization: The user is required to apply comprehensive analysis and select and prioritize models based on their performance and interpretability, and overall expert assessment, adding the most suitable models to the cart for final review. This cart acts as a staging area before deployment, where the user can make final adjustments by reiterating, going through the existing observation notes or update the notes with new insight as well (Fig. 4 left).
- Deployment Preparation: Once the model is selected, it will be ready for download and deployment. The system allows users to download a ZIP folder containing all relevant model artifacts along with comprehensive report by clicking on the Deploy button. (Figure 4 right).

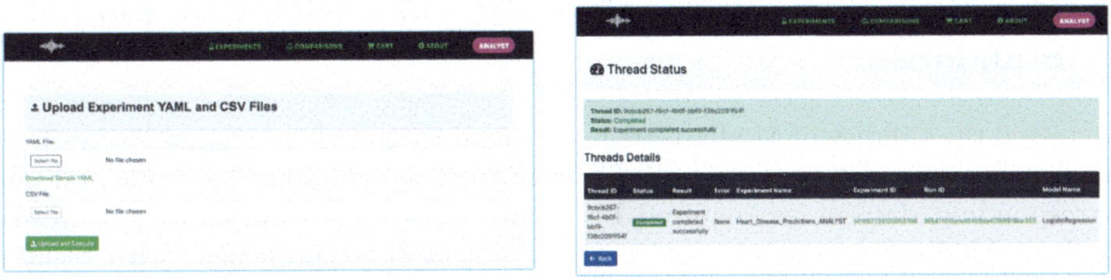

Fig. 2. Model-Cart user interface including Experiment submission interface (left) and and execution progress status report (right). Menu items leading to various modules of Model-Cart are shown at the top.

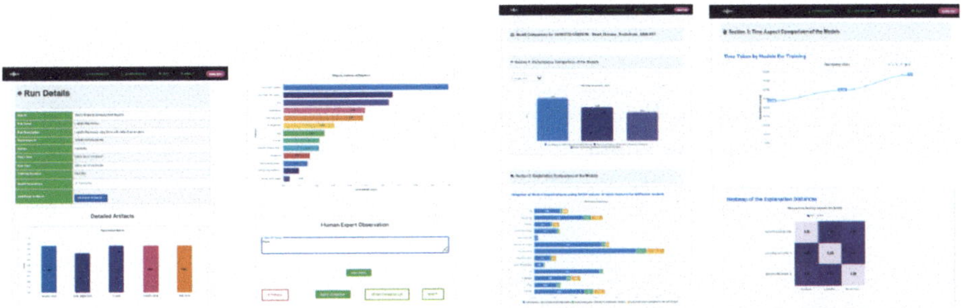

Fig. 3. Left: Experiment run analysis and model evaluation with form requiring users to write their analysis notes. Right: results of performance and explanation similarity metrics presented the user for their qualitative evaluation.

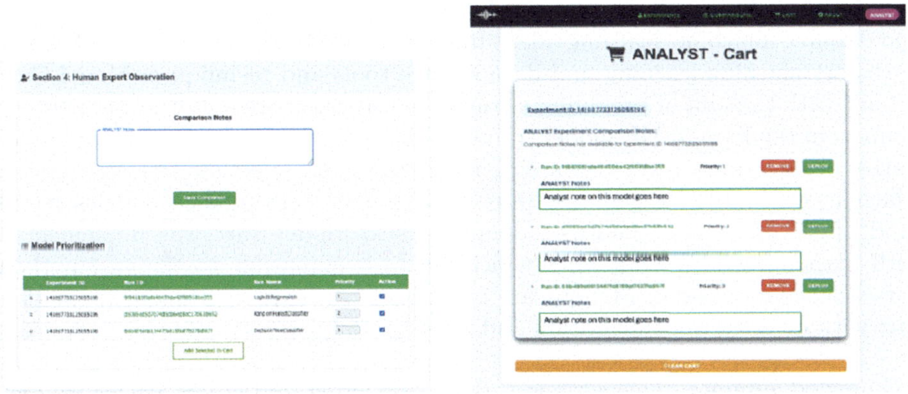

Fig. 4. The final stages of experiment execution: on the left model prioritization where the user, after thorough analysis, prioritizes the trained models while selecting most suitable models to added to the cart. On the right the prioritized list of selected ML models along with thorough model analysis and comparison notes are shown in the cart. Finally, users can choose a model from this cart and export it for deployment which includes the entire artifact along with model reports.

5 Benchmarking Model-Cart's Results with Community Standards

We tested the validity of Model-Cart results against Kaggle's community-verified benchmarks (generating the YAML file from the Kaggle public notebook and the corresponding dataset as required by Model-Cart). The comparative analysis confirms Model-Cart's performance scores are in close agreement with results from Kaggle notebooks 1. Details of comparisons of raw SHAP values and also results from other experiments are available on Model-Cart's Github page.

Table 1. Reproducing results from Kaggle notebooks on Model-Cart

	Model Cart		Kaggle	
	Accuracy	F1	Accuracy	F1
Logistic Regression	86.49	0.86	86.49	0.86
Random Forest	83.8%	0.83	83.8%	0.83
Decision Tree	82.4%	0.82	82.4%	0.82

6 Conclusion

Building machine learning systems that improve transparency, reproducibility, and dependability is crucial in this day and age where ML models are being integrated in high stakes settings. In this paper we present Model-Cart, an open source machine learning meta-framework that is built on top of state-of-the-art machine learning and data science tools and techniques. Model-Cart is built with the aim of providing a comprehensive user-friendly tool that brings humans in the loop of ML model training, incorporating their wisdom and qualitative observations as a determining factor in model selection and deployment which provides advantages over pure quantitative performance metrics based (and often automated) end-to-end approaches. Model-Cart was implemented with python on flask framework which makes it extensible to the vast python data science ecosystem. Model-Cart is available at https://github.com/model-cart with detailed documentation and a set of complete experiment examples.

Acknowledgements. This work is partially supported by the US Army Engineering Research and Development Center (ERDC) by contracts W912HZ21C0060 and W912HZ23C0005, and Award Number 1762034 from the National Science Foundation.

References

1. Asri, H., Mousannif, H., Al Moatassime, H., Noel, T.: Using machine learning algorithms for breast cancer risk prediction and diagnosis. Procedia Computer Science **83**, 1064–1069 (2016)
2. Beam, A.L., Manrai, A.K., Ghassemi, M.: Challenges to the reproducibility of machine learning models in health care. JAMA **323**(4), 305–306 (2020)
3. Chai, C., Li, G.: Human-in-the-loop techniques in machine learning. IEEE Data Eng. Bull. **43**(3), 37–52 (2020)
4. Gil, Y., Honaker, J., Gupta, S., Ma, Y., D'Orazio, V., Garijo, D., Gadewar, S., Yang, Q., Jahanshad, N.: Towards human-guided machine learning. In: Proceedings of the 24th international conference on intelligent user interfaces, pp. 614–624. ACM, New York, NY, USA (2019)
5. Gundersen, O.E., Shamsaliei, S., Isdahl, R.J.: Do machine learning platforms provide out-of-the-box reproducibility? Futur. Gener. Comput. Syst. **126**, 34–47 (2022)
6. Hastie, T., Tibshirani, R., Friedman, J.H., Friedman, J.H.: The elements of statistical learning: data mining, inference, and prediction, vol. 2. Springer, USA (2009)
7. Kale, A., Kassa, Y., Ricks, B., Gandhi, R.: A comparative assessment of bridge deck wearing surfaces: Performance, deterioration, and maintenance. Appl. Sci. **13**(19), 10883 (2023)
8. Kale, A., Kassa, Y., Ricks, B., Gandhi, R.: How to select simple-yet-accurate model of bridge maintenance? (2023)
9. Lundberg, S.M., Lee, S.I.: A unified approach to interpreting model predictions. Adv. Neural. Inf. Process. Syst. **30**, 4768–4777 (2017)
10. Mashrur, A., Luo, W., Zaidi, N.A., Robles-Kelly, A.: Machine learning for financial risk management: a survey. Ieee Access **8**, 203203–203223 (2020)
11. Min, J., Tu, J., Xu, C., Lukas, H., Shin, S., Yang, Y., Solomon, S.A., Mukasa, D., Gao, W.: Skin-interfaced wearable sweat sensors for precision medicine. Chem. Rev. **123**(8), 5049–5138 (2023)
12. Moscato, V., Picariello, A., Sperlí, G.: A benchmark of machine learning approaches for credit score prediction. Expert Syst. Appl. **165**, 113986 (2021)
13. O'neil, C.: Weapons of math destruction: How big data increases inequality and threatens democracy. Crown, USA (2017)
14. Pedregosa, F., Varoquaux, G., Gramfort, A., Michel, V., Thirion, B., Grisel, O., Blondel, M., Prettenhofer, P., Weiss, R., Dubourg, V., Vanderplas, J., Passos, A., Cournapeau, D., Brucher, M., Perrot, M., Duchesnay, E.: Scikit-learn: Machine learning in Python. J. Mach. Learn. Res. **12**, 2825–2830 (2011)
15. Ramdas, A., García Trillos, N., Cuturi, M.: On wasserstein two-sample testing and related families of nonparametric tests. Entropy **19**(2), 47 (2017)
16. Ribeiro, M.T., Singh, S., Guestrin, C.: " why should i trust you?" explaining the predictions of any classifier. In: Proceedings of the 22nd ACM SIGKDD international conference on knowledge discovery and data mining, pp. 1135–1144. ACM, New York, NY, USA (2016)
17. Sugimura, P., Hartl, F.: Building a reproducible machine learning pipeline. arXiv preprint arXiv:1810.04570 (2018)
18. Tsolaki, K., Vafeiadis, T., Nizamis, A., Ioannidis, D., Tzovaras, D.: Utilizing machine learning on freight transportation and logistics applications: A review. ICT Express **9**(3), 284–295 (2023)

19. Wang, M., Zheng, K., Yang, Y., Wang, X.: An explainable machine learning framework for intrusion detection systems. IEEE Access **8**, 73127–73141 (2020)
20. Wong, J., Murray Horwitz, M., Zhou, L., Toh, S.: Using machine learning to identify health outcomes from electronic health record data. Current epidemiology reports **5**, 331–342 (2018)
21. Wu, X., Xiao, L., Sun, Y., Zhang, J., Ma, T., He, L.: A survey of human-in-the-loop for machine learning. Futur. Gener. Comput. Syst. **135**, 364–381 (2022)
22. ydata: ydata profiling (2023). https://doi.org/10.5281/zenodo.1234. URL https://github.com/ydataai/ydata-profiling
23. Zaharia, M., Chen, A., Davidson, A., Ghodsi, A., Hong, S.A., Konwinski, A., Murching, S., Nykodym, T., Ogilvie, P., Parkhe, M., et al.: Accelerating the machine learning lifecycle with mlflow. IEEE Data Eng. Bull. **41**(4), 39–45 (2018)
24. Zainudin, S., Jasim, D.S., Bakar, A.A.: Comparative analysis of data mining techniques for malaysian rainfall prediction. Int. J. Adv. Sci. Eng. Inf. Technol **6**(6), 1148–1153 (2016)
25. Zytek, A., Liu, D., Vaithianathan, R., Veeramachaneni, K.: Sibyl: Understanding and addressing the usability challenges of machine learning in high-stakes decision making. IEEE Trans. Visual Comput. Graphics **28**(1), 1161–1171 (2021)

Cryptocurrency Price Forecasting: Insights from Deep Learning and Volatility Analysis

Rojin Manouchehri(✉), Ayesh Meepaganithage, Sergiu Dascalu, Mircea Nicolescu, Monica Nicolescu, and Lei Yang

Computer Science and Engineering, University of Nevada, Reno, NV, USA
{rmanouchehri,ameepaganithage,dascalus,mircea,monica,leiy}@unr.edu

Abstract. Cryptocurrency prices are highly volatile, making them a consistent target for machine learning models. This paper explores the application of advanced Recurrent Neural Networks (RNNs), specifically Long Short-Term Memory (LSTM) and Gated Recurrent Unit (GRU) networks, as well as traditional statistical and machine learning methods to predict the closing prices of major cryptocurrencies. Using daily price data sourced from public cryptocurrency exchange, the performance of RNN models with traditional predictive models was compared to determine the most effective approach. By evaluating model performance across cryptocurrencies with varying levels of volatility, it highlights how volatility influences predictive accuracy. Results indicate that both LSTM and GRU models outperform baseline statistical models with lower error rates. Additionally, traditional machine learning models, like decision trees and random forests, showed notable success with low-volatility assets, such as XRP, underscoring their suitability for stable data. These findings highlight the potential of deep learning techniques in capturing cryptocurrency market dynamics.

Keywords: Cryptocurrency · Deep Learning · Time-Series Forecasting · Long Short-Term Memory (LSTM) · Gated Recurrent Unit (GRU) · Volatility Effects · Machine Learning Models · Random Forests · Decision Trees · Statistical Forecasting

1 Introduction

Cryptocurrencies are considered important players in global financial markets. Unlike traditional currencies, they operate on decentralized networks without backing from central authorities. Their value is determined by market demand, leading to high volatility, influenced by factors such as market sentiment, regulatory news, technological advancements, and macroeconomic trends [9]. Among

R. Manouchehri and A. Meepaganithage—These authors contributed equally to this work.

S. Latifi (Ed.): ITNG 2025, AISC 1463, pp. 327–338, 2025.
https://doi.org/10.1007/978-3-031-89063-5_28

them, Bitcoin and Ethereum are two of the most influential and widely traded cryptocurrencies. Bitcoin, introduced in 2009, pioneered this decentralized financial model. This unique structure has spurred interest in predictive models due to their volatile and non-linear patterns [16].

Traditional financial forecasting methods often struggle to capture the complex dynamics and non-linear dependencies inherent in cryptocurrency price movements. In response, machine learning and deep learning techniques have been developed and employed [11]. Recurrent Neural Networks (RNNs) [17], particularly advanced architectures like Long Short-Term Memory (LSTM) [10] and Gated Recurrent Unit (GRU) [6] networks, have shown promising results in modeling sequential and time-series data. Unlike standard feedforward neural networks, these RNNs have feedback connections that enable them to process data sequences, particularly cryptocurrency time-series data [19]. Despite these advances, there remains a gap in research concerning the application of deep learning models to cryptocurrency predictions and their ability to adapt to sudden market shifts [4].

This paper aims to address these gaps by exploring the application of these advanced RNN architectures to predict the closing prices of selected cryptocurrencies due to their proven effectiveness in handling sequential data. Although more advanced architectures like Transformers have shown potential, they remain a promising avenue for future exploration beyond the scope of this study. Unlike prior studies, this work focuses on analyzing the performance of traditional and deep learning models across cryptocurrencies with varying volatility levels, analyzing their adaptability for diverse market conditions. Furthermore, it will compare the results of these models with traditional predictive models to determine the most effective approach, providing support for the conclusion that deep learning outperforms conventional methods.

The main contributions of this study are as follows:

- Extensive evaluation of the performance of traditional statistical methods, machine learning models, and modern deep learning models in cryptocurrency price prediction.
- Analysis of the performance of different models across various cryptocurrencies with varying volatility levels to evaluate each model's robustness and adaptability.

The paper is organized as follows: The Introduction section sets the stage by discussing the relevance and motivation behind the study. This is followed by the Related Work section, which reviews literature relevant to the methodologies and applications discussed. The Methodology section details the dataset preparation, model architecture, and the specifics of the LSTM and GRU models used. The Evaluation section describes how the models were tested and presents performance metrics. The Conclusion section reflects on the study's outcomes and potential areas for further research.

2 Related Work

Various predictive models, including statistical, machine learning, and deep learning techniques, have been explored for cryptocurrency price prediction. This section categorizes existing research into machine learning and deep learning approaches to illustrate their relative performance in cryptocurrency forecasting.

2.1 Statistical Machine Learning Models for Cryptocurrency Prediction

Cryptocurrency price forecasting has leveraged traditional models like ARIMA and GARCH, which perform well in stock markets but struggle with cryptocurrency volatility [13]. Machine learning approaches, including Support Vector Machines (SVM), Random Forest, and Neural Networks, have shown improved accuracy, particularly for stable assets [15].

Recent research expanding the variety of machine learning models used for cryptocurrency forecasting show that Random Forest often performs better for stable assets, and Gradient Boosting excels with volatile ones [18]. Some studies have further explored short-term forecasting, using models like Binary Autoregressive Trees (BART), Multilayer Perceptron (MLP), and Random Forest, and found that both BART and MLP can achieve lower error rates than Random Forest within a 90-day forecast horizon [8].

Researches also indicate that High-dimensional features show statistical methods like Logistic Regression perform well for daily predictions, while machine learning models excel with high-frequency data [5].

2.2 Deep Learning Models for Cryptocurrency Prediction

Recent studies have increasingly used deep learning models and Recurrent Neural Networks (RNNs) such as Long Short-Term Memory (LSTM) and Gated Recurrent Unit (GRU) networks due to their ability to capture temporal dependencies in highly volatile time series data [19,20]. For example, several studies employ LSTM architectures to forecast cryptocurrency prices, with Akila and Aditya demonstrating the model's effectiveness in handling long-term dependencies across various currencies [1,2]. These studies also highlight the potential for integrating additional techniques, such as Change Point Detection (CPD), to improve accuracy by adapting to structural shifts in the data [2].

Moreover, other studies have explored the application of Convolutional Neural Networks (CNN) and deep forward neural networks, either standalone or tuned with optimization techniques, for cryptocurrency price prediction. Oyedele's work, for instance, finds CNN models particularly effective for forecasting with smaller datasets [14]. Similarly, Bisvas examines LSTM networks trained with various market variables such as market capitalization and trading volume, highlighting the model's ability to capture complex, multi-dimensional patterns in digital currency valuation [3].

While prior research has explored different deep learning models for cryptocurrency prediction, few studies directly address the impact of volatility on model performance by a comprehensive examination of existing models, a key focus of our study.

3 Methodology

This section outlines the process followed in this study, from data preparation to model training and optimization. We begin by describing the dataset used and the initial data analysis conducted to understand its characteristics. The subsequent stages cover data preprocessing and splitting techniques employed. Finally, we detail the implemented models and the training and optimization strategies used to enhance predictive accuracy.

3.1 Dataset

For this study, we utilized a dataset comprising historical price information of various cryptocurrencies, recorded daily since April 28, 2013.[7] The dataset includes individual CSV files for each cryptocurrency, detailing metrics such as opening, high, low, and closing prices on a daily basis. Particular emphasis was placed on the Close prices, as these are most indicative of the market's daily valuation and important for the predictive modeling of future price movements.[9]

The following summary statistics provide an overview of the key metrics for selected cryptocurrencies (Table 1):

3.2 Initial Data Analysis

An initial analysis was conducted to calculate the volatility across all cryptocurrencies in the dataset by measuring the standard deviation of daily closing prices. Cryptocurrencies with higher standard deviations, such as Bitcoin and Ethereum, demonstrated more volatility, while stable coins like XRP showed lower volatility levels. This analysis helped distinguish highly volatile currencies from those with more stable price patterns, guiding the selection of currencies most suitable for this study.

Based on the initial volatility assessment, a selection of cryptocurrencies, including Bitcoin, Ethereum, Solana, Dogecoin, and XRP, were chosen for this study due to their diverse volatility profiles. This allowed for a comprehensive evaluation of model performance across cryptocurrencies with different levels of price fluctuation (Fig. 1).

Fig. 1. Cryptocurrency closing prices over time for selected currencies

3.3 Data Preprocessing

Data preprocessing involved several steps to transform raw cryptocurrency data into a format suitable for training deep learning models. The main feature engineering step involved creating lagged features to capture temporal dependencies essential for time series forecasting. Specifically, we prepared the dataset by creating lagged features for the Close prices, shifting them back by up to 9 days. This resulted in a series of new columns in the dataset, each representing the Close price from up to 9 days prior. We then structured the data using a 10-day window, where sequences of consecutive days' data (based on the lagged features) were grouped together as single training samples to predict the next day's price.

Table 1. Summary Statistics for Cryptocurrency Closing Prices

Statistic	Solana_Close	XRP_Close	Bitcoin_Close	Ethereum_Close
count	452.000000	2893.000000	2991.000000	2160.000000
mean	10.471388	0.234790	6711.290443	383.910691
std	14.114444	0.338629	11298.141921	601.078766
min	0.515273	0.002810	68.431000	0.434829
25%	1.424761	0.007346	430.569489	13.819200
50%	2.881389	0.182608	2286.409912	198.643691
75%	14.362851	0.312465	8576.238715	386.435272
max	55.910960	3.377810	63503.457930	4168.701049

The 10-day window size was selected through hyperparameter tuning, where we experimented with various window sizes (e.g., 5, 10, and 15 days) to determine the optimal length for capturing short-term patterns while maintaining computational efficiency. In addition, other hyperparameters such as the number of LSTM units, learning rate, and batch size were tuned through iterative testing to minimize validation loss and improve model performance. These experiments informed the final configuration of the model, ensuring a balance between accuracy and efficiency.

3.4 Data Splitting

Our data splitting strategy for better evaluations of the models included dividing the data into three sets: 70% of the data was used for initial training, 15% was allocated to the validation set for fine-tuning model parameters and preventing overfitting, and the remaining 15% was set aside as the test set to evaluate model performance and simulate real-world, unseen data conditions.

3.5 Implemented Models

In this study, we implemented three types of Recurrent Neural Networks (RNNs): Long Short-Term Memory (LSTM), Gated Recurrent Unit (GRU), and a basic RNN model. The following sections outline the architecture and operational details of these models.

Model Architecture:

The architectures of the LSTM, GRU, and basic RNN models consist of similar components but differ in their specific configurations:

1. Input Layer: The input to each model consists of the preprocessed lagged features of the Close prices.
2. Recurrent Layers:
   - LSTM Model: The core of the LSTM model includes two LSTM layers. The first LSTM layer has 128 hidden units, passing its outputs to a second LSTM layer with 64 hidden units to refine the information and extract deeper temporal relationships.
   - GRU Model: The GRU model also consists of two GRU layers with similar configurations to the LSTM model (128 and 64 hidden units, respectively), aimed at capturing dependencies with a different gating mechanism.
   - Basic RNN Model: The basic RNN model includes two RNN layers with 128 and 64 hidden units, respectively, which serves as a simpler recurrent architecture to compare with the more complex LSTM and GRU models.
3. Dropout Layers: Dropout layers were introduced following each recurrent layer to prevent overfitting by randomly omitting a portion of the features during each training pass.

4. Output Layer: The final part of each network is a dense layer with a single neuron and a linear activation function, outputting the predicted Close price for the following day based on the learned features and dependencies processed by the recurrent layers.

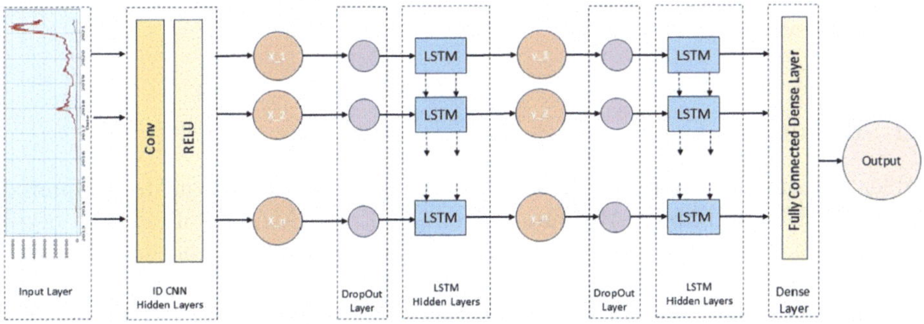

Fig. 2. The LSTM architecture for cryptocurrency price prediction, consisting of initial CNN layers followed by two LSTM layers, each paired with a dropout layer. A final dense layer produces the output.

Figure 2 shows the LSTM layers designed to process sequences of data.

Training and Optimization:

The training process for all models was standardized to ensure a fair comparison:

- Loss Function: The model was trained using the Mean Squared Error (MSE) loss function. MSE was chosen because it effectively penalizes large errors in prediction, making it ideal for ensuring accuracy in financial forecasts.
- Optimizer: The Adam optimizer was utilized for its computational efficiency, low memory requirements, dynamic learning rate adjustment, and faster convergence to the optimal set of weights [12].
- Epochs and Batch Size: Training was conducted over 30 epochs with a batch size of 16, to balance the speed of computation and model convergence quality.

4 Evaluation

4.1 Evaluation Metrics

To accurately assess the performance of the LSTM model in predicting cryptocurrency prices, the following metrics were used, which are standard in regression analysis:

- Root Mean Squared Error (RMSE): Represents the square root of the average of squared differences between prediction and actual observation. It offers a clear measure of the model's accuracy by showing how close the predicted values are to the actual values.

- Mean Absolute Error (MAE): Measures the average magnitude of the errors in a set of predictions, calculated as the mean of absolute differences between predictions and actual values.
- Mean Absolute Percentage Error (MAPE): Measures the average of the absolute percentage errors of forecasts, indicating model accuracy in percentage terms. It compares actual vs. predicted values, making it useful across datasets with different scales.

4.2 Results

Tables 2, 3 and 4 compare the Mean Absolute Error (MAE), Root Mean Squared Error (RMSE), and Mean Absolute Percentage Error (MAPE) of different cryptocurrencies when using various traditional machine learning and deep learning models, respectively.

From the results presented in these tables, it's clear that the GRU and Bi-GRU models outperformed all other deep learning and machine learning models. Both RNN and LSTM models also performed comparatively well; however, none of the machine learning models performed comparably to any of the deep learning models. When using "Bitcoin," "Ethereum," and "Solana," the Bi-GRU model performed best, with 179.80, 68.20, and 2.82 MAE, 264.28, 118.07, and 4.30 RMSE, and 2.69%, 4.30%, and 8.17% MAPE. When using "Dogecoin" and "XRP," the GRU model performed best, with 0.0086 and 0.0298 MAE, 0.0267 and 0.063 RMSE, and 5.65% and 4.70% MAPE. Figure 3 visually summarizes the comparative performance of these models, highlighting how Random Forest achieves lower errors for stable assets like XRP, which reflects its strength in low-volatility conditions.

A deeper analysis of these results shows that machine learning models, specifically decision trees and random forests, performed better when using the XRP cryptocurrency than with any other cryptocurrency. XRP's historical values were less volatile than those of other cryptocurrencies. It's easier for traditional machine learning models to predict when there's a stable pattern, hence their performance on stable XRP is better than on other highly volatile cryptocurrencies. For example, Random Forest outperformed GRU by approximately 10% in terms of MAPE for XRP, highlighting its suitability for low-volatility scenarios. The AutoRegressive Integrated Moving Average (ARIMA) model struggled with non-linear relationships, but both decision trees and random forests were better equipped to handle them.

Table 2. Mean Absolute Error (MAE) for Different Cryptocurrencies

Model	Bitcoin	Ethereum	Solana	Doge Coin	XRP
RNN	201.01	72.20	3.62	0.0811	0.0312
Bi-RNN	184.85	102.48	4.32	0.0249	0.0302
GRU	194.25	71.98	2.83	**0.0086**	**0.0298**
Bi-GRU	**179.80**	**68.20**	**2.82**	0.0103	0.0311
LSTM	194.71	86.44	2.99	0.0128	0.0304
Bi-LSTM	193.96	90.61	2.64	0.0131	0.0309
Decision Trees	5663.71	561.82	19.71	0.0774	0.0588
Random Forests	5611.29	569.59	22.51	0.0764	0.0344
ARIMA	5531.77	1231.27	22.97	0.0805	0.2583

Table 3. Root Mean Square Error (RMSE) for Different Cryptocurrencies

Model	Bitcoin	Ethereum	Solana	Doge Coin	XRP
RNN	276.29	120.24	5.04	0.1648	0.0649
Bi-RNN	264.13	170.51	5.92	0.0619	0.0620
GRU	277.28	120.69	4.26	**0.0267**	**0.0625**
Bi-GRU	**264.28**	**118.07**	**4.30**	0.0292	0.0639
LSTM	277.55	147.72	4.42	0.0364	0.0628
Bi-LSTM	274.45	160.10	4.05	0.0362	0.0625
Decision Trees	5896.46	866.44	20.63	0.1611	0.1136
Random Forests	5846.13	895.89	23.33	0.1599	0.0655
ARIMA	5769.84	1558.92	23.77	0.1643	0.4166

Table 4. Mean Absolute Percentage Error (MAPE) for Different Cryptocurrencies

Model	Bitcoin	Ethereum	Solana	Doge Coin	XRP
RNN	3.08	4.50	10.48	55.62	5.00
Bi-RNN	2.78	5.70	11.79	11.33	4.85
GRU	2.91	4.50	8.16	**5.65**	**4.70**
Bi-GRU	**2.69**	**4.30**	8.17	6.29	4.89
LSTM	2.85	5.05	8.74	7.58	4.73
Bi-LSTM	2.91	5.09	**7.69**	7.68	4.91
Decision Trees	81.41	27.31	51.95	38.89	9.36
Random Forests	80.59	25.61	59.73	36.67	5.85
ARIMA	79.34	78.07	61.00	50.36	42.68

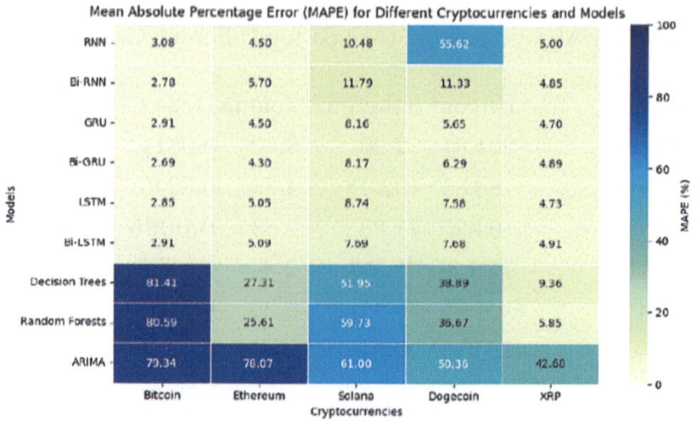

Fig. 3. Log-transformed Mean Absolute Error (MAPE) for different cryptocurrencies and models, transformed to the logarithmic scale to enhance comparability across a wide range of values

For highly volatile cryptocurrencies like Bitcoin and Ethereum, GRU and Bi-GRU demonstrated better performance, achieving an average MAPE improvement of 30% over Random Forest. These results showcase the robustness of GRU and Bi-GRU in capturing complex market dynamics. Among the deep learning models, GRU and Bi-GRU demonstrated the most consistent performance across different cryptocurrencies, which suggests their adaptability in handling both volatile and stable assets. This stability indicates that GRU-based models may be better suited for practical applications where asset volatility varies significantly.

This analysis highlights the distinct strengths of deep learning models in handling highly volatile assets. In contrast, traditional models like Random Forests excel in stable conditions. This volatility-focused comparison provides a novel perspective on selecting models tailored to specific market scenarios, underscoring how volatility influences model performance.

5 Conclusion

Overall, this study highlights the strength of RNN-based models, in handling the unique challenges of cryptocurrency price forecasting. Traditional machine learning models, performed relatively better for stable cryptocurrencies, emphasizing their utility in low-volatility scenarios. By emphasizing model performance across different volatility levels, this study provides practical insights for selecting prediction models tailored to specific market conditions.

These findings underscore the importance of model adaptability in financial forecasting and prepare the way for further exploration of hybrid architectures that integrate deep learning with real-world trading constraints. While this study focused on RNN architectures and closing price predictions, future work could incorporate additional market indicators, such as trading volume or macroeconomic factors, to further enhance performance. Exploring advanced architectures, like Transformers, or reinforcement learning for live trading applications, also represents promising directions for expanding the scope of this work. Finally, evaluating model robustness across broader datasets and diverse market conditions could provide deeper insights into their adaptability. Furthermore, the models employed in this study have potential applications beyond financial forecasting, such as analyzing and predicting user behavior patterns in VR systems. This could provide insights for optimizing user interactions and designing more effective VR experiences, which aligns with ongoing research in immersive technologies.

Acknowledgments. This material is based in part upon work supported by the National Science Foundation under grant IIS-2202640. Any opinions, findings, and conclusions or recommendations expressed in this material are those of the authors and do not necessarily reflect the views of the National Science Foundation.

References

1. Aditya Pai, B., Devareddy, L., Hegde, S., Ramya, B.: A time series cryptocurrency price prediction using lstm. In: Emerging Research in Computing, Information, Communication and Applications: ERCICA 2020, vol. 2, pp. 653–662. Springer (2022)
2. Akila, V., Nitin, M., Prasanth, I., Reddy, S., Kumar, A.: A cryptocurrency price prediction model using deep learning. In: E3S Web of Conferences, vol. 391, p. 01112. EDP Sciences (2023)
3. Biswas, S., Pawar, M., Badole, S., Galande, N., Rathod, S.: Cryptocurrency price prediction using neural networks and deep learning. In: 2021 7th international conference on advanced computing and communication systems (ICACCS), vol. 1, pp. 408–413. IEEE (2021)
4. Cao, J., Li, Z., Li, J.: Financial time series forecasting model based on ceemdan and lstm. Phys. A **519**, 127–139 (2019)
5. Chen, Z., Li, C., Sun, W.: Bitcoin price prediction using machine learning: an approach to sample dimension engineering. J. Comput. Appl. Math. **365**, 112395 (2020)
6. Chung, J., Gulcehre, C., Cho, K., Bengio, Y.: Empirical evaluation of gated recurrent neural networks on sequence modeling. arXiv preprint arXiv:1412.3555 (2014)
7. CoinMarketCap: cryptocurrency market data. https://coinmarketcap.com/. Accessed 04 Dec 2021
8. Derbentsev, V., Matviychuk, A., Soloviev, V.N.: Forecasting of cryptocurrency prices using machine learning. In: Advanced Studies of Financial Technologies and Cryptocurrency Markets, pp. 211–231 (2020)
9. Farell, R.: An analysis of the cryptocurrency industry. Wharton Res. Scholars **130**, 1–23 (2015)
10. Hochreiter, S., Schmidhuber, J.: Long short-term memory. Neural Comput. **9**(8), 1735–1780 (1997)
11. Khedr, A.M., Arif, I., El-Bannany, M., Alhashmi, S.M., Sreedharan, M.: Cryptocurrency price prediction using traditional statistical and machine-learning techniques: a survey. Intell. Syst. Account. Finan. Manage. **28**(1), 3–34 (2021)

12. Kingma, D.P.: Adam: a method for stochastic optimization. arXiv preprint arXiv:1412.6980 (2014)
13. Nayak, S.C.: Bitcoin closing price movement prediction with optimal functional link neural networks. Evol. Intel. **15**(3), 1825–1839 (2022)
14. Oyedele, A.A., Ajayi, A.O., Oyedele, L.O., Bello, S.A., Jimoh, K.O.: Performance evaluation of deep learning and boosted trees for cryptocurrency closing price prediction. Expert Syst. Appl. **213**, 119233 (2023)
15. Patel, J., Shah, S., Thakkar, P., Kotecha, K.: Predicting stock and stock price index movement using trend deterministic data preparation and machine learning techniques. Expert Syst. Appl. **42**(1), 259–268 (2015)
16. Rose, C.: The evolution of digital currencies: Bitcoin, a cryptocurrency causing a monetary revolution. Int. Bus. Econ. Res. J. (Online) **14**(4), 617 (2015)
17. Rumelhart, D.E., Hinton, G.E., Williams, R.J.: Learning representations by back-propagating errors. Nature **323**(6088), 533–536 (1986)
18. Samson, T.K.: Comparative analysis of machine learning algorithms for daily cryptocurrency price prediction. Inf. Dyn. Appl **3**(1), 64–76 (2024)
19. Yamak, P.T., Yujian, L., Gadosey, P.K.: A comparison between arima, lstm, and gru for time series forecasting. In: Proceedings of the 2019 2nd International Conference on Algorithms, Computing and Artificial Intelligence, pp. 49–55 (2019)
20. Zhao, Y., Yang, G.: Deep learning-based integrated framework for stock price movement prediction. Appl. Soft Comput. **133**, 109921 (2023)

An Application of Machine Learning in Computational Astrophysics Using Pre-trained Models

Joshua Ologbonyo[1], Reuben D. Budiardja[2], and Roger B. Sidje[1(✉)]

[1] Department of Mathematics, University of Alabama Tuscaloosa, Tuscaloosa, USA
jjologbonyo@crimson.ua.edu, roger.b.sidje@ua.edu

[2] National Center for Computational Sciences, Oak Ridge National Laboratory Oak Ridge, Oak Ridge, TN, USA
reubendb@ornl.gov

Abstract. Many problems in astrophysics are multi-physics and multi-scale in nature, involving the evaluation of nuclear equation of state (EOS) and opacity for radiation transport, which are often computationally expensive. Our study explores using machine learning techniques in one of the underlying computational building blocks of the general astrophysics simulation system (GenASiS). Specifically, we integrate a pre-trained neural network (NN) model into GenASiS, replacing a kernel process. Upon pre-training our experimental model with data generated by the Riemann problem, we tested it on a complex problem like the Kelvin-Helmholtz instability (KHI) problem. We observed the model's ability to accurately reproduce solutions without compromising runtime efficiency.

This success can be attributed to shared aspects between the two problems, including their similar nuclear EOS and fundamental physical principles. The study is therefore also an illustration that cross-over machine learning approaches can benefit problems with similar foundations.

Keywords: GenASiS · Neural Network (NN) · Pre-trained · Riemann Problem · Machine Learning · Astrophysics · Fortran · Core Collapse Supernova · nuclear equation of state (EOS) · Kelvin-Helmholtz instability (KHI)

1 Introduction

Understanding the universe and our place in it requires studying phenomena at every scale, from tiny quantum events to massive galactic processes. Each of these scales plays a role in shaping the broader cosmic picture. By examining the behavior of particles, the movement of celestial bodies, and the forces that govern both small and large systems, we gain a deeper insight into the nature of existence. Bridging the gaps between quantum mechanics, astrophysics, and

S. Latifi (Ed.): ITNG 2025, AISC 1463, pp. 339–353, 2025.
https://doi.org/10.1007/978-3-031-89063-5_29

everything in between is essential for unlocking the mysteries of the cosmos and understanding our role within it [1–3].

Many problems in astrophysics are multi-physics and multi-scale in nature, and often involve the evaluation of tabulated data obtained either from experiment or numerical modeling of those physical processes. Examples of these include the equation of state (EOS), opacity table for radiation transport, and the reaction rates of nuclear reaction network.

These problems are solved using numerical methods, interpolations and iterative non-linear solvers, which are often computationally expensive. In the last couple of decades, machine learning techniques have proven to be invaluable in completing tasks faster and accurately, with further improvements remaining possible through fine-tuning options.

In this study, we explore using machine learning techniques to mitigate the computational demands encountered when evaluating nuclear equation of state. Our objective is to leverage machine learning techniques in a computational building block within an existing astrophysics simulation code. We specifically applied a neural network (NN) architecture, with the goal of achieving better results in terms of both accuracy and runtime efficiency.

The rest of the work is subdivided as follows. Section 2 is a background about core collapse supernova simulations and their computational cost. Section 3 presents a short review on recent applications of machine learning in computational astrophysics, and how to mitigate the challenges arising there. Section 4 describes the general astrophysics simulation system (GenASiS), its properties and processes. Section 5 presents our methodology while in Sect. 6, we present the execution of our method on a process of GenASiS and the results. Finally in Sect. 7, we give our conclusion about this study and future work.

2 Core Collapse Supernova Simulation

In the vast expanse of the cosmos, the death of massive stars unfolds in a breathtaking display known as Core-Collapse Supernovae (CCSN). These cosmic explosions, typically involving stars with a mass greater than ten times that of our Sun, mark the end of a stellar life cycle and the birth of special objects like neutron stars and black holes.

CCSN is heralded as one of the most energetic phenomena in the cosmos. A staggering 10^{53} ergs of energy are released during the collapse, with the majority manifesting as elusive neutrinos. While neutrinos dominate the energy escape route, a significant amount (10^{51} ergs) is also released as visible electromagnetic radiation. This radiation, spanning various wavelengths, renders CCSN observable using telescopes, allowing scientists to capture the visual spectacle of these cosmic explosions [2,4,5].

Simulating supernovae stands as a critical endeavor in astrophysical research, providing invaluable insights into the universe and our place within it. Several key motivations drive the need for these simulations, these include understanding the formation of elements, stellar evolution and life cycle, cosmic chemical enrichment, neutrino physics, gravitational wave astronomy, gamma-ray bursts (GRBs), testing theoretical models, astrophysical phenomena on all scales, and advancing technology and computational astrophysics [6,7].

Understanding supernovae necessitates sophisticated numerical simulations incorporating equations from multiple physics branches like hydrodynamics, radiation transport, nuclear reactions, and gravity. Techniques like Adaptive Mesh Refinement (AMR) add complexity, challenging traditional numerical schemes to resolve sharp gradients and shock fronts. With grids of 1728×483, totaling 191 million cells, significant computational resources are needed. High-performance computing (HPC) resources become essential, with the simulations often pushing the boundaries of the available computational hardware. Effective data handling is crucial due to the vast data generated, maintaining numerical stability in extreme conditions. Access to cutting-edge computing infrastructure, including supercomputers and specialized hardware accelerators, is essential. As computational power advances, so does our ability to explore and understand supernova phenomena [4,5].

3 Machine Learning in Astrophysics

Machine-learning methods find diverse applications in the analysis of astronomical data, encompassing tasks like data interpretation, pattern recognition, prediction, classification, and compression [8,9]. A prominent approach is the use of artificial neural networks (ANNs), composed of interconnected nodes processing information and transmitting it through weighted connections. [10] highlights the debut of SkyNet, a generic NN training algorithm and its application in astronomical problem-solving, emphasizing its role in the BAMBI package for accelerated Bayesian inference in cosmology and the identification of gamma-ray bursters.

The landscape of computational algorithms is diverse, with many traditionally written in languages such as C, C++, and Fortran due to their efficiency and performance. On the other hand, the proliferation of machine learning tools has largely occurred in languages such as Python, Julia, and Matlab, which offer flexibility and ease of use. This disparity in languages necessitates a bridge, i.e., a means of connecting the robust computational algorithms written in lower-level languages with the wealth of machine learning tools available in higher-level languages. This bridge facilitates seamless integration, allowing researchers and practitioners to leverage the strengths of both worlds for comprehensive and efficient algorithmic solutions. Some examples include the Neural-Fortran package, a parallel Fortran framework for NN and deep learning by [11], Fortran-Keras Bridge (FKB) package, a Fortran-Keras deep learning bridge for Scientific Computing by [12] and Inference-Engine, by [13].

4 GenASiS

The General Astrophysics Simulation System, abridged GenASiS [14], is a platform meant for parallel, large-scale simulations. Tailored to cater to a spectrum of computational needs, ranging from weak-scale simulations to an epic

100 thousand MPI processes, GenASiS is a testament to the prowess of modern Fortran with some key features.

It is a multi-physics solver, strategically designed to excel in parallel computing environments. Its capabilities extend to orchestrating simulations with a magnitude of MPI processes, providing scalability for a diverse range of astrophysical scenarios. It is entirely written in modern Fortran, leverages the language's strengths in scientific and numerical computing [15,16]. GenASiS is modular and object-oriented allowing researchers to seamlessly integrate new features and modules to suit the evolving demands of astrophysical simulations as shown in Fig. 1. The inclusion of self-gravity in GenASiS enables the simulation of scenarios where gravitational interactions play a pivotal role. Additionally, the support for polytropic and nuclear EOS enhances the fidelity of simulations, capturing the nuanced behavior of astrophysical matter [14,17].

GenASiS stands as a versatile and powerful tool in the astrophysicist's arsenal, offering a high-performance computing solution for a diverse array of simulations. Its commitment to parallelism, modern Fortran principles, and a rich set of multi-physics solvers positions it as a valuable asset in the pursuit of unraveling the mysteries of the cosmos. As astrophysical simulations continue to evolve, GenASiS remains at the forefront, providing a reliable and flexible platform for pushing the boundaries of our understanding of the universe.

5 Method

The methodology adopted for this study follows the workflow shown in Fig. 2.

5.1 Data Extraction and Processes

This process begin with the extraction of relevant data from the solutions generated by GenASiS, and since these solutions are high dimensional data, they are written in silo files or hdf5 files. Hence the silo application programming interface (API) is used for extraction of data needed for building and training the model. The extracted data are then processed to ensure it meets the requirements of the NN model.

5.2 Neural Network Model

The foundation of every NN lies in its neurons, which closely resemble their biological counterparts. These neurons receive input values, calculate a weighted sum based on these inputs, incorporate a bias term, and ultimately produce an output by applying a nonlinear activation function, denoted as σ, to the computed sum. A feed-forward NN consists of multiple layers of such neurons, with the input and output layers referred to as the first and last layers, respectively. The intermediate layers are known as hidden layers. In the case of a fully connected NN, the output of each layer becomes the input for all neurons in the next layer. The output of the output layer serves as the overall output of the NN [18,19].

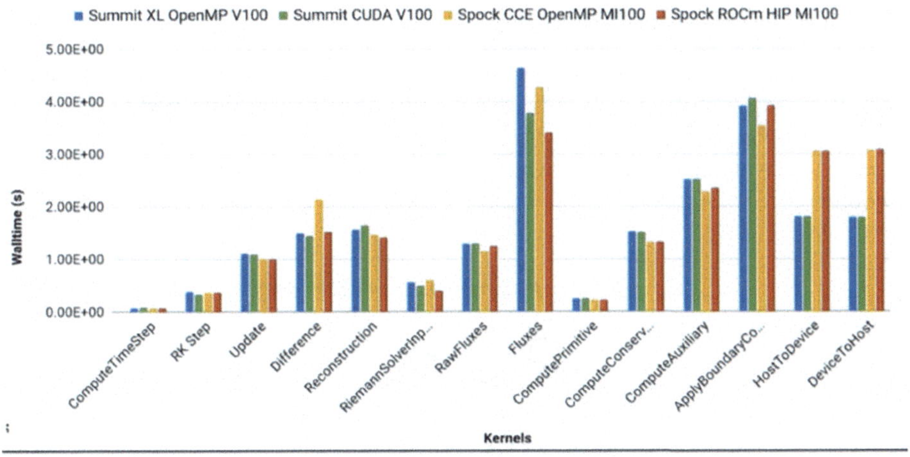

Fig. 1. GenASiS computational time for each kernels with different Compilers.

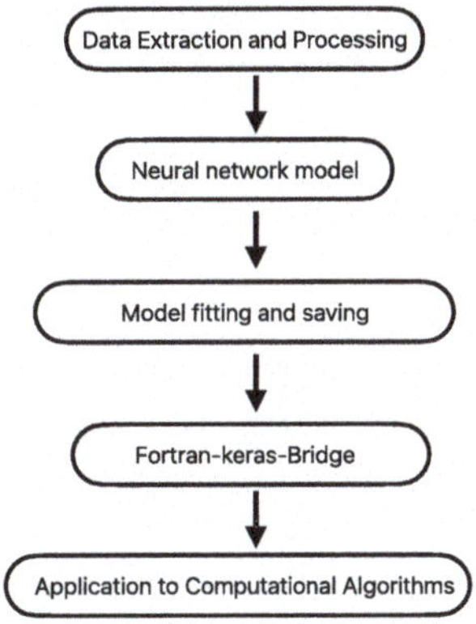

Fig. 2. Workflow

In this study, the NN employed are fully connected feed-forward architectures, featuring examples with one or two hidden layers. The NN parameters encompass weights and biases, denoted collectively as θ. During the training phase, these parameters are optimized by minimizing the disparity between the NN's output, $\hat{y}_i(\theta)$, and the precise values (y_i), known as "labels," from a training dataset [18]. This optimization process iteratively computes a loss function E, calculates the gradient of E with respect to all parameters using the chain rule (backpropagation step), and applies a gradient descent algorithm to adjust all parameters, aiming to minimize E. The loss function minimized in the training process is given by $E(\theta) = \frac{1}{n}\sum_{i=1}^{n}(\hat{y}_i(\theta) - y_i)^2$ and the Adam optimizer performed better compared other optimizers used for the gradient descent, for the purpose of this study, the NN model architecture is designed using Keras.

5.3 Model Fitting and Saving

We fit the model and explore a range of parameters, including the number of layers, layer sizes, and activation functions. Specifically, the Rectified Linear Unit (ReLU) function, defined as $\text{ReLU}(z) = \max(0, z)$, and some other functions were tested as activation functions. Among these, ReLU yields the most favorable outcomes.

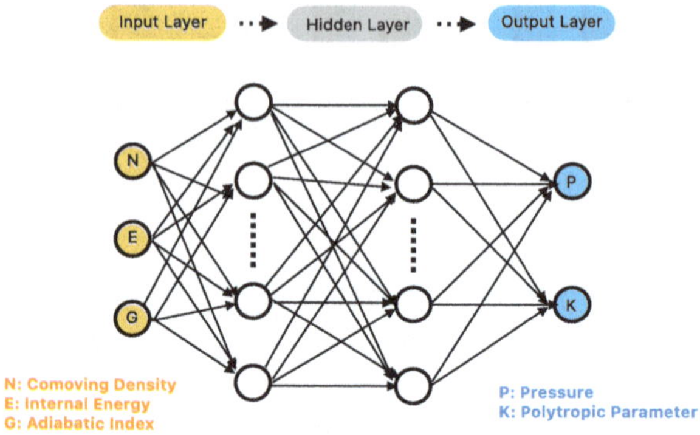

Fig. 3. Keras feed-forward fully connected architecture.

To assess the performance of the NNs, a separate test dataset is used, held back from the training process. Notably, the models consistently exhibit lower loss on this independent testing dataset compared to the training dataset, indicating a lack of overfitting. Consequently, dropout methods, aimed at mitigating overfitting, do not enhance the training process. The best fitted model is saved as a high dimensional hdf5 file which is the converted to txt file which is required by the Fortran-Keras Bridge (FKB) algorithm.

5.4 Fortran-Keras Bridge (FKB)

The Fortran-Keras Bridge (FKB) by [12] is a tool designed to seamlessly integrate the capabilities of Fortran, a long-standing language in large-scale scientific computation, with the versatility of Keras, a popular deep learning framework in Python. FKB serves as a bridge between Fortran and Keras, allowing researchers and developers to harness the power of Fortran's large-scale scientific computation capabilities while leveraging the ease and flexibility of Keras for deep learning tasks.

The FKB ecosystem encompasses various components, including FKB/F (Fortran), FKB/P (Python), and Neural Fortran. FKB/F represents the Fortran module or interface, providing the essential linkage to Fortran's computational prowess. On the other side, FKB/P encapsulates the Python module, facilitating seamless communication with Keras and enabling the integration of Fortran into the Python-based deep learning workflow.

This allow us to train our model in Python, save the model and import it into computational algorithm written in Fortran. This models then replace part of the computational process (subroutines) of the Algorithm.

The FKB comes equipped with a range of features that enhance its versatility for advanced scientific computations and machine learning tasks. Some of the notable features of FKB include: Fully connected layers for versatile connectivity within NN, batch normalization layers for stabilizing and accelerating training, dropout layers for regularization by randomly dropping connections during training, training in Fortran, custom loss functions and Ensembles.

5.5 Application to a Computational Building Block

We applied the integrated model to GenASiS, using the predictions generated by the NN into the existing GenASiS algorithm. The aim was to explore innovative ways in which the NN model can augment the efficiency and accuracy of the computational processes. This workflow underscores the collaborative forces of Fortran and Keras through the FKB, facilitating a seamless integration of traditional scientific computing and cutting-edge deep learning methodologies within a unified computational framework.

6 Experimentation in GenASiS: Fluid Dynamics

All software and experimentation done in this study are all carried out on Frontier, which is currently the world's fastest exascale supercomputer, and other Oak Ridge Leadership Computing Facility (OLCF) resources at the Oak Ridge National Laboratory, which is supported by the Office of Science of the U.S. Department of Energy under Contract No. DE-AC05-00OR22725.

GenASiS is made of the following processes (and subroutines): InputOutput, Computation, ComputeTimeStep, Communication, RK Step, Update, Difference, Reconstruction, RiemannSolverInput, RawFluxes, Fluxes, ComputePrimitive, ComputeConserved, ComputeAuxiliary, ApplyBoundaryConditions, DataTransfer to Device and Data Transfer to Host. Figure 1 shows the computational time for each kernels, our work aim to replace some of this kernels with a pre-trained model, thereby reducing computational time and resources while keeping the accuracy of the solutions produced by the algorithm intact.

6.1 The Riemann Problem

For the purpose of this study, we present two examples from categories of fluid dynamics problems built on GenASiS framework. The Riemann problem, extending beyond periodic advection to full fluid evolution with shocks and reflecting boundary conditions, in 1D, 2D, and 3D. For more specifics on the fluid dynamics classes and the numerical methods employed in GenASiS, we direct the reader to [14], offering a more systematic exploration of the functionalities available on GenASiS Basics.

We follow the process described in Sect. 5 as shown in the flow chart in Fig. 2, and replace the ComputeAuxiliary kernel process with a pre-trained NN model. The input variables for the ComputeAuxiliary are comoving density, internal energy and adiabatic while the process outputs pressure and polytropic parameter as show in the NN architecture in Fig. 3.

We ran GenASiS and GenASiS with ML with the following configuration: one thread of execution, 1 node, 7 logical cores per MPI task, 2 number of MPI tasks for 1D, 4 number of MPI tasks for 2D and 8 number of MPI tasks for 3D. To illustrate the results of the RiemannProblem for a resolution of 128 cells in 1D, 2D, and 3D, we present plots in Figs. 4, 5, 6, 7 and 8 respectively, showcasing the initial states at ($t = 0$) and final states of the fluid at ($t = 0.25$). The default initial conditions for the RiemannProblem for GenASiS and GenASiS with ML align with the well-known Sod shock tube problem for 1D, which has been extended to 2D and 3D.

We experimented three fully connected NN model; (i) model with two hidden layers, 1024 and 64 neurons respectively, (ii) model with one hidden layer of 32 neurons and (iii) model with one hidden layer of 2 neurons. NN model with two hidden layers performed abysmally in terms of computation time, NN model with two hidden layers of 32 neurons improved, also NN model with two hidden layers of 2 neurons' performance was shown in Fig. 9.

To improve the computation time for GenASiS ML, we attempt to use NN with no hidden layer which is also a multi-output linear regression model, the result was better in terms of computation time without altering the accuracy the solution as shown in Fig. 9, without compromising the accuracy of the solutions displayed in Figs. 4, 5, 6, 7 and 8.

Figure 10 shows the errors in the solutions of GenASiS with ML when compared with the solutions of GenASiS for each dimension, we utilize the L_1 norm to compute the error given by

$$L_1 \text{error} = \frac{1}{n} \sum_{i=1}^{n} |y_i - \hat{y}_i|,$$

and ran both algorithms with different cell resolutions, the error demonstrates improvement with higher dimensions as the number of cells increases.

6.2 Fluid Instability Tests: The Kelvin–Helmholtz Problem

We demonstrate the performance of GenASiS ML on instability of astrophysical relevance; the Kelvin–Helmholtz instability (KHI) problem. It describes the instability that occurs at the interface between two fluids or layers with different velocities or densities [20].

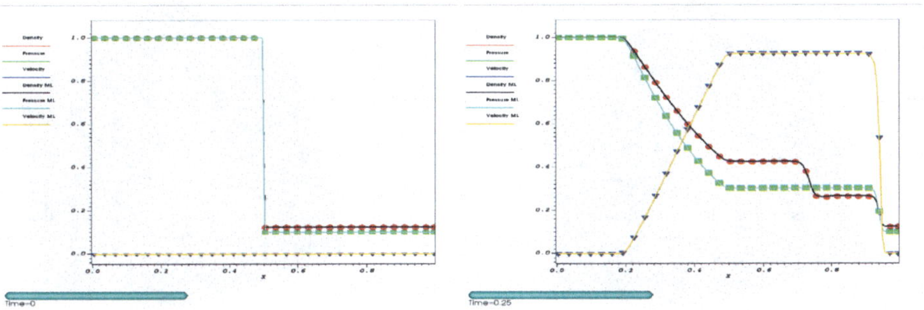

Fig. 4. Showing the plot of 1D version of Riemann problem at a resolution of 128 cells, for GenASiS [density (in red symbol), pressure (in green symbol) and velocity (in blue symbol)] and for GenASiS ML [density ML (in black line), pressure ML (in cyan line) and velocity (in yellow line)]. The initial solution at $t = 0$ is on the left, while the final solution at $t = 0.25$ on the right.

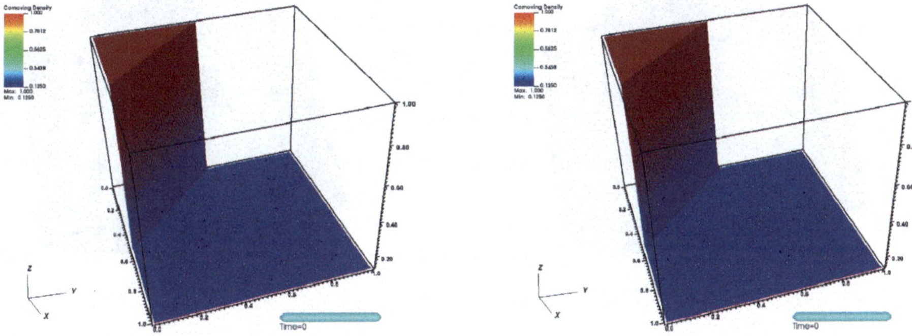

Fig. 5. Showing the plot of Density in 2D version of Riemann problem at $t =$ 0at a resolution of 128,128 cells. (a) GenASiS solution, (b) GenASiS ML solution.

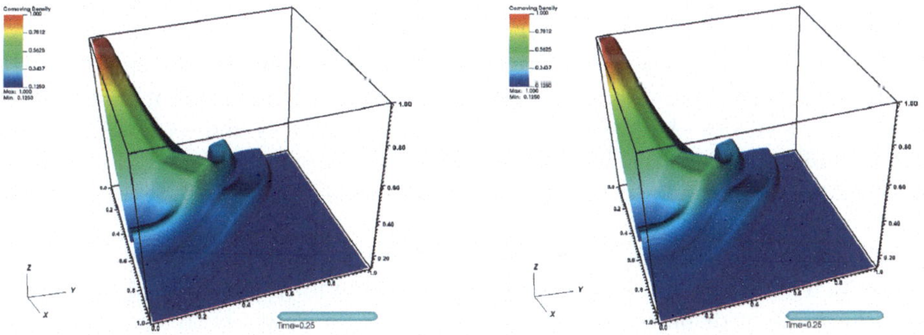

Fig. 6. Showing the plot of Density in 2D version of Riemann problem at $t = 0.25$ at a resolution of 128,128 cells. (a) GenASiS solution, (b) GenASiS ML solution.

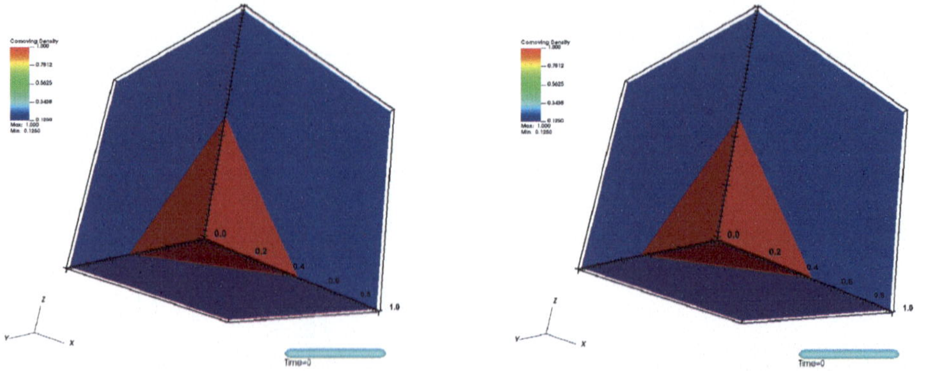

Fig. 7. Showing the plot of Density in 3D version of Riemann problem at $t = 0$ at a resolution of 128,128,128 cells. (a) GenASiS solution, (b) GenASiS ML solution.

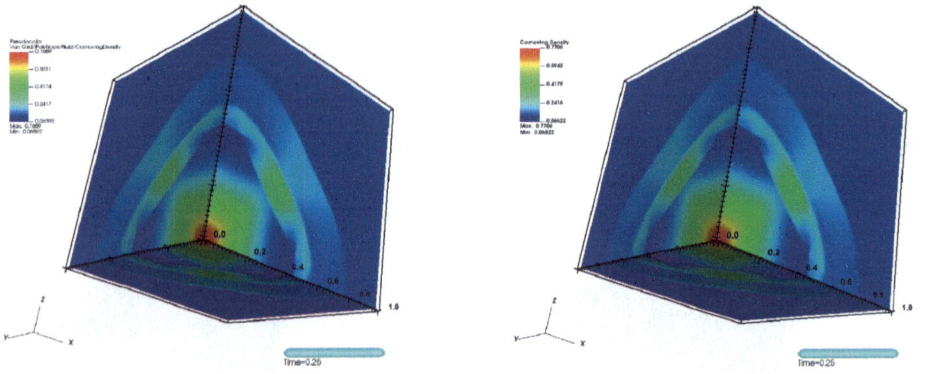

Fig. 8. Showing the plot of Density in 3D version of Riemann problem at $t = 0.25$ at a resolution of 128,128,128 cells. (a) GenASiS solution, (b) GenASiS ML solution.

Simulating KHI poses several challenges. Firstly, achieving sufficient resolution to capture fine-scale features near the shear layer, especially in three dimensions, can be computationally demanding. Secondly, maintaining numerical stability while accurately representing the growth of small perturbations over time is essential but challenging. Choosing appropriate boundary conditions to properly reflect inflow, outflow, and shear layer behavior is crucial [20,21]. For these reasons we decide to evaluate the performance of GenASiS ML on such complicated example.

The initial conditions for Kelvin-Helmholtz instability inherently involve stochasticity, which is the cause of GenASiS ML performance showed in Fig. 11, with the L_1 error displayed in Fig. 13. To address this issue, both algorithms are initialized with identical initial conditions, mitigating the stochasticity difference in their initial states. Figure 12 showed GenASiS ML solution as it compares to GenASiS solution while Fig. 13 displayed the L_1 error.

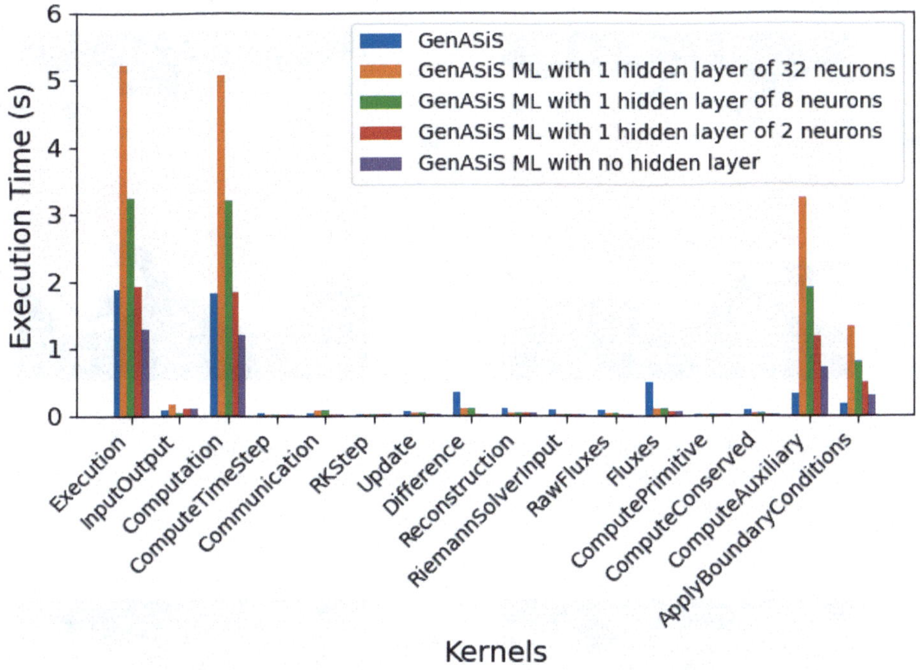

Fig. 9. Showing the computational time for each GenASiS kernels compared with GenASiS ML, using NN with one hidden layer and NN with no hidden layer.

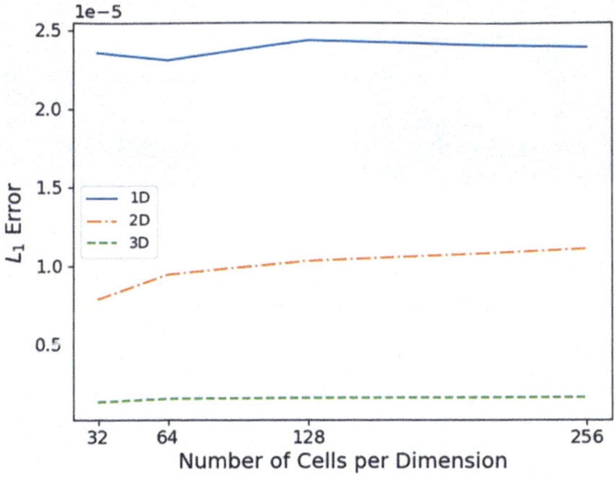

Fig. 10. Showing the L_1 error of GenASiS ML solution as compared with GenASiS solution for 1D, 2D and 3D.versions of Riemann problem.

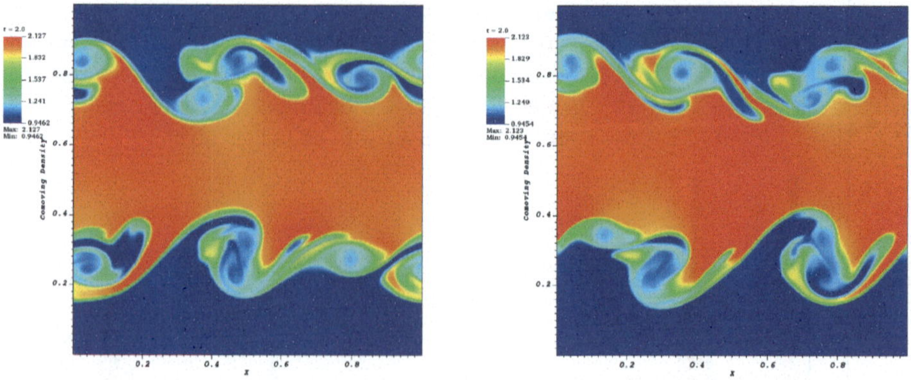

Fig. 11. Showing the plot of Density in 2D version of Kelvin–Helmholtz at $t = 2$ at a resolution of 512,512 cells, with stochasticity difference in initial state. (a) GenASiS solution, (b) GenASiS ML solution.

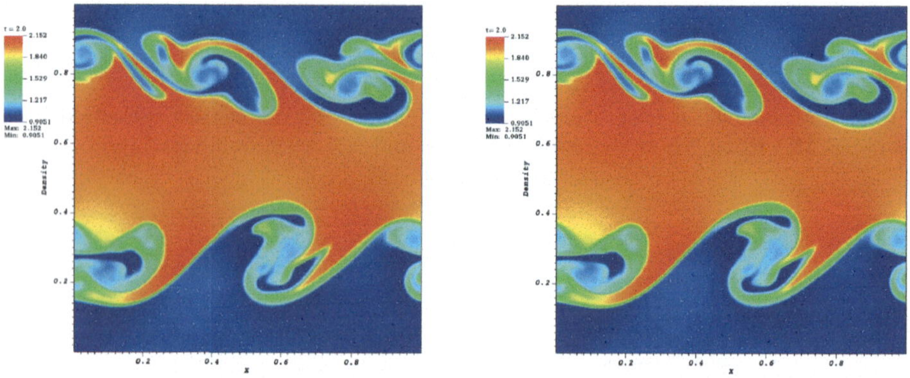

Fig. 12. Showing the plot of Density in 2D version of Kelvin–Helmholtz at $t = 2$ at a resolution of 512,512 cells, without stochasticity difference in initial state. (a) GenASiS solution, (b) GenASiS ML solution.

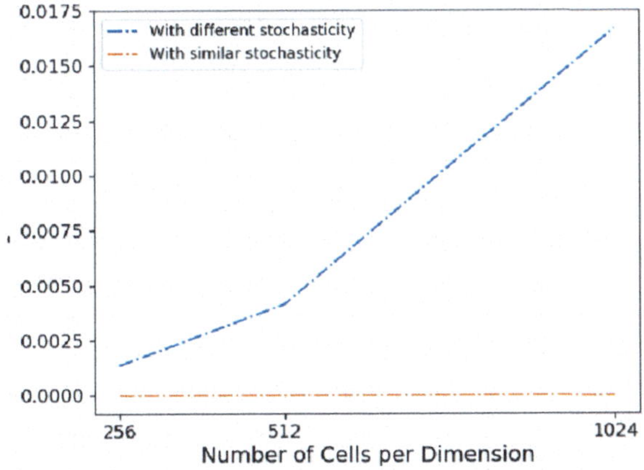

Fig. 13. Showing the L_1 error of GenASiS ML solution as compared with GenASiS solution for 2D version of Kelvin–Helmholtz problem, with different stochasticity and with similar stochasticity.

7 Conclusion

When using software like GenASiS to solve computationally intensive astrophysics problems, the size of the saved pre-trained model, as shown in Fig. 9, is crucial. This is because the model's size directly affects various aspects of its usage and deployment. A larger model size implies increased storage requirements and longer loading times, affecting the overall efficiency of the software. Furthermore, in scenarios where the model needs to be deployed to edge devices or distributed systems, a smaller model size becomes essential to minimize network bandwidth usage and ensure efficient deployment. Therefore, optimizing the size of the pre-trained model while preserving its performance is imperative for effective use of software like GenASiS in astrophysical computations.

By substituting one subroutine process with a pre-trained model, the algorithm consistently achieves accurate solutions. The pre-trained model was trained using data generated by the Riemann problem, and we applied the same model to the KHI problem, yielding accurate results. This means that the success in using the pre-trained model for both the Riemann problem and the KHI problem is facilitated by their shared aspects, such as using the same nuclear EOS and fundamental physical principles. Since the model was pre-trained with data generated by the Riemann problem, which encompasses these common aspects, it can effectively generalize to the KHI problem, resulting in accurate solutions.

Indeed, the model's ability to accurately generalize to the KHI problem suggests the potential applicability of the approach to other nuclear EOS that share similar fundamental physical principles. This highlights the versatility and broad utility of the pre-trained model in addressing computational challenges across various astrophysical simulations.

GenASiS ML With two hidden layer of 1024 and 64 neurons gave a fairly error of 5.575695×10^{-05} while GenASiS ML with one hidden layer and 8 neurons gave an error of 3.485631×10^{-05}, but the latter performed better in terms of computation time and storage. This is a trade-off situation of accuracy versus computation time, in which we try to find the middle ground. As the NN architecture deepens, the computational expense of the algorithm increases.

Our future work aim to explore the development of more optimized models that strike a balance between size and accuracy. There is the caveat that, while our methodology is effective, it depends on identify a standalone subroutine or kernel suitable for replacement with a pre-trained neural network model. We noted in GenASiS that, the ComputeAuxiliary kernel, though not highly computationally intensive, is the only one designed independently from other kernels. A future application of our technique could target a more computationally demanding building block of the software.

References

1. Kwok, S.: Our Place in the Universe: Understanding Fundamental Astronomy from Ancient Discoveries. Springer (2017)
2. Janka, H.-T., Langanke, K., Marek, A., Martínez-Pinedo, G., Müller, B.: Theory of core-collapse supernovae. Phys. Rep. **442**(1–6), 38–74 (2007)
3. Vainio, O.-P: Cosmology in theological perspective: understanding our place in the universe. Baker Academic (2018)
4. Branch, D., Wheeler, J.C.: Supernova explosions. Springer, vol. 4 (2017)
5. Arnett, D.: Supernovae and Nucleosynthesis: an Investigation of the History of Matter, from the Big Bang to the Present. Princeton University Press (1996)
6. Hempel, M., Fischer, T., Schaffner-Bielich, J., Liebendörfer, M.: New equations of state in simulations of core-collapse supernovae. Astrophys. J. **748**(1), 70 (2012)
7. Hopkins, P.F., et al.: How to model supernovae in simulations of star and galaxy formation. Mon. Not. R. Astron. Soc. **477**(2), 1578–1603 (2018)
8. Frank, M., Drikakis, D., Charissis, V.: Machine-learning methods for computational science and engineering. Computation **8**(1), 15 (2020)
9. Longo, G., Merényi, E., Tiňo, P.: Foreword to the focus issue on machine intelligence in astronomy and astrophysics. Publ. Astron. Soc. Pac. **131**(1004), 1–6 (2019)
10. Hobson, M., Graff, P., Feroz, F., Lasenby, A.: Machine-learning in astronomy. Proc. Int. Astron. Union **10**(S306), 279–287 (2014)
11. Curcic, M.: A parallel fortran framework for neural networks and deep learning. In: Acm sigplan fortran forum, vol. 38, no. 1. ACM New York, NY, USA, pp. 4–21 (2019)
12. Ott, J., Pritchard, M., Best, N., Linstead, E., Curcic, M., Baldi, P.: A fortran-keras deep learning bridge for scientific computing, arXiv preprint arXiv:2004.10652 (2020)
13. Rouson, D., Welsman, J., Nguyen, T.: Inference-engine v0. 1.0. Technical Report, Lawrence Berkeley National Laboratory (LBNL), Berkeley, CA, United States (2022)
14. Cardall, C.Y., Budiardja, R.D.: Genasis basics: object-oriented utilitarian functionality for large-scale physics simulations. Comput. Phys. Commun. **196**, 506–534 (2015)

15. Budiardja, R.D., Cardall, C.Y.: Targeting gpus with openmp directives on summit: a simple and effective fortran experience. Parallel Comput. **88**, 102544 (2019)
16. Budiardja, R.D.: Towards simulations of binary neutron star mergers and core-collapse supernovae with genasis (2010)
17. Budiardja, R.D., Cardall, C.Y., Endeve, E.: Accelerating our understanding of supernova explosion mechanism via simulations and visualizations with genasis. In: Proceedings of the 2015 XSEDE Conference: Scientific Advancements Enabled by Enhanced Cyberinfrastructure, pp. 1–8 (2015)
18. Géron, A.: Hands-on Machine Learning with Scikit-Learn, Keras, and TensorFlow. O'Reilly Media, Inc. (2022)
19. Dieselhorst, T., Cook, W., Bernuzzi, S., Radice, D.: Machine learning for conservative-to-primitive in relativistic hydrodynamics. Symmetry **13**(11), 2157 (2021)
20. Cardall, C.Y., Budiardja, R.D., Endeve, E., Mezzacappa, A.: Genasis: general astrophysical simulation system. i. refinable mesh and nonrelativistic hydrodynamics. Astrophys. J. Supplement Ser. **210**(2), 17 (2014)
21. Gramer, L.: Kelvin-helmholtz instabilities. J. Plasma Phys. **28**, 1–11 (2007)

Bug Triaging Based on Transformer Models Utilizing Commit Messages

Farhan Rahman Arnob(✉), Rubel Hassan Mollik, Pooja Goyal, and Renee Bryce

Department of Computer Science and Engineering, University of North Texas, Denton, TX, USA
{farhanrahmanarnob,rubelhassanmollik,poojagoyal}@my.unt.edu, renee.bryce@unt.edu

Abstract. Bug triaging is a crucial process in software maintenance that involves assigning bug reports to the appropriate developers for resolution. Automated bug triaging eliminates the need for manual assignment by leveraging machine learning (ML), information retrieval (IR), and graph neural network (GNN) techniques to classify bug reports based on extracted features. These approaches typically utilize bug report fields and metadata as training features. However, models often underperform when relying heavily on a few specific fields, such as the bug report summary and description. To address this issue, we integrate bug commit messages into the feature set to enhance the efficiency of bug-triaging models. Our study demonstrates significant improvements in triaging accuracy by incorporating commit messages. In the Firefox dataset, prediction accuracy increases from 23.28% to 69.45%. Similarly, in the Eclipse dataset, the accuracy increases from 70.01% to 89.28%. These results prove that including commit messages can significantly improve bug-triaging accuracy.

Keywords: Bug Triaging · Automated Bug Assignment · Commit Messages · Automation · Machine Learning · Natural Language Processing · Software Testing · Transformer Models · XLNet · Software Maintenance

1 Introduction

Since bugs are a pervasive part of software development, fixing them is essential to preserving and improving the quality of the product [1]. Frequently, bug reports are used to monitor the bug resolution process, as they provide thorough records of the problems found [2]. Effective bug-triaging and assigning these reports to the most suitable developer is crucial for quick and successful defect resolution throughout the software development lifecycle.

Triaging bugs takes a lot of effort and is difficult by nature. This duty has historically been overseen by senior team members and calls for long work hours,

S. Latifi (Ed.): ITNG 2025, AISC 1463, pp. 354–366, 2025.
https://doi.org/10.1007/978-3-031-89063-5_30

and traditionally, they often rely on bug summaries and descriptions. For example, Mozilla receives over 300 bug reports every day [3] and Eclipse reports around 37 per day [4]. The Mozilla code base has over 233 developers contributing to it [5]. It is difficult and error-prone to assign defects appropriately based on developer experience and bug requirements, causing delays and frequent reassignment of reports. One way to address this is by automating the bug triaging, which speeds up the assignment process while preserving human resources.

Recent work in automated bug triaging has integrated natural language processing (NLP) methods with machine learning (ML) models [3,6,7]. These techniques treat bug triaging as a problem of multi-class text classification. Conventional methods such as Word2Vec and TF-IDF (Term Frequency-Inverse Document Frequency) have been applied in multiple works [8,9]. Although Word2Vec's static embedding is unable to adjust to changing contexts, reducing their expressiveness and accuracy when capturing intricate textual patterns, TF-IDF assesses term frequency but lacks semantic context.

Even with these developments, performance with existing Machine Learning (ML) models has frequently been less than ideal. Significant improvement has been demonstrated by transformer-based approaches, such as BERT (Bidirectional Encoder Representations of Transformers) [5,10]. To capture complex word relationships and semantic context, these models use self-attention mechanisms [11]. Further improving this is XLNet (Generalized Autoregressive Pretraining for Language Understanding), which combines autoregressive and autoencoding techniques [12]. It does this by modeling complicated word dependencies and producing more accurate embedding by using permutation language modeling [13].

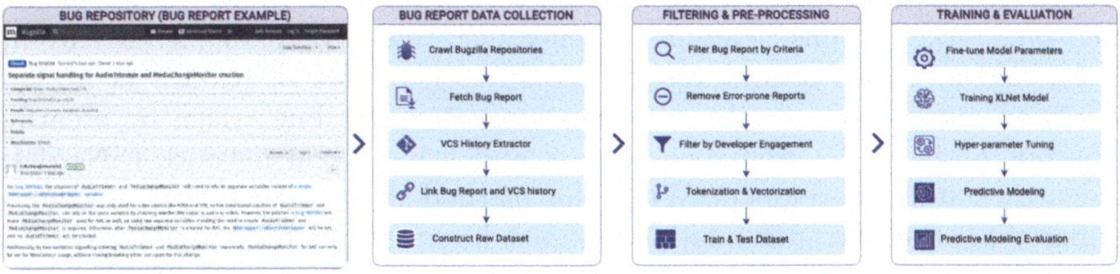

Fig. 1. An Overview of Data Collection, Training Transformer Model, and Evaluation

In this study, we propose a novel approach to enhance bug triaging by employing a feature engineering technique for improved dataset preparation and training. We use a state-of-the-art transformer-based neural network model to evaluate our methodology. We address the limitation by including commit messages into the feature set as commit messages align more closely with developers and provide critical insights into prior bug fixes. Our results demonstrate a clear advantage over existing models, that is, showing substantial improvements in performance metrics [10,12,14–17]. Specifically, XLNet improved accuracy from 23.28% to 69.45% for the Firefox dataset and from 70.01% to 89.28% for the

Eclipse dataset, underscoring its effectiveness in capturing nuanced contexts of bug reports.

This paper is organized as follows: Sect. 2 reviews background and related work on bug triaging using various graph neural networks, machine learning, and NLP techniques. Section 3 describes our infrastructure setup for data collection, preprocessing, and research methodology. Section 4 outlines our research prolongs and evaluation methods. Section 5 presents our results and highlights. Section 6 concludes with a detailed discussion of our findings and explores potential future work to enhance the process.

2 Background

Automated bug triaging is critical to addressing inefficiencies of manual defect assignment as software systems scale. Due to the growing number of bug reports, manual triaging is becoming increasingly problematic and prone to errors. Therefore, researchers are leveraging machine learning approaches to automate the triaging process to overcome these limitations. To address the issue in extensive open-source projects, this section examines the development of automated bug triaging, emphasizing advancements in ML and Natural Language Processing (NLP) that improve triaging efficiency and accuracy.

Conventional semi-automated methods for triaging bug reports use ML algorithms to examine past issue data and suggest developers who can address fresh reports. For instance, in the Eclipse and Firefox projects, prior techniques yielded precision levels of 57% and 64%, respectively [18]. Our method expands on this foundation by improving the dataset that models were trained on [19] to address the problem of "tossing" bug reports—where reports are reallocated to other developers, lengthening the time-to-correction—this work uses a graph model based on Markov chains. This model effectively reduced tossing events by up to 72% and increased accuracy by up to 23%. The work of Kukkar et al. [20] focuses on automating developer recommendations for bug triage by creating a concept profile (CP) from bug reports and using social network (SN) data to identify suitable developers. While this method shows improved performance compared to other recommendation systems, however, due to limited contextual information and the narrow scope of Bug it has its limitations [5,21]. The provided approach addresses bug assignments in large, long-lived software projects using a combination of ML techniques and probabilistic graph-based models. This approach aims to improve prediction accuracy and reduce software evolution costs by employing a diverse set of tools and evaluating their effectiveness on extensive datasets. However, the approach had a high dependence on Historical Data, the Complexity of the Implementation of the model as well as methodology Focused on Static Metrics. DevRec is another advanced study to recommend developers capable of resolving bugs [22], employing a dual-analysis framework that integrates bug report-based (BR-based) and developer-based (D-based). Although DevRec demonstrates superior performance compared to methods such as Bugzie [23] and DREX [24], particularly in terms of recall scores, it is not without limitations.

Although useful in many situations, the dual-analysis approach has certain built-in drawbacks. BR-based analysis might not adequately capture the changing context of current difficulties, but might offer insightful information based on previous bug reports. The D-based analysis can evaluate the skills and knowledge of developers, but it is not dynamic enough to account for changes in developer competence or recent modifications in the codebase, which can result in static developer affinity. As such, this may lead to a restricted contextual comprehension of the bug reports and may miss up-to-date, pertinent information, which could affect the accuracy of developer recommendations.

Applying machine learning techniques is the most popular approach in automated bug triaging. [3] first applied SVM (Support Vector Machine), a supervised machine learning algorithm that shows promise in automated bug triaging. Following the path, classical supervised ML algorithms such as Naive Bayes(NB) and Ensemble learning have been experimented with to produce better results [25]. Although traditional machine-learning approaches paved the way, the result was not satisfactory. With the evolution of Deep Learning (DL) [26], researchers started applying various DL methods like Convolutional Neural Networks(CNN) [27], Artificial Neural Networks(ANN) [28], and Deep bidirectional recurrent neural network with attention mechanism technique (DB-RNN-A) to advance the field. However, these techniques have limitations in grasping the contextual expressiveness of bug report features [29]. Studies with variants of BERT models that demonstrate unique abilities in terms of classification. Nevertheless, the bug report features used to train the BERT model were inadequate in achieving outstanding results.

Current automated approaches for triaging bugs usually concentrate on identifying the best developer by utilizing term selection strategies and allocating tasks according to developer experience. However, these methods have several limitations. For example, term selection methods often work on a restricted scope that can lead to the potentially omitted critical contextual information required for precise bug classification. The dynamic character of developer expertise and current workloads may not be taken into consideration by experience-based load balancing, resulting in recommendations that are not ideal. Furthermore, insufficient context in problem reports often affects these methodologies, which might reduce the precision of triaging results. However, the commit message can enhance the contextual efficiency. For instance, if a developer resolved a bug related to log in issue like 'fixed username validation', they are more likely to understand and effectively address a new bug like 'username not being added'.

3 Data Collection and Preprocessing

We use the well-documented Eclipse [30] and Firefox [31] bug repositories in this project. These large datasets include a wide variety of structured modules, which makes them an excellent foundation for evaluating our methods in a variety of software settings and bug-reporting situations.

3.1 Dataset Description and Characteristics

1) Eclipse Dataset: Eclipse is a feature-rich integrated development environment (IDE) with a broad range of plugins. We extracted 32,776 bug reports from four different modules - Core, Platform, UI, and IDE from the Eclipse Bugzilla repository. *2) Firefox Dataset:* Firefox is a popular web browser with copious features. We gathered 9,999 bug reports. These two large-scale systems have a broad range of features and developers which facilitates a unique perspective for our methodological assessment.

3.2 Data Preprocessing

Raw data often contains incomplete bug reports, missing metadata, or properties related to the bug that are not suitable for training models. We developed the following multi-step data filtering process to retain the quality and relevance of the dataset, as shown in Fig. 1.
1) Selection of Completed Bug Reports: To ensure that the dataset includes only those reports for which the resolution and developer expertise are well-documented, we focused exclusively on bug reports marked as "FIXED."
2) Exclusion of Incomplete Reports: As missing or incomplete information causes noise and undermines the training process we excluded the reports lacking crucial metadata, such as title, description, or summary.
3) Developer Engagement Filtering: We removed bug reports from developers who fixed fewer than ten or more than fifty problems to mitigate potential biases from developers who were either too involved or too little involved. This method guarantees that the model stays impartial and does not give preference to developers who have a negligible or significant impact.

4 Experimental Framework, Algorithm Settings and Evaluation Metrics

4.1 Experimental Framework

Our goal was to test how the inclusion of commit messages in bug reports affected the accuracy of developer assignments. We used well-documented datasets from Eclipse and Firefox bug repositories, which allowed us to evaluate our method across diverse software projects.
1) Data Processing: We utilized publicly available bug repositories from large-scale open-source projects to test the effectiveness of our model. After applying strict filtering criteria to ensure the data quality, we curved a subset of bug reports contain necessary metadata, including titles, descriptions and commit messages. The dataset was split into training, validations and test sets using a 80-10-10 ratio (Table 1).

Table 1. Dataset Description

Project	Time Interval	#Bug Reports	#Developers
Firefox	July 13, 2004 - January 06, 2024	9,999	635
Eclipse	October 19, 2001 - June 05, 2023	32,776	526

2) Text Processing: We preprocessed the text fields of the bug reports, like title, description, and commit messages to standardize our data. We converted all text to lowercase and removed punctuations and stopwords. Stemming is used to reduce words to their base forms, to reduce the dimensionality of the input data feature, and to focus on relevant features, as shown in Fig. 1.

3) Tokenization and Vectorization: XLNet[12], a state-of-the-art transformer model, is particularly suited for contextual relationships within text sequences, which is imperative for understanding the nuances of bug reports and commit messages. We used the model to tokenize and vectorize the text data. Each text field was transformed into a sequence of numerical tokens for features of the model.

4) Data Augmentation: We leveraged the data augmentation techniques to prevent overfitting by exposing the model to a broader range of input scenarios during training. By introducing minor perturbations in the text, we generated synthetic variations of existing bug reports.

5) Dataset Splitting: Finally, we divided the dataset into validation (10%), training (80%), and test (10%). The test set assesses the model's ability to generalize to new, untested data, guaranteeing reliable performance evaluation; the validation set is utilized for hyperparameter tuning and model optimization; and the training set allows the model to learn from a variety of examples.

Table 2. XLNet Metrics in Eclipse and Firefox

Features	Metrics	Eclipse (XLNet)				Firefox (XLNet)			
		Top-1	Top-3	Top-5	Top-10	Top-1	Top-3	Top-5	Top-10
Summary + Description	Accuracy	47.89%	64.44%	70.01%	77.26%	23.28%	40.78%	48.36%	62.89%
	F1 Score	64.77%	78.37%	82.36%	87.17%	37.77%	57.94%	65.19%	77.22%
Summary + Commit	Accuracy	54.96%	59.54%	63.07%	68.46%	64.14%	75.62%	80.16%	86.41%
	F1 Score	70.93%	74.64%	77.36%	81.28%	78.15%	86.12%	88.99%	92.71%
Summary + Description + Commit	Accuracy	**84.88%**	**88.29%**	**89.28%**	**90.89%**	**69.45%**	**81.48%**	**84.53%**	**89.14%**
	F1 Score	**91.82%**	**93.78%**	**94.34%**	**95.23%**	**81.97%**	**89.80%**	**91.62%**	**94.26%**

4.2 Algorithm Settings

1) XLNet Architecture: We used a custom `XLNetForMultiLabelSequenceClassification` model. The architecture leverages the pre-trained `xlnet-base-cased` model. We have built our model with the consist of the following components:

- **XLNet Model**: We build our model from the pre-trained `xlnet-base-cased` model to extract contextual embedding from the input text sequences. The model captures bidirectional dependencies in the text using a permutation-based language model.
- **Classifier Layer**: Our model is a fully connected linear layer that is equal to the number of labels (developers) that were added on top of XLNet.
- **Initialization**: We used `Xavier initialization` for weight initialization of the classifier for faster convergence by keeping the variance of the gradients consistent across layers.
- **Pooling Layer**: The hidden states produced by XLNet were pooled into a mean vector, reducing the sequence output to a single representation while retaining the most relevant information for classification.

2) Freezing and Unfreezing XLNet: During initial training, XLNet layers were frozen to prevent weight updates, allowing the classifier to train efficiently. Once stable, we unfroze the layers to allow fine-tuning, improving performance.
3) Optimization Details: We trained our XLNet-based model using the following optimization settings:

- **Optimizer**: We employed the `AdamW optimizer` for training, which is well-suited for transformer-based models. The optimizer updates the model weights while applying weight decay to prevent overfitting.
  - **Learning Rate**: The learning rate was set to $2e-5$, balancing rapid convergence and stable updates.
  - **Weight Decay**: We applied a weight decay of 0.01 to regularize the model.
- **Loss Function**: We used `Binary Cross-Entropy with Logits Loss (BCEWithLogitsLoss)` for our multi-label classification task. Thus, the loss function independently evaluates the probability of each developer being the correct assignee, treating each potential developer as a binary classification task.
- **Batch Size**: We have used a batch size of 16 to balance computational efficiency and learning performance.

4) Training and Validation: We trained the modelfor 20 epochs with an 80–20 training-validation split. Early stopping was applied to prevent overfitting, and the model was checkpointed based on validation loss. Training could be resumed from the latest checkpoint if interrupted.

4.3 Evaluation Metrics

We are evaluating our model performance of Firefox and Eclipse Dataset with different features based on the below evaluation metrics.
1. Acuracy: As a percentage of all bug reports, accuracy measures how many bug reports are accurately classified. It quantifies the frequency with which the model accurately designates a bug to the relevant developer based on the input features, including commit messages, summary, and description of the bug report.

2. F-1 Score: The harmonic mean of precision and recall is computed to account for both erroneous positives and false negatives.

Our experiments are guided by three research questions.

RQ1: How Model Performs in terms of Accuracy and F-1 Score for Eclipse and Firefox Dataset Respectively when using features separately and combined without Commit Messages?

RQ2: How does the Model perform in terms of Accuracy and F-1 Score for Eclipse and Firefox Dataset, Respectively, when using features separately and combined, including Commit Messages?

RQ3: When using Firefox and Eclipse datasets, how does our model perform in terms of accuracy and F-1 score as compared to existing implemented models?

The study questions examine how different feature combinations impact a bug-triaging model's performance. We concentrate on F1 scores and Accuracy on the Eclipse and Firefox datasets. Performance in the first question is evaluated using only the Summary + Description and Summary + Commit Messages from the bug report. The second question looks at the effects of using Summary + Description + Commit messages. The final question evaluates the performance of our transformer model with similar models implemented in the past.

5 Results and Discussion

We analyzed top-1, top-3, top-5, and top-10 prediction accuracies for each feature set across the developer pool to evaluate the model's predictive capability. Top-N metrics (e.g. Top-1, Top-3, Top-5, Top-10) are widely used in bug triaging literature [5], enabling fair comparison with prior studies.

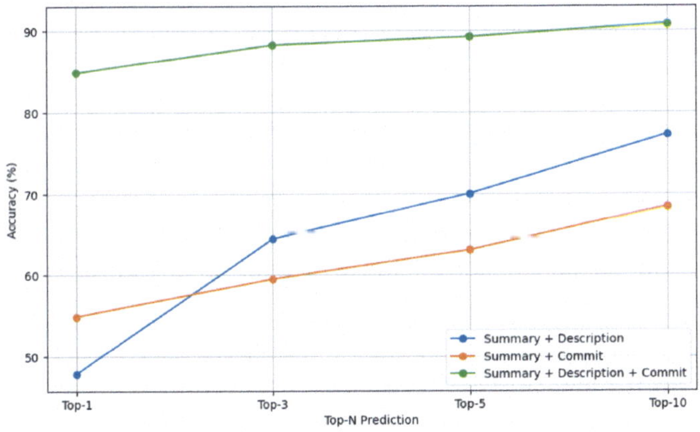

Fig. 2. Features Impact on Eclipse Triaging Result

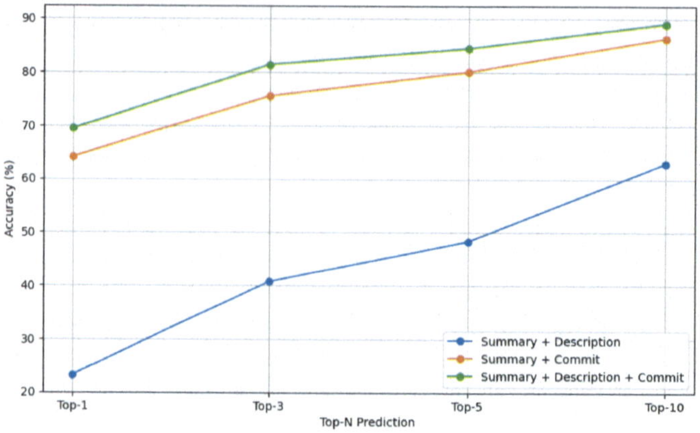

Fig. 3. Features Impact on Firefox Triaging Result

Table 3. XLNet vs. Transformer Models Accuracy on Eclipse and Firefox Datasets

Dataset	Model	Top-1	Top-5	Top-10
Eclipse	BERT	29.8%	66.3%	79.1%
	ALBERT	29.1%	67.3%	80.5%
	CodeBERT	32.3%	69.6%	81.2%
	DeBERTa	34.4%	73.1%	86.6%
	RoBERTa	32.1%	69.5%	81.6%
	XLNet	**84.88%**	**89.24%**	**90.89%**
Firefox	BERT	34.6%	46.7%	53.6%
	ALBERT	39.2%	50.9%	45.7%
	CodeBERT	33.1%	44.1%	50.6%
	DeBERTa	65.3%	79.5%	83.8%
	RoBERTa	35.4%	46.5%	52.5%
	XLNet	**69.45%**	**84.53%**	**89.14%**

RQ1: Table 2 compares the performance of Summary + Description and Summary + Commit messages in terms of F-1 score and accuracy for both data sets.

5.1 With Summary and Description

Firefox: Top-1 accuracy is 23.28%, top-10 accuracy is 23.28%, and the top-10 F-1 Score is 77.22%. This performs better over a wider range of predictions despite weaker initial accuracy, as shown in Table 2. Eclipse: The top-1 precision is 47. 89%, the top-10 precision is 77. 26%, and the top-10 F-1 score is 87. 17%. The model performs better initially and across broader predictions.

5.2 With Summary and Commit Messages

Firefox: Top-1 accuracy improves to 64.14%, top-10 accuracy is 86.41%, and the top-10 F-1 Score is 92.71%. Commit Messages to enhance precision and recall.

Eclipse: The top-1 precision is 54. 96%, the top-10 precision is 68. 46%, and the top-10 F-1 score is 81. 28%. Commit messages provide an improvement, though less pronounced. Higher top-10 F-1 Scores reflect the model's improved ability to balance precision and recall across multiple predictions. The inclusion of Commit Messages significantly enhances performance, as shown in Fig. 2 and 3.

RQ2: Table 2 shows the performance of all the combinations of features - Summary Description and Commit Messages. Seemingly, the result obtained incorporating the Commit messages outperforms the other.

Bugs are reported by end users and support engineers who do not have insights into the software systems but rather the features. Users' bug descriptions relate more to the features rather than the technicalities of the software system. On the contrary, the developers possess the ability to investigate the origin of the issue as a white-box view of the system. Traditionally, a developer commits an issue-fix with a proper message that expresses the root cause of the issue. Thus, commit messages with other bug report properties generate more likelihood of a developer relating with a new bug report. Moreover, commit messages to boost the learning for the model when the texts of summary and descriptions are insufficient. Figure 2 and 3 depict the positive impact on predictive modeling when Commit messages are included in the feature set.

Figure 2 and 3 depict the positive impact on predictive modeling when Commit messages are added to the feature set. The best overall performance yields for Firefox's Top-1 accuracy is 69. 45%, the top-10 accuracy is 89.14%, and the top-10 F-1 score is 94. 26% and for Eclipse: The top-1 precision is 84. 88%, the top-10 precision is 90. 89%, and the top-10 F-1 score is 95. 23%. The highest scores are achieved with all features combined, demonstrating significant improvements in all Top-N predictions.

RQ3: Our model outperforms current models, such as BERT, ALBERT, CodeBERT, DeBERTa, RoBERTa and XLNet [10,12,14–17] for the Eclipse dataset in all accuracy measures (Top-1, Top-5, and Top-10), as shown in Table 3. Similarly, our model outperforms other models significantly on the Firefox dataset, achieving the greatest accuracy across all criteria. Table 3 demonstrates our model's superior performance compared to existing models due to its integration of additional data sources, including system logs, developer history, and code snippets. This incorporation enhances the model's contextual understanding and significantly improves bug-triaging accuracy.

With the use of sophisticated contextual processing algorithms and sophisticated transfer learning architectures, our model grasps complicated language patterns with more accuracy and better F1 scores. These enhancements are validated by the empirical results from the Firefox and Eclipse datasets, which demonstrate our model's improved ability to efficiently and precisely categorize and prioritize issues.

6 Conclusion

In this project, we improved automated issue triaging by incorporating developer commit messages into our predictive models using the Eclipse and Firefox datasets. Utilizing the XLNet [12] model significantly enhances the performance of triaging. More specifically, for the Firefox dataset, the top-1 prediction accuracy went from 23.28% to 69.45%, and for the Eclipse dataset, it went from 70.01% to 89.28%. This large uplift shows how effective it is to include commit messages - which offer important context -in the triaging process. More insights are added to defect reports by the comprehensive commit data, which results in more precise developer assignments and fewer reassignment errors. Large-scale open-source projects like Mozilla Firefox and Eclipse, where the sheer number of bug reports and developers makes manual triaging difficult, benefit significantly from our method. Additionally, it can be optimized, and resolution times can shortened in various software development environments through automated triaging strategies.

7 Future Work

In the future, we aim to enhance our model to include data from other sources, such as system logs, developer histories, and code snippets. The subpar performance of summary and commit messages in the Eclipse dataset suggests that our first study may not be sufficient. Commit message opacity can negatively impact the quality of triaging, resulting in less accurate problem classification and prioritization.

We plan to include additional metrics such as recall, precision, and confusion matrices to provide a more comprehensive evaluation of our model's behavior. In our future work, we will include web service like AWS EC2 to handle analysis on larger datasets.

We suggest that future studies focus on creating systems to extract and use high-quality commit messages. There is potential to significantly improve bug triaging systems by looking into ways to make commit messages more helpful and thorough. A more comprehensive grasp of the development context may be possible by including real-time data from code snippets, developer histories, and system logs in the triaging framework. This would further optimize the triaging procedure. Subsequent investigations can expand on our discoveries to create more resilient and flexible bug classification systems by tackling these aspects. Providing insightful information and valuable enhancements for bug-triaging techniques will contribute to the advancement of automated code analysis and software maintenance.

References

1. Kim, S., Whitehead, E.J.: How long did it take to fix bugs? In: Proceedings of the 2006 International Workshop on Mining Software Repositories, ser. MSR 2006. Association for Computing Machinery, pp. 173–174 (2006)
2. Arcuri, A.: On the automation of fixing software bugs. In: Companion of the 30th International Conference on Software Engineering, pp. 1003–1006 (2008)
3. Anvik, J., Hiew, L., Murphy, G.C.: Who should fix this bug? In: Proceedings of the 28th International Conference on Software Engineering, pp. 361–370 (2006)
4. Anvik, J.: Automating bug report assignment. In: Proceedings of the 28th International Conference on Software Engineering, pp. 937–940 (2006)
5. Dipongkor, A.K., Moran, K.: A comparative study of transformer-based neural text representation techniques on bug triaging. In: 2023 38th IEEE/ACM International Conference on Automated Software Engineering (ASE), pp. 1012–1023. IEEE (2023)
6. Alenezi, M., Magel, K., Banitaan, S.: Efficient bug triaging using text mining. J. Softw. **8**(9), 2185–2190 (2013)
7. Bhattacharya, P., Neamtiu, I., Shelton, C.R.: Automated, highly-accurate, bug assignment using machine learning and tossing graphs. J. Syst. Softw. **85**(10), 2275–2292 (2012)
8. Aizawa, A.: An information-theoretic perspective of tf-idf measures. Inf. Process. Manage. **39**(1), 45–65 (2003)
9. Church, K.W.: Word2vec. Nat. Lang. Eng. **23**(1), 155–162 (2017)
10. Devlin, J.: Bert: pre-training of deep bidirectional transformers for language understanding, arXiv preprint arXiv:1810.04805 (2018)
11. Vaswani, A.: Attention is all you need, In: Advances in Neural Information Processing Systems (2017)
12. Yang, Z., Dai, Z., Yang, Y., Carbonell, J., Salakhutdinov, R.R., V. Le, Q.: Xlnet: Generalized autoregressive pretraining for language understanding. In: Advances in Neural Information Processing Systems, vol. 32, (2019)
13. Arabadzhieva-Kalcheva, N., Kovachev, I.: Comparison of bert and xlnet accuracy with classical methods and algorithms in text classification. In: 2021 International Conference on Biomedical Innovations and Applications (BIA), vol. 1, pp. 74–76. IEEE (2022)
14. Lan, Z.: Albert: a lite bert for self-supervised learning of language representations, arXiv preprint arXiv:1909.11942 (2019)
15. Feng, Z., et al.: Codebert: a pre-trained model for programming and natural languages, arXiv preprint arXiv:2002.08155 (2020)
16. He, P., Liu, X., Gao, J., Chen, W.: Deberta: decoding-enhanced bert with disentangled attention, arXiv preprint arXiv:2006.03654 (2020)
17. Tan, K.L., et al.: Roberta-lstm: a hybrid model for sentiment analysis with transformer and recurrent neural network. IEEE Access **10**, 21 517–21 525 (2022)
18. Guo, S., et al.: Developer activity motivated bug triaging: via convolutional neural network. Neural Process. Lett. **51**(3), 2589–2606 (2020)
19. Dong, H., Ren, H., Shi, J., Xie, Y., Hu, X.: Neighborhood contrastive learning-based graph neural network for bug triaging. Sci. Comput. Program. **C**, 103093 (2024)
20. Kukkar, A., et al.: Prore: An aco-based programmer recommendation model to precisely manage software bugs. J. King Saud Univ.-Comput. Inf. Sci. **35**(1), 483–498 (2023)

21. Zaidi, S.F.A., Awan, F.M., Lee, M., Woo, H., Lee, C.-G.: Applying convolutional neural networks with different word representation techniques to recommend bug fixers. IEEE Access **8**, 213 729–213 747 (2020)
22. Dedík, V., Rossi, B.: Automated bug triaging in an industrial context. In: 2016 42th Euromicro Conference on Software Engineering and Advanced Applications (SEAA), pp. 363–367. IEEE (2016)
23. Tamrawi, A., Nguyen, T.T., Al-Kofahi, J., Nguyen, T.N.: Fuzzy set-based automatic bug triaging (nier track). In: Proceedings of the 33rd International Conference on Software Engineering, pp. 884–887 (2011)
24. Zhang, T., Lee, B.: An automated bug triage approach: a concept profile and social network based developer recommendation. In: Intelligent Computing Technology: 8th International Conference, ICIC 2012 Huangshan, China, 25–29 July 2012, Proceedings 8. Springer, pp. 505–512 (2012)
25. Jeong, G., Kim, S., Zimmermann, T.: Improving bug triage with bug tossing graphs. In: Proceedings of the 7th Joint Meeting of the European Software Engineering Conference and the ACM SIGSOFT Symposium on The Foundations of Software Engineering, pp. 111–120 (2009)
26. LeCun, Y., Bengio, Y., Hinton, G.: Deep learning. Nature **521**(7553), 436–444 (2015)
27. O'Shea, K.: An introduction to convolutional neural networks, arXiv preprint arXiv:1511.08458 (2015)
28. Zupan, J.: Introduction to artificial neural network (ann) methods: what they are and how to use them. Acta Chim. Slov. **41**(3), 327 (1994)
29. Lee, S.-R., Heo, M.-J., Lee, C.-G., Kim, M., Jeong, G.: Applying deep learning based automatic bug triager to industrial projects. In: Proceedings of the 2017 11th Joint Meeting on Foundations of Software Engineering, pp. 926–931 (2017)
30. Bug triage datasets: Eclipse ide, eclipse core, eclipse platform, eclipse ui by eclipse foundation. Accessed 9 Sept 2024. https://bugs.eclipse.org/bugs/
31. Bug triage datasets: Mozilla firefox and mozilla core. Accessed 9 Sept 2024. https://bugzilla.mozilla.org/

CMA-ES with Radial Basis Function Surrogate for Black-Box Optimization

Farshid Farhadi Khouzani[1](✉), Abdolreza Mirzaei[2], Paul La Plante[1], and Laxmi Gewali[1]

[1] University of Nevada Las Vegas, Las Vegas, USA
farhadikhoozani@gmail.com
[2] Isfahan University of Technology, Isfahan, Iran

Abstract. Evolutionary optimization algorithms often face defects and limitations that complicate the evolution processes or even prevent them from reaching the global optimum. A notable constraint pertains to the considerable quantity of function evaluations required to achieve the intended solution. This concern assumes heightened significance when addressing costly optimization problems. However, recent research has shown that integrating machine learning methods, specifically surrogate models, with evolutionary optimization can enhance various aspects of these algorithms.

Among the evolutionary algorithms, the Covariance Matrix Adaptation Evolutionary Strategy (CMA-ES) is particularly favored. This preference is due to its use of Gaussian distribution for calculating evolution and its ability to adapt optimization parameters, which reduces the need for user intervention in adjusting initial parameters.

In this research endeavor, we propose the adoption of surrogate models within the CMA-ES framework called CMA-SAO to develop an initial surrogate model that facilitates the adaptation of optimization parameters through the acquisition of pertinent information derived from the associated surrogate model. Empirical validation reveals that CMA-SAO algorithm markedly diminishes the number of function evaluations in comparison to prevailing algorithms, thereby providing a significant enhancement in operational efficiency.

Keywords: Evolutionary Computation · Evolutionary Strategies · Optimization · Surrogate Models and Machine Learning

1 Introduction

Evolutionary algorithms are commonly used for optimization because they guide solutions to better positions through mutation and selection, adapting iteratively to find optimal outcomes. Despite their power, these algorithms can struggle when target functions or constraints are computationally expensive, as each evaluation may take significant time. In such cases, the computational load can extend processing times to hours or even days, making these algorithms less feasible for complex problems. To overcome this,

S. Latifi (Ed.): ITNG 2025, AISC 1463, pp. 367–379, 2025.
https://doi.org/10.1007/978-3-031-89063-5_31

researchers have integrated surrogate models—approximations of target functions—that reduce the need for direct, costly evaluations, allowing faster convergence and more efficient optimization.

Among evolutionary strategies, the Covariance Matrix Adaptation Evolutionary Strategy (CMA-ES) stands out in black-box optimization for its use of Gaussian distributions to sample and search the solution space. CMA-ES iteratively adjusts distribution parameters based on previous samples, refining the solution direction effectively. However, like other evolutionary algorithms, CMA-ES demands numerous function evaluations to reach optimality, which can be impractical in high-cost problems. Integrating surrogate models with CMA-ES can help minimize these evaluations, improving efficiency without compromising solution quality.

Several surrogate modeling approaches have been applied to enhance evolutionary algorithms:

Fitness Inheritance: This model allows individuals to inherit fitness scores from previous evaluations, reducing the need for repeated assessments. Introduced as a local model, fitness inheritance has evolved with variations, such as three types of inherited surrogate models to further streamline the process [1].

Gaussian Process Surrogates: These are global models that use dimensionality reduction to manage high-dimensional optimization problems [2]. However, due to the complexity of creating accurate models across dimensions, interest has grown in local models for their adaptability.

Neural Network Surrogates: The first surrogate model for evolutionary strategies used a neural network to evaluate structural optimization problems, retraining periodically with new samples around the solution mean [3].

nlmm-CMA-ES and sACM-ES Algorithms: nlmm-CMA-ES employs a quadratic local meta-model, showing 2-4x speedups for problems with quadratic landscapes. The sACM-ES algorithm, meanwhile, uses a Ranking SVM surrogate and introduces a life-length parameter to prevent premature convergence, marking a significant innovation [4].

Elite-driven Surrogate-assisted CMA-ES: It combines Gaussian Process surrogates with an adaptive ILCB method and chaotic operators, enhancing diversity and efficiency, achieving up to 7.7% performance gains in complex optimization tasks [5].

Multi Objective Surrogate-assisted KE-VFS-CMA-ES: It integrates a Knowledge-Extraction-based Variable-Fidelity Surrogate model and CMA-ES, leveraging low-fidelity data for exploration and high-fidelity data for exploitation. With an MHVI criterion and pre-screening strategy, it enhances convergence, diversity, and efficiency, excelling in benchmark tests and aerospace engineering optimization tasks [6].

This manuscript presents a new surrogate-assisted CMA-ES approach, enhancing evolutionary optimization by leveraging surrogate-derived data to guide the search process toward high-quality solutions. The paper begins with an overview of CMA-ES principles, explains the design of the new algorithm, and provides experimental results demonstrating its efficiency, concluding with a comprehensive evaluation of the algorithm's potential to advance surrogate-assisted optimization.

2 CMA-ES

CMA-ES represents a repetitive stochastic optimization methodology [7], wherein each iteration entails the sampling of a population comprising candidate solutions. In contrast to traditional population-based stochastic algorithms such as Genetic Algorithms (GAs), which delineate each step as the application of operators like mutation and crossover on the population, the CMA-ES framework is conceptualized as an evolutionary transition of probabilistic distribution, facilitating the sampling of novel individuals in every generation.

In fact, the operational protocol of CMA-ES is predicated upon the following systematic steps:

- Establish a population of λ candidate solutions.
- Compute λ candidate solutions with respect to the target function.
- Adapt the sampling distribution informed by feedback derived from the function evaluations conducted in the previous step.

3 Proposed Method

In this segment, we present our novel algorithm along with its operational mechanics; however, prior to this exposition, it is imperative to introduce the concept of model exploitation (Sect. 3.1)) as delineated by Loschilov in [4] and adaptive encoding (Sect. 3.2) as articulated by Hansen [8], given that these components are integral to our algorithmic framework.

This motivates us to define visibility aware polygonal chain approximation problem as follows.

3.1 Model Exploitation

Loshchilove introduced a surrogate-assisted Covariance Matrix Adaptation Evolution Strategy (CMA-ES) algorithm denoted as ACM-ES in [4]. A significant limitation of ACM-ES was its deficiency in surrogate control, prompting him to propose sACM-ES, which could ensure the efficacy of the surrogate model. This was achieved through the introduction of the lifelength parameter $\hat{n}$, which is adjusted in accordance with the error experienced by the surrogate $\hat{f}$ on the new test points evaluated on the function f.

He demonstrated that it is imperative to evaluate the quality of the surrogate model prior to its application. Upon the construction of the model, the quality is assessed on the $l_{test} = |\Lambda|$ points ordered subsequent to f, as measured in the following manner:

$$\mathrm{Err}\left(\hat{f}\right) = \frac{2}{|\Lambda|(|\Lambda| - 1)} \sum_{i=1}^{|\Lambda-1|} \sum_{j=i+1}^{|\Lambda|} 1_{\hat{f}_{i,j}} \tag{1}$$

For comprehensive information, please refer to [4].

3.2 Adaptive Encoding

Hansen introduced the Adaptive Encoding procedure in [8], which incrementally acquires a suitable coordinate system, demonstrating that the CMA-ES can be interpreted as a cumulative step-size algorithm executed within the modified coordinate system that is refined by the adaptive encoding procedure.

Kern et al. advanced the concept of coordinate transformation in [9], which is delineated from the existing Covariance matrix C and the prevailing mean value m:

$$x_j' = C^{-\frac{1}{2}}(x_j - m) \tag{2}$$

Loschilove evidenced in [4] that by transforming the training data into the newly established coordinate system as described by Eq. 2 and developing a surrogate model on these transformed training points, the surrogate model would exhibit an enhanced predictive capability regarding the target function. The author conducted an empirical investigation to substantiate this enhancement. We also replicate the same experimental procedure as depicted in Fig. 1.

It is evident that employing transformed data points facilitates the acquisition of more satisfactory contour representations of the surrogate model, which closely align with the contour lines of the rotated Ellipsoid function.

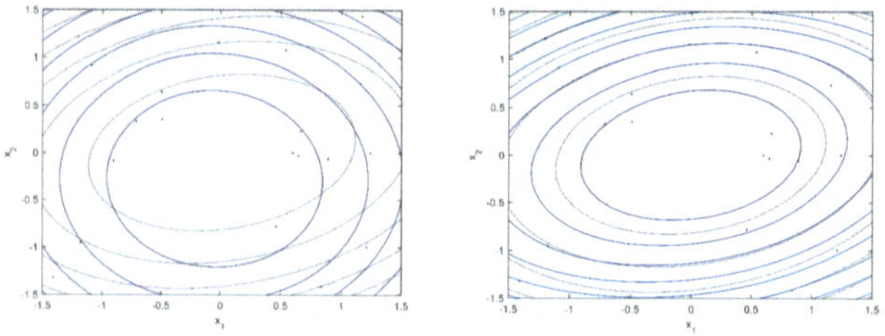

Fig. 1. Contour plots of rotated Ellipsoid function (dotted lines) and cubic RBF surrogates (solid lines) obtained with standard cubic RBF model (Left) or covariance matrix-based transformed cubic RBF model (Right). The transformed cubic RBF model is more appropriate than the standard one.

3.3 A New Surrogate-Assisted Optimization Method

In this manuscript, we present an innovative surrogate-assisted CMA-ES framework designated as CMA-SAO (CMA-ES surrogate-assisted optimization), which delineates a novel methodology aimed at maximizing the utilization of information derived from the surrogate models.

The utilization of existing data pertaining to surrogate models in order to accelerate the search for the optimal solution represents a novel concept that has garnered significant interest among contemporary researchers. This approach has the potential to decrease the requisite number of function evaluations required to attain optimality, as it facilitates the algorithm's progression toward the optimum more expeditiously.

In the context of CMA-SAO, subsequent to the development of the surrogate model $\hat{f}$, we identify the optimum of this model $\hat{f}$ within a relatively constrained region surrounding the mean of CMA-ES, followed by the evaluation of the optimum point within the target function f. Should the assessed point yield a superior target function value compared to the target function value associated with the mean of CMA-ES, it will be substituted for the optimum of CMA-ES, and the covariance matrix will be adjusted based on the evolutionary trajectory from the mean to the optimum of $\hat{f}$ as delineated in Algorithm 1.

Algorithm 1 CMA-SAO

given $g \leftarrow 0$; $Err \leftarrow 0.5$; $\mathcal{A}_g \leftarrow \phi$;
$\theta_{opt} \leftarrow$ InitializationCMA();
repeat
$\{\theta_{opt}, \mathcal{A}_{g+1}\} \leftarrow GenCMA(f, \theta_{opt}, \mathcal{A}_g\}$;
$g \leftarrow g + 1$;
until $g = g_{start}$;
repeat
$\hat{f} \leftarrow BuildSurrogateModel(\mathcal{A}_g, \theta_{opt})$;
$x_{mean} = \theta_{opt}.m$;
$x_{best} \leftarrow FindSurrogateMinimum(\hat{f}, [x_{mean} - \frac{\xi}{2}, x_{mean} + \frac{\xi}{2}]$;
if $f(x_{best}) < f(x_{mean})$ **then**
$\theta_{opt} \leftarrow$ UpdateCMAParameters(x_{best}, θ_{opt});
end if
for $i = 1, \ldots, \hat{n}$ **do**
$\{\theta_{opt}, \mathcal{A}_{g+1} = \mathcal{A}_g\} \leftarrow GenCMA(\hat{f}, \theta_{opt}, \mathcal{A}_g\}$;
$g \leftarrow g + 1$;
end for
$\{\theta_{opt}, \mathcal{A}_{g+1}\} \leftarrow GenCMA(f, \theta_{opt}, \mathcal{A}_g\}$;
$g \leftarrow g + 1$
$Err(\hat{f}) \leftarrow MeasureSurrogateError(\hat{f}, \theta_{opt})$;
$Err \leftarrow (1 - \beta_{Err})Err + \beta_{Err}Err(\hat{f})$;
$\hat{n} = \left\lfloor \frac{\tau_{err} - Err(\hat{f})}{\tau_{err}} \widehat{n_{max}} \right\rfloor$;
Until stopping criterion is met;

The CMA-SAO algorithm preserves a comprehensive global hyper-parameter vector $\theta = (\theta_{opt}, \hat{n}, \mathcal{A})$, where:

θ_{opt} denotes the optimization parameters of CMA-ES employed to ascertain the optimum of a costly target function $f(x)$.

$\hat{n}$ represents the number of generations for which the existing surrogate model is utilized.

$\mathcal{A}$ constitutes the repository of all point pairs employed for the training of the surrogate model.

In the CMA-SAO algorithm, $GenCMA(h, \theta_{opt}, \mathcal{A}_g\}$ signifies one generation of CMA-ES where h is the function designated for optimization (either the actual objective f or the surrogate $\hat{f}$), θ_{opt} denotes the current optimization parameters in CMA-ES (for instance, the CMA step-size and covariance matrix, among others) and $\mathcal{A}$ represents the repository of f.

Subsequent to each invocation of $GenCMA$, the CMA-ES parameters θ_{opt} are revised, and if the invocation was made with the true objective function, the archive $\mathcal{A}$ would consequently be updated; in other words, new points $(x, f(\mathrm{x}))$ would be incorporated into $\mathcal{A}$.

It should also be acknowledged that prior to each construction of the surrogate model, the current points x within $\mathcal{A}$ would be transformed into a new characteristic space predicated on adaptive encoding (refer to Sect. 3.2)).

During the initial phase of the CMA-SAO algorithm, θ_{opt} and $\mathcal{A}$ are regarded as the default parameters of CMA-ES and an empty set, respectively; thereafter, to establish the training set, *GenCMA* is executed with the target function f for g_{start} generations, during which all generated individuals x along with their corresponding target function values f (x) are incorporated into the training set $\mathcal{A}$.

After the initiation phase, the CMA-SAO engages in a repetitive six-step methodology:

Constructing a surrogate model $\hat{f}$ utilizing the existing archive $\mathcal{A}$ and the prevailing optimization parameters θ_{opt} (e.g., the matrix $C^{-\frac{1}{2}}$) to facilitate the transformation of training data into a novel characteristic space (procedure BuildSurrogateModel, line 9, Sect. (3.3)).

Determining the minimum of the surrogate model $\hat{f}$ within a relatively constrained domain (procedure Find Surrogate Minimum, line 10) employing the function and initializing it to x_{best}. The constrained domain for a minimization problem is defined as follows:

$$[x_{mean} - \frac{\xi}{2}, x_{mean} + \frac{\xi}{2}] \tag{3}$$

where x_{mean} denotes the current mean of the CMA-ES algorithm and ξ is defined as follows:

$$\xi = 0.1 \min_{1 \le i \le N} (b_i - a_i) \tag{4}$$

where N represents the dimensionality of the problem, b_i and a_i indicate the upper and lower bounds of the training set within the *ith* dimension.

If x_{best} exhibits superiority over the current mean of the CMA-ES, the CMA-ES parameters would be modified in accordance with Algorithm 2 (line 11–13).

Optimizing the surrogate model $\hat{f}$ for $\hat{n}$ generations. In this phase, $GenCMA(\hat{f}, \theta_{opt}, \mathcal{A}_g\}$ is consistently invoked for generations, thus θ_{opt} is revised while $\mathcal{A}$ remains static due to the absence of target function evaluations (line 14–17).

Enhancing the target function f for a singular generation. During this phase, λ individuals are assessed on f, the CMA-ES parameters θ_{opt} are revised, and the new points $(x, f$ (x)) are incorporated into $\mathcal{A}$ (line 18).

Assessing the surrogate model error $Err\left(\hat{f}\right)$ in accordance with Sect. 3.1 based on the λ currently evaluated points (line 20–21).

Modifying the lifelength $\hat{n}$ in accordance with Sect. 3.1 (line 22).

Algorithm 2 Adaptation of CMA-ES Parameters

$x_{mean-old} = x_{mean}$;

$x_{mean-new} = x_{best}$;

$P_\sigma \leftarrow (1 - c_{n\sigma})P_\sigma + \sqrt{c_{n\sigma}(2 - c_{n\sigma})\mu_{eff}}(C^{-\frac{1}{2}}(x_{mean-new} - x_{mean-old})/\sigma)$;

$P_c \leftarrow (1 - c_{cn})P_c + \sqrt{c_{cn}(2 - c_{cn})\mu_{eff}}(x_{mean-new} - x_{mean-old})/\sigma$;

$C \leftarrow (1 - c_{covn})C + c_{covn}P_cP_c^T$;

$\sigma \leftarrow \sigma\exp(\frac{c_{covn}}{d_{n\sigma}}(\frac{\|P_\sigma\|}{E\|N(0,1)\|} - 1))$;

In Algorithm 2, we modify the conjugate evolution path P_σ by incorporating the newly generated path $(x_{mean-new} - x_{mean-old})$ into it. As referenced in [7], we initialize μ_{eff} value to 1 and delineate the learning rate of the conjugate evolution path $c_{n\sigma}$ as follows:

$$c_{n\sigma} = (\mu_{eff} + 2)/(N + \mu_{eff} + 3) \tag{5}$$

where N represents the dimensionality of the problem.

The evolution path P_c is adjusted utilizing the new path $(x_{mean-new} - x_{mean-old})$. The rank-one learning rate c_{cn} is expressed as follows:

$$c_{cn} = (4 + \frac{\mu_{eff}}{N})/(N + 4 + 2\frac{\mu_{eff}}{N}) \tag{6}$$

Subsequent to the adaptation of the evolution path in the preceding step, the rank-one matrix $P_c P_c{}^T$ is integrated into the covariance matrix C, thereby facilitating its adaptation. The learning rate for the covariance matrix c_{covn} is articulated as follows:

$$c_{covn} = \frac{2}{(N + 1.3)^2 + \mu_{eff}} \tag{7}$$

The step-size σ is adjusted in accordance with the new conjugate evolution path. The $d_{n\sigma}$ parameter is represented as follows:

$$d_{n\sigma} = 1 + 2\max(0, \sqrt{\frac{\mu_{eff} - 1}{N + 1}} - 1) + c_{n\sigma} \tag{8}$$

3.4 Radial Basis Function Model

In our methodology, we implement the Radial Basis Function (RBF) model [10] due to its strong performance compared to other surrogate models like Support Vector Machine Regression and Neural Networks, and its successful application in prior optimization techniques [11, 12]. Unlike traditional RBF learning methods used in Machine Learning, our approach tailors the model specifically for optimization. We chose a cubic RBF model, as it has demonstrated success in previous studies and is considered more suitable for surrogate-based optimization than Gaussian RBF models [12, 13].

Furthermore, we conducted a comparative analysis of the cubic RBF model against thin plate spline and linear RBF models. For the purpose of this evaluation, we executed experiments on two widely recognized functions, namely the Sphere and Rosenbrock functions, utilizing the aforementioned RBF models. To evaluate the efficacy of these models, we employed Eq. 1, which was introduced in Sect. 3.1.

In our experimental framework, we constructed each model across various dimensions (2, 4, 8, 16, 32, 64, 128) utilizing a training set, and subsequently estimated the surrogate model error $Err\left(\hat{f}\right)$ based on a test set, repeating this process twenty times to plot the average errors in relation to the dimensionality of the problem.

The training and test sets are composed of uniformly distributed random points within the [−2,2] interval, which have been evaluated using the Sphere and Rosenbrock functions. The aggregate number of training points designated for the Sphere and Rosenbrock functions is equivalent to $2(d+1)$ and $5(d+1)$ respectively, with d denoting the dimensionality. The total number of test points is considered to be 10 times the total number of training points.

Figure 2 presents a graphical representation of the average evaluated errors across varying dimensions for both the Sphere and Rosenbrock functions. It is evident that for each function across all dimensions, the cubic RBF model exhibits the lowest average error when juxtaposed with the two other models.

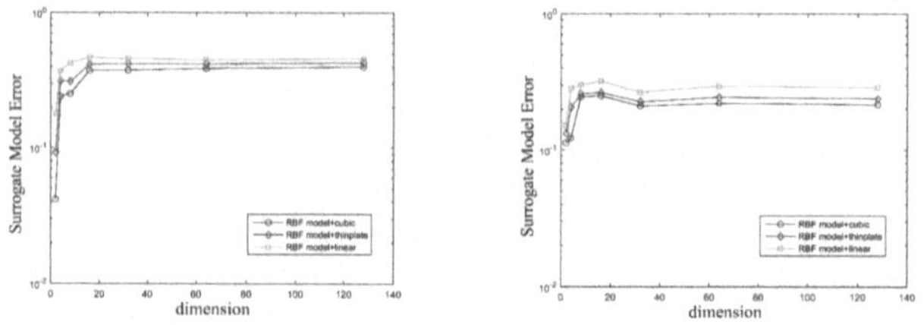

a: Sphere function b: Rosenbrock function

Fig. 2. The average errors of the surrogate model based on different functions for the RBF model concerning varying dimensions.

4 Experiments

Experiment 1: In the context of CMA-SAO, we commence g_{start} by assigning a value of 5, which corresponds to the number of evaluations for CMA-ES during the preliminary phase, and we designate the maximum lifespan of the surrogate model $\widehat{n}_{max}$ to be 20. The error threshold τ_{err} and the factor for error relaxation β_{Err} are set at 0.45 and 0.2, respectively.

We employ the fmincon function from the Matlab optimization toolbox for the purpose of optimizing the surrogate model. Following an examination of the CMA-SAO algorithm across various dimensionalities and differing sizes of training datasets, as suggested by [7], we ascertain that the most recent data points l are more advantageous for determining the number of training instances, l contingent upon the dimension n as delineated below:

$$l = 40 + 3n^{1.7} \tag{9}$$

Initially, we compare CMA-SAO with nlmm-CMA, sACM-ES, and the conventional CMA-ES across test functions (refer to Table 1), with regard to the number of function evaluations as documented in Table 2. These experiments were conducted under conditions that remain consistent for all algorithms:

- All functions considered possess an optimal value of zero.
- The results are derived from 20 independent trials. The termination criterion is the attainment of the target value 10^{-10}, with a cap placed on the number of $1000n^2$ function evaluations.

- We evaluate all algorithms across a specified dimension [22,40].

Our tables include a column *spu* designated as below:

$$spu = \frac{the\ number\ of\ function\ evalutions\ in\ CMA-ES}{the\ number\ of\ function\ evalutions\ in\ considered\ algorithm}$$

Table 1. Test Functions for experiments

Name	Function	Init	σ^0
Noisy Sphere	$f_{NoisySphere}(x) = (\sum_{i=1}^{n} x_i^2)\exp(\epsilon\mathcal{N}(0,1))$	$[-3,7]^n$	5
Ellipsoid	$f_{Elli}(x) = \sum_{i=1}^{n} 10^{6\frac{i-1}{n-1}} x_i^2$	$[1,5]^n$	2
Schwefel	$f_{Schwefel}(x) = \sum_{i=1}^{n} (\sum_{j=1}^{i} x_j)^2$	$[-10,10]^n$	10
Schwefel1/4	$f_{Schwefel^{1/4}}(x) = (f_{Schwefel}(x)^{1/4})$	$[-10,10]^n$	10
Rosenbrock	$f_{Rosenbrock}(x) = \sum_{i=1}^{n-1} (100(x_i^2 - x_{i+1})^2 + (x_i + 1)^2)$	$[-5,5]^n$	0.5
Ackley	$f_{Ackley}(x) =$ $-20\exp(-0.2\sqrt{\frac{1}{n}\sum_{i=1}^{n} x_i^2}) + \exp(\frac{1}{n}\sum_{i=1}^{n}\cos(2\pi x_i))$	$[1,30]^n$	14.5
Rastrigin	$f_{Rastrigin}(x) = 10d + \sum_{i=1}^{n} (x_i^2 - 10\cos(2\pi x_i))$	$[1,5]^n$	2
Sphere	$f_{Sphere}(x) = \sum_{i=1}^{n} x_i^2$	$[-10,10]^n$	10
Cigar	$f_{Cigar}(x) = x_1^2 + 10^6 \sum_{i=2}^{n} x_i^2$	$[-5,5]^n$	0.5
Bohachevsky	$f_{Bohachevsky}(x) =$ $\sum_{i=1}^{n-1} (x_i^2 + 2x_{i+1}^2 - 0.3\cos(3\pi x_i) - 0.4\cos(4\pi x_{i+1}) + .07)$	$[1,15]^n$	7

An examination of Table 2 reveals that CMA-SAO exhibits significantly superior performance in terms of reducing the number of function evaluations when juxtaposed with two alternative algorithms, particularly in the context of high-dimensional objective functions. It is also pertinent to note that (n)lmm-CMA demonstrates enhanced efficacy exclusively on functions characterized by quadratic landscapes. Furthermore, it is essential to highlight that CMA-SAO performs commendably on multimodal functions such as f_{Ackley}, $f_{Rastrigin}$, and $f_{Bohachevsky}$. This observation underscores the robustness of CMA-SAO when confronted with multimodal challenges.

In Experiment 2, Loschilove conducted investigations aimed at estimating the empirical complexity associated with the training and testing of Ranking SVM, utilizing training points $100\sqrt{n}$, ceasing after $50000\sqrt{n}$ iterations, and evaluating the surrogate model on 500 test points (refer to Fig. 3-Left). We replicated this experiment employing identical values for both training and test points to facilitate a comparative analysis of the complexities inherent to these two surrogate models (refer to Fig. 3-Right).

For both models, the empirical complexity concerning dimension n is slightly super-linear; however, it is evident that the cubic RBF model necessitates a shorter learning time in comparison to Ranking SVM. Specifically, for dimensions ranging from 100 to 200, the learning time for the cubic RBF model remains below 1 s, whereas this metric for Ranking SVM exceeds 2 s.

Table 2. The mean frequency of function evaluations subsequent to twenty iterations of speedup performance (spu) for CMA-SAO, sACM-ES, (n)lmm-CMA, and CMA-ES is analyzed. The finding presented in the (n)lmm-CMA section represents the optimal values derived from the sources [5] (designated with a preliminary "n:" for the latter). The variable ϵ represents the intensity of noise (when applicable)

Function	*n*	λ	ϵ	*CMA-SAO*	*Spu*	*sACM-ES*	*spu*	*(n)lmm-CMA*	*spu*	*CMA-ES*
$f_{schwefel}$	2	6		86	4.3	187	1.9	**81**	4.5	370
	4	8		147	5.9	305	2.8	**145**	6.0	879
	8	10		366	5.4	650	3.0	**282**	7.1	2010
	16	11		1117	4.6	1823	2.8	**627**	8.2	5156
	20	12		**1635**	4.3	3500	2.0			7042
	32	14		**3741**	4.0	9689	1.5			15072
	40	15		**5985**	3.7	16771	1.3			22400
$f_{schwefel^{1/4}}$	2	6		**398**	3.8	532	2.8	n:413	3.7	1527
	4	8		**835**	3.4	889	3.2	n:971	2.9	2847
	8	10		1959	3.0	**18525**	3.1			5882
	16	11		**4604**	2.6	4733	2.6			12411
	20	12		**6353**	2.4	7761	2.0			15600
	32	14		**12349**	2.3	18635	1.6			29378
	40	15		**20550**	2.0	27365	1.5			41534
$f_{Rosenbrock}$	2	6		480	1.4	508	1.4	**n:252**	2.8	700
	4	8		779	2.8	826	2.6	**n:719**	3.0	2187
	8	10		2168	2.6	**1802**	3.2	2494	2.3	5769
	16	11		5618	2.9	**5417**	3.0	7299	2.2	16317
	20	12		7673	2.8	**7218**	3.0			21749
	32	14		**20781**	2.5	24326	2.1			52671
	40	15		**32102**	2.5	41003	2.0			82043
$f_{NoisySphere}$	2	6	.35	214	1.9	419	-	**n:109**	3.7	407
	4	8	.25	327	2.5	864	-	**n:236**	3.6	844
	8	10	.18	**587**	2.8	1232	1.3	n:636	2.6	1663
	16	11	.13	**935**	3.3	2089	1.5	n:2156	1.4	3120

(continued)

Table 2. *(continued)*

Function	n	λ	ϵ	*CMA-SAO*	*Spu*	*sACM-ES*	*spu*	*(n)lmm-CMA*	*spu*	*CMA-ES*
	20	12	.11	**1215**	3.1	2753	1.4			3777
	32	14	.09	**1616**	3.5	4607	1.2			5767
	40	15	.08	**2173**	3.2	6217	1.1			7023
f_{Ackley}	2	6		**197**	3.7	269	2.7	n:227	3.2	735
	4	8		**407**	3.8	421	3.7			1577
	8	10		**689**	4.4	1141	2.6			3066
	16	11		**1145**	4.9	1451	3.9			5672
	20	12		**1497**	4.4	3299	2.0	8150	0.8	6641
	32	14		**2521**	3.9	5921	1.7			10063
	40	15		**3533**	3.4	8456	1.4			12084
f_{Elli}	2	6		**174**	4.4	349	2.2			774
	4	8		**391**	4.3	569	2.9			1688
	8	10		**1045**	4.3	1121	4.0			4542
	16	11		3688	3.5	**2351**	5.6			13177
	20	12		5311	3.5	**4653**	4.0			19060
	32	14		**12015**	3.7	14304	3.1			44562
	40	15		**19201**	3.6	22101	3.1			69642
$f_{Rastrigin}$	2	50		621	3.1	1402	1.4	**n:528**	3.6	1970
	5	140		**3783**	3.2	5121	2.4	n:4037	3.0	12310
	10	500		**27105**	1.4	32501	1.2	29250	1.3	40152
f_{Sphere}	2	6		**81**	4.1	167	2			334
	4	8		**129**	5.9	284	2.7			770
	8	10		**251**	6.2	589	2.7			1579
	16	11		**557**	5.2	1196	2.4			2942
	20	12		**733**	4.8	1790	2.0			3542
	32	14		**1786**	3.0	3212	1.6			5378
	40	15		**1986**	3.3	4247	1.5			6722
f_{Cigar}	2	6		**198**	4.1	274	2.9			818
	4	8		**373**	4.5	470	3.6			1704
	8	10		**796**	4.7	1138	3.3			3766
	16	11		**1511**	4.8	3089	2.3			7343
	20	12		**1761**	5.1	4262	2.1			8989

(continued)

Table 2. (*continued*)

Function	n	λ	ϵ	*CMA-SAO*	*Spu*	*sACM-ES*	*spu*	*(n)lmm-CMA*	*spu*	*CMA-ES*
	32	14		**3373**	4.0	8290	1.6			13660
	40	15		**3703**	4.6	12237	1.3			17039
$f_{Bohachevsky}$	2	6		**133**	3.1	260	1.5			415
	4	8		**415**	3.1	1082	1.2			1301
	8	10		**675**	5.1	2192	1.5			3451
	16	11		**1450**	6.2	4251	2.1			9001
	20	12		**2282**	5.5	6282	2.0			12734
	32	14		**4832**	4.5	8812	2.4			21762
	40	15		**5383**	5.6	14102	2.1			30302

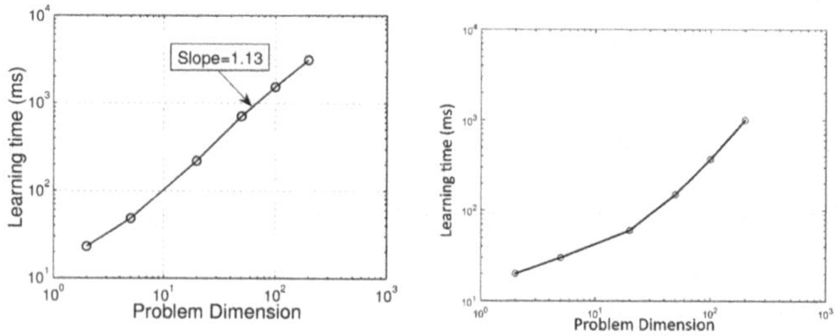

Fig. 3. The cost of model learning/testing w.r.t problem dimension for Ranking SVM model (Left), cubic RBF model (Right).

5 Summary and Conclusion

We have introduced a novel algorithm for a surrogate-assisted Covariance Matrix Adaptation Evolution Strategy (CMA-ES), designated as CMA-SAO, with the objective of minimizing the number of function evaluations required for computationally intensive black-box optimization tasks.

The CMA-SAO algorithm employs an innovative strategy to maximize the informational yield from the surrogate model, distinguishing it from existing surrogate-assisted CMA-ES methodologies documented in the academic literature, and rendering it particularly adept for addressing high-dimensional optimization challenges. The approach adopted by CMA-SAO aims to enhance the mean position of the CMA-ES by refining the surrogate model within a relatively confined region surrounding the CMA-ES mean; subsequent to the identification of a promising point, the parameters of CMA-ES are iteratively updated to expedite the convergence toward the optimal solution of the optimization problem. Furthermore, our implementation of CMA-SAO incorporates a cubic Radial Basis Function (RBF) model. Empirical findings across ten benchmark functions demonstrate a significant advancement over the traditional CMA-ES, outperforming two additional surrogate-assisted CMA-ES algorithms, namely sACM-ES and (n)lmm-CMA.

References

1. Fonseca, L.G., Lemonge, A.C.C., Barbosa, H.J.C.: A study on fitness inheritance for enhanced efficiency in real-coded genetic algorithms. IEEE Cong. Evol. Comput. **2012**, 1–8 (2012)
2. Liu, B., Zhang, Q., Gielen, G.G.E.: A Gaussian process surrogate model assisted evolutionary algorithm for medium scale expensive optimization problems. IEEE Trans. Evol. Comput. **18**(2), 180–192 (2014)
3. Papadrakakis, M., Lagaros, N.D., Tsompanakis, Y.: Structural optimization using evolution strategies and neural networks. Comput. Methods Appl. Mech. Eng. **156**(1), 309–333 (1998)
4. Loshchilov, I., Schoenauer, M., Sebag, M.: Self-adaptive surrogate-assisted covariance matrix adaptation evolution strategy. In: Proceedings of the 14th Annual Conference on Genetic and Evolutionary Computation, Philadelphia, Pennsylvania, USA (2012). https://doi.org/10.1145/2330163.2330210
5. Li, Z., Gao, T., Tian, K., Wang, B.: Elite-driven surrogate-assisted CMA-ES algorithm by improved lower confidence bound method. Eng. Comput. **39**(1), 2543–2563 (2023)
6. Li, Z., Tian, K., Zhang, S., Wang, B.: Efficient multi-objective CMA-ES algorithm assisted by knowledge-extraction-based variable-fidelity surrogate model. Chin. J. Aeronaut. **36**(6), 213–232 (2023). https://doi.org/10.1016/j.cja.2022.09.020
7. Hansen, Ostermeier, A: Completely derandomized self-adaptation in evolution strategies. Evol. Comput. **9**(2), 159–195 (2001)
8. Hansen, N.: Adaptive encoding: how to render search coordinate system invariant. In: Parallel Problem Solving from Nature – PPSN X, pp. 205–214. Springe, Berlin, Heidelberg (2008)
9. Kern, S., Hansen, N., Koumoutsakos, P.: Fast quadratic local meta-models for evolutionary optimization of anguilliform swimmers (2004)
10. Powell, M.J.D.: The theory of radial basis function approximation in 1990. In: Advances in Numerical Analysis, pp. 105–210 (1992)
11. Gutmann, H.M.: A radial basis function method for global optimization. J. Global Optim. **19**(3), 201–227 (2001)
12. Regis, R.G., Shoemaker, C.A.: Local function approximation in evolutionary algorithms for the optimization of costly functions. IEEE Trans. Evol. Comput. **8**(5), 490–505 (2004)
13. Wild, S.M., Shoemaker, C.: Global convergence of radial basis function trust region derivative-free algorithms. SIAM J. Optim. **21**(3), 761–781 (2011)

Predictive Modeling of Crime Trends Using Time-Series Analysis

Opeoluwa Williams and Natarajan Meghanathan(✉)

Department of Electrical and Computer Engineering and Computer Science, Jackson State University, Jackson, MS, USA
opeoluwa.d.williams@students.jsums.edu,
natarajan.meghanathan@jsums.edu

Abstract. The goal of this research is to explore and analyze historical crime data in order to develop predictive models that can forecast future crime trends. By leveraging time-series analysis techniques, this study aims to uncover temporal patterns and identify key trends in criminal activities. Specifically, we use a variant (SARIMAX) of the well-known ARIMA (AutoRegressive Integrated Moving Average) model for time-series forecasting and analyze the prediction accuracy. The crime dataset (for the city of Chicago) used for this project contains detailed records of criminal incidents, including variables such as time, location, and crime type. The SARIMAX model helps to identify trends, seasonality, and periodic fluctuations in crime rates, while other machine learning techniques could be employed to optimize predictions based on additional factors such as geographic hotspots and socio-temporal variables. The relative error in the prediction is observed to be 2.88% to predict the number of crimes in a month for the city of Chicago whose average monthly crime rate is 28,677.

Index Terms: Time-series analysis · Predictive modeling · Crime data · ARIMA · Temporal pattern · Crime prediction · Machine learning · Parameter analysis · Univariate data · Seasonality

1 Introduction

In urban environments, predicting crime trends is a critical challenge for law enforcement agencies tasked with maintaining public safety. Criminal activities often exhibit complex patterns that evolve over time and space, influenced by various socio-economic and environmental factors [1]. Traditional crime analysis methods are reactive, focusing on responding to crimes after they occur rather than predicting and preventing them [2]. The inability to accurately forecast crime trends hinders proactive resource allocation, efficient law enforcement, and strategic urban planning. We address the problem of predicting crime trends through the use of historical crime data. By leveraging time-series analysis, this study aims to provide law enforcement agencies with the ability to forecast when and where crimes are likely to occur [3, 4]. Identifying these trends is essential for enhancing urban safety, preventing crime before it happens, and optimizing the deployment of resources (like police officers, surveillance systems, etc.) in high-risk areas [5].

S. Latifi (Ed.): ITNG 2025, AISC 1463, pp. 380–388, 2025.
https://doi.org/10.1007/978-3-031-89063-5_32

Predictive modeling has the potential to revolutionize the way law enforcement agencies operate by shifting from reactive crime response strategies to proactive crime prevention efforts. By analyzing historical crime data, predictive models can identify recurring patterns and forecast future criminal activities [6]. This allows for the strategic deployment of police forces and other resources, optimizing patrol routes and schedules, and reducing response times to high-risk areas [7].

The significance of this study lies in its ability to provide actionable insights that enhance crime prevention and resource allocation [2, 3]. Law enforcement agencies often face budget constraints and personnel limitations, necessitating efficient resource allocation. Predictive modeling offers a data-driven solution to these challenges by identifying patterns in crime occurrences and suggesting the best strategies for intervention [8]. Predictive insights can inform policy-making, allowing urban planners and local governments to implement long-term safety initiatives aimed at reducing crime rates in most vulnerable communities [5].

Machine learning and time-series analysis represent cutting-edge approaches to solving the challenges associated with predicting crime trends [9]. Traditional crime analysis tools are often limited by their inability to account for complex temporal relationships within large datasets. In contrast, machine learning algorithms are capable of analyzing vast amounts of data and identifying subtle patterns that may be missed by traditional methods [4, 10]. Time-series analysis has become a valuable tool in crime prediction, allowing researchers and practitioners to model temporal dependencies and forecast future crime occurrences based on historical patterns. Through time-series models, patterns like seasonality, trends, and occasional irregularities can be analyzed to make informed predictions about crime rates.

Time-series analysis models such as ARIMA (AutoRegressive Integrated Moving Average) offer a robust framework for understanding the temporal aspects of crime data [6]. The ARIMA (Auto-Regressive Integrated Moving Average) model is one of the most widely applied time-series models in crime forecasting due to its strength in modeling linear patterns and capturing short-term dependencies. ARIMA excels in handling data with strong temporal correlations, making it well-suited for crime data with consistent seasonal trends. Studies demonstrate that ARIMA can successfully model crime rates by using historical crime counts and seasonal variations. For example, higher crime rates during summer or on certain holidays can be effectively anticipated by ARIMA models [4, 12]. However, ARIMA's effectiveness diminishes when applied to highly non-linear data, as it requires the data to be stationary and assumes a linear relationship between time points. Some researchers have argued that while ARIMA performs well with univariate data, it lacks flexibility for multivariate analyses where multiple variables (like weather, population density, or economic indicators) affect crime rates [7]. To address these limitations, other methods, such as Seasonal ARIMA (SARIMAX), incorporate seasonality adjustments that allow for periodicity, though it still encounters challenges with non-linear trends.

In this paper, we explore the use of SARIMAX on more than 20 years of crime dataset for the City of Chicago to evaluate its ability to capture seasonality, trends, and fluctuations over time, providing accurate forecasts for future crime occurrences [7]. This study aims to offer an advanced, comprehensive solution to predicting crime trends, providing law enforcement with powerful tools for crime prevention and resource optimization [11]. The rest of the paper is organized as follows: Sect. 2 reviews related literature on time-series analysis for crime forecasting and highlights the gaps in the research in this topic. Section 3 presents the data pre-processing steps used in this research. Section 4 presents and discusses the results of time-series analysis of the crime dataset for the city of Chicago using SARIMAX. Section 5 concludes the paper and outlines plans for future work.

2 Literature Review

Time-series analysis has become a valuable tool in crime prediction, allowing researchers and practitioners to model temporal dependencies and forecast future crime occurrences based on historical patterns. Through time-series models, patterns like seasonality, trends, and occasional irregularities can be analyzed to make informed predictions about crime rates. Techniques such as ARIMA, Prophet [13] and Exponential Smoothing [14] have been widely applied in crime forecasting due to their ability to capture trends, seasonality, and fluctuations over time [3][4]. Each of these time-series models offers distinct advantages depending on the characteristics of the crime data and the forecasting horizon. ARIMA is best suited for linear, short-term patterns with strong seasonality, while Prophet provides flexibility for handling irregular seasonal changes and long-term trends. Emerging studies have explored combinations of these models to leverage their individual strengths. For example, hybrid models that combine ARIMA with machine learning algorithms such as LSTM (Long Short-Term Memory) networks have been developed to improve the accuracy of crime predictions [4]. LSTM, a type of recurrent neural network, excels in handling sequential data and capturing non-linear patterns, making it highly compatible with traditional time-series methods in hybrid applications [8].

Time-series forecasting has found substantial applications in urban planning and public safety management. Beyond predicting monthly or yearly crime counts, it is used to guide strategic resource allocation, optimize police patrol schedules, and design intervention programs. For instance, many urban areas have adopted ARIMA-based forecasting to deploy law enforcement resources to high-risk areas during predicted peak crime periods. Similarly, the flexibility of Prophet to adjust for holidays and unusual events provides actionable insights for event-based policing [10].

While time-series forecasting has shown considerable promise in crime prediction, there remain several gaps in existing research. Many studies have focused primarily on predicting crime frequency over time but have paid less attention to the spatial distribution of crime, such as identifying hotspots or regional variations within a city [6]. Integrating spatial analysis with time-series forecasting could offer a more holistic view of crime trends, especially when combined with GIS (Geographic Information System) tools. Moreover, few studies have examined the impact of socio-economic variables (e.g., unemployment rates, income levels) or external events (e.g., policy changes, economic downturns) in time-series crime forecasting models. Incorporating these factors could improve the model's robustness and enhance interpretability, addressing the root causes of crime trends rather than solely predicting occurrences.

3 Data Pre-processing

Our original dataset is the Chicago crime dataset from January 2001 to September 2024, and can be accessed at https://data.cityofchicago.org/Public-Safety/Crimes-2001-to-Present/ijzp-q8t2/data_preview. The dataset contains 8.18 million rows and 22 columns, recording critical information about reported crimes, including details such as the date and time, crime type, and location. The "Date" column records the exact date and time of each crime. For our time-series analysis, we will use this column to aggregate the number of crimes per month from January 2001 to September 2024. This aggregation serves as the foundation for building our forecasting models and identifying long-term crime trends. Given that the *Date* column holds crucial information about when each crime occurred, it was converted into 'datetime' format to allow seamless time-based operations such as extracting months and years. This transformation was critical, as it allowed us to focus our analysis on the temporal dimension of crime patterns. From the *Date* column, we extracted a 'YearMonth' feature, reducing the granularity of the data from individual crime incidents to monthly crime counts. This step ensured that each row now represented the total number of crimes for a given month, making it easier to detect long-term trends and seasonal patterns. After the necessary filtering and aggregation, the cleaned data was saved as 'crimes.csv,' ensuring that it was ready for use in the next stages of the project. This structured dataset, now organized by monthly crime counts (see Fig. 1 for a snapshot), lays a solid foundation for building time-series forecasting models. For data preprocessing, we used Pandas, a powerful Python library widely used for data manipulation and analysis. With Pandas, we can clean, transform, aggregate, and export data, ensuring it is ready for further analysis using ARIMA and Prophet. Any bias in the dataset is fixed using the SMOTE technique [15].

	YearMonth	Crimes
0	2001-01	38122
1	2001-02	33788
2	2001-03	40571
3	2001-04	40094
4	2001-05	41843
..	...	...
280	2024-05	22841
281	2024-06	22971
282	2024-07	23808
283	2024-08	22555
284	2024-09	22186

[285 rows x 2 columns]

Fig. 1. A snapshot of the crimes.csv file

We then perform Exploratory Data Analysis (EDA) on the Chicago crime dataset to gain insights into its temporal characteristics. We will use essential libraries like *Matplotlib*, and *Statsmodels* to carry out this analysis, focusing specifically on ETS (Error-Trend-Seasonality) decomposition to understand the underlying trend, seasonality, and residuals in the data. The ETS decomposition technique splits the time-series data into three components: (1) *Trend*: The long-term movement in the data shows whether crime rates are increasing or decreasing over time. (2) *Seasonality*: The recurring patterns or cycles in the data which can indicate periodic spikes in crime during specific times of the year, like holidays or weekends. (3) *Residuals*: The random noise or irregular component left after removing the trend and seasonal patterns. Using the *Statsmodels* library, we applied multiplicative ETS decomposition to the crime dataset. Multiplicative decomposition assumes that the seasonal variation depends on the trend, which makes it suitable for data that shows changing seasonal effects over time, as is typically with crime data. Figure 2 presents the Trend, Seasonality and Residuals plot for the Chicago dataset.

The trend plot shows a general decline in crime rates from 2001 to around 2016, with a slight uptick post-2018. This suggests that overall crime rates have decreased over the years, but more recent data shows a potential increase. The seasonal plot shows a clear recurring pattern with peaks and troughs over the year. Crime tends to spike during certain months, likely summer months, and decline during others, reflecting seasonal variations in criminal activity. The residual plot shows the random noise remaining after removing trend and seasonality. The randomness in the residuals suggests that other unknown factors may influence crime rates, or it could indicate areas for further investigation to improve the model's accuracy.

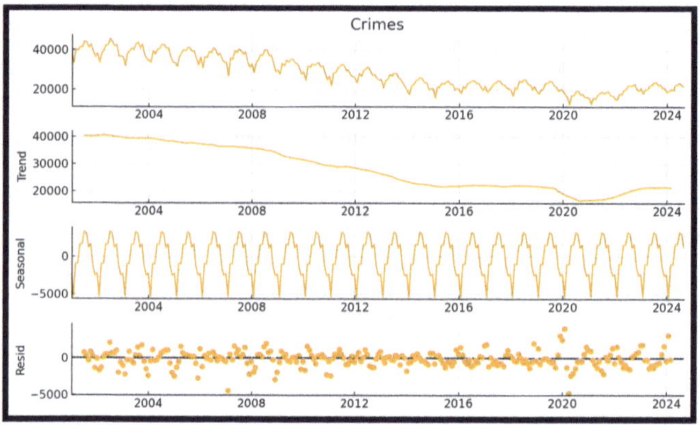

Fig. 2. Trend, seasonality and residuals plots for the Chicago crime dataset

4 Time-Series Forecasting with ARIMA

We use a variant of the ARIMA model, called the Seasonal ARIMA with Exogenous Variables (SARIMAX) for predicting crime trends. In the context of crime data, exogenous variables might include factors like weather conditions or economic indicators, but for our study, we'll focus on the time-series components alone. SARIMAX is a powerful extension of the ARIMA model that accounts for seasonality and allows for the inclusion of exogenous variables, which can improve prediction accuracy by incorporating external factors influencing crime rates. SARIMAX is specified by the following parameters: (1) p, d, q: representing respectively the autoregressive (the number of previous observations to include along with the current observation), differencing (the number of times the data needs to be differenced to remain constant over time), and moving average orders (the number of lagged residuals to include) for the non-seasonal part; (2) *P*, *D*, *Q*: The seasonal counterparts of *p*, *d*, *q*, representing the seasonal autoregressive, seasonal differencing, and seasonal moving average orders. (3) *m*: The length of the seasonal cycle. For our crime data, since we are dealing with monthly data, the seasonal period is 12 months. The SARIMAX model allows us to capture both the short-term dependencies and seasonal patterns in the crime data. To accurately do this we will perform parameter analysis using essential libraries like *pmdarima* (helps in parameter tuning as well as in performing stepwise searches to minimize AIC: Akaike Information Criterion) and *warnings* (to filter and suppress harmless warnings).

```
Best model:  ARIMA(1,1,1)(2,1,1)[12]
Total fit time: 26.781 seconds
```

SARIMAX Results

Dep. Variable:	y	**No. Observations:**	285
Model:	SARIMAX(1, 1, 1)x(2, 1, 1, 12)	**Log Likelihood**	-2328.312
Date:	Mon, 14 Oct 2024	**AIC**	4668.624
Time:	11:32:17	**BIC**	4690.258
Sample:	01-01-2001	**HQIC**	4677.309
	- 09-01-2024		
Covariance Type:	opg		

	coef	**std err**	**z**	**P>|z|**	**[0.025**	**0.975]**
ar.L1	0.3507	0.138	2.537	0.011	0.080	0.622
ma.L1	-0.5468	0.131	-4.170	0.000	-0.804	-0.290
ar.S.L12	0.2345	0.050	4.716	0.000	0.137	0.332
ar.S.L24	-0.1158	0.021	-5.400	0.000	-0.158	-0.074
ma.S.L12	-0.6338	0.060	-10.611	0.000	-0.751	-0.517
sigma2	1.433e+06	9.82e+04	14.584	0.000	1.24e+06	1.63e+06

Ljung-Box (L1) (Q):	5.36	**Jarque-Bera (JB):**	65.64
Prob(Q):	0.02	**Prob(JB):**	0.00
Heteroskedasticity (H):	0.61	**Skew:**	-0.70
Prob(H) (two-sided):	0.02	**Kurtosis:**	4.96

Fig. 3. Results of parameter analysis

The auto_arima() function is used to determine the best parameters for the ARIMA model. We specify a few key arguments: (1) start_$p = 1$, start_$q = 1$, max_$p = 3$, max_$q = 3$; (2) $m = 12$; (3) start_$P = 0$, seasonal = True (allows the model to account for seasonal patterns by adding the seasonal ARIMA component); (4) trace = True (allows us to see the stepwise search for the best model configuration); (5) error_action = 'ignore', suppress_warnings = True: (to ensure that any failed configurations or minor issues are handled without interrupting the fitting process) and (6) stepwise = True (to enable stepwise fitting, optimizing the model selection process by trying combinations of parameters iteratively). After fitting the ARMIA model, the summary() function outputs the best configuration of the model based on AIC and other evaluation metrics.

The ARIMA(1,1,1)(2,1,1) [12] model was selected (see Fig. 3) as the optimal model based on the stepwise search process. This means that the model includes: (1) $p = 1$: An autoregressive component using one lag; (2) $d = 1$: First-order differencing to ensure stationarity; (3) $q = 1$: A moving average component with one lag; (4) $P = 2$, $D = 1$, $Q = 1$: These represent the seasonal autoregressive, differencing, and moving average components, with a seasonal period of 12 months (capturing annual seasonality). The AIC (Akaike Information Criterion) value for the best model is 4668.624, indicating a well-fitting model. A lower AIC suggests a model that explains the data well with fewer parameters. From the SARIMAX results image, the coefficients of the AR and MA components are statistically significant (p-values < 0.05), which indicates that these components contribute to the model's predictive power. The results of the Ljung-Box test (p-value of 0.02) suggest that there may still be some autocorrelation in the residuals, indicating room for further model refinement or improvement.

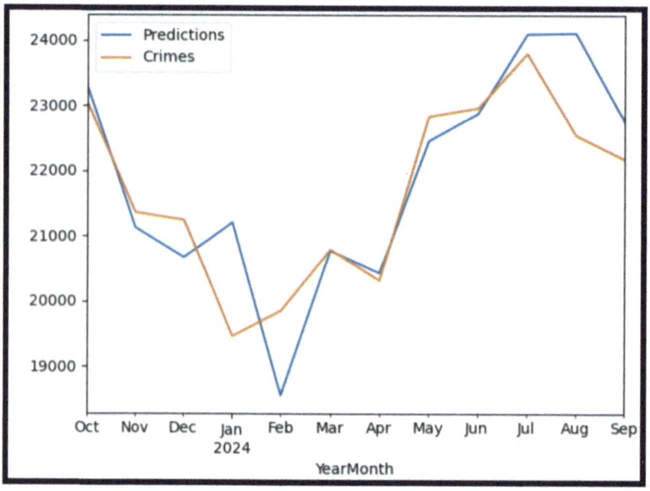

Fig. 4. Predicted crime count vs. Actual crime count for one year of test data

We then fit the SARIMAX model to our Chicago crime dataset based on the results of our parameter analysis, make predictions, and evaluate the model's performance using appropriate error metrics. The following libraries will be used: statsmodels (provides the SARIMAX function) and sklearn.metrics (to compute the Root Mean Squared Error: RMSE of the predictions). The dataset is then split into train and test sets, with the last 12 months (1 year) reserved for testing the model. The train set is used to fit the SARIMAX model, and the test set is used to evaluate the model's predictive performance. Figure 4 presents the crime predictions (blue) vs. actual number of crimes (orange) for months ranging from October 2023 to September 2024. There are some discrepancies during certain periods, such as late winter and early spring, where the model slightly underestimates crime rates. However, overall, the model captures the seasonality and general trends effectively, especially during the summer months when crime spikes are more pronounced. The RMSE value is 826.67, implying that on average, the model's predictions deviate from actual crime counts by about 827 crimes per month. With a mean monthly crime count of 28,677, the relative error (827/28,677) of approximately 2.88% of the typical monthly crime count is a reasonable margin for error given the real-world complexity and variability of crime data.

5 Conclusion and Future Work

Given the scale of the crime data - where monthly crime counts range between 12,938 and 46,014 - and the standard deviation of 8,276, an RMSE of 826.67 indicates that the SARIMAX model performs well in predicting future crime trends. The 2.88% error margin is considered reasonable for time-series forecasting, especially when dealing with real-world crime data, which is inherently noisy and difficult to predict with complete accuracy. The model also revealed a strong seasonal component, with crime rates showing distinct patterns over time. The analysis also highlighted a downward trend in crime from 2001 to 2021, followed by a period of stabilization. Understanding these patterns can help law enforcement agencies optimize resource allocation during periods of increased crime activity. The research results offer significant implications for crime prevention, resource allocation, and policy-making. Predictive insights generated by these models can enable law enforcement agencies to anticipate crime surges and optimize the deployment of resources, such as adjusting patrol schedules during high-risk periods or focusing attention on emerging crime hotspots. T results of this study can assist policymakers in creating data-driven strategies to improve urban safety and reduce criminal activities. As part of future work, we plan to conduct a comparative time-series analysis of the crime datasets with two other models such as Prophet [13] and Exponential Smoothing [14] and evaluate the prediction accuracy of these models with that of SARIMAX. Also, there is room to apply more advanced machine learning techniques such as LSTMs (Long Short-Term Memory Networks) or GRUs (Gated Recurrent Units) that could improve accuracy by capturing more complex temporal dependencies in the data. Hybrid models that combine traditional statistical methods with machine learning approaches may also enhance performance.

Acknowledgement. Research was sponsored by the Army Research Office and was accomplished under Grant Number W911NF-21-10264. The views and conclusions contained in this document are those of the authors and should not be interpreted as representing the official policies, either expressed or implied, of the Army Research Office or the U.S. Government. The U.S. Government is authorized to reproduce and distribute reprints for Government purposes notwithstanding any copyright notation herein.

References

1. Yuki, J., Md. Sakib, M.Q., Zamal, Z., Habibullah, K.M., Das, A.K.: Predicting crime using time and location data. In: Proceedings of the 7th International Conference on Computer and Communications Management, pp. 124–128. Bangkok, Thailand (2019)

2. Kamalanabhan, E., Shylaja, J., Navinesha, M., Kaviya, S.: Chicago crime: a novel perspective on forecasting and evaluation. Int. J. Emerg. Technol. Innov. Res. **11**(5), 96–103 (2024)
3. Palanivinayagam, A., Gopal, S.S., Bhattacharya, S., Anumbe, N., Ibeke, E., Biamba, C.: An optimized machine learning and big data approach to crime detection. Wirel. Commun. Mob. Comput. **2021**(5291528), 1–10 (2021)
4. Vimala Devi, J., Kavitha, K.S.: Automating time series forecasting on crime data using RNN-LSTM. Int. J. Adv. Comput. Sci. Appl. **12**(10), 458–463 (2021)
5. Jha, N., Paranjape, V., Sharma, S., Hasan, Z.: Machine learning for crime analysis and prediction. Int. J. Recent Dev. Eng. Technol. **11**(9), 35–40 (2022)
6. Shah, N., Bhagat, N., Shah, M.: Crime forecasting: a machine learning and computer vision approach to crime prediction and prevention. Visual Comput. Ind. Biomed. Art **4**(9), 1–14 (2021)
7. Nenga, K., Catal, C., Kar, K.: Machine learning in crime prediction. J. Ambient. Intell. Humaniz. Comput. **14**, 2887–2913 (2023)
8. Sinha, B.B., Biswas, T.: An efficient framework for forecasting of crime trend using machine learning technique. In: Proceedings of International Conference on Data Science and Applications, LNNS, vol. 552, pp. 741–755 (2023)
9. Kalyan Sagar, M., Gopala Krishna, P.: Crime data analysis using machine learning. RB J. Libr. Inf. Sci. **12**(12), 44–51 (2022)
10. Aggarwal, K., et al.: Implementing machine learning algorithms on criminal databases to develop a criminal activity index. J. Emerg. Invest. **6**(1), 1–5 (2023)
11. Dakalbab, F., Talib, M.A., Waranga, O.A., Nassif, A.B., Abbas, S., Nasir, A.: Artificial intelligence & crime prediction: a systematic literature review. Soc. Sci. Hum. Open **6**(1), 100342, 1–23 (2022)
12. Borowik, G., Zbigniew, M.W., Cichosz, P.: Time series analysis for crime forecasting. In: Proceedings of the 26th International Conference on Systems Engineering, pp. 1–10, Sydney, Australia (2018)
13. Prophet, https://facebook.github.io/prophet/
14. Holt, C.C.: Forecasting trends and seasonal by exponentially weighted averages. Int. J. Forecast. **20**(1), 5–10 (2004)
15. Chawla, N.V., Bowyer, K.W., Hall, L.O., Kegelmeyer, W.P.: SMOTE: synthetic minority over-sampling technique. J. Artif. Intell. Res.Intell. Res. **16**(1), 321–357 (2002)

aSFD: An Autoencoder-Based Streaming Fraud Detection in Supervised Learning

Luis Olmos and Rashida Hasan(✉)

Department of Computer Science, California State University, Northridge, Northridge, CA, USA
luis.olmos.563@my.csun.edu, rashida.hasan@csun.edu

Abstract. The increasing volume of financial transactions and the dynamic nature of fraud patterns pose significant challenges for real-time fraud detection. Traditional methods often struggle to detect anomalies in streaming data due to their static nature and inability to adapt to evolving patterns. To address these challenges, we propose a novel hybrid model, 'aSFD' (Autoencoder-based Streaming Fraud Detection), which integrates an autoencoder for latent feature extraction with an ExtraTrees classifier, specifically tailored for streaming learning and real-time performance. The autoencoder is trained on non-fraudulent data to effectively identify outliers in real time, enhancing fraud detection capabilities. A sliding window mechanism is employed to enable continuous updates, ensuring adaptive decision-making and efficient management of concept drift, which distinguishes our model from static alternatives. Our main contribution is a hybrid model combining an autoencoder and ExtraTrees classifier, specifically adapted for streaming data environments with a sliding window mechanism to handle concept drift effectively. Experimental evaluations on the Kaggle and PaySim fraud datasets showed improved performance, with precision, recall, and F1-scores of 0.95, 0.88, 0.92 (Kaggle) and 0.93, 0.80, 0.86 (PaySim), surpassing state-of-the-art algorithms.

Keywords: Fraud detection · Credit card fraud · Anomaly detection · Streaming fraud detection · Autoencoder · Concept drift · Online detection · Extra tree classifier · Supervised Learning · Data stream mining

1 Introduction

In the rapidly evolving landscape of online transactions, they have become an integral part of daily life for individuals and businesses worldwide. While this growth brings convenience and efficiency, it also invites a surge in fraudulent activities, such as unauthorized access, identity theft, phishing, and illicit money transfers [4]. Fraudsters continuously adapt their tactics, leveraging advanced

R. Hasan—These authors contributed equally to this work.

S. Latifi (Ed.): ITNG 2025, AISC 1463, pp. 389–400, 2025.
https://doi.org/10.1007/978-3-031-89063-5_33

technologies and exploiting vulnerabilities in digital platforms to bypass security measures [7]. These fraudulent activities not only lead to significant financial losses—costing businesses and consumers billions annually—but also erode customer trust and compromise the integrity of financial systems. As a result, enhancing fraud detection mechanisms is a critical imperative to safeguard assets, maintain customer confidence, and ensure the stability of the financial ecosystem [4].

Traditional fraud detection methods, such as decision trees, random forests, and neural networks, work well on static datasets but struggle with the evolving nature of fraud in real-time systems. Their reliance on periodic retraining delays responses to new fraud patterns, increasing the risk of undetected fraudulent transactions [3,11]. While batch models like random forests and boosting have been explored [13], they are not suitable for real-time detection. To address this, researchers have explored streaming methods. Carcillo et al. introduced an incremental learning framework that updates models with new transactions, reducing retraining needs, but scalability remains a challenge [2].

To tackle these issues, we proposed a novel hybrid algorithm that adapts a traditionally batch-oriented autoencoder and ExtraTrees classifier for real-time fraud detection on streaming data. The autoencoder, trained on non-fraudulent data, identifies anomalies by extracting key features, while the ExtraTrees classifier leverages these features to detect fraud in real-time. The method integrates continuous updates via a sliding window mechanism, allowing the model to adapt to evolving fraud patterns without the need for frequent retraining.

Our contributions are summarized as follows:

- **Algorithm Design:** Development of a hybrid model combining autoencoder feature extraction with ExtraTrees classification for real-time fraud detection.
- **Streaming Data Handling:** Adapting both models for streaming environments with continuous updates and minimal retraining.
- **Concept Drift Handling:** Incorporating a sliding window mechanism to adapt to evolving fraud patterns and concept drift.

The rest of the paper is structured as follows: Sect. 2 reviews related work on fraud detection, focusing on the limitations of existing models in streaming environments. Section 3 details our methodology, including the modifications made to the autoencoder and ExtraTrees classifier for streaming data. Section 4 presents experimental results, evaluating the performance of our approach on both datasets and Sect. 5 concludes with future research directions.

2 Related Work

As digital payments grow, real-time fraud detection is critical to mitigate evolving fraud tactics. Traditional machine learning models, such as support vector machines (SVMs) and decision trees, have been widely used but suffer from limitations like reliance on static datasets and periodic retraining, which slow adaptation to new fraud patterns. Hussain et al. demonstrated the effectiveness of SVMs and decision trees but highlighted the challenge of frequent retraining for dynamic fraud environments [6]. Dal Pozzolo et al. addressed class imbalance in fraud detection using undersampling and SMOTE, though their methods remained suited to batch processing environments [13].

Ensemble methods like Random Forests and ExtraTrees have gained popularity for handling high-dimensional, imbalanced data. Geurts et al. introduced ExtraTrees [5], which has outperformed simpler models in complex fraud scenarios, but these models also rely on static data and regular updates. Deep learning models, like Recurrent Neural Networks (RNNs) and Long Short-Term Memory (LSTM) networks, capture temporal patterns but face scalability challenges in real-time environments due to high computational costs [1,8,16].

To address the need for real-time detection, Carcillo et al. proposed incremental learning to enable continuous updates without frequent retraining, but scalability challenges remained [2]. Rajeshwari and Babu introduced a sliding window approach for near real-time fraud detection, though its effectiveness was constrained by window size [15].

More recently, Lin and Jiang applied the AERFAD framework, integrating autoencoders with Random Forest for fraud detection [12]. While accurate, Random Forest's computational demands make it unsuitable for real-time use. In contrast, our model leverages the ExtraTrees classifier, offering comparable accuracy with greater computational efficiency, making it better suited for real-time fraud detection where rapid decision-making is crucial.

Our research advances real-time fraud detection by developing a hybrid model that combines autoencoders for anomaly detection with an ExtraTrees classifier for fraud classification, both optimized for streaming data. Utilizing a sliding window approach, our model continuously learns from incoming transactions, improving detection accuracy while maintaining scalability. By integrating autoencoders with ExtraTrees, the model effectively handles class imbalance, reducing false positives and ensuring real-time detection. Furthermore, our model addresses the computational challenges of deploying deep learning in real-time by implementing an optimized retraining mechanism. This allows the model to adapt to evolving fraud patterns efficiently, making it suitable for high-volume financial institutions.

3 Methods

This section details the approach we used to develop a real-time fraud detection model capable of handling large-scale streaming data. Our model was designed with a focus on computational efficiency, adaptability, and robustness, to ensure high performance across multiple fraud datasets without being overly complex or specific to a particular dataset. Figure 1 shows the general framework of our proposed model for streaming fraud detection.

3.1 Autoencoder Architecture and Pretraining

The autoencoder consists of an encoder that compresses input data into a latent representation and a decoder that reconstructs the input from this compressed form. The encoder progressively reduces the data's dimensionality, from 100 units to 50 units, and finally to 10 units in the bottleneck layer. The decoder mirrors this architecture. The latent features generated by the autoencoder provide a low-dimensional representation of the input data, capturing its most essential characteristics.

Although the reconstruction error

$$L_{\text{recon}}(X_i) = \left\|X_i - \hat{X}_i\right\|^2 \tag{1}$$

is commonly used for anomaly detection, we found that leveraging latent features for downstream classification yielded more robust and consistent performance. These latent features encapsulated richer, more informative aspects of the data. To promote the learning of a sparse and meaningful representation, we incorporated L_1 regularization with a penalty term of 10^{-5}. The model was trained for 50 epochs using 0.001 learning rate of the Adam optimizer, which further enhanced stability during training.

Our experiments showed that relying solely on reconstruction error did not produce optimal results. In contrast, the latent features from the bottleneck layer were particularly effective at capturing complex relationships in the data, significantly improving classification performance when combined with the ExtraTrees algorithm.

Once trained, the autoencoder is used in a simulated online setting for real-time fraud detection. To ensure adaptability to evolving data distributions, we propose a novel mechanism for incrementally updating the encoder's weights without full retraining. This is particularly useful in scenarios where the characteristics of non-fraudulent transactions shift over time, a phenomenon known as concept drift.

Instead of retraining the autoencoder entirely, we update the encoder incrementally using a sliding window $\mathcal{W}(t)$ that captures the most recent non-fraudulent observations at time t. The encoder's parameters θ are updated as follows:

$$\theta \leftarrow \theta - \eta \nabla_\theta L_{\text{auto}}(\theta; \mathcal{W}(t)), \tag{2}$$

where η is the learning rate, and $\mathcal{W}(t)$ represents the data within the sliding window. This approach ensures that the autoencoder remains up-to-date and adapts to changes in the data without the computational overhead of full retraining. Crucially, this incremental update focuses on recent trends in non-fraudulent transactions, maintaining efficiency in real-time operations.

3.2 ExtraTrees Classifier for Latent Feature Classification

The autoencoder's latent features are classified using ExtraTrees, an ensemble method that balances accuracy and efficiency. ExtraTrees is well-suited to handling imbalanced datasets and high-dimensional data, making it effective for fraud detection.

In our model, the ExtraTrees classifier is initialized with 200 decision trees, each of which is trained on a random subset of both features and samples. This feature randomization ensures that the classifier is less prone to overfitting, while the ensemble of trees provides a strong generalization capability. To balance the class distributions, we set 'class_weight='balanced", which automatically adjusts the weights of the classes based on their frequencies.

The prediction for a new instance Z_j is computed by averaging the predictions across all trees in the ensemble:

$$\hat{y}_j = \frac{1}{M} \sum_{m=1}^{M} T_m(Z_j) \tag{3}$$

where M is the number of trees, and $T_m(Z_j)$ is the prediction of the m-th tree for instance Z_j.

The adaptability of ExtraTrees, combined with its ability to handle class imbalance through weighted sampling, made it the best option for our fraud detection task. Furthermore, since ensemble methods typically offer better generalization, using ExtraTrees allowed us to deploy a model capable of detecting subtle fraud patterns in real-time.

3.3 Streaming Adaptation via Sliding Window and Concept Drift Handling

The novel adaptation of the ExtraTrees classifier for online learning involves periodic updates to the decision boundaries of the trees. Rather than fully retraining the model with each batch of new data, we selectively update a subset of the trees with recent transactions while retaining older trees to preserve knowledge of historical patterns. This technique is essential in combating catastrophic forgetting, where the model would otherwise forget earlier learned patterns when exposed to new data.

The sliding window mechanism also plays a key role in handling concept drift, which refers to shifts in the underlying data distribution over time. Our model continuously monitors model performance and if performance degrades beyond a predefined threshold τ, we assume that concept drift has occurred and trigger a model update.

During retraining, a window $\mathcal{W}(t)$ of size w is used to capture the most recent observations. The model is then optimized using the combined loss function:

$$L_{\text{total}}(\theta, \alpha) = \lambda_1 L_{\text{auto}}(\theta) + \lambda_2 L_{\text{clf}}(\alpha) \tag{4}$$

where $L_{\text{clf}}(\alpha)$ represents the classification loss of the ExtraTrees classifier, and λ_1, λ_2 are hyperparameters controlling the relative importance of the autoencoder and the classifier. This dynamic balance between the autoencoder and the classifier ensures that the model remains both adaptive and computationally efficient. The details of the algorithm are presented in Algorithm 1

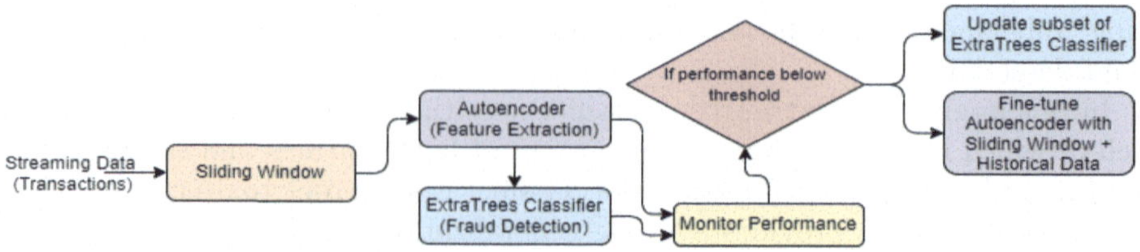

Fig. 1. Framework of the Proposed Algorithm aSFD

Algorithm 1. Sliding Window for Concept Drift Handling and Model Update

1: **Input:** Sliding window $\mathcal{W}(t)$ of size w, Autoencoder A, ExtraTrees Classifier C, Performance threshold τ, Metrics (precision, recall, F1-score), Hyperparameters λ_1, λ_2
2: **Output:** Updated classifier C, updated autoencoder A, drift_detected
3: **while** new data arrives in $\mathcal{W}(t)$ **do**
4: Monitor model performance using precision, recall, and F1-score
5: **if** performance degrades below threshold τ **then**
6: Detect concept drift
7: Update a subset of trees in C using recent data from $\mathcal{W}(t)$
8: Incrementally update autoencoder A with a combined dataset of recent data from $\mathcal{W}(t)$ and a subset of historical data
9: Use a gradual fine-tuning approach for A to retain prior knowledge while adapting to new patterns
10: Optimize the model using the combined loss function as defined in Eq. 1
11: Retain older trees in C to preserve historical knowledge
12: **end if**
13: **end while**
14: **Return** Updated classifier C, updated autoencoder A, drift_detected

3.4 Streaming Simulation and Model Evaluation

To simulate real-time fraud detection, we divided the test data into batches of 1000 transactions, each representing a stream of financial transactions. For each batch, the model extracted latent features using the pretrained autoencoder, then passed these features to the ExtraTrees classifier for prediction.

After processing each batch, the model's performance was evaluated and monitored. If a significant drop in performance was detected, the model was retrained using a combination of recent and historical data to ensure adaptability without sacrificing memory retention.

4 Results

This section presents an analysis of the performance of our model using the Kaggle Credit Card Fraud and Paysim datasets. We provide detailed evaluations of the model's precision, recall, and F1-score, comparing these metrics with those of state-of-the-art methods in the literature. Furthermore, we examine the time complexity of our model, illustrating its ability to balance computational efficiency with effective fraud detection performance in real-time environments.

Algorithm 2. Fraud Detection with Autoencoder and ExtraTrees Classifier

---

Input: X_{train}, y_{train}, X_{stream}, Autoencoder A, ExtraTrees Classifier C, threshold τ, sliding window $\mathcal{W}(t)$
Output: y_{pred}, drift_detected
Offline Pretraining:
Train Autoencoder A on non-fraudulent transactions from X_{train}
Extract latent features Z_{train} from X_{train}
Train ExtraTrees classifier C using Z_{train} and y_{train}
while streaming data X_{stream} **do**
    Extract latent features Z_{batch} from X_{stream} using autoencoder A
    Predict fraud using classifier C: $y_{\text{pred}} = C(Z_{\text{batch}})$
    Monitor model performance (precision, recall, F1-score)
    if performance drops below threshold τ **then**
        Detect concept drift
        Apply the concept drift handling and model update as described in **Algorithm 1**
    end if
end while
Return y_{pred}, drift_detected

4.1 Datasets

We evaluated our fraud detection methods using two datasets: the PaySim mobile money fraud simulation and the Kaggle Credit Card Fraud dataset [9,10]. PaySim simulates mobile money transactions based on real-world logs from an African service, containing around 24 million records with various transaction types (CASH-IN, CASH-OUT, etc.) [10]. Key challenges in using this dataset is that certain balance-related features must not be used for fraud detection, as these values are altered when fraud is detected and transactions are canceled, it is also highly imbalanced with fraudulent transactions making up less than 1% of the total data.

The Kaggle Credit Card Fraud dataset consists of 284,807 European cardholder transactions collected over two days, with only 492 fraudulent cases (0.172% of the total) [9]. Most features have been anonymized via Principal Component Analysis (PCA), except for 'Time' and 'Amount'.

Both datasets present challenges related to class imbalance and feature limitations, making them ideal for testing fraud detection methods across various financial services.

4.2 Performance Analysis on Kaggle and PaySim Datasets

Our hybrid model, aSFD, was tested on both the Kaggle Credit Card Fraud dataset and the PaySim dataset. The performance metrics for the minority class (fraudulent transactions) are displayed in Table 1.

In both datasets, the majority class (non-fraudulent transactions) achieved a perfect F1-score of 1.00, indicating that the model was able to classify all non-fraudulent transactions correctly.

Table 1 compares the performance metrics for the minority class (fraudulent transactions) across the two datasets. The model achieves slightly higher precision and recall on the Kaggle dataset (0.95 and 0.88) compared to PaySim (0.93 and 0.80). These results indicate that the model effectively minimizes false positives while maintaining high detection rates, especially in the Kaggle dataset.

Table 1. Comparison of Precision, Recall, and F1-score for the Minority Class (Fraudulent) on Kaggle and PaySim Datasets.

Metric	Kaggle Dataset	PaySim Dataset
Precision	0.95	0.93
Recall	0.88	0.80
F1-score	0.92	0.86

Our model was pre-trained offline using historical data to avoid the cold start problem, which typically arises in streaming data environments. This step ensured that the Autoencoder and ExtraTrees classifier started with sufficient information to make accurate predictions from the onset of real-time data flow.

4.3 Performance Comparison: Our Proposed Model aSFD vs. State-of-the-Art Models

We compared our model against several state-of-the-art online learning approaches in fraud detection. These methods vary in complexity, ranging from deep learning architectures to traditional machine learning techniques.

Both the T2 Control Chart and OCSVM methods reported perfect performance, but these results were obtained using a balanced test set, which does not reflect the imbalanced nature of fraud detection datasets [14]. In contrast, our model operates in an imbalanced setting, making our metrics more representative of real-world scenarios.

The graphs present the precision, recall, and F1-score of various fraud detection models tested on the Kaggle Credit Card Fraud dataset. As shown in Fig. 2, traditional models like Linear SVM and Logistic Regression struggle with low precision (0.139 and 0.190), indicating high false positive rates [1]. Deep Neural Networks and Non-linear Auto Regression also perform poorly, with precision values below 0.20 [1]. In contrast, our hybrid model (Autoencoder-ExtraTrees)

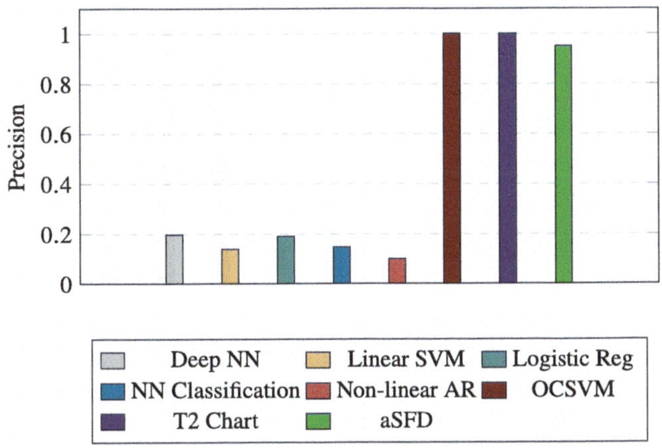

Fig. 2. Precision for Various Fraud Detection Models

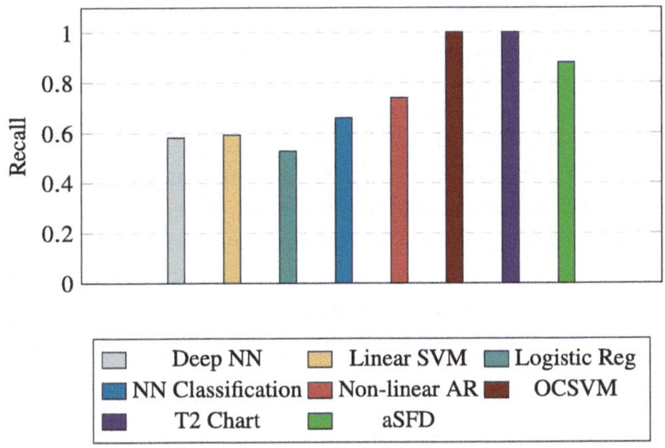

Fig. 3. Recall for Various Fraud Detection Models Tested on Kaggle Dataset

achieved much higher precision, as detailed in the Results section. In terms of recall (Fig. 3), models like Non-linear AR report high recall (0.740), but at the cost of poor precision, indicating a high rate of false positives. Our approach provides a better balance between recall and precision, which is crucial for real-time fraud detection.

Our approach outperforms 5 out of 7 models, providing a better balance between recall and precision, which is crucial for real-time fraud detection. Figure 4 shows that most models fail to deliver strong F1-scores, with almost none exceeding 0.30, while our hybrid model consistently outperforms them with much stronger F1-scores, highlighting its ability to manage the trade-off between precision and recall.

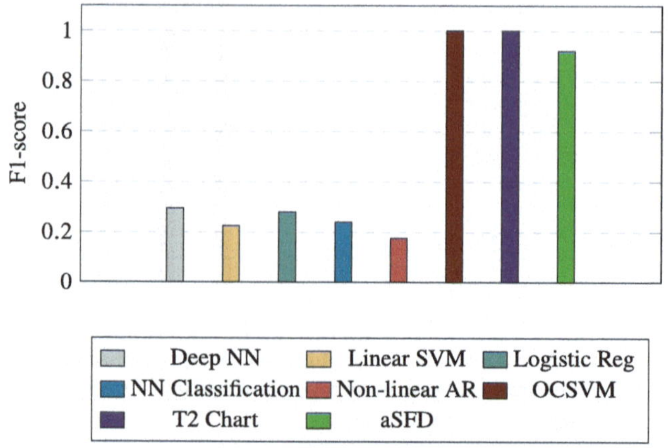

Fig. 4. F1-score for Various Fraud Detection Models Tested on Kaggle Dataset

4.4 Time Complexity Analysis

The overall time complexity of our model is determined by the combined operations of the Autoencoder and the ExtraTrees classifier. The Autoencoder's complexity depends on the number of layers and neurons, which is roughly $O(n)$ where n is the number of transactions processed. The ExtraTrees Classifier has a complexity of $O(T \log n)$, where T is the number of trees and n is the number of samples.

The final time complexity for training the model offline is approximately $O(n + T \log n)$, which is highly efficient compared to more complex deep learning models like CNNs or RNNs. During real-time processing, the Autoencoder extracts latent features with constant time complexity per batch, while the ExtraTrees classifier processes these features in $O(\log n)$ time for classification.

Our proposed algorithm aSFD is designed to handle streaming data efficiently. By leveraging the Autoencoder for feature extraction and the ExtraTrees classifier for rapid decision-making, the overall inference time per transaction remains low, enabling the algorithm to operate effectively in a high-frequency environment without significant delays.

5 Conclusion

We developed a hybrid model for real-time fraud detection that integrates an autoencoder for anomaly detection and an ExtraTrees classifier for classification, optimized for streaming data. We addressed the challenge of evolving fraud patterns by employing a sliding window mechanism, enabling continuous learning and efficient adaptation without full retraining. Our proposed model handled imbalanced datasets by using the autoencoder to capture key transaction features and the sliding window to mitigate catastrophic forgetting for continuous adaptation Our model aSFD demonstrated superior performance on the Kaggle Credit Card Fraud and PaySim datasets, particularly excelling in detecting minority-class fraud.

The sliding window's sensitivity to size highlights the need for further refinement, and developing an adaptive sliding window will be a focus of future work to better handle evolving data patterns. Additionally, future research will explore advanced deep learning models, like transformers, to enhance system adaptability.

References

1. Abakarim, Y., Lahby, M., Attioui, A.: An efficient real time model for credit card fraud detection based on deep learning. In: Proceedings of 12th International Conference on Intelligent Systems: Theories and Applications (SITA 2018), pp. Art. no. 30, 1–7. Rabat, Morocco (2018). https://doi.org/10.1145/3289402.3289530
2. Carcillo, F., Pozzolo, A.D., Borgne, Y., Caelen, O., Mazzer, Y., Bontempi, G.: Scarff: a scalable framework for streaming credit card fraud detection with spark. Inf. Fusion **41**, 182–194 (2018). https://doi.org/10.1016/j.inffus.2017.09.005
3. Effectiv: Predictive analytics for fraud detection. https://effectiv.ai/resources/predictive-analytics-for-fraud-detection/
4. Fortinet: what is internet fraud? types of internet fraud. https://www.fortinet.com/resources/cyberglossary/internet-fraud
5. Geurts, P., Ernst, D., Wehenkel, L.: Extremely randomized trees. Mach. Learn. **63**(1), 3–42 (2006). https://doi.org/10.1007/s10994-006-6226-1
6. Hussain, S.K.S., Reddy, E.S.C., Akshay, K.G., Akanksha, T.: Fraud detection in credit card transactions using svm and random forest algorithms. In: 2021 Fifth International Conference on I-SMAC, pp. 1013–1017. Palladam, India (2021). https://doi.org/10.1109/I-SMAC52330.2021.9640631
7. Identity, P.: Online fraud: defined, types, methods & prevention. https://www.pingidentity.com/en/resources/blog/post/everything-need-know-online-fraud.html
8. Jurgovsky, J., et al.: Sequence classification for credit-card fraud detection. Expert Syst. Appl. **100**, 234–245 (2018). https://doi.org/10.1016/j.eswa.2018.01.037
9. Kaggle: credit card fraud detection dataset. https://www.kaggle.com/datasets/mlg-ulb/creditcardfraud. Accessed: [Insert Date]
10. Kaggle: Paysim: a financial mobile money simulator for fraud detection. https://www.kaggle.com/datasets/ealaxi/paysim1. Accessed: [Insert Date]
11. Lebichot, B., Verhelst, T., Borgne, Y., He-Guelton, L., Oblé, F., Bontempi, G.: Transfer learning strategies for credit card fraud detection. IEEE Access **9**, 114754–114766 (2021). https://doi.org/10.1109/ACCESS.2021.3104472
12. Lin, T.H., Jiang, J.R.: Anomaly detection with autoencoder and random forest. In: Proceedings of 2020 International Computer Symposium (ICS), pp. 96–99 (2020). https://doi.org/10.1145/3097983.3098052
13. Pozzolo, A.D., Boracchi, G., Caelen, O., Alippi, C., Bontempi, G.: Credit card fraud detection: a realistic modeling and a novel learning strategy. IEEE Trans. Neural Netw. Learn. Syst. **29**(8), 3784–3797 (2018). https://doi.org/10.1109/TNNLS.2017.2736643

14. Tran, P.H., Tran, K.P., Huong, T.T., Heuchenne, C., Tran, P.H., Huong, T.M.: Real-time data-driven approaches for credit card fraud detection. In: Proc. 2018 International Conference on E-Business and Applications, pp. 6–9. Da Nang, Viet Nam (2018). https://doi.org/10.1145/3194188.3194196
15. U, R., Babu, B.S.: Real-time credit card fraud detection using streaming analytics. In: 2016 2nd International Conference on Applied and Theoretical Computing and Communication Technology, pp. 439–444. Bangalore, India (2016). https://doi.org/10.1109/ICATCCT.2016.7912039
16. Zhou, C., Paffenroth, R.C.: Anomaly detection with robust deep autoencoders. In: Proceedings of the 23rd ACM SIGKDD International Conference on Knowledge Discovery and Data Mining, pp. 665–674. ACM, New York, USA (2017). https://doi.org/10.1145/3097983.3098052

ML Classification of Cancer Types Using High Dimensional Gene Expression Microarray Data

Dwaipayan Mukhopadhyay[1(✉)], Rohan J. Dalpatadu[2], Laxmi P. Gewali[3], and Ashok Singh[4]

[1] Department of Epidemiology and Biostatistics, University of Nevada Las Vegas, Las Vegas, USA
dwaipayan.mukhopadhyay@unlv.edu

[2] Department of Mathematical Sciences, University of Nevada Las Vegas, Las Vegas, USA
rohan.dalapatadu@unlv.edu

[3] Department of Computer Science, University of Nevada Las Vegas, Las Vegas, USA
laxmi.gewali@unlv.edu

[4] Department of Resorts Gaming and Golf Management, University of Nevada Las Vegas, Las Vegas, USA
ashok.singh@unlv.edu

Abstract. Cancer is a disease caused by the abnormal growth of cells in different parts of the body and is one of the top causes of death globally. Microarray gene expression data plays a critical role in the identification and classification of cancer tissues. Due to recent advancements in Machine Learning (ML) techniques, researchers are analyzing gene expression data using a variety of such techniques to model the progression rate & treatment of cancer patients with great effect. But high dimensionality alongside the presence of highly correlated columns in gene expression datasets leads to computational difficulties. This paper aims to propose the use of ML classification techniques- Linear Discriminant Analysis (LDA) & Random Forest (RF) for classifying five types of cancer (breast cancer, kidney cancer, colon cancer, lung cancer and prostate cancer) based on high dimensional microarray gene expression data. Principal component analysis (PCA) was used for dimensionality reduction, and principal component scores of the raw data for classification. Six distinct categorization performance measures were used to evaluate these approaches; RF method provided us with higher accuracy than LDA method. The method and results of this article should be helpful to researchers who are dealing with many genes in microarray data.

Keywords: Principal Components Analysis · Linear Discriminant Analysis · Random Forest · Precision · Recall · F1 · AUC · macro-averaged AUC · micro-averaged AUC

1 Introduction

Cancer is a disease which can start almost anywhere in the human body, in which some of the body's trillion cells grow uncontrollably and spread to other parts of the body. There are over 200 types of cancer such as colon, liver, ovarian and breast etc. [1,

S. Latifi (Ed.): ITNG 2025, AISC 1463, pp. 401–407, 2025.
https://doi.org/10.1007/978-3-031-89063-5_34

2]. In 2023, 1,958,310 new cancer cases and 609,820 cancer deaths were projected in the United States. [3]. This prompts a clear understanding of the underlying mechanism and characteristics of this potentially fatal disease alongside identifying the most significant genes responsible for it. Microarray data analysis has been a popular approach for diagnosing cancer, and DNA microarray is a technology used to collect data on large numbers of various gene expressions at the same time [4–6]. Analysis of gene expression data involves the identification of informative genes, [7] and [8] demonstrate that cancer classification can be improved by identifying informative genes which in turn can be used to accurately predict new sample classes.

Machine learning (ML) is a branch of artificial intelligence (AI) that enables computers to "self-learn"; develop their own predictions, improving over time without being explicitly programmed [9]. ML techniques are highly suited for analyzing massive volume of data, being generated in the medical industry [10, 11]. However, there are several difficulties like (a) Many highly correlated genes for just a small sample size. (b) Various uncertainties for example, fabrication, image processing etc., resulting in unexplained fluctuation in the data. [12].

The earliest detection of cancer is among the most efficient approaches to reduce cancer-related death [13]. Due to overestimation & various linearity issues it is difficult to categorize high-dimensional microarray data ($p > n$) using statistical approaches [14, 15]. Various supervised and unsupervised ML techniques have also been adopted to identify the most significant genes [16, 17]. In microarray gene expression analysis, gene selection or feature selection (FS) is utilized to improve cancer classification performance by countering the curse of dimensionality by identifying the most informative genes to enhance disease prediction accuracy [18].

The remainder of this paper is structured in the following manner. Section 2 presents the materials and methods. In Sect. 3, we present the experimental results. Finally, in Sect. 4, we provide a conclusion. Respecting the rules, we provide a very brief version of this paper. Detailed works are provided in the full version.

2 Material and Methods

In this article, we have used the Linear Discriminant Analysis (LDA) classifier [19] and the random forest (RF) classifier [20] on an 801 rows x 20531 columns (genes) dataset of patients with five cancer types: BRCA, KIRC, COAD, LUAD and PRAD; the dataset has no missing values. Variables in this dataset are RNA-Seq gene expression levels measured by **illumina HiSeq** platform. The variables are dummy named gene XX. This dataset (gene expression cancer RNA-Seq) was downloaded from the UCI Machine Learning Repository [21]. The statistical software package R (2023) was used for all data analyses and visualizations [22].

A dataset of 801 rows and 25 PC-scores was created, and then this dataset of PC-scores was randomly split into an 80% training set and 20% test set. All accuracy measures are computed from the confusion matrix which is a cross-tabulation of observed Y and predicted Y values. The following accuracy measures are computed for each level by calculating the one vs all binary confusion matrices: Area Under the Curve (AUC) & Macro- and micro-averages of AUC.

3 Results

PCA was run on the entire 801 rows × 20531 genes data set, and trial-and-error showed that just the first two principal components were sufficient for classification purposes. The genes with highest absolute loadings are shown in Fig. 6.

A scatterplot of the first two PC-scores for the entire dataset is shown in Fig. 1.

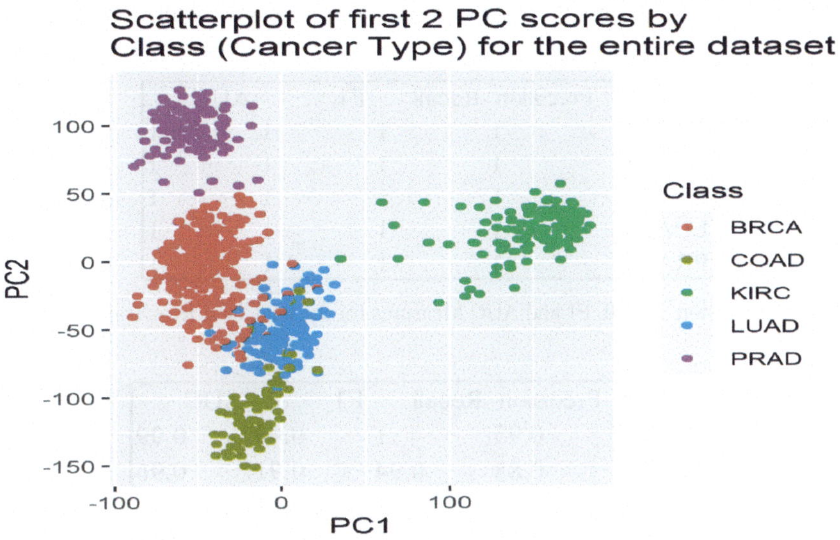

Fig. 1. Scatterplot of PC2 vs PC1 of the Entire Dataset

A clear separation between BRCA and KIRC cancer sub-types with some overlap between COAD, LUAD and PRAD is seen in Fig. 1.

Figures 2, 3, 4, 5 and 6 show that all measures of multi-level accuracy are high for both training and test datasets and both LDA and RF methods.

3.1 Accuracy Measures for the LDA Classifier for Training and Test Data

	Precision	Recall	F1	AUC
BRCA	0.97	0.93	0.95	0.96
COAD	0.96	0.84	0.9	0.92
KIRC	1	0.97	0.98	0.98
LUAD	0.77	0.95	0.85	0.95
PRAD	1	0.97	0.99	0.99

Fig. 2. Precision, Recall, F1 and AUC Measures for the LDA Classifier – Training Data

	Precision	Recall	F1	AUC
BRCA	0.98	0.97	0.97	0.98
COAD	1	0.94	0.97	0.97
KIRC	1	1	1	1
LUAD	0.92	1	0.96	0.99
PRAD	1	0.96	0.98	0.98

Fig. 3. Precision, Recall, F1 and AUC Measures for the LDA Classifier – Test Data

	Precision	Recall	F1	AUC
BRCA	1	1	1	1
COAD	1	1	1	1
KIRC	1	1	1	1
LUAD	1	1	1	1
PRAD	1	1	1	1

Fig. 4. Precision, Recall, F1 and AUC Measures for the RF Classifier – Training Data

	Precision	Recall	F1	AUC
BRCA	0.95	1	0.98	0.99
COAD	0.88	0.94	0.91	0.96
KIRC	1	1	1	1
LUAD	0.97	0.88	0.92	0.94
PRAD	1	0.96	0.98	0.98

Fig. 5. Precision, Recall, F1 and AUC Measures for the RF Classifier – Test Data

3.2 Accuracy Measures for the RF Classifier for Training and Test Data

In Fig. 6 we provide the variables (genes) with high absolute loadings on the first two PC-scores; such a table can be very useful for selection of features (genes).

PC1	PC2	PC1	PC2	PC1	PC2
gene_3439	gene_9176	gene_16379	gene_1073	gene_14818	gene_8597
gene_6733	gene_9175	gene_16449	gene_4178	gene_2639	gene_10620
gene_439	gene_3540	gene_16155	gene_12848	gene_19160	gene_3440
gene_219	gene_3541	gene_7489	gene_11012	gene_13507	gene_15668
gene_1510	gene_9177	gene_18042	gene_11249	gene_9226	gene_3849
gene_16132	gene_12995	gene_7649	gene_14386	gene_17906	gene_2404
gene_16169	gene_12069	gene_3921	gene_5667	gene_8988	gene_2507
gene_220	gene_12568	gene_7964	gene_15437	gene_18108	gene_10646
gene_19153	gene_18135	gene_13818	gene_6594	gene_4223	gene_9232
gene_19159	gene_3737	gene_10950	gene_1482	gene_172	gene_6361
gene_6593	gene_17664	gene_2774	gene_5009	gene_8348	gene_5829
gene_16392	gene_11250	gene_4442	gene_3523	gene_11250	gene_4422
gene_16342	gene_1189	gene_16133	gene_7395	gene_13497	gene_13076
gene_16246	gene_11355	gene_5657	gene_7896	gene_5600	gene_11409
gene_11566	gene_11910	gene_16337	gene_19760	gene_13084	gene_17145
gene_3461	gene_18745	gene_16130	gene_4247	gene_2288	gene_15865
gene_8801	gene_4456	gene_14114	gene_2639	gene_12808	gene_7417
gene_17109	gene_6720	gene_2129	gene_7234	gene_5836	gene_17166
gene_1858	gene_203	gene_5199	gene_6937	gene_11713	gene_5539
gene_19151	gene_7113	gene_628	gene_6160	gene_17585	gene_4031
gene_19236	gene_6584	gene_16377	gene_17168	gene_3860	gene_7965
gene_2844	gene_19373	gene_16118	gene_399	gene_19201	gene_11107
gene_3843	gene_18753	gene_3862	gene_5691	gene_15736	gene_4866
gene_450	gene_11388	gene_18	gene_14623	gene_2879	gene_10402
gene_7421	gene_18383	gene_440	gene_3542	gene_7234	gene_11259
gene_7490	gene_148	gene_6935	gene_8050	gene_7625	gene_15453
gene_12078	gene_11019	gene_1410	gene_1201	gene_553	gene_19296
gene_7116	gene_13004	gene_5442	gene_1554	gene_4737	gene_6723
gene_6890	gene_15898	gene_18676	gene_17949	gene_9177	gene_7933
gene_16402	gene_13976	gene_545	gene_9529	gene_134	gene_7992
gene_7965	gene_9626	gene_16156	gene_4464	gene_13493	gene_9184
gene_19148	gene_13111	gene_19914	gene_5752	gene_14467	gene_19193
gene_14503	gene_5017	gene_7896	gene_218	gene_12977	gene_510
gene_5729	gene_10141	gene_17920	gene_6784	gene_742	gene_11449
gene_13916	gene_7238	gene_3861	gene_4170	gene_14427	gene_863
gene_7792	gene_2506	gene_16088	gene_12881	gene_16363	gene_18650
gene_6816	gene_14199	gene_4046	gene_15301	gene_3369	gene_1336
gene_180	gene_11762	gene_4587	gene_16372	gene_1427	gene_5050
gene_6734	gene_9075	gene_16105	gene_3730	gene_18282	gene_1448
gene_16259	gene_15894	gene_3541	gene_7178	gene_9711	gene_12245

Fig. 6. Significant genes with highest absolute loadings on the first two PC-scores

4 Conclusion

PCA results showed that the first 50 components (PC) cumulatively explained 71% of all variability present in the 801 $\times$ 20532 gene expression data, with the first two PC's explaining only 26% of total variability. The first two PC's, however, were sufficient for classification of cancer sub-types with high accuracy. LDA was able to classify each of the five cancer-subtypes with high accuracies except for LUAD which had a precision of 77% for the training set. The RF method was able to classify each sub-type with very high accuracy. The PCA loadings on 20532 genes were sorted in order of magnitude and genes (features) important for classification were identified.

Acknowledgement. This manuscript contains content from the first author's previously completed dissertation (Title: An Application of Machine Learning Algorithms in Healthcare, completed in 2024 at University of Nevada, Las Vegas), which has been adapted and summarized for this publication. Any reuse of the first author's prior work is clearly cited and referenced in the manuscript. This work has not been published elsewhere in its entirety and that proper attribution has been made to my thesis as the original source.

References

1. Alladi, S.M., Ravi, V., Murthy, U.S.: Colon cancer prediction with genetic profiles using intelligent techniques. Bioinformation **3**(3), 130 (2008). https://doi.org/10.6026/97320630003130

2. Alon, U., et al.: Broad patterns of gene expression revealed by clustering analysis of tumor and normal colon tissues probed by oligonucleotide arrays. Proc. Natl. Acad. Sci. **96**(12), 6745–6750 (1999). https://doi.org/10.1073/pnas.96.12.6745
3. Siegel, R.L., Miller, K.D., Wagle, N.S., Jemal, A.: Cancer statistics, 2023. Ca Cancer J. Clin. **73**(1), 17–48 (2023). https://doi.org/10.3322/caac.21763
4. Slonim, D.K.: From patterns to pathways: gene expression data analysis comes of age. Nat. Genet. **32**(4), 502–508 (2002). https://doi.org/10.1038/ng1033
5. Harrington, C.A., Rosenow, C., Retief, J.: Monitoring gene expression using DNA microarrays. Curr. Opin. Microbiol. **3**(3), 285–291 (2000). https://doi.org/10.1016/S1369-5274(00)00091-6
6. Schena, M., Shalon, D., Davis, R.W., Brown, P.O.: Quantitative monitoring of gene expression patterns with a complementary DNA microarray. Science **270**(5235), 467–470 (1995). https://doi.org/10.1126/science.270.5235.467
7. Alpaydin, E.: Introduction to Machine Learning. MIT Press, Cambridge (2020)
8. Erickson, B.J., Korfiatis, P., Akkus, Z., Kline, T.L.: Machine learning for medical imaging. Radiographics **37**(2), 505–515 (2017). https://doi.org/10.1148/rg.2017160130
9. Mahmood, N., Shahid, S., Bakhshi, T., Riaz, S., Ghufran, H., Yaqoob, M.: Identification of significant risks in pediatric acute lymphoblastic leukemia (ALL) through machine learning (ML) approach. Med. Biol. Eng. Comput. **58**, 2631–2640 (2020). https://doi.org/10.1007/s11517-020-02245-2
10. Kononenko, I.: Machine learning for medical diagnosis: history, state of the art and perspective. Artif. Intell. Med. **23**(1), 89–109 (2001). https://doi.org/10.1016/S0933-3657(01)00077-X
11. Kourou, K., Exarchos, T.P., Exarchos, K.P., Karamouzis, M.V., Fotiadis, D.I.: Machine learning applications in cancer prognosis and prediction. Comput. Struct. Biotechnol. J. **13**, 8–17 (2015). https://doi.org/10.1016/j.csbj.2014.11.005
12. Sung, H., et al.: Global cancer statistics 2020: GLOBOCAN estimates of incidence and mortality worldwide for 36 cancers in 185 countries. CA Cancer J. Clin. **71**(3), 209–249 (2021). https://doi.org/10.3322/caac.21660
13. Reid, A., de Klerk, N., Musk, A.W.: Does exposure to asbestos cause ovarian cancer? A systematic literature review and meta-analysis. Cancer Epidemiol. Biomarkers Prev. **20**(7), 1287–1295 (2011). https://doi.org/10.1158/1055-9965.EPI-10-1302
14. Brahim-Belhouari, S., Bermak, A.: Gaussian process for nonstationary time series prediction. Comput. Stat. Data Anal. **47**(4), 705–712 (2004). https://doi.org/10.1016/j.csda.2004.02.006
15. Bray, F., et al.: Global cancer statistics 2018: GLOBOCAN estimates of incidence and mortality worldwide for 36 cancers in 185 countries. CA Cancer J. Clin. **68**(6), 394–424 (2018). https://doi.org/10.3322/caac.21492
16. Jain, A., Zongker, D.: Feature selection: evaluation, application, and small sample performance. IEEE Trans. Pattern Anal. Mach. Intell. **19**(2), 153–158 (1997). https://doi.org/10.1109/34.574797
17. Wang, Y., et al.: Gene selection from microarray data for cancer classification—a machine learning approach. Comput. Biol. Chem. **29**(1), 37–46 (2005). https://doi.org/10.1016/j.compbiolchem.2004.11.001
18. Liu, K.H., Tong, M., Xie, S.T., Yee Ng, V.T.: Genetic programming based ensemble system for microarray data classification. Comput. Math. Methods Med. **2015** (2015). https://doi.org/10.1155/2015/193406
19. Johnson, R.A., Wichern, D.W.: Applied multivariate statistical analysis (2002). https://books.google.com/books?id=gFWcQgAACAAJ
20. Genuer, R., Poggi, J.M.: Random Forests. Springer International Publishing, Cham (2020). https://doi.org/10.1007/978-3-030-56485-8_3

21. Frank, A.: UCI machine learning repository (2010). http://archive.ics.uci.edu/ml. https://doi.org/10.24432/C5R88H
22. Team, R.C.: R: a language and environment for statistical computing. R Foundation for Statistical Computing. (No Title) (2013). https://www.R-project.org/

Generative Artificial Intelligence and Its Applications

A Robust Adversarial Attack Framework for Enhanced Software Vulnerability Detection Systems

Biswash Nursingh Kunwar and Prasanthi Sreekumari(✉)

Department of Computer Science, University of Louisiana at Monroe, Monroe, LA, USA
kunwarbn@warhawks.ulm.edu, sreekumari@ulm.edu

Abstract. Deep learning has made detecting software vulnerabilities faster and more accurate. However, these models can still be misled by adversarial attacks. Such attacks cause the models to make incorrect predictions, which threatens system security. This study presents a new framework to address these problems. The framework combines Shapley Additive Explanations (SHAP), Support Vector Machines (SVM), and a Bidirectional Long Short-Term Memory (BiLSTM) network with an attention mechanism. Each component helps identify and rank important features, allowing the model to recognize patterns linked to vulnerabilities. In addition, Foolbox generates adversarial examples to test the model's resilience. This approach strengthens both the accuracy and reliability of vulnerability detection models, building a stronger defense against adversarial attacks.

Keywords: Adversarial Attack · Software Vulnerability · Deep Learning · Foolbox · Security · Framework · Machine Learning · Defense · Attackers · Model

1 Introduction

A software vulnerability is a weakness in a system. Attackers can use this weakness to break into the system or disrupt it. Vulnerabilities often happen because of mistakes in coding, and it can also occur when the system is set up incorrectly [1]. If important security steps are skipped during development, vulnerabilities can appear as well. When attackers find these weaknesses, they can steal data or cause the system to fail, and this can lead to financial losses for the company. It can also hurt the company's reputation and people may lose trust if a company's system is not secure [2]. As software becomes more common in everyday life, finding and fixing these weaknesses is very important. It helps keep systems safe and protects sensitive information.

Research on software vulnerability detection uses a mix of advanced techniques to boost both security and clarity. These include machine learning, graph-based models, and Explainable AI (XAI). Traditional machine learning methods, such as Support Vector Machines (SVM), Decision Trees, and Random Forests, analyze data from software

S. Latifi (Ed.): ITNG 2025, AISC 1463, pp. 411–417, 2025.
https://doi.org/10.1007/978-3-031-89063-5_35

processes and code text to spot vulnerabilities at different stages, like code commits or releases [3]. Graph-based approaches, such as Abstract Syntax Trees (ASTs) and graph neural networks, help capture the structure of code, making it easier to find patterns linked to vulnerabilities in complex systems [4]. Transformer models and Tree-LSTM networks add even more detail, examining code line by line or by function to identify specific areas that might be at risk. Real-time methods, like Just-in-Time (JIT) vulnerability detection, focus on each code commit. They give developers instant feedback on any new vulnerabilities that might come up with code changes [5].

To make machine learning predictions easier to understand, XAI techniques like SHAP show how each feature influences the model's decisions. Adversarial attack simulations, such as EaTVul, test how strong vulnerability detection models are by creating adversarial data through ChatGPT [14]. This reveals any weaknesses in the model against attacks designed to trick it. Together, these techniques create a complete framework for software vulnerability detection, combining accuracy, real-time feedback, and protection against adversarial threats.

Although software vulnerability detection has improved, these techniques still have important limitations. One major issue is the high rate of false positives and false negatives. This means that models might mistakenly label safe code as vulnerable or miss actual vulnerabilities. This can overwhelm developers and lower trust in the system. The accuracy of these models also heavily relies on the quality of their training data. However, many datasets are imbalanced or lack a variety of vulnerability types, which limits how well the models work. In addition, these advanced models need high computational resources and specialized knowledge to set up and maintain, which can be challenging for many organizations.

These limitations show that while current techniques are helpful, they still need to be more accurate, understandable, and reliable for real-world use. To address these issues, this paper introduces a new approach called SHAP-Aware Adversarial BiLSTM (SAAB) for strengthening software vulnerability detection. SAAB combines SHAP, SVM, and BiLSTM network with an attention mechanism [15]. Together, these tools help clearly identify which features are most important for detecting vulnerabilities. SAAB also uses Foolbox, a Python library that generates adversarial examples to test the model's resilience, providing a systematic way to make software security stronger against sophisticated attacks.

2 Related Work

Recent advancements in software vulnerability detection use both machine learning and adversarial methods. These methods help improve detection accuracy and test how well models can resist attacks. This section reviews only the studies most relevant to our research.

Lin et al. [6] introduced CatCode, a framework that uses category theory to evaluate large language models (LLMs) on various code-related tasks, like code generation, translation, and explanation. Yiyang Li et al. [7] proposed Semantic-Preserving Adversarial Code Embeddings (SPACE), a technique to improve the robustness and accuracy of pre-trained language models for code comprehension tasks. Unlike traditional methods that must balance performance with robustness, SPACE uses adversarial training in the embedding space to achieve both goals. Srikant et al., [8] presented adversarial techniques for fooling machine learning models that analyze computer programs. Zhou Yang et al., [9] introduced ALERT (Naturalness Aware Attack), a black-box adversarial attack designed to mislead code models like CodeBERT and GraphCodeBERT. Noam Yefet et al., [10] introduced the Discrete Adversarial Manipulation of Programs (DAMP), a novel method for generating adversarial examples to test neural models for code. Huangzhao Zhang et al., [11] proposed the Metropolis-Hastings Modifier (MHM) algorithm, a technique for generating adversarial examples specifically for deep learning models used in source code analysis. Nils Loose et al., [12] presented Madvex, a novel adversarial attack framework targeting machine learning-based malware detection systems, particularly those used for WebAssembly (Wasm) binaries. Nguyen et al., [13] proposed GraphCodeAttack, an adversarial framework targeting pre-trained language models for code, such as CodeBERT and GraphCodeBERT. Shigang Liu et al., [14] introduced EaTVul, a novel adversarial attack framework targeting machine learning-based software vulnerability detection systems. EaTVul employs a systematic approach, starting with SVM to identify critical non-vulnerable samples, followed by an attention mechanism to highlight important features within those samples. Using these features, EaTVul leverages ChatGPT to generate adversarial code snippets that subtly modify vulnerable code without altering functionality. These adversarial examples are further optimized through a fuzzy genetic algorithm to enhance their effectiveness. This work highlighted the vulnerabilities of current machine learning-based detection models to evasion attacks, emphasizing the need for improved defense mechanisms to protect software systems from such adversarial threats.

Although advanced software vulnerability detection techniques face key limitations, such as high false positives or negatives due to challenges in capturing subtle differences in code functionality, many models remain vulnerable to adversarial attacks, which exploit minor code changes to bypass detection without altering functionality, indicating a lack of robustness. Moreover, methods like MHM and DAMP are computationally intensive and difficult to scale, highlighting the need for more efficient, scalable, and interpretable solutions in real-world applications.

3 Overview of SAAB

To tackle challenges in software vulnerability detection, this study introduces a framework called SAAB. SAAB builds on techniques like EaTVul to make adversarial attacks on detection systems more effective. The framework is designed to find and use weaknesses in these models with advanced tools. First, SHAP identifies and ranks the most important features in the code. This step focuses on the features that have the biggest impact on model predictions, helping choose features that might get past detection. Then, SVM filter and highlight key features that influence the model's decisions. This process creates a refined list of features, making adversarial samples more precise and effective. Next, the framework uses BiLSTM network with an attention mechanism. This setup reads code sequences in both directions, capturing the full context and relationships within the code. The attention mechanism then focuses on the most important parts, which helps make the adversarial samples more precise.

Moreover, SAAB uses Foolbox, a Python tool that creates adversarial examples, to systematically generate input samples to trick the vulnerability detection model. Foolbox provides a more focused and dependable way to produce adversarial data compared to earlier methods, keeping the attack effective. The new method SAAB combines SHAP, SVM, BiLSTM with attention mechanism, and Foolbox to create a method that accurately selects and interprets features and produces stronger, more reliable adversarial examples. This approach overcomes limitations of earlier models, offering a clear, precise, and resilient framework for evasion attacks on software vulnerability detection systems.

The proposed method builds upon the existing EaTVul attack framework, addressing its limitations by incorporating advanced components. As shown in Fig. 1, the framework SAAB is designed to improve code vulnerability detection by addressing limitations in existing methods. The different components include input phase, feature analysis, deep learning, adversarial components and enhanced output. The different components of SAAB framework are presented below:

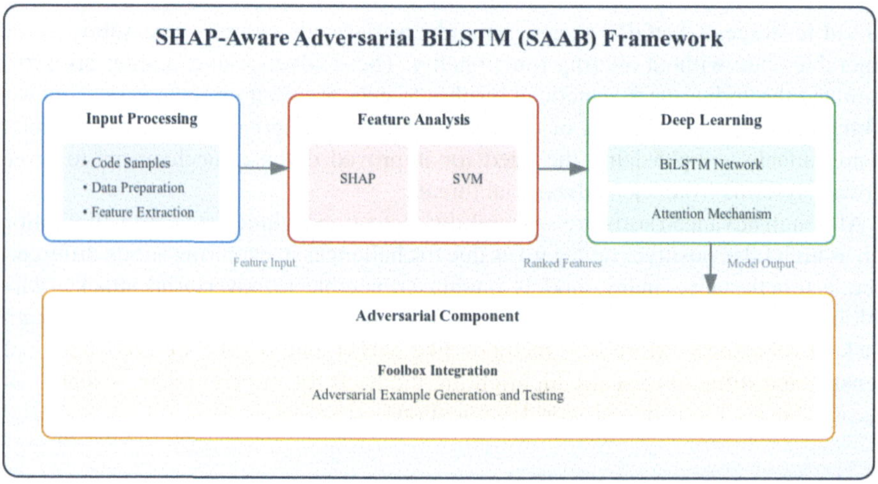

Fig. 1. Framework of SAAB

Input Processing: The process starts by organizing and preparing the raw data. This includes labeled code samples, both vulnerable and non-vulnerable, for training and testing. At this stage, the data is cleaned to make it consistent and ready for the models. This involves breaking the code into tokens, standardizing its structure, and removing unnecessary details. The result is a set of key features for effective vulnerability detection. The goal of this phase is to prepare the data in a structured, standardized format for further analysis.

Feature Analysis: This component focuses on identifying the most important features in the data, helping the model concentrate on relevant aspects while reducing noise. The key features include SHAP and SVM. SHAP calculates importance scores for each feature, showing which features, most influence the model's predictions. SHAP ranks these features by their scores, allowing the model to focus on the most relevant data and reduce complexity. This makes the detection process more accurate and easier to interpret. SHAP is available as a form of Python library which can be imported in popular open-source platform for machine learning like Tensorflow.

The dataset in the input phase is first run through a SVM model, which does an initial classification of the code samples. SVM helps identify patterns and organize features, laying the groundwork for deeper analysis.

Deep Learning Component: This component processes sequential data to understand its structure and relationships. After feature selection, the chosen features are sent to a BiLSTM model. This model processes code sequences in both forward and backward directions. This helps capture important context and relationships in the code. It is essential for accurately detecting vulnerabilities. Furthermore, an attention mechanism is added to the BiLSTM to focus on the most critical parts of each code sequence. By giving more importance to certain sections, the attention mechanism directs the model's focus to areas most likely to contain vulnerabilities. This improves overall detection accuracy.

Table 1. Comparison between SAAB and EaTVul

Features	EaTVul	SAAB
Feature Importance	Uses SVM for feature importance	Incorporates SHAP with SVM for feature importance providing a theoretically grounded and precise measure of feature contributions
Adversarial Attack	Uses unreliable, inconsistent and non-code method ChatGPT for generation of adversarial examples	Uses a dedicated python library, FoolBox, which generates targeted adversarial examples with high precision
Pipeline	Uses SVM, BiLSTM with attention mechanism and ChatGPT	Combines the strengths of SVM, BiLSTM with attention mechanism, SHAP, and Foolbox into a cohesive framework.

Adversarial Component: To ensure the model can handle real-world challenges, the SAAB framework incorporates adversarial testing. The SAAB framework includes Foolbox, a library that generates adversarial examples. Foolbox makes small changes to the code input that keep its original function but are designed to trick the model. This testing challenges the model, helping ensure it can handle real-world evasion tactics and remain strong against attacks.

Table 1 presents a comparison between SAAB and EaTVul. EaTVul's approach leverages BiLSTM for impotant feature analysis. However, our proposed method enhances this further as follows: SAAB incorporates SHAP for feature importance which provides a theoretically grounded and precise measure of feature contributions. Foolbox generates targeted adversarial examples with high precision. It ensures scalability and reproducibility, which are not achievable with inconsistent, non-code-based methods like ChatGPT. The proposed method combines the strengths of SVM, BiLSTM with attention mechanism, SHAP, and Foolbox into a cohesive framework. It enhances both interpretability and robustness, providing a more comprehensive solution compared to EaTVul.

4 Discussion

The proposed SAAB framework shows promise for improving software vulnerability detection. The different components of SAAB create a layered, understandable approach that detects vulnerabilities and resists adversarial attacks. Each part of SAAB plays a key role. SHAP helps the model focus on the most relevant features. SVM provides an initial classification. BiLSTM with attention mechanism captures important code patterns. Foolbox tests the model's ability to handle attacks. These components make SAAB a stronger, more reliable model than previous methods. However, to fully understand SAAB's potential, the future work includes testing on a variety of real-world datasets.

5 Conclusion

In this research, we propose the SAAB framework that provides a new approach to software vulnerability detection that is both accurate and resilient to attacks. By using SHAP, SVM, BiLSTM, and Foolbox, SAAB not only improves detection accuracy but also makes the model more secure against adversarial threats. This framework is a step forward in creating reliable and safe software vulnerability detection systems, addressing both the need for precise detection and protection against attacks. SAAB offers a strong foundation for future research and development in the field of cybersecurity, bringing us closer to more effective ways to keep software systems secure.

References

1. Guo, Y., Bettaieb, S., Casino, F.: A comprehensive analysis on software vulnerability detection datasets: trends, challenges, and road ahead. Int. J. Inf. Secur. **23**(3311–3327) (2024). https://doi.org/10.1007/s10207-024-00888-y
2. Lomio, F., Iannone, E., De Lucia, A., Palomba, F., Lenarduzzi, V.: Just-in-time software vulnerability detection: are we there yet? J. Syst. Softw. **188**, 111283 (2022). https://doi.org/10.1016/j.jss.2022.111283
3. Li, D., Liu, Y., Huang, J.: Assessment of software vulnerability contributing factors by model-agnostic explainable AI. Mach. Learn. Knowl. Extr. **6**(1087–1113) (2024). https://doi.org/10.3390/make6020050
4. Biggio, B., Nelson, B., Laskov, P.: Poisoning attacks against support vector machines. In: ICML, 2012

5. Chen, X., Liu, C., Li, B., Lu, K., Song, D.: Targeted backdoor attacks on deep learning systems using data poisoning, 2017. arXiv:1712.05526
6. Lin, Z., Yao, Y., Yuan, Y.: CatCode: a comprehensive evaluation framework for LLMs on the mixture of code and Text J-GLOBAL (2024). jst.go.jp. https://jglobal.jst.go.jp/en/detail?JGLOBAL_ID=202402213837839346
7. Li, Y., Wu, H., Zhao, H.: Semantic-preserving adversarial code comprehension. In: International Conference on Computational Linguistics, 2022
8. Srikant, S., et al.: Generating Adversarial Computer Programs using Optimized Obfuscations (2021). ArXiv abs/2103.11882
9. Yang, Z., Shi, J., He, J., Lo, D.: Natural attack for pre-trained models of code. In: Proceedings of the 44th International Conference on Software Engineering (ICSE '22). Association for Computing Machinery, New York, NY, USA, pp. 1482–1493 (2022). https://doi.org/10.1145/3510003.3510146
10. Yefet, N., Alon, U., Yahav, E.: Adversarial examples for models of code. Proc. ACM Program. Lang. **4**(OOPSLA), Article 162, 30 (2020). https://doi.org/10.1145/3428230
11. Zhang, H., Li, Z., Li, G., Ma, L., Liu, Y., Jin, Z.: Generating adversarial examples for holding robustness of source code processing models. In: Proceedings of the AAAI Conference on Artificial Intelligence, vol. 34, no. 01, pp. 1169–1176 (2020). https://doi.org/10.1609/aaai.v34i01.5469
12. Loose, N., Mächtle, F., Pott, C., Bezsmertnyi, V., Eisenbarth, T.: Madvex: instrumentation-based adversarial attacks on machine learning malware detection. In: Gruss, D., Maggi, F., Fischer, M., Carminati, M. (eds.) Detection of Intrusions and Malware, and Vulnerability Assessment: 20th International Conference, DIMVA 2023, Hamburg, Germany, 12–14 July 2023, Proceedings, LNCS, vol. 13959, pp. 69–88. Springer-Verlag, Berlin, Heidelberg (2023). https://doi.org/10.1007/978-3-031-35504-2_4
13. Nguyen, T.-D., et al.: Adversarial Attacks on Code Models with Discriminative Graph Patterns (2023). https://doi.org/10.48550/arXiv.2308.11161
14. Liu, S., et al.: EaTVul: ChatGPT-based Evasion Attack Against Software Vulnerability Detection (2024). https://doi.org/10.48550/arXiv.2407.19216
15. Moharana, P.: Guide to Bi-LSTM with attention Mechanism. Analytics India Magazine, 1 August 2024. https://analyticsindiamag.com/topics/bi-lstm-with-attention-mechanism/

Artificial Intelligence Evolution: The Rise of Generative AI

Amber Williams, Dextany Hatfield, and Bharat S. Rawal(✉)

Computer and Digital Technologies, Grambling State University, Grambling, USA
brawal@gram.edu

Abstract. Utilizing machine learning techniques and models, AI enables computers to learn from data and enhance their performance over time. Deep learning, a subset of machine learning, employs networks with layers to uncover complex patterns from extensive datasets, transforming tasks such as image recognition, natural language processing, and self-driving cars. The advent of Generative AI marks a significant milestone in the evolution of Artificial Intelligence. Building upon the foundations of traditional AI, Generative AI has revolutionized the landscape by enabling machines to create, innovate, and learn from vast amounts of data. Generative AI explores the use of generative adversarial networks to create content like images, music, and text. Applications such as language translation, sentiment analysis, and chatbots rely on natural language processing (NLP) to help machines understand, interpret, and generate language effectively. This survey paper provides an overview of AI concepts and underlines their significance in various domains.cy. This article investigates the security and privacy implications of AI in our daily lives.

Keywords: artificial intelligence · machine learning · deep learning · generative artificial intelligence · natural language processing · adversarial attack · federated learning · security · privacy

1 Introduction

AI has made advancements in industries such as healthcare and banking. The goal of AI is to equip machines with intelligence to humans enabling them to perceive, analyze and make decisions autonomously. Machine learning techniques and models play a role in AI by enabling computers to learn from data and enhance their capabilities [1]. One popular method within machine learning is Deep Learning, which utilizes networks with multiple layers to automatically learn patterns from data. Deep neural networks have shown performance in tasks like processing audio, images, natural language as well as enabling advancements in autonomous driving technology [1]. A growing field within AI is Generative AI, which focuses on creating systems that can produce content. Technologies like Generative Adversarial Networks (GANs) and Variational Autoencoders (VAEs) empower machines to generate visuals, music compositions, written content and more. Generative AI finds applications in areas such as art creation, content generation and enhancing data sets across industries [2].

S. Latifi (Ed.): ITNG 2025, AISC 1463, pp. 418–432, 2025.
https://doi.org/10.1007/978-3-031-89063-5_36

Natural Language Processing (NLP) plays a role in helping machines understand, interpret, and generate language. It serves as the backbone for applications involving language like language translation, sentiment analysis, chatbots and virtual assistants [3]. Recent advancements in NLP through transformer-based models such as BERT and GPT have significantly improved language processing capabilities.

Generative AI has made drastic strides within the world in recent years. Generative AI can create realistic pictures like humans. As technology becomes more advanced, several areas of concern have grown. Mode collapse, training instability, and ethical concerns have become problem areas with the advancements within Generative AI. This research aims to explore the uses, challenges, and ethics of generative AI models.

During the 1950s neural network research began. These networks can mimic human brain activity. Perception research also began during this time. This area grew in popularity because of the need for a biological process to solve problems. However, scientists face challenges with limited data availability and computational power. Hopfield networks emerged by the 1980s. These networks could recall patterns in partial and noisy inputs. Also, Boltzmann machines became better "stochastic" [4] neural networks. The machines could know how to learn data patterns. Generative AI has been useful in other industries as well. For instance, generative models create "synthetic data" [5, 6] to train and test intrusion systems. This process allows companies to ensure their defense plans can prepare them for cyber threats. While Generative AI has provided many advancements, there have also been several challenges with technology. The "reliability and consistency" [7] of outputs generated has been a concern for many. Since computers do not understand semantics like humans, they can often misinterpret the inputs given to them. The interpretability of generative AI models is another area of concern. Knowing why a generative model produces an output is "crucial for trust and accountability" [6] in industries like healthcare and criminal justice. Since users are unable to see the decision-making process of Generative AI models, it is difficult for users to interpret the outputs provided. Protecting sensitivity is important to promote equity. Framework protocol and regulations must be placed to address the ethical usage of Generative AI. Another area of concern with Generative AI is the rate at which fake information can travel across the Internet. If AI-generated content can "mimic reality" [7] traditional authentication and verification methods are threatened. The democratization of content created by generative AI blurs the lines between fact and fiction.

The rest of this research is as follows. Section 2 describes related work. Section 3 Describes artificial intelligence AI. Section 4 presents machine learning (ML). Section 5 describes the Deep Learning concept. Section 6 talks about natural language processing. Section 7 describes generative artificial intelligence. Section 8 talks about adversarial attacks. Section 9 discusses Federated learning and homomorphic encryption, and Sect. 10 concludes the paper.

2 Related Works

Yann LeCun is a prominent figure in artificial intelligence, widely recognized for his groundbreaking contributions to machine learning and Deep Learning. ("What resources are available in French regarding Yann LeCun's work in…") This review explores

LeCun's research, innovative models, and profound impact on the field, particularly in areas such as unsupervised learning and convolutional neural networks (CNNs), which have significantly advanced domains like computer vision, natural language processing, and robotics [8].

LeCun's influence on computer vision is particularly noteworthy due to his pioneering work on CNNs. In 1989, he and his team introduced the LeNet-5 architecture, specifically designed for digit recognition. This architecture demonstrated the effectiveness of layered structures, including pooling layers and fully connected layers, in extracting hierarchical features from images. LeNet-5 laid the groundwork for modern CNNs and has inspired countless subsequent architectures that dominate the field today [8].

In addition to his work on CNNs, LeCun has made significant strides in advancing self-supervised and unsupervised learning techniques. He proposed "energy-based models" in the early 2000s, which provide a framework for learning representations from data by assigning lower energy to known data points and higher energy to outliers [9]. This approach has opened new avenues for model training without the need for extensive labeled datasets.

LeCun also introduced the concept of autoencoders—a type of neural network capable of learning efficient representations by preserving the local structure of input data [10]. Recently, he has been a strong advocate for self-supervised learning methods, which enable models to predict features of input data without human intervention. This paradigm has shown remarkable success in various applications, including image classification, speech recognition, and natural language processing [11].

Beyond these contributions, LeCun's work has profoundly shaped the theoretical foundations of Deep Learning. He has elucidated key principles such as the significance of network depth, the use of nonlinear activation functions, and the importance of gradient-based optimization techniques. In the seminal paper "Efficient BackProp," LeCun and his collaborators introduced the backpropagation algorithm, which remains a cornerstone of neural network training [12]. Furthermore, he advocated for the use of rectified linear units (ReLUs) as activation functions to mitigate the vanishing gradient problem, thereby enhancing convergence speed in deep networks [13]. Through his extensive research and innovative approaches, LeCun has not only advanced the capabilities of AI but has also inspired a new generation of researchers and practitioners in the field.

3 Artificial Intelligence

Impact of Artificial Intelligence Across Industries: Artificial Intelligence (AI) is transforming various sectors, including healthcare, banking, transportation, and entertainment. This rapid advancement has garnered significant attention from scholars, policymakers, and the public alike. This paper examines AI's evolution, practical applications, challenges, and prospects.

History and Background

The development of AI can be understood as a progression through distinct stages, each marked by significant achievements. In the 1950s and 1960s, researchers primarily focused on exploring methods for reasoning and problem-solving. This foundational work laid the groundwork for future advancements in the field [1].

The 1970s saw the introduction of expert systems, which enabled the implementation of rule-based AI applications across various domains, such as healthcare and finance [14]. However, progress during this time was often hampered by limitations in computational capabilities and data availability.

The 1980s marked a turning point with the advent of machine learning techniques, which empowered computers to analyze data, extract insights, and improve their performance over time [15]. This period also witnessed advancements in neural networks, genetic algorithms, and other methodologies that would become essential to modern AI.

The 21st century has been characterized by unprecedented growth in AI, driven by the proliferation of data, advancements in hardware, and significant breakthroughs in Deep Learning technologies [16]. As a result, AI has gained widespread popularity and attracted increasing investments year after year.

Figure 1 This scatterplot represents the funding in the U.S. throughout the years starting from 2011 to 2023. Funding for AI increased dramatically from 2020 to 2021 and slowed down towards 2022.

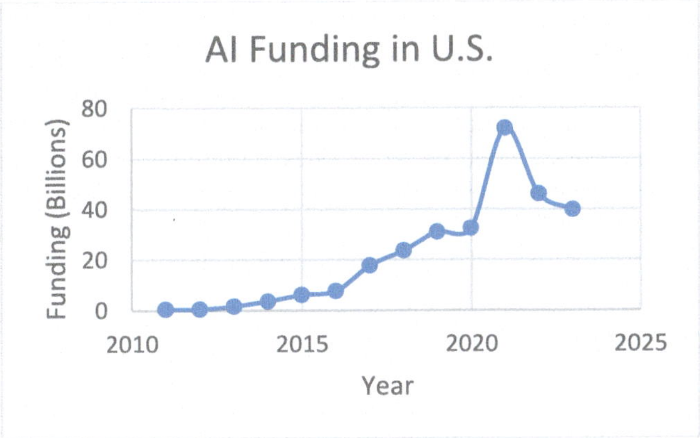

Fig. 1. Represents AI funding in the U.S.

3.1 Implementation of Artificial Intelligence

Various industries are leveraging AI to transform their operations and enhance productivity. In healthcare, AI algorithms analyze medical images to provide diagnoses and customize treatment plans based on patient data [17]. In finance, AI-driven trading systems effectively manage investment portfolios, detect anomalies in real-time, and predict market trends [18].

The automotive industry is also undergoing significant changes with the adoption of AI technologies for navigation, obstacle detection, and decision-making in vehicles [19]. Additionally, the widespread use of AI-powered virtual assistants has simplified tasks such as scheduling appointments, managing emails, and controlling smart home devices [20].

3.2 Challenges and Limitations

Despite significant advancements, the adoption of AI faces several challenges. One major concern revolves around the implications of AI, including algorithmic bias, privacy breaches, and potential job displacement [21]. Addressing these issues requires collaboration across disciplines such as engineering, policymaking, and ethics to ensure that AI benefits society positively.

Another critical obstacle is the lack of transparency in AI models, particularly neural networks, which operate as complex systems. This complexity makes it difficult to understand their decision-making processes [22]. Such opacity poses risks in sectors like healthcare and criminal justice, where transparency and accountability are essential.

3.3 AI's Potential in the Future

Looking ahead, the potential opportunities presented by AI are immense. Ongoing developments in AI research, particularly in areas like lifelong learning, have the potential to enhance the transparency, flexibility, and human-centered aspects of AI technologies [22, 23]. Fields such as cybersecurity, supply chain management, and environmental monitoring stand to benefit from leveraging AI in conjunction with advanced technologies like blockchain, the Internet of Things (IoT), and quantum computing [24]. Thanks to open-source frameworks, cloud-based services, and online learning platforms, individuals and businesses without specialized expertise will gain access to AI capabilities [25]. However, maximizing the benefits of AI requires a comprehensive approach that simultaneously addresses societal and technological challenges.

4 Machine Learning: An Overview

In today's data-driven landscape, **Machine Learning (ML)** has emerged as a transformative tool that empowers computers to autonomously analyze information, recognize patterns, and draw conclusions. By leveraging algorithms, ML allows systems to make predictions and decisions without explicit programming for each task. This paper provides an in-depth exploration of ML, covering its foundational concepts, various algorithms, real-world applications, challenges, and potential future developments. By effectively harnessing the capabilities of Machine Learning, we can address complex issues and significantly enhance the well-being of humanity across diverse sectors.

5 Deep Learning

The advent of Deep Learning has significantly transformed the field of artificial intelligence, driving remarkable advancements in areas such as computer vision, natural language processing, and reinforcement learning. This section provides a comprehensive exploration of Deep Learning, covering its fundamental concepts, frameworks, training techniques, real-world applications, challenges, and potential future developments.

In the 1980s, researchers developed the backpropagation algorithm, a breakthrough in training neural networks that effectively minimized errors during the learning process [26]. This algorithm has enabled a range of applications, allowing farmers to make informed decisions based on predictive analytics and helping healthcare professionals analyze X-rays and MRIs more efficiently for faster diagnosis [27]. In the automotive sector, self-driving cars utilize Deep Learning to interpret their surroundings, identifying pedestrians, traffic signals, and other vehicles to navigate safely [27].

Deep Learning, a subset of machine learning, empowers computers to comprehend and make decisions based on complex data patterns, mimicking the human brain's neural architecture. Its applications are diverse and impactful. For instance, Deep Learning has enhanced multimedia systems and security technologies [50]. It has also improved predictive models in various domains, such as using Deep Learning for REM prediction to create more accurate and efficient alternatives [51].

In healthcare, Deep Learning models have proven effective in analyzing chest X-ray images for diagnostic purposes [52]. Additionally, these models can process various satellite images to analyze environmental features, such as dune crests [53]. Furthermore, Deep Learning algorithms are being increasingly utilized for medical evaluations, such as stroke assessment [54]. Numerous studies affirm that Deep Learning continues to drive significant technological advancements across multiple sectors.

6 Natural Language Processing

Natural language processing (NLP) is a discipline within artificial intelligence and computer science that applies machine learning techniques to enable computers to understand and communicate in human languages. NLP encompasses various computational methods for text analysis and has been implemented in numerous applications. For instance, it has been utilized to assist perioperative pain providers by leveraging technology to enhance patient care [56]. Additionally, there have been instances where NLP has enabled software to perform artifact extraction from natural language requirements without human intervention [57]. Moreover, the use of machine learning algorithms and large language models has facilitated the extraction of information from client records, demonstrating the potential for data analytics in the criminal justice system [58]. Studies have also explored various data augmentation (DA) methods, including back translation, text generation, and paraphrasing [59].

Despite these advancements, computers still struggle to fully comprehend human language. They often find it challenging to grasp the semantics and sentiment behind the inputs provided, which can lead to inaccurate outputs. To address this issue, it is essential to integrate language models and knowledge graphs so that NLP systems [28] can understand and apply the context of inputs in their generated outputs. This integration helps improve the performance of voice assistants and other audio technologies.

Non-negative matrix factorization (NMF) is a useful technique that aids in making models more interpretable. NMF assists computers in breaking sounds into individual components, making it easier to understand the nuances within audio data [29]. As advancements in NLP continue, communication between humans and computers is expected to become smoother and more effective.

Early NLP systems have paved the way for current capabilities, but they had significant limitations, such as restricted vocabulary, difficulties with spelling mistakes, and challenges in interpreting sentences. These limitations hindered the seamless integration of technology into daily life. Today, several improvements in NLP have been integrated into our everyday experiences. For instance, the Natural Language Toolkit (NLTK) [30] enhances computers' ability to comprehend sentences, manage grammatical errors, and interpret the sentiment conveyed by the words used. Figure 2. describe.

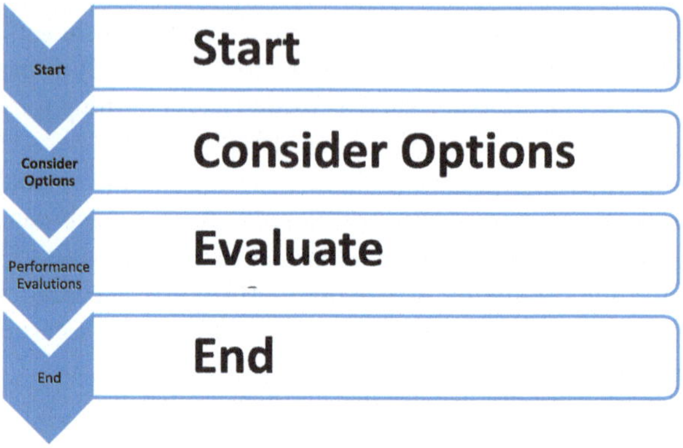

Fig. 2. Is a simplified flow chart for the "Generative AI [2].

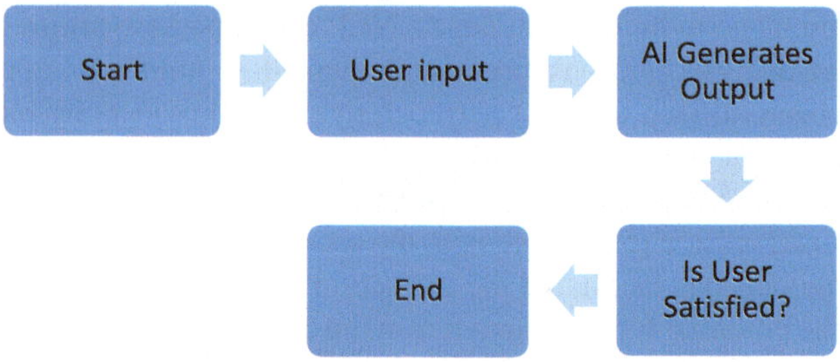

Fig. 3. Describes the operational steps for Gen AI.

Figure 3 is a simplified chart for the "Harnessing the power of LLMS in practice" passage [19].

6.1 Applications and Implementations NLP

NLP has a vast array of applications spanning different domains. The interactions between various media processes through NLP aim to understand and interpret human languages, facilitating the generation of human-like responses in the future. Some notable applications include:

- **Chatbots and Virtual Assistants**: These systems utilize NLP to engage in human-like conversations and provide customer support.
- **Sentiment Analysis**: Businesses employ NLP to analyze customer feedback and social media interactions, gaining insights into public sentiment regarding products or services [64]. ("Information Retrieval Vs NLP Insights - Restackio")
- **Language Translation**: NLP technologies power translation services, allowing for real-time conversion of text across different languages.
- **Text Summarization**: Automated summarization tools leverage NLP to condense lengthy documents into concise summaries. ("Natural Language Understanding Automated Solutions")

6.2 Challenges and Limitations

Despite significant advancements in NLP, several challenges persist. One major issue is the inherent complexity and ambiguity of human language, which can lead to errors and inaccuracies within NLP systems. Addressing ambiguity and understanding context remain ongoing research challenges in the field of NLP [32].

Another obstacle is the availability of training data for specific tasks and domains. NLP models often require substantial amounts of data for effective training, a process that can be both expensive and time-consuming to acquire. To tackle this challenge, researchers are exploring transfer learning and semi-supervised learning methods, which aim to reduce the data burden while maintaining model performance [33].

6.3 Potential Future

The future of NLP looks promising, thanks to advancements in machine learning, transfer learning, and multimodal approaches [34]. Progress in few-shot learning and language comprehension offers exciting opportunities for enhancing NLP capabilities. IT experts are increasingly adopting NLP models that integrate text with visuals and audio, as these models provide a more nuanced understanding of language [35].

However, the reliance on large data volumes raises significant data privacy concerns. Regulations such as the General Data Protection Regulation (GDPR) in Europe, which includes the "right to explanation," may impact AI systems, especially those utilizing complex algorithms, posing challenges for communication policies [44, 45]. Furthermore, AI-powered technologies, like deepfakes, have the potential to disrupt communication, particularly on social media platforms. To combat misinformation, existing regulations may require updates, or new rules may need to be enacted [46, 47].

6.4 Applications and Implementations

AI is rapidly evolving and becoming increasingly integrated into daily life through various electronic devices, programs, and vehicles. It is trained in extensive pools of collected data to enhance the efficiency and accuracy of its outputs [36, 37]. However, the incorporation of AI also presents potential risks that can be significant, particularly in critical applications such as autonomous vehicles. The primary concern in training AI is to ensure that the data used is consistent, accurate, and relevant to the specific tasks it is designed to perform [38]. Figure 4 represents the simplified version of NLP architecture.

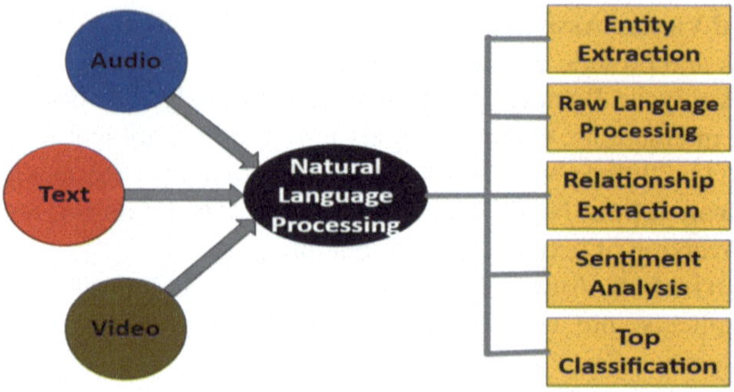

Fig. 4. NLP architecture

7 Generative Artificial Intelligence

Generative Artificial Intelligence (AI) has emerged as a powerful tool for creating various forms of content, including images, text, music, and videos. This study explores the core concepts, architecture, training methods, applications, challenges, and future developments in generative AI. At the heart of generative AI lies probability modeling, which involves understanding the probability distributions that govern the generation of observed data. These models aim to capture the underlying patterns within the data to produce instances that closely resemble the training set.

Neural networks play a crucial role in generative modeling, with techniques such as Generative Adversarial Networks (GANs) and Variational Autoencoders (VAEs) being widely adopted. GANs consist of two networks: a generator and a discriminator. These networks are trained simultaneously, where the generator creates data examples while the discriminator distinguishes between real and generated examples [30]. VAEs, on the other hand, learn to encode data examples into a continuous latent space, allowing for the generation of new examples by sampling from the learned distribution [31].

In the coming years generative AI shows promise with advancements, in learning, reinforcement learning and neuromyotonic AI. These breakthroughs are expected to push the boundaries of modeling and open up possibilities for uses, in creativity, design and entertainment [1]. By tackling obstacles and venturing into research avenues we can unleash the complete capabilities of generative AI and develop smarter and more expressive technologies.

8 Adversarial Attack

AI adversarial attacks represent a unique category of cybersecurity threats that manipulate AI systems' behavior. Unlike traditional cyberattacks, these attacks exploit inherent vulnerabilities within AI algorithms, which are often difficult to rectify [39]. Adversarial attacks can occur in various forms, including the physical manipulation of objects—such as placing stickers on stop signs to mislead self-driving cars—or feeding the AI misleading data. These actions raise significant security concerns and societal questions.

Adversarial attacks can target a range of systems, including content filters on social media platforms, where AI decides what content appears in users' feeds [40]. The dissemination of false information or harmful content can erode user trust in these platforms. The implications extend beyond social media; AI systems are also deployed in military and law enforcement contexts. If adversaries compromise these systems, the consequences could be dire, leading to decisions based on inaccurate information and disrupting critical operations. Many organizations rely on AI to automate repetitive tasks, making them vulnerable to such attacks. Disruptions in supply chain processes, for instance, could result in significant economic losses and operational challenges.

A common misconception is that all AI attacks are detrimental to society [42]. While some nations, such as China, may employ AI as a tool for surveillance and control, adversarial attacks can also serve as a protective measure against governmental oppression. In this context, citizens may engage in what could be termed "defensive AI attacks" [43] to safeguard their civil liberties and evade detection. However, employing these tactics can escalate tensions between individuals and governments, resembling a digital arms race where parties strive to outmaneuver one another using technology.

This dynamic raises important questions about the regulation of offensive AI capabilities. To mitigate the risks associated with AI adversarial attacks, the implementation of "AI Security Compliance" [38], programs is essential. These programs would establish guidelines and frameworks to bolster the security of AI systems against attacks. They would enable stakeholders to assess attack risks, implement IT reforms to make attacks more challenging, and develop actionable response plans in case of an incident. Furthermore, regulatory measures should be established for governmental and high-risk environments to guide both nations and citizens in maintaining peace and security.

9 Federated Learning and Homomorphic Encryption

Federated learning and homomorphic encryption are two rapidly advancing fields in machine learning and cryptography that hold great promise for enhancing privacy and security in distributed data processing. Federated learning is a decentralized machine learning paradigm where multiple devices or nodes collaboratively train a shared model while keeping their data local. This approach is different from traditional machine learning, where data is centralized on a server, posing privacy risks and challenges related to data sovereignty. In Federated learning, instead of sending raw data to a central server, each device computes model updates locally, which are then aggregated at the server to refine the global model. This process allows organizations to build machine learning models while respecting user privacy, which is especially important in industries such as healthcare, finance, and mobile computing, where sensitive data is generated. Researchers have tried to apply transformers to federated learning [62]. It is very important to utilize user-generated data for the improvement of Tactile Internet without interfering with user privacy.[63]. Federated learning is a type of machine learning framework that deals with communication constraints [64].

However, while Federated learning addresses privacy concerns by ensuring that raw data never leaves individual devices, it introduces new privacy challenges during the aggregation process. For example, the model updates shared between devices and the central server might still reveal information about the underlying data. This issue motivates the integration of homomorphic encryption into federated learning frameworks. Homomorphic encryption is a form of encryption that allows computations to be performed on encrypted data without needing to decrypt it first. In other words, it enables data to remain encrypted during the entire processing cycle, ensuring that privacy is maintained even when computations are performed on the data. Federated algorithms are biased toward certain models that are trained by native federated algorithms datasets. They are not independent and are identically distributed, which is known as fairness problem [65]. However other researchers believe that storing sensitive data as a source in the cloud produces the risk of being hijacked. To compensate for this, homomorphic encryption is utilized to encrypt and store the sensitive data, and can safely operate in an encrypted state, is being studied [66]. Homomorphism algorithms can carry out multi-level homeomorphism addition under specific parameters [67]. Homomorphism encryption lets the users use mathematical operations on open data without knowing the decryption key [68].

In conclusion, Federated learning and Homomorphism encryption are fast growing topics in the world today. They have been used in doing many important things which have great benefit for the world. Homomorphic encryption is a form of encryption that allows computations to be performed on encrypted data without needing to be decrypted first. Federated learning is a machine learning form where multiple devices can collectively train a shared model while keeping their data native. Federated learning and Homomorphism encryption are important to fields like healthcare, banking etc.

10 Conclusion

Natural language processing (NLP) has appeared as a prominent area within artificial intelligence, finding applications across various domains, including healthcare, social media analysis, customer service, and information retrieval. Despite its advancements, NLP faces challenges such as data availability and linguistic ambiguity, which can hinder its effectiveness.

Generative AI, another significant side of artificial intelligence, offers solutions to complex problems but also raises critical ethical questions about reliability, interpretability, and privacy. To ensure the ethical use of these technologies, robust policies must be established to protect privacy, ensure factual accuracy, promote equity, and enhance transparency.

As we look to the future of artificial intelligence, it is essential for governments to strike a balance between fostering innovation and safeguarding public interests. This balanced approach will help mitigate risks and ensure that the benefits of AI technologies are realized while protecting individuals and society.

References

1. Wick, C.: Deep learning. Informatik-Spektrum **40**(1), 103–107 (2016). https://doi.org/10.1007/s00287-016-1013-2

2. Goodfellow, I.J., et al.: Generative adversarial nets. In: Advances in Neural Information Processing Systems, pp. 2672–2680, 2014
3. Devlin, J., et al.: BERT: pre-training of deep bidirectional transformers for language understanding. In: Proceedings of the 2019 Conference of the North American Chapter of the Association for Computational Linguistics: Human Language Technologies, vol. 1, pp. 4171–4186, 2019
4. Michel-Villarreal, R., et al.: Challenges and opportunities of generative AI for higher education as explained by ChatGPT. Educ. Sci. **13**(9), 856 (2023)
5. Bandi, A., Adapa, P.V.S.R., Kuchi, Y.E.V.P.K.: The power of generative ai: a review of requirements, models, input-output formats, evaluation metrics, and challenges. Future Internet **15**(8), 260 (2023)
6. Fui-Hoon Nah, F., Zheng, R., Cai, J., Siau, K., Chen, L.: Generative AI and ChatGPT: applications, challenges, and AI-human collaboration. J. Inf. Technol. Case Appl. Res. **25**(3), 277–304 (2023). https://doi.org/10.1080/15228053.2023.2233814
7. Ratten, V., Jones, P.: Generative artificial intelligence (ChatGPT): implications for management educators. Int. J. Manag. Educ. **21**(3), 100857 (2023)
8. LeCun, Y., et al.: Gradient-based learning applied to document recognition. Proc. IEEE **86**(11), 2278–2324 (1998)
9. LeCun, Y., et al.: A Tutorial on Energy-Based Learning," Predicting Structured Data, MIT Press, Cambridge (2006)
10. Lee, H., et al.: Sparse deep learning for robust multiclass object detection. In: Proceedings of the IEEE Conference on Computer Vision and Pattern Recognition, 2012
11. LeCun, Y.: Self-Supervised Learning: The Dark Matter of Intelligence? (2019)
12. LeCun, Y., et al.: Efficient BackProp, Neural Networks: Tricks of the Trade, Springer, 1998
13. Glorot, X., et al.: Deep sparse rectifier neural networks. In: Proceedings of the 14th International Conference on Artificial Intelligence and Statistics, 2011
14. Silver, D., et al.: Mastering the game of go without human knowledge. Nature **550**(7676), 354–359 (2017)
15. Esteva, A., et al.: Dermatologist-level classification of skin cancer with deep neural networks. Nature **542**(7639), 115–118 (2017)
16. Silver, D., et al.: Mastering the game of go with deep neural networks and tree search. Nature **529**(7587), 484–489 (2016)
17. Kohavi, R., Provost, F.: Glossary of terms. Mach. Learn. **30**(2–3), 271–274 (1998)
18. Domingos, P.: A few useful things to know about machine learning. Commun. ACM **55**(10), 78–87 (2012)
19. Mullainathan, S., Spiess, J.: Machine learning: an applied econometric approach. J. Econ. Perspect. **31**(2), 87–106 (2017)
20. Bostrom, N.: Superintelligence: Paths, dangers, strategies. Oxford University Press, Oxford (2014)
21. Floridi, L., Cowls, J.: A unified framework of five principles for AI in society. Harv. Data Sci. Rev. **1**(1) (2019)
22. Jordan, M.I., Mitchell, T.M.: Machine learning: trends, perspectives, and prospects. Science **349**(6245), 255–260 (2015)
23. Marcus, G., Davis, E., Shrobe, H.: The next decade in AI: four steps towards robust artificial intelligence (2018). arXiv preprint arXiv:1803.08099
24. Schölkopf, B., Smola, A.J.: Learning with Kernels: Support Vector Machines, Regularization, Optimization, and Beyond. MIT press, Cambridge (2002)
25. Bishop, C.M.: Pattern recognition and machine learning. Springer (2006)
26. Sejnowski, T.J.: The Deep Learning Revolution. MIT Press, Cambridge (2018)
27. Pan, S., et al.: Natural language processing: a historical review. Comput. Linguist. **19**(4), 409–431 (1993)

28. Bird, S., Klein, E., Loper, E.: Natural Language Processing with Python: Analyzing Text with the Natural Language Toolkit. O'Reilly Media, Inc., Boston, 2009
29. Zhang, X., et al.: Unleashing the power of edge-cloud generative AI in mobile networks: a survey of AI-GCS services. IEEE Commun. Surv. Tutor. **23**(4), 2952–2975 (2021)
30. Goodfellow, I., et al.: Generative adversarial nets. Adv. Neural Inf. Process. Syst. **27**, 2672–2680 (2014)
31. Kingma, D.P., Welling, M.: Auto-Encoding Variational Bayes, 2013. arXiv:1312.6114 [stat.ML]
32. Mihalcea, R., Radev, D.: Graph-based Natural Language Processing and Information Retrieval. Cambridge University Press, Cambridge, 2011
33. Turian, J., Ratinov, L., Bengio, Y.: Word representations: a simple and general method for semi-supervised learning. In: Proceedings of the 48th Annual Meeting of the Association for Computational Linguistics, pp. 384–394, 2010
34. Devlin, J., et al.: BERT: Pre-training of Deep Bidirectional Transformers for Language Understanding, 2018. arXiv:1810.04805 [cs.CL]
35. A. Vaswani et al., "Attention Is All You Need," Advances in Neural Information Processing Systems 30, 2017
36. Rajagopal., A., Nirmala, V.: Strategies to Architect AI Safety: Defense to Guard AI from Adversaries, 2019. https://search.ebscohost.com/login.aspx?direct=true&db=edsarx&AN=edsarx.1906.03466&site=eds-live
37. Zhang, T., Chen, X., Qu, C., Yuille, A., Zhou, Z.: Leveraging AI Predicted and Expert Revised Annotations in Interactive Segmentation: Continual Tuning or Full Training? 2024. https://search.ebscohost.com/login.aspx?direct=true&db=edsarx&AN=edsarx.2402.19423&site=eds-live
38. Bogani, R., Dr, A.T., Arnaboldi, L., Wortham, R.H.: Garbage in, toxic data out: a proposal for ethical artificial intelligence sustainability impact statements. AI and Ethics **3**(January), 1135–42 (2023). https://doi.org/10.1007/s43681-022-00221-0
39. Belfer Center for Science and International Affairs, Harvard Kennedy School. (2023). Artificial Intelligence Systems Can Be Attacked
40. Bengio, Y., et al.: Generating Sequences with Recurrent Neural Networks, 2013. arXiv:1308.0850, [cs.NE]
41. IBM. (2022). Understanding Adversarial Attacks on AI
42. MIT Technology Review. (2023). How to protect AI systems from adversarial attacks
43. National Institute of Standards and Technology. (2022). AI Risk Management Framework
44. Chesney, R., Citron, D.: Deep Fakes: A Looming Challenge for Privacy, Democracy, and National Security. Social Science Research Network (2018). https://papers.ssrn.com/sol3/papers.cfm?abstract_id=3213954
45. Graefe, A.: Guide to Automated Journalism. Tow Center for Digital Journalism (2016). https://towcenter.org/research/guide-to-automated-journalism/
46. Goodman, B., Flaxman, S.: European Union regulations on algorithmic decision-making and a "right to explanation." AI Mag. **38**(3), 50–57 (2016). https://doi.org/10.1609/aimag.v38i3.2741
47. West, D.M.: The future of work: robots, AI, and automation. Brookings Institution Press (2018)
48. Mitchell, R., Frank, E.: Accelerating the XGBoost algorithm using GPU computing. PeerJ Comput. Sci. **3**, e127 (2017)
49. https://video.search.yahoo.com/search/video?fr=mcafee&p=logistic+regression+explained&type=E210US739G0#id=1&vid=bb8e12762217e8b8c75d5fdbdc3792e6&action=click

50. Shabbir, M., Kamal, M., Ullah, Z., Khan, M.M.: Securing autonomous vehicles against GPS spoofing attacks: a deep learning approach. IEEE Access **11**(January), 105513–26 (2023). https://doi.org/10.1109/ACCESS.2023.3319514
51. Sallouha, H., Sarkar, S., Krijestorac, E., Cabric, D.: REMU-Net: deep learning based agile rem prediction with energy-efficient cell-free use case. IEEE Open J. Signal Process. **5**(January), 750–65 (2024). https://doi.org/10.1109/OJSP.2024.3378591
52. Hussain, B.Z., et al.: Wasserstein GAN based chest X-Ray dataset augmentation for deep learning models: COVID-19 detection use-case. In: Annual International Conference of the IEEE Engineering in Medicine and Biology Society. IEEE Engineering in Medicine and Biology Society. Annual International Conference 2022, no. July, pp. 2058–61, 2022. https://doi.org/10.1109/EMBC48229.2022.9871519
53. Nowatzki, M.: Application of a deep learning framework to explore transfer learning for dune mapping across regions. In: European Geosciences Union General Assembly 2023 25 (January): @Abstract EGU23-16209, 2023. https://doi.org/10.5194/egusphereegu23-16
54. Luvizutto, G.J., et al.: Use of artificial intelligence as an instrument of evaluation after stroke: a scoping review based on international classification of functioning, disability and health concept: AI applications for stroke evaluation. Top. Stroke Rehabil. **29**(5), 331–346 (2022)
55. Maciejewski, C., et al.: Practical use case of natural language processing for observational clinical research data retrieval from electronic health records: AssistMED project. Pol. Arch. Internal Med. **134**(5) (2024). https://doi.org/10.20452/pamw.16704
56. Seng, E.C., Mehdipour, S., Simpson, S., Gabriel, R.A.: Tracking persistent postoperative opioid use: a proof-of-concept study demonstrating a use case for natural language processing. Reg. Anesth. Pain Med. **49**(4), 241–47 (2024). https://doi.org/10.1136/rapm-2023-104629
57. Imtiaz Malik, M., Azam Sindhu, M., Ayaz Abbasi, R.: Extraction of use case diagram elements using natural language processing and network science. PloS One **18**(6): e0287502 (2023). https://doi.org/10.1371/journal.pone.0287502
58. Elyazori, H., Thurman, T., Lybarger, K., Taxman, F.S.: Exploring probation and parole records using natural language processing: a case study of supervisory condition notes. Fed. Probat. **87**(3), 19–26 (2023)
59. Campbell, J.: Generative Artificial Intelligence. Salem Press Encyclopedia of Science, September 2023. https://research-ebsco-com.grambling.idm.oclc.org/linkprocessor/plink?id=e9dca70d42f4-35b2-acd5-933f6dbf8457
60. Reddy, G.P., Shrutika, S., Soo-Hyun, P.: Generative AI for the maritime environments. In: 2024 Fifteenth International Conference on Ubiquitous and Future Networks (ICUFN), Ubiquitous and Future Networks (ICUFN), 2024 Fifteenth International Conference On, pp. 618–23, , July 2024. https://doi.org/10.1109/ICUFN61752.2024.10625100
61. Holtz, N., Sven, W., Jorge Marx, G.: The new era of knowledge retrieval: multi-agent systems meet generative AI. In: 2024 Portland International Conference on Management of Engineering and Technology (PICMET), Management of Engineering and Technology (PICMET), 2024 Portland International Conference On, 1–10 August 2024. https://doi.org/10.23919/PICMET64035.2024.10653018
62. Wang, D., Guo, B., Shen, Y., Cheng, S.-J., Lin, Y.-H.: A faster fully homomorphic encryption scheme in big data. In: Proceedings of the 2017 IEEE 2nd International Conference on Big Data Analysis (ICBDA), Beijing, China, pp. 345–349, 2017. https://doi.org/10.1109/ICBDA.2017.8078836
63. Wen, J., Dai, H., He, J., Xi, M., Xiao, S., Yang, J.: Federated offline reinforcement learning with multimodal data. IEEE Trans. Consum. Electron. **70**(1), 4266–4276 (2024). https://doi.org/10.1109/TCE.2023.3330943
64. Huang, H., et al.: Active client selection for clustered federated learning. IEEE Trans. Neural Netw. Learn. Syst. **35**(11), 16424–16438 (2024). https://doi.org/10.1109/TNNLS.2023.3294295

65. Ju, L., Zhang, T., Toor, S., Hellander, A.: Accelerating fair federated learning: adaptive federated Adam. IEEE Trans. Mach. Learn. Commun. Netw. **2**, 1017–1032 (2024). https://doi.org/10.1109/TMLCN.2024.3423648
66. Ryu, J., Kim, K., Won, D.: A study on partially homomorphic encryption. In: Proceedings of the 2023 17th International Conference on Ubiquitous Information Management and Communication (IMCOM), Seoul, South Korea, pp. 1–4, 2023. https://doi.org/10.1109/IMCOM56909.2023.10035630
67. Zhang, Z., Li, Z.: A new LWE-based homomorphic encryption algorithm over integer. In: Proceedings of the 2021 International Conference Computer Information Science and Artificial Intelligence (CISAI), Kunming, China, pp. 521–525, 2021. https://doi.org/10.1109/CISAI54367.2021.00106
68. Kadykov, V., Levina, A.: Homomorphic properties within lattice-based encryption systems. In: Proceedings of the 2021 10th Mediterr. Conference Embedded Computer (MECO), Budva, Montenegro, pp. 1–4, 2021. https://doi.org/10.1109/MECO52532.2021.9460165

The Use of GenAI in Graph Based Unit Testing

Abubakr S. Masood, Mir H. Ali, Mohammed W. Amair, Ziad A. Al-Sharif(✉), and Safwan Omari

Department of Engineering, Computing and Mathematical Sciences (ECaMS), College of Aviation, Science and Technology (CoAST), Lewis University, 1 University Parkway, Romeoville, IL 60446, USA
{abubakrsyedmasood,mirhyderali,mohammedwaheeduddi,zalsharif, omarisa}@lewisu.edu

Abstract. Software testing verifies that the software is free of defects and meets its requirements. This process includes various levels, one of which is unit testing, where developers create test cases alongside their regular code, and use frameworks, such as JUnit for Java, to enable a frequent automated execution of these test cases. However, designing test cases remains a significant challenge. Graph-based testing offers a solution by representing units in the source code as graphs, with nodes representing basic code blocks and edges representing transitions or interactions between these nodes. Additionally, modern Generative AI (GenAI) models, including ChatGPT, Gemini, and Copilot, present new opportunities for enhancing the software testing process. This paper investigates the potential of using GenAI models to automate and improve unit testing, particularly through graph-based methods. Experiments are designed to evaluate these models, assessing their ability to reduce manual effort while improving test coverage, efficiency, and code quality. The results reveal that GenAI models can streamline test generation and execution, but their effectiveness heavily relies on prompt quality and they lack an inherent understanding of program logic. In contrast, traditional graph-based unit testing ensures comprehensive coverage through systematic exploration of control flow paths but is resource-intensive. Therefore, this paper recommends a hybrid approach that combines the automation capabilities of GenAI with the rigor of traditional methods to achieve robust and efficient software testing.

Keywords: Large Language Models · LLM · Generative AI · GenAI · ChatGPT · Gemini · Copilot · Unit Testing · Graph Based Testing · Software Testing

A. S. Masood, M. H. Ali, M. W. Amair, Z. A. Al-Sharif, and S. Omari—These authors contributed equally to this work.

S. Latifi (Ed.): ITNG 2025, AISC 1463, pp. 433–444, 2025.
https://doi.org/10.1007/978-3-031-89063-5_37

1 Introduction

Software testing is a critical process that ensures the functionality, performance, and reliability of software applications. It helps identify defects, validate requirements, and improve the overall quality of software systems. This testing process can be approached by different means, one of which is the Model-Based Testing (MBT). MBT consists of a number of methods and approaches, with the aim of improving the quality, efficiency, and effectiveness of test processes, tasks, and test cases. MBT helps design test cases by using models to represent the desired behavior of the system. By leveraging models of the System Under Test (SUT), MBT allows for the systematic generation of test cases, ensuring comprehensive test coverage and early detection of defects. This approach enhances the reliability of the software [1,2].

Graph-based testing is an approach within MBT, where the software's *control flow* or *data flow* is represented as an abstract graph (model). In this method, the SUT is represented as a graph, with *nodes* typically denoting states or specific code blocks (sequences of statements with a single entry and exit point), and *edges* representing transitions or interactions between these nodes. This graphical representation facilitates the systematic generation of test cases that cover various paths within the graph, ensuring thorough exploration of the control flow and/or data flow. By leveraging the graph model, testers can identify critical transitions and states that require testing, thereby enabling comprehensive test coverage and early defect detection [3,4]. This approach aligns with the principles of MBT, as it utilizes a model (the graph) to automate and guide the testing process, utilizing graph-based coverage criteria, and ultimately enhancing the efficiency and effectiveness of software testing efforts [5].

A key advantage of graph-based testing is its ability to systematically cover different parts of the software's control flow. To achieve this, various coverage criteria can be employed, such as Node Coverage (NC), Edge Coverage (EC), Edge-Pair Coverage (EPC), and Prime Path Coverage (PPC). These criteria define the extent to which the graph is traversed and help guide the design of effective test cases to ensure comprehensive testing [6].

On the other hand, Generative AI (GenAI) is revolutionizing the field of software engineering and software testing by automating and enhancing various aspects of these processes. GenAI tools can generate code snippets, refactor existing code, and translate code between different programming languages, significantly accelerating development and improving code quality. Additionally, GenAI can automate the creation of documentation, ensuring consistency and adherence to standards across projects, and generate test cases to streamline the creation of realistic scenarios for testing. Moreover, GenAI can automatically generate various test cases based on specifications and requirements, create synthetic test data that mimics real-world scenarios, and predict potential bugs and vulnerabilities by analyzing patterns in code and test results. Furthermore, GenAI can allow to automate repetitive tasks such as test script creation and execution [7,8]. Therefore, GenAI tools, including OpenAI's ChatGPT, Gemini, and GitHub Copilot, leverage machine learning models to automate various software development tasks, from code generation and debugging to software testing and quality assurance [9,10].

This paper focuses on exploring the use of GenAI in graph-based unit testing, specifically in the context of Control Flow Testing. In this approach, test cases are designed to evaluate the control structures of software code, with test scenarios derived from the program's control flow graph. We investigate how GenAI can aid in generating test cases that meet specific graph-based coverage criteria, helping to ensure that the software performs correctly across a wide range of scenarios.

To illustrate this, we examine a Java program designed to compute statistical metrics, such as Sum, Length, Median, Mean, Variance, and Standard Deviation on a *List* of integer numbers. By generating test cases informed by specific coverage criteria such as NC, EC, EPC, and PPC, we assess the robustness and reliability of the program, particularly its handling of different inputs and special cases. This paper aims to demonstrate the potential of GenAI-driven test case generation for improving software testing efficiency and effectiveness, while also highlighting the challenges and future directions in the field. As a result, GenAI has the potential to enhance the testing process by automatically generating diverse test cases based on coverage criteria, such as the aforementioned NC, EC, EPC, and PPC coverage criteria.

The remaining parts of this paper are organized as follows. Section 2 sheds some light on the background research. Section 3 presents our research methodology. Section 4 describes our experimentation environment. Section 5 presents our finding results from the three experimented models. Section 6 sheds more light on the results. Finally, Sect. 7 concludes our work, presents its limitations, and highlights the potential for future studies.

2 Background and Related Work

Software testing verifies that software applications fulfill their requirements and operate as intended in various scenarios. Graph-based testing is one of the crucial tools for ensuring comprehensive test coverage. These graph-based models represent the software's behavior as a graph, with nodes representing states, statements, or sets of statements called basic blocks, and edges representing transitions between these nodes [11]. Control Flow Graphs (CFGs) and Data Flow Graphs (DFGs) are particularly useful for illustrating a program's behavior during execution, aiding in the identification of potential testing areas [12]. Coverage criteria such as statement coverage, branch coverage, and path coverage are essential for measuring the completeness of test cases. These criteria ensure that all lines of code (*statement coverage*), and all decision points (*branch coverage*), are tested.

For example, a CFG can help visualize the logical flow of a program, mapping out the different paths the program may take. This representation allows for the application of various coverage criteria, such as NC, which ensures that all code statements are tested. On the other hand, DFG can help identify how data moves through a program, highlighting where variables are defined, used, and modified, thereby ensuring that critical data manipulations are tested [7,8].

However, GenAI has emerged as a transformative tool in the field of software engineering, including software testing. Unlike traditional testing approaches [13–17], GenAI has the capability to advance software development and quality assurance (QA). The ability of GenAI models, such as ChatGPT, Gemini, and Copilot, to comprehend and produce software code, documentation, and automated testing has motivated researchers to investigate how these new GenAI models might improve the software testing process.

The integration of GenAI in software testing has gained considerable attention due to its potential to automate and enhance various processes. GenAI models have been used to generate unit test cases, optimize test suites, and predict potential issues, thereby improving test coverage and efficiency [18]. Several tools and frameworks facilitate the integration of GenAI in software testing, such as Tata Elxsi's GenAI based unit testing framework [19]. Despite advancements, challenges remain in ensuring the reliability and accuracy of GenAI generated test cases. Researchers are focusing on improving AI models to better understand software logic and enhancing the ability to handle complex testing scenarios to further automate and optimize the testing process. [20].

This paper explores the incorporation of GenAI models for generating graph-based test cases, which could lead to significant improvements in the overall quality of software products. By employing graph-based representations of the program's structure, this paper investigates how GenAI models, including ChatGPT, Gemini, and Copilot, can assist in automatically generating test cases that cover specific graph-based testing criteria, ensuring that key logic and data flows are thoroughly tested.

3 Methodology

Our methodology builds on the use of CFGs and DFGs as fundamental tools in unit testing. This approach facilitates comprehensive testing strategies that ensure all possible execution paths are examined [21]. Coverage criteria based on CFGs can ensure thorough testing of the SUT by focusing on different aspects (test requirements) of the program's execution paths. These criteria help in creating robust test cases that thoroughly examine the software's control flow and data flow, leading to more reliable and fault-resistant software [22].

This paper explores and evaluates the use of GenAI and its ability to design test cases that satisfy some of the coverage criteria related to CFG testing, see Fig. 1. Our experimentation uses one of the most common code examples in software testing, which is well-known in software testing literature [3]. First, we manually analyzed the program unit, built a CFG, and designed a set of test cases based on various CFG-based testing criteria. We then provided the same program unit to each of the investigated GenAI models. Our methodology begins by designing a set of prompts for each GenAI model. These prompts are crafted to enable the GenAI to produce a set of test cases that satisfy one of the CFG

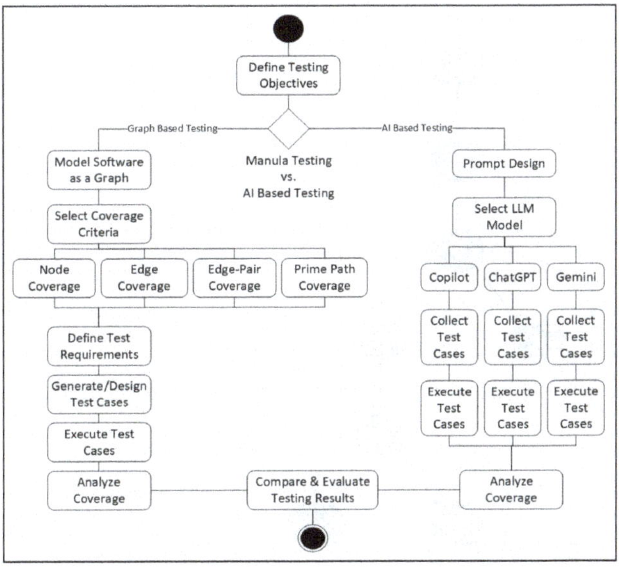

Fig. 1. Research Methodology. It employs a sample program to generate test cases based on various graph-based testing coverage criteria. It evaluates whether any of the GenAI models, including ChatGPT, Gemini, and Copilot, can produce similar test cases that meet the same test requirements. The coverage criteria used in this evaluation include NC, EC, EPC, and PPC coverage.

```
public static void computeStats (int [ ] numbers)
{
   int length = numbers.length;
   double med, var, sd, mean, sum, varsum;

   sum = 0.0;
   for (int i = 0; i < length; i++)
   {
      sum += numbers [ i ];
   }
   med  = numbers [ length / 2 ];
   mean = sum / (double) length;

   varsum = 0.o;
   for (int i = 0; i < length; i++)
   {
      varsum = varsum  + ((numbers [ i ] - mean) * (numbers [ i ] - mean));
   }
   var = varsum / ( length - 1 );
   sd  = Math.sqrt ( var );

   System.out.println ("length:              " + length);
   System.out.println ("mean:                " + mean);
   System.out.println ("median:              " + med);
   System.out.println ("variance:            " + var);
   System.out.println ("standard deviation: " + sd);
}
```

Fig. 2. The sample simple Java method used in our experimentation. It calculates various statistical measures for a *List* of integer numbers; taken from [3]. This method is utilized to construct the CFG presented in Fig. 3. Additionally, it is provided to each of the three GenAI models in our evaluation to generate test cases that meet common graph-based coverage criteria.

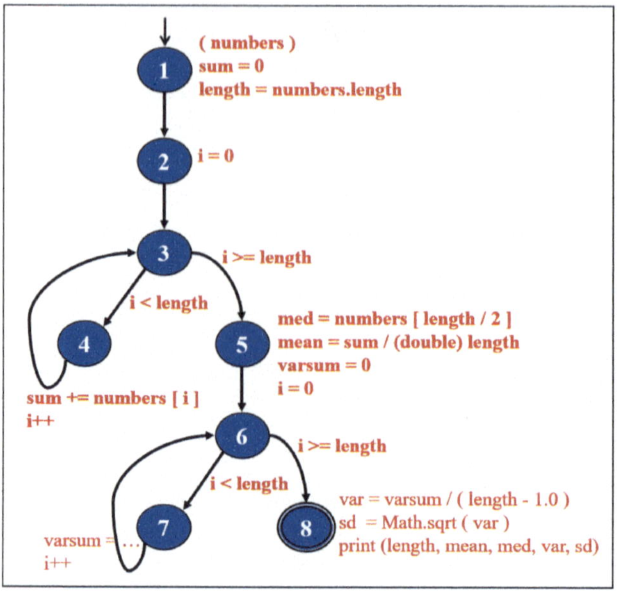

Fig. 3. A CFG that captures all possible paths presented in the simple sample Java method given in Fig. 2. The nodes are numbered from 1 to 8 and annotated with the related variable names from the corresponding basic block represented in each node. The edges are unidirectional, representing the potential control flow that can be taken by a test case.

criteria, including NC, EC, EPC, and PPC criteria. All prompts are applied to all the GenAI models, and the resulting test cases are compared between the models and against the manually derived test cases from the CFG (see Fig. 3) of the sample program presented in Fig. 2.

4 Experimentation

In this study, we evaluate the efficacy of three GenAI models—ChatGPT-4.0, Gemini, and Microsoft Copilot—in generating test cases for the sample Java program presented in Fig. 2. Our experimentation process and its detailed steps is presented in Fig. 1. The selected program is a commonly used Java program from a textbook [3], which performs statistical operations on a list of numbers. We carefully create a set of six targeted prompts, strictly tailored to test different aspects of the program and satisfy some of the graph-based coverage criteria including NC, EC, EPC, and PPC. For each GenAI model, the sample program and the six prompts are provided one at a time. Each model is used to generate a set of test cases based on the crafted prompts. The test cases are collected from each model and grouped based on the used prompt. During the analysis process, the collected test cases from each model are executed on the Java program and traced in the CFG that is created manually. This is used to measure the extent to which they satisfy predefined graph-based coverage criteria. The outputs of the three models are systematically compared against each other and against the manual testing.

4.1 Prompts Design

After giving the sampling method presented in Fig. 2 to each of the evaluated models, we used the following set of six prompts in the same order, one prompt at a time, evaluated the recommended test cases, and compared against the test cases that can be used based on various test requirements related to graph-based testing.

P_1: *Give me a set of test cases to test this program*
P_2: *Give me an optimal set of test cases*
P_3: *Give me a set of test cases that can satisfy the node coverage criterion*
P_4: *Give me a set of test cases that can satisfy the edge coverage criterion*
P_5: *Give me a set of test cases that can satisfy the edge-pair coverage criterion*
P_6: *Give me a set of test cases that can satisfy the prime-path coverage criterion*

5 Results

Using the targeted prompts presented in Sect. 4.1 that resemble the graph based testing scenarios, this sections presents and evaluates the results generated by each of the GenAI models. It evaluates the model effectiveness in designing the test cases based for the given sample program. It evaluates each model's outputs' accuracy, performance, and usability by examining them on same sample program.

5.1 ChatGPT

This AI module produced an average total of 8 test cases, primarily focusing on fundamental array properties. These test cases covered scenarios such as empty arrays, single-element arrays, odd and even numbers, negative values, repeated elements, arrays with zeroes, and mixed numbers. ChatGPT's approach is straightforward and aims to address common edge cases and array characteristics that can affect the functionality of a program. While it efficiently covers essential testing scenarios, it lacks more complex conditions and special cases.

However, experimenting with the graph-based testing, we find that this model generates 5 test cases for the NC, 6 test cases for the EC, 7 test cases for the EPC, and 6 test cases for the PPC. Comparing these results with the manual testing, it is clear that manual testing outperforms these models in terms of the number of test cases and the efficiency of the testing process, see Table 1.

5.2 Gemini

In contrast, Gemini generated around 14 test cases with a more varied and in-depth approach. In addition to covering basic scenarios like odd and even numbers, Gemini introduces complex test cases such as arrays with duplicate elements, arrays with large numbers, and median-based considerations (e.g., arrays where the median is the first or last element). It also includes cases for arrays with non-integer elements, providing a more comprehensive evaluation. By offering a wider range of test cases, Gemini allows for better testing of edge cases and complex input data, making it a more robust option for thorough program testing.

However, experimenting with the graph-based testing, we find that this model generates 11 test cases for the NC, 18 test cases for the EC, 17 test cases for the EPC, and 21 test cases for the PPC. Comparing these results with the manual testing, it is clear that manual testing outperforms these models in terms of the number of test cases and the efficiency of the testing process, see Table 1.

5.3 Copilot

Copilot has generated near to 9 different test cases, striking a balance between simplicity and core functionality. Its test cases cover a mix of basic array characteristics, including positive and negative numbers, large elements, and odd-even combinations. While not as extensive as Gemini's set, Copilot effectively addresses fundamental scenarios, making it useful for quick testing of core functionalities without delving too deeply into more complex edge cases.

However, experimenting with the graph-based testing, we find that this model generates 6 test cases for the NC, 7 test cases for the EC, 8 test cases for the EPC, and 8 test cases for the PPC. Comparing these results with the manual testing, it is clear that manual testing outperforms these models in terms of the number of test cases and the efficiency of the testing process, see Table 1.

5.4 Performance Evaluation in Code Execution

ChatGPT, Gemini, and Copilot offer varying levels of test case generation and coverage for software testing. ChatGPT focuses on structured input-output mappings, covering basic cases such as empty arrays, single elements, and mixed inputs, including odd and even-length scenarios. However, it is limited to basic functional nodes and lacks attention to edge cases, performance, and error handling. Gemini provides comprehensive coverage, addressing basic, edge cases, performance testing, and error scenarios. It effectively tests empty arrays, boundary values, large arrays, and error-handling paths, achieving full coverage of nodes, edges, edge pairs, and prime paths. In contrast, Copilot is primarily focused on basic test cases like empty arrays, single elements, and homogeneous arrays, with limited coverage of functional nodes and missing critical edge cases and non-functional aspects like performance testing and error handling. In terms

Table 1. Coverage metrics for ChatGPT, Gemini, and CoPilot. It summarizes the number of test cases designed and generated based on each criterion. It presents the number of test cases designed based on the used coverage criteria and compares the manual CFG-based testing against the AI-based testing obtained from three LLM models.

Prompt#	Criterion	Manual	ChatGPT	Gemini	Copilot
P_1	–	–	8	14	9
P_2	–	–	7	12	6
P_3	NC	1	5	11	6
P_4	EC	1	6	18	7
P_5	EPC	3	7	17	8
P_6	PPC	5	6	21	8

of coverage criteria, ChatGPT partially satisfies edge-pair and prime path coverage, while Gemini fully meets all criteria, including robust error handling and performance testing. Copilot, however, falls short in edge-pair and prime path coverage due to its limited scope. Gemini emerges as the most comprehensive solution, effectively addressing functional and non-functional testing needs, while ChatGPT and Copilot are better suited for simpler testing scenarios (Table 2).

Table 2. Comparison of Testing Criteria Across Tools

Criteria	ChatGPT	Gemini	Copilot
Basic Test Cases	✓	✓	✓
Node Coverage	✓	✓	✓
Edge Coverage	✓	✓	✗
Edge-Pair Coverage	✗	✓	✗
Prime Path Coverage	✗	✓	✗

6 Discussion

The evaluation of the three AI models—ChatGPT, Gemini, and Copilot—across various test cases reveals distinct approaches to handling input scenarios. ChatGPT consistently focuses on foundational cases, addressing essential scenarios such as empty arrays, single elements, and varying lengths (both odd and even), along with mixed positive and negative values. This targeted approach effectively covers key edge cases and basic paths through the code. Gemini, on the other hand, adopts a more comprehensive strategy with a broader range of test cases. It not only includes various array lengths but also examines conditions involving duplicates, zeros, non-integer elements, and even edge cases like overflow and underflow. This extensive set of 21 test cases allows for a deeper exploration of complex input scenarios and ensures thorough coverage of all potential paths within the program. Copilot provides a focused set of test cases, emphasizing core functionalities while still addressing fundamental scenarios such as basic arrays, even and odd numbers, and both positive and negative values. With its eight test cases, Copilot offers a streamlined approach that, while effective, may not delve as deeply into edge cases as Gemini does. Overall, while ChatGPT covers essential scenarios effectively and Copilot emphasizes simplicity and core functionalities, Gemini stands out for its extensive and detailed examination of diverse input cases, making it particularly robust in testing the models across a wide array of conditions.

7 Conclusion

GenAI testing and traditional graph-based testing exhibit complementary strengths and weaknesses. GenAI models, particularly Gemini, excel in automating test case generation, significantly reducing the time and effort required for test design while achieving considerable coverage across NC, EC, EPC, and PPC criteria. They offer diversity in test cases, including performance and error-handling scenarios, and are highly efficient and scalable. However, GenAI models rely on human-designed prompts and often lack a deep understanding of program logic, which can result in incomplete coverage of complex decision paths and specific program behaviors.

In contrast, traditional graph-based testing provides systematic and rigorous coverage, guaranteeing full exploration of all nodes, edges, and paths while tailoring test cases to specific program logic. This approach ensures comprehensive error handling and robust boundary testing, but it is time-intensive, resource-heavy, and prone to human error. Combining the strengths of both approaches is recommended, with GenAI models generating a wide array of test cases quickly, and traditional methods ensuring thorough path coverage and addressing gaps. For the 'computeStats' function, Gemini from GenAI closely approximates traditional testing, but full PPC coverage would benefit from supplemental manual testing. This hybrid approach balances efficiency and effectiveness for robust software testing.

References

1. Ahmad, T., Iqbal, J., Ashraf, A., Truscan, D., Porres, I.: Model-based testing using uml activity diagrams: a systematic mapping study. Comput. Sci. Rev. **33**, 98–112 (2019)

2. Garousi, V., Keleş, A.B., Balaman, Y., Güler, Z.Ö., Arcuri, A.: Model-based testing in practice: an experience report from the web applications domain. J. Syst. Softw. **180**, 111032 (2021)
3. Ammann, P., Offutt, J.: Introduction to Software Testing, 2 ed. Cambridge University Press, Cambridge (2016)
4. Bernardino, M., Rodrigues, E.M., Zorzo, A.F., Marchezan, L.: Systematic mapping study on mbt: tools and models. IET Softw. **11**(4), 141–155 (2017)
5. Utting, M., Legeard, B.: Practical Model-Based Testing: A Tools Approach. Elsevier, Amsterdam (2010)
6. Alshareef, A., Keller, N., Carbo, P., Zeigler, B.P.: Generative ai with modeling and simulation of activity and flow-based diagrams. In: International Conference on Simulation Tools and Techniques, pp. 95–109, Springer, 2023
7. Nguyen-Duc, A., et al.: Generative artificial intelligence for software engineering–a research agenda, *arXiv preprint*arXiv:2310.18648, 2023
8. Russo, D.: Navigating the complexity of generative ai adoption in software engineering. ACM Trans. Softw. Eng. Methodol. (2024)
9. Li, S., Cheng, Y., Chen, J., Xuan, J., He, S., Shang, W.: Assessing the performance of ai-generated code: a case study on github copilot. In: 2024 IEEE 35th International Symposium on Software Reliability Engineering (ISSRE), pp. 216–227, IEEE, 2024
10. Gonçalves, C.A., Gonçalves, C.T. : Assessment on the effectiveness of github copilot as a code assistance tool: an empirical study. In: EPIA Conference on Artificial Intelligence, pp. 27–38, Springer, 2024
11. Jha, P., Sahu, M., Isobe, T.: A uml activity flow graph-based regression testing approach. Appl. Sci. **13**(9), 5379 (2023)
12. Elasri, C., Kharmoum, N., Saoiabi, F., Boukhlif, M., Ziti, S., Rhalem, W.: Applying graph theory to enhance software testing in medical applications: a comparative study. In: International Conference on Advanced Intelligent Systems for Sustainable Development, pp. 70–78, Springer, 2023
13. De Soto, B.G., Adey, B.T.: Preliminary resource-based estimates combining artificial intelligence approaches and traditional techniques. Procedia Eng. **164**, 261–268 (2016)
14. Al-Sharif, Z.A., Jeffery, C.L.: Abstracttrace: the use of execution traces to cluster, classify, prioritize, and optimize a bloated test suite. Applied Sciences (2076-3417) **14**(23) (2024)
15. Al-Sharif, Z.A., Abdalrahman, W.F., Jeffery, C.L.: Encoding test cases using execution traces. In: 2021 12th International Conference on Information and Communication Systems (ICICS), pp. 239–244. IEEE (2021)
16. Al-Sharif, Z.A., Jeffery, C.L., Said, M.H.: The use of dynamic temporal assertions for debugging. Embed. Softw. Verif. Debugging 47–65 (2017)
17. Al-Sharif, Z.A., Jeffery, C.L., Said, M.H.: : Debugging with dynamic temporal assertions. In: 2014 IEEE International Symposium on Software Reliability Engineering Workshops, pp. 257–262. IEEE (2014)
18. Rajbhoj, A., Somase, A., Kulkarni, P., Kulkarni, V.: Accelerating software development using generative ai: Chatgpt case study. In: Proceedings of the 17th Innovations in Software Engineering Conference, pp. 1–11, 2024
19. Elxsi, T.: Accelerating unit testing using generative ai, Tata Elxsi Insights, 2024
20. Hnatushenko, V., Pavlenko, I.: The use of generative artificial intelligence in software testing. Syst. Technol. **2**(151), 113–123 (2024)

21. Kauffman, S., Moreno, C., S. Fischmeister, S.: Annotating control-flow graphs for formalized test coverage criteria. In: 2024 IEEE International Conference on Software Testing, Verification and Validation Workshops (ICSTW), pp. 55–62, IEEE, 2024
22. Yuan, X., Bo, H., Lina, Z., Chengyang, L.: A control flow graph optimization method for enhancing fuzz testing. IEEE Access (2024)

A Study of Prompt Engineering Techniques for Code Generation: Focusing on Data Science Applications

Dileon Saint-Jean[1,2(✉)], Baker Al Smadi[2], Syed Raza[1], Sari Linton[2], and Ugochukwu Igweagu[2]

[1] Department of Computer Science and Data Science, American Public University System, 111 W. Congress Street, Charles Town, WV 25414, USA
dileon.saintjean@gmail.com, syed.raza@mycampus.apus.edu
[2] Department of Computer Science and Digital Technologies, Grambling State University, 403 Main Street, Grambling, LA 71245, USA
alsmadib@gram.edu

Abstract. The increasing demand for data scientists has highlighted the need for up-skilling and efficient tools in the field. Generative AI offers a promising so-lution for automating coding tasks and enhancing productivity. This paper investigates the effectiveness of prompt engineering techniques for code generation in data science applications using Google's Gemini large language model (LLM). By examining the quality of AI-generated code for tasks such as data cleaning, exploratory data analysis (EDA), and machine learning model building, this study aims to identify best practices for prompt engi-neering and contribute to a better understanding of how to effectively utilize generative AI models for code generation in the data science field. The study analyzes the accuracy and efficiency of the generated code, as well as its ability to handle complex data science tasks. Findings suggest that prompt engineering plays a crucial role in shaping the quality and effectiveness of AI-generated code, and that generative AI has the potential to significantly accelerate data science workflows.

Keywords: Generative AI · Prompt Engineering · Code Generation · Large Language Models (LLM's) · Data Science · Deep Learning · Data Cleaning · Exploratory Data Analysis (EDA) · Machine Learning · Predictive Modeling

1 Introduction

The field of data science is experiencing explosive growth, driven by the increasing availability of data and the rising demand for insights across various industries. According to the U.S. Bureau of Labor Statistics, employment of data

---

B. Al Smadi, S. Raza, S. Linton, and U. Igweagu—These authors contributed equally to this work.

S. Latifi (Ed.): ITNG 2025, AISC 1463, pp. 445–453, 2025.
https://doi.org/10.1007/978-3-031-89063-5_38

scientists is projected to grow 36% from 2023 to 2033, far outpacing other occupations (4%) and Mathematical Science Occupations as a whole (28%) [11]. This surge in demand highlights the critical need for skilled data scientists who can effectively extract knowledge and value from data. However, the complexity and breadth of tasks involved in data science, from data cleaning and exploration to model building and visualization, often require significant coding expertise. Finding individuals with the necessary skills to meet this increasing demand is a significant challenge.

A recent study by LinkedIn revealed a substantial shortage of data science talent, with over 150,000 data scientist positions unfilled in the United States alone [6]. The dynamic nature of data science necessitates continuous learning and adaptation, making it challenging for traditional education models to fully equip individuals with the diverse and evolving skillset required for long-term success in this field [8]. Furthermore, the interdisciplinary nature of data science necessitates a diverse skillset encompassing programming, statistics, domain knowledge, and communication, which can be challenging to acquire through conventional educational approaches [7].

Considering these challenges, generative AI offers a promising avenue for upskilling aspiring and existing data science practitioners. By automating certain aspects of code generation, generative AI can lower the barrier to entry for individuals who may not have extensive programming experience, while also em-powering experienced professionals to work more efficiently and effectively. Recent research has demonstrated the potential of generative AI models, such as large language models (LLMs), to generate code in various programming languages, including Python, R, and SQL [1,2] These models can assist with a wide range of coding tasks.

The effectiveness of generative AI in code generation depends on the quality of the interaction between the human user and the AI model. This interaction is facilitated through prompt engineering, which involves creating effective input prompts to guide the AI model toward generating the desired output, code, in this case. The structure of a prompt is vital in shaping the output of these models, as it influences components of the response like accuracy, efficiency, style and security. Understanding the usage of prompts for Data Science Applications is there-fore important to produce well-structured and efficient code.

This research aims to address the following question: how do different prompt engineering techniques affect the quality and effectiveness of AI generated code for data science applications? To address this question, this study will examine the effectiveness of prompts, the quality of generated code, and identify best practices for AI code generation in data science.

In particular, this paper makes the following contributions:

- This study identifies best practices for prompt engineering with Gemini to generate high-quality code for data science tasks.
- It provides insights into the strengths and limitations of Gemini for code generation in data science, contributing to a better understanding of how to effectively utilize this powerful AI model.

– It highlights the potential of generative AI to accelerate data science workflows by automating certain aspects of code generation.

2 Literature Review

Recent advances in deep learning, particularly in natural language processing (NLP), have paved the way for significant progress in the field of code generation [5]. This progress is fueled by the increasing availability of large code corpora and the development of sophisticated language models (LMs) capable of capturing complex code structures and semantics. Deep learning models have demonstrated remarkable success in various code-related tasks, including code completion, code summarization, and cross-language code translation [5]. This success is attributed to the ability of deep learning models to automatically learn representations from data, capture long-range dependencies in code, and generalize to new and unseen code patterns.

One of the challenges in code generation is the accurate representation of code and the effective control of the generation process. Traditional approaches, such as rule-based systems or statistical language models, often struggle to capture the nuances of code and generalize to new scenarios. Deep learning models, particularly recurrent neural networks (RNNs) and transformers, have emerged as powerful tools for code generation due to their ability to learn complex patterns and dependencies in code. Svyatkovskiy et al. (2020) demonstrate the effectiveness of transformer models in their work on IntelliCode Compose, a code generation tool that can predict entire lines of syntactically correct code [10]. Their approach leverages a variant of the GPT-2 model trained on a massive dataset of code, achieving impressive results in terms of code completion accuracy and fluency.

The effectiveness of deep learning models in code generation is closely tied to the quality of the input prompts used to guide the generation process. Prompt engineering, which involves crafting effective input prompts to elicit desired outputs from the model, plays an important role in shaping the generated code [3]. Factors such as code tokenization, vocabulary size, and the inclusion of relevant context information in the prompt are important for improving the quality of the generated code [3].

The application of generative AI in data science is another area of growing interest. The ability to generate code for data science tasks, such as data cleaning, exploratory data analysis (EDA), and machine learning model building, has the potential to significantly accelerate data science workflows and empower data scientists to focus on higher-level tasks. Danilchenko (2011) explores the use of case-based reasoning (CBR) in combination with routine design and template-based programming to generate code for routine data manipulation tasks [9]. While this approach focuses on a specific domain, it highlights the potential of AI to automate code generation for data science applications.

However, the use of generative AI in data science also raises important considerations regarding the soundness and completeness of the generated code

[9]. Existing code quality metrics may have limitations, and improved metrics are needed to accurately capture the intended quality aspects of code. Factors such as code cohesion, coupling, and complexity are important when evaluating the quality of AI-generated code. Machine learning-based approaches to code quality assessment can potentially offer more accurate and nuanced evaluations compared to traditional metrics [9].

A comparative analysis of popular code generation tools, including GitHub Copilot, Amazon CodeWhisperer, and ChatGPT, evaluated their performance on the HumanEval dataset [12]. These tools can generate functionally correct code, but they may also produce code with potential security vulnerabilities or maintainability issues [12]. This highlights the need for careful evaluation and refinement of AI-generated code, as well as the importance of ongoing research to improve the quality and reliability of code generation tools.

The concept of "one-shot correction" leverages user feedback to improve the accuracy of code generation models without requiring retraining [4]. This approach involves decomposing the code generation task into sub-problems and utilizing user feedback to refine the generated code snippets [4]. This highlights the potential of incorporating human expertise and feedback into the code generation process to improve the quality and usability of AI-generated code.

3 Methodology

3.1 General Methodology

This study employs Gemini, Google's general-purpose large language model (LLM), as the generative AI model for code generation. Gemini offers several advantages that make it suitable for this research.

Gemini is trained on a massive dataset of text and code, enabling it to generate high-quality code in multiple programming languages, including Python, which is commonly used in data science. Gemini exhibits a strong ability to understand the context of the code and generate code that is relevant to the specific task and data at hand.

While Gemini offers significant potential for code generation, it is important to acknowledge its limitations. As with any LLM, Gemini may exhibit biases present in the training data, which could lead to biased or unfair code generation. The quality of the generated code is highly dependent on the quality of the input prompts. While Gemini can generate code based on patterns and examples, its reasoning abilities are still limited. It may struggle with complex or abstract coding tasks that require deeper understanding and problem-solving skills.

This research aims to address these limitations by systematically investigating the effectiveness of guiding Gemini to generate high-quality code for specific data science tasks. By understanding the strengths and limitations of Gemini and employing effective prompt engineering strategies, we aim to maximize the potential of this powerful AI model for code generation in the data science domain.

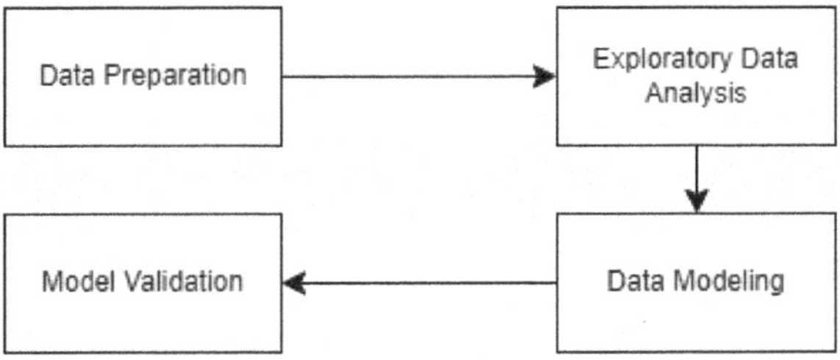

Fig. 1. Data Analysis Pipeline.

The diagram below outlines the specific four data science tasks that have been selected for this study.

The dataset in use for this research was the well-known iris dataset that is publicly available online.

3.2 Evaluation Metrics

To assess the quality of the code generated by Gemini, this study will look at two primary components: code correctness and model accuracy. By analyzing the results based on these metrics, this study aims to identify best practices for prompt engineering and contribute to a better understanding of how to effectively utilize generative AI models for code generation in the data science field. The following prompt will be used to start the process with Gemini:

Using the attached file, perform the following tasks:

1. **Generate histograms for:**
   (a) SepalLength
   (b) SepalWidth
   (c) PetalLength
   (d) PetalWidth
2. **How many unique species are in the dataset?**
3. **Generate a scatter plot for:**
   (a) SepalWidth vs. SepalLength
   (b) PetalWidth vs. SepalLength
4. **Split the data into training and testing datasets (80/20).**
5. **Create a predictive regression model that can classify the Species based on all combinations of SepalLength, SepalWidth, PetalLength, and PetalWidth using the training dataset.**
6. **Report the model accuracies on the testing dataset.**

4 Results

The following figures and table represent the results from the generative AI assisted Data Analysis.

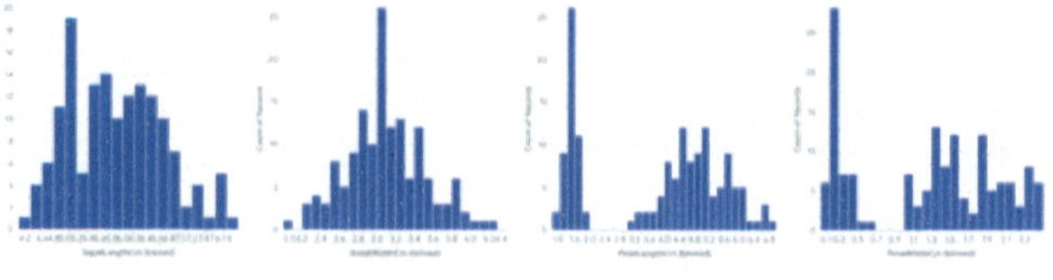

Fig. 2. Histograms generated by Gemini.

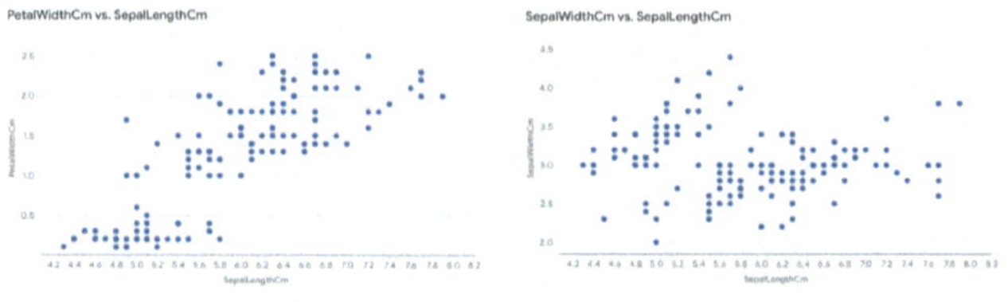

Fig. 3. Scatter plots generated by Gemini.

The results indicate that Gemini can generate accurate and efficient code for various data science tasks. The generated histograms and scatter plots were consistent with those produced using traditional coding methods. Additionally, the predictive models created by Gemini achieved high accuracy on the testing dataset. While some limitations were identified the overall results suggest that Gemini has the potential to be a valuable tool for data scientists.

5 Discussion and Conclusions

From the results, the generative AI was able to provide functionally accurate code that generated accurate plots. Additionally, while there is some room for improvement in their labeling, this can be addressed by modifying follow-up prompts to customize the look and feel of the plots. Table 1 also highlights the accuracy of the different predictive models created by Gemini, with a combination of predictors used ranging from a single one, to combinations of predictors that are equally accurate. Identical results were received by a data scientist working on the same problem using R and Python.

While this provides some evidence that Gen AI assisted data analytics is possible, a deeper exploration of how far we can leverage Gen AI is needed. One drawback of the data set used is that it was properly preprocessed, with no missing values, or any invalid values, therefore the capability of Gen AI to handle such a situation was not tested.

Based on the preceding, there is some evidence to suggest that Gen AI can be used to perform data analysis. While this use case employed some of the fundamentals, the accuracy is undeniable, as well as the formatting of the results. With some basic training in prompt engineering, and data analysis, a slightly trained individual will be able to perform complex data analysis.

Table 1. Accuracy values for predictor variables.

Predictor Variables	Accuracy
SepalLengthCm	0.83333
SepalWidthCm	0.60
PetalLengthCm	1
PetalWidthCm	1
SepalLengthCm, SepalWidthCm	0.90
SepalLengthCm, PetalLengthCm	1
SepalLengthCm, PetalWidthCm	1
SepalWidthCm, PetalLengthCm	1
SepalWidthCm, PetalWidthCm	1
PetalLengthCm, PetalWidthCm	1
SepalLengthCm, SepalWidthCm, PetalLengthCm	1
SepalLengthCm, SepalWidthCm, PetalWidthCm	1
SepalLengthCm, PetalLengthCm, PetalWidthCm	1
SepalWidthCm, PetalLengthCm, PetalWidthCm	1

5.1 Handling Complexity and Ethical Considerations

While our study demonstrates the potential of Gemini for code generation in data science, it is essential to acknowledge its limitations and the ethical considerations associated with its use. Our study primarily focused on the preprocessed Iris dataset, which may not sufficiently challenge Gemini's capabilities in handling real-world complexities like missing or noisy data. Future work should explore the effectiveness of prompt engineering in guiding Gemini to address such complexities. This could involve testing different prompt structures or levels of detail to identify best practices for handling complex data scenarios.

Furthermore, it is important to recognize that the quality of AI-generated code, including code produced by Gemini, is highly dependent on the quality of the input prompts and the training data. Biases present in the training data could lead to biased or unfair code generation, and Gemini's reasoning abilities are still limited, particularly for complex or abstract coding tasks.

From an ethical standpoint, the use of generative AI in data science raises concerns about the potential displacement of data scientists, the perpetuation of biases, and the misuse of AI-generated code. It is essential to address these concerns by promoting responsible AI practices, ensuring transparency in AI-generated code, and providing clear guidelines for the ethical use of generative AI in data science.

5.2 Strengths and Limitations

Our analysis reveals that Gemini possesses notable strengths in code generation for various data science tasks, including data cleaning, exploratory data analysis

(EDA), and machine learning model building. It effectively generates accurate and efficient code for tasks such as creating histograms, scatter plots, and predictive models, demonstrating its capability to automate these aspects of data science workflows. Additionally, Gemini exhibits a good understanding of code context, generating code relevant to the specific task and data. However, Gemini also presents certain limitations.

The quality of the generated code is heavily reliant on the quality of the input prompts, emphasizing the importance of effective prompt engineering. Furthermore, while Gemini can generate code based on patterns and examples, its reasoning abilities are limited, particularly for complex or abstract coding tasks that demand deeper understanding and problem-solving. As with other large language models, Gemini may exhibit biases present in its training data, potentially leading to biased code generation. Lastly, our study primarily focused on a preprocessed dataset, leaving its capabilities in handling real-world data complexities, such as missing or noisy data, less explored.

5.3 Future Work

Future work will focus on addressing the limitations identified in this study and further exploring the potential of Gemini for code generation in data science. Specifically, we plan to investigate more complex datasets and tasks to assess Gemini's capabilities in handling real-world complexities. We will evaluate its performance on more complex datasets with features such as missing or noisy data. This will involve exploring a wider range of data science tasks, including those that require more advanced coding techniques and problem-solving skills, such as hyperparameter tuning of deep learning models.

A comprehensive evaluation of Gemini's performance will be conducted by comparing it with other code generation tools like GitHub Copilot and Amazon CodeWhisperer using standardized datasets. This will provide a clearer understanding of Gemini's strengths and weaknesses relative to other available tools. To improve the presentation of results and facilitate a clearer understanding of the findings, we will incorporate additional visualizations. This will include exploring different types of visualizations to represent the performance of Gemini under various conditions and prompt engineering techniques. We will conduct a more systematic exploration of different prompt engineering techniques to identify best practices for generating high-quality code with Gemini. This will involve testing various prompt structures, levels of detail, and formats to determine their impact on the generated code's accuracy, efficiency, and maintainability.

To assess Gemini's ability to generalize and adapt, we will design experiments to evaluate its performance on unseen or novel coding problems. This will help determine how well Gemini can extrapolate from its training data and apply its knowledge to new and unfamiliar tasks. We will incorporate fuzzy datasets to evaluate how well Gemini performs data cleaning tasks within the data analysis pipeline. This will provide insights into Gemini's ability to handle real-world data imperfections and inconsistencies, further evaluating its robustness and practicality for data science applications. By addressing these areas in future research, we aim to contribute to a more comprehensive understanding of the potential and limitations of generative AI for code generation in data science.

Acknowledgements. The authors would like to acknowledge the financial support of the Department of Education under grant funding from Grambling State University. The authors also express their sincere gratitude to Dr. Stacey Duhon of their invaluable support through this research Project. Support from the LIDSS grant at Grambling State University is gratefully acknowledged.

References

1. Austin, J., et al.: Program synthesis with large language models. arXiv:2108.07732 (2021)
2. Chen, M., et al.: Evaluating large language models trained on code. arXiv:2107.03374 (2021)
3. Dehaerne, E., Dey, B., Halder, S., Gendt, S.D., Meert, W.: Code generation using machine learning: a systematic review. IEEE Access **10**, 82434–82455 (2022)
4. Le, K.T., Andrzejak, A.: Rethinking ai code generation: a one-shot correction approach based on user feedback. Autom. Softw. Eng. **31**(60), 1–42 (2024)
5. Le, T., Chen, H., Babar, M.A.: Deep learning for source code modeling and generation: models, applications, and challenges. ACM Comput. Surv. **53**(3), 1–38 (2020)
6. LinkedIn: The skills companies need most in 2023—and how to learn them (2023). https://www.linkedin.com/pulse/linkedin-2023-most-in-demand-skills-learn-companies-need-mrinmoy-paul
7. National Academies of Sciences, Engineering, and Medicine: Data Science for Undergraduates: Opportunities and Options. National Academies Press, Washington (2018)
8. Santos, B.L., Carvalho, L.: Educational data science: essentials, approaches, and tendencies. Springer Nature (2023)
9. Sharma, T., Spinellis, D.: Do we need improved code quality metrics? arXiv preprint arXiv:2012.12324 (2020)
10. Svyatkovskiy, A., Deng, S.K., Fu, S., Sundaresan, N.: Intellicode compose: code generation using transformer. In: Proceedings of the 28th ACM Joint European Software Engineering Conference and Symposium on the Foundations of Software Engineering, pp. 1433–1443 (2020)
11. U.S. Bureau of Labor Statistics: Data scientists. Occupational Outlook Handbook. https://www.bls.gov/ooh/math/data-scientists.htm
12. Yetiştiren, B., Özsoy, I., Ayerdem, M., Tüzün, E.: Evaluating the code quality of ai-assisted code generation tools: an empirical study on github copilot, amazon codewhisperer, and chatgpt. arXiv preprint arXiv:2304.10778 (2023)

Comparative Analysis of Pretrained Machine Learning Models in Cross Domain Applications

Dileon Saint-Jean[1,2](✉), Baker Al Smadi[2], Syed Raza[1], Sari Linton[2], and Ugochukwu Igweagu[2]

[1] Department of Computer Science and Data Science, American Public University System, 111 W. Congress Street, Charles Town, WV 25414, USA
dileon.saintjean@gmail.com, syed.raza@mycampus.apus.edu

[2] Department of Computer Science and Digital Technologies, Grambling State University, 403 Main Street, Grambling, LA 71245, USA
alsmadib@gram.edu

Abstract. The proliferation of pretrained machine learning models has facilitated their application across a variety of domains, from finance to social media analysis. However, these models often exhibit domain-specific biases, limiting their cross-domain performance. This study investigates the accuracy and reliability of various sentiment analysis models when applied to datasets outside of their primary training domain. It looks at eight different pretrained models, two from the financial domain, and seven from general sentiment analysis models. By analyzing the performance of these models on both finance-based datasets and social media-based datasets, we highlight the need for techniques like Structural Correspondence Learning (SCL) to bridge the domain gap and improve model generalization. The study also discusses the importance of domain-specific fine-tuning, data augmentation, transfer learning techniques, and hybrid approaches to further enhance the performance of pretrained models in cross-domain sentiment analysis tasks.

Keywords: Sentiment Analysis · Cross Domain Applications · Pretrained Models · Financial Data · Text Data · Cross-Domain · Domain Adaptation · Machine Learning · Predictive Modeling · Deep Learning

1 Introduction

Pretrained models on Hugging Face, such as BERT and RoBERTa variations, have become popular due to their ease of use and adaptability. While these

B. A. Smadi, S. Raza, S. Linton, and U. Igweagu—These authors contributed equally to this work.

S. Latifi (Ed.): ITNG 2025, AISC 1463, pp. 454–460, 2025.
https://doi.org/10.1007/978-3-031-89063-5_39

models have demonstrated impressive performance on a wide range of natural language processing tasks, their effectiveness can be hindered by domain shifts. When applied to domains significantly different from their training data, these models may struggle to capture domain-specific nuances and subtleties.

This study aims to explore how well sentiment analysis models, pretrained on datasets related to social media or general textual data, perform when used on financial data versus social media reviews. By understanding these performance variations, we can identify the limitations of direct model application and highlight the need for techniques like Structural Correspondence Learning (SCL) to enhance cross-domain performance. SCL can help bridge the domain gap by aligning the semantic and syntactic structures of the source and target domains, leading to improved model generalization and accuracy.

By investigating the performance of pretrained models on diverse domains and advocating for techniques like SCL, this study contributes to the advancement of robust and adaptable machine learning models capable of handling cross-domain challenges.

2 Literature Review

The need for domain adaptation in machine learning arises when models trained on data from a specific source domain perform poorly when applied to a different target domain [1]. This challenge is particularly pronounced in natural language processing (NLP), where domain-specific language and context can impact model performance [1]. Discriminative learning methods, while effective within consistent datasets, often suffer in cross-domain applications due to differences in data distribution.

Structural Correspondence Learning (SCL) addresses this issue by creating a shared feature representation between source and target domains [1]. By learning feature relationships, SCL enhances the generalizability of models, allowing them to better adapt to target domains with minimal labeled data. Although SCL shows promising results, it is not without disadvantages such as its reliance on pivot features which may not be readily identifiable [6]. Recent advancements in deep learning have led to the development of more sophisticated domain adaptation techniques, such as adversarial networks and transformers, which have shown impressive performance in various NLP tasks [4,7].

Research has shown that this correspondence-driven adaptation significantly boosts performance, particularly in low-resource environments [2,3]. For example, in tasks like part-of-speech tagging, SCL demonstrated improved accuracy in target domains by effectively modeling feature correlations [1]. Additionally, careful consideration of data selection is important for successful domain adaptation. For example, selecting source data that is closely related to the target do-main can significantly improve performance [5].

Closing the gap between source and target domains using SCL and other domain adaptation techniques provides a possible solution to enhance the robustness and generalizability of NLP models, especially in low resources use cases.

3 Methodology

This study evaluates the performance of pretrained language models on two distinct datasets for cross-domain sentiment analysis. To ensure a robust and comprehensive assessment, a structured evaluation methodology was employed, encompassing data preprocessing, model inference, and performance evaluation.

3.1 Data Preprocessing

Raw text data inherently requires preprocessing to be compatible with machine learning models. First, tokenization was performed, where text was segmented into individual units, either words or subwords, depending on the specific architecture of each pretrained model. This step is crucial as it allows the model to process the input text effectively. Next, truncation and padding were applied to standardize input sequence lengths. This ensures consistency and prevents issues arising from varying input sizes during model training and inference.

3.2 Model Inference

To assess the inherent cross-domain capabilities of pretrained models, a zero-shot learning approach was adopted. This means the models were directly applied to the target datasets without any fine-tuning or additional training on domain-specific data. This approach provides valuable insights into the generalization ability of these models. Each model then generated sentiment predictions for every sample in the datasets.

3.3 Performance Evaluation

Model performance was rigorously evaluated using multiple metrics. Accuracy, calculated as the ratio of correct predictions to total predictions, served as the primary metric for overall performance assessment. However, to gain a more nuanced understanding of model behavior, additional metrics were employed. Precision, recall, and F1-score were computed for each sentiment class (e.g., positive, negative, neutral), providing a detailed analysis of class-wise performance and revealing potential biases or weaknesses. Furthermore, confusion matrices were generated to visualize the distribution of correct and incorrect predictions, offering a clear picture of the types of errors made by each model.

This multi-faceted evaluation methodology allows for a thorough analysis of the strengths and weaknesses of different pretrained models when applied to the challenging task of cross-domain sentiment analysis. By combining overall accuracy with class-specific metrics and visualization tools, this approach provides a comprehensive understanding of model performance and generalization capabilities.

4 Results

The results of the various models' performance are outlined below in Table 1.

Table 1. Performance metrics for different models on Financial Data.

Model	Accuracy	Precision	Recall	F1-Score
distilbert-base-uncased-finetuned-sst-2-english	16.00%	5.33%	33.33%	9.20%
bert-base-multilingual-uncased-sentiment	16.00%	5.33%	33.33%	9.20%
sentiment-roberta-large-english	16.00%	5.33%	33.33%	9.20%
ProsusAI/finbert	74.00%	71.92%	73.72%	72.25%
cardiffnlp/twitter-roberta-base-sentiment-latest	62.00%	67.61%	55.53%	57.22%
finiteautomata/bertweet-base-sentiment-analysis	16.00%	5.33%	33.33%	9.20%
aychang/bert-base-cased-trec-coarse	16.00%	5.33%	33.33%	9.20%
bhadresh-savani/distilbert-base-uncased-emotion	16.00%	5.33%	33.33%	9.20%

The Bar charts below show the performance of these models on the two datasets, with Fig. 1 representing the financial dataset in blue and Fig. 2 the yelp review dataset in green.

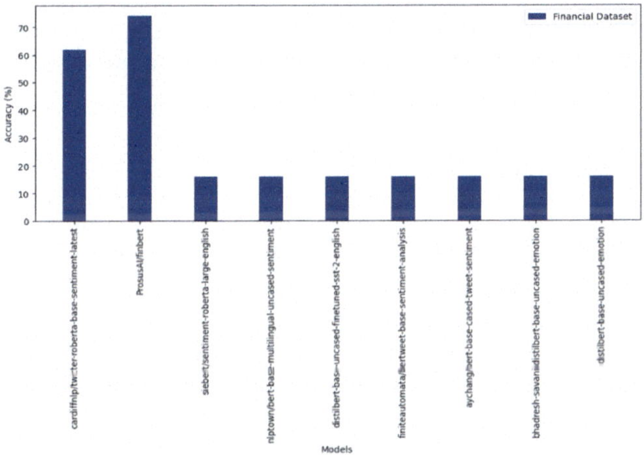

Fig. 1. Comparison of model accuracy on Financial Dataset.

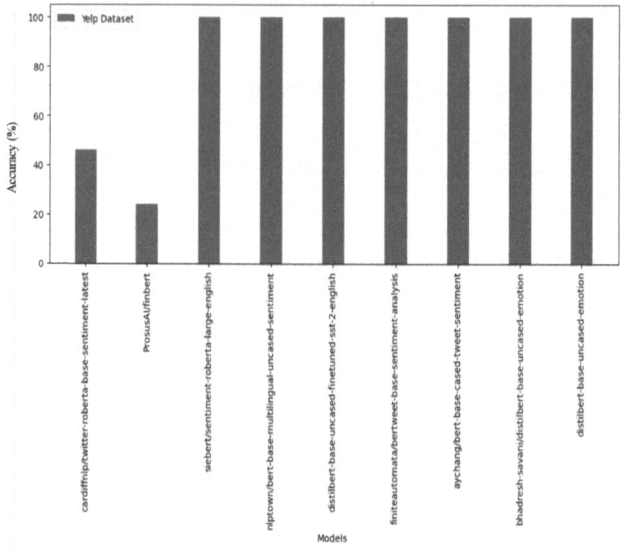

Fig. 2. Comparison of model accuracy on Text Dataset.

5 Analysis and Discussion

The study reveals that the cross-domain performance of pretrained sentiment analysis models is influenced by several factors, including the domain specificity of training data, model architecture and size, data preprocessing and feature engineering, and target domain characteristics.

5.1 Domain Specialization

Models trained on domain-specific data exhibit superior performance on datasets from the same domain, while the architecture and size of the pretrained model can affect its ability to generalize to new domains. The choice of data preprocessing techniques and feature engineering methods can impact the model's ability to capture relevant information from the text data, and the characteristics of the target domain, such as the type of language used, the level of formality, and the presence of domain-specific jargon, can affect the model's performance.

5.2 Transferability

The observed performance variations across different models and datasets can be attributed to several reasons, including domain shift, model bias, and lack of domain-specific knowledge. The mismatch between the training data and the target domain data can lead to a significant drop in performance, and pretrained models may exhibit biases toward the specific characteristics of their training data, which can hinder their performance on data from different domains.

General-purpose models may lack the domain-specific knowledge necessary to accurately analyze sentiment in specialized domains.

Specific characteristics of the models contribute to their success or failure in cross-domain sentiment analysis tasks. For example, ProsusAI/finbert's superior performance on the financial dataset can be attributed to its specific training on financial data, enabling it to capture the nuances of financial language. The success of general-purpose models on the Yelp Review dataset suggests their ability to handle informal, subjective language. However, their lower performance on the financial dataset highlights their limitations in generalizing to domain-specific language patterns.

5.3 Model Limitations

To gain a deeper understanding of the results, further analysis could explore error analysis, feature importance, and model interpretability. Examining the types of errors made by each model can reveal specific areas of weakness and provide insights for improvement. Identifying the features that contribute most to the model's predictions can help understand the factors driving its performance. Utilizing techniques to interpret the model's decision-making process can provide insights into its behavior and potential biases. By conducting a more in-depth analysis of the results, researchers can gain a deeper understanding of the factors influencing cross-domain performance and develop more effective strategies for addressing the challenges of domain adaptation in sentiment analysis.

6 Conclusions and Next Steps

This research demonstrated the effectiveness of pretrained language models in achieving high accuracy on various NLP tasks but also revealed limitations in cross-domain adaptation due to domain shift. To address this, the next steps involve expanding the scope of the study and implementing Structural Correspondence Learning (SCL).

The current study is limited to a small set of datasets and models. Future research should include a wider range of datasets from diverse domains, such as legal, medical, and scientific texts. Additionally, the study should evaluate more pretrained language models, including those with different architectures and training data.

SCL offers a promising approach by identifying and aligning shared semantic and syntactic structures between source and target domains, ultimately enhancing model generalization and accuracy. The implementation of SCL will involve a three-step process. First, pivot features will be identified, encompassing semantic features like word and phrase similarity, syntactic features such as common syntactic structures, and domain-specific features relevant to the task. Second, correspondences between these features will be learned through feature mapping, employing similarity measures like word embeddings, and utilizing alignment techniques such as canonical correlation analysis or deep learning methods. Finally, the pretrained model will be adapted by fine-tuning it on target domain data, incorporating learned correspondences, and applying transfer learning based on identified correspondences. Additionally, domain-specific regularization will be introduced to encourage the model to learn domain-invariant features.

By expanding the scope of the study and implementing SCL, we expect to gain a deeper understanding of the factors influencing cross-domain performance and develop more effective strategies for addressing the challenges of domain adaptation in sentiment analysis. This approach is expected to improve the model's ability to generalize across different domains and enhance its overall accuracy.

Acknowledgements. The authors would like to acknowledge the financial support of the Department of Education under grant funding from Grambling State University. The authors also express their sincere gratitude to Dr. Stacey Duhon of their invaluable support through this research Project. Support from the LIDSS grant at Grambling State University is gratefully acknowledged.

References

1. Blitzer, J., McDonald, R., Pereira, F.: In: Proceedings of the 2006 Conference on Empirical Methods in Natural Language Processing, pp. 120–128 (2006)
2. Devlin, J., Chang, M.W., Lee, K., Toutanova, K.: Bert: pre-training of deep bidirectional transformers for language understanding. In: Proceedings of the 2019 Conference of the North American Chapter of the Association for Computational Linguistics: Human Language Technologies, Volume 1 (Long and Short Papers), pp. 4171–4186 (2019)
3. Ganin, Y., et al.: Domain-adversarial training of neural networks. J. Mach. Learn. Res. **17**(1), 2096–2030 (2016)
4. Glorot, X., Bordes, A., Bengio, Y.: Domain adaptation for large-scale sentiment classification: a deep learning approach. In: Proceedings of the 28th International Conference on Machine Learning (ICML-11), pp. 513–520 (2011)
5. Kamanth, U., Liu, J., Whitaker, J.: Transfer Learning: Domain Adaptation. Springer, Cham (2019)
6. Li, S., Hu, X., Lin, L., Liu, A., Wen, L., Yu, P.S.: A multi-level supervised contrastive learning framework for low-resource natural language inference. IEEE/ACM Trans. Audio Speech Lang. Process. **31**, 1771–1783 (2023). https://doi.org/10.1109/TASLP.2023.3270771
7. Lo, P.C., Lee, Y.Y., Chen, H.H., Kwee, A.T., Lim, E.P.: Contrastive learning approach to word-in-context task for low-resource languages. In: Proceedings of the 37th AAAI Workshop on Knowledge Augmented Methods for Natural Language Processing, pp. 1–8. Washington, DC (2023)

Can ChatGPT Serve as a Cognitive Stimulus? A Proposed Model for Assessing the Impact on Graduate Students' Perceptions of Cognitive and Developmental Skills

Hamdan Alabsi[1(✉)], Majed Almotairi[2], Mohammed Alyami[3], Yahya Alqahtani[4], Rand Obeidat[1], and Faris Almansour[5]

[1] Management Information Systems Department, Bowie State University, Bowie, MD 20715, USA
{halabsi,robeidat}@bowiestate.edu
[2] Information Systems Department, Indiana Institute of Technology, Fort Wayne, IN 46803, USA
malyami@coloradomesa.edu
[3] Computer Information Systems Department, Colorado Mesa University, Grand Junction, CO 81501, USA
MMAlmotairi@indianatech.edu
[4] Computer Science Department, Jazan University, Jazan, Saudi Arabia
yalqahtani@jazanu.edu.sa
[5] Information Technology Department, Saudi Electronic University, Riyadh, Saudi Arabia
f.almansour@seu.edu.sa

Abstract. This research explores whether ChatGPT – as one significant contributor to the integration of Artificial Intelligence (AI) into daily life- can act as a cognitive stimulus to improve graduate students' learning experiences in the classroom. This study proposes a theoretical model to assess graduate students' perceptions of ChatGPT to explore whether its use positively or negatively affects stimulating their cognitive skills. Cognitive Load Theory was used in this research to examine how ChatGPT can be a mediator factor impacting cognitive load reduction, critical thinking, creativity, and problem-solving abilities. It is worth noting that this model has not been tested yet; future research will focus on data collection and hypothesis testing.

Keywords: ChatGPT · Artificial Intelligence (AI) · Cognitive Skills · Cognitive Load · E-learning

---

M. Almotairi, M. Alyami, Y. Alqahtani, R. Obeidat, and F. Almansour—These authors contributed equally to this work.

S. Latifi (Ed.): ITNG 2025, AISC 1463, pp. 461–468, 2025.
https://doi.org/10.1007/978-3-031-89063-5_40

1 Introduction

Technological advancements, particularly in Automation, Machine Learning (ML), and AI, have profoundly influenced many aspects of daily life, reshaping industries such as healthcare, business, and education. As AI applications continue to grow at a rapid pace, debates about its capabilities and limitations have surfaced, involving not only experts in academia and industry but also the general public. These debates often center on AI's intelligence, capabilities, and scalability. Many individuals maintain conservative views on AI, seeing it as a technology with limited capabilities and specific intended functions. In contrast, others view it as highly proficient at efficiently performing repetitive tasks and replicating human physical and cognitive processes. A more optimistic perspective sees AI's potential as virtually limitless, exceeding expectations. Despite these differing perspectives, it is evident that we are entering what could be called an "AI-driven world," signaling a new phase in human progress. A key player in this transformation is ChatGPT, the widely recognized chatbot developed by OpenAI. Since its launch on November 30, 2022, ChatGPT rapidly gained popularity, reaching an estimated 100 million monthly active users within just two months [18,25]. Given the growing impact of AI tools like ChatGPT, academic institutions must acknowledge the significant increase in their usage among students globally and begin incorporating them into educational practices. This paper proposes a model to explore graduate students' perceptions of using ChatGPT to complete university assignments. Furthermore, it aims to investigate whether ChatGPT can serve as a cognitive stimulus, reduce cognitive overload, and ultimately enhance problem-solving, critical thinking, and creativity, leading to improved learning outcomes. This paper is structured as follows: the Research Motivation, Research Gap and Questions, followed by the Literature Review, Theory & Hypotheses, Methodology, and Conclusion sections.

1.1 Research Motivation

The adoption of ChatGPT among college students is rapidly growing, with an increasing number already integrating its features into their academic routines [10]. A survey of 1,000 undergraduate and graduate students found that 51% use ChatGPT to help complete coursework more effectively [24]. Another study of 588 students revealed that 37% actively use ChatGPT [20]. Moreover, ChatGPT has been shown to enhance productivity significantly. An MIT study found that it led to a 40% increase in productivity and a 20% improvement in quality across various tasks, such as writing cover letters, drafting emails, and conducting cost-benefit analyses, benefiting professionals like data analysts and HR specialists [15,29]. Recent research also suggests that the growing use of ChatGPT is contributing to improvements in students' performance and the quality of their work. Our study proposes a model to explore whether ChatGPT helps reduce cognitive load and enhance cognitive skills, such as problem-solving, critical thinking, and creativity, particularly among university students. Although a wide variety of e-learning tools are available, generative AI tools like ChatGPT have become particularly impactful for students. Many e-learning platforms have already incorporated ChatGPT in roles such as virtual tutoring, helping students tackle academic challenges. Consequently, further integration of such tools in classroom environments is expected in the near future, particularly for customized learning [3,5,6].

1.2 Research Gap and Questions

Despite extensive research on the ethical implications of AI adoption in academia and its effects on students' experiences and outcomes, a significant gap remains in the literature regarding the impact of AI on students' cognitive processes, particularly in graduate-level academic settings [2,19]. Existing studies largely focus on undergraduate students and their interactions with AI technologies like ChatGPT, emphasizing concerns that such tools may hinder cognitive development at this early stage due to gaps in foundational knowledge and understanding [11,22,28]. In contrast, our research shifts the focus to graduate students, who have already completed foundational courses—typically referred to as liberal arts, core, and elective courses—and often possess professional experience or hold current professional positions. As such, they have developed a more advanced knowledge base compared to undergraduates. This study aims to explore the following question:

Graduate-level university students experience a reduced cognitive load that enhances cognitive abilities when they use AI-supported tools such as ChatGPT to complete their academic work.

2 Literature Reviews

The launch of ChatGPT by OpenAI prompted numerous researchers to explore its functionality, ethical implications, and effects on education and cognitive skills. Bai et al. suggest that educators can leverage ChatGPT to foster critical thinking, problem-solving, and personalized learning. However, they also caution against the risks of excessive dependence on AI tools, which could undermine critical thinking and memory retention [7]. Ethical concerns surrounding ChatGPT have been examined in several studies. For example, Farhi et al. analyzed students' perspectives on the ethical implications of ChatGPT, finding that while the chatbot can be useful, it raises concerns about academic integrity. The researchers advocate for establishing guidelines to support policymaking in educational institutions [13]. Similarly, a study on university students' perceptions of ChatGPT found that while students appreciate its time-saving capabilities, subject coverage, and personalized feedback, they also express concerns about the reliability and accuracy of its sources and citations [23]. On the one hand, Holmes et al. argue that AI in education can be integrated into curricula to enhance student outcomes [17]. On the other hand, Saritha et al. warn that incorporating AI technologies in education demands careful consideration, highlighting a decline in critical thinking and problem-solving skills following the adoption of AI tools like ChatGPT [26]. Similarly, Fulbright and Morrison's research on cognitive augmentation through ChatGPT highlights that the tool does not always yield positive cognitive outcomes. They noted instances where ChatGPT provided misleading information, potentially harming cognitive development [14]. In contrast, Mollick suggested that AI could play a crucial role in facilitating learning, particularly in terms of knowledge transfer [22]. One possible reason some students do not experience cognitive enhancement is a lack of proper training in how to interact with ChatGPT. Without sufficient guidance on how to use the tool effectively and evaluate the quality of its output, students may fail to fully benefit from its potential. Therefore, implementing comprehensive training programs that teach students how to engage with ChatGPT and assess its information is essential for maximizing its educational impact.

3 Theory and Hypotheses

3.1 Connection to Relevant Theory

This research builds on John Sweller's (1988) Cognitive Load Theory, which posits that working memory has a limited capacity and emphasizes minimizing extraneous cognitive load to optimize learning. ChatGPT can help reduce cognitive load by providing quick access to vast amounts of information, enabling students to expand their knowledge base. This allows learners to conserve cognitive resources, directing their focus toward more complex tasks rather than those requiring less cognitive effort [4,9,27]. Cognitive load may also act as a mediating factor that enhances cognitive performance when using ChatGPT. This concept can be linked to the NASA Task Load Index (TLX), a tool designed to assess subjective workload. The TLX enables users to evaluate the cognitive load experienced while interacting with various human-machine interface systems [1].

3.2 Hypothesis Evaluation and Testing

This research will evaluate the following hypotheses to determine whether there is a correlational relationship or not:

Hypothesis (H1): Utilizing AI tools like ChatGPT to complete university assignments positively correlates with enhancing problem-solving skills through reducing cognitive load.
Hypothesis (H2): Utilizing AI tools like ChatGPT to complete university assignments positively correlates with enhancing Creativity through reducing cognitive load.
Hypothesis (H3): Utilizing AI tools like ChatGPT to complete university assignments positively correlates with enhancing critical thinking through reducing cognitive load.

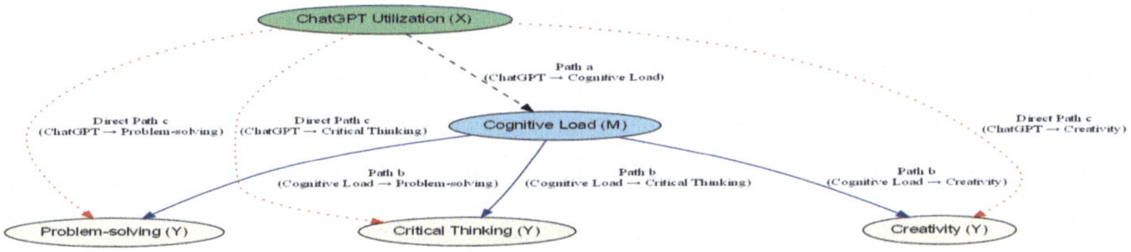

Fig. 1. The relationship among the variables in this proposed model.

3.3 Operationalization of Key Variables

Figure 1 presents a conceptual model illustrating the potential impact of ChatGPT (X) on cognitive abilities. Specifically, it suggests that the integration of ChatGPT into cognitive processes may alleviate Cognitive Load (M), thus fostering improvements in Problem-Solving (Y1), Creativity (Y2), and Critical Thinking (Y3).

4 Methodology

For this research, the primary unit of analysis will be the "individual," with a focus on graduate students at universities. We aim to explore whether these students experience cognitive stimuli when using AI-assisted tools, such as ChatGPT, to complete their academic work. The main data collection method will be a survey, with the questionnaire adapted from existing studies to align with the specific objectives of this research. Although we will draw on similar methodologies from prior research, we will tailor the approach to test our hypotheses [12,16,30].

4.1 Framework Process for Mediation Model

Figure 2 outlines the process flow of the analysis. The study begins by collecting student data on ChatGPT usage, cognitive load, problem-solving, creativity, and critical thinking. After preparing the data, a mediation model is set up to explore the relationships between these variables, with ChatGPT usage as the independent variable, cognitive load as the mediator, and cognitive outcomes as the dependent variables. Regression analyses are conducted to assess the effects of ChatGPT usage on cognitive load (Path a) and cognitive outcomes through cognitive load (Path b). The results are extracted and interpreted, focusing on direct, indirect, and total effects (Path c). Finally, the findings are visualized and summarized to understand how ChatGPT usage impacts cognitive skills comprehensively [8,21].

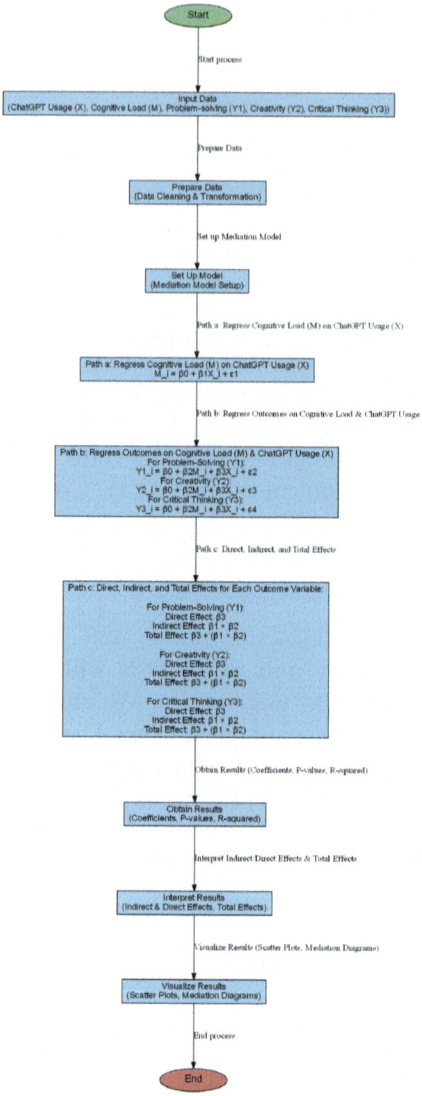

Fig. 2. Framework for Mediation Analysis of ChatGPT Usage on Cognitive Abilities

5 Conclusion

The main goal of this study is to develop a model to explore the potential relationship between ChatGPT usage and cognitive stimulation among graduate students working on their university assignments. The results are expected to support the proposed hypotheses by implementing and testing this model. A key limitation of this study is the absence of real-world data to test the proposed hypotheses objectively. In the future, we will collect data and test this proposed model. Furthermore, students might be affected by various biases, including the tendency to favor more accurate or socially acceptable responses, as well as the inclination to avoid answering certain questions altogether. These factors could introduce distortions, undermining the reliability and overall validity of the study's findings.

References

1. Tlx @ nasa ames - home. https://humansystems.arc.nasa.gov/groups/tlx/. Accessed December 2024
2. Academic Senate: Initial thoughts and suggestions regarding implications of large language models (llms) (e.g. chatgpt) for fall final assessments. https://academic-senate.berkeley.edu/implications-of-llms-chatgpt-for-final-assessments
3. Alabsi, H.Z., Almotairi, M., Alqahtani, Y., Alyami, M.A.: Enhancing students' job seeking process through a digital badging system. In: 2023 IEEE/ACIS 21st International Conference on Software Engineering Research, Management and Applications (SERA), pp. 276–281. IEEE (2023)
4. Ali, A.: Chat GPT: increasing or decreasing cognitive skills. Medium. https://medium.com. Accessed 29 Apr 2024
5. Almotairi, M., Alabsi, H.Z., Alqahtani, Y., Alyami, M.A., Aljazaeri, M.M., Song, Y.T.: Improving students' readiness toward the labor market through customized learning. In: International Conference on Software Engineering Research and Applications, pp. 169–181. Springer (2022)
6. Ayoubi, K.: Adopting chatgpt: pioneering a new era in learning platforms. Int. J. Data Netw. Sci. **8**(2), 1341–1348 (2024)
7. Bai, L., Liu, X., Su, J.: Chatgpt: the cognitive effects on learning and memory. Brain-X **1**(3), e30 (2023)
8. Bevans, R.: Multiple linear regression — a quick guide (examples). https://www.scribbr.com/statistics/multiple-linear-regression/
9. Clark, C., Kimmons, R.: Cognitive load theory. EdTechnica, pp. 109–115 (2023)
10. D'Agostino, S.: Chatgpt advice academics can use now. Inside Higher Ed. https://www.insidehighered.com
11. Essel, H.B., Vlachopoulos, D., Essuman, A.B., Amankwa, J.O.: Chatgpt effects on cognitive skills of undergraduate students: receiving instant responses from ai-based conversational large language models (llms). Comput. Educ. Artif. Intell. **6**, 100198 (2024)
12. Tyson, A.F.M.: What the data says about Americans' views of artificial intelligence. https://www.pewresearch.org
13. Farhi, F., Jeljeli, R., Aburezeq, I., Dweikat, F.F., Al-shami, S.A., Slamene, R.: Analyzing the students' views, concerns, and perceived ethics about chat gpt usage. Comput. Educ. Artif. Intell. **5**, 100180 (2023)
14. Fulbright, R., Morrison, M.: Does using chatgpt result in human cognitive augmentation? (2024). ArXiv:2401.11042
15. Gesikowski, C.: Chatgpt boosts productivity by 40% and quality by 20%, says mit study. Medium. https://bootcamp.uxdesign.cc. Accessed 19 Apr 2024
16. Guo, Y., Lee, D.: Leveraging chatgpt for enhancing critical thinking skills. J. Chem. Educ. **100**(12), 4876–4883 (2023)
17. Holmes, W., Bialik, M., Fadel, C.: Artificial Intelligence in Education. Promise and Implications for Teaching and Learning (2019)

18. Hu, K., Hu, K.: Chatgpt sets record for fastest-growing user base - analyst note. Reuters (2023). https://www.reuters.com
19. Huallpa, J.J., Al, E.: Exploring the ethical considerations of using chat gpt in university education. Period. Eng. Nat. Sci. (PEN) **11**(4), Art. no. 4 (2023)
20. Intelligent.com: 4 in 10 college students are using chatgpt on assignments. Intelligent. https://www.intelligent.com
21. Lee, H., et al.: A guideline for reporting mediation analyses of randomized trials and observational studies: the agrema statement. JAMA **326**(11), 1045 (2021)
22. Mollick, E.R., Mollick, L.: New modes of learning enabled by ai chatbots: three methods and assignments (2022). Rochester, NY: 4300783
23. Ngo, T.: The perception by university students of the use of chatgpt in education. Int. J. Emerg. Technol. Learn. (Online) **18**(17), 4–19 (2023)
24. Nietzel, M.T.: More than half of college students believe using chatgpt to complete assignments is cheating. Forbes. https://www.forbes.com. Accessed 19 Apr 2024
25. Ortiz, S.: What is chatgpt and why does it matter? here's what you need to know. ZDNET. https://www.zdnet.com
26. Saritha, D.D., et al.: Exploring chat gpt: a comprehensive analysis of performance, user perception, and satisfaction. IJO - Int. J. Bus. Manag. **7**(02) (2024). Art. no. 02
27. Sweller, J.: Chapter two - cognitive load theory. In: Mestre, J.P., Ross, B.H. (eds.) Psychology of Learning and Motivation, vol. 55, pp. 37–76. Academic Press (2011)
28. Tang, C., Mao, S., Naumann, S.E., Xing, Z.: Improving student creativity through digital technology products: a literature review. Thinking Skills Creat. **44**, 101032 (2022)
29. Winn, Z.: Study finds chatgpt boosts worker productivity for some writing tasks. MIT News — Massachusetts Institute of Technology. https://news.mit.edu
30. Zawiah, M., al.: Chatgpt and clinical training: perception, concerns, and practice of pharm-d students. J. Multidiscip. Healthc. **16**, 4099–4110 (2023)

Entity Resolution with Household Movement Discovery Using Google Generative AI

Muzakkiruddin Ahmed Mohammed(✉), John R. Talburt, Altaf Mohammed, and Khizer Syed

Information Quality, University of Arkansas at Little Rock, Little Rock, USA
{mmohammed6,jrtalburt,amohd6,skdaniyal}@ualr.edu

Abstract. This paper presents a prototype system for discovering patterns of household movements using the Google Generative AI (Gemini-1.5) large language model (LLM) and implemented through the Flask web framework. The system processes unstandardized name and address data to identify instances where two or more individuals share addresses at different points in time and subsequently move together to another location. The key advantage of using LLMs is their ability to understand data regardless of syntax quality, variations in address formats, and differences in how names or personal information are recorded. This research shows that an LLM is capable of detecting these patterns and provides a roadmap for automating the extraction of complex relationships between household members. This will improve entity resolution (ER) for address-based movements, and explore the efficacy of generative models in handling unstructured, low-quality data.

Keywords: Unstructured Data Handling · Address Discrepancies · Data Standardization · Semantic Matching · Relationship Inference · Data Preprocessing · Record Linkage · Household Dynamics · Address Variability · Migration Patterns

1 Introduction

Entity resolution (ER) is essential in applications where it is necessary to identify and link multiple references to the same entity. This can be challenging in scenarios with missing and poor-quality data. An additional challenge with demographic data is that over time, the same person may have different addresses. Without explicit knowledge of these changes of address or other connecting information, it may not be possible to make a direct link by matching two references to the same person at different addresses. In a large population, the probability of two different persons having the same name is too high to establish a confident match.

However, a method of indirectly linking persons between two different addresses is sometimes possible when more than one member of a household participates in the same move. These situations are referred to as Household Moves. By determining that two different addresses are associated with the same group of names, it is possible to infer

S. Latifi (Ed.): ITNG 2025, AISC 1463, pp. 469–481, 2025.
https://doi.org/10.1007/978-3-031-89063-5_41

links between the references by name even though the addresses are different. While the probability that the same name at two different addresses refers to the same person is low, the probability that the same two names at two different addresses are the same two persons is much higher, and for three names, even higher still.

Standard ER blocking processes are primarily focused on finding direct, pairwise matches, i.e., two references that have enough similarity to reach a match threshold for linking. Because linking inferred from household moves involves finding a specific pattern of at least four related references, traditional blocking methods do not help. The challenge of tracking individuals or households based on their address data is compounded by the frequent occurrence of misspellings, inconsistent address formats, and incomplete records. Identifying household move patterns is especially difficult due to the inherent variability in how addresses and names are recorded, often leading to errors in traditional methods.

In this paper, we propose a novel approach for household movement discovery using Large Language Models (LLMs) to automate the detection of individuals or groups of individuals who share addresses at different points in time and move together to another location. Household movement discovery refers to the process of identifying when groups of two or more individuals relocate together between addresses. This type of movement analysis can not only improve recall rates in standard linking processes, but it can also provide insights into migration patterns, urban planning, and even fraud detection. Tracking these movements, however, is a difficult task when working with unstandardized and heterogeneously structured references.

Advanced techniques leveraging AI and LLMs, such as Google's Generative AI (Gemini-1.5), significantly enhance our ability to process unstructured data by understanding language, semantics, and syntax. LLMs excel at grasping the nuances of language, allowing them to interpret variations in naming conventions and address formats effectively. Their natural language understanding (NLU) capabilities enable them to handle poor-quality data, decipher contextual meaning, and resolve inconsistencies. This comprehensive understanding of language helps LLMs navigate the complexities of unstructured data, which traditional ER techniques often struggle to address due to their reliance on clean, structured data and extensive preprocessing.

The system described in this paper integrates Gemini-1.5 with the Flask web framework, creating an interactive tool that automatically processes raw address data. By leveraging LLMs, our system automates entity resolution, simplifies the identification of household movements, and provides an accurate, scalable solution for working with complex data.

2 Related Work

Household discovery within entity resolution (ER) has become an increasingly important area, especially with advancements in generative AI and large learning models, which offer new ways to address complex data challenges. Traditionally, household discovery relied on deterministic matching, where attributes like addresses and surnames were used to link records. However, these methods often struggled with inconsistencies and variations in data, particularly when records contained typos or alternative spellings [1].

Probabilistic methods later emerged, improving accuracy by assigning likelihood scores to determine household membership and accounting for minor inconsistencies. These methods laid the groundwork for more sophisticated models, especially with the rise of machine learning (ML) techniques, which have been used to enhance clustering and classification in household discovery [2].

The introduction of generative AI, particularly through Generative Adversarial Networks (GANs) and large learning models like BERT and GPT, has further revolutionized household discovery. GANs, for example, generate synthetic data that mimic real records, providing more data for model training and improving ER accuracy without requiring extensive labeled datasets [7]. This is especially useful in household discovery, where labeled data can be scarce and costly to produce. Synthetic data generated by GANs can simulate common variations in household records, such as slight discrepancies in names or addresses, making the model more robust and adaptable to real-world conditions. In addition, large language models (LLMs) such as GPT-4 are increasingly used to augment household data by generating alternative representations of attributes, like different address formats or family name spellings, that occur naturally in datasets. These models capture subtle variations, allowing for more nuanced matching within household groups [3].

Graph-based approaches in household discovery have also benefited from large learning models. In graph-based methods, individuals are represented as nodes, and their relationships, such as shared addresses, are represented as edges. With the advent of transformers and attention mechanisms, these graph representations can be more accurately analyzed, as attention layers allow the model to dynamically weigh important attributes when making household matching decisions. This dynamic approach is particularly effective in managing complex household relationships within large datasets, as illustrated by Khan Mohammed et al. [4], where the model could detect relationships based on subtle data connections that traditional approaches might overlook. Moreover, Bayesian frameworks incorporating LLMs have enabled more flexible probabilistic approaches, leveraging prior knowledge about household structures alongside observed data to assign probabilities to household associations. This is particularly useful for situations where household compositions are uncertain or evolve over time [5].

Generative AI also aids in data privacy—a critical concern in household discovery due to the sensitive nature of household data. Techniques such as differentially private GANs have been developed to generate synthetic data while ensuring no specific individual's data is retained, thus safeguarding privacy. This privacy-preserving synthetic data allows researchers to train and share ER models without risking the exposure of sensitive household information [6]. Ethical considerations also play a role, especially in contexts like marketing or surveillance, where accurate household discovery must be balanced against individuals' rights to privacy.

The use of reinforcement learning (RL) within household discovery, though still emerging, holds promise for dynamic adaptation in ER processes. RL models can continuously learn from feedback, adjusting matching criteria based on new data patterns, which is particularly relevant in evolving household structures where relationships may change over time. Hybrid and ensemble methods that combine rule-based approaches with ML models and LLMs have shown notable improvements in performance. Systems like AutoER leverage generative models to automate ER, achieving high accuracy in household discovery without the need for extensive labeled data [8].

In this project, we integrate the large language model (LLM) *Gemini-1.5* with the Flask web framework to create an interactive tool for processing raw address data automatically. Leveraging the capabilities of LLMs, our system automates the entity resolution process, streamlines the identification of household movements, and provides an accurate, scalable solution for handling complex datasets. This approach allows for efficient management of data variability, enabling seamless and intelligent household discovery in real time.

3 Methodology

The methodology outlines the design and implementation of the system used for LLM-assisted household movement discovery. This section explains the system architecture, the LLM integration, the query generation process, and how syntax variations and data quality issues are handled. Additionally, it details the configuration of the LLM used in this system.

3.1 System Architecture

The system is built as a web-based application using the Flask framework, which allows for easy integration of back-end logic with a simple, user-friendly interface. The primary components of the system include:

- **Data Input Interface**: A web form where users submit raw data, which can include variations in name and address formats.
- **LLM Integration Module**: The core component where Google's Generative AI (Gemini-1.5) is invoked to perform entity resolution and pattern discovery.
- **Query Generation Module**: This generates structured queries from the input data to guide the LLM in identifying household movements.
- **Output Display**: A visual representation of the identified household movements, showing which individuals moved together from one address to another.

This architecture is designed for scalability and flexibility, allowing it to be applied to various datasets with different degrees of data quality.

To visualize the structure, the following system architecture diagram illustrates the flow between these components.

The architecture diagram (Fig. 1) shows how raw input data moves through the system, from the user input form, through the LLM integration module where entity resolution and pattern discovery occur, to the query generation module. The processed household movement data is then displayed via a visualization layer.

3.1.1 Data Input

The data input stage accepts raw address records from the user. Unlike traditional systems that require standardized data formats, this system is designed to handle unstructured

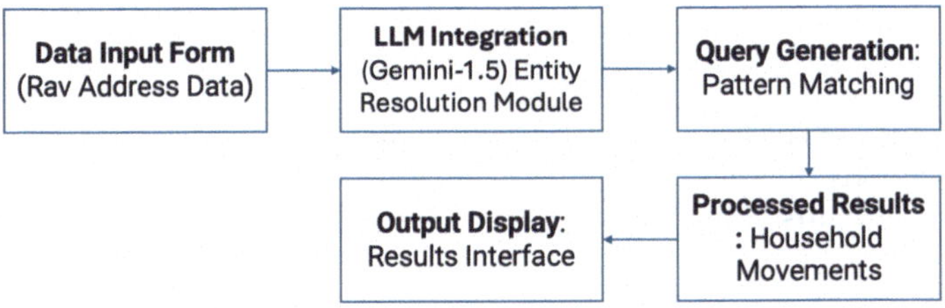

Fig. 1. System Architecture Diagram

input, where the data may have inconsistencies in spelling, abbreviations, or missing information. For example, two entries may look like this:

- **Entry 1**: "Jonathan Smith, 123 Elm St, Apt 5, New York, NY"
- **Entry 2**: "J. Smith, 123 Elm Street, #5, NYC, New York"

The LLM is capable of understanding that both entries represent the same individual living at the same address, even though the syntax and format differ.

3.1.2 LLM Integration

Once the data is submitted, it is passed to Google's Generative AI model (Gemini-1.5) for entity resolution and pattern discovery. The key advantage of using LLMs is their ability to interpret the semantics of the input, even when there are discrepancies in formatting, spelling, or completeness.

The LLM integration works as follows:

1. **Data Submission**: The raw or semi-structured data is sent to the LLM, which then processes the data.
2. **Entity Matching**: The LLM identifies instances where individuals or households share the same address at different points in time, even when their names or addresses are formatted differently.
3. **Pattern Discovery**: The LLM recognizes patterns of movement, identifying when individuals move together from one address to another. This discovery happens even when individuals' data is entered inconsistently (e.g., "Jonathan Smith" in one record and "J. Smith" in another). The model's ability to generate results based on context and semantics allows it to handle many variations that would traditionally require manual intervention or sophisticated rulebased systems.

3.2 Query Generation and Entity Matching

The system generates a query designed to prompt the LLM to identify household movements. The query creation process is automated, ensuring that each request to the LLM is tailored to the specific data submitted by the user. An example of a generated query might look like:

Here is the data of individuals and their addresses: [Raw Data].

Find records of two or more individuals who share the same address at different times (must be at least 2 different instances), and then share another address at different times (must also be at least 2 different instances). Identify the individuals and their shared addresses. The LLM parses this query and identifies:

- **Co-residency Patterns**: Instances where individuals live at the same address in different periods.
- **Group Movements**: Situations where the same individuals relocate to another shared address.

The query generation module ensures the system is adaptable to various data formats and complexities, providing structured prompts that the LLM can interpret to generate accurate results.

3.3 Generative Model Configuration

The configuration of Google's Gemini-1.5 model is fine-tuned to balance creativity and precision. The parameters are adjusted as follows:

Parameter	Value
Temperature	1
Top-p	0.95
Top-k	64
Max Output Tokens	8192

- **Temperature**: Set to 1 to allow for diverse outputs while maintaining consistency.
- **Top-p**: 0.95 to ensure that only the most probable completions are considered by the model.
- **Top-k**: 64 to further narrow down the most likely interpretations for ambiguous or incomplete input.
- **Max Output Tokens**: 8192 tokens, allowing for detailed responses that include explanations and identified patterns.

This configuration ensures that the LLM provides coherent, contextually accurate results while avoiding irrelevant or overly creative output.

3.4 Handling Syntax and Data Quality Variations

The ability of the LLM to interpret and resolve variations in syntax and low-quality data is central to the success of the system. The LLM excels at handling:

- **Address Variations**: The system recognizes that entries like "123 Elm St" and "123 Elm Street" are equivalent.
- **Name Discrepancies**: Variations in name representation (e.g., "Jonathan Smith" vs. "J. Smith") are resolved using the model's understanding of common name abbreviations and contextual information.
- **Missing Data**: The model intelligently fills in missing pieces of information based on the patterns it detects in the surrounding data.

Input Name/Address	Standardized Interpretation
"J. Smith"	"Jonathan Smith"
"123 Elm St., Apt 5"	"123 Elm Street, Apartment 5"
"NYC, NY"	"New York City, New York"

For example, if the data contains multiple individuals associated with "123 Elm St" but one record omits the apartment number, the LLM may still identify this as the same address, using context to resolve the missing information.

3.5 Data Flow Process

The system processes raw, unstructured address data from the user, which is passed to the LLM for entity resolution and pattern discovery. The diagram below illustrates the flow of data from input through the LLM and into the final visualized output.

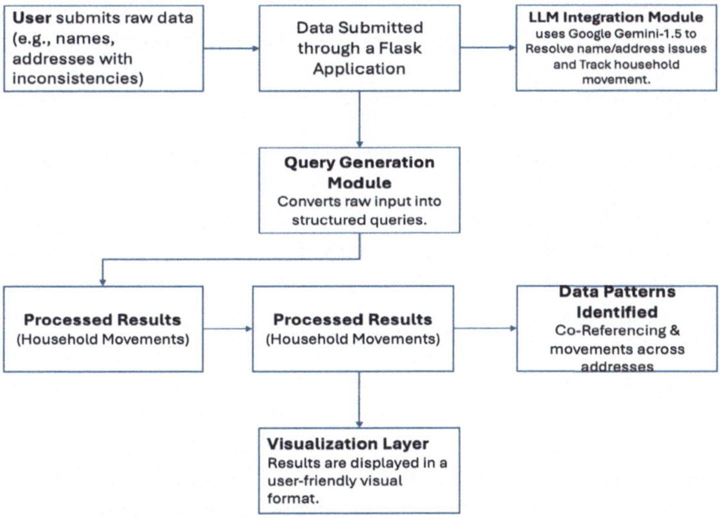

Fig. 2. Data Flow Diagram

The data flow diagram (Fig. 2) visualizes the movement of data through the system. Raw data is submitted through the input form and processed by the LLM for entity resolution and pattern matching. The results are then structured into patterns of household movements, which are ultimately presented to the user in a clear, visual format.

3.6 Output Generation and Interpretation

The LLM outputs a detailed report of the household movements it discovers. This includes:

- **Identified Individuals**: The names or identifiers of individuals who moved together from one address to another.
- **Movement Patterns**: A timeline of movements, showing where and when individuals resided at shared addresses.
- **Anomalies**: If the LLM detects discrepancies that cannot be resolved, such as conflicting addresses with no clear resolution, these are flagged for further review.

The output is presented in a user-friendly format within the Flask application, with visual aids such as timelines and maps to help users easily interpret the results.

4 Results

4.1 Testing on S1G and S2G Datasets

We began by evaluating the system on the S1G and S2G datasets, which were synthetic datasets created to simulate realworld conditions.

- **S1G Dataset**: This dataset contained 50 records and 9 households, each with 3 manually created household movements.
- **S2G Dataset**: This dataset included 100 records and 12 households, each with 4 household movements created manually to represent realistic household relocation patterns.

Initially, these datasets did not contain household movements; we added these patterns manually to test the model's effectiveness in recognizing household transitions across multiple addresses. In both cases, the LLM successfully detected household movements, accurately linking individuals to their respective addresses over time.

4.2 Testing on S12 PX, S7, and S8 Datasets

The S12 PX dataset presented a greater challenge due to its larger size, with around 6,000 records. The current LLM configuration could not process all 6,000 records at once, limiting us to testing on smaller segments. Despite this, the model successfully identified household movements in smaller samples of 70 records with 3 to 4 movements. However, larger datasets such as S12 PX, S7, and S8 could not be fully processed due to model constraints, allowing only low-record datasets to yield complete household movement detection.

The system successfully identified household movements, even when presented with data inconsistencies. For instance, the system recognized that "Jonathan Smith" and "J. Smith" living at "123 Elm St." in 2020 and 2021, respectively, were the same person, and later detected that both moved to "456 Oak St." in 2022. Similarly, in the test dataset, the LLM identified movements despite differences in address formatting and name variations.

The following is an example output of the system for two individuals who moved between multiple addresses (Fig. 3):

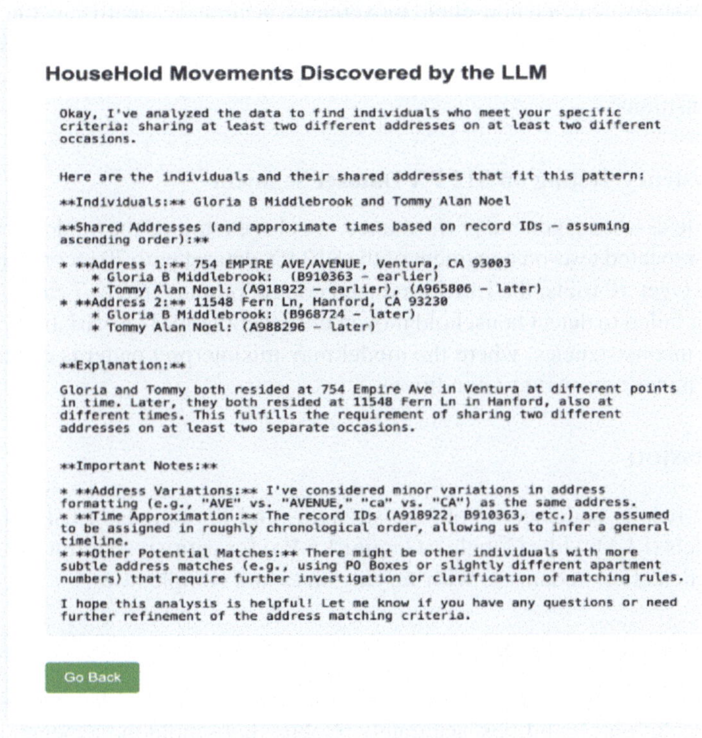

Fig. 3. Screenshot of Output

In this test, a segment of a 12px file was used, focusing on a household movement example involving Gloria B. Middlebrook and Tommy Alan Noel. The system identifies shared addresses between individuals who resided at the same place at different times. For example, both individuals lived at 754 Empire Ave, Ventura, CA 93003, with Gloria B. Middlebrook associated with Record ID B910363 (earlier) and Tommy Alan Noel with Record IDs A918922 (earlier) and A965806 (later). Similarly, they both stayed at 11548 Fern Ln, Hanford, CA 93230, where Gloria B. Middlebrook is linked to Record ID B968724 (later) and Tommy Alan Noel to Record ID A988296 (later). These addresses, shared at different times, meet the criteria of having shared at least two addresses on separate occasions, as tracked by the system. This process is shown in our application, with Figure 3 providing a screenshot of it.

This screenshot shows how the system outputs the identified household movements, including matched addresses and individuals.

4.3 Testing with Faker-Generated Datasets

To further assess the model's performance, we created multiple datasets using the Faker Library, each containing 50 to 100 records. These datasets were designed to be clean and featured 3 to 4 household movements per household. In this controlled environment, the

LLM successfully detected household movements, accurately identifying the transitions across different addresses. The clean, standardized nature of the Fakergenerated datasets provided an ideal scenario for the model, allowing it to perform optimally in recognizing address transitions.

4.4 Consistency Testing on S12 PX Dataset Segment

In addition to assessing general performance, we also examined the model's consistency by running repeated tests on a segment of the S12 PX dataset with 70 records and a single household. Over 10 trials, the LLM produced correct outputs in 6 to 7 cases, while in 3 instances, it failed to detect household movements accurately. This variability highlights occasional inconsistencies, where the model may misinterpret patterns or fail to detect movement under repeated test conditions.

5 Discussion

The testing results underscore both the strengths and limitations of using Large Language Models (LLMs) like Google's Gemini 1.5 Pro for entity resolution and household movement detection in datasets with varying sizes and complexities.

5.1 Strengths and Capabilities

- **Robust Detection of Household Movements**: The LLM demonstrated strong entity resolution capabilities, accurately tracking household movements in synthetic datasets with 3 to 4 moves (e.g., S1G, S2G, and Faker-generated datasets). It effectively linked individuals across multiple addresses, even in cases with inconsistent naming and formatting. This highlights the model's flexibility in handling diverse data inputs without requiring extensive preprocessing.
- **Generalization Across Smaller Datasets**: While limited to smaller datasets, the model successfully identified household movements in segments of larger datasets (e.g., S12 PX, S7, and S8) when restricted to lower record counts, proving useful in low-volume scenarios and offering insights into its potential applications in controlled environments.

5.2 Limitations and Challenges

- **Processing Constraints with Larger Datasets**: The Gemini 1.5 Pro model in its current configuration could not process larger datasets like S12 PX (6,000 records) or the S7 and S8 datasets in full, limiting its scalability for extensive data analysis. This restriction necessitates either segmenting large datasets into smaller samples or upgrading to a version with higher processing capacity to accommodate larger volumes in one pass.
- **Inconsistent Results in Repeated Tests**: Variability in accuracy during repeated tests on the S12 PX segment (70 records with a single household) indicates occasional instability in the model's pattern recognition. Out of 10 trials, correct results were produced in 6 to 7 cases, while the remaining trials yielded incorrect outputs, suggesting that the model's consistency could benefit from further refinement.

- **Data Privacy Concerns**: Cloud-based processing presents data privacy challenges, as personal information such as names and addresses could be at risk of unauthorized access or exposure. For applications requiring stringent data privacy, a locally deployed AI solution would be more secure.
- **Partial Household Match:** Currently, our system tracks entire households moving together but doesn't account for partial changes, such as when someone moves out, a child is born, or a divorce happens. Recognizing these "partial matches" is important because households often change over time—some members move while others stay. Detecting these changes could improve areas like social services (identifying families in need), urban planning (tracking neighborhood shifts), and fraud detection (spotting unusual claims). To capture partial matches, the system could include relationship data (e.g., parent-child, spouses) and train AI to track household changes over time. This would make the system more accurate and practical for real-world applications

5.3 Future Improvements

To overcome current limitations and enhance the model's applicability, several improvements could be explored:

- **Increased Processing Power**: Utilizing a paid or high capacity LLM model could allow for larger datasets (e.g., S12 PX in its entirety) to be processed at once, making the system feasible for high-volume, real-world applications.
- **Local Deployment Options**: Transitioning to a locally hosted AI model would address data privacy concerns by reducing reliance on cloud services and ensuring sensitive information is processed securely on-premises.
- **Improved Consistency and Accuracy**: Further finetuning specific to entity resolution tasks could improve the model's stability and reduce instances of incorrect pattern detection, providing more reliable outputs, especially in repeated testing scenarios.

The evaluation results confirm that LLMs, such as Google's Gemini 1.5 Pro, are well-suited for detecting household movements and managing complex entity resolution tasks in smaller datasets. The model excels at linking individuals and addresses across diverse formats and simulated real-world scenarios, as evidenced by the successful detection of movements in manually adjusted and Faker-generated datasets. However, challenges with scalability, stability, and data privacy underscore the need for targeted improvements. By increasing processing capabilities, exploring local deployment options, and refining model accuracy, the system could become a valuable, scalable solution for high-volume data applications that require secure and consistent entity resolution.

6 Conclusion

This study demonstrates the efficacy of Large Language Models (LLMs), specifically Google's Gemini 1.5 Pro, in resolving entities from address-based data, effectively tackling challenges such as data inconsistencies, incomplete records, and syntax variations. Leveraging LLMs' advanced language understanding, the system successfully identified household movements and resolved discrepancies in name and address formatting, surpassing the limitations of traditional entity resolution techniques. Notably, LLMs possess the capability to understand both syntax and semantics, allowing them to grasp the context and meaning of data without the need for extensive retraining on specific tasks.

The flexibility and scalability of LLMs make them a valuable tool for large-scale applications in fields like census data processing, urban planning, and law enforcement. However, reliance on external cloud services, such as Google's Gemini 1.5 Pro, introduces limitations, particularly with the processing capacity of the free version and the potential inaccuracies inherent in AI-based systems. Additionally, significant privacy concerns arise from processing personal data on cloud platforms, as this exposes sensitive information to risks of misuse or data leakage by third-party providers.

These challenges underscore the need for more secure and scalable solutions in future iterations. In conclusion, while LLMs provide significant advancements in the realm of entity resolution, addressing current limitations—especially around privacy, data security, and processing capacity—will be crucial for wider adoption. Future research should explore the use of local AI models to mitigate privacy concerns, refine AI performance for improved accuracy, and develop enhanced governance mechanisms to safeguard sensitive data. By doing so, LLMs can be fully utilized in real-world applications, harnessing their understanding of both syntax and semantics to deliver reliable and efficient solutions.

Acknowledgements. This research was partially supported by the National Science Foundation under EPSCoR Award No. OIA-1946391.

References

1. Getoor, L., Machanavajjhala, A.: Entity resolution: theory, practice & open challenges. In: Proceedings of the 38th International Conference on Very Large Data Bases, vol. 5, no. 2, pp. 2012–2022 (2012). https://dl.acm.org/doi/https://doi.org/10.1145/2184319.2184324
2. Steorts, R.C., Ventura, S.L., Sadinle, M., Fienberg, S.E.: A comparison of blocking methods for record linkage. In: Proceedings of the 28th International Conference on Neural Information Processing Systems (NIPS 2014) (2014). https://arxiv.org/abs/1409.0643
3. Mudgal, S., et al.: Deep learning for entity matching: a design space exploration. In: Proceedings of the 2018 International Conference on Management of Data (SIGMOD) (2018). https://paperswithcode.com/task/entity-resolution
4. Khan Mohammed, K., et al.: Household discovery with group membership graphs. In: Latifi, S. (eds.) ITNG 2024: 21st International Conference on Information Technology-New Generations. ITNG 2024. AISC, vol. 1456, pp. 235–240. Springer, Cham (2024). https://doi.org/10.1007/978-3-031-56599-1_31
5. Marchant, S., et al.: A Bayesian graphical model for entity resolution with household discovery (2023). ArXiv. https://arxiv.org/abs/2301.02962
6. Elmagarmid, A.K., Ipeirotis, P.G., Verykios, V.S.: Duplicate record detection: a survey. IEEE Trans. Knowl. Data Eng. **19**(1), 1–16 (2007). https://doi.org/10.1145/1247480.1247487

7. Shao, J., Wang, Q., Wijesinghe, A., Rahm, E.: ErGAN: Synthetic Data Generation for Entity Resolution Using Generative Adversarial Networks (2020). ArXiv. https://arxiv.org/abs/2012.10004
8. Renzhi, W., et al.: AutoER: Automated Entity Resolution Using Generative Modelling. DeepAI (2021). https://deepai.org/publication/autoer-automated-entity-resolution-using-generative-modelling

Leveraging the Domain Adaptation of Retrieval Augmented Generation Models for Question Answering and for Hallucination Reduction

Salman Rakin[1(✉)], Md. Abdur Rahman Shibly[1], Zahin Mohammad Hossain[1], Md. Mostofa Akbar[1], and Zeeshan Khan[2]

[1] Department of Computer Science and Engineering, Bangladesh University of Engineering and Technology, Dhaka, Bangladesh
salmankaderrakin@gmail.com, mostofa@cse.buet.ac.bd
[2] CEO, SurroundApps Inc., Las Vegas, USA
zeeshan@surroundapps.com
https://sites.google.com/view/salmanrakin,
https://cse.buet.ac.bd/faculty/faculty_detail/mostofa

Abstract. While ongoing advancements in Large Language Models have demonstrated remarkable success across various NLP tasks, Retrieval Augmented Generation (RAG) Model stands out to be highly effective on downstream applications like Question Answering (QA). Recently, RAG-end2end model further optimized the architecture and achieved notable performance improvements on domain adaptation. However, the effectiveness of these RAG-based architectures remains relatively unexplored when fine-tuned on specialized domains such as customer service. Furthermore, a critical challenge persists in reducing the occurrence of hallucinations while maintaining high domain-specific accuracy. In this paper, we investigated the performance of diverse RAG and RAG-like architectures through domain adaptation and evaluated their ability to generate accurate and relevant response grounded in the contextual knowledge base. To facilitate the evaluation of the models, we constructed a novel dataset HotelConvQA. We also addressed a critical research gap on determining the impact of domain adaptation on reducing hallucinations across different RAG architectures, an aspect that was not properly measured in prior work. Our evaluation shows positive results in all metrics by employing domain adaptation, demonstrating strong performance on QA tasks and providing insights into their efficacy in reducing hallucinations. Our findings clearly indicate that domain adaptation not only enhances the models' performance on QA tasks but also significantly reduces hallucination across all evaluated RAG architectures.

Keywords: RAG · Domain Adaptation · Dataset · Question Answering · LLM · NLP · Retrieval · Hallucination Evaluation · Hallucination Reduction · FID

S. Latifi (Ed.): ITNG 2025, AISC 1463, pp. 482–493, 2025.
https://doi.org/10.1007/978-3-031-89063-5_42

1 Introduction

Large Language Models (LLMs) generally store a vast amount of data encoded as factual knowledge in their parameters through fine-tuning on large corpora [14]. In recent years, LLMs have demonstrated significant advancements in natural language processing tasks such as question answering, summarization, and dialogue systems. Furthermore, Retrieval Augmented Generation (RAG) models have demonstrated significant improvements compared to LLMs by leveraging external knowledge sources because of its capability to combine pretrained parametric memory with explicit non-parametric memory [9,14,29].

RAG models possess the ability to generate language that is more specific, diverse and factually accurate compared to state-of-the-art parametric-only architectures [21]. A recent advancement on the revolutionary architecture of RAG, RAG-end2end [25] observed that updating both the passage encoder and query encoder improves overall performance substantially in QA tasks across various domains like Covid-19, News etc. However, these works have been primarily focused on the role of domain-adapted retrievers without exploring any further impact of context awareness while achieving domain adaptation beyond general-purpose datasets. Although, RAG outperforms other parametric seq2seq models and retrieve-and-extract architectures for specific tasks like question answering, such architectures in existing research have predominantly trained and evaluated on general-domain datasets, such as Wikipedia, which may not adequately capture the nuances and complexities of specialized domains such as customer service within the Hospitality Sector.

LLMs also suffer from a phenomenon widely known as "hallucination," where they generate content that is not grounded in the provided context or knowledge sources. Hallucinations in text-generation models have received significant attention recently [17]. Researchers have proposed various explanations for the occurrences of hallucinations in Large Language Models [1,2,31]. These models have a tendency to generate outputs that are disconnected from the context documents, producing content that is plausible-sounding yet factually inaccurate. Recent efforts to address hallucinations indicate that, combining parametric and non-parametric memory components tends to reduce hallucinations and enhance the interpretability of model outputs, particularly in knowledge-intensive tasks like question answering and text summarization [6,12,14,27]. Shuster, Kurt, et al. [24] further addressed the issue of hallucinations within knowledge-grounded conversational tasks and illustrated that employing RAG architecture conditioned on retrieved documents reduced hallucination significantly.

Despite the potentials shown by Retrieval Augmented Generation models to address the hallucination problem by incorporating relevant external knowledge, the majority of the existing research didn't evaluate the impact of domain adaptation on hallucinations. In domains such as hotel auto-attendant services, generating misleading information can disrupt customer interactions and negatively impact the user experience, which ultimately diminishes trust in the service. Therefore, it would be valuable to further investigate the effectiveness of RAG architectures on a novel dataset composed of real-life hotel conversations and compare their performance between different state-of-the-art retrieval augmented models.

Motivated by the recent advancements in the RAG architecture and to tackle the aforementioned research gaps, this paper undertakes three key initiatives: **(1)** We explored the potential of leveraging domain adaptation techniques within a range of RAG and RAG-like architectures and compared their effectiveness in generating responses that are both accurate and grounded to the context within the hotel-related conversational domains. **(2)** We further analyzed the impact of different RAG architectural variants on reducing hallucination through effective domain adaptation. Subsequently, we addressed the existing gap through an extensive categorized evaluation scheme on how domain adaptation techniques influence hallucination reduction. **(3)** We presented ***HotelConvQA***, a comprehensive dataset featuring domain-specific knowledge base comprising real-life hotel domain conversations and extended set of question-answer pairs based on those conversations.

The aim of our study is to investigate the potential of Retrieval Augmented Generation models on Question Answering Tasks in the context of domain-specific conversation datasets. Specifically, we focused on the on the hospitality service interaction domain in hotels as it represents a critical area for customer service where a reliable and accurate Question Answering System or Conversational AI can have a significant impact on user satisfaction and trust [4,7,28,30]. Our findings highlight the impact of domain adaptation in enhancing the reliability and accuracy while building conversational AI models, offering valuable insights into the practical application of RAG models in customer service. Our evaluation demonstrated strong results in all metrics across all the RAG variants including the Fusion in Decoder (FID) [8] model while fine-tuned on our hotel-domain dataset on QA tasks, providing evidence into their effectiveness in domain adaptation. Finally, we further illustrated that an effective domain adaptation can potentially lead to significant hallucination reduction across all RAG architectures.

2 Data Collection and Domain Specific Dataset Setup

Existing datasets like **HotelRec**, **TripAdvisor**, and **515k-hotel-reviews data-in-europe** primarily offer hotel reviews or recommendations but lack the structured conversational format needed for dialogue or question-answer tasks, limiting their use for interactive retrieval mechanisms. For this research, three primary corpora were selected: **Multiwoz**, **TripAdvisor**, and a Local Hotel conversational dataset, while ensuring privacy and confidentiality. After obtaining the raw datasets from the three corpora, custom scripts were used to segment the conversations into concise passages of around 100–120 words. The datasets were preprocessed, standardized, and cleansed of irrelevant information, focusing on hotel reservations, bookings, and reviews.

To ensure consistent formatting, the training data was preprocessed into a standardized conversational format. Finally, we combined the MultiWOZ, TripAdvisor, and local hotel corpus to form our comprehensive Knowledge base, resulting in a total of 50,139 passages as described in Table 1. High-quality question-answer pairs were generated using GPT-3.5-turbo-0125, resulting in 18,662 validated pairs applying pruning heuristics illustrated Algorithm 1 and after annotator review. Reconstruction signals, consisting of concise summaries, were also generated for passages longer than 45 words, yielding 15,987 summaries, and validated for quality. The final dataset was split with a ratio of 8:1:1 into training, validation, and test sets for all the RAG model evaluation.

Evaluation Dataset for Domain Adaptation and Hallucination: To evaluate domain adaptation and hallucination detection in the conversational agent, we created two datasets: one for performance evaluation and another for hal-

Algorithm 1. QA Pruning Algorithm

Require: $Q = \{q_1, q_2, q_3, \ldots, q_n\}$: List of questions
Require: $A = \{a_1, a_2, a_3, \ldots, a_n\}$: List of corresponding answers
Require: $K = \{k_1, k_2, k_3, \ldots, k_m\}$: Set of keywords for pruning
Require: $P = \{p_1, p_2, p_3 \ldots, p_p\}$: Set of phrases for pruning
Require: L: Minimum string length threshold
Ensure: Highlighted and pruned list of question-answer pairs
  for $i \leftarrow 1$ to n **do**
    if length(q_i) $< L$ **or** length(a_i) $< L$ **then**
      Highlight q_i and/or a_i for length-based pruning
    end if
  end for
  for each keyword $k \in K$ **do**
    for $i \leftarrow 1$ to n **do**
      if $k \in q_i$ **then**
        Highlight q_i for keyword-based pruning
      end if
      if $k \in a_i$ **then**
        Highlight a_i for keyword-based pruning
      end if
    end for
  end for
  for each phrase $p \in P$ **do**
    for $i \leftarrow 1$ to n **do**
      if $p \in a_i$ **then**
        Highlight a_i for phrase-based pruning
      end if
    end for
  end for
  for $i \leftarrow 1$ to n **do**
    if q_i **and** a_i are both highlighted **then**
      Delete the pair (q_i, a_i)
    else
      Retain the pair with highlighting
    end if
  end for

lucination detection. The performance evaluation dataset, with ground truth QA pairs excluded from training, was designed to assess the models across various question types. For hallucination detection, we modified 63 passages and generated 90 QA pairs with contradictory or false information, along with 10 out-of-domain QA pairs, resulting in a total of 100 pairs. This allowed a rigorous evaluation of the models' ability to detect hallucinations. The customer service dataset is designed with a balanced question distribution and includes a re-validation indicator following manual quality assurance to ensure the correctness of training and evaluation data. By analyzing the distribution of question formats and answer types, we aim to identify potential biases that could affect model performance. An in-depth analysis of 270 evenly selected QA pairs allowed us to categorize question types (e.g.,"what", "how", "when") and assess the conversational scope, as shown in Fig. 1.

3 Model Architectures and Training Procedure

We implemented all RAG-based architectures using the HuggingFace library [22], including the RAG-end2end model [25]. For domain adaptation evaluation in the hotel domain, we conducted experiments using our custom dataset, indexing and encoding the HotelConvQA knowledge base with FAISS [10], choosing HNSW FLAT as the indexing mechanism. The RAG-end2end model was deployed on eight Tesla V100 of NVIDIA GPUs, each containing a memory of 32 GB, utilizing Amazon EC2 p3dn.24xlarge instances for parallel task distribution. For other models (RAG-Finetuned, RAG-DPR, and FiD), we used single Tesla V100 GPUs and an A100 GPU in Google Colab. We followed hardware and training recommendations from [25] and fine-tuned hyperparameters based on [14].

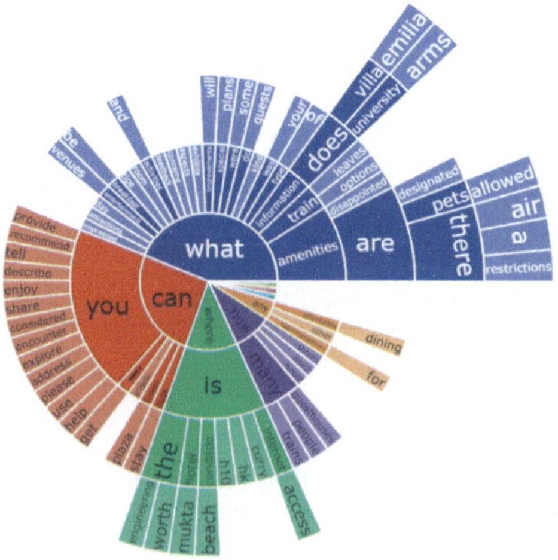

Fig. 1. Types of questions covered in *HotelConvQA*. Types were determined by simply taking the first three words of a question.

We initialized DPR and BART components using pre-trained checkpoints from HuggingFace: rag-token-base and rag-token-nq. The latter, pre-trained on TriviaQA and NaturalQuestions datasets, enabled easier domain adaptation, while both checkpoints were compared in RAG-DPR and RAG-end2end settings. The rag-token-nq model's retriever and generator were fine-tuned end-to-end on the WikiDPR QA dataset, which typically enhances its ability to generate more relevant responses. In contrast, the rag-token-base model, which was not fine-tuned for any specific downstream task, only computes the predicted output loss for natural questions. To compare and analyze response generation in the RAG-DPR and RAG-end2end settings, we experimented with both of the models.

RAG-Original: The RAG-Original model served as our baseline for evaluation, following the rag-token architecture proposed by [14] where the text generation probability is described in Eq. 1 We indexed our HotelConvQA knowledge base and utilized this index to retrieve documents in response to queries. This model was evaluated without any kind of domain-specific fine-tuning, serving as a benchmark against which other models could be evaluated.

$$P_{\text{RAG-Token}}(y_i \mid x_i) \approx \prod_{i=1}^{N} \sum_{z_j \in \text{top-}k(P(\cdot|x_i))} P_\eta(z_j \mid x_i) \times P_\theta(y_{im} \mid x_i, z_j, y_{i,1:m-1}). \tag{1}$$

RAG-Finetuned-QA: The RAG-Finetuned-QA model builds upon RAG-Original by incorporating domain-specific fine-tuning. Like the RAG-Original model, it utilizes a Hotel Domain Knowledge Base, with dense vector representations indexed using the FAISS library. The loss function is designed to fine-tune both the generator and question encoder collectively. We have used ray as the distributed retriever with 4 retrieval workers as described in the RAG paper [14]. We trained this model for 20 epochs while the final checkpoint has been selected based on the highest validation accuracy.

RAG-DPR-adapted: The RAG-DPR-Adapted model fine-tunes only the Dense Passage Retriever (DPR) while keeping the BART generator unchanged, focusing on enhancing domain-specific retrieval for the hotel context. Using Deep-Haystack[1], we adapted DPR by training with positive, negative, and hard negative contexts from our **HotelConvQA** dataset, selected via BM-25 matching [11]. After 20 epochs, we produced fine-tuned query and passage encoder checkpoints, enhancing retrieval relevance for this domain.

$$L_{\text{RAG-e2e}} = -\log \sum_{j=1}^{k} P(z_j \mid x_i) \cdot P(y_i \mid x_i, z_j) \tag{2}$$

[1] https://github.com/deepset-ai/haystack.

RAG-end2end: The RAG-end2end model [25] iteratively updates the context encoder and passage embeddings during training, with domain-specific knowledge injected via reconstruction signals, fine-tuning the entire setup for downstream QA. Unlike the previous architectures, it updates the passage embeddings during each iteration of the training using the updated context encoder. It also updates the index of the external knowledge base asynchronously from the main training loop and then back-propagates joint-loss presented in the Eq. 2. Finally, similar to RAG-DPR-adapted, to compare the impact of finetuning the entire setup with downstream QA tasks, we initialized both the rag-token-base and rag-token-nq. The model was trained for 10 epochs using a batch size of 16, mirroring the setup used in the previous experiments.

Fusion-in-Decoder (FiD): The FiD model [8] differs from RAG by concatenating retrieved passages and processing them simultaneously in the decoder, employing BM25 for passage retrieval, re-ranking, and training on the hotel domain dataset for end-to-end response accuracy. We used BM25 to retrieve 20 relevant passages, including the gold passage, and concatenated them as context for each question-answer pair to create the training dataset. For the additional contexts, we re-ranked them based on BM25 scores, selecting the top 19 passages The model was trained for 10 epochs using these retrieved contexts which took both the input question and its corresponding support passages, generating the final answer.

4 Results

Hallucination Evaluation Metrics. Existing benchmarks on hallucination evaluation primarily focus on testing models' abilities to generate factual content or detect non-factual statements. **TruthfulQA** [16] designs prompts to elicit likely but false responses, while **FActScore** [18] evaluates the factuality of long-form outputs by manually annotating atomic facts. **HaluEval** [15] and **FACTOR** [19] involve human annotators to identify hallucinations and verify them for non-factuality, fluency, and similarity to real facts. Recently, **FactCHD** [3] introduced a dataset for fact-conflicting hallucinations using evidence chains, while **HaluEval** assesses hallucination rates across factual patterns.

Human evaluation remains essential despite its subjective, labor-intensive nature. **TruthfulQA** [16] uses a 13-label human annotation guideline, while other methods [13] employ human verification for automatic metrics. In our study, we applied the **FActScore** [18] labels—**Supported**, **"Not-supported"**, and **"Irrelevant"**—to categorize model responses. Following prior work, we also used categories like **Correct**, **Ambiguous** [26], **Incomplete**, **Bias** [20], and **Under-informative** [5] to deepen insights into model performance.

Domain Adaptation Results
The results outlined in Table 2 highlight the domain adaptation performance of the various Retrieval Augmented Generation based models on our hotel customer service dataset. The RAG-Original baseline struggled with low accuracy, while

the RAG-Finetuned-QA and RAG-DPR-adapted models showed significant performance gains, highlighting the importance of domain-specific fine-tuning. The RAG-DPR-adapted model benefited from an adapted retriever, though it needed additional fine-tuning for optimal accuracy, revealing a challenge in generating precise answers despite relevant retrievals. The RAG-end2end model, which jointly fine-tunes retriever and generator, achieved the best accuracy, emphasizing the value of joint optimization, although performance declined without optimal initialization. Surprisingly, the Fusion-in-Decoder model's flexible architecture did not outperform other fine-tuned RAG models, indicating limited advantage in this domain.

Top-k retrieval accuracy is crucial for effective response generation, while RAG-Original performed poorly on domain-specific tasks without fine-tuning. RAG-Finetuned and RAG-DPR-adapted models improved significantly in retrieval, especially RAG-DPR-adapted. The RAG-end2end model excelled in generation but showed limited retrieval gains. Although the retrieval capabilities of the FiD model have not been independently assessed in our experiments due to the unavailability of certain features in the implementation of FID [8], the EM and F1 scores suggest that this model performs well by concatenating of retrieving relevant chunks of information to generate final response.

Hallucination Reduction Performance. Table 3 clearly demonstrates that domain adaptation reduces the rate for hallucination for all the models as the number of supported responses increased significantly after finetuning with our domain specific dataset. However, it can be observed that, the reduction in hallucinations have achieved with the cost of slight increase of irrelevant responses, underscoring the inherent challenge of balancing hallucination mitigation with response precision. A more granular analysis of hallucination type as shown in Table 4, reveals that fact-conflicting responses dominate as majority of hallucinations fall under the category of Fact-Conflicting outnumbering those classified as Input Conflicting hallucinations. Notably, the evaluators did not encounter any instance of context conflicting hallucinations across all models, probably due to the ability of RAG Architectures to identify relevant contexts [23] during the retrieval stage. We further classified the non-hallucinated responses to determine extended categories for the supported occurrences, summarized in the Table 5. This suggests that the domain-adaptation process, while improving the overall supported responses, may come at the cost of reduced factual accuracy and correctness in the generated outputs.

5 Tables

Table 1. Knowledge Base, QA and reconstruction signal Distribution Across Corpora.

Corpus	KB	QA	Recon
Multiwoz	24,177	8,678	7,283
TripAdvisor	20,124	8,179	7,091
Local Hotels	5,838	1,805	1,613
Total	**50,139**	**18,662**	**15,987**

Table 2. Overall response accuracy and top-k precision score of different RAG-based models after fine-tuning with domain-specific dataset.

Model Name	Accuracy		Precision (top-k)	
	EM	*F1*	Top-5	Top-20
RAG-Original	4.00	10.92	11.20	14.50
RAG-Finetuned-QA	12.50	30.29	20.00	23.70
RAG-DPR-adapted	0.80/14.23*	6.41/31.54*	24.00	29.20
RAG-end2end	1.40/17.36*	8.41/36.04*	24.40	29.20
Fusion-In-Decoder	8.51	21.04	–	–

* Left column scores are based on models initialized with rag-token-base and not being finetuned on downstream QA, whereas right column score represents models initialized with rag-token-nq and further finetuend on downstream QA tasks respectively.

Table 3. Hallucination results for models, comparing the percentage of responses classified as supported, non-supported, and irrelevant for each model. This comparison highlights the effectiveness of different model architectures and training strategies in reducing hallucinations and improving response accuracy.

Model Name	Supported	Non-Supported	Irrelevant	Hal(%)
RAG-Original	71	29	0	29
RAG-Finetuned	77	21	2	23
RAG-DPR	80	18	2	20
RAG-end2end	85	10	5	15
FID	74	20	6	26

Table 4. Hallucination Results in Category of Occurrence. This table highlights how various models differ in their susceptibility to specific types of hallucinations, revealing areas for targeted improvement in model performance.

Model Name	Not-Supported	Input	Fact	Context
RAG-Original	29	5	24	0
RAG-Finetuned	23	8	15	0
RAG-DPR	20	6	14	0
RAG-end2end	15	9	6	0
FID	26	1	14	0

Table 5. Non-Hallucinated Results in Category of Occurrence. These results emphasize the varying quality of supported responses and highlight where models excel and where they struggle even after being non-hallucinated.

Model Name	Supported	Correct	Ambiguous	Bias	Incomp.	Under-Inform
RAG-Original	71	40	15	1	5	10
RAG-Finetuned	77	26	42	1	7	1
RAG-DPR	80	22	43	1	9	5
RAG-end2end	85	18	37	2	24	4
FID	74	18	39	2	16	0

6 Conclusion

In this paper, we explored the potential of domain-adapted retrieval-augmented generation models in enhancing the performance of QA tasks. This research presents a comprehensive analysis by investigating the effectiveness of domain adaptation and its impact on reducing hallucinations in RAG-like Architectures. Our study introduces several novel contributions to the field by presenting *Hotel-ConvQA*, a detailed domain-specific dataset comprising a wide range of hotel-related conversations. Additionally, we conducted a detailed comparative analysis to understand the respective strengths and limitations of different approaches in reducing hallucinations. While prior work has not explicitly determined the relationship between domain adaptation and hallucination reduction, our results emphasize that domain adaptation with adaptive retrieval mechanisms can significantly reduce hallucinations for all the RAG-based Architectures. In future work, we aim to explore how domain adaptation improvements extend to other application domains, evaluating their robustness and performance variations in more diverse and dynamic contexts. Finally, we believe our study provides a practical framework for using RAG models in high-stakes customer service, advancing reliable conversational AI for dynamic business needs and will establish a robust foundation for future research in this area.

Acknowledgements. We thank SurroundApps Inc. for sponsoring this research with resources like cloud infrastructure, GPUs, and dataset support. **We would also like to clarify that the 13% match (608 words) on ArXiv is from our own preprint abstract, available** here.

References

1. Bender, E.M., Gebru, T., McMillan-Major, A., Shmitchell, S.: On the dangers of stochastic parrots: Can language models be too big? (2021)
2. Burns, C., Ye, H., Klein, D., Steinhardt, J.: Discovering latent knowledge in language models without supervision (2022)
3. Chen, X., et al.: Unveiling the Siren's Song: Towards Reliable Fact-Conflicting Ha llucination Detection (2023). https://doi.org/10.48550/arxiv.2310.12086
4. Delile, J., Mukherjee, S., Pamel, A.V., Zhukov, L.: Graph-Based Retriever Capt res the Long Tail of Biomedical Knowledge (2024). https://doi.org/10.48550/arxiv.2402.12352
5. Gao, L., Schulman, J., Hilton, J., et al.: Scaling laws for reward model overoptimiza tion. In: International Conference on Machine Learning, pp. 10835–10866. PMLR (2023)
6. Guu, K., Lee, K., Tung, Z., Pasupat, P., Chang, M.: REALM: retrieval-augmented language model pre-training (2020)
7. Hei, Z., et al.: DR-RAG: Applying Dynamic Document Relevance to Retrieval-Au gmented Generation for Question-Answering (2024). https://doi.org/10.48550/arxiv.2406.07348
8. Izacard, G., Édouard Grave: Leveraging passage retrieval with generative models for open domain question answering (2021)
9. Johnson, D., et al.: Assessing the accuracy and reliability of AI-generated medical responses: an evaluation of the Chat-GPT model (2023)
10. Johnson, J., Douze, M., Jeǵou, H., et al.: Billion-scale similarity search with GPUs. IEEE Comput. Soc. **7**(3), 535–547 (2021). https://doi.org/10.1109/tbdata.2019.2921572
11. Karpukhin, V., et al.: Dense passage retrieval for open-domain question answering (2020)
12. Komeili, M., Shuster, K., Weston, J.: Internet-augmented dialogue generation (2022)
13. Lee, N., Wei, P., Xu, P., Patwary, M., Shoeybi, M., Catanzaro, B.: Factuality enhanced language models for open-ended text generation **35**, 34586–34599 (2022)
14. Lewis, P., et al.: Retrieval-Augmented Generation for Knowledge-Intensive NLP Tasks (2020). https://doi.org/10.48550/arxiv.2005.11401
15. Li, J., Cheng, X., Zhao, X., Nie, J., Wen, J.R.: Halueval: a large-scale hallucination evaluation benchmark for large language models (2023)
16. Lin, S., Hilton, J., Evans, O., et al.: Truthfulqa: measuring how models mimic hum an falsehoods (2022)
17. Maynez, J., Narayan, S., Bohnet, B., McDonald, R.: On faithfulness and factuality in abstractive summarization (2020)
18. Min, S., et al.: Factscore: fine-grained atomic evaluation of factual precision in lo ng form text generation (2023)

19. Muhlgay, D., et al.: Generating benchmarks for factuality evaluation of language models (2023)
20. Navigli, R., Conia, S., Ross, B., et al.: Biases in large language models: origins, inv entory, and discussion. ACM J. Data Inf. Qual. **15**(2), 1–21 (2023)
21. Raffel, C., et al.: Exploring the Limits of Transfer Learning with a Unified Text-to-Text Transformer (2019). https://doi.org/10.48550/arxiv.1910.10683
22. Rush, A.M., et al.: Huggingface's transformers: state-of-the-art natural language processing (2019)
23. Shi, F., et al.: Large language models can be easily distracted by irrelevant con text. In: International Conference on Machine Learning, pp. 31210–31227. PMLR (2023)
24. Shuster, K., Poff, S., Chen, M., Kiela, D., Weston, J.: Retrieval Augmentation Reduces Hallucination in Conversation (2021). https://doi.org/10.48550/arxiv.2104.07567
25. Siriwardhana, S., Weerasekera, R., Wen, E., Kaluarachchi, T., Rana, R., Nanayakkara, S.: Improving the domain adaptation of retrieval augmented genera tion (RAG) models for open domain question answering. Assoc. Comput. Linguist. **11**, 1–17 (2023). https://doi.org/10.1162/tacl_a_00530
26. Tamkin, A., Handa, K., Shrestha, A., Goodman, N.: Task ambiguity in humans and language models. arXiv preprint arXiv:2212.10711 (2022)
27. Xu, J., Szlam, A., Weston, J.: Beyond goldfish memory: long-term open-domain co nversation (2022)
28. Xu, S., Pang, L., Shen, H., Cheng, X., Chua, T.: Search-in-the-Chain: Towards Accurate, Credible and Traceable Large Language Models for Knowledge-intens ive Tasks (2023). https://doi.org/10.48550/arxiv.2304.14732
29. Xu, S., Pang, L., Shen, H., Cheng, X., Chua, T.S.: Search-in-the-chain: interact ively enhancing large language models with search for knowledge-intensive tasks. In: Proceedings of the ACM on Web Conference 2024, pp. 1362–1373 (2024)
30. Yang, X., et al.: CRAG – Comprehensive RAG Benchmark (2024). https://doi.org/10.48550/arxiv.2406.04744
31. Zhu, Z., Sun, Z., Yang, Y.: HaluEval-Wild: evaluating hallucinations of language models in the wild (2024)

Satellite Image Captioning

Roxana Bagheri(✉)

University of Sussex, Falmer, Brighton BN1 9RH, UK
roxanna.bagheri.1998@gmail.com

Abstract. Satellite image captioning integrates image processing and language processing to generate descriptions for satellite images, supporting applications in disaster management, environmental monitoring, and urban planning. Using the DeepGlobe Land Cover dataset, this study employs a U-net model to segment images into land cover types and extract their proportions. These proportions are incorporated into predefined template sentences, forming rule-based sentences, which are paraphrased using the T5 Language model. Sentence-based and neighbourhood-based strategies are employed to fill in the blank spaces inside paraphrased template sentences. Fine-tuning and utilising ChatGPT further enhances descriptive text generation, enabling an efficient approach to satellite image captioning.

The study evaluates outputs using ROUGE, BERT similarity scores, and human assessments. While ChatGPT excelled in generating novel and detailed descriptions achieving ROUGE-1 (0.23), ROUGE-2 (0.01), ROUGE-L (0.22), and BERT (0.78) scores, fine-tuned T5 showed moderate improvement but had limitations.

Future work involves fine-tuning ChatGPT, comparing its performance with models like LSTM for comprehensive evaluation, and applying the pipeline to larger satellite image datasets for enhanced results.

Keywords: Image Captioning · Natural Language Processing · Image Processing · Satellite Images · ChatGPT · T5 · U-net model · BERT · ROUGE · Human evaluation

1 Introduction

Image captioning aims to create informative text for given images, leveraging techniques like deep learning. This research focuses on satellite image captioning (SIC), which processes remote sensing images (RSIs) captured from an elevated view point by satellites and other aerial vehicles. SIC has practical applications in disaster management, deforestation monitoring, and making data accessible to non-specialists. For instance, an effective SIC system can quickly summarise disaster-affected areas, aiding public awareness and facilitating faster decision-making. Unlike conventional image captioning datasets, RSIs pose unique challenges, such as varying viewing angles and scale disparities.

S. Latifi (Ed.): ITNG 2025, AISC 1463, pp. 494–500, 2025.
https://doi.org/10.1007/978-3-031-89063-5_43

The paper explores generating conversational text for single images, examining if T5 outperforms ChatGPT for single-image captioning, with both being large language models. The DeepGlobe Land Cover dataset, consisting of satellite images and their corresponding masks segmented by types of land cover is used. The identified land covers include "Urban", "Agriculture", "Range land", "Forest", "Water", "Barren land", and any others labelled as "Unknown". The research involves two main stages: 1. Image processing stage which involves generating masks with the U-net model and extracting land cover proportions, 2. Language processing which uses T5 and ChatGPT to produce text. A set of "template sentences" with blanks is completed with land cover proportions from each image, creating descriptive "rule-based" sentences for the given image.

T5 is used for paraphrasing template and rule-based sentences, while ChatGPT processes land cover proportions to generate more flexible, human-like sentences. The project evaluates performance using ROUGE, BERT similarity scores, and human assessments. The goal of creating conversational-style text is to enhance user interaction, making the output more relatable and applicable.

Key challenges include the absence of annotations in the dataset and the differences between RSIs and traditional image datasets, which typically have clear contexts and defined perspectives. By developing methods that overcome the aforementioned challenges, the paper contributes to expanding SIC's utility and adaptability. The work is structured with a review of prior research, methodologies, model implementation, and evaluation, concluding with future work suggestions.

2 Background and Related Work

In the context of image processing, Qu et al. [1], introduced an encoder-decoder model for high spatial resolution remote sensing image captioning, inspired by neural language models for conventional image captioning [2, 3] using CNN for feature extraction, RNN and LSTM for sentence generation, tested on remote sensing images, the model's performance was evaluated with BLEU, METEOR and CIDEr scores.

BLEU, METEOR, ROUGE, CIDEr, and SPICE are prominent evaluation metrics, where BLEU, METEOR and ROUGE were introduced for natural language processing and machine translation tasks, but CIDEr and SPICE are specifically designed for image captioning evaluation. BLEU [4] focuses on n-gram precision, while METEOR [5] incorporates precision, recall of unigrams, penalises exact word matches and rewards partial matches. ROUGE [6] evaluates n-gram overlaps using precision and recall with variants capturing the longest common subsequences. CIDEr [7] applies TF-IDF weighting to capture consensus in image captions. SPICE [8] evaluates semantic content using scene graphs to assess meaning beyond n-gram overlap.

Early studies in RSIC utilised RNN and LSTM models, with LSTM showing superior performance [1, 9]. By 2020, transformer-based models became favoured for their parallel training capabilities and time efficiency [10]. Template-based [11] and retrieval-based [12] language models offer well-structured captions but lack flexibility and variety, limiting their performance. RNNs, particularly LSTM models, gained attention for improving caption quality, but transformer-based models now surpass them in performance and efficiency. Enhancements like active attention [13], attribute relation encoding [14], and persistent memory mechanisms [15] further refine image captioning. To address overfitting in small datasets [16], cross-entropy-based modified loss functions [17] and reinforcement learning strategies [10] have been explored.

Transfer learning which leverages prior knowledge from one task to improve performance on another, has significantly improved language models' performance. Early models like Word2vec [18] provided basic word embeddings but lacked contextual understanding, leading to ELMO [19], which introduced contextual embeddings using biLSTM. Transformers [20] innovated with self-attention mechanisms, forming the basis for models like GPT (decoder-only) and BERT (encoder-only) with enhancements like masked language modelling [21]. S-BERT reduced computation for sentence-level tasks using siamese networks and triplet loss. The T5 model [22], using the same transformer architecture at its both parts of encoder and decoder, further refined transformer efficiency by reconstructing the original text.

3 Methods and Materials

3.1 Dataset

The pipeline utilises the DeepGlobe Land Cover dataset, comprising 803 satellite images with masks. The masks are RGB images that semantically segment its corresponding satellite image into seven land cover types, namely: "Urban", "Agriculture", "Range land", "Forest", "Water", "Barren land", and "Unknown". The dataset is divided into training, validation, and testing sets, each pair of mask and image has a unique image ID. The test and validation images lack corresponding masks. For this study, 750 images are used for training, 53 images for validation, and 343 for testing.

3.2 Method

The pipeline has two main stages, consisting of an image processing stage using the U-net model for semantic segmentation and a language processing stage for text generation. The U-net model produces image masks. Text generation is performed using T5 and ChatGPT, covering individual image caption generation.

Image Processing. The U-net architecture, introduced by Ronnerberg et al. (2015) [23], is widely used for semantic segmentation tasks, featuring U-shaped network with encoder and decoder parts, comprising four blocks of convolution, ReLu activations, and pooling or upsampling steps. For each block inside the extracting part, the output is concatenated to its corresponding block in the contracting part and finally followed by two convolutions and a ReLu. At the final layer, a 1×1 convolution is done to map each feature vector to the desired number of classes.

In this study, U-net was adapted for satellite image segmentation, with images resized from (2448, 2448) to (256, 256) to reduce computational complexity. A keras-based random search tuner optimised the learning rate and batch size by running the U-net model and evaluating its accuracy on the validation set for 5 times for each combination randomly selected from the predefined search spaces of {0.001, 0.0001} for learning rate and {16, 32, 48, 64} for batch size, ultimately selecting 0.001 and 16 respectively resulting in the highest accuracy. The final model generated 1146 masks, calculating pixel counts for 7 categories per image to form a table with 1146 rows and 7 columns of land cover proportions, which were used for subsequent text generation.

Text for Single Images T5. The T5_base model, known for "text-to-text" framework, which can be fine-tuned, was used for generating image descriptions. "Template sentences" and "Rule-based sentences" were passed to T5 in various ways. Template sentences, predefined with blank spaces, are converted into rule-based sentences by filling the blanks with land cover proportions extracted from masks. Both types aid in providing structured and informative descriptions for corresponding images.

First, template sentences with blank spaces paraphrased with T5, and blank spaces were filled using two methods: A sentence-based method analysing all tokens excluding stopwords, while a neighbourhood-based method focusing on non-stopword tokens near the blanks. In both approaches, the selected words and land cover names are encoded into vectors using "word2vec-google-news-300" [18], and the most relevant land cover name to the selected sentence tokens is identified using cosine similarity. However, these methods often generated less informative sentences and required reordering and filtering based on land cover proportions to improve relevance.

The other approach directly passed rule-based sentences to T5 for paraphrasing, bypassing blank-filling procedures. This approach produced more fluent and precise summaries but still lacked domain-specific adaptability. To address this, the T5 model was fine-tuned on rule-based sentences. To fine-tune T5, numeric proportions of train images with brief descriptions are provided as source text, while rule-based sentences from the train images serve as target sentences. Fine-tuning improved the coherence and relevance of generated text, demonstrating the model's adaptability when tailored to specific data and tasks.

Text for Single Images ChatGPT. ChatGPT_3.5 turbo, based on a generative pre-trained transformer architecture, is utilised for text generation for single images without needing predefined templates or rule-based sentences. Trained on extensive internet data, it can dynamically interpret and generate human-like responses. In this approach, numerical data representing land cover proportions are passed to ChatGPT, and a carefully designed prompt structure is used to elicit informative, concise descriptions for each image, highlighting significant categories while disregarding zero values.

4 Results

The results for text generation for single images were evaluated through human scoring, ROUGE, and BERT similarity scores. In the evaluation tasks employing ROUGE and BERT, the source text is the rule-based sentence with which the generated text will be compared. Human evaluation focused on four aspects: category coverage, sentence coherence, relevance to the context, and accuracy of numeric information. For a given sentence each fulfilled aspect added 1 point and each unmet aspect subtracted 1 point. According to Table 1, human evaluation found that despite occasional mismatches between numbers and the category labels and a smaller average of 3.4 sensible sentences, ChatGPT outperformed other models, with the highest average score of 4.3725, excelling in reporting multiple categories and accurate numeric information. The fine-tuned T5 generated texts discussing an average 4.26 categories among all 7 categories per image, with at most 2 categories per sentence and around 2.17 categories represented accurately by numbers per text. However, issues remain, including altered numeric values, sentence replications and inconsistent descriptions of the same categories. Other methods, such as passing sentences with numbers to T5 and neighbourhood-based approaches, had lower scores. Table 2 demonstrates ROUGE and BERT scores confirm the human evaluations, with ChatGPT and fine-tuned T5 as the two best performing models in novelty (low ROUGE scores) and contextual relevance (high BERT scores).

Table 1. Average human evaluated scores for different aspects of all proposed models

	Model				
Average Numbers of	ChatGPT	fine-tuned T5	Passing Sentences with Numbers	Sentence-Based	Neighbourhood-Based
Mentioned Categories	6.8	4.26	4.4	2.66	3.1
Sensible Sentences	3.39	3.11	2.26	2.66	0.81
Relevant Sentences	3.4	3.99	2.42	1.39	1.68
Accurate Numbers	3.9	2.17	1.59	−0.47	−0.36

Table 2. ROUGE-1, ROUGE-2, ROUGE-L, and BERT scores on top human-evaluated sentences

	ChatGPT	fine-tuned T5	Passing Sentences	Sentence-Based	Neighbourhood-Based
R1	0.23	0.17	0.59	0.37	0.29
R2	0.01	0.05	0.44	0.15	0.06
RL	0.22	0.17	0.58	0.36	0.28
BERT	0.78	0.85	0.85	0.83	0.81

5 Conclusion

This paper explored generating descriptions for satellite images using T5 and ChatGPT. The results showed that ChatGPT outperformed T5 in text generation for satellite images. ChatGPT's larger number of parameters enabled it to handle language complexities better than T5, which requires fine-tuning for specific tasks. However, both models had limitations, such as errors in reporting land cover proportions and category labels.

Future work could include improved evaluation methods beyond human evaluation scores discussed in this paper, incorporating approaches that assess various textual aspects. Additionally, it would be valuable to evaluate efficiency of T5 and ChatGPT in generating descriptions for satellite image clusters or devise models to identify specific images or clusters of images based on real-time descriptive text. For instance, rule-based sentences or template sentences corresponding to each image within a cluster could be provided to T5 or ChatGPT to produce a unified description, enabling a performance comparison.

Furthermore, employing LSTM-based image captioning models could broaden this comparison, providing deeper insights into the relative performance of ChatGPT and T5 in satellite image captioning tasks. Future work also includes applying the pipeline to larger satellite image datasets to achieve improved results.

SIC plays a crucial role in disaster management by automating the analysis of satellite imagery for tasks such as damage assessment, real-time monitoring of events like floods or wildfires, resource allocation, and supporting search and rescue operations. Future work could also focus on fine-tuning ChatGPT to improve its accuracy and domain specificity, enhancing its ability to generate precise, context aware descriptions.

Designing a mobile app capable of generating detailed, accurate captions about affected areas, infrastructure damage, and accessible routes could further enhance situational awareness and decision-making. Moreover, integrating the proposed model with Geographic Information Systems (GIS) and AI-driven analytics, would enable scalability and preparedness, ensuring timely and effective resource distribution while reducing human error.

References

1. Qu, B., Li, X., Tao, D., Lu, X.: Deep semantic understanding of high resolution remote sensing image. In: Proceedings of the International Conference on Computer, Information and Telecommunication Systems (CITS), pp. 1–5 (2016)
2. Vinyals, O., Toshev, A., Bengio, S., Erhan, D.: Show and tell: a neural image caption generator. In: Proceeding of the IEEE Conference on Computer Vision and Pattern Recognition (2015)
3. Karpathy, A., Fei Fei, L.: Deep visual-semantic alignment for generating image descriptions. In: Proceedings of the IEEE Conference on Computer Vision and Pattern Recognition (2015)
4. Papineni, K., Roukos, S., Wars, T., Zhu, W.: Bleu: a method for automatic evaluation of machine translation. In: 40th Annual Meeting of the Association for Computational Linguistics, Philadelphia, Pennsylvania, USA, pp. 311–318 (2002)
5. Banerjee, S., Lavie, A.: METEOR: an automatic metric for MT evaluation with improved correlation with human judgments. In: ACL Workshop on Intrinsic and Extrinsic Evaluation Measures for Machine Translation and/or Summarization, pp. 65–72. Association for Computational Linguistics, Ann Arbor (2005)
6. Lin, C.: ROUGE: a package for automatic evaluation of summaries. In: Text Summarization Branches Out, pp. 74–81. Association for Computational Linguistics, Barcelona (2004)
7. Vedantam, R., Zitnick, C.L., Parikh, D.: Consensus-based image description evaluation. ArXiv. /abs/1411.5726 (2014)
8. Anderson, P., Fernando, B., Johnson, M., Gould, S.: SPICE: semantic propositional image caption evaluation. ArXiv. /abs/1607.08822/ (2016)

9. Lu, X., Wang, B., Zheng, X., Li, X.: Exploring models and data for remote sensing image caption generation. IEEE Trans. Geosci. Remote Sens. **56**(4), 2183–2196 (2017)
10. Shen, X., Liu, B., Zhou, Y., Zhao, J.: Remote sensing image caption generation via transformer and reinforcement learning. Multimedia Tools Appl. **79**(35), 26661–26682 (2020)
11. Shi, Z., Zou, Z.: Can a machine generate humanlike language description for a remote sensing image? IEEE Trans. Geosci. Remote Sens. **55**(6), 3623–3634 (2017)
12. Wang, B., Lu, X., Zheng, X., Li, X.: Semantic description of high-resolution remote sensing images. IEEE Geosci. Remote Sens. Lett. **16**(8), 1274–1278 (2019)
13. Lu, X., Wang, B., Zheng, X.: Sound active attention framework for remote sensing image captioning. IEEE Trans. Geosci. Remote Sens. **58**(3), 1985–2000 (2020)
14. Yuan, Z., Li, X., Wang, Q.: Exploring multi-level attention and semantic relationship for remote sensing image captioning. IEEE Access **8**, 2608–2620 (2020)
15. Fu, K., Li, Y., Zhang, W., Yu, H., Sun, X.: Boosting memory with a persistent memory mechanism for remote sensing image captioning. Remote Sens. **12**(11), 1874 (2020)
16. Li, X., Zhang, X., Huang, W., Wang, Q.: Truncation cross entropy loss for remote sensing image captioning. IEEE Trans. Geosci. Remote Sens. **59**(6), 5246–5257 (2021)
17. Zhang, X., Wang, X., Tange, X., Zhou, H., Li, C.: Description generation for remote sensing images using attribute attention mechanism. Remote. Sens. **11**(6), 612 (2019)
18. Mikolov, T., Chen, K., Corrado, G., Dean, J.: Efficient estimation of word representation in vector space. arXiv preprint arXiv:1301.3781 (2013)
19. Peters, M.E., et al.: Deep contextualized word representation. arXiv:1802.05365v2 (2018)
20. Vaswani, A., et al.: Attention is all you need. arXiv:1706.03762v7 (2017)
21. Devlin, J., Chang, M., Lee, K., Toutanova, K.: BERT: pre-training of deep bidirectional transformers for language understanding. arXiv:1810.04805 (2019)
22. Raffel, C., et al.: Exploring the limits of transfer learning with a unified text-to-text transformer. J. Mach. Learn. Res. **21**(140), 1–67 (2020)
23. Ronneberger, O., Fischer, P., Brox, T.: U-net convolutional networks for biomedical image segmentation. In: Medical Image Computing and Computer-Assisted Intervention-MICCAI 2015: 18th International Conference, pp. 234–241. Springer, Munich (2015)

Software Engineering

AI-Powered Semantic Search for Historical Documentation: A Collaborative Research with Hitachi Energy

Ivan Kelly Maranan Hansson, Edvin Wiklund, Alessio Bucaioni(✉), and Luciana Provenzano

Mälardalen University, Västerås, Sweden
{alessio.bucaioni,luciana.provenzano}@mdu.se

Abstract. Companies with long operational histories often face the challenge of managing vast repositories of documentation, which hold critical knowledge needed for maintaining ongoing projects. Retrieving relevant information from these extensive archives is a time-consuming and complex task, requiring specialized expertise and familiarity with outdated terminology. Semantic search has emerged as a promising technology to address these issues by improving the precision and efficiency of information retrieval. In this paper, we present our collaborative research with Hitachi Energy, exploring the development of a semantic search engine based on existing open-source solutions to assist practitioners in searching large industrial historical document repositories. We first analyzed available No-SQL databases with search-engine interfaces, followed by an evaluation of pre-trained semantic transformers to determine which offers the best balance of accuracy and speed for semantic search. Our research identified OpenSearch as the most suitable No-SQL database due to its flexibility, free usage, and support for semantic transformers. After evaluating various pre-trained semantic transformers, we found all-MiniLM-L6-v2 to offer the best balance of accuracy and speed for semantic search. Based on the findings, we developed a prototype AI-powered semantic search tool, which was tested in a workshop involving Hitachi Energy professionals. Our findings demonstrate the feasibility and effectiveness of AI-powered semantic search for handling historical documentation, offering significant potential for industries tasked with managing large legacy archives.

Keywords: Semantic search · pre-trained semantic transformers · documentation · AI

1 Introduction

Corporations with long histories, such as Hitachi Energy, manage vast repositories of documentation spanning decades, often critical for maintenance, operations, and customer support. For instance, Hitachi's High Voltage Direct Current

S. Latifi (Ed.): ITNG 2025, AISC 1463, pp. 503–516, 2025.
https://doi.org/10.1007/978-3-031-89063-5_44

(HVDC) division still relies on projects initiated in the 1950s. However, these documents, often stored in paper-based or poorly digitized formats, are difficult to search efficiently due to inconsistent structures, heterogeneous formats, and irrelevant content like headers and footers. Searching these archives is time-consuming and error-prone. Traditional keyword-based search engines struggle with variations in terminology, misspellings, paraphrased concepts, and outdated terminology, making it harder to find relevant information. Recent advancements in semantic search, powered by AI [1,2] and technologies like BERT and SBERT, address these challenges by understanding query intent and retrieving results based on semantic meaning, even in unstructured datasets [3].

This paper presents a collaborative effort with Hitachi Energy to develop an AI-powered semantic search solution. We evaluated No-SQL databases and pre-trained semantic transformers, selecting OpenSearch for its flexibility and free functionality and all-MiniLM-L6-v2 for its superior balance of accuracy and speed. Using the MS MARCO dataset to ensure data confidentiality, we measured performance with metrics such as Mean Reciprocal Rank ($MRR@k$) and Normalized Discounted Cumulative Gain ($nDCG@k$). The resulting prototype, tested with Hitachi Energy professionals, demonstrated the feasibility and effectiveness of AI-powered semantic search for managing large industrial archives, offering significant potential for similar use cases.

The remainder of this paper is structured as follows. Section 2 details the research process adopted for this work. Section 3 describes the design and execution of the experiments to evaluate pre-trained semantic transformers in conjunction with a No-SQL database. Section 4 introduces the prototype of the AI-powered semantic search tool developed for Hitachi Energy. Section 5 presents insights from the validation workshop at Hitachi Energy, discussing the strengths and weaknesses of the solution, along with threats to validity. Section 6 reviews related works from the literature, and Sect. 7 concludes the paper with final remarks and possible future research directions.

2 Research Process

In this work, we employed a research process that is an adaptation of two methodologies: the engineering method by Basili [4] and the one by Gorschek et al. [5]. Our process focuses on implementing and connecting existing solutions [4] while also maximizing the transfer of technology between academia and industry through a multi-step validation process [5]. The first step is the elicitation of industrial needs, which was initiated by the Hitachi Energy representatives involved in this research. Starting from the identified need, we derived a generalized Research Goal (RG) that is:

RG: to investigate the feasibility of implementing an AI-powered solution to search historical documentation.

To comply with the engineering method, we began by building a candidate solution. This process started with collecting and reviewing existing solutions through a comprehensive literature review. To strike a balance between rigor and agility[1], we did not opt for a full-fledged systematic review. We searched three scientific databases: ACM Digital Library[2], IEEE Explore[3], and Google Scholar[4], with the aim of identifying existing No-SQL databases with search-engine interfaces, pre-trained semantic transformers, and their benefits. In doing so, we excluded studies that were not written in English, were not peer-reviewed, or were shorter than 4 pages. We conducted experiments on the No-SQL databases with search-engine interfaces and pre-trained semantic transformers identified during the literature review. The goal was to select the most performant sentence transformer for implementing a prototype AI-powered semantic search tool for Hitachi Energy. In particular, we evaluated all the collected pre-trained semantic transformers using Mean Reciprocal Rank (MRR), Normalized Discounted Cumulative Gain (nDCG), and search speed (milliseconds) on the MS MARCO Top 1000 Evaluation set. Section 3 provides further details on these activities. Building a candidate solution was an iterative step coupled with a preliminary validation step called validation in academia, where each iteration of the candidate solution was evaluated in academic settings to assess its initial feasibility and applicability. To further validate the candidate solution and measure its relevance and applicability in industrial settings, we conducted a workshop with Hitachi Energy employees, including three global managers, one line manager, one pre-eminent engineer, and a team of seven engineers. During this workshop, we allowed participants to test the solution with a small sample dataset from Hitachi Energy. Section 4 provides further details on the workshop.

3 Evaluating Semantic Transformers with No-SQL Databases: The Experiment

This section describes the experiment to evaluate pre-trained semantic transformers with a No-SQL database using three key metrics: Mean Reciprocal Rank (MRR), Normalized Discounted Cumulative Gain (nDCG), and search speed in milliseconds. The evaluation was conducted on the first 100,000 pairs of the MS MARCO Top 1000 Evaluation dataset, as the full 8.8 million query-answer pairs were impractical to process due to computational constraints. We focused on $k = 10$ and $k = 100$ for MRR and nDCG, assessing how often relevant answers appeared in the top results. High scores at $k = 10$ indicate that relevant answers are frequently found within the top 10 results, while high scores at $k = 100$ suggest relevance within a broader set of results.

3.1 No-SQL Database Selection

Handling unstructured data is a major challenge in industrial environments. While databases like SQL and No-SQL manage data efficiently, searching for

[1] For the sake of space, we omit all the details of the literature review.
[2] https://dl.acm.org/.
[3] https://ieeexplore.ieee.org/Xplore/home.jsp.
[4] https://scholar.google.com/.

relevant information remains difficult. Traditional keyword-based methods often fail to handle the nuances of unstructured data. The state-of-the-art approach uses No-SQL databases with search-engine interfaces and semantic transformers [6]. From our literature review, we identified three No-SQL databases with built-in search interfaces as potential candidates for our solution.

- ElasticSearch: Built on the Lucene library, ElasticSearch is a full-text search engine with a distributed search and analysis engine at the core of the Elastic Stack. It operates under the Elastic License.
- OpenSearch: Originally a fork from version 7.10.2 of ElasticSearch, OpenSearch also builds on the Lucene library and supports full-text search. It is distributed under the Apache License Version 2.0 (ALv2).
- Apache Solr: Another search engine built on the Lucene library, Apache Solr supports full-text indexing and searching. Its operations follow three main steps: indexing, querying, and ranking. Solr is licensed under ALv2.

All three search engines are built on Apache Lucene, an open-source library for indexing and managing documents in No-SQL databases, leading to similar baseline performance. Table 1 compares their key benefits based on six categories identified in the literature, referencing studies such as [3,7–14].

- Document storage: How data is stored in the database.
- Free of charge: Whether the database is entirely free to use without additional charges for full functionality.
- Dense vector search: Whether the database supports semantic search using dense vector techniques.
- Built-in sentence transformers: Whether the database has native support for sentence transformers.
- Support for third-party embedding models: Whether the system can integrate with external embedding models.
- Pre-embedding requirement: Whether sentences must be pre-embedded before insertion into the database.

Table 1. Comparison of No-SQL databases based on their capabilities.

Capability	OpenSearch	ElasticSearch	Apache Solr
No-SQL storage type	Document	Document	Document
Free of charge	X		X
Dense vector search (Semantic search)	X	X	X
Built-in sentence transformers	X	X	
Supports third-party embedding models	X	X	
Pre-embedding of sentences before insertion			X

From Table 1 and the literature review, OpenSearch emerged as the best choice for our implementation. It meets all criteria, including being open-source, supporting dense vector searches, sentence transformers, and third-party models, and offering a free license. Its flexibility and modifiability make it ideal for both academic research and industrial use, such as at Hitachi Energy.

3.2 Dataset

The experiment used the first 100,000 lines of the MS MARCO Top-1000 Evaluation dataset [15], chosen for its similarity to Hitachi Energy's data and query patterns. A Python script automated the splitting and uploading of documents, creating passages of varying lengths, mirroring those in MS MARCO. The dataset's structure, with short queries matching both long and short passages, aligned with Hitachi Energy's typical queries. MS MARCO was also selected for its public availability, ensuring repeatability and generalizability for other researchers. It consists of question-answer pairs, each with a unique ID.

3.3 Evaluation Metrics

We used three evaluation metrics, where the first two metrics were used to evaluate the accuracy of the sentence transformers, while the final metric assesses the speed of the search.

1) Accuracy Metrics. We define accuracy based on how well the results of a query match the user's intent, which can be measured by the ranking of the retrieved results. To assess accuracy, we used two metrics that consider the ranking: MRR@k and nDCG@k.

MRR@k measures how quickly a user can find a relevant answer. Equation 1 shows how MRR is calculated by summing the reciprocal rank of the first relevant result for each query, where the rank of the correct answer is used.

$$\text{MRR@}k = \frac{1}{j}\sum_{j=1}^{j}\frac{1}{rank_j} \tag{1}$$

The result is an average across all queries, and the MRR value ranges between 0 and 1, with a higher value indicating better accuracy (i.e., relevant results appearing higher in the list of retrieved answers). The k value limits how many results are included in the retrieved list, and if no relevant result is found within the top k results, the query receives a reciprocal rank of zero. The MRR@k metric is well-suited for our goal, as it reflects how efficiently the system retrieves relevant answers, providing insight into whether the implementation is viable in real-world applications.

To further support our evaluation, we utilized the nDCG metric. Unlike MRR, nDCG takes into account not just the first correct result, but also the ordering of subsequent relevant results. nDCG compares the actual ranking of relevant documents to an ideal ranking. Equation 2 shows the formula for nDCG, which divides the Discounted Cumulative Gain (DCG) by the Ideal Discounted Cumulative Gain (IDCG).

$$\text{nDCG@}k = \frac{DCG@k}{IDCG@k} \tag{2}$$

The DCG at rank k is calculated using Eq. 3, where rel_i is the relevance of the passage at rank i. A relevant result is assigned a value of 1, while an irrelevant result is assigned 0.

$$\text{DCG@}k = \sum_{i=1}^{k} \frac{rel_i}{\log_2(i+1)} \tag{3}$$

IDCG is computed by ordering the relevant passages in descending order and calculating the DCG for the ideal ranking (Eq. 4).

$$\text{IDCG@}k = \sum_{i=1}^{k} \frac{rel_i}{\log_2(i+1)} \tag{4}$$

Both MRR and nDCG yield results between 0 and 1, with higher scores indicating that correct answers are ranked higher in the list of results. Together, these metrics provide a comprehensive view of the accuracy and ranking performance of the semantic transformers.

2) Speed Metric. Speed is defined as how quickly the system returns an answer to the user. To evaluate speed, we measured the execution time for each query in milliseconds, ensuring precise timing for each search operation. The total execution times for all queries were summed and averaged, as shown in Eq. 5, where k represents the number of queries and t_i is the time in milliseconds for each query.

$$\text{Average Speed} = \frac{\sum_{i=1}^{k} t_i}{k} \tag{5}$$

This metric allowed us to assess whether the search system could return results quickly enough to be practical in real-world scenarios.

3.4 Experimental Process

The experiment process, shown in Fig. 1, was conducted on a laptop with an AMD Ryzen 7 4800U CPU and 16 GB of RAM using a Docker environment. A cluster of four OpenSearch nodes (version 2.13.0) was set up: two for data storage, one for machine learning tasks, and one for the dashboard interface. After initializing the cluster, a sentence transformer was uploaded (step 1, Fig. 1), and an index containing answers, IDs, and sentence embeddings was created. The MS MARCO dataset answers were embedded and inserted into the database via a pipeline (step 2, Fig. 1). Semantic searches were conducted using dataset queries (step 3, Fig. 1). Queries were embedded and compared against the indexed data using k-NN search and cosine similarity to rank semantically closest matches (step 4, Fig. 1). Execution times were averaged across all queries, and results were evaluated using nDCG and MRR metrics. Steps two through four were automated with Python to ensure accuracy and consistency, and the experiment was repeated for all selected sentence transformers.

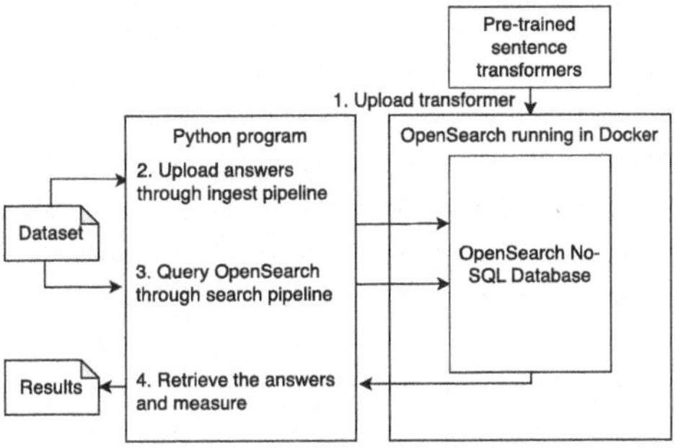

Fig. 1. Experimental process.

3.5 Results

The metrics used during the experiments are *MRR* and *nDGG* for accuracy of the search, and *average speed* for the time to provide the response to the search, as explained in Sect. 3.3. A low score in MRR and nDCG indicates that the correct answers was ranked lower than expected, while a low average time per query suggests that the transformer executes queries quickly. From Table 2, four sentence transformers stand out, demonstrating better performance across the different metrics.

Upon analysis, it became evident that the fluctuations in results between most transformers were minimal. However, two models, distiluse-base-multilingual-cased-v1 and paraphrase-multilingual-MiniLM-L12-v2, performed significantly worse in both MRR and nDCG, indicating that these transformers ranked the correct answers much lower in the list of retrieved answers.

No clear correlation was observed between embedding dimensions and accuracy, nor did optimization for semantic search guarantee superior performance. Notably, paraphrase-mpnet-base-v2, with 512 dimensions, performed poorly across all metrics, particularly in terms of speed. The worst transformer in terms of speed was alldistilroberta-v1, which took 203ms per query, 92ms longer than the average. This shows that while some models may excel in certain areas, they may fall behind significantly in others.

Table 2. Result from the experiment

Sentence transformer	$MRR@10$	$MRR@100$	$nDCG@10$	$nDCG@100$	Avg. $ms/query$
alldistilroberta-v1	0.3377	0.3424	0.2546	0.4039	203
all-MiniLM-L6-v2	0.3617	0.3654	0.2876	0.4484	73
all-MiniLM-L12-v2	0.3572	0.3611	0.2824	0.4422	53
all-mpnet-base-v2	0.3470	0.3514	0.2640	0.4220	162
msmarco-distilbert-base-tas-b	0.3540	0.3582	0.2650	0.3867	94
multi-qa-MiniLM-L6-cos-v1	0.3566	0.3605	0.2810	0.4360	67
multi-qa-mpnet-base-dot-v1	0.3223	0.3278	0.2341	0.3722	169
paraphrase-MiniLM-L3-v2	0.3580	0.3618	0.2817	0.4192	74
paraphrase-multilingual-MiniLM-L12-v2	0.3060	0.3115	0.2202	0.3458	55
paraphrase-mpnet-base-v2	0.3341	0.3389	0.2459	0.3829	177
distiluse-base-multilingual-cased-v1	0.2531	0.2617	0.1648	0.2930	94
Average	0.3352	0.3401	0.2528	0.3957	111

4 Hitachi AI-Powered Semantic Search Tool Prototype

Based on previous section findings, OpenSearch was selected as the most suitable No-SQL database for its flexibility, open-source nature, and support for semantic transformers. The "all-MiniLM-L6-v2" transformer was chosen for its balance of accuracy and speed, forming the foundation of an AI-powered semantic search prototype for Hitachi Energy. The tool's architecture integrates OpenSearch with a PyQt6-based GUI, enabling intuitive user interaction. The opensearch-py library connects the UI to OpenSearch, which runs in a Docker containerized setup with three nodes: two data nodes for storing documents and embeddings and one node hosting the sentence transformer for efficient semantic processing. The GUI allows users to set query parameters, including the search text, the number of answers to retrieve, and the k-value for top-k results. Results are displayed with details such as scores, documents, and retrieved answers, while additional metrics on document retrieval are shown for user insight. This setup provides a seamless and efficient semantic search experience.

4.1 Evaluating the Hitachi AI-Powered Semantic Search Tool Prototype

We evaluated the prototype in a workshop with 12 Hitachi Energy employees, including managers and engineers who regularly work with historical documents and face challenges in retrieving information. The workshop had three parts: a 15-min introduction presenting the tool's purpose, functionality, and architecture; a 35-min hands-on session where participants tested the prototype using domain-specific queries and discussed its limitations and improvements; and a 30-min discussion on the tool's broader applicability and potential integration within the company. Insights and feedback from the workshop are detailed in Sect. 5.

5 Discussion

In this section, we reflect on the results of our study and the development of the AI-powered semantic search prototype. We evaluate the insights gained from both the technical experiment and the practical validation workshop conducted with Hitachi Energy. Additionally, we discuss the strengths and limitations of our approach, while also addressing potential threats to the validity of our findings.

5.1 On the Results and Prototype

As discussed in Sect. 3, we identified three search engines that met the needs of our solution. Ultimately, OpenSearch was chosen due to its flexibility and the fact that it offers all functionalities free of charge. While we focused on feasibility rather than optimization (refer to Sect. 5.2 for more details on this topic), we acknowledge that the other solutions, such as ElasticSearch, could potentially perform better in terms of speed, memory usage, and redundancy, as some papers suggest ElasticSearch has faster execution. However, since all the identified search engines are built on the Apache Lucene library, we believe they would perform similarly and be relatively easy to switch between due to their shared underlying architecture. This assumption is further supported by CERN's successful transition from ElasticSearch to OpenSearch [7]. Our experiments revealed that the all-MiniLM-L6-v2 sentence transformer was the best fit for our solution, outperforming other transformers across most of our chosen metrics. While it was marginally slower in query time compared to the second-best model, the difference of 20ms was considered negligible. We prioritized accuracy metrics, which showed that all-MiniLM-L6-v2 consistently returned correct answers higher in the result list, thereby improving the overall search speed by minimizing user effort. Interestingly, transformers optimized specifically for semantic search did not outperform the others in terms of accuracy; only multi-qa-MiniLM-L6-cos-v1 made it into the top four. However, the performance gap between the top four transformers was minimal, suggesting that any of them could work well in our solution. It's important to note that the experiment was limited to 100,000 query-answer pairs due to time constraints. While this limitation may have affected the results, we argue that our sample size was sufficient to draw meaningful conclusions. Another observation is that sentence transformers producing larger embedding dimensions performed worse in terms of speed, which aligns with the expectation that k-NN search in OpenSearch performs less efficiently with larger embeddings. Surprisingly, larger embeddings did not correlate with higher accuracy, which seems counter-intuitive, as we expected larger dimensions to capture more nuanced differences in passages and sentences. Our findings align with the work of Xian et al. [3], demonstrating that existing tools can be combined easily to create a powerful vector search solution. We expanded on their work by testing some of the AI tools provided by OpenSearch, and our validation activities at Hitachi Energy confirm that it is feasible to achieve AI-powered semantic search for historical documentation. Lastly, our study did not focus on optimizing the performance of the tested sentence transformers. A more in-depth fine-tuning process, specifically tailored to a dataset like Hitachi Energy's, could yield better results. During the validation process, we observed indications that such customization would be highly beneficial, as a model trained on domain-specific data would likely outperform the pre-trained models we used.

5.2 Insights from the Validation Workshop

The practitioners at Hitachi Energy were generally satisfied with the potential offered by our prototype for searching large historical documentation. One key observation was that search speed was prioritized over accuracy. In the current process at Hitachi Energy, searching for information in historical archives involves locating documents by title and manually sifting through an average of 40 to 50 pages containing varied information, such as installed equipment details, user guides, and release notes. In this context, the primary requirement for a semantic search engine is that it completes searches swiftly, ideally within seconds, and without exceeding a 30-second threshold. A faster search engine mitigates the need for perfect accuracy, as engineers can refine their queries and quickly repeat searches until they obtain more satisfying results. This balance between speed and accuracy aligns with the primary operational needs at Hitachi Energy. Another strength of our solution was the use of a GUI. The 12 practitioners who tested the tool found the GUI intuitive and easy to interact with. Despite the difference in the way the tool operates—where the search process does not require document titles—the new search method was quickly understood and adopted. This ease of use is an encouraging result that may positively influence the future adoption of the tool within the company. Finally, we believe that the decision to use off-the-shelf and open-source solutions to develop an AI-powered semantic search tool offers a cost-effective and feasible approach for companies looking to introduce semantic search into their workflows. By leveraging readily available, continuously updated technologies and existing know-how, companies can significantly reduce the cost and time associated with developing and maintaining such tools. This approach provides flexibility and scalability without requiring the heavy investment associated with proprietary solutions.

5.3 Threats to Validity

We identified several possible threats to the validity that may have affected this study [16]. The first threat concerns the metrics we used to evaluate the accuracy of the search engine, namely MRR and nDCG. Since there are no universally recommended metrics for evaluating accuracy in the literature, the choice of these metrics may affect the construct validity of the experiment. However, we selected MRR and nDCG because they are widely used in similar studies, as explained in Sect. 3. Furthermore, MRR is the standard metric used for the MS MARCO dataset, which we used in our experiment. The choice of the sentence transformers and dataset may represent a further threat to the validity of our work because they can introduce unknown factors that affect the causal relationship we are investigating. We limited the number of evaluated sentence transformers to those already provided in OpenSearch, the No-SQL database chosen for our experiment (Sect. 3). However, the selected transformers are among the most popular and well-regarded multilingual and English transformers on the HuggingFace community website[5], making them appropriate for our study. A further threat is related to the computational environment used for our experiment. We ran the prototype on a laptop with limited computational resources, which may have impacted the performance of the evaluated transformers. Although this setup reflects realistic constraints in terms of hardware for some users, the system might perform differently in a high-performance computing environment, potentially yielding faster results or better scalability. Another threat concerns the confidentiality of our test data, which may limit the generalizability of our findings. To maintain the confidentiality of Hitachi Energy's documents, we used the publicly available MS MARCO dataset to simulate their documentation, as previously explained. While this approach enhances the relevance of our work for other research studies, the solution might perform differently when applied to proprietary or domain-specific datasets. Finally, the validation process used in our study may introduce another threat to external validity. Although we conducted a workshop with Hitachi Energy employees to evaluate the usability and effectiveness of the tool, the feedback was limited to a small number of participants within a specific organizational context. Broader evaluations involving different industries, datasets, and end-users may reveal additional insights or limitations.

[5] https://huggingface.co/models?pipeline_tag=sentence-similarity&sort=likes.

6 Related Work

In this section, we present state-of-the-art research that aligns with our work.

6.1 About Sentence Transformers

The use of sentence transformers has expanded significantly, with key contributions from Devlin et al., who introduced BERT [17], and Reimers and Gurevych, who developed SBERT [18]. SBERT addressed BERT's limitations for Semantic Textual Similarity (STS) tasks by using a Siamese network, improving processing speed while retaining accuracy. These foundational works underpin our research, which evaluates pre-trained transformers based on BERT and SBERT, focusing on performance in speed and accuracy on external test data. Zhang et al. introduced the Hybrid List Aware Transformer Reranking (HLATR) module, outperforming BERT and RoBERTa in efficiency and robustness for document retrieval [19]. Using the MS MARCO dataset and Mean Reciprocal Ranking (MRR), their study aligns with ours but differs in scope. While they used the full MS MARCO dataset, we focused on the top 1000 passages, resembling Hitachi Energy's data. Additionally, our study evaluates off-the-shelf transformers in OpenSearch, offering insights into practical, ready-to-use tools.

6.2 About AI-Powered Search in Large-Scale Database

The challenge of retrieving relevant information from large-scale databases has gained attention with AI advancements, though the field remains relatively new and underexplored. Few papers provide comprehensive evaluations, and ongoing developments make it difficult to identify the most feasible solutions. Andre et al. explored semantic search for COVID-19 publications using the TREC-COVID and CORD-19 datasets [20]. Their evaluation employed metrics like nDCG, top-N precision, and MAP, but their CO-search engine relied on complete relevance judgments, highlighting a limitation. While their work developed a new system, our study focuses on evaluating existing sentence transformers within a search engine context rather than creating a system from scratch. Xian et al. examined vector search technologies, comparing Lucene (used in OpenSearch and ElasticSearch) with Facebook's Faiss [3]. They identified differences in speed, query latency, and throughput, emphasizing the importance of dense vector manipulation for modern search. Their findings provided a foundation for our work, which leverages Lucene for dense vector searches. We extend their approach by evaluating pre-trained sentence transformers integrated with Lucene in OpenSearch. Our study builds on these works by focusing on practical applications, evaluating pre-trained models, and integrating them into real-world systems like OpenSearch to improve information retrieval in unstructured datasets.

7 Conclusion and Future Work

This study examined the feasibility of using AI-powered semantic search to retrieve historical documentation in an industrial setting, focusing on Hitachi

Energy. We identified OpenSearch as the most suitable No-SQL database and selected the all-MiniLM-L6-v2 transformer for its balance of accuracy and speed. A prototype integrating these components into a graphical user interface was developed and evaluated by 12 Hitachi Energy employees using real-world data. The tool delivered accurate results in under one second, meeting the company's primary requirements. These findings demonstrate the potential of AI-powered semantic search for managing extensive legacy archives and contribute valuable insights to this emerging field.

Future work could include training a custom sentence transformer on organization-specific data, enhancing scalability for larger user loads, refining search parameters, and improving the GUI. Testing with larger datasets and across industries would further validate and expand the tool's applicability.

References

1. Musabimana, B.B., Bucaioni, A.: Integrating AiaaS into existing systems: the gokind experience. In: International Conference on Information Technology-New Generations, pp. 417–426. Springer, Cham (2024)
2. Nazir, R., Bucaioni, A., Pelliccione, P.: Architecting ml-enabled systems: challenges, best practices, and design decisions. J. Syst. Softw. **207**, 111860 (2024)
3. Xian, J., Teofili, T., Pradeep, R., Lin, J.: Vector search with openai embeddings: Lucene is all you need. In: Proceedings of the 17th ACM International Conference on Web Search and Data Mining, ser. WSDM 2024, pp. 1090–1093. Association for Computing Machinery, New York (2024)
4. Basili, V.R.: The experimental paradigm in software engineering. In: Experimental Software Engineering Issues (1992). https://api.semanticscholar.org/CorpusID:14982809
5. Gorschek, T., Garre, P., Larsson, S., Wohlin, C.: A model for technology transfer in practice. IEEE Softw. **23**(6), 88–95 (2006)
6. Ni, C., Wu, J., Wang, H., Lu, W., Zhang, C.: Enhancing cloud-based large language model processing with elasticsearch and transformer models (2024)
7. Papadopoulos, S., Saiz, P., Ulrich, S.: Architecting the opensearch service at CERN. In: 26th International Conference on Computing in High Energy and Nuclear Physics (CHEP 2023) (2023)
8. Parker, Z., Poe, S., Vrbsky, S.V.: Comparing NoSQL MongoDB to an SQL DB. In: Proceedings of the 51st ACM Southeast Conference, ser. ACMSE 2013. Association for Computing Machinery, New York (2013)
9. Hansen, J., Porter, K., Shalaginov, A., Franke, K.: Comparing open source search engine functionality, efficiency and effectiveness with respect to digital forensic search (2018). https://ntnuopen.ntnu.no/ntnu-xmlui/bitstream/handle/11250/2584227/577-Article%2BText-1167-1-10-20181009.pdf?sequence=2&isAllowed=y
10. Chang, X.: The analysis of open source search engines. Highlights Sci. Eng. Technol. **32**, 32–42 (2023)
11. Kathare, N., Reddy, O.V., Prabhu, V.: A comprehensive study of elasticsearch. Int. J. Sci. Res. (IJSR) (2020)
12. Gupta, S., Rani, R.: A comparative study of elasticsearch and couchdb document oriented databases. In: 2016 International Conference on Inventive Computation Technologies (ICICT), vol. 1, pp. 1–4 (2016)

13. Liu, G., Li, C., Tian, W., Li, Z.: Distributed geospatial data service based on opensearch. In: 2016 2nd IEEE International Conference on Computer and Communications (ICCC), pp. 100–104 (2016)
14. LeVan, R.: Opensearch and SRU: a continuum of searching. Inf. Technol. Libr. **25**(3), 151–153 (2006)
15. Bajaj, P., et al.: MS MARCO: a human generated machine reading comprehension dataset (2018)
16. Runesson, P., Höst, M.: Guidelines for conducting and reporting case study research in software engineering. Emperical Softw. Eng. 131–164 (2009)
17. Devlin, J., Chang, M., Lee, K., Toutanova, K.: BERT: pre-training of deep bidirectional transformers for language understanding, CoRR, vol. abs/1810.04805 (2018)
18. Reimers, N., Gurevych, I.: Sentence-bert: sentence embeddings using siamese bert-networks. In: Proceedings of the 2019 Conference on Empirical Methods in Natural Language Processing. Association for Computational Linguistics, Stroudsburg (2019). The 2018 Conference on Empirical Methods in Natural Language Processing. https://tubiblio.ulb.tu-darmstadt.de/117723/
19. Zhang, Y., Long, D., Xu, G., Xie, P.: HLATR: enhance multi-stage text retrieval with hybrid list aware transformer reranking (2022)
20. Esteva, A., Kale, A., Paulus, R., et al.: Covid-19 information retrieval with deep-learning based semantic search, question answering, and abstractive summarization. NPJ Digit. Med. **4**, 68 (2021)

The Use of Call Graphs and Deep Learning to Improve Software Testing

Hemanth G. Chintala, Ziad A. Al-Sharif(✉), and Safwan Omari

Department of Engineering, Computing and Mathematical Sciences (ECaMS), College of Aviation, Science and Technology (CoAST), Lewis University, 1 University Parkway, Romeoville, IL 60446, USA
{hemanthganeshchint,zalsharif,omarisa}@lewisu.edu

Abstract. Software testing is a critical part of software development, it is essential for preventing failures and enhancing software quality attributes. However, the testing process can be costly and time-consuming, often involving a large number of test cases. Over time, the accumulation of redundant and overlapping test cases can complicate and lengthen the testing time. To address these challenges, this paper utilizes graph similarity and deep learning techniques to optimize test suites. It uses call graphs from test cases to identify redundant and similar test cases. A machine learning model is used to calculate and predict the similarity scores between these call graphs, helping to classify and prioritize the test cases. This helps rank test cases based on their similarity scores, with lower scores indicating higher priority due to their unique code coverage. This approach allows test engineers to focus on the most diverse set of test cases, ensuring comprehensive code coverage and efficient testing. By reducing the number of redundant test cases, this method aims to streamline the testing process, reduce costs, and maintain high software quality standards. Ultimately, this paper seeks to provide a systematic framework for test engineers to determine the optimal amount of testing needed to effectively meet the software quality objectives.

Keywords: Software Testing · Regression Testing · Redundant Test Cases · Call Graph · Graph Similarity · Similarity Scores · Test Case Prioritization · Runtime Monitoring · Execution Trace

1 Introduction

Software testing is defined as the process of executing a program or system with the intent of finding errors and verifying that the software behaves as expected [1]. This involves applying a set of criteria to a structure or model of the software to ensure it meets specified requirements and performs its intended

H. G. Chintala, Z. A. Al-Sharif and S. Omari—These authors contributed equally to this work.

S. Latifi (Ed.): ITNG 2025, AISC 1463, pp. 517–527, 2025.
https://doi.org/10.1007/978-3-031-89063-5_45

functions. Others define software testing as a set of activities intended to find errors in software so that it can be corrected before the product is released to the end user [2]. It involves the execution of a software component or system component to evaluate one or more properties of interest. Nevertheless, all definitions emphasize that software testing is an indispensable part of the Software Development Life Cycle (SDLC), and it is essential for ensuring high software quality and reliability.

However, traditional software testing methods, while effective, often become cumbersome and resource-intensive, especially as software systems increase in complexity. The urgency for more efficient testing methodologies is more pronounced in today's fast-evolving technology landscape. Advances in machine learning provide promising avenues to revolutionize these testing processes, particularly through the optimization of test suites and the automation of test case generation and prioritization [3,4].

Additionally, one of the foremost challenges in software testing is the effective management of test suites. As projects evolve, test suites can become bloated with redundant or irrelevant test cases, leading to wasted resources and increased testing time without proportional gains in fault detection or coverage [5,6]. Nevertheless, analyzing the test suite can be achieved by different means [7]. Therefore, machine learning algorithms are increasingly employed to analyze patterns in data derived from test executions, helping to identify and eliminate redundant test cases, thereby enhancing the testing process's efficiency [8].

In this paper, we make use of the abstract call graph as a visual representation for the test case. These abstract graphs can help visually capture the internal behavior of the test case and lead to identifying the similarities between various test cases. However, the possibility of having hundreds or thousands of these graphs can make this visual analysis challenging. Therefore, this paper is employing a machine learning model to identify the similarity between various graphs can facilitate an automated process to help the test engineer identify, classify, and rank these test cases into levels or clusters. However, the results from this paper are preliminary. As part of our future goals, we aim to apply this methodology to real-life software projects.

The remaining parts of this paper are organized as follows. Section 2 sheds some light on the background research. Section 3 highlights the goals and objectives of this project. Section 4 presents our research methodology. Section 5 describes our experimentation environment. Section 6 presents our results. Finally, Sect. 7 concludes our work, and highlights the potential for future studies.

2 Background and Related Work

Graphs play a crucial role in software testing by providing visual representations of data, workflows, and system states. They are used to visualize data, making it easier to interpret and understand software structure, and relationships between entities. Graph-based testing involves representing software components or states as nodes and their interactions or transitions as edges, which helps in testing different scenarios. This includes state transition testing, data flow testing, control flow testing, and interaction flow testing [9]. Program graphs represent the flow of a program, aiding testers in visualizing and understanding the program's structure and behavior [10]. Overall, graphs enhance the clarity and effectiveness of software testing by providing a visual approach to understanding complex data and workflows [11].

2.1 The Use of AI in Software Testing

Many researchers have highlighted the importance of using modern AI in reducing redundant and similar test cases to save time and resources in software testing [12]. Techniques such as K-Nearest Neighbors (KNN) [3], Naïve Bayes [5], K-Means clustering [5], and Dynamic Domain Reduction (DDR) [6] have been employed to streamline test suites. The advancements in Machine Learning and Deep Learning offer promising approaches to tackling these challenges [4]. For instance, Yuan et al. [4] emphasized the capacity of deep learning models to automate and improve the identification and elimination of redundant test cases, leading to more efficient software development cycles [13,14].

2.2 Graph Similarities

Over the years, different methods for measuring how similar two graphs are implemented and researched. Depending on the type of graphs and the application. Some methods uses basic edit operations such as deleting, adding, changing vertices and edges, while other methods uses a partial matching of two graphs with an inexact matching relation [15]. However, techniques for graph matching and similarity assessments have evolved, with recent studies focusing on improving the accuracy and efficiency of these methods [8].

On the other hand, graph-based representations of software execution traces provide a novel approach to identifying redundant test cases by analyzing the similarities in their execution paths [16]. Researchers have proposed algorithms that adjust graph matching techniques to account for variations in graph structure, thereby enhancing the detection of similarities [17]. These methods have been crucial in reducing the number of test cases required for thorough software testing without compromising quality. Furthermore, the deployment of graph-based techniques to represent and analyze software behavior has opened new avenues for identifying similarities between different test cases. These innovative methodologies facilitate the visualization of test case behavior, enabling testers to prioritize or deprioritize test cases based on their uniqueness or similarity [18]. This approach not only streamlines the testing process but also ensures comprehensive coverage, thereby enhancing the software's overall robustness [19].

Moreover, the integration of machine learning into software testing represents a transformative shift towards more data-driven, intelligent testing strategies. This evolution significantly enhances testing efficiency, cost-effectiveness, and overall efficacy, offering substantial benefits across all phases of software development [20]. By harnessing the power of advanced analytics and machine learning, developers and testers can better navigate the complexities of modern software requirements, ensuring that the products not only meet but exceed the rigorous standards of quality and functionality expected in today's competitive markets [21].

3 Project Goals and Objectives

This project aims to leverage machine learning techniques to enhance the effectiveness and efficiency of software testing processes. By integrating advanced machine learning models, the project seeks to automate the identification, classification, and prioritization of test cases based on their execution traces and derived graph representations. The objectives are:

1. **Automate Test Case Analysis:** Develop machine learning algorithms capable of automatically identifying redundant or irrelevant test cases to reduce the manual effort required in the testing process [5,6].
2. **Enhance Test Case Prioritization:** Implement models that prioritize test cases based on their potential to detect faults, thereby optimizing the testing cycle and resource allocation [3].
3. **Improve Fault Detection Capabilities:** Utilize predictive analytics to improve the detection of software faults earlier in the development cycle, increasing the reliability and stability of software products [22].
4. **Optimize Test Coverage:** Ensure comprehensive test coverage with minimal test suites by employing graph-based models to analyze the execution paths and behavioral similarities of test cases [17].
5. **Graph-Based Test Case Representation:** Create detailed graph-based representations of test cases to visually and analytically assess the complexity and coverage of test scenarios [8].
6. **Streamline Regression Testing:** Develop methods to streamline regression testing processes by continuously integrating and assessing the impact of new or changed code on existing functionalities [4].

By achieving these objectives, the project will deliver a sophisticated framework that supports test engineers in executing more effective and efficient software testing strategies, fundamentally transforming the approach to software quality assurance.

4 Methodology and Approach

In this project, we employ and evaluate various Machine Learning models to analyze call graphs of various test cases. Our methodology includes building a dataset of call graphs. This involves combining a dataset of execution traces for various test cases, automatically building call graphs for each of the given test

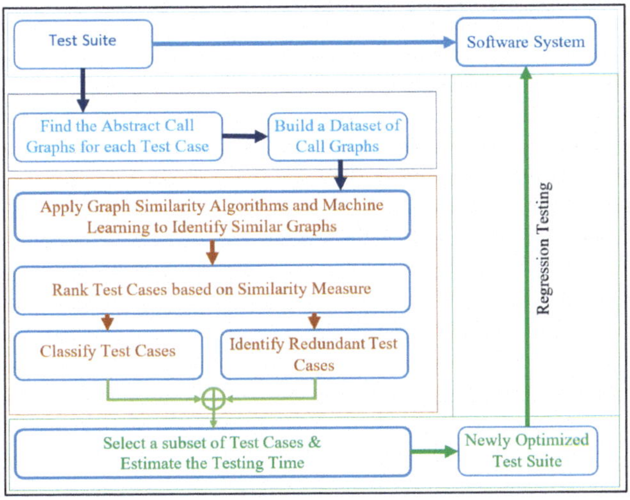

Fig. 1. Proposed methodology and approach

cases, combining these call graphs into a dataset, and evaluating some of the machine learning models on those call graphs to find the similarity measures.

We will utilize these similarities to identify the degree of uniqueness for each of the undertaken test cases. This will assist the test engineer to rank the test cases from the most important to least important based on the similarity measures reflected in their code coverage and execution behavior. Figure 1 in the document illustrates the proposed methodology and approach. Algorithm 1 shows our algorithmic approach. This algorithm analyzes test cases by generating call graphs for each test case and calculating similarity scores between them. It starts by iterating through test cases to create a graph representing the sequence of actions within each test case, connecting actions as nodes with directed edges. These graphs are subsequently compared in pairs using a similarity metric (including edge, node similarity and overall similarity). The similarity scores indicate the level of similarity between test case graphs. This process facilitates the identification of patterns and relationships among the test cases which helps in eliminating redundant test cases, and significantly reduces software testing time and resources.

Algorithm 1. Analyze Banking Test Cases

```
Input: Set of banking test cases
Output: Similarity scores between test cases

procedure GenerateCallGraphs(TestCases)
    for each testCase in TestCases do
        graph ← new Graph()
        for each action in testCase do
            node ← CreateNode(action)
            graph.AddNode(node)
            if previous node exists then
                graph.AddEdge(previousNode, node)
            end if
            previousNode ← node
        end for
        SaveGraph(testCase, graph)
    end for
end procedure

procedure CalculateSimilarity(Graphs)
    for each pair of Graphs do
        similarityScore ← CompareGraphs(G1, G2)
    end for
end procedure

TestCases ← LoadTestCases()
Graphs ← GenerateCallGraphs(TestCases)
CalculateSimilarity(Graphs)
```

5 Experiments

5.1 Data Representation and Execution

This project uses the runtime execution traces and runtime events to build call-graphs, each of which represents a test case. The analysis of the dataset of call graphs will be reflected in the analysis of their corresponding test cases. For example, clustering of various call graphs and ranking their similarity measures will be reflected in clustering the runtime execution traces and their corresponding test cases. These results will be reflected in optimizing the testing process for the software under test. This will result in helping the test engineer better manage and distribute the resources.

For example, functions and methods calls and returns can be embedded in the execution behavior of the program or its test cases. To gain insights into the distribution of these embedding spaces such as the distribution of these events inside each test case, we may need to find and cluster similar functionalities, test cases with similar execution patterns, runtime events, or even program behaviors that may share the same logic and internal design, or the same part of code, for the code coverage criteria.

To demonstrate this study, here we employed a graph-based approach to model and analyze transaction workflows in banking systems. Each workflow representing a distinct banking operation, such as a balance inquiry or fund transfer, is modeled as a directed graph with nodes representing transaction steps and edges representing transitions. The methodology involves several phases: designing test cases, constructing directed graphs for each transaction, performing similarity analysis between workflows, identifying bottlenecks, and proposing optimizations. The following sections detail each step in the methodology, from graph representation to the export and analysis of results.

5.2 Test Cases Description

Each scenario is carefully crafted to simulate real-world transactions.

1. **Balance Inquiry (Savings Account):** Represents the user checking their balance in a savings account to verify available funds and retrieve account information securely.
2. **Internal Transfer (Savings to Checking):** Simulates an intra-bank transfer from the user's savings to their checking account, including identity verification and fund movement steps.
3. **Utility Bill Payment (Checking Account):** Models the process of paying a utility bill from a checking account, involving user identity verification, bill validation, and funds check.
4. **External Transfer (Checking to External Account):** Simulates transferring funds from the user's checking account to an external bank account, with multiple security verifications.
5. **Statement Generation (Last 6 Months):** Emulates the generation of an account statement for the past six months, detailing the transaction history for user records.
6. **Balance Inquiry (Checking Account):** Similar to the first test case but for a checking account, to demonstrate the functionality and flexibility of balance inquiries.
7. **Transfer to Another User's Savings Account:** Represents a user-initiated transfer to another individual's savings account within the same bank, including security checks and transfer processes.
8. **Business to Savings Transfer:** Models a business transaction transferring funds from a business account to a savings account, focusing on business-specific transaction requirements.
9. **Loan Balance Inquiry:** Simulates a loan balance inquiry, focusing on retrieving outstanding balance and due payment details for loan management.
10. **Business Utility Bill Payment:** Similar to Utility Bill Payment but from a business account, addressing operations specific to business banking needs.

5.3 Graph-Based Representation of Test-Cases

In *Directed Graph Model*, each banking transaction workflow is modeled as a directed graph where nodes represent transaction stages such as "Initiate Transfer" or "Validate Account," and directed edges depict the actions between stages, such as transfers or verifications. Attributes such as edge weights indicate the frequency of transitions, and colors differentiate transaction types, enhancing the workflow's visualization and analysis.

In *Graph Construction*, graphs are constructed by mapping each transaction step to a node and transitions to directed edges, with attributes such as weight and color providing additional context. This method visually represents the transaction sequence, aiding in the understanding and analysis of banking operations.

For example, the internal fund transfer might include nodes such as *Initiate Transfer* and *Verify User* with edges indicating actions between these nodes. Edge attributes such as weights and colors help distinguish the types of transactions and highlight the most frequented paths, simplifying complex banking workflows into understandable diagrams (Fig. 2).

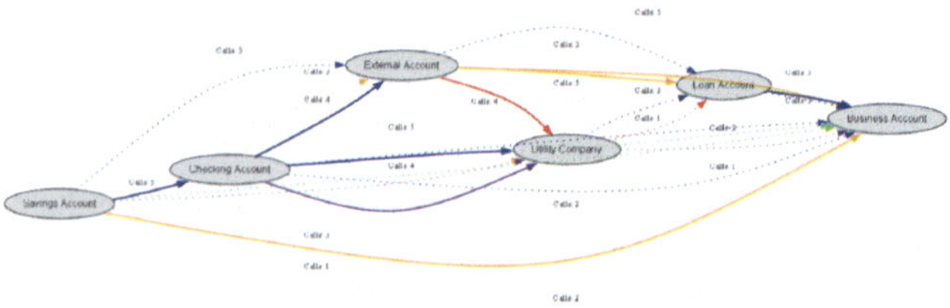

Fig. 2. Sample Call Graph from Test Case 6

This call graph for Test Case 6 showcases a network of financial transactions and queries between various account types, including a Savings Account, Checking Account, External Account, Business Account, Loan Account, and a Utility Company. The graph visually represents the flow and frequency of interactions, marked by "calls", with thicker, more vividly colored lines indicating more frequent transactions, highlighting key pathways such as frequent checks or transfers to and from the Business and External Accounts.

5.4 Graph Similarity Analysis

Define Similarity Metrics: We employ several metrics to quantify the similarity between transaction workflows represented as directed graphs. These metrics facilitate the comparison of structural and operational similarities across workflows.

- **Node Similarity:** Measures similarity based on common transaction states shared between two graphs. It is defined as:

$$\text{Node Similarity} = \frac{\text{Number of Common Nodes}}{\text{Total Unique Nodes}}$$

- **Edge Similarity:** Assesses transitions between states by comparing both the presence and weights of edges, calculated as:

$$\text{Edge Similarity} = \frac{\text{Number of Matching Edges}}{\text{Total Unique Edges}}$$

- **Overall Similarity Score:** A composite metric that averages node and edge similarities to provide a holistic view:

$$\text{Overall Similarity} = \frac{\text{Node Similarity} + \text{Edge Similarity}}{2}$$

Rationale for Similarity Analysis: Analyzing graph similarities aids in workflow optimization, security vulnerability identification, and efficient resource allocation by identifying structural and sequential commonalities in banking transactions.

6 Results

The results shown in the Table 1 represent the similarity scores between 10 test cases based on node similarity, edge similarity, and overall similarity. Here is detailed interpretation of the findings:

Table 1. Sample similarity scores between test cases.

Test Suite A	Test Suite B	Node Similarity	Edge Similarity	Overall Similarity
TC_4	TC_6	1	0.725	0.862
TC_3	TC_4	1	0.700	0.850
TC_1	TC_7	1	0.683	0.841
TC_2	TC_6	1	0.677	0.839
TC_5	TC_7	1	0.677	0.839
TC_5	TC_9	1	0.677	0.839
TC_4	TC_9	1	0.671	0.836

6.1 Node Similarity

In the Banking test cases, all pairs have a node similarity score of 1, except for a few pairs (i.e. TC_1 with TC_7, TC_5 with TC_7). A node similarity of 1 indicates that both test cases involve exactly the same set of nodes, meaning they use the same set of account types in their transactions. Lower values (i.e. 0.6774 or 0.6825) indicate slight variations in the nodes used between the test cases.

6.2 Edge Similarity

For the Banking test cases, this varies more significantly between pairs, ranging from about 0.6721 to 0.7246. Edge similarity takes into account not only the presence of similar actions (*edges*) between nodes but also the frequency of these actions as depicted by the weights of the edges. Lower scores indicate differences in how transactions are processed between the cases, either by different sequences of actions or different frequencies of certain transactions.

6.3 Overall Similarity

This metric averages the node and edge similarities to provide a comprehensive measure of similarity. The overall similarity scores reflect a mix of both node and edge characteristics. Scores closer to 1 (i.e. 0.8623 between TC_4 and TC_6) suggest highly similar workflows, whereas lower scores (i.e. 0.8356 between TC_4 and TC_9) suggest more variation.

The key insights from the test case similarities demonstrate potential areas for optimization by identifying redundant tests and suggesting the consolidation of similar processes. They also highlight the need for differentiation in test cases where transactions differ significantly, which may be crucial for adhering to business rules or enhancing security measures. This analysis facilitates targeted improvements in testing strategies, promoting efficiency while ensuring that critical functionalities are thoroughly tested.

7 Conclusion

The test case similarity results help achieve the project's objectives by identifying and prioritizing test cases that exhibit redundancy or are significantly similar. This allows for streamlining the software testing test suite, which directly aligns with the project's goal of optimizing the software testing process using machine learning models. By recognizing which test cases cover similar functionalities or execute similar sequences, the project can suggest a more efficient test suite that reduces redundancy without compromising the quality of the software. Furthermore, this efficiency reduces software testing time and enables better resource allocation, ultimately enhancing the effectiveness of the Software testing process.

References

1. Pressman, R.S.: Software Engineering: A Practitioner's Approach. McGraw-Hill (2005)
2. Sommerville, I.: Software Engineering, 9th edn. Pearson (2011)
3. Bajaj, A., Sangwan, O.P.: Test case prioritization using bat algorithm. Recent Adv. Comput. Sci. Commun. **14**(2), 593–598 (2021)
4. Ma, G., Ahmed, N.K., Willke, T.L., Yu, P.S.: Deep graph similarity learning: a survey. Data Min. Knowl. Disc. **35**, 688–725 (2021)

5. Gokilavani, N., Bharathi, B.: Test case prioritization to examine software for fault detection using PCA extraction and k-means clustering with ranking. Soft. Comput. **25**(7), 5163–5172 (2021)
6. Panday, A., Gupta, M., Singh, M.K., Ali, N.: Test case redundancy detection and removal using code coverage analysis. MIT Int. J. Comput. Sci. Inf. Technol. **3**(1), 6–10 (2013)
7. Al-Sharif, Z.A., Abdalrahman, W.F., Jeffery, C.L.: Encoding test cases using execution traces. In: 2021 12th International Conference on Information and Communication Systems (ICICS), pp. 239–244. IEEE (2021)
8. Liu, C., Niu, D., Yang, X., Zhao, X.: Graph matching based on feature and spatial location information. Vis. Comput. **39**(2), 711–722 (2023)
9. Ammann, P., Offutt, J.: Introduction to Software Testing, 2 edn. Cambridge University Press (2016)
10. Gold, R.: Decision graphs and their application to software testing. Int. Sch. Res. Not. **2013**(1), 432021 (2013)
11. Elasri, C., Kharmoum, N., Saoiabi, F., Boukhlif, M., Ziti, S., Rhalem, W.: Applying graph theory to enhance software testing in medical applications: a comparative study. In: International Conference on Advanced Intelligent Systems for Sustainable Development, pp. 70–78. Springer, Cham (2023)
12. Al-Sharif, Z.A., Jeffery, C.L.: Abstracttrace: the use of execution traces to cluster, classify, prioritize, and optimize a bloated test suite. Appl. Sci. (2076-3417) **14**(23) (2024)
13. Al-Sharif, Z.A., Jeffery, C.L., Said, M.H.: The use of dynamic temporal assertions for debugging. In: Lettnin, D., Winterholer, M. (eds.) Embedded Software Verification and Debugging. ES, pp. 47–65. Springer, New York (2017). https://doi.org/10.1007/978-1-4614-2266-2_3
14. Al-Sharif, Z.A., Jeffery, C.L., Said, M.H.: Debugging with dynamic temporal assertions. In: 2014 IEEE International Symposium on Software Reliability Engineering Workshops, pp. 257–262. IEEE (2014)
15. Bougleux, S., Brun, L., Carletti, V., Foggia, P., Gaüzère, B., Vento, M.: Graph edit distance as a quadratic assignment problem. Pattern Recogn. Lett. **87**, 38–46 (2017)
16. Bousse, E., Mayerhofer, T., Combemale, B., Baudry, B.: Advanced and efficient execution trace management for executable domain-specific modeling languages. Softw. Syst. Model. **18**(1), 385–421 (2019)
17. Wang, R., Yan, J., Yang, X.: Unsupervised learning of graph matching with mixture of modes via discrepancy minimization. IEEE Trans. Pattern Anal. Mach. Intell. **45**(8), 10500–10518 (2023)
18. Romanov, V., Ivanov, V., Succi, G.: Approaches for representing software as graphs for machine learning applications. In: 2020 International Computer Symposium (ICS), pp. 529–534. IEEE (2020)
19. Müller, R., Fischer, M.: Graph-based analysis and visualization of software traces (2019)
20. Ramadan, A., Yasin, H., Pektas, B.: The role of artificial intelligence and machine learning in software testing, arXiv preprint arXiv:2409.02693 (2024)
21. Groumpos, P.P.: A critical historic overview of artificial intelligence: issues, challenges, opportunities, and threats. Artif. Intell. Appl. **1**, 197–213 (2023)
22. Hemmati, H.: How effective are code coverage criteria? In: 2015 IEEE International Conference on Software Quality, Reliability and Security, pp. 151–156. IEEE (2015)

State of Test Optimization for Variability in Industry

Muhammad Abbas[1,2](✉), Mehrdad Saadatmand[1], Eduard Paul Enoiu[2], and Bernd-Holger Schlingloff[3]

[1] RISE Research Institutes of Sweden, Västerås, Sweden
mehrdad.saadatmand@ri.se
[2] Mälardalens University, Västerås, Sweden
muhammad.abbas@ri.se, eduard.paul.enoiu@mdu.se
[3] Fraunhofer FOKUS, Berlin, Germany
holger.schlingloff@fokus.fraunhofer.de

Abstract. This paper presents an exploratory survey on test optimization practices for variant-intensive systems in the industry, focusing on understanding current challenges and prospects. Testing variant-intensive systems, particularly in domains like automotive, embedded systems, and telecom, involves considerable complexity due to the high number of configurations and variability in hardware and software. The survey responses reveal varying adoption of test categorization, prioritization, and selection techniques across different domains, with some employing feature-based testing approaches and others relying on more ad-hoc methods. Challenges identified include difficulties with change impact analysis, the scalability of test optimization in systems with many configurations, and the ineffective reuse of test cases across product variants. The study highlights the need for automated tools to support test categorization, prioritization, and coverage-based test selection to address these challenges. The findings highlight the importance of developing more scalable solutions tailored to the needs of specific domains for optimizing tests in variant-intensive systems.

Keywords: Test Optimization · Variant-Intensive Systems · Test Reuse · Coverage Criteria · Product line engineering · Variability · Test Selection · Test Categorization · Coverage · Survey

1 Introduction

In engineering industrial systems, the testing process introduces complexities far beyond those seen in traditional software testing. Industrial systems must account for an extensive range of configurations, enclosing not only different hardware components but also variant software versions and their integration into a functional system [2,19]. These diverse configurations, combined with the constrained testing infrastructure, present significant challenges for efficient

S. Latifi (Ed.): ITNG 2025, AISC 1463, pp. 528–542, 2025.
https://doi.org/10.1007/978-3-031-89063-5_46

testing. Traditional testing techniques [10] cannot manage the combinatorial explosion of variants required by industrial systems testing. As industrial systems are increasingly deployed in mission-critical domains, ensuring that all possible configurations are tested effectively and efficiently becomes essential.

Several research efforts have highlighted the complexities involved in testing industrial systems due to the variability in hardware and software configurations. For instance, Golagha et al. [21] emphasize the need for test selection techniques that intelligently and automatically determine which system variant to test and where to execute these tests, considering the limitations of hardware testbeds. Variants of industrial systems, whether in hardware tools such as those in controllers for industrial machines, require careful optimization and selection of test cases that can cover the breadth of system behaviors while minimizing the cost and time spent on exhaustive testing [9].

Recent advances in test optimization, particularly in industries such as automotive software development, have shed light on methods to reduce test analysis effort. Golagha et al. [11] present a clustering approach that groups failing test cases based on non-code features to reduce manual analysis time. This kind of failure clustering highlights the importance of optimizing test efforts when multiple configurations are involved, as seen in regression testing environments where large tests may fail due to a few core issues [1,11].

Contributions. While studies focus on test optimization and selection in the presence of variability, it is still unclear if they are standard industry practices. This paper reviews the state of practice in industry in variability management, test optimization, and test selection in the presence of variability. The paper further reports perceived practical challenges in the area.

Structure. The rest of the paper is organized as follows. Section 2 introduces related work briefly. Section 3 presents the method, Sect. 4 presents and discusses the obtained qualitative results. Section 5 presents the potential threats to validity and limitations. Finally, Sect. 6 concludes the paper with future directions.

2 Related Work

Researchers have addressed some issues of variant selection and test optimization for industrial software systems. How to handle variability testing in product lines has been surveyed by Sinnema et al. [19], Lamancha et al. [14] and Lee et al. [15]. These surveys indicate little work exists on variant handling and selection during software testing compared to other aspects of software development. In recent years, some contributions [4,6–8,18] have proposed applications of combinatorial testing concepts for variant and product line handling as well as for variability testing and test optimization.

Engstrom et al. [10] provides a comprehensive view of variability testing and identifies several challenges regarding the testing of variant-intensive software. This survey particularly mentions i) how to handle many tests, ii) how to balance the effort between reusable components and concrete products, and iii) how to manage variability. Additionally, there is a lack of substantial empirical studies on actual industrial systems and the applicability of these optimization techniques for handling variability in industrial scenarios.

Test case selection and prioritization play a critical role in test optimization and regression testing. Previous research has explored various approaches for selecting and prioritizing the most effective test cases, especially when executing the entire test suite is impractical [3]. These approaches typically use data from variants or prior system executions, including fault history, execution time, test coverage, and previous failing tests [17,21]. Yoo and Harman [22] provide a comprehensive review of regression test minimization, selection, and prioritization strategies, underscoring their importance in test optimization. Despite over 40 years of maturing research, the industrial application of these techniques still needs to be improved. Several surveys on regression test selection methods [12,22] highlight that most techniques lack industrial application. As the complexity of software systems increases, particularly with the introduction of variability, there is a growing demand for advanced test optimization techniques that can efficiently handle both the scale and variety of industrial systems.

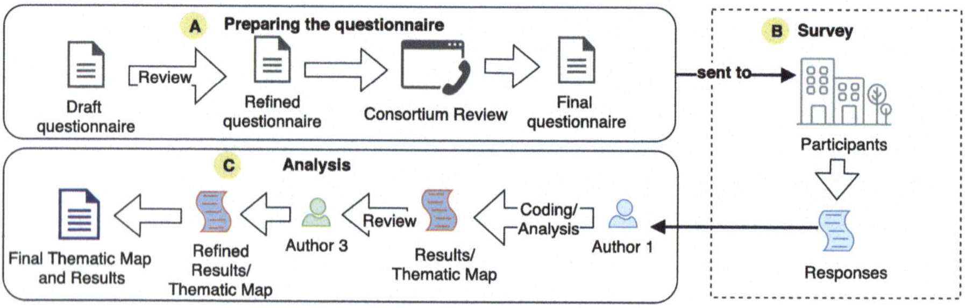

Fig. 1. Data collection and analysis procedure

3 Method

The survey conducted for this study was designed to assess the current state of test optimization practices in variant-intensive systems.

Context and Research Questions: This work can be classified as an *exploratory* questionnaire-based case study that focuses on documenting the current state of practices in effectively testing variant-intensive systems in industry. We define the following research questions (RQs) that guide our data collection for the study.

RQ1: What is the current state-of-practice in test optimization for variant-intensive systems in industry?

This RQ collects general data about the nature of the system under test (SUT),

the test suite, and the process for testing variants in the industry.
RQ2: What challenges exist in testing and test optimization for variant-intensive systems?
In this RQ, we aim to gather qualitative insights on the perceived challenges in testing and test optimization of variant-intensive systems.

We collect data for both research questions using a questionnaire containing closed and open-ended questions. For RQ1, we analyze the responses and provide an overview of the results directly based on them. For RQ2, we employ qualitative data analysis (thematic analysis) to identify themes and sub-themes regarding perceived challenges from the reactions.

Our method was based on guidelines for surveys, case studies, and ethical data collection [13,16,20]. The rest of this section details the data collection procedure.

Data Collection Procedure: As shown in Fig. 1, the procedure is divided into three main phases, i.e., Ⓐ, Ⓑ, and Ⓒ, focusing on the preparation of the questionnaire, conducting the survey, and analysis of the collected data, respectively. Below, we provide a brief overview of each phase.

A. Questionnaire: We followed a team-based approach to design and execute the questionnaire-based survey. The team included subject matter experts from industry and researchers involved in a European project. During various meetings with the experts, an initial questionnaire was drafted with open and close-ended questions, as shown in phase Ⓐ of Fig. 1. After an internal review, the questions were refined, re-ordered, and simplified. Following the guidelines for survey instrument evaluation [16], we used expert reviews to evaluate our refined questionnaire. The consortium reviewed the refined questionnaire in an online meeting, and based on the feedback, a final questionnaire was obtained. Following this collaborative approach, the survey questions were formulated based on discussions with project stakeholders, including use case providers and technology/research partners, to address key aspects of test management, test reuse, prioritization, and optimization for product variants. The questionnaire development process was iterative, involving internal reviews by industry experts and stakeholders within the consortium to ensure the relevance of the questions to practical, real-world testing scenarios.

The final questionnaire included open and close-ended questions covering the following main areas.

1. General questions about the nature of SUT, test suite, variability in the SUT, and the current test process (RQ1).
2. Questions to collect data about the state of test optimization, test optimization for variability, test reuse, and test categorization and prioritization (RQ1).
3. Questions to collect data about perceived challenges and areas of improvement (RQ2).

The first group of questions aims to document the demographics of the SUT, the testing process, and the test suite. The second group focuses on collecting practices in test optimization for variability and related topics that enable test optimization.

B. Survey and Participants: After drafting the final questionnaire, we selected representatives from various companies in diverse domains based on convenience. Therefore, we chose representative industrial professionals involved in the European project. Some of these participants also helped review and refine the initial version of the questionnaire. Every participant had a representative role in their company and was distributed across Europe and North America. The survey was distributed to the participants using a collaborative word-processing tool, and they were encouraged to answer all the questions. We also encouraged the participants to add comments while answering the questions they thought could be necessary for the survey.

Ethical Considerations: The survey targets representative software development and testing professionals to gather data about the current state of practices in test optimization for variant-intensive systems. Via a European project, we accessed seven representative companies from which we selected one participant each. However, participation in the survey was strictly voluntary. Furthermore, we ensured that the data collected during the survey was only used to draw general conclusions about the topic and that the confidentiality of the data and anonymity of the participants were respected.

C. Analysis and Interpretation: The collected data is analyzed, reported, and discussed in Sect. 4. As is typical for surveys, we report and discuss the participants' responses to close-ended questions, supporting the discussions with statistics extracted from the responses. On the other hand, for open-ended questions, we use qualitative content analysis primarily to document and report high-level themes about the topic and the perceived challenges. In particular, the responses were analyzed, and open-ended questions were coded by Author 1 (as shown in step Ⓒ of Fig. 1), and a document reporting the results and thematic map was produced. Author 3 of this paper reviewed and refined the results and thematic map, producing a refined document for the results with a thematic map. Finally, all authors reviewed the analysis and the thematic map reported in this paper.

Table 1. Background on representative companies

Domain	# of Part.	Region	Nature
Automotive	2	Germany, Canada	Safety/Security-critical (1/2)
Embedded Observability	1	Sweden	N/A
Production & Manufacturing	1	Germany	Safety-critical
Railway control	1	Sweden	Safety-critical
Telecom	1	Türkiye	N/A
TV OS	1	Türkiye	N/A

4 Results and Discussions

In this section, we present and discuss the results of the analysis and interpretation of the responses. Where needed, we also support the results with supporting quotes— rewritten for better readability—from the responses presented in "*italic text*" for emphasis.

4.1 Background on SUT and Test Suite (RQ1)

Table 1 summarizes the background of the representative companies and their domains. As seen in Table 1, the participants ranged from various domains, such as automotive, embedded software observability, and TV OS, to name a few. The `# of Part.` column shows the total number of participants from the domain and the region column shows the country of operation of the representative company. Finally, the `Nature` column shows the nature of the SUT under consideration for the survey. One of the representative companies' products for the automotive domain is safety and security-critical. The other automotive SUT (from Germany) provides supporting software to enable the engineering of automotive systems and is, therefore, not a safety or security-critical product. Finally, the SUTs in production, manufacturing, and the railway domain are classified as safety-critical.

The questionnaire also covered data collection for existing test suites for the products. When asked if the SUT has an existing test suite and its size, all the respondents highlighted that they have a test suite at the product level. The size of the test suite varies across all domains. In the automotive domain, one respondent exemplified infotainment systems with a test suite that captures different parts of the road map data; thus, the test suite captures around 220000 mi of road. In addition, radar and stereo camera data are also used for testing, which results in additional tests. In the railway domain, the system contains over 180 sub-components, each of which is tested with at least four to five test scenarios. In addition, there are 20 test scenarios in the railway domain at the integration level. A respondent from the telecom domain highlighted that their product has a test suite of 1500 test scenarios for their product, where each test scenario might contain multiple test cases. The TV OS domain has a test suite of 101K test scenarios, 100K requiring manual execution. The respondents from embedded observability and production domains did not specify the size of their test suites.

4.2 Variability, Testing and Test Optimization Practices (RQ1)

The survey responses provided valuable insights into the current practices in testing variant-intensive systems across different domains. The results revealed significant variability in the adoption of test optimization techniques. While some domains, such as automotive and railway, have well-established processes for feature-based testing, others use specific practices due to the nature of their systems.

Feature-Based Testing and Reuse A key finding was the widespread use of feature-based testing in safety-critical domains, where testing is closely tied to documented features and system requirements. Automotive and railway domains, in particular, showed a reliance on structured requirements and models to guide their testing efforts. However, the survey also revealed inconsistent test reuse across variants, with some domains successfully reusing test cases. In contrast, others, such as embedded observability, struggle due to the high variability of their platforms. Test reuse is often not feasible in these cases due to the dynamic configuration changes introduced at runtime.

As most of the SUTs have to work with varying hardware configurations and regional regulations, we also asked questions to collect data about the variability in the systems and how they are handled. We asked how new features are

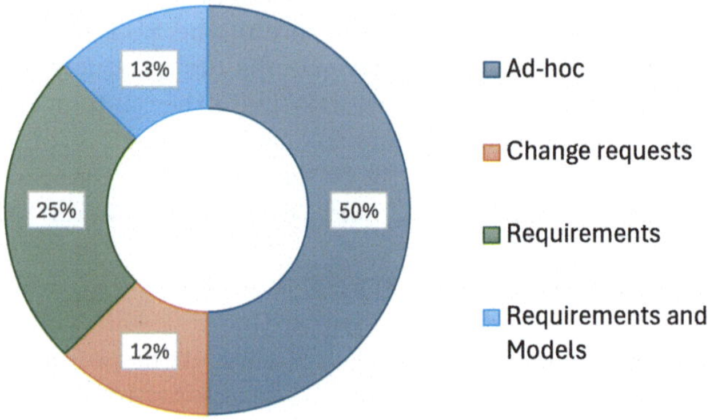

Fig. 2. New feature introduction process

documented and introduced to the systems, if they are modeled, and how. In addition, we asked if existing artifacts such as requirements, tests and source code is being reused when deriving new variants. As shown in Fig. 2, ad-hoc approaches for documenting features are widespread. The most representative respondents of the domains define an ad hoc process that can vary. They can use requirements, feature requests, or other documentation sources when introducing new features or modeling variability. Following ad-hoc feature modeling and documentation, requirements are also a primary source for introducing and documenting new features. In some cases (13%), requirements are combined with models to document new features and their variability. Finally, change requests are more frequently used in the automotive domain when new features are introduced. For artifact reuse, except the embedded observability domain, all other domains reuse existing artifacts when deriving new variants. The representative expert from the embedded observability domain cited that their SUT is one big platform that has to work with varying hardware configurations, often at runtime. Due to the nature of the SUT, they do not derive their systems' variants, making reuse irrelevant.

Test Prioritization and Selection. Classifying the test cases into different categories might be necessary to effectively apply test selection for test optimization. For example, one classification would be whether or not a test case is testing a particular product configuration. Such categorizations and classifications improve the test selection process and, in turn, enable test optimization. In addition, coverage criteria for test cases also play a crucial role in test selection. Therefore, we asked our respondents if they categorize test cases. We also asked about the categorization scheme used, whether they keep track of the variants and features the test cases are testing, and whether they use coverage criteria.

Five of the seven representative companies from their domains categorize test cases based on various classification schemes. In addition, another company plans to introduce test categorization into their process. Only one of the seven companies does not have test categorization in place and has no plans for it in the near future. When asked what classification schemes are being used for test categorization, a respondent from the automotive domain mentioned that tests are classified based on their perceived difficulty levels. Respondents from the embedded observability and telecom domain use functional and various non-functional quality labels to categorize test cases. In the production and manufacturing domain, the mode of operation of the tests is used to categorize them (e.g., automatic or human input required). Finally, the respondent in the TV domain responded that the company uses a classification scheme unknown to the respondent.

We also asked if separate test suites exist for the variants and the standard product. We found that only the automotive and railway domains have separate test suites for the standard product and the derived variants. When asked if the test cases could be traced to the standard product and its product variants, we found that all domains except for two (production and one representative company from the automotive domain) have mechanisms to trace test cases to variants and standard products.

When new features are introduced or changes are made to the system under development, regression testing is common. In such cases, redundant and repeated testing of shared features can make the testing process inefficient. Furthermore, not all the tests in a suite are the same in effectiveness and might need ranking, prioritization, and selection. In addition, the execution of all test suites might not be possible in some cases. For example, one billion test cases may be needed to test all the possible variants of the product in the embedded observability domain. This is due to the varying configurations that could be enabled at compile time. "*A single test program, where 10–20 sanity checks are applied on the output. This, however, needs to be executed on every combination of 30–50 static build flags that affect the code in undocumented ways, meaning at least 2^30 = 1 billion tests if using a 'brute force' approach with no test case prioritization.*" Different test criteria are used to select a subset of tests from the test suite in such cases. We asked if the shared code and features are tested across different variants and versions. In addition, we asked for test optimization enablers. In particular, we asked if they order, prioritize, or categorize tests based on some criteria before execution in regression scenarios.

Four of the seven representative companies from their domains order test cases. In the automotive domain, the representative companies do not order test cases for execution, but one representative company does select sub-sets based on the test suite size. For example, due to time constraints, they might select a subset of map data from a large test suite to test the navigation system. In the embedded observability domain, engineers could order the test randomly or systematically based on exploring various build flags. In addition, they only test the critical sub-set of build configurations—identified manually—and select a sub-set from the test suite that targets the most critical build configurations.

On the other hand, all tests are executed at least once in the representative company from the production and manufacturing domain. In the rail domain, the test cases are independent and, thus, do not require any ordering. However, test selection is performed to target the changes in the code. Nevertheless, due to the safety-critical nature of the product, the entire test suite is once again executed before the release of the software. In the telecom domain, test cases are ordered, but no test selection is made, and all the tests are executed. Finally, test cases might depend on each other in the TV domain, and thus, ordering is required based on their dependency. Executing the entire test suite is preferred, but it is sometimes not possible, and an ordered subset is selected to meet time constraints.

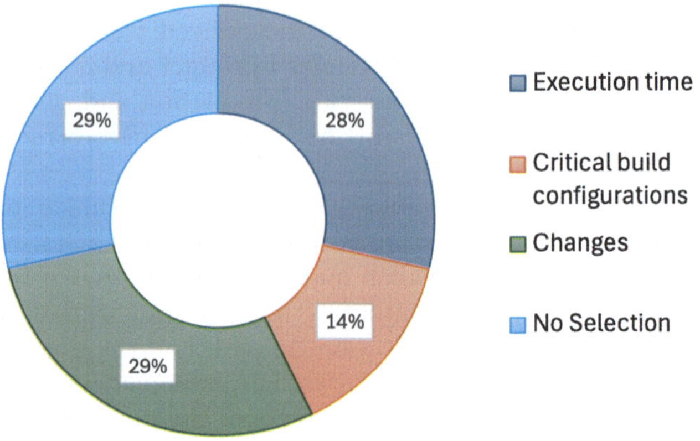

Fig. 3. Test selection criteria

In summary, four representative companies (57%) do not order the test cases for execution. The ones that do order tests are often based on test dependencies. Also, five of the representative companies (70%) have some test selection in place with varying selection criteria, shown in Fig. 3. The automotive and TV domains select tests that could finish execution in a pre-defined time window.

The representative company from the embedded observability domain identifies critical build configurations based on experience and selects a subset of tests targeting those configurations. Finally, the telecom and railway domains select a subset of tests targeting changes.

Use of Coverage Criteria. The survey also explored the use of coverage criteria in test optimization. Results indicated that coverage criteria, such as feature coverage, code coverage, and combinatorial interaction testing, are applied across domains. Automotive and railway control systems, in particular, utilize feature coverage to ensure that new variants or features are sufficiently tested without redundant testing of common components.

When asked what coverage metrics are recorded that can be used in test selection, we found various traditional and variability-aware metrics used to record coverage, summarized in Fig. 4. In particular, one respondent from the automotive domain and one from the railway domain mentioned the use of code, feature, and change coverage as recorded metrics for optimization. One of the partners in the automotive domain plans to have these metrics in place in the near future. The coverage of static build flags is of more importance for test optimization for the embedded systems observability domain. Furthermore, the TV and production domains use feature coverage as a metric that can enable test optimization. Finally, the representative company from the telecom domain uses the traditional code coverage metric for test optimization.

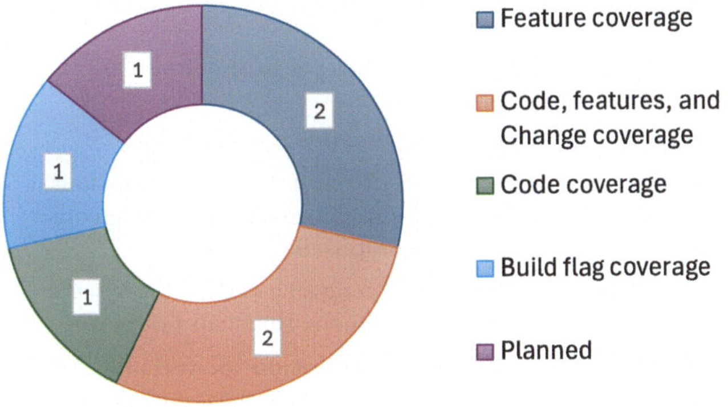

Fig. 4. Prominent coverage metrics recorded

> **Takeaway: RQ1**
>
> In practice, variability in systems are still captured informally, often with natural language requirements. The number of variants and their combinations can explode and test selection and optimization for variability is a practical concern. Most of the studied representative companies do use test categorization

and a variety of coverage metrics to enable test optimization for variability. Nevertheless, the respondents still see this as a challenging problem.

4.3 Challenges and Improvement Opportunities (RQ2)

Our RQ2 focuses on collecting and reporting potential areas of improvement and opportunities for research by collecting practical challenges in test optimization for variability. In this regard, we asked the representative companies from their domains how the current testing process can be improved. As mentioned in Sect. 3, we use a qualitative data analysis method known as thematic analysis to collect and document the findings for RQ2. Table 2 reports the area's high-level themes and concrete challenges (sub-themes and codes), briefly summarized as follows.

Table 2. Perceived challenges in the area (RQ2)

Theme	Sub-theme and codes
1. Configurations to (not) test	1.1 Detection of redundant configurations 1.2 Effect of configuration change on code 1.3 Testing composed features
2. Impact analysis	2.1 Flagging changes that can break tests 2.2 Flagging changes that can fail tests
3. Variance and testing	3.1 Test script derivation for variants 3.2 Test selection for changes and variants 3.3 Test classification based on criticality 3.4 Critical scenarios identification
4. Coverage	4.1 Dead code effects coverage 4.2 Measuring feature coverage from component-level coverage

The thematic analysis of the survey responses highlighted several recurring challenges in test optimization. One major issue was the difficulty of effectively dealing with configuration parameters and variability. (1) Respondents found it challenging to detect redundant configurations of their product that could result in equivalent behavior. "*Many configurations have been tested, but many of these are believed to produce identical code (the same variant) as the relevance of many parameters depends on other configuration parameters. These dependencies are not documented. It could be useful for finding which configuration combinations are redundant.*" The industry could benefit from approaches and tooling that can extract and capture configuration parameter dependencies. This will help enable the detection of equivalent variants and, in turn, optimize variability testing processes. In addition, the uncertain effects of a small change to configuration parameters on the resultant software hinder the maintenance and evolution of both the software and its tests. Finally, the thematic analysis also highlighted the challenges with test optimization in the presence of large and composed features (features that use other features). "*Some features are reused [...] The granularity of features is very low for us, and we don't model; we re-test [...]*"

(2) Another reoccurring theme in the survey responses was the challenges in applying change impact analysis to identify breakages. Most companies reported that while they could identify changes in code or requirements, they lacked tools to trace these changes to specific test cases, making it harder to prioritize tests. The analysis further highlighted the need for the detection of small changes that may break or fail tests. Additionally, the manual nature of test categorization and selection in some domains, such as embedded observability, was a barrier to optimizing test efforts. "*We don't have a tool for change management of requirements, features, and test cases. [...] We don't know anything about new features before we see failing test cases. Also, we have repeating scenarios with one parameter/flag change in tests. If we can analyze the effects of these parameters and remove repetition, we can save a lot of time.*"

(3) Another challenge was the scalability of test optimization in systems with high variability. In this regard, the analysis highlights the difficulty in reusing test cases across product variants, particularly in domains where system architectures or configurations vary significantly between product variants. In embedded systems, testing becomes complex due to dynamic configurations, making test reuse ineffective, manual, and time-consuming. Therefore, test scripts for individual variants have to be manually adapted and evolved for variants. While there are design-time approaches for deriving tests for variants [5], there is still a lack of tooling for automated test evolution for product variants. Furthermore, some participants, particularly from the embedded systems domain, reported that the number of potential test cases could reach billions due to varying configurations. In such cases, effective test selection based on coverage criteria becomes critical but is often not fully automated. This highlights the challenge of balancing exhaustive testing against constraints, such as tests that cover changes. Finally, automated critical scenario identification from configuration could help avoid exhaustive testing. "*The existing product tests are used and modified as needed [for variants] [...] We don't have a tool for change management of requirements, features, and test cases. By identifying only the changes performed, then the tester can target only needed areas. [...] We don't know the combination of configurations that are used in practice by our customers. We naturally test the default settings and other 'likely' settings, and then as many other variants as possible.*"

Test prioritization is widely used, but determining the most effective criteria for test case selection and ordering is challenging. In safety-critical domains like railway, although test selection is performed to target changes in code, the entire test suite still needs to be executed before release, making testing inefficient. "*[We target] the changes performed within the components, we only perform tests which affect those components. However, running all tests before releasing software is a common practice.* "

(4) Finally, the analysis highlighted several challenges on the integration between code coverage and variability. In particular, variability in systems can introduce dead code that handles a configuration that is not relevant in a particular variant, resulting in low code coverage. "*Testers need to know what is possible given the current design [...] is the dead code really dead, or is it a bad test case?*" In this regard, participants also highlighted the lack of tooling that can link component/variant-level coverage results to derive or compute feature/product-level code coverage. "*[It is challenging] to feed test results from variants into the product.*"

The survey results indicate significant interest in further developing automated tools to support test optimization. Many participants highlighted the need for tools to automatically categorize and prioritize test cases based on coverage, risk, and criticality.

Takeaway: RQ2

A major issue is the lack of tools to trace code changes to specific test cases, hindering test prioritization. Manual test categorization and scalability are also problematic, especially in embedded systems with high variability, where test selection is often not automated. Reusing test cases across product variants is difficult due to dynamic configurations. While test prioritization is common, selecting the most effective criteria remains challenging, especially in safety-critical embedded domains.

5 Validity Threats

Participant selection for the survey posed a concern, as using non-probability sampling could introduce selection bias. Although a diverse range of industrial practitioners was included, they may not fully represent the broader population of industry professionals. This limitation was acknowledged, and efforts were made to include participants from different domains and regions to enhance the survey's representativeness.

Another potential threat to internal validity when using questionnaires is the risk of respondents misinterpreting the questions. However, this risk was significantly minimized since all industrial participants were involved in the same research project and shared a similar understanding of the terminology used.

6 Conclusion and Future Directions

In conclusion, this study reveals that while test optimization practices for variant-intensive systems are common in some domains, such as automotive and railway systems, current practices often need to address the scalability and complexity of high variability, especially in domains like embedded systems. Many companies use manual processes for test categorization, and there is a need for more tooling to trace changes in code to specific test cases, which hampers efficient test prioritization. Reusing test cases across product variants remains challenging, particularly in systems with dynamic configurations. Future work should enhance automation for test categorization, prioritization, and selection based on coverage and risk, explicitly scaling these solutions for highly configurable systems.

References

1. Abbas, M., Hamayouni, A., Moghadam, M.H., Saadatmand, M., Strandberg, P.E.: Making sense of failure logs in an industrial devops environment. In: Latifi, S. (ed.) ITNG 2023 20th International Conference on Information Technology-New Generations, pp. 217–226. Springer, Cham (2023)
2. Abbas, M., Jongeling, R., Lindskog, C., Enoiu, E.P., Saadatmand, M., Sundmark, D.: Product line adoption in industry: an experience report from the railway domain. In: Proceedings of the 24th ACM Conference on Systems and Software Product Line: Volume A - Volume A, SPLC '20. ACM, New York (2020)
3. Ali, N., et al.: On the search for industry-relevant regression testing research. J. Empir. Softw. Eng. (2018)
4. Bryce, R.C., Colbourn, C.J.: Prioritized interaction testing for pair-wise coverage with seeding and constraints. Inf. Softw. Technol. **48**(10), 960–970 (2006). https://doi.org/10.1016/j.infsof.2006.03.004. Advances in Model-based Testing
5. Bucaioni, A., Di Silvestro, F., Singh, I., Saadatmand, M., Muccini, H.: Model-based generation of test scripts across product variants: an experience report from the railway industry. J. Softw. Evol. Process **34**(11), e2498 (2022)
6. Cohen, M.B., Dwyer, M.B., Shi, J.: Interaction testing of highly-configurable systems in the presence of constraints. In: Proceedings of the 2007 International Symposium on Software Testing and Analysis, ISSTA '07, pp. 129–139. Association for Computing Machinery, New York (2007). https://doi.org/10.1145/1273463.1273482
7. Cohen, M.B., Snyder, J., Rothermel, G.: Testing across configurations: implications for combinatorial testing. SIGSOFT Softw. Eng. Notes **31**(6), 1–9 (2006). https://doi.org/10.1145/1218776.1218785
8. Colbourn, C.J., Cohen, M.B., Turban, R.: A deterministic density algorithm for pairwise interaction coverage. In: IASTED Conference on Software Engineering, pp. 345–352 (2004)
9. Eljasik-Swoboda, T., Demuth, W.: Leveraging clustering and natural language processing to overcome variety issues in log management. In: ICAART (2), pp. 281–288 (2020)
10. Engström, E., Runeson, P.: Software product line testing-a systematic mapping study. Inf. Softw. Technol. **53**(1), 2–13 (2011)
11. Golagha, M., Lehnhoff, C., Pretschner, A., Ilmberger, H.: Failure clustering without coverage. In: Proceedings of the 28th ACM SIGSOFT International Symposium on Software Testing and Analysis, pp. 134–145 (2019)

12. Hao, D., Zhang, L., Mei, H.: Test-case prioritization: achievements and challenges. Front. Comp. Sci. **10**(5), 769–777 (2016). https://doi.org/10.1007/s11704-016-6112-3
13. Kasunic, M.: Designing an effective survey. Carnegie Mellon University, Software Engineering Institute Pittsburgh, PA (2005)
14. Lamancha, B.P., Usaola, M.P., de Guzman, I.G.R.: Model-driven testing in software product lines. In: International Conference on Software Maintenance, pp. 511–514. IEEE (2009)
15. Lee, J., Kang, S., Lee, D.: A survey on software product line testing. In: International Software Product Line Conference, pp. 31–40. ACM (2012)
16. Linåker, J., Sulaman, S.M., Höst, M., de Mello, R.M.: Guidelines for conducting surveys in software engineering v. 1.1. Technical report, Lund University, Sweden (2015)
17. Memon, A., et al.: Taming Google-scale continuous testing. In: International Conference on Software Engineering, pp. 233–242. IEEE Press (2017)
18. Petke, J., Cohen, M.B., Harman, M., Yoo, S.: Practical combinatorial interaction testing: empirical findings on efficiency and early fault detection. IEEE Trans. Softw. Eng. **41**(9), 901–924 (2015). https://doi.org/10.1109/TSE.2015.2421279
19. Sinnema, M., Deelstra, S.: Classifying variability modeling techniques. Inf. Softw. Technol. **49**(7), 717–739 (2007)
20. Strandberg, P.E.: Ethical interviews in software engineering. In: 2019 ACM/IEEE International Symposium on Empirical Software Engineering and Measurement (ESEM), pp. 1–11. IEEE (2019)
21. Strandberg, P.E., Ostrand, T.J., Weyuker, E.J., Sundmark, D., Afzal, W.: Automated test mapping and coverage for network topologies. In: Proceedings of the 27th ACM SIGSOFT International Symposium on Software Testing and Analysis, pp. 73–83 (2018)
22. Yoo, S., Harman, M.: Regression testing minimization, selection and prioritization: a survey. Softw. Test. Verif. Reliab. **22**(2), 67–120 (2012)

Dynamic Event Processing Framework for API Gateways: Optimizing Real-Time Performance

Mohammadmahdi Ghobadi(✉) and Renee Bryce

University of North Texas, Denton, TX, USA
mohammadmahdighobadi@my.unt.edu, renee.bryce@unt.edu

Abstract. Modern event-driven microservices require API gateways capable of handling diverse and dynamic workloads with minimal latency. Traditional gateways lack mechanisms for real-time event prioritization and dynamic batching. We introduce a **Dynamic Event Processing Framework (DEPF)** for API gateways that processes high-priority events immediately and batches low-priority events with dynamically adjusted batch sizes and processing intervals based on system metrics. DEPF achieves significant latency reduction for high-priority events up to **43.65%** compared to Traefik and over **99%** compared to NGINX while maintaining high throughput and resource efficiency for low-priority workloads. Evaluations demonstrate DEPF's improved performance in real-time, event-driven architectures, achieving latency reductions ranging from approximately **7%** to over **43%** compared to Traefik and from **86%** to over **99%** compared to NGINX across different event types.

Keywords: Event-driven microservices · API gateway · Real-time prioritization · Dynamic batching · Optimizing

1 Introduction

The rise of microservices architecture has transformed the way applications are developed and deployed, enabling scalability, flexibility, and rapid deployment [7]. Within this paradigm, event-driven microservices have gained prominence due to their ability to decouple services and promote asynchronous communication [9].

Central to microservices architectures are API gateways, which act as intermediaries between clients and backend services, handling tasks such as request routing, load balancing, authentication, and rate limiting [15]. Traditional API gateways like NGINX and Traefik are well-suited for stateless request-response patterns but face challenges in event-driven environments with mixed-priority workloads and real-time processing requirements.

R. Bryce—This author contributed equally to this work.

S. Latifi (Ed.): ITNG 2025, AISC 1463, pp. 543–553, 2025.
https://doi.org/10.1007/978-3-031-89063-5_47

Real-time systems demand low latency for critical events while ensuring efficient throughput for non-urgent traffic [3]. In domains such as finance, healthcare, and IoT, the timely processing of high-priority events is crucial. Existing API gateways lack mechanisms for dynamic prioritization and adaptive batching, often leading to suboptimal performance under fluctuating workloads [16].

1.1 Motivation and Challenges

Microservices often experience variable traffic patterns due to user behavior, scaling operations, and the deployment of new services [8]. Traditional gateways are typically configured with static parameters, which makes them poorly suited to handle fluctuating loads efficiently. This can result in performance degradation, increased latency, and resource bottlenecks during peak demand periods.

There is a trade-off between optimizing for low latency essential to process critical high-priority events quickly and maximizing throughput, important to handle large volumes of low-priority events [2]. Existing gateways lack mechanisms to adjust processing parameters dynamically based on real-time system metrics like CPU and memory usage [1]. Without the ability to adapt to current system conditions, gateways may either underutilize available resources or become overwhelmed under high load, affecting overall system reliability.

1.2 Contributions

We introduce the **dynamic event processing framework (DEPF)** for API gateways, designed for real-time event-driven architectures. Key contributions:

1. **Hybrid Processing Model**: Combines immediate processing for high-priority events with adaptive batching for low-priority events, optimizing both latency and throughput.
2. **Dynamic Event Classification**: Implements flexible event prioritization based on predefined criteria, such as event type and source.
3. **Adaptive Batching Strategy**: Adjusts batch sizes and processing intervals based on system metrics, enabling real-time adaptation to changing load conditions.
4. **Empirical Evaluation**: Demonstrates DEPF's performance improvements, highlighting significant latency reductions ranging from 7% to over 99% compared to traditional gateways.

2 Related Work

2.1 API Gateways in Microservices

API gateways simplify client interactions in microservices architectures by providing a single entry point [15]. NGINX and Traefik are popular gateways known for their scalability and configurability [11,13], but they lack support for event-driven processing and prioritization [16]. They are primarily designed for stateless, synchronous request-response patterns and do not offer mechanisms for differentiating between high and low-priority events.

Our work differs by integrating real-time event prioritization and adaptive batching directly into the API gateway, addressing the limitations of traditional gateways in handling mixed-priority, event-driven workloads.

2.2 Event-Driven Architectures

Event-driven architectures decouple event producers from consumers, allowing services to react to events asynchronously [2]. Messaging systems such as RabbitMQ and Apache Kafka provide robust event streaming capabilities [6]. RabbitMQ supports priority queues [12], enabling message prioritization at the messaging layer.

However, these systems require additional middleware for prioritization before events reach backend services. We integrate prioritization directly into the API gateway, eliminating the need for extra middleware layers and reducing latency by processing high-priority events immediately at the gateway level.

2.3 Prioritization Techniques

Quality of Service (QoS) mechanisms prioritize network traffic based on predefined policies [5], but do not address application-level priorities. Middleware solutions can intercept and prioritize events before they reach back-end services [14], but they add complexity and potential latency.

In contrast, our framework operates at the application layer within the API gateway, providing application-level prioritization without the overhead of additional middleware. This approach simplifies the architecture and improves the performance of high-priority events.

2.4 Dynamic Batching Strategies

Dynamic batching optimizes resource utilization in machine learning systems by adjusting batch sizes based on workloads [4]. Clipper introduces an adaptive batching strategy for prediction-serving systems.

Our work extends this concept to API gateways that handle heterogeneous event types. Unlike Clipper, which focuses on machine learning workloads, our adaptive batching algorithm adjusts batch sizes and processing intervals based on real-time system metrics (CPU and memory usage), catering to the unique demands of event-driven microservices.

2.5 Performance Engineering in Dynamic Systems

Nambiar et al. [10] highlight the need for real-time adaptation mechanisms in software performance engineering, citing the lack of adaptive solutions as a future research direction.

Our framework addresses this gap by providing a real-time adaptive batching strategy that responds to system performance metrics, enhancing resource utilization and performance in dynamic microservices environments.

3 System Architecture

3.1 Overview

DEPF integrates real-time event prioritization and adaptive batch processing directly into the gateway layer. The architecture consists of modular components that work together to manage event flow efficiently, ensuring that high-priority events are processed with minimal latency while maintaining throughput for low-priority events.

3.2 Components

Ingress Handler: Acts as the unified entry point for all incoming events. It receives events over HTTP, performs initial validation and parsing, and dispatches events to the Event Prioritization Module.

Event Prioritization Module: Classifies events into high and low priority queues based on predefined criteria, such as event type (e.g., "transaction," "log") and source. This ensures that critical events receive immediate attention.

High-Priority Event Processor: Processes high-priority events immediately by dispatching them to backend services without delay, minimizing latency for critical operations.

Low-Priority Batch Processor: Handles low-priority events by aggregating them into batches of adjustable size. It processes these batches at intervals determined by the Parameter Adjustment Module, optimizing throughput and resource utilization.

Parameter Adjustment Module: Dynamically adjusts the batch size and processing interval for low-priority events based on current system metrics, such as CPU and memory usage. This allows the system to adapt in real time to changing load conditions, preventing resource exhaustion and maintaining optimal performance.

Monitoring and Metrics Collector: Collects system metrics and provides data for parameter adjustments. By monitoring resource utilization and performance metrics, it enables the gateway to make informed decisions about processing parameters.

3.3 Event Flow

Incoming events are received by the Ingress Handler and passed to the Event Prioritization Module for classification. High-priority events are sent to the High-Priority Event Processor and dispatched immediately, while low-priority events are queued in the Low-Priority Batch Processor. The Parameter Adjustment Module adjusts batch size and processing interval based on system metrics collected by the Monitoring and Metrics Collector, ensuring optimal processing of low-priority events.

4 Hybrid Batching Strategy

4.1 Conceptual Framework

The Hybrid Batching Strategy optimizes event processing by balancing latency and throughput while efficiently utilizing system resources. It involves three key components:

1. **Dynamic Event Classification**: Separating events into high and low priority based on predefined criteria.
2. **Adaptive Batch Processing**: Adjusting batch size and processing interval for low-priority events in response to system metrics.
3. **Real-Time Parameter Adjustment**: Modifying processing parameters dynamically to reflect current CPU and memory usage.

4.2 Dynamic Event Classification

Events are classified using a priority function $P(E)$:

$$P(E) = \begin{cases} 1, & \text{if } E \text{ is high priority} \\ 0, & \text{if } E \text{ is low priority} \end{cases} \tag{1}$$

Criteria for classification may include event type (ET) and source. For example, transaction events from trusted sources may be deemed high priority, while log events may be considered low priority.

4.3 Adaptive Batch Processing

4.3.1 Retrieve Metrics Current CPU usage (CPU) and memory usage (MEM) are retrieved at regular intervals from the Monitoring and Metrics Collector.

4.3.2 Adjust Parameters The batch size (B) and processing interval (I) are adjusted based on the following conditions:

- **Both CPU and Memory Low**:

$$B \leftarrow B + \delta_B, \quad I \leftarrow I - \delta_I \tag{2}$$

  Increase batch size to improve throughput and decrease processing interval to process batches more frequently.
- **Both CPU and Memory High**:

$$B \leftarrow B - \delta_B, \quad I \leftarrow I + \delta_I \tag{3}$$

  Decrease batch size and increase processing interval to reduce system load.

- **CPU High, Memory Low**:

$$B \leftarrow B + \delta_B, \quad I \leftarrow I + \delta_I \tag{4}$$

Adjust parameters to balance CPU load without overusing memory.

- **Memory High, CPU Low**:

$$B \leftarrow B - \delta_B, \quad I \leftarrow I - \delta_I \tag{5}$$

Reduce batch size to lower memory usage and decrease processing interval to prevent backlog.

Here, δ_B and δ_I are adjustment increments. Constraints are applied to maintain parameters within acceptable ranges.

4.4 Algorithms

Parameter Adjustment Algorithm

Algorithm 1. Adjust Batch Parameters

Require: Current batch size B, processing interval I
Ensure: Adjusted batch size B, processing interval I
Retrieve CPU usage CPU and memory usage MEM
if CPU is low and MEM is low **then**
    $B \leftarrow B + \delta_B$
    $I \leftarrow I - \delta_I$
else if CPU is high and MEM is high **then**
    $B \leftarrow B - \delta_B$
    $I \leftarrow I + \delta_I$
else if CPU is high and MEM is low **then**
    $B \leftarrow B + \delta_B$
    $I \leftarrow I + \delta_I$
else if CPU is low and MEM is high **then**
    $B \leftarrow B - \delta_B$
    $I \leftarrow I - \delta_I$
end if
Apply constraints:
$B \leftarrow \max(B_{\min}, \min(B, B_{\max}))$
$I \leftarrow \max(I_{\min}, \min(I, I_{\max}))$

4.5 Advantages

1. **Latency Reduction**: Immediate processing of high-priority events minimizes delay, crucial for real-time systems.

2. **Throughput Optimization**: Adaptive batching maximizes throughput for low-priority events, improving overall system efficiency.
3. **Resource Efficiency**: Dynamic adjustments ensure optimal use of CPU and memory, preventing resource bottlenecks.
4. **Scalability**: The modular architecture facilitates horizontal scaling, allowing the system to handle increased loads.

5 Implementation and Evaluation

5.1 Implementation Details

The Dynamic Event Processing Framework was implemented using Golang due to its performance and concurrency capabilities. The language's support for goroutines and channels allowed efficient handling of concurrent event streams. The development environment consisted of four virtual machines (VMs) within the same network:

- **VM 1**: Hosted Apache Kafka, serving as the message broker.
- **VM 2**: Deployed producer microservices that generated events.
- **VM 3**: Deployed consumer microservices that processed events.
- **VM 4**: Ran the API gateways Traefik, NGINX, and our DEPF for comparative evaluation.

An application was developed to simulate HTTP requests with five types of events: "transaction," "log," "notification," "command," and "query." Each type included a mix of high and low priority events.

5.2 Evaluation Setup

The evaluation focused on measuring latency and throughput. The following metrics were collected:

- **Event Count**: Total number of events processed per event type.
- **High-Priority Count**: Number of high-priority events per event type.
- **Low-Priority Count**: Number of low-priority events per event type.
- **Average Event Duration**: Average time taken to process an event, measured from ingestion to completion.

The system was subjected to varying load conditions to assess its performance under different scenarios.

5.3 Results

The performance of DEPF was evaluated in comparison with Traefik and NGINX. Table 1 summarizes the collected metrics across five different event types.

Table 1. Performance Metrics Comparison

Gateway	Event Type	Event Count	High-P Count	Low-P Count	Avg Duration (ms)
DEPF	transaction	100,000	90,049	9,951	29.05
	log	25,000	2,521	22,479	55.20
	notification	42,000	8,390	33,610	45.20
	command	100,000	65,020	34,980	28.05
	query	64,000	31,976	32,024	41.39
Traefik	transaction	100,000	90,106	9,894	47.32
	log	25,000	2,581	22,419	59.42
	notification	42,000	8,239	33,761	52.02
	command	100,000	64,980	35,020	49.80
	query	64,000	31,938	32,062	51.12
NGINX	transaction	100,000	89,912	10,088	3,143.37
	log	25,000	2,502	22,498	416.16
	notification	42,000	8,587	33,413	403.63
	command	100,000	65,092	34,908	3,055.39
	query	64,000	31,954	32,046	1,796.01

5.4 Analysis of Results

The performance metrics indicate that DEPF consistently outperforms Traefik and NGINX across all event types in terms of average event duration.

For **"transaction"** events, typically high priority, DEPF achieved an average duration of 29.05 ms, compared to Traefik's 47.32 ms and NGINX's 3,143.37 ms. This represents a latency reduction of approximately **38.6%** over Traefik and over **99%** over NGINX, highlighting DEPF's efficiency in handling high-priority events promptly.

For **"command"** events, DEPF processed events in an average of 28.05 ms, outperforming Traefik's 49.80 ms and NGINX's 3,055.39 ms, indicating latency reductions of about **43.65%** and over **99%**, respectively.

For low-priority events like **"log"** and **"notification"**, DEPF also showed improved performance. The average duration for "log" events was 55.20 ms with DEPF, compared to 59.42 ms with Traefik and 416.16 ms with NGINX, translating to latency reductions of approximately **7%** and over **86%**, respectively. Although the improvement over Traefik is modest, it indicates that DEPF's adaptive batching does not negatively impact low-priority event processing.

For "notification" events, DEPF achieved an average duration of 45.20 ms, about **13%** faster than Traefik's 52.02 ms and significantly faster than NGINX's 403.63 ms (over **88%** reduction).

Regarding **"query"** events, DEPF's average duration was 41.39 ms, compared to Traefik's 51.12 ms and NGINX's 1,796.01 ms, resulting in latency reductions of approximately **19%** and over **97%**, respectively.

Overall, DEPF consistently demonstrated lower average durations across all event types. The most significant improvements were observed against NGINX, particularly for high-priority events, where DEPF achieved latency reductions exceeding **99%**. Improvements over Traefik ranged from approximately **7%** to **43.65%**, indicating that DEPF provides substantial performance benefits even compared to more modern gateways.

These results validate the effectiveness of DEPF's hybrid processing model, which prioritizes high-priority events for immediate processing while efficiently batching low-priority ones. The adaptive batching mechanism enables DEPF to adjust processing parameters in real time based on system metrics, ensuring optimal performance under varying load conditions.

The consistent performance gains across different event types suggest that DEPF is well-suited for environments with diverse workloads. The ability to reduce latency for high-priority events without compromising the throughput of low-priority events makes DEPF advantageous for applications requiring both real-time responsiveness and high throughput.

6 Discussion

DEPF's improved performance is attributed to its hybrid processing model, which reduces overhead by immediately processing high-priority events and efficiently batching low-priority ones. The adaptive batching mechanism dynamically adjusts to system load, preventing resource saturation and ensuring consistent performance even under fluctuating workloads.

The nature of the input events, with a high proportion of critical "transaction" and "command" types, showcased DEPF's ability to prioritize effectively. The gateway's responsiveness to system metrics allowed it to maintain optimal processing parameters throughout the experiments.

These results imply that DEPF is highly suitable for time-sensitive applications such as financial transaction processing, real-time monitoring systems, and emergency response coordination. By prioritizing critical events and adapting to system load, DEPF ensures reliable and efficient operation in environments where performance is paramount.

Moreover, the framework's ability to maintain high throughput for low-priority events without compromising the processing of high-priority events demonstrates its effectiveness in balancing the trade-off between latency and resource utilization. This balance is crucial in modern microservices architectures that must handle diverse workloads with varying priority levels.

The successful implementation and evaluation of DEPF suggest that integrating dynamic prioritization and adaptive batching directly into API gateways can significantly enhance performance in event-driven systems. This approach eliminates the need for additional middleware layers, simplifying the architecture and reducing potential points of failure.

7 Threats to Validity

Several threats to validity were identified in our study. Regarding internal validity, the experiments were conducted in a controlled environment using virtual machines within a single network. This setup may not fully capture the complexities and unpredictability of real-world distributed systems, such as network latency variability and hardware heterogeneity.

For external validity, the event types and priority distributions used in our evaluation reflect common scenarios but may differ in other domains or applications. Consequently, the results may not generalize to all event-driven architectures without further validation.

To address reliability, experiments were repeated multiple times, and average values were used in the analysis to mitigate anomalies and ensure consistent results. However, the lack of detailed system metrics like CPU and memory usage limits the depth of resource utilization analysis.

8 Conclusion

The Dynamic Event Processing Framework (DEPF) provides an effective solution for managing mixed-priority workloads in real-time event-driven architectures. By integrating dynamic prioritization and adaptive batching based on system metrics, DEPF optimizes both latency and throughput while efficiently utilizing resources.

Our implementation and evaluation demonstrate that DEPF achieves significant latency reductions across various event types. Specifically, DEPF reduces latency for high-priority events by up to **43.65%** compared to Traefik and over **99%** compared to NGINX. For low-priority events, DEPF maintains high throughput and resource efficiency, with latency reductions ranging from approximately **7%** to over **88%** compared to traditional gateways.

These performance gains are attributed to DEPF's hybrid processing model, which processes high-priority events immediately and batches low-priority events with dynamically adjusted parameters based on real-time system metrics. This approach ensures that critical events receive prompt attention without compromising the efficient handling of less urgent workloads.

Potential areas for further research include advanced parameter tuning using machine learning techniques for predictive adjustments, extending support to additional protocols like WebSocket or MQTT to enhance applicability in IoT and real-time communication systems, and integrating enhanced security features such as anomaly detection and automated threat response mechanisms to improve resilience against attacks.

In summary, DEPF represents a significant advancement in API gateway design for event-driven microservices, offering quantifiable improvements in performance that make it highly suitable for time-sensitive applications requiring both low latency and high throughput.

Acknowledgements. The authors thank all contributors who supported this research. Special thanks to the development team for their valuable insights and assistance.

References

1. Balalaie, A., Heydarnoori, A., Jamshidi, P.: Microservices architecture enables devops: migration to a cloud-native architecture. IEEE Softw. **33**(3), 42–52 (2016)
2. Bellemare, A.: Building Event-Driven Microservices: Leveraging Organizational Data at Scale. O'Reilly Media Inc, Sebastopol (2020)
3. Coulouris, G., Dollimore, J., Kindberg, T., Blair, G.: Distributed Systems: Concepts and Design. Pearson Education, London (2011)
4. Crankshaw, D., Wang, X., Zhou, H., Franklin, M.J., Gonzalez, J.E., Stoica, I.: Clipper: a low-latency online prediction serving system. In: 14th USENIX Symposium on Networked Systems Design and Implementation (NSDI 17), pp. 613–627 (2017)
5. Floyd, S., Jacobson, V.: Link-sharing and resource management models for packet networks. IEEE/ACM Trans. Networking **3**(4), 365–386 (1995)
6. Kreps, J., Narkhede, N., Rao, J.: Kafka: a distributed messaging system for log processing. In: Proceedings of the NetDB, vol. 11, pp. 1–7 (2011)
7. Li, S., et al.: Understanding and addressing quality attributes of microservices architecture: a systematic literature review. Inf. Softw. Technol. **131**, 106449 (2021)
8. Mazzara, M., Dragoni, N., Bucchiarone, A., Giaretta, A., Larsen, S.T., Dustdar, S.: Microservices: migration of a mission critical system. IEEE Trans. Serv. Comput. **14**(5), 1464–1477 (2018)
9. Michelson, B.M.: Event-driven architecture overview. Patricia Seybold Group **2**(12), 10–1571 (2006)
10. Nambiar, M., Kattepur, A., Bhaskaran, G., Singhal, R., Duttagupta, S.: Model driven software performance engineering: current challenges and way ahead. ACM SIGMETRICS Perform. Eval. Rev. **43**(4), 53–62 (2016)
11. NGINX, Inc.: Nginx: High performance load balancer, web server, & reverse proxy (2024). https://www.nginx.com/. Accessed 25 Nov 2024
12. RabbitMQ Documentation Team: Classic queues support priorities (2024). https://github.com/rabbitmq/rabbitmq-website/blob/main/versioned_docs/version-4.0/priority.md. Accessed 25 Nov 2024
13. Traefik Labs: Traefik: The cloud-native application proxy (2024). https://traefik.io/. Accessed 25 Nov 2024
14. Vinoski, S.: Where is is middleware. IEEE Internet Comput. **6**(2), 83–85 (2002)
15. Zhao, J., Jing, S., Jiang, L.: Management of API gateway based on micro-service architecture. J. Phys: Conf. Ser. **1087**(3), 032032 (2018)
16. Zhelev, S., Rozeva, A.: Using microservices and event driven architecture for big data stream processing. In: AIP Conference Proceedings, vol. 2172. AIP Publishing (2019)

Enhancing IoT Edge Platforms: Selecting an MQTT-Compatible Broker for Kubernetes Environments

Bahareh Aghajanpour, Alessio Bucaioni(✉), and Gabriele Capannini

Mälardalen University, Västerås, Sweden
{bahareh.aghajanpour,alessio.bucaioni,gabriele.capannini}@mdu.se

Abstract. This paper aims to enhance IoT communication in edge platforms by identifying suitable MQTT-compatible solutions. The primary focus is on investigating two critical quality attributes of message brokers: data throughput and latency. To guide this research, we address two questions: (1) How do protocols like MQTT and their architecture contribute to messaging in IoT communications? and (2) How can MQTT architecture be optimized to meet the requirements of the ABB IoT Edge Platform? A mixed-methods research approach was applied, encompassing both quantitative and qualitative phases. The quantitative phase involves experimental evaluation of data throughput and latency among various message brokers, while the qualitative phase includes stakeholder interviews and literature review. Results indicate that event-driven architecture, publisher-subscriber design patterns, and the NATS message broker provide effective support for MQTT-based IoT edge platforms. Notably, NATS demonstrated high throughput and low latency compared to alternative brokers, making it a strong candidate for use in IoT edge environments. These findings offer valuable insights into best practices for MQTT-compatible message brokers in IoT platforms, specifically benefiting software architects developing for the ABB IoT edge platform.

Keywords: IoT · Edge · MQTT · Kubernetes · NATS

1 Introduction

The Internet of Things (IoT) refers to a network of interconnected, intelligent objects capable of autonomously transferring data [1]. At its core, IoT involves the seamless integration of physical objects with the internet, creating vast networks that serve applications ranging from household devices to industrial systems [2]. Regardless of scale, efficient data transfer is essential for IoT devices. Typically, these devices transmit telemetry data to edge gateways and cloud platforms, often relying on the Message Queuing Telemetry Transport (MQTT) protocol. Known for its publisher-subscriber design pattern, MQTT facilitates

S. Latifi (Ed.): ITNG 2025, AISC 1463, pp. 554–567, 2025.
https://doi.org/10.1007/978-3-031-89063-5_48

machine-to-machine communication, making it particularly suited for IoT applications [3]. Consequently, MQTT has emerged as a key protocol in supporting reliable communication and data exchange within IoT platforms.

Message brokers act as intermediaries in these systems, translating and routing messages to meet the requirements of different software components, services, or systems. They leverage the MQTT protocol to bridge applications and middleware, creating seamless connections between disparate components [4]. However, the abundance of broker implementations presents software architects with the challenge of selecting a suitable broker, especially in enterprise IoT contexts where data throughput and latency are critical considerations [5]. In clustered environments, factors such as capacity planning, configuration, and system availability add further complexity [6].

Initial research by ABB has highlighted limitations with certain brokers, such as Mosquitto, especially concerning latency and data throughput. For instance, on ABB's IoT edge platform, Mosquitto has exhibited persistence issues and struggles to handle the high data throughput required by enterprise-level IoT deployments. As IoT platforms expand, the need for scalable brokers with robust performance becomes increasingly important.

In this paper, we aim to enhance IoT communication in edge platforms by identifying an MQTT-compatible broker that aligns with the requirements of an event-driven, publisher-subscriber architecture. Experimental evaluation of various MQTT-compatible brokers, including NATS, indicates that NATS is a promising candidate to replace Mosquitto in Kubernetes-orchestrated edge environments, particularly for ABB's IoT edge platform.

The remainder of this paper is structured as follows. Section 2 outlines the research process. Section 3 details the experimental setup and results. Section 4 presents the findings derived from experimental and research data. Section 5 discuss the findings of this research, while Sect. 6 compares this study with relevant works. Finally, Sect. 7 concludes the paper and suggests avenues for future research.

2 Research Process

The primary Research Goal (RG) was to identify an architectural solution, including a message broker, that would function effectively for an IoT application on an edge-based platform with Kubernetes orchestration. We broke down the overarching RG into the following Research Questions (RQs):

RQ1: How do protocols like MQTT and their architecture contribute to messaging in IoT communications?

RQ2: How can the MQTT architecture be enhanced to meet the requirements of the ABB IoT edge platform?

To provide answers to these questions, we employed a mixed-methods research approach, inspired by [7,8], combining experimental analysis and literature review to gain insights into MQTT broker limitations and to identify best practices for IoT edge platforms on Kubernetes clusters. The research process began with a quantitative phase, where we conducted experiments to evaluate the data throughput and latency of different message brokers under various conditions. This phase helped us in answering to RQ1. The qualitative phase followed, involving an in-depth literature review to contextualize these findings within broader architectural considerations and industry best practices. This phase helped us in answering to RQ2 and, partially, RQ1. In the final interpretation phase, we integrated results from both quantitative and qualitative methods, creating a comprehensive perspective that addresses existing limitations and offers informed recommendations for enhancing IoT edge platforms.

2.1 Literature Review

To address our research questions, particularly RQ2, we conducted comprehensive literature reviews as the qualitative phase of our study. Striving to maintain a balance between rigor and agility, we adopted a streamlined protocol for our literature review, structured as follows:

Search Strategy: our primary academic databases were IEEE Xplore and ScienceDirect, with the majority of literature sourced from IEEE. We conducted searches using keywords such as "MQTT," "Kubernetes clustering," "IoT architecture," and "message brokers." Additionally, we employed Boolean operators to combine these keywords (e.g., "MQTT AND IoT architecture," "MQTT OR IoT architecture," "MQTT AND message broker"). To ensure comprehensive coverage, we also included relevant conference proceedings, official website documentation, and books.

Inclusion and Exclusion Criteria: given that IoT is a rapidly evolving field, we focused on articles and conference papers published within the last decade. We prioritized publications that concentrated on MQTT architecture, IoT edge computing, and message brokers, selecting only those written in or translated into English. Publications not directly relevant to our main focus were excluded.

Paper Selection: an initial screening of titles and abstracts led to the selection of 25 relevant papers for full-text review. Following a detailed examination, three papers were excluded due to lack of relevance.

Data Extraction: to directly address RQ2, we systematically reviewed nine of these papers, while the remaining studies provided a broader understanding of the topic. Key findings and insights were extracted and synthesized to answer the research questions comprehensively throughout the report.

2.2 Experiment

Our selection of message brokers was guided by specific criteria aligned with the requirements of this study, informed by relevant literature [3,9]. The key features considered were as follows:

- Latency: time taken for each broker to deliver a message from publisher to subscriber.
- Throughput: rate of messages sent per second.
- Ease of setup.
- Compatibility with MQTT protocol standards.

These criteria were chosen to ensure that the selected message broker could meet the real-time communication needs of our IoT system while adhering to industry standards and minimizing implementation complexity. Initially, five message brokers were considered—HiveMQ, VerneMQ, RabbitMQ, NATS, and Apache Kafka—which are discussed further in Sect. 3. Due to time constraints, only one of these brokers was benchmarked in this study. Although Apache Kafka was considered initially, challenges in setup led to an unsuccessful attempt at benchmarking. Kafka's complex deployment requirements were deemed unsuitable, given the priority for ease of setup in this study. As a result, we relied on existing research data for Apache Kafka [10,11]. Similarly, we used benchmarking data from existing studies for HiveMQ and VerneMQ [12], which are also discussed in Section 3. After further consideration, NATS was selected over RabbitMQ for benchmarking due to its simple deployment, efficient performance characteristics, and its event-driven architecture [13].

Experimental Environment: The test environment consisted of a Linux Ubuntu 20.04.6 LTS Virtual Machine with a 16-core CPU, 128 GB RAM, and 250 GB HDD. Docker version 26.0.0 was installed to facilitate deployment. To benchmark the performance of NATS, we used JetStream [14], a built-in distributed system within the NATS ecosystem that enhances functionality by operating atop the core NATS services and enabling accessibility from all client applications. JetStream was used to conduct both benchmark and stress tests, simulating high loads and evaluating data throughput and latency within the NATS messaging system. The tests focused on both publishers and subscribers under various load scenarios, enabling a detailed examination of system performance under different conditions.

Experimental Setup: to work around security restrictions limiting internet access, we pre-downloaded the NATS image for deployment. The NATS server was started using Docker, which allowed for simplified deployment. Although Docker can support clustered configurations with multiple NATS servers, our focus was on benchmarking a single broker instance to isolate and measure performance in a controlled environment. JetStream was activated on top of the NATS server via the command `nats-server -m 8222 -js`, providing confirmation of server readiness through a "Server is ready" status message. Due to network restrictions, the Docker command for running JetStream included network proxy credentials to facilitate connectivity.

Test Scenarios for Throughput and Latency: we structured the benchmarking tests around six throughput scenarios and eight latency scenarios to assess NATS under different conditions. For the throughput Scenarios, configurations varied by message size, number of publishers and subscribers, and message volume. Each scenario was designed to test the broker's performance under progressively increased load and message size variations. For the latency scenarios, we tested NATS in request-reply mode using various combinations of publishers, subscribers, and message counts. Each latency scenario simulated different usage patterns to evaluate system responsiveness under varying loads. Latency was measured by averaging the time taken for each request-response cycle, allowing for comparative analysis across scenarios. This procedural setup allowed for a structured assessment of NATS under diverse workload scenarios, providing insights into the broker's capacity to handle the demands of an IoT edge environment. A summary of the specific experimental results is presented in Sect. 3.

3 How Do Protocols Like MQTT and Their Architecture Contribute to Messaging in IoT Communications? (RQ1)

As outlined earlier, to address RQ1, we evaluated five message brokers: HiveMQ, VerneMQ, RabbitMQ, NATS, and Apache Kafka. Due to time constraints, direct benchmarking was conducted only for NATS. For the remaining brokers, we relied on benchmarking data from prior studies, as detailed below.

3.1 Experimental Results for NTS Industrial Benchmarking

Due to security restrictions limiting internet connectivity, the NATS image was pre-downloaded for deployment. The remainder of the setup was conducted via Docker. A NATS server was initiated using the `docker run` command, though it is possible to create clusters of multiple NATS servers via Docker [13]. However, for this study, we aimed to benchmark NATS in a controlled, single-server environment to accurately measure performance for individual publisher and subscriber pairs. JetStream, an integral feature of NATS for benchmarking, was activated using the command `nats-server -m 8222 -js`, with the server readiness confirmed by the `Server is ready` status.

For the experimental work, network proxy credentials were incorporated into the Docker command to overcome connection restrictions. The performance of NATS was then tested across six throughput scenarios and eight latency scenarios to evaluate its capacity in handling varied workloads.

A set of tests was conducted by varying the number of publishers in the set $\{1, 4, 16\}$ and the number of subscribers in the set $\{1, 4, 16, 64\}$, across different scenarios with message sizes of 2 and 20 kilobytes. The results in Table 1 indicate that as the number of publishers and subscribers increases, the overall bandwidth also increases until the load on the system exceeds its capacity, then it becomes congested and the bandwidth degrades.

To evaluate latency, NATS was tested in eight request-reply scenarios with varied publisher, subscriber, and message configurations, simulating different system loads. In each scenario, average latency was calculated as the inverse of the publishing rate.

Table 1. Bandwidth results for messages of 2 kB and 20 kB in Messages per Second and GB/s.

# Pubs	# Subs	Msg/s (GB/s)	
		2 kB	20 kB
1	1	568820 (1.08)	86408 (1.65)
1	4	804433 (1.53)	137822 (2.63)
1	16	1026244 (1.96)	162124 (3.09)
1	64	903633 (1.72)	156847 (2.99)
4	1	579655 (1.11)	73812 (1.41)
4	4	963657 (1.84)	112364 (2.14)
4	16	1391364 (2.65)	218907 (4.18)
4	64	1233010 (2.35)	182787 (3.49)
16	1	562086 (1.07)	76624 (1.46)
16	4	1056885 (2.02)	125987 (2.40)
16	16	1306175 (2.49)	213806 (4.08)
16	64	1152619 (2.20)	118377 (2.26)

- Scenario 1: configured with 1 publisher, 1 subscriber, and 1,000 messages, this setup resulted in an average latency of approximately 0.4325 ms.
- Scenario 2: with 20 publishers and 20 subscribers for 1,000 messages, latency decreased significantly to 0.0321 ms, highlighting how increasing publishers and subscribers can reduce latency.
- Scenario 3: with 50 publishers, 50 subscribers, and 1,000 messages, latency dropped further to 0.0260 ms.
- Subsequent scenarios: increasing the message count to 10,000 across different publisher-subscriber configurations resulted in latency values as low as 0.0173 ms in high-load conditions, showing an inverse relationship between message volume and latency.

These latency results, shown in Table 2, indicate that higher numbers of publishers, subscribers, and messages generally lead to reduce the latency, showing NATS efficiency under high-load conditions. The explanation could be that, when more publishers and subscribers are present in the system, multiple messages can be processed concurrently. Applying batching and pipelining techniques, NATS is able to optimize the network usage more effectively by reducing the average latency per message.

These findings demonstrate a trade-off between latency and throughput in NATS. Increasing system complexity—such as by adding more publishers and subscribers—can improve performance but also requires additional resources, impacting overall system design. However, by further increasing the number of publishers and subscribers, we expect that the broker and the network will become saturated, thereby increasing the latency due to resource contention.

Table 2. Latency results NATS benchmark

# Pubs	# Subs	Messages	Latency (ms)
1	1	1000	0.4325
20	20	1000	0.0321
50	50	1000	0.0260
1	1	10000	0.3950
20	20	10000	0.0299
50	50	10000	0.0173

3.2 Comparison of NAS with Other Brokers

We selected other four brokers, encompassing both MQTT-native and MQTT-compatible options. VerneMQ and HiveMQ are MQTT-native brokers, while the others—RabbitMQ, Apache Kafka are compatible with the MQTT protocol. For all these brokers, we relied on benchmarking data from prior studies.

VerneMQ: VerneMQ is a high-performance MQTT-native broker designed for scalability and robust messaging applications [15]. According to Koziolek et al. [12], VerneMQ was developed to overcome scalability limitations in brokers like RabbitMQ, which is based on the AMQP protocol. VerneMQ is lightweight yet powerful, capable of managing millions of device connections [15].

HiveMQ: Another MQTT-native broker, HiveMQ is known for its reliability and scalability, making it well-suited for mission-critical IoT and messaging applications [16]. HiveMQ offers strong security features, including TLS encryption, authentication, and integration with third-party security systems, making it ideal for applications with stringent security requirements. Performance tests conducted by Koziolek et al. [12] (see Table 3) reveal key distinctions between VerneMQ and HiveMQ. VerneMQ achieved a throughput of 10,000 messages per second (msg/s) with a latency of 8.7 ms, delivering messages quickly and efficiently for applications that require low latency. In contrast, HiveMQ achieved a throughput of 8,000 msg/s but with a higher latency of 119.4 ms, making it less suitable for time-sensitive applications, despite its strengths in reliability and message handling. Both brokers offer scalability through multi-threaded architectures and support horizontal scaling, with the test environment limited to eight CPU cores. This suggests that performance could potentially improve with additional resources.

RabbitMQ: an open-source broker implementing the Advanced Message Queuing Protocol (AMQP), RabbitMQ offers flexibility and reliability in message delivery [17]. Supporting both point-to-point and pub-sub architectures, RabbitMQ is versatile for diverse application needs. However, it may not be ideal for applications requiring guaranteed precise delivery of each message, as it ensures delivery of at least one message among several. Despite this, RabbitMQ performs well in complex routing scenarios, which makes it valuable for certain applications.

Apache Kafka: Kafka is a distributed streaming platform designed for high-throughput data processing, commonly used in real-time analytics and event-driven architectures [10]. Kafka's high throughput and low latency provide strong message delivery and ordering guarantees, making it suitable for applications requiring strict sequencing. However, Kafka's complexity in setup and configuration may present challenges for simpler IoT implementations. Performance analyses by GCORE [11] and benchmarking by Confluent [18] (Table 3) indicate that Kafka achieves a throughput of up to 2,000,000 msg/s with a latency of 5 ms, ideal for high-speed data streams. RabbitMQ, in comparison, manages up to 60,000 msg/s with a latency of around 10 ms, a figure that may vary based on message routing complexity, system configuration, and hardware resources. RabbitMQ's emphasis on reliability may contribute to slightly higher latency than other platforms.

Table 3. Performance comparison of HiveMQ, VerneMQ, RabbitMQ, and Kafka

System	Throughput (msg/s)	Latency(ms)
HiveMQ	8000	119.4
VerneMQ	10000	8.7
RabbitMQ	60000	10.0
Kafka	2000000	5.0

In summary, based on the requirements of this work—including high throughput, low latency, MQTT compatibility, and ease of deployment—NATS emerges as the most suitable message broker for the application discussed in this study. Its strong performance metrics, combined with MQTT compatibility and a focus on simplicity, make it an ideal choice for enhancing IoT communication within the ABB IoT edge platform.

4 How Can the MQTT Architecture Be Enhanced to Meet the Requirements of the ABB IoT Edge Platform? (RQ2)

To strengthen the MQTT protocol for the ABB IoT Edge Platform, we conducted a systematic literature review to inform architectural improvements. Al-Awami et al. [1] emphasize the importance of redundancy and scalability in IoT architectures, highlighting the need for robust messaging systems to mitigate potential failures. Al-Fuqaha et al. [2] identify MQTT's lightweight advantages for IoT applications but note its limitation in relying on a single broker, thus supporting the enhancement of MQTT with multiple brokers for greater resilience. Johari et al. [19] underscore the need for efficient routing within MQTT to ensure reliable message delivery, suggesting a more resilient broker architecture. Additionally, Shahri et al. [20] propose real-time and reliability enhancements to MQTT, such as message prioritization and deadlines, aligning with the demands of dependable IoT communication. Further studies reinforce this perspective. Koziolek et al. [12] evaluate various MQTT brokers, emphasizing performance and scalability in distributed IoT environments. Deploying each broker instance on a separate cluster node allows for future scalability to handle increased demands. Bandai et al. [21] advocate for geographically distributed brokers to boost performance and redundancy, with Kubernetes assigning brokers to distinct cluster nodes for continued stability in case of failures. Grüner et al. [6] highlight how Kubernetes orchestration can improve system stability by distributing MQTT brokers across nodes, thus minimizing single points of failure and enhancing system availability.

Drawing on these insights, we propose an enhanced MQTT architecture that integrates an event-driven model with a pub/sub pattern to support real-time communication in IoT systems. The MQTT architecture includes clients that publish or subscribe to messages, with a central broker acting as an intermediary to deliver these messages to subscribers. To support real-time communication, we propose an event-driven architecture, where events are generated by an event producer, transmitted to an event handler, and distributed among event consumers [22]. This real-time event handling is well-suited to IoT applications, where timely communication is essential. To enhance this architecture, we couple the event-driven model with the pub/sub (publisher-subscriber) design pattern, wherein a publisher sends messages to a broker that then distributes these messages to subscribers via output channels. This design supports real-time data handling and aligns with MQTT's architecture for IoT systems. Additional features, such as client properties and a centralized controller, are introduced to enable effective message delivery by prioritizing attributes like deadlines. A database aids in analyzing these real-time attributes to ensure prompt communication. In edge-platform IoT environments, Kubernetes orchestration provides a robust framework for distributed systems by organizing containers into Pods, which are deployed on virtual or physical nodes based on system requirements [23]. Running system modules within a Kubernetes cluster offers benefits such as container management, service discovery, and self-healing, although it also introduces some complexities [24]. To further enhance system stability and redundancy in MQTT-based IoT architectures, we propose three solutions:

- Node-based broker deployment: each broker instance is deployed on a separate node within a single system cluster, allowing the system to scale as needed. Instances synchronize to transfer messages and related data. If an instance fails, others can continue to operate, thus increasing system availability and stability. This setup uses the pub/sub design pattern, allowing applications, such as mobile apps, to subscribe to topics published by device broker clients, resulting in a more resilient messaging framework.
- Clustered broker deployment without queue mirroring: this solution involves deploying multiple broker instances across different hardware nodes. Although

the instances operate independently, each can handle client connections and process messages, ensuring system resilience even if one broker fails [6].

- Geographical broker deployment: this solution involves deploying brokers based on geographic locations to enhance system stability. Clients subscribe to topics based on their location, and Kubernetes assigns each broker to a distinct cluster node. This approach distributes workload across brokers, allowing the system to maintain operations if a broker fails, as other brokers take over responsibilities.

Each of these solutions enhances MQTT's performance, redundancy, and scalability in IoT systems by leveraging Kubernetes orchestration and clustered broker setups.

5 Discussion

In this study, we set out to enhance IoT communications on edge platforms by analysing the MQTT protocol, architectural considerations, and broker selection. Our focus on message brokers was guided by performance evaluations centred around the key quality attributes of throughput and latency within MQTT-based IoT architectures. Additionally, we explored the role of orchestration tools like Kubernetes in optimizing message broker performance within distributed systems.

As described in Sect. 3, our experimental analysis primarily concentrated on benchmarking the NATS message broker. The detailed results, presented in Tables 1 and 2, offer insights into metrics such as the number of messages sent, average message rate, and message size, providing a nuanced understanding of throughput for individual publishers and subscribers. Specifically, our tests on the NATS broker with 8 publishers and 4 subscribers, using message sizes of 2 kB and 20 kB, yield valuable findings for applications with similar configurations. However, this setup may not fully capture performance characteristics for larger-scale implementations with an extended number of clients.

Our focus throughout this study was on the MQTT protocol, which is widely recognized as a suitable choice for IoT environments due to its lightweight design. While our work did not assess protocol alternatives, as such considerations fell outside the scope of this work, discussions on protocol selection are well-covered in works such as [25]. Instead, our analysis concentrated on choosing an appropriate message broker, evaluating both MQTT-native and MQTT-compatible options. This narrowed focus provided clarity in broker selection but may limit some alternative solutions. It is also important to note that this work was conducted within a company context, where the decision to use the MQTT protocol and specific brokers had been made prior to our involvement. Nonetheless, selecting a manageable and efficient broker like NATS offers considerable benefits, particularly for software architects seeking ease of deployment and flexibility across varied applications where setup simplicity is essential.

Our findings also suggest that MQTT, when combined with an event-driven architecture and pub/sub design pattern, is effective in supporting real-time communication and message handling in IoT systems. While using MQTT as the primary messaging protocol may restrict some architectural flexibility, adopting this protocol along with a pub/sub design pattern can significantly improve the efficiency of real-time request-reply interactions in IoT communications.

Additionally, Kubernetes plays a crucial role in managing brokers within an IoT edge platform. Its capabilities extend beyond basic management; Kubernetes enables clustering of message brokers, allowing flexible allocation and failover support. In scenarios where one or more brokers fail, Kubernetes can automatically reassign responsibilities among the remaining brokers, ensuring system continuity. As discussed in Sect. 4, this clustering feature helps maintain the functionality of other application components even if specific brokers experience failures.

In conclusion, our proposed solution integrates the MQTT protocol with an event-driven architecture and a pub/sub design pattern within a clustered Kubernetes environment, where each node hosts a dedicated message broker. Based on our findings, we recommend NATS as the most suitable message broker due to its compatibility with MQTT and its strong performance in applications characterized by a moderate number of publishers and subscribers and message sizes between 2 kB and 20 kB. This configuration offers a viable model for system architects designing similar IoT edge platforms that require scalability, reliability, and real-time communication.

6 Related Work

In this study, our goal was to propose an efficient and reliable message broker architecture for IoT applications, specifically identifying a broker compatible with an IoT platform that utilizes the MQTT protocol for message queuing services. Previous research has proposed various architectures for client-broker communication. For example, [21] outlines two architectures: one where brokers broadcast messages to all clients and another where clients subscribe to specific subjects, allowing publisher clients to share information only with relevant subscribers. Several studies, including [19] and [26], present the basic MQTT architecture consisting of publishers, subscribers, and a broker. In this setup, publishers send messages to the broker, which then delivers these messages to subscribers based on their subscribed topics. This architecture aligns with our approach, where we adopt a pub/sub design pattern with a central broker that facilitates communication between publisher and subscriber clients.

Additionally, message brokers vary from native MQTT solutions, such as HiveMQ [16] and VerneMQ [15], to alternatives that are compatible with MQTT but not exclusively MQTT-based, like Amazon Kafka [10], RabbitMQ [17], and NATS [13]. Various studies have compared these brokers based on different quality attributes. For instance, [12] conducted experiments focusing on throughput and latency across three MQTT-based brokers, revealing that factors such as the underlying programming language affect performance. Erlang-based brokers, such as EMQX and VerneMQ, demonstrated superior performance compared to the Java-based HiveMQ in this study, indicating that EMQX and VerneMQ achieved better results in terms of both throughput and latency.

In this work, we propose NATS as the preferred message broker due to its MQTT compatibility, despite not being natively MQTT-based. According to our experiments, NATS achieved higher throughput and lower latency compared to HiveMQ, VerneMQ, and RabbitMQ. A study by [6] also evaluated MQTT brokers, finding that EMQX had the highest throughput, followed by VerneMQ and HiveMQ. However, our results with NATS demonstrated significantly higher throughput and lower latency than these MQTT-native brokers. Additionally, NATS exhibited an estimated average latency of approximately 238.6 ms per request, underscoring its suitability for high-performance IoT applications.

To ensure a robust evaluation of message brokers, we followed a structured benchmarking approach, building on the results of studies by [12] and [6]. These studies conducted comprehensive performance tests on HiveMQ and VerneMQ using a test infrastructure with Dual Intel Xeon E5-2640 v3 CPUs (32 threads), 128 GB RAM, and Gigabit connectivity, with StarlingX v3.0 on CentOS 7.6 supporting the virtualization environment. MZBench was employed for benchmarking. In comparison, our test environment consisted of a Linux Ubuntu 20.04.6 LTS Virtual Machine with a 16-core CPU, 128 GB RAM, and a 250 GB HDD. Docker version 26.0.0 was used, with JetStream [14] deployed to benchmark NATS performance.

Comparisons of broker architectures have revealed that protocol efficiency, as well as scalability, is crucial for high-performance IoT and automotive systems [27].

7 Conclusion and Future Work

This work enhanced IoT communication on edge platforms by analysing the MQTT protocol. Our findings indicate that an event-driven architecture, pub/sub design pattern, and NATS message broker are effective choices for MQTT-based IoT systems, achieving high throughput and low latency. Benchmarking results showed NATS handling over 1 million messages per second, with data transfer rates up to 1.22 GB/sec, highlighting its suitability for high-performance IoT applications. Additionally, Kubernetes was found to improve system scalability and availability by effectively managing message brokers.

For future work, expanding the scope to include additional transport protocols beyond MQTT could reveal more options for IoT communication. Investigating the newer MQTT version 5, which offers improved scalability and error handling, is also recommended for applications requiring enhanced performance in large-scale IoT systems.

References

1. Al-Awami, S.H., Al-Aty, M.M., Al-Najar, M.F.: Comparison of IoT architectures based on the seven essential characteristics. In: 2023 IEEE 3rd International

Maghreb Meeting of the Conference on Sciences and Techniques of Automatic Control and Computer Engineering (MI-STA). IEEE (2023)
2. Al-Fuqaha, A., Guizani, M., Mohammadi, M., Aledhari, M., Ayyash, M.: Internet of Things: a survey on enabling technologies, protocols, and applications. IEEE Commun. Surv. Tutorials, **17**(4) (2015)
3. Ansyah, A.S.S., et al.: MQTT broker performance comparison between AWS, Microsoft azure and google cloud platform. In: 2023 International Conference on Recent Trends in Electronics and Communication (ICRTEC), IEEE (2023)
4. Apukhtin, V., Shirokopetleva, M., Skovorodnikova, V.: The relevance of using message brokers in robust enterprise applications. In: IEEE International Scientific-Practical Conference Problems of Infocommunications, p. 2019. IEEE, Science and Technology (PIC S&T) (2019)
5. Bucaioni, A., Pelliccione, P., Mubeen, S.: Modelling centralised automotive e/e software architectures. Adv. Eng. Inform. **59**, 102289 (2024)
6. Gruener, S., Koziolek, H., Rückert, J.: 2021 IEEE 18th International Conference on Software Architecture (ICSA), pp. 69–79 (2021)
7. Creswell, J.W., Plano Clark, V.L.: Designing and Conducting Mixed Methods Research. Sage Publications, Thousand Oaks (2017)
8. Bucaioni, A., Di Salle, A., Iovino, L., Malavolta, I., Pelliccione, P.: Reference architectures modelling and compliance checking. Softw. Syst. Model. **22**(3), 891–917 (2023)
9. Mishra, B., Mishra, B., Kertesz, A.: Stress-testing MQTT brokers: a comparative analysis of performance measurements. Energies, **14**(18) (2021). https://www.mdpi.com/1996-1073/14/18/5817
10. "Apache Kafka." https://kafka.apache.org/
11. "Gcore Compare." https://gcore.com/learning/nats-rabbitmq-nsq-kafka-comparison/
12. Koziolek, H., Grüner, S., Rückert, J.: A comparison of MQTT brokers for distributed IoT edge computing. In: Jansen, A., Malavolta, I., Muccini, H., Ozkaya, I., Zimmermann, O. (eds.) ECSA 2020. LNCS, vol. 12292, pp. 352–368. Springer, Cham (2020). https://doi.org/10.1007/978-3-030-58923-3_23
13. "NATS." https://nats.io/
14. "Jetstream benchmark." https://docs.nats.io/nats-concepts/jetstream/
15. "VerneMQ." https://vernemq.com/
16. "HiveMQ." https://www.hivemq.com/
17. "RabbitMQ." https://www.rabbitmq.com/
18. "Confluent.io." https://developer.confluent.io/learn/kafka-performance/
19. Johari, R., Bansal, S., Gupta, K.: Routing in IoT using MQTT protocol. In: 2020 12th International Conference on Computational Intelligence and Communication Networks (CICN). IEEE (2020)
20. Shahri, E., Pedreiras, P., Almeida, L.: Enhancing MQTT with real-time and reliable communication services. In: 2021 IEEE 19th International Conference on Industrial Informatics (INDIN), pp. 1–6 (2021)
21. Kawaguchi, R., Bandai, M.: Edge based MQTT broker architecture for geographical IoT applications. In: 2020 International Conference on Information Networking (ICOIN). IEEE (2020)
22. Event-driven architecture, wikipedia. https://en.wikipedia.org/wiki/Event-driven_architecture/
23. "redhat." https://www.redhat.comentopicscontainerswhat-is-kubernetespod#:~:text=A%20Kubernetes%20pod%20is%20a,a%20more%20common%20use%20case)

24. Kebbani, N., Tylenda, P., McKendrick, R.: The Kubernetes Bible: The Definitive Guide to Deploying and Managing Kubernetes Across Major Cloud Platforms (2022)
25. Naik, N.: Choice of effective messaging protocols for IoT systems: MQTT, COAP, AMQP and HTTP. Defence Sch. Commun. Inf. Syst. Year of publication
26. Jukić, O., Filjar, R., Heđi, I., Ciriković, E.: MQTT-like network management architecture. In: 44th International Convention on Information, p. 2021. IEEE, Communication and Electronic Technology (MIPRO) (2021)
27. Kukulicic, N., Samardzic, D., Bucaioni, A., Mubeen, S.: Automotive service-oriented architectures: a systematic mapping study. In: 2022 48th Euromicro Conference on Software Engineering and Advanced Applications (SEAA), pp. 459–466. IEEE (2022)

Virtual Reality Fire Drill for Campus Evacuation

Danah Omary[1](✉), Sai Karthik Gamineedi[2], Adnan Shareef[2], and Sharad Sharma[2]

[1] Electrical Engineering, University of North Texas, 3940 N Elm St, Denton, TX 76207, USA
danahomary@my.unt.edu

[2] University of North Texas, 3940 N Elm St, Denton, TX 76207, USA

Abstract. Evacuation drills are needed for campus safety. There is a need to prepare for fire drills on campus as emergencies cannot be predicted. Virtual reality (VR) offers an experimental set-up to conduct fire evacuation drills in a collaborative VR environment. This paper presents a VR simulation of a campus fire evacuation scenario to educate and train individuals on proper evacuation procedures. The simulation includes a realistic virtual environment of the campus building and surrounding areas, interactive elements to guide users through the evacuation process, and AI-controlled characters to simulate various behaviors and responses. The simulation provides feedback and evaluation mechanisms to assess users' performance and understanding of evacuation procedures. The implementation of the VR campus fire evacuation environment was done using Unity 3D gaming engine. The VR environment incorporates detailed modeling, sound effects, animations, interactivity, and AI functionality.

Keywords: XR · VR · Simulation · Training · Evacuation Drill · Educational Training · 3D Modeling · GUI design · Unity · Interaction

1 Introduction

College institutions have a need to prepare for various emergency scenarios in an effective manner. One of the significant emergency situations universities should prepare and train for is for evacuations when there are fire hazards that the campus occupants must escape. However, training people with just the campus buildings and general escape protocol does not give a realistic sense of the fire hazard. It is also very difficult to include realistic fires without actually having fires, which would risk the safeties of the participants of training exercises. Another limitation in implementing evacuation training exercises is that there are large amounts of involved parties of various roles that should be trained to handle the evacuation scenario, not just emergency responders. However, even just basic training for all the people involved would be difficult to implement

S. Latifi (Ed.): ITNG 2025, AISC 1463, pp. 568–579, 2025.
https://doi.org/10.1007/978-3-031-89063-5_49

on a large scale. One possible way to handle training to account for all of these existing limitations is by leveraging Virtual Reality (VR) to make a realistic simulation for college campus fire evacuation training.

Using VR for campus fire evacuation training can provide many advantages. Firstly, VR can provide a safe mechanism to simulate various aspects of the hazards of fires without risking the safeties of participants. Effective representation of the fire hazards will allow for people to be familiar with how fires would be seen and experienced along with the actual risks of fire hazards rather than just the evacuation procedure. However, the VR application can still provide the evacuation protocol in an effective manner as well. To increase the realism of the evacuation, a realistic and accurate model of the campus environment should be developed and used for the VR simulation. With realistic, immersive and high quality simulation of a fire evacuation scenario, VR can be used to effectively familiarize and train users on how to not only evacuate but to also handle the fire. In the case that a user does get virtually harmed, they could also be trained on what actions to take when some they or anyone else in the evacuation need medical attention.

Another advantage of using VR for campus fire evacuation training is that development and deployment of the application would be of much lower cost than organizing in person training drills, especially over several years. An application could also much more easily be modified and adapted than a training drill. In addition to saving users from having to wait for scheduled, formal training, using an application - especially a mobile application - would enable users to train whenever they would like to. Due to the application-based nature of the training, the college would still have the option to offer both the application for use at any time and official, planned instruction. Another advantage of application-based training is that the VR application can integrate user feedback forms and tools so that the institution may also more quickly assess the training's efficacy and experience and obtain feedback.

Another advantage of using a VR application for fire evacuation is that it makes training on a greater scale possible. Training a large number of people using a downloadable application is far simpler than training them all in person. VR also has a greater capacity to accommodate various types of participant roles and their responsibilities. For fire evacuations, the primary roles that should be accounted for are evacuees and emergency responders. Within those roles, there are many audiences that could be targeted by a VR campus fire evacuation training application. Firstly, because they are often present on campuses and for extended periods of time, students, staff, and faculty at educational institutions should receive training on handling fire evacuations. They should be equipped to face the scenario of having to evacuate the establishments they frequently visit for their own safety and well-being. Even visitors and university guests should also have the opportunity to get familiar with evacuation protocol in the unlikely scenario that an evacuation is necessary. Secondly, university security personnel in particular should be trained on how to oversee the evacuation of all building occupants and ensure that they are all safe and receive medical attention if they are injured. Lastly, the emergency responders who work with colleges should also be trained in managing the evacuation's aftermath as well as how to put out fires and deal with any damage or emergencies they may have caused. Emergency responders that have received training should work more effectively.

In this paper, we present a VR application we developed for fire evacuation training for our campus environment specifically. For this work, we have developed a high-quality simulation of a fire and its risks and hazards, modeled our campus building and its interior furnishings and surroundings accurately and realistically, developed an evacuation protocol with guiding Graphical User Interface (GUI) elements, and used Artificial Intelligence (AI) controlled agents to simulate a user guiding evacuees to safety and providing them with medical attention.

The rest of the paper is organized as follows. Section 2 for evacuation simulations. Section 3 discusses and demonstrates how we implemented the realistic campus modeling and VR simulation application, including the evacuation protocol and the design of the GUI. Lastly, Sect. 4 presents our conclusion.

2 Related Work

This section presents recent studies and advancements in the area of VR fire evacuation simulations, emphasizing the contributions, approaches, and results of these studies. This section offers insights into the development of VR-based fire evacuation training and highlights significant trends and advances by using a wide range of research. Using an evolutionary virtual reality platform, Lorusso et al. demonstrated a revolutionary method to fire emergency evacuation, demonstrating how evolutionary algorithms may optimize evacuation techniques inside a virtual setting [5]. Their work aids in the creation of cutting-edge virtual reality instruments that enhance safety precautions and evacuation procedures. With the use of an online serious game, Yang et al. investigated virtual fire evacuation exercises and provided a novel viewpoint on the use of web-based platforms for fire safety instruction [10]. Their research highlights the adaptability of VR technology in providing educational material by using gamification aspects that improve user engagement and learning results. A mixed reality-based fire evacuation practice system was presented by Sakaguchi et al., combining components of augmented and virtual reality to improve training [6]. Their hybrid method offers a complete disaster preparation training solution by fusing the immersive experiences of VR with the contextual information of Augmented Reality (AR).

The efficacy of evacuation procedures was the main focus of simulation-based evaluations of fire emergency preparation and response carried out in virtual reality by Kwegyir-Afful et al. [2]. Through its ability to improve safety precautions, their research highlights the value of using VR simulations to evaluate and improve emergency response plans. To emphasize the need of cooperation and coordination in emergency response situations, Sharma et al. suggested a collaborative virtual reality environment for building evacuation training [7]. The research improves communication and cooperation during evacuation exercises, resulting in more efficient emergency responses, by enabling user participation inside a shared virtual reality environment. Virtual reality was first used in fire evacuation research by Kinateder et al., who emphasized the technology's ability to produce lifelike simulations that may be used to examine how people behave in emergency situations [4]. The virtual reality system for fire evacuation training was presented by Sookhanaphibarn et al. It allows users immersive training experiences to improve their grasp of emergency response methods and evacuation processes in a 3D virtual environment[9]. The value of realistic training scenarios and collaborative learning experiences were pointed out by Sharma et al. when they built an immersive collaborative virtual environment of a university campus for conducting virtual campus evacuation drills and tours for campus safety[8]. A fire evacuation and management model based on virtual reality and building information modeling was introduced by Gao et al. It integrates cutting-edge simulation methods to simulate fire incidents and optimize evacuation routes in intricate building environments[3]. Cal et al. studied AR's application in evacuation training games. They explored how AR optimized evacuation training participant engagement and learning. The researchers added AR strengths to the training game to make it more immersive and engaging, help users realize evacuation and emergency response processes[1]. Their research established the foundation for further investigations by proving that virtual reality can accurately simulate evacuation situations and evaluate evacuation plans. From modeling evacuation situations to improving evacuation techniques and increasing user engagement, these studies together show the many uses of virtual reality in fire evacuation training. Researchers are making communities safer and more resilient by using VR technology to further their innovative work in the area of preparing for emergencies.

3 Implementation

In this work, we specifically integrated a model of our campus building to provide a sense of realism and familiarity to the users. Using realistic modeling for the virtual environment will better prepare the users we train to handle evacuations as they will be practicing evacuating with a model similar to the building they frequently occupy. The implementation of an extensive protocol for handling the evacuation also helps increase the realism of the VR training simulation by emulating essential elements the experience of evacuating involves. Lastly, we developed a robust GUI for the user to be well engaged when in the simulation. Each of these aspects will be covered in detail in the following subsections.

3.1 Modeling

The virtual environment of our simulation, shown in Fig. 1, includes realistic, textured models of the engineering campus layout with customized, furnished building interiors, terrain and trees, surrounding areas, evacuation routes and fire hazards (Fig. 2).

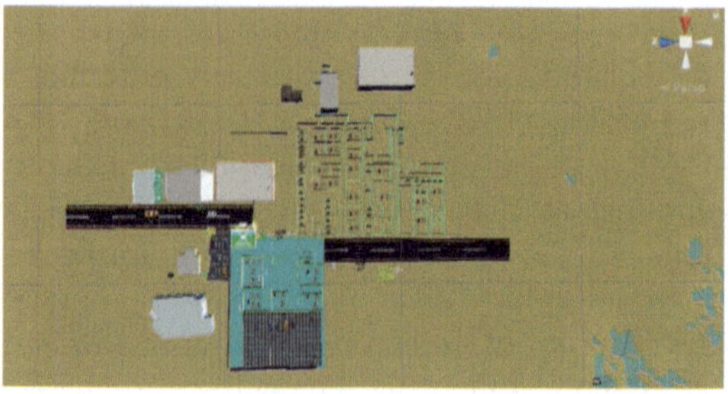

Fig. 1. Full scene model.

We converted a 2D Drawing (DWG) model of our university's engineering campus, which is named Discovery Park, that included classrooms, offices, walkways and more. Using Sketchup, we extruded the faces in order to create a 3D model of the campus layout for our virtual environment. The model was then imported into Unity for use in our VR application. Within Unity, to increase the sense of realism and immersion, we furnished the campus layout environment. We imported and placed assets of a variety of models for different types of items typically found in a college campus such as desks, chairs, computers, briefcases and more. By including furniture, obstructions, and moving parts, the simulated setting becomes more complicated and replicates the difficulties of a real-world evacuation. In addition to the modeling of the campus buildings, our virtual environment modeling also includes surrounding areas such as parking lots, other buildings and transportation roads. The target evacuation area is also within the surrounding area.

3.2 Evacuation Protocol

The objective of the evacuation is to get to the evacuation area goal without either the player or AI agents losing all health, which virtually would represent dying. The AI agents in the simulation represent other people in evacuation scenarios as evacuations are rarely done in isolation or with only one person. The AI agents specifically follow the user, so the user is serving as an evacuation guide in the implemented protocol. This trains the user to get familiar with guiding others in fire evacuation emergencies rather than just escaping by themselves, giving motivation to users to learn how to lead and take consideration to other people and their safety (Fig. 3).

The evacuation protocol and the fires do not immediately begin once the user starts the application. Rather, the user has the option of either pressing a keyboard key or a GUI button specifically for starting the drill. However, the user can still move around the environment with either the joysticks or the keyboard movement keys (W, A, S, D) both before the drill and during the drill. Once the user triggers the drill to start, several things occur to indicate the drill is active. Firstly and most noticeably, a fire alarm sound plays and continues as long as

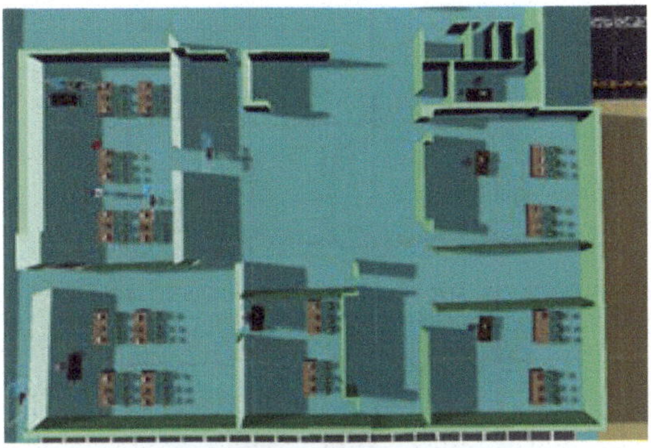

(a) Virtual Campus Building 1.

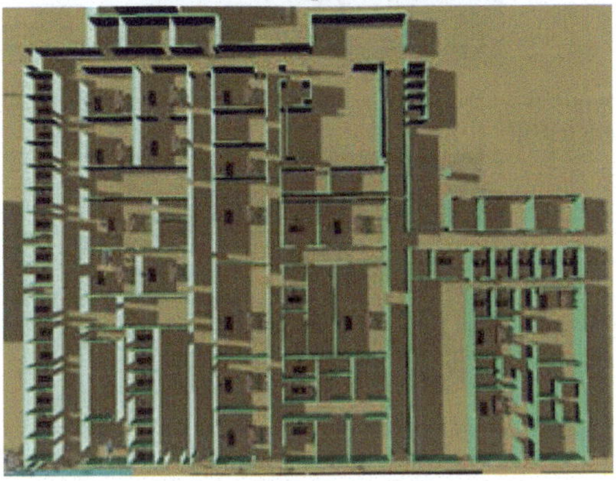

(b) Virtual Campus Building 2.

Fig. 2. Campus 3D Model.

the drill is active. Secondly, fires become "active" in that they are not only visible but also cause damage to the user and AI agents when entered. Lastly, GUI elements specific to the evacuation procedure are displayed, one of which is a stopwatch that only starts after the evacuation is triggered. Additionally, though not analogous to real life evacuations but necessary for simulation user experience, the evacuation can be paused. The evacuation continues until the user reaches the evacuation area goal.

Once the user reaches the evacuation goal area, the health statuses of both the users and the AI agents are checked. Firstly, they are checked to see if anyone lost all health as that represents dying and is an unacceptable result as that would be analogous to letting someone die in evacuations in real life. The user gets informed of failing the training through both a failure sound as well as a displayed message. In the other case that everyone still has some health left after completing the evacuation, the amount of health all parties have is checked. If anyone at all lost any health, medical help must be called for. In that situation, a GUI button to call emergency services becomes visible and the user is informed

(a) Initial VR simulation environment state.

(b) VR simulation environment in the middle of the evacuation drill.

(c) VR simulation environment when the user and AI agent have reached the evacuation target area.

Fig. 3. VR simulation environment at different evacuation drill states.

to call emergency services. After the user presses the button or an assigned keyboard key, then a call sound plays and an ambulance near the evacuation goal area becomes visible. This is analogous to how in real evacuations, medical first responders need to actively be called in order for them to come; they are not just immediately present after evacuating.

3.3 Graphical User Interface and Interactions

In order to allow effective user agency in the training, a GUI must be designed to display everything the user needs to know as well include sufficient interfacing to interact with the simulation components. Thus, we created a robust GUI to properly be able to simulate fire evacuation training. The GUI should also be dynamic and reflect the current state of the simulation and the user's status rather than just include a fixed set of interface elements. However, there are some GUI elements that are necessary to keep throughout all evacuation stages, though their contents and values might change. In our GUI design, the GUI elements that are present throughout include a status panel, a status display message, an exit button, and some joysticks for movement and looking around. Especially for a high-emotion and stress environment of a fire emergency, the GUI should include guidance for the user as the danger could cloud their judgement and lead them to panic instead of acting. The status panel should be present throughout the simulation in order to help users know what actions to take, not only for the simulation but also to be trained on how to handle actual fire evacuations at various points in the process. Though not analogous to actual fire evacuations, having a way to exit the simulation at any instance is essential, so the exit button is available for the entire duration of the simulation. Lastly, the user should be able to move and look around at any point in the simulation just like they would be able to in real life, so the joysticks are also available throughout the simulation. The joysticks also allow for movement and looking otherwise not possible on mobile applications, which can be more widely distributed than personal computer (PC) applications for training purposes. Consideration for both PC and mobile application roll-out has been done throughout our GUI design by assigning actions for multiple different possible input devices such as PC keyboards and computer mouses as well as mobile touchscreens.

(a) Default closed state of the door.

(b) The door being open after the user moved into close proximity to it.

Fig. 4. Door interaction.

In addition to the GUI buttons, there are 2 other main types of objects that can be interacted with: the doors and the fires. As shown in Fig. 4, when the user's avatar gets close enough to any of the doors in the VR environment, the door plays an opening animation and plays a door opening sound when the user enters. This is analagous to how users will need to open doors when there is a fire evacuation as it cannot be assumed there is a completely empty path towards the area to evacuate to in the case of actual emergency situations.

Figure 5 shows the GUI in the state before the drill starts. Though there is no fire or anything to cause damage before the drill starts, the health status within a GUI panel is still displayed to the user to show that they start off with full health. Instructions are displayed in the status panel on what inputs should be used to start the drill either through keyboard input or GUI buttons, and to start the drill there is a dynamic GUI button dedicated to starting the drill (Fig. 6).

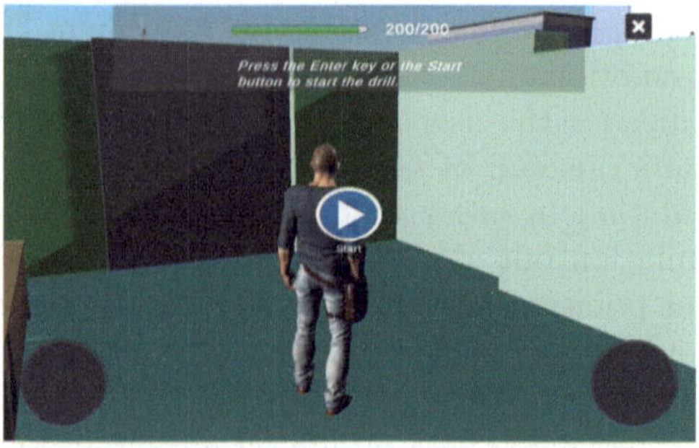

Fig. 5. The GUI elements before the drill is started.

Fig. 6. The GUI during the drill.

During the drill itself, there are several changes to the GUI. Some of the contents are changed for the constant GUI elements, such as displaying a different instruction message. The user is no longer waiting for the drill to start, so the message was changed to indicate that the evacuation has started and that the user needs to go to the evacuation area. Another change is that previous GUI elements that were available before become unavailable as they are no longer needed, such as the start button as the drill is already happening and the user no longer needs to be able to start the drill. Lastly, new GUI elements become available. Firstly, as there is time-sensitive environment activity and procedure happening, a pause button becomes available as a user should be able to pause in the case that something briefly needs their attention and they need to be able to get back to their current level of progress in the evacuation after addressing it. Additionally, a stopwatch icon and a stopwatch time display are added in order for the user to know how long they are taking to complete the evacuation.

Including a stopwatch display for the user specifically is beneficial for evacuation training for several reasons. Firstly, the fires themselves, and thus their hazard, could grow as time moves forward. Growth of fire could also cause more damage to both the area and its occupants. If the user gets harmed through fire, they could also lose their ability to move, especially their ability to move quickly, which thus impairs their ability to evacuate as well as lead others to evacuate. It is thus very important for the users to feel the urgency of the fire emergency and try to escape as quickly as possible. However, a user should not be pressured to rush as it would risk not only their safety, but also the safeties of those following them, represented in simulation through the health of the AI agents following them. Including a stopwatch display that shows how much time has passed keeps users aware of how long they are taking which should help them balance the need to evacuate quickly while also proceeding with caution to not unnecessarily risk their own safety or the safeties of those being guided to evacuate (Fig. 7).

Fig. 7. Fire damage interaction.

Regarding the fires, as long as the user or an AI agent is in any of the fires, the user takes damage and a fire sound is also played. The behavior continues until the user is sufficiently away from the fire. This is similar to how people can hear and feel fire in real life when close to it. The combination of the visuals, sound, and damage of the fire in the VR simulation makes the hazard of fire more tangible to the user to take it more seriously and gives a more immersive experience. Whenever the user is harmed, not only does the health bar visual decrease, the value decreases. Whenever a user gets harmed from actual fire, due to the nature of fire damage, the user will require medical assistance and treatment. Actual harm from fire cannot be left untreated. In order to emphasize the urgent need for medical attention, as soon as the user gets damaged from fire, a representative health icon starts flashing to indicate the need for medical attention and motivate the user to move towards the evacuation goal area to be able to call for help to get medical treatment.

In the case that evacuation was done successfully in that both the user and the AI agents received no damage to their health, the a message display as well as a sound indicating success will be shown, as seen in Fig. 8. As no one needed medical attention due to no loss in health, the GUI button to call for help would not be available (Fig. 9).

In the case that either the user, the AI agents or both were harmed once everyone reached the evacuation goal area, an emergency phone call button

Fig. 8. GUI when drill is complete.

(a) GUI after the drill in the case that someone needs medical assistance.

(b) Drill complete, ambulance.

Fig. 9. GUI in both scenarios after the drill.

becomes available. The icon indicating medical assistance is needed still flashes until the ambulance "arrives," i.e. until the ambulance model becomes visible.

In both the case of successful evacuation and evacuation requiring medical assistance at the end, the final values of the stopwatch and user health are left for the user to see and evaluate their simulated evacuation performance.

4 Conclusion

In conclusion, using VR to simulate fire evacuations can be more effective than physical, in person training. VR allows for risk-free immersion that is not possible with in person training and also can help people become more familiar with the fire evacuation experience how to handle university campus fire evacuations. In this work, we successfully implemented a VR application for campus fire evacuation with realistic modeling and immersive, high-quality user experience that has an extensive protocol and robust interactions.

Acknowledgements. The authors wish to thank the other classmates who contributed on the initial class project this was based upon.

References

1. Catal, C., Akbulut, A., Tunali, B., Ulug, E., Ozturk, E.: Evaluation of augmented reality technology for the design of an evacuation training game. Virtual Reality **24**, 359–368 (2020)
2. Ebo Kwegyir-Afful, T.O.H., Kantola, J.I.: Simulation-based assessments of fire emergency preparedness and response in virtual reality. Int. J. Occup. Saf. Ergon. **28**(2), 1316–1330 (2022). https://doi.org/10.1080/10803548.2021.1891395. PMID: 33591217
3. Gao, R., Wang, K., Lai, X., Hung, W.: Fire evacuation and management model based on building information modeling and virtual reality. In: ISARC. Proceedings of the International Symposium on Automation and Robotics in Construction, vol. 39, pp. 291–296. IAARC Publications (2022)
4. Kinateder, M., et al.: Virtual reality for fire evacuation research. In: 2014 Federated Conference on Computer Science and Information Systems, pp. 313–321 (2014). https://doi.org/10.15439/2014F94
5. Lorusso, P., De Iuliis, M., Marasco, S., Domaneschi, M., Cimellaro, G.P., Villa, V.: Fire emergency evacuation from a school building using an evolutionary virtual reality platform. Buildings **12**(2) (2022). https://doi.org/10.3390/buildings12020223
6. Sakaguchi, S., Makino, M.: A mixed reality-based fire evacuation drill system. In: Nakajima, M., Kim, J.G., Lie, W.N., Kemao, Q. (eds.) International Workshop on Advanced Imaging Technology (IWAIT) 2021, vol. 11766, p. 1176620. International Society for Optics and Photonics, SPIE (2021). https://doi.org/10.1117/12.2591015
7. Sharma, S.: Improving emergency response training and decision making using a collaborative virtual reality environment for building evacuation. In: Stephanidis, C., Chen, J., Fragomeni, G. (eds.) HCI International 2020 - Late Breaking Papers: Virtual and Augmented Reality, pp. 213–224. Springer, Cham (2020)
8. Sharma, S., Rajeev, S.P., Devearux, P.: An immersive collaborative virtual environment of a university campus for performing virtual campus evacuation drills and tours for campus safety. In: 2015 International Conference on Collaboration Technologies and Systems (CTS), pp. 84–89 (2015). https://doi.org/10.1109/CTS.2015.7210404
9. Sookhanaphibarn, K., Choensawat, W., Paliyawan, P., Thawonmas, R.: Virtual reality system for fire evacuation training in a 3D virtual world. In: 2016 IEEE 5th Global Conference on Consumer Electronics, pp. 1–2 (2016). https://doi.org/10.1109/GCCE.2016.7800430
10. Yang, Y., Xu, Z., Wu, Y., Wei, W., Song, R.: Virtual fire evacuation drills through a web-based serious game. Appl. Sci. **11**(23) (2021). https://doi.org/10.3390/app112311284. https://www.mdpi.com/2076-3417/11/23/11284

IoT and Networking

Towards Model-Based Assessment of Trustworthiness in Autonomous Cyber-Physical Production Systems

Maryam Zahid[1(✉)], Alessio Bucaioni[2], and Francesco Flammini[1]

[1] School of Innovation, Design and Engineering, Division of Product Realisation, Mälardalens Universitet, Västerås, Sweden
{maryam.zahid,francesco.flammini}@mdu.se

[2] School of Innovation, Design and Engineering, Division of Computer Science and Software Engineering, Mälardalens Universitet, Västerås, Sweden
alessio.bucaioni@mdu.se

Abstract. The latest industrial revolution has introduced autonomous cyber-physical production systems, integrating machine learning into smart manufacturing to optimize production and resource management. However, this integration impacts trustworthiness due to less predictable and explainable behaviors. This paper presents a novel model-based methodology for evaluating the trustworthiness of such systems. A study was conducted to explore the potential and limitations of model-based assessment, categorizing limitations into structural, behavioral, and resource-related aspects. The findings highlight inadequate risk identification and assessment of ML components in these systems and the constraints of single modeling approaches. Based on these insights, we propose a new methodology to address these limitations and improve the risk assessment of ML components in autonomous production systems.

Keywords: Model-based risk assessment · Limitations · Machine Learning · ISO/IEC 61508 Standard · ISO 31010 Standard · Autonomous Cyber Physical Production Systems · Autonomous Cyber Physical Manufacturing Systems · Architecture · Software System Behavior · Trustworthiness

1 Introduction

Autonomous Cyber-Physical Production Systems (ACPPS) is a product of having artificial intelligence (AI) based solutions such as the machine learning (ML) algorithms integrated into the cyber layer of the traditional cyber-physical production/manufacturing systems (CPPS/CPMS), or the involvement of Collaborative Robots (Cobots) [6], making it intelligent and independent [23].

A. Bucaioni and F. Flammini—These authors contributed equally to this work.

S. Latifi (Ed.): ITNG 2025, AISC 1463, pp. 583–594, 2025.
https://doi.org/10.1007/978-3-031-89063-5_50

Utilizing ML algorithms, ACPPS autonomously refines its predictive models, optimizing decision-making for parameters, and conducts thorough root cause analysis on detected anomalies [20].

Although attractive, such autonomy in ACPPS also raises concerns regarding its trustworthiness [13]. According to a taxonomy, the word trustworthiness here can be defined as a collection of attributes such as robustness, data governance, lawful, ethics, human agency and oversight, and societal and environmental well-being [14]. The Artificial Intelligence Act (AI-ACT) proposed by the European Union classifies AI-based systems by the defined risks and, therefore, mandates the implementation of trustworthiness-related requirements [32]. The scope of such requirements spans the entire ACPPS where, e.g., robustness is crucial for maintaining the system's overall stability, efficient and safe decision-making requires trust, human agency, and oversight is essential to gain trust among its human operators [33]. Moreover, for a product such as vehicles, medical equipment or even cobots to be marketed, it must undergo an extensive risk assessment process per the International Standardization Organization (ISO) standards. Model-based risk assessment (MBRA) techniques have been preferred for detection, and accurate and comprehensive assessment of risks at an early stage of the development life-cycle [29]. However, existing techniques focus on general/traditional CPS or evaluate specific risks related to single trustworthiness attributes [35]. Moreover, since ACPPS is a new concept introduced, no studies have been conducted on assessing trustworthiness-related risks, especially the ones originating from the dynamic behavior of the ML component of ACPPS.

As a main contribution, this paper presents an overview of the proposed risk assessment methodology titled *"Model-based Assessment for Trustworthiness of Industrial Cyber-physical Systems"* (MATrICS), aiming to identify and evaluate trustworthiness-related risks originating from the dynamic behavior of ML component in ACPPS [25]. To develop the new methodology, we conducted a preliminary literature review of the existing MBRA techniques used to assess trustworthiness-related risks in ACPPS and their limitations. The limitation identified became the basis for the developed methodology.

The rest of the paper is structured as follows. Section 2 presents a preliminary review of the existing state-of-the-art MBRA techniques and their limitations, and Sect. 2.1 discusses the adopted research process. Section 3 presents the new methodology proposed to overcome the limitations highlighted in the preliminary review. Finally, Sect. 4 summarizes the study's findings and presents plans for future work.

2 Preliminary Review

This section discusses the review process and presents the findings of a preliminary literature review on the model-based techniques used to assess the trustworthiness-related risks in ACPPS and their reported limitations. To have a discussion, the techniques studied were categorized as shown below for presentation purposes and, therefore, are not semantic in any way.

2.1 Preliminary Review Process

The MATrICS methodology was developed using a cyclic research process following Crnkovic's guidelines [7].

It began with a systematic literature review of MBRA techniques for assessing trustworthiness-related risks in ACPPS, aiming to identify their limitations concerning ML components. The preliminary studies followed Kitchenham et al.'s guidelines [19], using a search string with five parts to include all possible term synonyms and ensure comprehensive coverage, as follows: *("Cyber-Physical System" OR CPS) AND ("Manufacturing" OR "Industry 4" OR "Production") OR (CPPS OR CPMS) AND ("Smart" OR "Artificial Intelligent" OR "Self-Sustain" OR "Autonomous") AND ("Safety" OR "Security" OR "Trust" OR "Dependability" OR "Resilience" OR "Robust" OR "Self-Heal" OR "Self-Repair") AND ("Risk") AND ("Model-Based") AND ("Assessment" OR "Analysis").* The search was conducted on four major databases in software engineering and computer science [19], with details and findings available in Sect. 2.1 and the replication package[1]. The initial search yielded 959 articles, but only peer-reviewed, English-language studies focusing on model-based risk techniques for trustworthiness in ACPPS ML components were included. Studies addressing MBRA limitations were prioritized, while tertiary studies were excluded, resulting in a final set of 44 relevant articles. A data extraction form was created to address the research objective. Extracted data was analyzed, stored in the form, and clustered into two groups: MBRA techniques (Sect. 2.2) and their limitations (Sect. 2.3). Relevant information not initially captured was reviewed and, if necessary, added to the form after re-analysis against the updated requirements.

During the problem formulation step, the literature review data was analyzed and synthesized following Crtuzes et al.'s guidelines [8]. Initially, we identified techniques for assessing trustworthiness-related risks in ACPPS, evaluating their coverage, benefits, and challenges. Vertical and orthogonal analyses were conducted to understand the relationship between the techniques and their characteristics. ACPPS, like traditional CPPS, includes three layers: cyber, network, and physical. However, ACPPS differs due to the ML component in its cyber layer. Findings revealed vulnerabilities from integrating heterogeneous components without proper configuration and risk mitigation, particularly with software patches. Behavioral and structural models were commonly used for risk assessment. Further analysis highlighted that while ACPPS systems rely on intelligent components, risks associated with the ML component and its integration with other ACPPS elements are largely overlooked. These insights informed the formulated problem: *Existing solutions proposed lacked focus on the trustworthiness-related risks, possibly resulting due to the dynamic nature of the machine learning component of ACPPS.*

In the solution proposal step, an iterative approach was used to develop the solution, starting with defining requirements to address the formulated problem. This involved further analysis and a survey of ISO recommendations for risk management in ACPPS, ensuring the proposed solution aligned with these standards. Due to space constraints, process details are omitted. Using the survey data as a foundation, existing techniques were re-examined to identify those

[1] https://github.com/PhDResearcher/ACM2025.

meeting the defined requirements and ISO standards. Based on this analysis, the proposed solution, the MATrICS methodology, is detailed in Sect. 3.

2.2 Model-Based Risk Assessment Techniques

Integrating digital twins in production sections has introduced many issues as they are heavily reliant on data and, therefore, prone to attacks. Based on the review results, identified MBRA techniques used can be clustered into two main categories, namely: *quantitative techniques* providing numerically the probability of occurrence and the severity of the identified risk, and *qualitative techniques* highlighting the risks along with their propagation paths, control strategies and seriousness in terms of high, medium, and low [37]. Among the most reported MBRA techniques, Attack Trees, Fault Tree Analysis (FTA), Event Tree Analysis (ETA), Probability Risk Assessment (PRA), stochastic modeling, Graphs, Decision Trees and Binary Trees, and Bayesian Networks (BN) were the quantitative techniques. Similarly, most reported qualitative techniques included the bow-tie Diagram (BTD). BTD, HAZOP, Petri Nets, and Markov Chain Model (MCM) [36].

In addition, model-checking techniques have been reported for verifying the ML models in ACPPS, ensuring they meet safety and performance requirements [11]. For instance, a model-checking approach was applied using an ML-based fault-diagnostic tool to verify reliability requirements in an automated hydraulic press [34]. Unlike MBRA techniques, which focus on identifying and managing risks [16], model-checking is used for formal system verification [12].

2.3 Limitations of Existing MBRA Techniques

Limitations reported against the proposed MBRA techniques can be categorized into three main types (see Fig. 1).

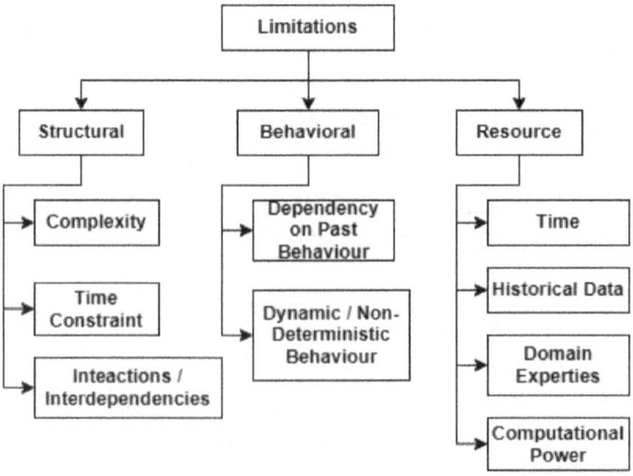

Fig. 1. Identified limitations of MBRA techniques

Structural limitations is associated with the ACPPS's architecture, including complex models with tightly coupled components, numerous interdependencies, and resource-intensive requirements. According to the findings obtained, such limitations are mainly due to the networked nature of ACPPS, where MBRA models fail to capture the tight interactions between the components within a single layer or between multiple layers of ACPPS [10]. This is evident in probabilistic models used for risk assessment, often overlooking real-time requirements and the heterogeneous nature of ACPPS, leading to incomplete risk assessments [11].

ML components' dynamic and periodically evolving nature poses additional challenges for the existing MBRA techniques. In the case of ML components within ACPPS, the traditional MBRA techniques may not account for continuous learning and adaptation capabilities, leading to risk evaluation gaps such as overlooking data quality risks, model bias, and adversarial attacks [22].

Scalability is another significant limitation. The increase in system size leads to an increase in the complexity of the probabilistic model, as evident with the use of probabilistic models. To manage this, studies often focus on a single MBRA technique or apply it to a specific layer of the ACPPS rather than the entire system [28].

Behavioral limitations are tied to the dynamic nature of systems, such as complex behaviors or dependency on past actions [9]. Influenced by concurrency, asynchrony, and distribution, ACPPS operates in variable environments [33]. The reliance of ACPPS on real-time adaptation, data processing, high-speed communication, autonomous decision-making, decentralized control, scalability, flexibility, and robust fault tolerance complicates implementing these requirements [22]. Failing to adhere to such dynamic behavior is why most MBRA techniques, like FTAs, ETAs, BN, Petri Nets, and HAZOP, provide incomplete or incorrect risk assessment of ML models in ACPPS. This can further result in poor decisions, affecting production. Complex behaviors, such as accident sequences or resilience requirements, complicate risk assessment model construction [1]. Studies often overlook device dependencies and complex network topologies in ACPPS [9]. Techniques treating cyber and physical layers separately fail to model risk propagation accurately. MBRAs are generally unsuitable for dynamic risks in ACPPS [15]. As systems grow, techniques like PNs struggle with concurrent processing [3]. Dependence on past behavior documentation limits current MBRA techniques' accuracy [27]. Assessment of Trustworthiness-related risks is even more complex due to rapid changes in the trust metric, complex interdependencies, lack of real-time modeling, overlooked interactions, scalability issues, and reliance on historical data. These limitations highlight the need for advanced methodologies to better address dynamic, interdependent, and real-time trustworthiness risks in ACPPS.

Resource-related limitations refer to the need for high computational power, extensive historical data, and specialized expertise [4]. MBRA techniques such as BN, FTA, ETA, and MCM's dependency on domain-specific historical data for model construction and risk calculations can be challenging and expensive due to privacy concerns and system size [38]. Real-time risk assessment demands significant computational power, often unavailable [27]. For example, continuously monitoring nodes to identify malicious activity increases complexity and requires high computational power and memory [2]. MBRA techniques (e.g., FTA, HAZOP, PRA, ETA) need deep domain knowledge, adding to development and maintenance costs [30]. Domain-specific data may include complete system source code, requiring domain experts [5].

Evaluating trustworthiness-related risks of ML components within ACPPS adds further limitations as it requires extensive datasets for training and validation, which can be challenging to compile. Moreover, training complex models requires substantial computational power, often needing specialized hardware like GPUs or TPUs, leading to limitations associated with the availability of adequate resources. Developing, training, and validating these models requires highly specialized expertise, adding to the overall cost and complexity [24].

2.4 Mapping of Limitations Onto MBRA Techniques for ACCPS

Mapping of the identified MBRA techniques and the constraints limiting their applicability can be seen in Table 1. Looking into the limitations of the model-based techniques, studies have reported overlapping of these limitations such that some techniques showed structural and behavioral limitations. At the same time, some had structural and resource-related limitations or, in other cases, had reported limitations falling into all three of these categories. Although Stochastic models allow the identification of critical vulnerabilities by graphically modeling the system behavior, it is assumed that the system is secure from the start, leading to possible underestimation of risks and misallocation of resources. As for Model-checking techniques, they are more prone to all three types of limitations due to the possibility of a state explosion or a scenario explosion in autonomous systems, making it challenging to handle concurrency or other complex interactions within ACPPS [26].

Table 1. Limitations of Model-Based Assessment Techniques

Type of Limitations	MBRA Technique
Structural	BTD, FTA, ETA, BN, PNs, Graphs, Trees, Stochastic Modelling
Resource Consumption	BTD, FTA, PRA, ETA, BN, HAZOP, MCM
Behavioral	BTD, Stochastic Modelling, FTA, ETA, HAZOP, BN, MCM, HRD, safety use cases, models constructed using OUM, Component Diagram

Current studies focus more on limited system areas, lacking discussions on probable scalability and reliability-related issues of the applied MBRA techniques. A significant gap identified is the lack of focus on trustworthiness-related risks possibly emerging from the evolving nature of ML components within ACPPS and how they can impact the overall system. Moreover, there exists a lack of information on the mitigation measures to counter risks associated with data availability, interpretability, model simplification, expert reliance, and computational costs.

3 The MATrICS Methodology

This section presents an overview of the proposed MATrICS methodology and how it implements the findings of the review.

Although proven helpful in assessing risks, MBRA techniques such as Fault-/Attack Trees, variants of HAZOP, and Markov Models have limitations with multiple layers of ACPPS, ML components, and various aspects of trustworthiness. Researchers, therefore, often focus on a specific set of trustworthiness-related attributes (e.g., safety and security) on a single layer of ACPPS, mainly

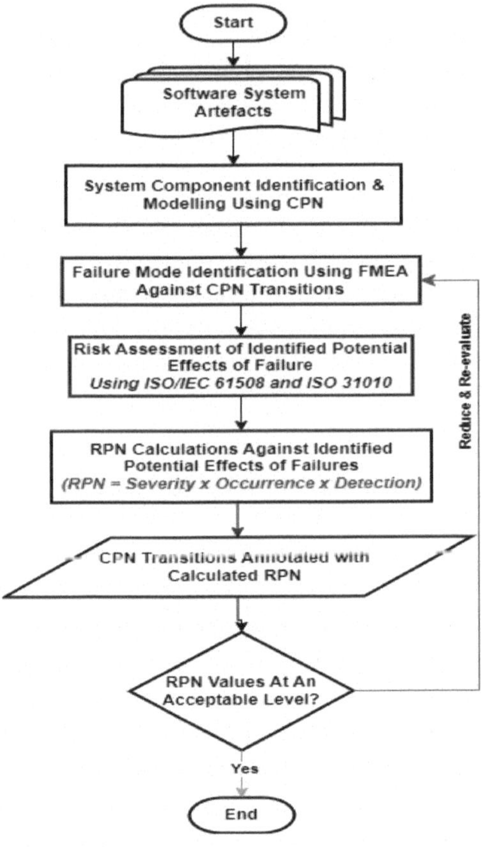

Fig. 2. The MATrICS methodology

the network layer. In the case of the cyber layer, the risk assessment has been neglecting ML-related risks like opacity and bias, which require a multi-faceted evaluation. Here, a hybrid MBRA technique can help capture system complexities, provide a more comprehensive risk assessment, and optimize resources.

The proposed hybrid MBRA methodology, named MATrICS (see Fig. 2), uses a combination of Colored Petri Nets (CPN) with Failure Model and Effect Analysis (FMEA) for better trustworthiness-related risk evaluation of ML components within ACPPS. MATrICS, when applied incrementally, makes the risk evaluation process more suitable and scalable for complex systems. Here, CPN's dynamic modeling capabilities can help capture the concurrent, asynchronous, distributed, non-deterministic, and stochastic behavior of ML components in ACPPS [18]. At the same time, FMEA applied to generated CPN can systematically identify and address multiple failure modes [21], offering a comprehensive, systematic, and practical framework for assessing trustworthiness-related risks in ACPPS. Moreover, the outcomes of mitigation measures can be observed by looping the proposed risk assessment technique.

3.1 MATrICS Requirements

The requirements for the proposed MATrICS methodology were driven based on the findings obtained from the review (as seen in Table 2). CPNs dynamically illustrate the interactions between components and how a failure propagates through the system, providing a clear structural representation of the system. Combining the FMEA analysis with the CPN's graphical depiction can allow for a better understanding of how the structural changes affect the system and help identify the critical areas for improvement. CPNs can model the flow of processes, highlighting possible bottlenecks or inefficiencies. At the same time, a targeted FMEA analysis identifies the potential failure points within the constructed model, improving the overall accuracy of the risk assessment process and optimizing the resource allocation. Moreover, CPNs simulating scenarios can help visualize the impact of identified failure modes, offering a more com-

Table 2. Driven Requirements for MATrICS Methodology

Type of Requirement	Requirement
Structural	Risk Assessment at Design Stage
Structural	Coverage of Tight Interactions or Inter-Dependencies Between Components
Behavioral	Coverage of System's Dynamic Behavior
Behavioral	Coverage of System's Asynchronous, Concurrent, and Dynamic Requirements
Resources	Optimal Resource Allocation
Resources	Identification of Bottlenecks or Inefficiencies

prehensive view of the system's performance and the impact risks can have on it.

3.2 MATrICS Steps

MATrICS methodology is composed of the following seven steps (see Fig. 2):

1. Take as an input the available software system design artifacts to construct a CPN.
2. For each Transition in CPN, conduct a Process-focused risk assessment using FMEA, and the ISO/IEC 61508 standard [31] and ISO 31010 standard [17].
3. Calculate Risk Priority Number (RPN) for each of the identified and assessed potential effects of failure (see Eq. 1).
4. Annotate transitions in CPN associated with relevant failure modes using the calculated RPN values.
5. Check if the calculated RPN value for a particular CPN transition is acceptable.
   - If acceptable, end execution of the methodology.
   - If unacceptable, implement required mitigation measures to reduce the severity of the identified risks and repeat the process starting from step 2.

Here, step 1 is responsible for mapping the components of the system under evaluation, and steps 2–5 evaluate the system for possible risks, their severity, possibility of occurrence, and probability of detection.

$$RPN = \text{Severity} \times \text{Occurrence} \times \text{Detection} \tag{1}$$

Here, RPN is a product of three factors associated with the identified risk [21]: the seriousness of the consequences if the failure occurs *(Severity)*, the likelihood of a failure to occur *(Occurance)* and probability of the failure being identified before the occurrence of the resultant problem *(Detection)*.

Values for RPN's above-stated factors are ranked on a scale of *1–10* with '1' being the lowest and '10' being the highest. Overall, the value of a calculated RPN can range between *1–1000* [21]. Once the RPN values are calculated against each failure mode, they are summed up based on the failure modes involved against each transition of the CPN. The higher the accumulated RPN (ARPN) value, the higher the priority of the studied risk to be mitigated.

Moreover, the methodology allows for evaluating the effects an applied risk mitigation measure can have on the system via the implementation of step 7. Since a threshold RPN value generally does not exist, the risk management team will decide the threshold RPN value [21]. However, the scope of this study is limited only to assessing the identified risks.

Considering the ML components, the proposed methodology aims to provide a structured and systematic risk assessment methodology addressing all trustworthiness-relevant aspects of ACPPS. It supports a detailed understanding of ML components' dynamic and adaptive aspects and their failure modes. This enables prioritization and mitigation of trustworthiness-related risks during ACPPS development and validation.

4 Conclusion and Future Work

Considering the new risks and complexity introduced by ML components, we moved further toward model-based assessment of trustworthiness in ACPPS in this paper. We developed an abstract of the novel hybrid methodology combining several techniques. Based on the results of the preliminary studies, we found that stochastic modeling formalisms, such as extensions of Petri Nets and Fault Trees, are the most reported techniques for risk assessment, while metrics such as MTTF and intrusion probabilities are the most commonly used for evaluating and prioritizing the identified risks. Model-checking has also been reported as a means to evaluate trustworthiness in ACPPS. However, none of the existing studies addresses the application of those techniques on the ML component of ACPPS. The reported limitations of the existing techniques can be categorized as structural, behavioral, and resource-related. The most reported limitations were scalability-related issues, failure to capture the system's dynamic behavior, and a lack of resources such as documented data, domain experts, and hardware capable of handling complex computations. Based on the current limitations of model-based techniques and tools in assessing the risks associated with the ML components of ACPPS, our methodology – named MATrICS – aims to comprehensively and systematically evaluate relevant trustworthiness attributes. The MATrICS methodology is hybrid since it combines Colored Petri Nets to ensure high expressive power to model any dynamic and adaptive behavior due to ML and FMEA for modular, iterative, and incremental bottom-up modeling to ensure scalability to complex (i.e., large, distributed, and heterogeneous) production systems.

In the future, we plan to further develop the MATrICS methodology and empirically evaluate its effectiveness and performance regarding accurate ACPPS risk identification.

Acknowledgements. This work has been partially funded by the European HORIZON-KDT-JU research project MATISSE "Model-based engineering of Digital Twins for early verification and validation of Industrial Systems", HORIZON-KDT-JU-2023-2-RIA, Proposal number: 101140216-2, KDT232RIA_00017, the Swedish Knowledge Foundation through the MoDEV project (20200234), and the Sweden's innovation agency Vinnova through the project iSecure (202301899).

References

1. Evrin, E., et al.: Risk assessment and analysis methods: qualitative and quantitative. ISACA J. **28** (2021)

2. Ambore, B., et al.: Novel model for boosting security strength and energy efficiency in Internet-of-Things using multi-staged game. Int. J. Electri. Comput. Eng. (2088-8708) **9**(5) (2019)
3. Berger, S., et al.: Modelling availability risks of it threats in smart factory networks–a modular petri net approach. In: European Conference on Information Systems (2021)
4. Burkart, N., Huber, M.F.: A survey on the explainability of supervised machine learning. J. Artif. Intell. Res. **70**, 245–317 (2021)
5. Castellanos, J.H., et al.: Finding dependencies between cyber-physical domains for security testing of industrial control systems. In: Proceedings of the 34th Annual Computer Security Applications Conference, pp. 582–594 (2018)
6. Čelebić, V., Bucaioni, A.: A systematic mapping study on the role of software engineering in enabling society 5.0. In: 2023 IEEE International Smart Cities Conference (ISC2), pp. 1–8. IEEE (2023)
7. Crnkovic, G.D.: Constructive research and info-computational knowledge generation. In: Model-Based Reasoning in Science and Technology: Abduction, Logic, and Computational Discovery, pp. 359–380. Springer (2010)
8. Cruzes, D.S., Dyba, T.: Recommended steps for thematic synthesis in software engineering. In: 2011 International Symposium on Empirical Software Engineering and Measurement, pp. 275–284 (2011)
9. Darandale, S., Mehta, R.: Risk assessment and management using machine learning approaches. In: 2022 International Conference on Applied Artificial Intelligence and Computing (ICAAIC), pp. 663–667 (2022)
10. Deng, S., et al.: Security risk assessment of cyber physical power system based on rough set and gene expression programming. IEEE/CAA J. Autom. Sin. **2**(4), 431–439 (2015)
11. DeSmit, Z., et al.: An approach to cyber-physical vulnerability assessment for intelligent manufacturing systems. J. Manuf. Syst. **43**, 339–351 (2017)
12. Embedded.com: An introduction to model checking (2023). https://www.embedded.com/an-introduction-to-model-checking
13. Euro News: The enemy is false information': world leaders and businesses take on cybersecurity in riyadh (2023). https://www.euronews.com/2023/11/09/
14. Flammini, F., Alcaraz, C., Bellin, E., Marrone, S., Lopez, J., Bondavalli, A.: Towards trustworthy autonomous systems: a survey of taxonomies and future perspectives. J. Artif. Intell. Res. **70**, 245–317 (2021)
15. Geng, S., et al.: Web application architecture security evaluation method based on AADL. In: 2015 20th International Conference on Engineering of Complex Computer Systems (ICECCS), pp. 186–189 (2015)
16. Gran, B.A., Fredriksen, R., Thunem, A.P.J.: An approach for model-based risk assessment. In: International Conference on Computer Safety, Reliability, and Security, pp. 311–324 (2004)
17. ISO International Organization for Standardization: ISO 31010 - risk management (2020). https://practicalrisktraining.com/iso31010
18. Jitmit, C., Vatanawood, W.: Simulating artificial neural network using hierarchical coloured petri nets. In: Proceedings of the 2021 6th International Conference on Machine Learning Technologies, pp. 127–131 (2021)
19. Kitchenham, B., Brereton, P.: A systematic review of systematic review process research in software engineering. Inf. Softw. Technol. **55**(12), 2049–2075 (2013)
20. Lazaroiu, G., Androniceanu, A., Grecu, I., Grecu, G., Neguriţă, O.: Artificial intelligence-based decision-making algorithms, Internet of Things sensing networks,

and sustainable cyber-physical management systems in big data-driven cognitive manufacturing. Oeconomia Copernicana **13**(4), 1047–1080 (2022)
21. Li, J., Chignell, M.: FMEA-AI: AI fairness impact assessment using failure mode and effects analysis. AI Ethics **2**(4), 837–850 (2022)
22. Li, X., Kang, K., Hu, S.: Real-time requirements for cyber-physical systems: a survey. IEEE Trans. Industr. Inf. **13**(6), 2817–2826 (2017)
23. Lichte, D., Wolf, K.D.: Use case-based consideration of safety and security in cyber physical production systems applied to a collaborative robot system. In: Safety and Reliability–Safe Societies in a Changing World, pp. 1395–1401. CRC Press (2018)
24. Mattioli, J., et al.: An overview of key trustworthiness attributes and KPIs for trusted ML-based systems engineering. AI Ethics **4**(1), 15–25 (2024)
25. Nazir, R., Bucaioni, A., Pelliccione, P.: Architecting ml-enabled systems: challenges, best practices, and design decisions. J. Syst. Softw. **207**, 111860 (2024)
26. Pappagallo, A., et al.: Monte Carlo based statistical model checking of cyber-physical systems: a review. Information **11**(12), 588 (2020)
27. Patel, A.R., Liggesmeyer, P.: Machine learning based dynamic risk assessment for autonomous vehicles. In: 2021 International Symposium on Computer Science and Intelligent Controls (ISCSIC), pp. 73–77 (2021)
28. Peng, Y., et al.: Cyber-physical system risk assessment. In: 2013 Ninth International Conference on Intelligent Information Hiding and Multimedia Signal Processing, pp. 442–447 (2013)
29. Rathore, L.K., Sao, N.: An integrated model based test case prioritization using UML sequence and activity diagram. Int. J. Res. Comput. Appl. Robot. **3**(12), 31–41 (2015)
30. Tam, K., Jones, K.: Cyber-risk assessment for autonomous ships. In: 2018 International Conference on Cyber Security and Protection of Digital Services (Cyber Security), pp. 1–8 (2018)
31. TÜV SÜD: Functional safety - IEC 61508. https://www.tuvsud.com/en-us/services/functional-safety/iec-61508. Accessed Apr 2024
32. Union, E.: The enemy is false information': world leaders and businesses take on cybersecurity in riyadh (2023). https://www.weforum.org/agenda/2023/06/european-union-ai-act-explained/
33. Wan, J., Yi, M., Li, D., Zhang, C., Wang, S.: Data-driven and autonomous manufacturing control in cyber-physical production systems. J. Manuf. Syst. **54**, 97–107 (2020)
34. Yang, R., Zhong, M.: Machine Learning-Based Fault Diagnosis for Industrial Engineering Systems. CRC Press (2022)
35. Zahid, M., Bucaioni, A., Flammini, F.: Trustworthiness-related risks in autonomous cyber-physical production systems-a survey. In: 2023 IEEE International Conference on Cyber Security and Resilience (CSR), pp. 440–445. IEEE (2023)
36. Zahid, M., Bucaioni, A., Flammini, F.: Model based trustworthiness evaluation of autonomous cyber-physical production systems: a systematic mapping study. ACM Comput. Surv. (2024)
37. Zahid, M., Inayat, I., Deneva, M.: Security risks in cyber physical systems-a systematic mapping study. J. Softw. Evol. Process **33**(9), e2346 (2021)
38. Zografopoulos, I., et al.: Cyber-physical energy systems security: threat modeling, risk assessment, resources, metrics, and case studies. IEEE Access **9**, 29775–29818 (2021)

Context-Aware Event Sequence Prediction for Sustainable Mobile Computing Using LSTM: A Deep Learning Approach

Pooja Goyal[1(✉)], Caiden Chadwick[2], Melody Jing[3], Samuel Williams[4], and Renee Bryce[1]

[1] Computer Science and Engineering, University of North Texas, Denton, TX, USA
PoojaGoyal@my.unt.edu, Renee.Bryce@unt.edu
[2] Arkansas State University, Jonesboro, AR, USA
caiden.chadwick@smail.astate.edu
[3] Rice University, Houston, TX, USA
mj56@rice.edu
[4] Texas Christian University, Fort Worth, TX, USA
s.a.williams1@tcu.edu

Abstract. As smartphone applications become increasingly integrated into daily life, ensuring their reliability in diverse contexts is crucial. This study investigates how various contextual factors—such as Wi-Fi connectivity, screen activity, and battery status—impact application behavior and device performance. Using the CM-SPAM algorithm from the SPMF library to preprocess context event data from Android smartphones, we apply machine learning models i.e., Support Vector Machine (SVM), Random Forest (RF), and Long Short-Term Memory (LSTM) to predict application usage and behavior. Our results show that one-minute time interval yields the best prediction accuracy, with RF achieving the highest performance (F1-score of 0.842), outperforming both SVM and LSTM. These findings emphasize the importance of context-aware event data for improving the reliability and accuracy of smartphone applications, suggesting that context-driven models like RF can significantly enhance app performance across varying conditions.

Keywords: Smartphone Application Testing · Context-Aware Applications · Episode Mining · Sequential Pattern Mining · Deep Learning · Machine Learning

1 Introduction

The huge network of connected smart gadgets that can interact is known as the Internet of Things (IoT). These devices, which span applications in transportation, security, healthcare, smart homes, and industrial sectors, have become integral to modern society [1,2]. For instance, in healthcare, IoT devices like smartwatches and biometric trackers monitor vital signs, while smart homes employ

S. Latifi (Ed.): ITNG 2025, AISC 1463, pp. 595–608, 2025.
https://doi.org/10.1007/978-3-031-89063-5_51

IoT-enabled systems such as lighting, thermostats, and security cameras [3]. Of particular importance, smartphones—especially those running the Android operating system—are central to the IoT ecosystem, facilitating interaction with a wide range of other devices [4,5]. The widespread adoption of smartphones calls for rigorous testing of their software applications to ensure both reliability and robustness in diverse environments.

Smartphone applications are influenced by various environmental factors, such as screen orientation, battery level, Wi-Fi connectivity, and more [6,7]. Collectively, these factors are referred to as context events. Context-aware applications, designed to adapt to such environmental variables, predict these events which can enhance user experience and device functionality [8]. As a result, comprehensive testing of applications in relation to potential context events is essential [9]. Recent research suggests that identifying patterns in context event sequences can offer valuable insights, which can be leveraged to improve application testing, user experience and application performance [10].

A key approach in analyzing time-series context event data is to predict the next event in a sequence based on prior events [11]. Machine learning techniques are increasingly being used to model such event prediction tasks in context-aware applications [12]. The primary objective of this study is to investigate machine learning methods for efficiently predicting subsequent events in context event sequences, specifically in the context of Android smartphones.

The structure of this document is as follows: Sect. 2 reviews background and related work, Sect. 3 outlines the data collection, preprocessing, and research methodology. Section 4 presents the research questions and evaluation metrics. Section 5 discusses the results, and Sect. 6 concludes with an analysis of the findings and discusses future directions for applications of this research.

2 Background and Related Work

This section provides an overview of smartphone context events, the SPMF library, and the machine learning techniques employed in our experiments as well as previous research works in the similar domain.

2.1 Background and Related Work

Several previous works have explored approaches for predicting smartphone context events, contributing to the growing field of context-aware systems. One such study utilized sequence rule mining algorithms, specifically TKE-Rules and POERMH [13] [14], indicated that the model incorporating these algorithms was notably successful in predicting individual events [8].

In a similar vein, another study introduced a novel variant of the Compact Prediction Tree (CPT) method, known as the Context Data Compact Prediction Tree (CD-CPT), for the prediction of real-world context events. While the CD-CPT outperformed other models in predicting full sequences of context events, it was found to be less effective in predicting individual events compared to more accurate alternatives. This approach demonstrated promising results when applied to predicting complete sequences of context event data, independent of real-world context data [10].

Additionally, a recent study applied High-Utility Itemset (HUI) mining algorithms to context data, aiming to improve context-aware software testing. The results of this work suggested that HUI-based models could achieve high predictive accuracy for significant context events, whether they occur individually or simultaneously, thereby enhancing the robustness of context-aware applications [15].

Previous research on sequence rule mining (TKE-Rules, POERMH) and prediction trees (CD-CPT) has advanced smartphone context event prediction but falls short in capturing long-term temporal dependencies and handling dynamic time-series data. For instance, TKE-Rules excelled in event prediction but lacked sequence modeling, while CD-CPT performed better for full sequences than individual events. Additionally, HUI mining showed promise but did not fully address the complexities of real-time context data. Building on these limitations, our work explores CM-SPAM and integrates LSTM models to better capture long-range dependencies in time-series data, aiming for more accurate predictions of context events.

2.2 SPMF

The Sequential Pattern Mining Framework (SPMF) [16] is an open-source Java-based tool that specializes in discovering patterns in large datasets, offering around 254 algorithms. Due to the robust pattern mining capabilities in SPMF, we implemented the CM-SPAM algorithm to identify frequent event sequences, applying machine learning to enhance predictive accuracy and improve context event forecasting

2.3 CM-SPAM

CM-SPAM, an optimized version of the SPAM algorithm [17], efficiently mitigates performance bottlenecks in earlier implemented SPMF algorithms by using the CMAP data structure to prune the search space through co-occurrence-based mapping. This enhances its scalability in large-scale sequence mining tasks.

2.4 Random Forest

Random forest classification [18] as an ensemble method, it integrates several decision trees that were trained using arbitrary selections of the characteristics and data that enhances model accuracy and generalization. This diversity among trees allows the random forest to capture complex relationships and handle large, noisy datasets effectively, making it a robust and reliable choice for classification tasks as well as best method address the instability and over-fitting issues of individual decision trees. [19].

2.5 Support Vector Machine

Support Vector Machine (SVM) [20]are proven to be effective in handling both linear and nonlinear classification tasks, even in the presence of noise or class imbalance. By mapping data to a high-dimensional space and constructing hyperplanes that maximize the margin between classes, SVM enhances classification performance. Additionally, SVM performance on sparse data makes it a robust choice for our dataset [21][22].

2.6 LSTM

Long Short-Term Memory (LSTM) networks [23] are best choice suited for this research due to their ability to capture long-term dependencies in sequential data, a critical feature for time series forecasting. Unlike basic Recurrent Neural Networks (RNNs) [24], LSTMs use specialized memory cells with gates (input, forget, and output) to manage and retain relevant information over longer sequences. This makes LSTMs particularly well-suited for tasks involving complex temporal relationships, such as predicting smartphone context events, where long-range dependencies are very critical [25].

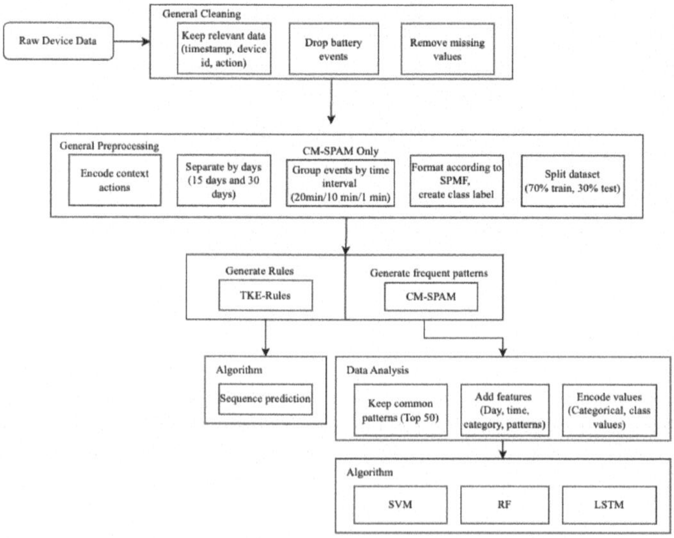

Fig. 1. Overview of the Research Methodology

3 Research Methodology

This research employs below steps in the methodology: Data Collection and Preprocessing, CM-SPAM rule generation, and comparison of various machine learning and deep learning models as an Overview presented in Fig. 1

3.1 Data Collection

Dataset used in this study is from real world users i.e. using the ContextMon Android app [26]. The app operates in the background, continuously monitoring context events and keeping them on the user's device in a SQLite database. The application uploads the device ID, timestamp, and related context event action to a remote MySQL database every 15 min. The research's dataset, which comes from 146 different Android smartphone users, covers the period from October 2020 to December 2023. It includes 86 distinct kinds of context events that reflect different facets of consumers' regular smartphone use.

3.2 Data Preprocessing

For experiments we have divided the experiments into two classes i.e., category and event classes. For both categories, we followed a common set of steps, outlined as follows:

3.2.1 Data Cleaning We retained only the relevant columns for this research: timestamp, action, and device ID, discarding all other columns. To address the bias introduced by battery events, which accounted for over half of the records, we removed all battery-related events. This reduced the dataset from 30,828,454 instances to 14,281,063, and the number of unique context events from 86 to 83. We also removed records with invalid or missing values. The dataset was then mapped to numeric representations and split into two categories: users with 15 continuous days of data and users with 30 continuous days. Data from devices lacking at least 15 days of continuous data were excluded.

3.2.2 CM-SPAM Preprocessing To extract meaningful patterns using CM-SPAM [16], we grouped events by timestamp into parallel events, then formed sequences based on time intervals (ranging from 0.5 to 30 min). Specifically, 1, 10, and 20-minute intervals were used for category and event classification, resulting in six datasets for each 15-day and 30-day periods, in total 12 datasets.

For preprocessing, we created a separate column for the last event in each sequence to mark the sequence's end and removed instances with empty sequences after excluding the target. For event classification, labels were unchanged, while for category classification, events were mapped to their categories, discarding multi-category labels. The datasets were split into two training files and one test file 70:30 split, with devices having fewer than 500 instances excluded. One training file contained all columns for feature engineering, and the other included only the sequence column for CM-SPAM processing.

3.3 Feature Engineering for CM-SPAM

The output files were used to implement features across the entire dataset, ensuring pattern extraction occurred only from the training set to prevent data leakage. Frequent sequential patterns identified in the training set were applied to both the training and test files.

For category classification, we used frequency features for the top 50 patterns, two categorical features (day and time), and a frequency feature for each category. Event classification included additional features: frequency features for each of the 83 unique context events, as well as proportionality features for serial and parallel events in a sequence.

The top 50 patterns, derived from the training set using CM-SPAM, were used to generate 50 frequency features based on their occurrence, considering overlapping patterns. Time of day was divided into eight intervals e.g., Early Morning: 12:00 AM–6:00 AM and the day of the week into seven categories (Monday to Sunday), both of which were one-hot encoded.

For both tasks, we created frequency features for each event category (media, net, wifi, power, location, bluetooth, screen, and other) by counting event occurrences within each sequence. Event classification also included frequency features for each context event and additional features for serial and parallel event occurrences, along with a proportion feature for serial events.

4 Empirical Study and Research Questions

4.1 Evaluation Metrics

We evaluate the performance of our classification models using three primary metrics, each scaled with weighted averaging:

1) **Precision**: The ratio of true positives to the sum of true and false positives. Weighted precision averages across classes, with individual scores weighted by class instance frequency.
2) **Recall**: The ratio of true positives to the sum of true positives and false negatives. Weighted recall averages across classes, weighted by class instance frequency.
4) **F1-score**: The harmonic mean of precision and recall. The weighted F1-score is the average F1-score across classes, weighted by the number of instances in each class.

This research explores four key questions based on above metrics:

RQ1: How does integrating CM-SPAM affect the precision, recall, and accuracy of RF, SVM, and LSTM classifiers in predicting smartphone context events using temporal and sequential patterns?

RQ2: To what extent does CM-SPAM enhance the classification performance of RF, SVM, and LSTM for diverse event types, based on feature extraction and pattern recognition?

RQ3: Which classifier (RF, SVM, or LSTM) performs best in predicting smartphone context events and categories, with and without CM-SPAM features?

RQ4: How do RF, SVM, and LSTM compare for nuanced event patterns, before and after incorporating CM-SPAM features, in real-time Android applications?

4.2 Experimental Setup

For our experiments each device's data was processed separately through the relevant SPMF algorithm, generating a single output file per device, which is considered in the context of its respective device.

Since this iteration of the context event dataset has not been previously used in research, the model parameters are independent of prior work, including earlier versions of this dataset. The current dataset includes additional features and newly collected data.

4.2.1 CM-SPAM The implementation of CM-SPAM requires an input file and a minimum support (minsup) parameter, which defines the minimum percentage of instances in the dataset that must contain a given pattern for it to be considered frequent [16]. Through empirical testing, we determined that a minsup value of 0.1 (10%) yielded a reasonable number of patterns across most device files. However, setting minsup to 0.1 alone resulted in an excessive number of patterns for certain devices, while increasing the minsup parameter reduced the number of patterns across all devices. Instead of adjusting minsup, we employed additional parameters: minimum and maximum pattern lengths set to 2 and 10 respectively, to avoid overemphasizing single or parallel events. Serial and parallel events were treated as single elements in patterns [16], resulting in a more balanced number of patterns across devices. All other optional parameters were left at their default values.

Table 1. CM-SPAM Example Inputs and Outputs

CM-SPAM Input	CM-SPAM Output
1 -1 1 2 3 -1 1 3 -1 4 -1 3 6 -1 -2	1 -1 2 -1 #SUP: 4
1 4 -1 3 -1 2 3 -1 1 5 -1 -2	2 -1 1 -1 #SUP: 2
5 6 -1 1 2 -1 4 6 -1 3 -1 2 -1 -2	5 -1 1 -1 3 -1 #SUP: 2
5 -1 7 -1 1 6 -1 3 -1 2 -1 3 -1 -2	2 -1 3 -1 #SUP: 3

Table 1 provides an example of the output generated by CM-SPAM for each device, which consists of frequent sequential patterns containing both serial and parallel events. Parallel events within the same time step are separated by spaces, while partially ordered serial and parallel events are delimited by the special character "−1". Each pattern concludes with "−1 −2" and a support value indicating the number of sequences in which the pattern appeared [16].

4.2.2 Random Forest We found that our data was not highly sensitive to the maximum tree depth or the number of decision trees in the random forest model. Although we trained the model separately for each user device, the performance variations across devices with identical parameters were minimal. As a result, we opted to use consistent parameters across all devices for efficiency.

The tuned parameters included the number of estimators, maximum tree depth, and the minimum samples required to split a node. Ultimately, we selected 100 estimators, a maximum depth of 20, and a minimum of 5 samples per split to balance model performance, minimize over-fitting, and maintain manageable runtime.

4.2.3 Support Vector Machine Our dataset is sparse and imbalanced, characteristics that align well with the strengths of support vector machines (SVM). Motivated by this, we applied SVM to the data and tested various penalty types, iteration limits, and C values to optimize performance. The final configuration used was $C=0.5$, no iteration limit, and an L1 penalty, based on performance evaluation on both training and test datasets.

4.2.4 Long Short-Term Memory Due to time constraints, we limited hyper-parameter tuning to epoch count, batch size, and LSTM layers. After empirical testing, we settled on a model with 80 LSTM layers, trained for 50 epochs, and using a standard batch size of 32, recognizing that further experimentation could potentially improve performance.

4.3 Time Interval Testing

For sequence derivation using CM-SPAM, we grouped context events into time intervals of varying lengths, including 1, 10, and 20 min. Figures 2 and 3 display the average F1-scores for all models across these time intervals.

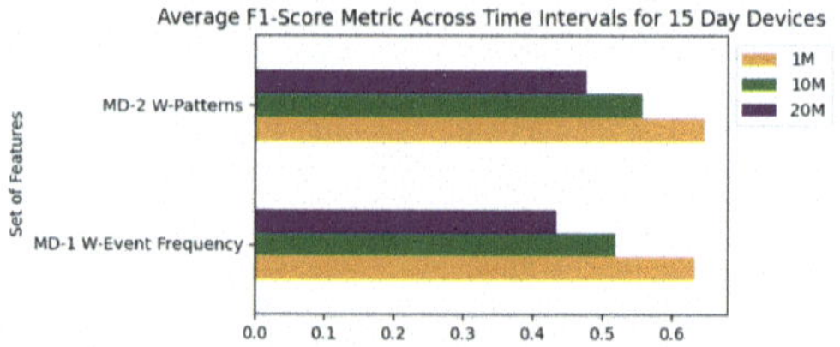

Fig. 2. F1-score for event-based prediction across time intervals of 1 min, 10 min, and 20 min.

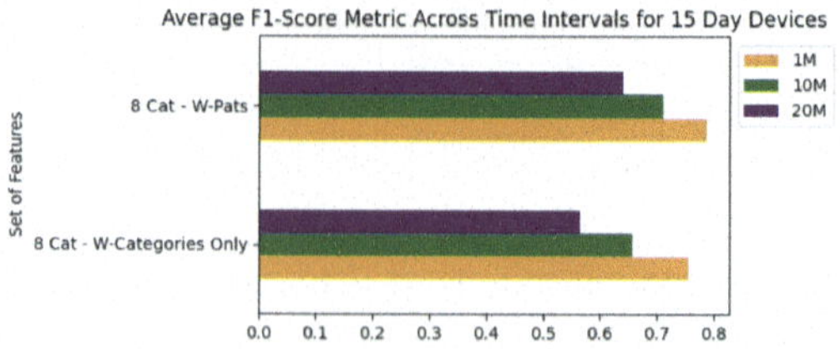

Fig. 3. F1-score for category-based prediction across time intervals of 1 min, 10 min, and 20 min.

As the time interval for grouped context-event sequences increased towards 20 min, model performance decreased. Conversely, reducing the interval to 1 min led to significantly higher scores across all models, classification tasks, and device sets. This trend is clearly illustrated in Figs. 2 and 3, which show the relationship for each classification task and model, based on the 15-day device sets.

Table 2. Event classification averages for 15 and 30 day devices

Model	Day Set	Set Name	Precision	Recall	F1-Score
LSTM	15	MD-1 W-Event Frequency	51.47%	60.88%	53.74%
LSTM	15	MD-2 W-Patterns	54.28%	62.71%	56.06%
RF	15	MD-1 W-Event Frequency	67.95%	72.93%	69.39%
RF	15	MD-2 W-Patterns	68.03%	73.08%	69.51%
SVM	15	MD-1 W-Event Frequency	65.39%	70.47%	66.70%
SVM	15	MD-2 W-Patterns	67.58%	72.06%	68.81%
LSTM	30	MD-1 W-Event Frequency	49.31%	57.86%	51.52%
LSTM	30	MD-2 W-Patterns	51.83%	59.72%	53.78%
RF	30	MD-1 W-Event Frequency	62.37%	67.32%	63.47%
RF	30	MD-2 W-Patterns	62.94%	67.49%	63.96%
SVM	30	MD-1 W-Event Frequency	60.16%	65.20%	60.85%
SVM	30	MD-2 W-Patterns	61.38%	66.19%	62.16%

5 Results and Discussion

Results and performance for CM-SPAM, RF, SVM, and LSTM are discussed below:

RQ1: Table 2 shows that LSTM benefited the most from CM-SPAM-derived features, with an average score improvement of 2.32% for 15-day devices and 2.21% for 30-day devices, across precision, recall, and F1-score. The SVM showed a smaller improvement of 1.96% for 15-day devices and 1.17% for 30-day devices. The RF classifier showed the least improvement, with a score increase of 0.12% for 15-day devices and 0.41% for 30-day devices. These results are further illustrated in Fig. 5, showing average weighted F1-score improvements.

RQ2: Single-category classification yielded better results than event classification across all models, with CM-SPAM-derived features having a greater impact. As shown in Figs. 5 and 4, the LSTM model saw a significant improvement, with an average score increase of 5.17% for 15-day devices and 4.59% for 30-day devices (Table 3). The SVM also benefitted, with a 2.52% improvement for 15-day devices and 4.71% for 30-day devices. The RF model showed the least benefit, with a score increase of 1.11% for 15-day devices and 1.17% for 30-day devices.

Table 3. Category classification averages for 15 and 30 day devices

Model	Day Set	Set Name	Precision	Recall	F1-Score
LSTM	15	8 Cat - W-Categories Only	66.55%	72.38%	66.40%
LSTM	15	8 Cat - W-Pats	72.15%	76.59%	72.11%
RF	15	8 Cat - W-Categories Only	82.83%	84.40%	83.16%
RF	15	8 Cat - W-Pats	84.04%	85.45%	84.22%
SVM	15	8 Cat - W-Categories Only	76.04%	81.06%	77.03%
SVM	15	8 Cat - W-Pats	79.15%	82.65%	79.89%
LSTM	30	8 Cat - W-Categories Only	65.37%	70.21%	65.23%
LSTM	30	8 Cat - W-Pats	70.96%	73.82%	69.79%
RF	30	8 Cat - W-Categories Only	81.45%	81.95%	81.13%
RF	30	8 Cat - W-Pats	82.44%	83.18%	82.41%
SVM	30	8 Cat - W-Categories Only	72.83%	77.14%	73.88%
SVM	30	8 Cat - W-Pats	78.60%	80.63%	78.76%

RQ3: As shown in Tables 2 and 3, Figs. 5, and 4, the RF classifier outperforms both SVM and LSTM models in terms of precision for both category and event classification tasks. Although the integration of CM-SPAM-derived features has a modest effect on the RF model, it leads to a larger performance gap for the SVM and LSTM, which still fall short of the RF scores even with CM-SPAM features.

RQ4: As shown in Tables 2 and 3, RF consistently outperforms both SVM and LSTM across all evaluation metrics and devices for both classification tasks, while LSTM scores remain the lowest. The impact of CM-SPAM features is noticeable but relatively small, as observed in RQ3.

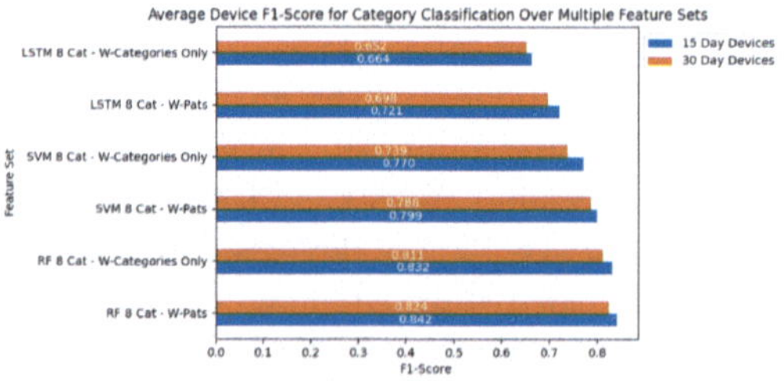

Fig. 4. F1-score for category-based prediction at a 1 min time interval for 15 and 30 day devices across models.

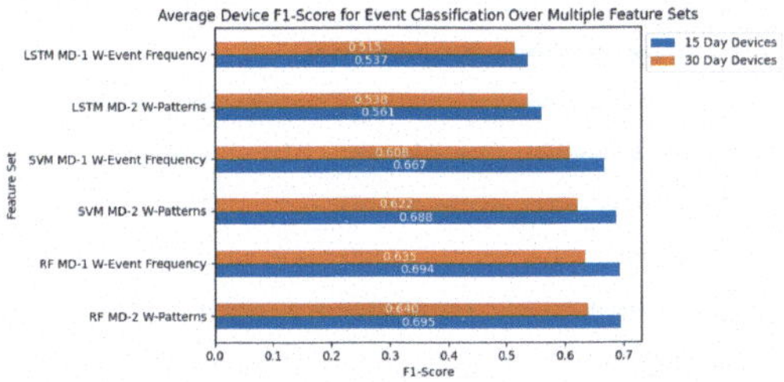

Fig. 5. F1-score for event-based prediction at a 1 min time interval for 15 and 30 day devices across models.

In summary, RF consistently outperformed SVM and LSTM in both predictive accuracy and computational efficiency. It demonstrated superior performance for real-time optimization in context-aware smartphone applications, with notably lower time costs compared to LSTM, which had the longest processing times.

While SVM performed slightly worse than RF, it holds potential for improvement through hyperparameter tuning and testing with shorter time intervals, such as 30 s. Additionally, increasing the number of top pattern features derived from CM-SPAM could enhance its performance.

Although LSTM produced the lowest scores and took the most time, its results indicate promise for event and category classification tasks. Its performance could improve with fine-tuning of hyperparameters like the number of layers, epochs, and batch sizes, which were not extensively explored in this study. Further experimentation, including adjustments to data preprocessing and model training, could yield a more effective LSTM model for these prediction tasks.

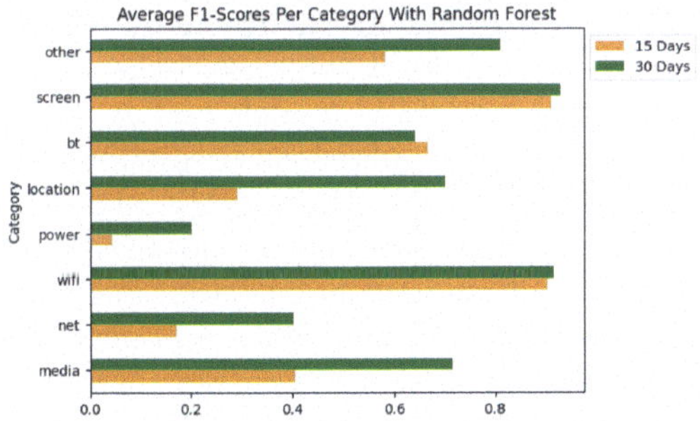

Fig. 6. F1-score for each category of context event with random forest.

For category classification, Figs. 6 and 7 display the unweighted average F1-scores for RF and SVM across the 15-day and 30-day datasets. A strong correlation exists between category frequency and model performance: more frequent events yield higher prediction accuracy, while less frequent events are harder

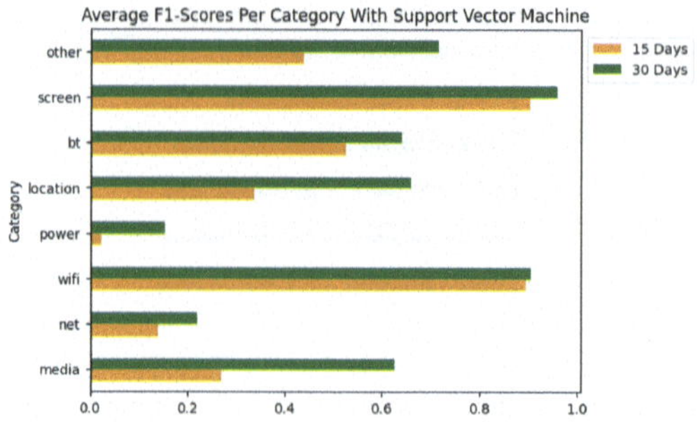

Fig. 7. F1-score for each category of context event with support vector machines.

to predict. Although the 30-day set showed slightly lower weighted scores than the 15-day set, the models exhibited greater consistency in category prediction with the larger dataset, as evidenced by Figs. 6 and 7 the increase in unweighted averages from 15 days to 30 days.

6 Conclusion

To optimize context-aware smartphone applications in the IoT landscape, it's crucial to improve their ability to predict user behavior. Using the CM-SPAM algorithm, we extracted features from context data and applied three classification models—Random Forest (RF), Support Vector Machine (SVM), and Long Short-Term Memory (LSTM). Our results show that CM-SPAM slightly improved classification scores, with LSTM benefiting the most (2.32% improvement for 15-day devices). However, RF outperformed both SVM and LSTM in terms of precision and overall performance, consistently yielding the best results (RQ3, RQ4). These findings highlight RF as the most reliable model.

Future research could focus on improving user behavior prediction in smartphone applications by enhancing LSTM models and exploring other RNN variants like GRU for better long-event sequence predictions. Experimenting with different CM-SPAM parameters and other SPMF algorithms, such as those for sequential or periodic pattern mining, could boost classification performance. Finally, exploring continuous context-based authentication could open new research avenues, improving the reliability and accuracy of context-aware apps in IoT environments.

References

1. Kumar, S., Tiwari, P.: Internet of Things is a revolutionary approach for future technology enhancement: a review. J. Big Data **6**(1), 111 (2019). https://doi.org/10.1186/s40537-019-0268-2
2. Mouha, R.A.: Internet of things (IoT). J. Data Anal. Inf. Process. **9**(2), 77–101 (2021). number: 2 Publisher: Scientific Research Publishing. https://www.scirp.org/journal/paperinformation?paperid=108574
3. Nižetić, S., Šolić, P., López-de-Ipiña González-de Artaza, D., Patrono, L.: Internet of Things (IoT): opportunities, issues and challenges towards a smart and sustainable future. J. Clean. Prod. **274**, 122877 (2020). https://www.ncbi.nlm.nih.gov/pmc/articles/PMC7368922/
4. El Khaddar, M.A., Boulmalf, M.: Smartphone: the Ultimate IoT and IoE device, In: Smartphones from an Applied Research Perspective. IntechOpen (2017). https://www.intechopen.com/chapters/56113
5. Planet of the phones. The Economist (2015). https://www.economist.com/leaders/2015/02/26/planet-of-the-phones
6. Dybå, T., Sjøberg, D., Cruzes, D.: What works for whom, where, when, and why? On the role of context in empirical software engineering. In: Proceedings of the 2012 ACM-IEEE International Symposium on Empirical Software Engineering and Measurement, pp. 19–28. IEEE, Lund, Sweden (2012)
7. Do, T.M.T., Gatica-Perez, D.: Where and what: using smartphones to predict next locations and applications in daily life. Perv. Mob. Comput. **12**, 79–91 (2014). https://www.sciencedirect.com/science/article/pii/S1574119213000576
8. Goyal, P., Khan, M.K., Steil, C., Martel, S.M., Bryce, R.: Smartphone context event sequence prediction with POERMH and TKE-rules algorithms. In: 2023 IEEE 13th Annual Computing and Communication Workshop and Conference (CCWC), pp. 0827–0834 (2023). https://ieeexplore.ieee.org/document/10099166
9. Amalfitano, D., Fasolino, A., Tramontana, P., Amatucci, N.: Considering context events in event-based testing of mobile applications. In: IEEE 6th International Conference on Software Testing, Verification and Validation Workshops, ICSTW 2013. IEEE, Luxembourg, Luxembourg, pp. 126–133 (2013)
10. Goyal, P., Khan, M., Teshome, N., Geary, B., Bryce, R.: Context data compact prediction tree (CD-CPT): transforming user experience through predictive analysis, vol. 2, pp. 166–173. SCITEPRESS (2024). http://www.scitepress.org/Papers/2024/126158
11. Letham, B., Rudin, C., Madigan, D.: Sequential event prediction. Mach. Learn. **93**(2), 357–380 (2013). https://doi.org/10.1007/s10994-013-5356-5
12. Sarker, I.H.: Context-aware rule learning from smartphone data: survey, challenges and future directions. J. Big Data **6**(1), 95 (2019). https://doi.org/10.1186/s40537-019-0258-4
13. Fournier-Viger, P., Yang, Y., Yang, P., Lin, J.C.-W., Yun, U.: TKE: mining top-k frequent episodes. In: Trends in Artificial Intelligence Theory and Applications. Artificial Intelligence Practices: 33rd International Conference on Industrial, Engineering and Other Applications of Applied Intelligent Systems, IEA/AIE 2020, Kitakyushu, Japan, September 22-25, 2020, Proceedings, vol. 33, pp. 832–845. Springer (2020)
14. Chen, Y., Fournier-Viger, P., Nouioua, F., Wu, Y.: Mining partially-ordered episode rules with the head support. In: Golfarelli, M., Wrembel, R., Kotsis, G., Tjoa, A.M., Khalil, I. (eds.) DaWaK 2021. LNCS, vol. 12925, pp. 266–271. Springer, Cham (2021). https://doi.org/10.1007/978-3-030-86534-4_26

15. Goyal, P., Seale, A., Charubin, K., Bennett, Q., Bryce, R.: AI-powered smartphone context and high-utility itemset mining for enhanced app testing and personalization. In: Latifi, S. (ed.) ITNG 2024: 21st International Conference on Information Technology-New Generations, pp. 427–435. Springer Nature Switzerland, Cham (2024)
16. SPMF: a Java open-source data mining library. https://www.philippe-fournier-viger.com/spmf/index.php?link=documentation.php
17. Fournier-Viger, P., Gomariz, A., Campos, M., Thomas, R.: Fast vertical mining of sequential patterns using co-occurrence information. In: Tseng, V.S., Ho, T.B., Zhou, Z.-H., Chen, A., Kao, H.-Y. (eds.) PAKDD 2014. LNCS (LNAI), vol. 8443, pp. 40–52. Springer, Cham (2014). https://doi.org/10.1007/978-3-319-06608-0_4
18. Breiman, L.: Random forests. Mach. Learn. **45**(1), 5–32 (2001). https://doi.org/10.1023/A:1010933404324
19. Biau, G., Scornet, E.: A random forest guided tour. TEST: Off. J. Span. Soc. Statist. Oper. Res. **25**(2), 197–227 (2016). https://ideas.repec.org/a/spr/testjl/v25y2016i2d10.1007_s11749-016-0481-7.html
20. Cortes, C., Vapnik, V.: Support-vector networks. Mach. Learn. **20**(3), 273–297 (1995). https://doi.org/10.1007/BF00994018
21. What Are Support Vector Machines (SVMs)? (2023). https://www.ibm.com/topics/support-vector-machine
22. Nalepa, J., Kawulok, M.: Selecting training sets for support vector machines: a review. Artif. Intell. Rev. **52**(2), 857–900 (2019). https://doi.org/10.1007/s10462-017-9611-1
23. Hochreiter, S., Schmidhuber, J.: Long short-term memory. Neural Comput. **9**, 1735–80 (1997)
24. Sherstinsky, A.: Fundamentals of recurrent neural network (RNN) and long short-term memory (LSTM) network. Physica D **404**, 132306 (2020)
25. Siami-Namini, S., Tavakoli, N., Namin, A.S.: The performance of LSTM and BiLSTM in forecasting time series. In: 2019 IEEE International Conference on Big Data (Big Data), pp. 3285–3292. IEEE (2019). https://ieeexplore.ieee.org/document/9005997
26. Piparia, S., Khan, M., Bryce, R.: Discovery of real world context event patterns for smartphone devices using conditional random fields. In: ITNG 2021 18th International Conference on Information Technology-New Generations, pp. 221–227. Springer (2021)

The Importance of Network Neutrality

Victor Clincy(✉) and Lama Zaabab

Computer Science Department, Kennesaw State University, Marietta, GA 30060, USA
vclincy@kennesaw.edu

Abstract. Network neutrality (net neutrality) is the idea that all information on the Internet should be treated equally. It is an essential component to the Internet. The repeal of net neutrality, the philosophy of equal treatment of all data on the web by Internet Service Providers (ISPs), was met with much disappointment, and the fight for net neutrality has been a long battle. With its recent restoration, and in a world with an increasing use of technology, cyber laws are relevant now more than ever. The Federal Communications Commission (FCC) was in charge of getting rid of net neutrality in the first place, and also reinstating it as of recent. The fairness of net neutrality is ensuring ISPs are not bought out is essential because it promotes the success of small businesses and access to information for all. COVID-19 gave a taste of the Internet with and without net neutrality. The historical context with Ajit Pai's hand in the fight is necessary to highlight. The history of the fight for and against net neutrality gives a taste of the ever-changing landscape of the Internet and how network slicing, discrimination, and regulations play into each other.

Keywords: Network neutrality · Net neutrality · ISP · FCC

1 Introduction

Network (net) neutrality mandates Internet Service Providers (ISPs) to treat all data on the web equally. It is an Obama-era regulation put on the Internet in order to prevent ISPs from speeding up, slowing down, or blocking any content, websites, or applications (Hildebrandt & Wiewiorra, 2023). Net neutrality regulations were put into place in 2015, repealed in 2017, and put into place in 2018 (Jamison, 2018). The Federal Communications Commission (FCC) was responsible for approving the plan to get rid of net neutrality. The Internet broke out against this vote, and people began getting acquainted with the idea of an Internet without the net neutrality regulations—a concept so foreign yet so familiar. Eventually, the FCC reinstated net neutrality regulations with the rules coming into effect as of recent in July 2024.

The journey was a long hard fought battle. Supporters of the repeal of net neutrality, however, argue that the Internet was better before the fairly recent additions of net neutrality regulations, and that the new way of calculating Internet usage through bundles is, in fact, fairer and more convenient. Although at the surface an Internet without net neutrality may seem to have these qualities, the sheer power Internet Service Providers

S. Latifi (Ed.): ITNG 2025, AISC 1463, pp. 609–614, 2025.
https://doi.org/10.1007/978-3-031-89063-5_52

gain with its removal wipe out any hope of a truly fair Internet. The ability to access information is a right guaranteed by the First Amendment, but the removal of the regulation shifts that right to a luxury.

2 Internet Service Providers

Regardless, it is important to note just how necessary net neutrality is. Without net neutrality, ISPs have too much power. Their ability to slow down, speed up, and block content turns into injustice when they can essentially control what users view. The lack of net neutrality is a censorship which directly places the United States with the lot of authoritarian countries which manage what their citizens are able to access. With this lack of net neutrality, the ISPs can promote media associated with their partisan views by slowing or blocking content they disagree with (Hildebrandt & Wiewiorra, 2023). In an already polarized society, the idea of people only being able to access information associated with one particular political view is frightening. With the repeal of net neutrality allowing Internet Service Providers to slow down or block certain content, it gives rise to a new partisan-based issue—they can now block content they disagree with. Save The Internet cautions that depending on the politics of the Internet Service Provider one is associated with, the consumers would have specific news content faster or more widely available to them (Hildebrandt & Wiewiorra, 2023). This issue brings about an ethical dilemma in that information would no longer be truly widely available to everyone. The United States would become increasingly polarized as citizens would be more inclined to hold the views of the political party their Internet Service Provider is associated with, and as such, would have more access to information aligning with only their party and a lack of information about the other side/sides. Net neutrality is crucial for minorities, especially communities of color who are able to share their stories of racial injustices they encounter. This demise of net neutrality means that the fundamental rights of Americans guaranteed by the first amendment of the United States Constitution—freedom of speech, freedom of the press—are no longer protected.

With the power ISPs gain with the repeal of net neutrality, small businesses are at stake. They are now put at a large disadvantage to large businesses merely due to their size and wealth. ISPs are able to speed up the content of the largest provider of money, which boots small businesses to the dugout considering their websites and content will be slower and last (Greenstein, Peitz, & Valletti, 2016). This repeal essentially wipes out real competition which is one of the reasons people come to this country – in search of something better, something hopeful. Net neutrality ensures an equal platform for all businesses, but without it, already-large businesses are able to get ahead and make it close to impossible for small businesses to succeed. The demise of net neutrality means that Internet Service Providers can make a lot of money off large business' payments, and that small businesses will struggle. This is because these large corporations can now pay these Internet Service Providers, such as AT&T, Comcast, and Verizon, to make their content faster. This puts the small businesses at a disadvantage considering consumers are enticed to follow the fastest content available. This grows especially important when faced with the fact that Netflix itself was a small business that has grown to be one of the most widely used television services today. Without net neutrality, which levels the initial playing field for all businesses, Netflix would never have grown to the business it is today. Even Netflix highlighted how they would not have made it to where they are today without net neutrality regulations considering they themselves started out as a small business (Wygorski, 2019). The repeal of net neutrality puts limits to creativity and innovation that small businesses are able to bring about. With the new possibility of large corporations paying off Internet Service Providers to make their content faster than their competitors', money becomes too large of a factor in whether a business will succeed.

3 Federal Communications Commission

The use of home broadband increased between twenty percent and forty percent. It was such an increase in Internet traffic, especially video streaming, that ISPs utilized the lack of net neutrality to promote certain traffic over others (Trivino, Franco, & Ochoa, 2021). The pandemic highlighted "Network discrimination is rewidening the gap for access to educational opportunities," (Yamagata-Lynch et al., 2017). The lack of net neutrality unfortunately widened the gap for educational opportunities in COVID considering underprivileged families did not have the same internet access. Network slicing played a role in the changing Internet to come with regulations thrown in the air (Yoo, 2024).

Back in mid-2017, the FCC issued its comment period for its proposal on the riddance of net neutrality. The FCC had to extend the end of the comment period, August 16th, by two weeks, to August 30th, due to the numerous complaints filed (they did not have enough time to wade through them all.) With over 20 million filings, the proposal has been the most commented entry in FCC's history, and it can be compared with the 3.7 million filings received for the proposal to create net neutrality rules (Novak & Hajibayova, 2018). The comment even became inaccessible at one point due to a spike in traffic after a John Oliver segment on net neutrality, but the FCC claimed that this inaccessibility was due to a DDoS (Distributed Denial of Service) attack. The sheer difference in the number of filings for the enactment and repeal of net neutrality shows the difference in the appeal.

The former director of the FCC is Ajit Pai, a former Verizon lawyer chosen to head the executive agency, and his stance on net neutrality grows concerning when considered with the events that followed. He is the one who released the false statement claiming that the FCC comments system's shutdown was due to a DDoS attack. Many were doubtful of these claims, but Pai stuck to his story and added that the same attack happened back in 2014 when net neutrality laws were being put in place. Later evidence coming from senior officials confirmed the falsity of the 2014 claims. It was not until a series of internal emails containing information about the FCC's manufacture of the DDoS attack were exposed that Ajit Pai finally spoke up, attributing his lies to, "a flawed comment system we inherited from the prior Administration" (Trivino, Franco, & Ochoa, 2021). Pai's shift of the blame was especially alarming considering the series of lies told to prevent the American people from knowing something they already knew: people want net neutrality. How can he be trusted when relaying information the American people do not already know? He conveniently had members of Congress fooled for seven months in terms of the FCC comment traffic concerning net neutrality regulations. It is disappointing to ponder how the December 2017 vote could have been different. Pai, compelled to tell the "truth," could not admit to what he did, but the overall coverup illustrates the FCC's knowledge of how unpopular the removal of net neutrality regulations is.

Ajit Pai's fabrication of the DDoS attack is not the only thing he is guilty of, though, considering the allegations regarding his association with Sinclair Broadcasting. The alleged coordination was called to be investigated by Representatives Pallone and Cummings. Sinclair Broadcasting, the largest broadcaster in the United States, wanted to merge with Tribune Media, and Pai was investigated for changing rules in order to allow for this $3.9 billion merge to happen. The 72% postmerger reach to United States homes would well surpass the Congress-imposed cap of 39% reach (Jamison, 2018). The issue with the merge surrounded the excessive national power Sinclair would gain, which discourages locality. The clear violation of the Congress cap combined with the myriad of enacted FCC rules that have benefited the broadcasting company give cause for questions concerning the timing of the rules. His previous ties to Verizon add on to suspicions of "perfect timing."

The Sinclair-Tribune merger, whether despite or because of the allegations against Pai, has thankfully been denied. Even with the deregulating proprietorship rules from Pai, Sinclair was mandated to sell some stations as a condition to be able to merge with Tribune. Sinclair, however, in its documents highlighting who would acquire the stations, planned to sell them to companies where Sinclair still had power—directly and indirectly. The FCC attributed their unanimous decision to decline the merger based on "misrepresentation or lack of candor" from Sinclair Broadcasting (Hildebrandt & Wiewiorra, 2023). The conflict of interest concerning Sinclair and its relentless efforts to hold onto any power they can acquire was enough to sway the FCC vote to turn down the merge deal. Tribune is also spurring up a legal case against Sinclair for breach of contract. President Trump came out against the result of the vote in a tweet declaring the decision incredibly "sad and unfair" (Trivino, Franco, & Ochoa, 2021). It is no surprise Trump would want Sinclair to increase their coverage to virtually the entire country. President Trump's son-in-law, Jared Kushner, declared he had an arrangement with Sinclair in early 2016 for advantageous coverage of Trump during the presidential campaign (Trivino, Franco, & Ochoa, 2021). Sinclair's close ties with the Trump administration give all the more reason to decline the merger. Sinclair already has the most widespread coverage in the United States over any other broadcasting company. Another increase, for almost double the coverage it has now, would have the potential to become an issue of controlled partisan-based political knowledge, and would mean the potential creation of an illegal media monopoly.

Critics of net neutrality argue that the Internet operated superbly without the regulations beforehand, and that it would be better for consumers to have no net neutrality. The Internet became widely available to the public in 1990, and until the Obama-era legislation in 2015, there was no net neutrality (Hildebrandt & Wiewiorra, 2023). After all, the Internet has existed without net neutrality regulations far longer than it has with them. What is important to note, however, is that that is an extremely small portion of time for a concept so new to the public to truly develop. The development of the Internet brought about issues that made the placement of net neutrality regulations necessary in the first place. Verizon, for example, had blocked PayPal for its users for a lengthy period of time before net neutrality, perhaps because of their own new version of the payment system, "Isis" (Trivino, Franco, & Ochoa, 2021). Internet Service Providers had not been widely introduced to the idea, and now that lots of money could be made, they had new plans thanks to the help of the FCC. This opportunity of blocking opponents' information provided a huge incentive—money. The net neutrality regulations became necessary due to the unfair treatment of information on the Internet by the Internet Service Providers.

Another point opponents of net neutrality argue is that perhaps the bundle way of calculating Internet usage costs is far better. This would put those who do not use numerous forms of media at an advantage as they would see significant decrease in their Internet data (Trivino, Franco, & Ochoa, 2021). In theory, the idea that those who use more should pay more, and those who use less should pay less seems like a reasonable ask. However, the bundles would be split based on sites that people of all demographics use. For example, one bundle could include Gmail, and the other would include USNews or Netflix. Users would have to split their costs up and, because it would become a burden, most would eventually be obliged to limit their information usage and access. This limit would most probably create an advantage to information for the wealthy and an increased gap in educational value for the poor.

4 Conclusion

It is hard to grasp what the lack of net neutrality means with mere speculation, but countries such as Guatemala and Portugal with a great lack of net neutrality regulations can provide a clearer picture. In Guatemala, numerous people own more than one SIM card because each will provide a different app for free (Yoo, 2024). For example, one card will allow for one to access GroupMe for free, while another gives free access to Skype. These apps can still be used when their Internet data runs out, but if one wants to access any other website, they would need to pay. SIM cards are not as easily transferrable in the United States, but this showcases the hassle of the bundle calculations. In Portugal, although the country adheres to net neutrality rules enacted by the European Union, it permits bundle-pricing Internet calculations. Users have standard data based on the initial payment but are required to pay an additional monthly fee for any of the other five packages distinguished by messaging, social, video, music, and email and cloud (Yoo, 2024). By this method, if a user wanted to use Instagram, Apple Music, and WhatsApp they would be obligated to pay for three extra monthly bundles. Not only does this method of Internet usage cost calculation raise a financial concern, but it raises a concern in terms of fair competition for upstart apps. For example, if a user loves listening to music, but the bundle includes Apple Music without Spotify, then the user will have more of an incentive to use Apple Music considering that Spotify would be more expensive. Competition becomes hard for apps/businesses without a great pool of money to start with.

The reinstatement of net neutrality was a long hard-fought battle at both the state and federal level. The repeal of net neutrality brought about many concerns in terms of the degree to which ISPs control users' access. The ability of ISPs to block, slow down, and speed up content limits competition, small businesses, and access to information while allowing the ISP to acquire large sums of money from large businesses and data users.

Net neutrality allows for a leveled initial playing field for all businesses, users to pay smaller amounts for more information, and better options in general. With the increase in Internet traffic and concepts such as network slicing, there are many nuances to net neutrality which are still necessary to explore.

References

Greenstein, S., Peitz, M., Valletti, T.: Net neutrality: a fast lane to understanding the trade-offs. J. Econ. Perspect. **30**(2), 127–150 (2016). https://doi.org/10.1257/jep.30.2.127

Hildebrandt, C., Wiewiorra, L.: The past, present, and future of (net) neutrality: a state of knowledge review and research agenda. J. Inf. Technol. **39**(1), 167–193 (2023). https://doi.org/10.1177/02683962231170891

Jamison, M.A.: Net neutrality policies and regulation in the United States. Rev. Netw. Econ. **17**(3), 151–173 (2018). https://doi.org/10.1515/rne-2018-0041

Novak, A., Hajibayova, L.: Net neutrality in the context of provision of fair and equitable access to information sources and services. Pub. Libr. Quart. **38**(1), 19–33 (2018). https://doi.org/10.1080/01616846.2018.1524703

Trivino, R.D., Franco, A.A., Ochoa, R.L.: Internet and net neutrality in the time of COVID-19: a global overview. In: 2021 Eighth International Conference on eDemocracy & eGovernment (ICEDEG), pp. 133–138 (2021). https://doi.org/10.1109/icedeg52154.2021.9530952

Wygonski, J.: Net neutrality enforcement is necessary to provide small businesses with the opportunity for fair competition. Ohio State Bus. Law J. **13**(2), 403–432 (2019)

Yamagata-Lynch, L.C., Despande, D.R., Do, J., Garty, E., Mastrogiovanni, J.M., Teague, S.J.: Net neutrality and its implications to online learning. Int. Rev. Res. Open Distrib. Learn. **18**(6) (2017). https://doi.org/10.19173/irrodl.v18i6.3129

Yoo, C.S.: Network slicing and net neutrality. Telecommun. Policy **48**(2) (2024). https://doi.org/10.1016/j.telpol.2023.102619

Online Advertising is a Regrettable Necessity: On the Dangers of Pay-Walling the Web

Yonas Kassa(✉)

School of Data Intelligence & Technology, Robert Morris University, Pittsburgh, PA, USA
ykassa@acm.org

Abstract. The exponential growth of the web and its benefits can be attributed largely to its open model where anyone with internet connection can access information on the web for free. This has created unprecedented opportunities for various members of society including the most vulnerable, as recognized by organizations such as the UN. This again can be attributed to online advertising, which has been the main financier to the open web. However, recent trends of paywalling information and services on the web are creating imminent dangers to such open model of the web, inhibiting access for the economically vulnerable, and eventually creating digital segregation. In this paper, we argue that this emerging model lacks sustainability, exacerbates digital divide, and might lead to collapse of online advertising. We revisit the ad-supported open web business model and demonstrate how global users actually pay for the ads they see. Using data on GNI (gross national income) per capita and average paywall access costs, we established a simple income-paywall expenditure gap baseline. With this baseline we show that 135 countries with a total population estimate of 6.56 billion people cannot afford a scenario of a fully paywalled web. We further discuss how a mixed model of the so-called "premium services" creates digital segregation and poses danger to online advertising ecosystem. Finally, we call for further research and policy initiatives to keep the web open and more inclusive with a sustainable business model.

1 Introduction

The advent of the World Wide Web has revolutionized the way we access and share information, leading to an enormous amount of accessible knowledge and opportunities for individuals and societies globally. The open model of the web which allows anyone with an internet connection to access and contribute information for free, has been instrumental in creating a global information ecosystem that has benefited billions of users worldwide [6]. This open model of the web has provided unprecedented opportunities for the most vulnerable members of society, enabling them to access educational resources, connect with communities, and participate in the digital economy and social development [9]. It is important

S. Latifi (Ed.): ITNG 2025, AISC 1463, pp. 615–624, 2025.
https://doi.org/10.1007/978-3-031-89063-5_53

to note that getting access and staying connected to the internet by itself is an expensive challenge for most parts of the world [27]. Recognizing this challenge (also referred to as first level digital divide) various efforts have been made by governments, industry, and research bodies to bring internet connection widely accessible[36]. The UN has identified ICT access as one of sustainable development goals [9]. Researchers have also studied approaches to address the second challenge, such as creating low-cost ICT devices, creating innovative solutions to make internet reachable to farthest places, and lowering the user cost of web traffic and improving user experience with low-speed internet[18]. Despite the many challenges such as high cost of connecting to internet [27], online privacy [28] and security issues [34], the ad-supported web has generally been free (arguably) and open without requiring payment beforehand, until recently.

The primary financier of this open web has been online advertising [17]. As such, online advertising has enabled content creators and service providers to monetize their offerings and sustain their businesses, providing global internet users with a wealth of free information and online services. However, as we will see in the rest of the paper, recent trends towards paywalling information and services on the web pose imminent dangers to this open model.

The multifaceted benefits of the open web is enormous and the advantages it has been offering to the least privileged has been unprecedented. Multiple reports have underlined its transformative power such as its effect on bridging employment inequality [19], reducing the proportion of households(particularly women) below the poverty line [2], and helping eradicate illiteracy [39].

The emergence of paywalling an information content or part of it as a business model can be linked to several factors, including the high costs of producing and operating a large-scale information content and service, and the ad-tech shift towards highly targeted advertising and entailed user technologies to circumvent it [26]. While these emerging paywalls may seem to provide short-term financial gains for content platforms and service providers, they exacerbate the digital divide and lack sustainability as a business-model in the long term.

In this paper, we raise concerns about the sustainability and inclusivity of the emerging trend towards paywalled content and services on the web. We highlight dangers of this emerging challenge contending that a fully paywalled web is unsustainable and poses significant risks to entire advertising-supported ecosystem. After covering some of important related work in the area, we begin with revisiting the ad-supported open web business model and demonstrate how global users actually pay for the ads they see. Using results from previous research [29] and the GNI per capita data, we then discuss how cost of accessing a small portion of content on internet is impractical for most part of the world. Next, we discuss how a mixed model of premium services creates digital segregation and poses a danger to the online advertising ecosystem. With these points we aim to shed light on how this emerging business model exacerbates existing inequalities and creates new ones. Then, we proceed to call for research and policy efforts towards creating ethical and desirable advertising systems that can keep the web open and more inclusive. Finally, we conclude the paper pointing limitations of our work and suggesting potential directions for future work.

2 Related Work

While online advertising has been widely acknowledged as backbone of the open web [14,20] enabling free global information dissemination, its modus operandi has always been the subject of user dissatisfaction [8]. The dissatisfaction primarily emerges from embedded adverts between user interactions that degrade user experience [25,40], and the increasing incorporation of sophisticated tracking mechanisms. Advertisers use tracking techniques to collect user activity data with the aim of bringing most relevant ads to the users [4,5]. In relation to this, multiple open platforms serving billions of users have been a subject of several scrutiny [24,31]. A particular example is the Cambridge Analytica case [15], that attempted to influence US presidential election analysing user behavior, which was widely covered by mainstream media resulting in global awareness on privacy. Even though the Cambridge Analytica study was performed on Facebook, it is worth to note that the method can theoretically be replicated to other social platforms such as X (formerly Twitter) and TikTok. Privacy risks associated with online advertising have been widely documented [16,22]. The advertising monetary value of user data has also been documented in multiple studies [23,28]. While these studies showed the risks and undesirable sides of online advertising and collecting user data, other results have shown how proper handling of user data such as via anonymization and aggregation can effectively be used for societal benefit [13,32,33]. Furthermore, multiple works have proposed to improve user interaction with online advertising with approaches such as bringing transparency and improving user privacy [10,30], and ad profit sharing with users [7]. A few research works have explored impact of advertising on ad-supported open web [4,29,35]. Varian [38] studied interplay between advertising costs and product prices and has shown that when advertisers cut back on ad spending, they do not necessarily change product prices accordingly. A paper by Shiller et al. [35] studied the compounded impact of ad-blocking on revenue concluding that ad blocking poses a substantial threat to the ad-supported web. A more recent article of high relevance to our discussion is a paper by Papadopoulos et al. [29] which studies prevalence of paywalls and discussed their impacts. Their findings show that paywall use has been increasing, and the median cost of an annual paywall access is 108 USD per site.

3 Revisiting Ad-Supported Internet

A well-known architectural presentation to discuss modern day online advertising systems involves putting advertisers and end users on the two extremes that are connected via online advertising system as intermediary. In this popular scenario the online advertising system composed of parties like supply side platform, ad exchange, and demand side platforms tracks user activities to gather

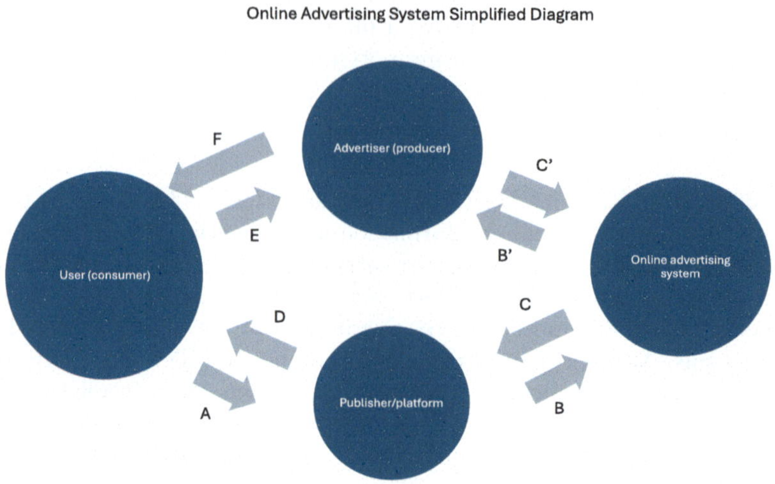

Fig. 1. Diagram Revisiting ad-supported Web: First, advertiser assigns a total cost (TC) to a product/service by adding the advertising dollar (Ca) to the initial cost of a product/service. When a user visits ad-supported publisher/platform (arrow A), the platform contacts ad-system (arrow B), which in turn performs RTB (arrow B') that leads to winning advertiser sending back ads and a dollar amount (arrow C'). the advertising system then sends the winning ad along with a fraction of the dollar amount to the ad-slot on the platform (arrow C). The ad is presented to the user (arrow D). Up on conversion, user eventually pays the dollar amount for ad they saw (arrow E) plus Cps, finally product/service is delivered to user (arrow F).

data that can be used during Realtime Bidding (RTB). When a user visits ad-supported platform RTB takes place, where potential advertisers bid for ad-slot to be displayed as part of the content presented to the end user [41]. To build the most optimal matching between advertisers and ad-slots to each individual user advertising system uses a variety of sophisticated techniques that heavily make use of personal data [37]. Though it is widely known that user personal data is a tradable commodity, the fact that end users actually pay for the ads they see is often overlooked.

Online advertising is a multibillion-dollar business. This multibillion-dollar budget comes from advertisers which are producers of goods and services that aim to reach their potential clients or customers. Companies spend a significant portion of their revenue to advertising, the average ad spending budget of small businesses ranges between 4–6% of their gross revenue, while ad spending by some companies rises up to 20% [12].

Let's take a scenario of company X getting its customers exclusively via online advertising. If gross revenue of company X (i.e. total sum of products and services sold) is R, then a portion of this revenue goes to online advertising Ca. Let X's target cost after advertising be Cps.

Then,

$$Cps = R - Ca \tag{1}$$

This gross revenue, R, is the sum of total cost (TC) paid by X's clients/customers.

$$TC = Cps + Ca \tag{2}$$

$$TC = \sum_{i \epsilon items} \sum_{j \epsilon users} \left(Cps_{(j)}^{(i)} + Ca_{(j)}^{(i)} \right) \tag{3}$$

With this second equation, we can say that the advertising budget is practically paid by end users. We can diagrammatically see this model on Fig. 1. The diagram can be summarized as follows: initally advertiser assigns a total cost (TC) to a product/service by adding the advertising cost (Ca) to the initial target cost of a product/service. The advertiser places this ad with advertising dollar Ca on RTB. The advertiser then sends the winning ad through the publisher/platform paying a fraction of Ca to the platform. Up on conversion, user eventually pays for ad they saw. This visualization demonstrates that the advertising cycle can sustainably continue as long as the system is healthy and end users find products/services advertised to them relevant and appealing. It also suggests us to postulate that end users practically pay for the ads they are exposed to.

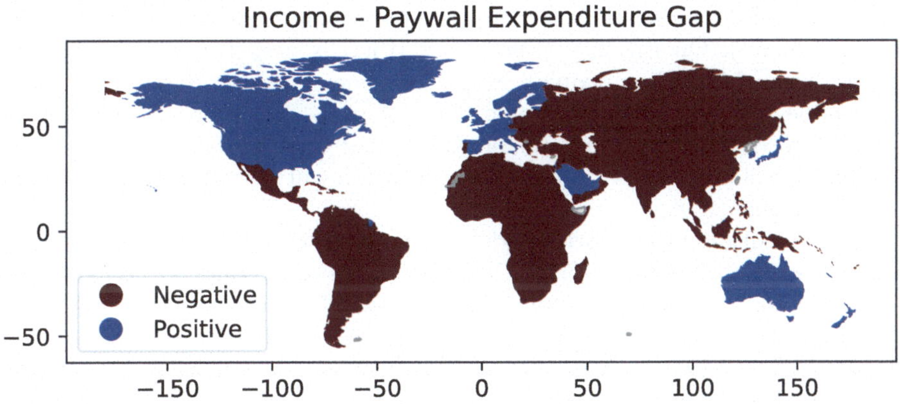

Fig. 2. IPEG, income paywall expenditure gap (blue indicates IPEG value > 0, dark red indicates IPEG < 0) demonstrating majority of countries having negative IPEG value indicating that paywall budget cost is already higher than the average before tax income of a country's citizens.

4 Pay-Walling the Internet Exacerbates Digital Segregation

Given there are over a billion websites in the world today, users are faced with multiple alternatives of varying degrees of utility for their digital needs. This information overload also exposes users to multiple challenges such as disinformation [21], and malware [34]. Services such as Google AdWords and SimilarWeb provide categorization of websites into interest groups such as books, music, business services, adult, games, etc. SimilarWeb[1] for instance provides 250 such categories. It is challenging to analyze and categorize each individual website to decide which platform can be used as a replacement (representative) for other websites in its category. For example, CNN and Fox News fall under news and media category, but it is known that one cannot replace the other [11]. An interesting example of website uniqueness is the case when the Seattle times temporarily had to lift its paywalls during the 2014 Oso mudslide recognizing that it was the main platform covering the disaster on its website [1].

However, for our discussion we will assume that a user can satisfy their information needs with one website from each category (i.e. user can live with only 250 websites). According to results from Papadopoulos et al. [29], median annual paywall access is 108 per site, this gives that in a fully paywalled scenario a user will need to budget 27,000 USD (i.e. 108 USD*250) just to get past paywalled services. Taking the GNI (Gross national income) per capita, GNIpc, of countries from world bank, we can estimate a baseline income-paywall expenditure gap, IPEG, for users per country.

$$IPEG = GNIpc - PWAC \tag{4}$$

For paywall business model to be sustainable, the IPEG has to be positive (bare minimum). On Fig. 2, we show that IPEG is negative on 135 countries (amounting to 6.56 billion people). Note that some of those countries don't even have online payment system even if a few of users can afford to pay for the paywalls. Also, it is important to remember that even in those countries with blue IPEG, a significant portion of the population will have difficulty allocating paywall budget.

Now let's discuss the hybrid model which is spearheaded by large platforms such as YouTube, Twitter (X), and TikTok[2]. Data on of some of the most popular platforms and their premium fees is presented on appendix A. One of the main incentives to go premium on such platforms is ad-free experience or a service with fewer ads. Hybrid model gives rich people option to avoid watching ads, get higher quality information and experience. However, as we saw previously, targeted advertising via RTB works by finding potential customers for each advertiser. If certain groups of advertisers can't find their target users they will avoid investing in online advertising (e.g. advertisers of cars, luxury watches, villas, high end phones will pull their advertising budget for lack of reaching potential customer) and hence reducing the market capitalization of total advertising ecosystem. In other words, if we rearrange Eq. 3 based on income, rich and affluent groups will disappear from Eq. 5.

$$TC = (Cps + Ca)^{rich} + (Cps + Ca)^{affluent} + (Cps + Ca)^{poor} \tag{5}$$

[1] https://support.similarweb.com/hc/en-us/articles/213856729-Top-Websites.

[2] https://techcrunch.com/2023/02/27/social-media-apps-adopting-subscription-models/.

Note that advertisers pulling from platforms has immediate effect with potential to kill the company (advertisers vs X (formerly Twitter), for example[3]). Reduced advertising economy will discourage content generators and online service providers from serving the poor, specifically people in least developed regions will be at a high risk. On the other hand, economically vulnerable people in developed regions will be unable to access paywalled content and hence will be digitally segregated getting less quality digital content, making them less informed, and less competent in general.

5 Conclusion

5.1 We Need More Desirable and Ethical Ads

So far we have discussed the important role ads play in keeping the web open, and the impracticality of the emerging premium-paywall business models. We have also raised some of the concerns that make advertising undesirable part of the web. However, there is not alternative business model that is viable and as scalable as online advertising that can drive an open and inclusive global scale information exchange. If we address its undesirable issues, Online advertising can be the most viable information super-highway that connects producers with consumers, or act as a mechanism to deliver timely information to the masses as it was proven during COVID-19 pandemic [3]. Our position is that with enough research and strong policy, there is a possibility to device an advertising framework that balances the needs of different stakeholders while ensuring the sustainability and inclusivity of the web. Rather than shifting towards paywalled services, our focus should be on fixing and refining this incumbent infrastructure to make advertising a desirable feature of the web.

6 Limitations and Future Work

While we cover several aspects of this emerging issue in this short paper, our approach has several limitations. First, in our IPEG calculation we didn't take into account other expenses which made our baseline highly generous. Second, we relied on SemanticWeb categorization, a more detailed and sophisticated content and traffic analysis could result in a more accurate number of representative websites. Finally, we have not included the possible solutions to improve online advertising. Future work could capitalize on these limitations.

Acknowledgments. Author(s) would thank reviewers for their valuable feedback and insightful comments. preprint version of this paper was shared on arXiv at: https://arxiv.org/pdf/2409.00026.

A Appendix A: Premium Costs of Some of the Popular Platforms

(See Table 1).

[3] https://time.com/6340905/elon-musk-advertiser-boycott-x/.

Table 1. Paywall costs of some of the popular websites from top 50 websites based on similarweb.com. Note that some of the platforms offer discounts when purchased in bulk or if paid for the full year plan.

Platform	*Monthly*	*Annual*	*Name*
youtube.com	13.99	167.88	YouTube premium
instagram.com	11.99	143.88	verified
twitter.com	16	192.0	premium+
xvideos.com	9.99	119.88	premium
tikTok.com	4.99	59.88	premium
reddit.com	5.99	71.88	premium
openai.com	20	240	plus
netflix.com	22.99	275.88	premium
office.com	6.99	83.88	personal
Zoom.us	13.33	159.96	pro
Total	*126.26*	*1515.12*	

References

1. Ananny, M., Bighash, L.: Why drop a paywall? Mapping industry accounts of online news decommodification. Int. J. Commun. **10** (2016). 3359–3380
2. Bahia, K., et al.: The welfare effects of mobile broadband internet: evidence from Nigeria. J. Dev. Econ. **170**(2024), 103314 (2024)
3. Breza, E., et al.: Effects of a large-scale social media advertising campaign on holiday travel and COVID-19 infections: a cluster randomized controlled trial. Nat. Med. **27**, 9 (2021). 1622–1628
4. Budak, C., Goel, S., Rao, J., Zervas, G.: Understanding emerging threats to online advertising. In: Proceedings of the 2016 ACM Conference on Economics and Computation. ACM, USA, pp. 561–578 (2016)
5. Castelluccia, C., Olejnik, L., Minh-Dung, T.: Selling off privacy at auction. In: Network and Distributed System Security Symposium (NDSS), NDSS, USA, p. 16 (2014)
6. Chaqfeh, M., et al.: Towards a world wide web without digital inequality. Proc. Natl. Acad. Sci. **120**, 3 (2023), e2212649120
7. Chen, Y., Ghosh, A., McAfee, R.P., Pennock, D.: Sharing online advertising revenue with consumers. In: International Workshop on Internet and Network Economics, pp. 556–565. Springer, USA (2008)
8. Chua, M.Y.K., Yee, G.O., Gu, Y.X., Lung, C.H.: Threats to online advertising and countermeasures: a technical survey. Digit. Threats: Res. Pract. **1**, 2 (2020). 1–27
9. Tshilunda, M.A., et al.: Towards a New Social Contract: reducing inequalities through digital public goods and youth collaboration for the Sustainable Development Goals (SDGs). UN Digit. Librar. **1**(1), 15 (2022). http://digitallibrary.un.org/record/4014948
10. Fiesler, C., Hallinan, B.: We are the product public reactions to online data sharing and privacy controversies in the media. In: Proceedings of the 2018 CHI Conference on Human Factors in Computing Systems. CHI, USA, pp. 1–13 (2018)

11. Flamino, J., et al.: Political polarization of news media and influencers on Twitter in the 2016 and 2020 US presidential elections. Nat. Hum. Behav. **7**(6), 904–916 (2023)
12. Frenkiel, T.: Digital marketing budget allocation in six steps, 2023. https://funnel.io/blog/allocate-digital-marketing-budget
13. Garcia, D., et al.: Analyzing gender inequality through large-scale Facebook advertising data. Proc. Natl. Acad. Sci. **115**(27), 6958–6963 (2018)
14. Glenday, J.: One in four Brits don't realise the internet is funded by advertising, 2020. https://www.thedrum.com/news/2020/10/19/one-four-brits-don-t-realise-the-internet-funded-advertising. Accessed 04 Apr 2024
15. González, F., Yu, Y., Figueroa, A., López, C., Aragon, C.: Global reactions to the Cambridge analytica scandal: a cross-language social media study. In: Companion Proceedings of the 2019 World Wide Web Conference. World Wide Web Conference, USA, pp. 799–806 (2019)
16. González-Cabañas, J., Cuevas, Á., Cuevas, R., López-Fernández, J., García, D.: Unique on Facebook: formulation and evidence of (nano) targeting individual users with non-PII data. In: Proceedings of the 21st ACM Internet Measurement Conference. ACM, USA, pp. 464–479, 2021
17. Greenwood, J., Ma, Y., Yorukoglu, M.: 'You will:'A macroeconomic analysis of digital advertising. Technical report, National Bureau of Economic Research (2021)
18. Habib, R., et al.: A framework for improving web affordability and inclusiveness. In: Proceedings of the ACM SIGCOMM 2023 Conference. SIGCOMM, USA, pp. 592–607, 2023
19. Hjort, J., Poulsen, J.: The arrival of fast internet and employment in Africa. Am. Econ. Rev. **109**(3), 1032–1079 (2019)
20. Kassa, Y.M., Cuevas, R., Cuevas, Á.: A large-scale analysis of Facebook's user-base and user engagement growth. IEEE Access **6**(2018), 78881–78891 (2018)
21. Kumar, S., West, R., Leskovec, J.: Disinformation on the web: impact, characteristics, and detection of wikipedia hoaxes. In: Proceedings of the 25th International Conference on World Wide Web. WWW, USA, pp. 591–602, 2016
22. Leon, P.G., Rao, A., Schaub, F., Marsh, A., Cranor, L.F., Sadeh, N.: Privacy and behavioral advertising: towards meeting users' preferences. In: Symposium on Usable Privacy and Security (SOUPS), pp. 22–24. SOUPS, USA (2015)
23. Marciel, M., Gonzalez, J., Kassa, Y.M., Gonzalez, R., Ahmed, M.: The value of online users: empirical evaluation of the price of personalized ads. In: 2016 11th International Conference on Availability, Reliability and Security (ARES), pp. 694–700. IEEE, USA (2016)
24. Matz, S.C., Appel, R.E., Kosinski, M.: Privacy in the age of psychological targeting. Curr. Opin. Psychol. **31**(2020), 116–121 (2020)
25. Miroglio, B., Zeber, D., Kaye, J., Weiss, R.: The effect of ad blocking on user engagement with the web. In: Proceedings of the 2018 World Wide Web Conference. WWW, USA, pp. 813–821, 2018
26. Nithyanand, R., et al.: Adblocking and counter blocking: a slice of the arms race. In: 6th USENIX Workshop on Free and Open Communications on the Internet (FOCI 16). USENIX, USA, p. 12, 2016
27. Papadopoulos, P., Kourtellis, N., Markatos, E.P.: The cost of digital advertisement: comparing user and advertiser views. In: Proceedings of the 2018 World Wide Web Conference. WWW, USA, pp. 1479–1489, 2018

28. Papadopoulos, P., Kourtellis, N., Rodriguez, P.R., Laoutaris, N.: If you are not paying for it, you are the product: how much do advertisers pay to reach you? In: Proceedings of the 2017 Internet Measurement Conference. IMC, USA, pp. 142–156, 2017
29. Papadopoulos, P., Snyder, P., Athanasakis, D., Livshits, B.: Keeping out the masses: understanding the popularity and implications of internet paywalls. In: Proceedings of the Web Conference 2020. The Web Conference, USA, pp. 1433–1444, 2020
30. Parra-Arnau, J., Achara, J.P., Castelluccia, C.: Myadchoices: bringing transparency and control to online advertising. ACM Trans. Web (TWEB) **11**(1), 1–47 (2017)
31. Patnaik, S., Litan, R.E.: TikTok shows why social media companies need more regulation. Brook. Inst. Policy Brief May **1**(1), 1 (2023)
32. Piaggesi, S., Giurgola, S., Karsai, M., Mejova, Y., Panisson, A., Tizzoni, M.: Mapping urban socioeconomic inequalities in developing countries through Facebook advertising data. Front. Big Data **5**(2022), 1006352 (2022)
33. Rama, D., Mejova, Y., Tizzoni, M., Kalimeri, K., Weber, I.: Facebook ads as a demographic tool to measure the urban-rural divide. In: Proceedings of The Web Conference 2020. The Web Conference, USA, pp. 327–338, 2020
34. Sakib, M.N., Huang, C.-T.: Automated collection and analysis of malware disseminated via online advertising. In: 2015 IEEE Trustcom/BigDataSE/ISPA, vol. 1, pp. 1411–1416. IEEE, USA (2015)
35. Shiller, B., Waldfogel, J., Ryan, J.: Will ad blocking break the internet? Technical Report, National Bureau of Economic Research (2017)
36. Singh, S., Nanda, V., Sen, R., Sengupta, S., Kumaraguru, P., Gummadi, K.P.: Leveraging facebook's free basics engine for web service deployment in developing regions. In: Proceedings of the Ninth International Conference on Information and Communication Technologies and Development. ICTD, USA, pp. 1–11, 2017
37. Speicher, T., et al.: Potential for discrimination in online targeted advertising. In: Conference on Fairness, Accountability and Transparency, pp. 5–19. PMLR, USA (2018)
38. Varian, H.: Advertising costs and product prices. J. Law Econ. **65**(S2), S419–S431 (2022)
39. West, M., Ei, C.H.: Reading in the mobile era: a study of mobile reading in developing countries. UNESCo, USA (2014)
40. Zeng, E., Kohno, T., Roesner, F.: What makes a bad ad? User perceptions of problematic online advertising. In: Proceedings of the 2021 CHI Conference on Human Factors in Computing Systems. CHI, USA, pp. 1–24, 2021
41. Zhang, W., Yuan, S., Wang, J.: Optimal real-time bidding for display advertising. In: Proceedings of the 20th ACM SIGKDD International Conference on Knowledge Discovery and Data Mining. ACM, USA, pp. 1077–1086, 2014

Potpourri

Decision Support Framework for Precision Agriculture Using Multimedia and Open Data

Amira Kerkad[1](✉) and Rabah Gouri[2]

[1] IECPS Lab – Department of Industrial Engineering & Maintenance, ENSTA, Algiers, Algeria
akerkad@usthb.dz

[2] Ecole Nationale Supérieure des Technologies Avancées – ENSTA, Algiers, Algeria
rabah.gouri@ensta.edu.dz

Abstract. This research introduces a new framework using Internet of Things (IoT) and data-warehousing to accurately identify the most suitable crops for specific regions worldwide. The first step of our approach consists of a soil-sampling protocol used with IoT sensors to accurately identify soil properties. Secondly, a data-warehouse is constructed based on web-scraping that collects relevant agricultural information (crops, regions, climate…) sourced from online databases and websites. A specialized tool is developed to convert multimedia content into exploitable datasets. The data-driven model considers a wide range of factors, including soil nutrient levels (NPK), temperature, pH, conductivity, wind sensitivity, and crop-specific requirements. The third step is to filter the data-warehouse content by the soil sample properties obtained from step one to get the most relevant crops for the selected field. A predicted yield is associated to each crop using Machine Learning model trained on historical production data. By providing farmers with informed decisions, our framework aims to enhance agricultural productivity, sustainability, and resilience in the face of resource scarcity. Finally, we show how a large-scale sampling operation can provide a precise map of agricultural lands that allows national crop cultivation and rotation planning. While the experimental study was conducted in Algeria to validate the framework's effectiveness, its principles remain applicable to diverse agricultural environments worldwide.

Keywords: Decision Support Systems · Open Data · Data Analytics · Data-Warehousing · Multimedia · Data Visualization · Maps · Precision Agriculture · Soil Sampling · Sustainability

1 Introduction

The global agricultural landscape faces increasing pressure from factors such as population growth, climate change, and resource scarcity. To address these challenges, Internet of Things (IoT) technology, coupled with advanced data-warehousing, has emerged as a powerful tool for decision support in precision agriculture. IoT devices equipped with sensors can collect real-time data on soil conditions, weather patterns, and crop health. This data, combined with analytical information stored in data-warehouses, can be analyzed to provide valuable insights for decision-making.

S. Latifi (Ed.): ITNG 2025, AISC 1463, pp. 627–633, 2025.
https://doi.org/10.1007/978-3-031-89063-5_54

This paper proposes a new framework for crop recommendation that uses IoT data from soil samples to search for suitable crops from a dedicated data warehouse. The conception is done on three main steps: (1) soil sampling and IoT sensor measurement, (2) creating a data-warehouse from scraped multimedia and tabular open data, and (3) yield prediction model trained on historical data.

To illustrate the potential of this global framework, we will examine a case study in Algeria. With its diverse agricultural landscapes and its unique climate and resource challenges, Algeria provides a compelling example of how precision agriculture can be applied to improve crop yields and sustainability. Additionally, this work aligns with several of the United Nations Sustainable Development Goals (SDGs), particularly:

- *SDG 2 (Zero Hunger):* Reducing hunger through improved agricultural productivity.
- *SDG 12 (Responsible Consumption and Production):* Fostering responsible resource management.
- *SDG 13 (Climate Action):* Enhancing climate resilience in agriculture.

This research has the potential to optimize agricultural practices in the country, leading to increased food production, improved environmental sustainability, and enhanced economic growth.

The paper begins by presenting related work in Sect. 2, followed by the proposed framework in Sect. 3. Section 4 is dedicated to the experimental study. We conclude by discussing the results, the limits and further enhancements of our approach.

2 Related Work

The latest statistical data of land usage in Algeria reveals a disparity between the total land area and the portion dedicated to agriculture (18%). This Total Agricultural Surface (TAS) contains 6% of unproductive lands, and 74.5% of wastelands and pastures. This imbalance is due to predominantly semi-arid, arid, and hyper-arid climate. Moreover, suboptimal land use hinders agricultural potential. IoT technology combined with sampling protocols may ensure precise field-level data on soil potential.

When conducting environmental and agricultural monitoring, soil sampling is a crucial step. Soil sampling references the collection of small soil material extracted from a soil pit [1–3]. Many approaches are defined based on the soil's condition and the desired state of the collected material [4]. Depending on the targeted measures, we can make [1]: (a) loose samples to get texture, nutrients and pollutants etc. [5–8], (b) volume-controlled samples for bulk density [9], or (c) undisturbed samples for morphological properties [10]. For crop cultivation, we need to perform basic loose samples since we need to measure soil texture, pH, electrical conductivity, nutrients etc.

IoT technology enables farmers to monitor various aspects of their operations through connected devices and sensors. This leads to increased efficiency and data-driven decision-making [11–13]. Combining a well-designed soil sampling strategy with relevant IoT sensors is essential to obtain accurate data for a specific field.

Agricultural yields do not depend only on soil, but also on climate, and crop requirements. Integrating heterogenous data like weather forecasts and crop properties, allows to unlock soil potential and optimize agricultural yields. Business Intelligence offers valuable tools for data-driven decision-making across various disciplines [14]. Data-warehousing enables to store and analyze large datasets, facilitating comparisons and insights. While data-warehousing holds significant promise, its widespread adoption is still in its early stages in agricultural sector [11]. The studies on data-warehousing in agriculture using IoT technology consider storing measurement data for analysis, without integrating multimedia data, nor a sampling strategy [15].

Our study introduces a new decision support approach using data-warehousing to provide recommendations based on integrated climate, soil, and crop data, that fit the IoT provided measurements. Our methodology allows performing a nationwide sampling which enables the development of a holistic national precision agriculture map.

3 Framework

To analyze soil potential, including unproductive lands, we propose a new framework (Fig. 1) that we detail in the following sections.

3.1 Soil Sampling and IoT Device

Different soil sampling strategies are defined worldwide [1]. To ensure representative soil samples, we adhere to a rigorous sampling protocol adapted to our study:

- Timing: Ideal sampling occurs in the fall.
- Sampling density: The number of samples per acre varies based on soil heterogeneity, typically ranging from 1 to 4 per acre.
- Sampling technique: A zigzag pattern is followed to collect approximately 15 subsamples per sample, with a depth of 25–30 cm [3].
- Sampling tool: Different tools can be used in sampling such as blades, tubes or augers. It has to be uncontaminated and provide similar sampling units.
- Sample preparation: Collected soil subsamples are thoroughly mixed to achieve a homogeneous mixture that represents a soil sample.

Soil sensor needles are inserted into the soil sample to measure several key properties:

- Nutrient Levels: Nitrogen (N), Phosphorus (P), and Potassium (K) levels are assessed, which are essential for plant growth and development.
- Soil pH: This measurement determines the soil's acidity or alkalinity, significantly influencing nutrient availability for plants.
- Electrical Conductivity (EC): EC measures the salt content in the soil solution, which can impact plant water uptake and overall health.
- Soil Temperature: Soil temperature plays a crucial role in microbial activity and influences plant growth rates.
- Soil Moisture: Accurate measurement of soil moisture content is vital for optimizing irrigation practices and preventing both overwatering and drought stress.

This data helps farmers make informed decisions about fertilization, irrigation, and other crop management practices, leading to improved yields and overall plant health.

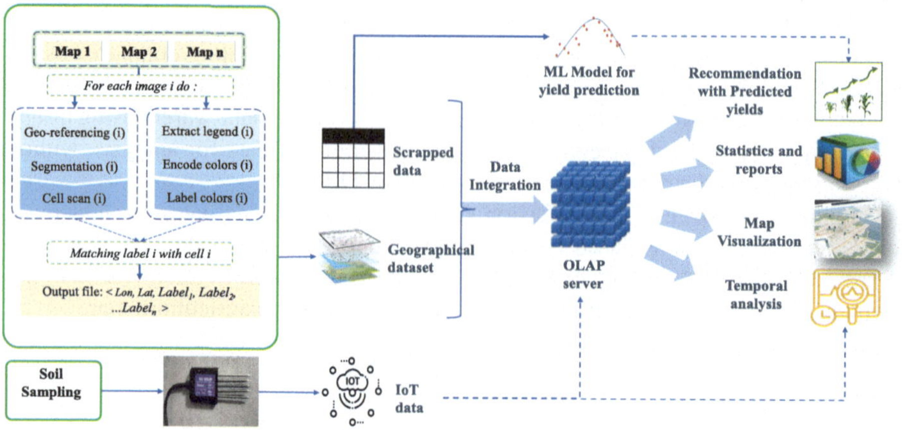

Fig. 1. Framework for extracting heterogeneous data (map images, IoT and tabular data) to create a decision support system for crop recommendation, empowered by a yield prediction model.

3.2 Data Warehousing

Creating the data-warehouse involves transforming source datasets into a target schema, enabling reporting based on defined analytical dimensions as follows:

Step 1: Define target schema with three dimensions: (1) Climate: province, month, temperature intervals, precipitation, wet days, and wind speed; (2) Crop: crop name, temperature intervals, nutrients intervals, wind sensitivity, ideal months, climate and soil type; (3) Area: latitude, longitude, city, province, climate type, and soil type. A recapitulative fact table 'Recommendation' is derived from dimensional joins and filtering, and no additional measures are required in our use case.

Step 2: Define data sources: As open data sources have heterogenous formats (PNG maps, GeoJson, tables…), we used web-scraping tools and APIs to collect data. Climate statistics dataset, scraped from the FAO/AquaStat website, provide monthly climate information (precipitation, temperature, wind speed etc.) for each province. Crop properties and Algerian cities datasets have also been created from open data. To extract soil type and bioclimatic stage, we created a tool that converts static maps (PNG from British Geological Survey website) into CSV (latitude, longitude, soil type, and bioclimatic stage) using georeferencing and image processing scripts.

Step 3: Data integration: The collected data is cleaned and structured to align with the target schema. The ETL jobs connect the source files to their corresponding tables.

3.3 Yield Prediction

Based on historical data from governmental annual reports, we compiled a comprehensive dataset encompassing various crop types through a list of most important crops, their corresponding production yields (measured in quintals and Kg/ha) for the last decade. A polynomial regression model was developed and trained on this dataset to predict future crop yields for each specific region within Algeria. The polynomial regression is used to provide: (1) the national yield per crop type, and (2) the local yield for each pair (crop, province) for the next years.

The predicted yield for each crop in a given region is then integrated into the recommended crop list, providing farmers with valuable insights into potential production outcomes.

3.4 Deployment

Once populated, the data-warehouse provides recommendations as follows: A farmer prepares a soil sample following our protocol, then measures its properties using the IoT. The measurements used to filter the data-warehouse, considering the location climate and soil type to identify suitable crops for the field. A list of recommended crops is provided, along with the predicted yield (Kg/ha) and other insights that help optimize fertilization, planting periods, and crop rotations.

Nationwide soil sampling using our approach provides a robust dataset for constructing a detailed precision agriculture map. The provided map of productive lands will consequently enlarge the existing TAS (Total Agricultural Surface), and make strategic crop planning and rotation. This granular-level information empowers decision-makers to select a periodic national crop plan and optimize resource utilization.

4 Experimental Study

To evaluate the effectiveness of our proposal, we conducted an experimental study to assess the accuracy of crop recommendation based on soil sample analysis and climate properties (Fig. 2).

Crop dataset contains properties of 105 crops. The climate dataset contains 696 entries (12 months x 58 provinces), areas dataset contains 1,541 cities across 58 provinces. To complement the data sources, the map extraction process generated respectively 95,583 and 39,140 entries for soil types and bioclimatic stages to cover the national territory. The discrepancy in cells number is due to the finer granularity required for accurate soil type mapping compared to bioclimatic stage. Indeed, in a small surface, we can find many soil types and few or one single bioclimatic stage. For this reason, we defined smaller cells to scan our static map during georeferencing process for soil types extraction. In contrast, we defined larger cells to scan the map for bioclimatic stage extraction.

To facilitate ongoing analysis and tracking, soil samples collected from farmers are stored in a dedicated database. The experimental study involved the analysis of 178 samples using the IoT device. Due to logistical and time constraints, the measurements were limited to controlled laboratory conditions. Yet, the simulation shows how data-warehouse efficiently interacts with the soil sensor to provide coherent results.

While this approach allowed for a focused evaluation of the system's core functionalities, it's important to acknowledge that real-world field conditions can introduce additional variables that may influence our recommendation and measuring accuracy.

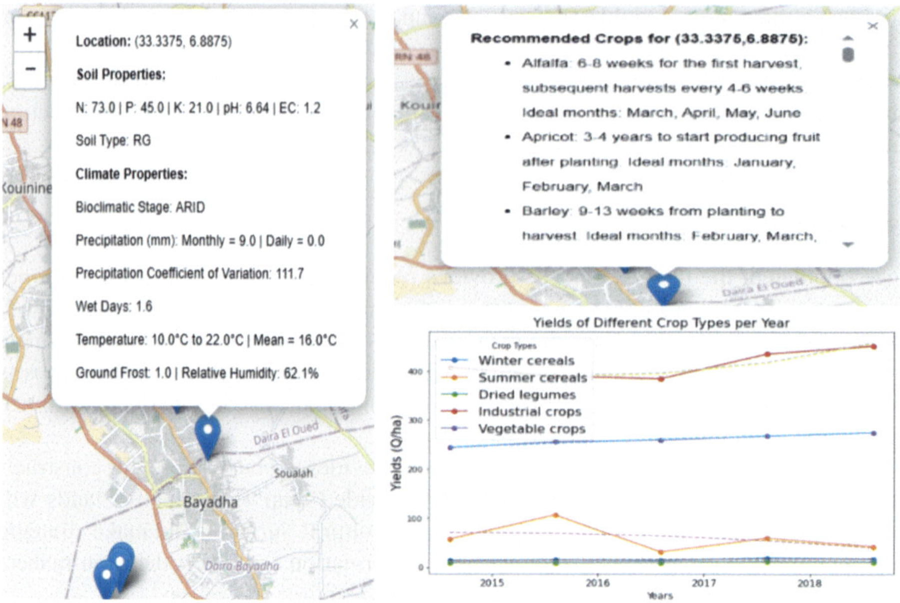

PROVINCE	POTATO	TOMATO	OIGNON	GARLIC	WATERMELON	CARROT	PEPPER	BELL-PEPPER	CUCUMBERS	ZUCCHINI	...	IND. TOMATO	TOBACCO
1 ADRAR	195.0	253.2	120.0	55.0	98.7	115.2	117.3	150.7	422.5	90.6	...	376.5	31.9
2 CHLEF	301.1	634.2	450.8	79.0	559.7	415.3	91.0	209.8	533.8	242.7	...	770.0	0.0
3 LAGHOUAT	282.9	321.4	239.4	80.0	203.9	210.0	210.0	300.0	219.4	380.0	...	0.0	0.0
4 O.E.BOUAGHI	261.5	300.0	451.3	276.7	491.6	301.4	200.0	200.0	340.2	294.4	...	450.0	17.1
5 BATNA	297.8	218.2	192.6	158.9	234.1	197.9	143.7	163.5	187.2	173.4	...	0.0	17.7

Fig. 2. Obtained recommendations and yields from the data-warehouse. The upper left image shows the soil and climate properties of the soil sample. The upper right image shows a scrolling list of recommended crops. The chart (center) shows the predicted yields per crop category. The lower table provides yield estimation of each crop in each province.

5 Conclusion

This paper presents a crop recommendation framework combining IoT technology and data-warehousing. Core components of this approach are: (1) a soil sampling protocol and a soil sensor device, (2) a data warehouse constructed from web scrapped data, using a new process for multimedia content integration, and (3) a ML model for crop yield prediction.

While our initial experimental study demonstrated the efficiency of the soil samples measuring and the data-warehouse deployment, it was limited to controlled environments. For further work, we aim to:

- Conduct extensive soil sampling across various regions in Algeria to identify potential limitations and refine the approach accordingly,
- Enhance data quality for both crop property datasets and climate forecasts to bolster the accuracy of recommendations.

This research represents a significant step towards realizing a national-scale precision agriculture map. By employing our innovative methodology, we aim to empower agricultural planning to effectively address both food security and industrial needs.

To achieve this ambitious goal on a national level, we will leverage the power of cloud computing platforms such as AWS EC2 or MS Azure. These platforms will enable us to process and analyze large-scale, real-time datasets, providing valuable insights and facilitating dynamic decision-making.

References

1. Mabit, L., Fulajtar, E., Toloza, A., Ochoa, V., Maestroni, B.: Implementation and Optimization of Soil Sampling: Some Practical Guidance and Considerations. Academic Press, pp. 47–63 (2018)
2. Vitti, G.C., Luz, P.C., Altran, W.S.: Nutrition and fertilization. In: Book Sugarcane (2015)
3. Huo, Y., Dijkstra, F.A., Possell, M., Singh, B.: Plastics in soil environments: all things considered. In: Advances in Agronomy (2022)
4. Loveland, P.J., Bellamy, P.H.: Environmental monitoring. In: Encyclopedia of Soils in the Environment, pp. 441–448 (2005)
5. Schoeneberger, P.J., Wysocki, D.A., Benham, E.C.: Soil Survey Staff, Field Book for Describing and Sampling Soils, Version 3.0. Natural Resources Conservation Service, National Soil Survey Center, p. 300. Lincoln, NE (2012)
6. Kemper, W.D., Koch. E.J.: Aggregate stability of soils from western United States and Canada. Technical Bulletin 1355. Agricultural Research Service, US Department of Agriculture, Beltsville, MD, p. 52 (1966)
7. Walworth, J.I.: Soil drying and rewetting, or freezing and thawing, affects soil solution composition. Soil Sci. Soc. Am. J. **56**, 433437 (1992)
8. Ritz, K.: Views of the underworld: in situ visualisation of soil biota. In: Ritz, K., Young, I. (Eds.), The Architecture and Biology of Soils: Life in Inner Space, p. 12. CABI, Wallingford/Cambridge (2011)
9. Zapata, F.: Handbook for the Assessment of Soil Erosion and Sedimentation Using Environment Radionuclides. Kluwer Academic Publishers, Dordrecht, p. 219 (2002)
10. Kulmatiski, A., Vogt, D.J., Siccama, T.G., Beard, K.H.: Detecting nutrient pool changes in rocky forest soils. Soil Sci. Soc. Am. J. **67**, 12821286 (2003)
11. Xu, J., Gu, B., Tian, G.: Review of agricultural IoT technology. Artif. Intell. Agric. **6**, 10–22 (2022)
12. Adamchuk, V.I., Hummel, J.W., Morgan, M., Upadhyaya, S.D.: On-the-go soil sensor for precision agriculture. Comput. Electron. Agric. **44**(1), 71–91 (2004)
13. Farooq, M.S., Riaz, S., Abid, A., Abid, K., Naeem, M.A.: A survey on the role of IoT in agriculture for the implementation of smart farming, vol. 7. IEEE (2019)
14. Jarke, M., et al.: Fundamentals of Data Warehouses. Springer Science Business Media (2013)
15. Rahman, A., Ermatita, Budianta, D.: Data warehouse design for soil nutrients with IoT based data sources. In: International Conference on Informatics, Multimedia, Cyber and Information System (2019)

Jiapi: A Pluggable Type Checker Generator for Statically-Typed JVM-Based Languages

Benjamin Cisneros(✉) and Jan Bækgaard Pedersen

Department of Computer Science, University of Nevada Las Vegas, Las Vegas, NV 89154, USA
cisneo1@unlv.nevada.edu, matt.pedersen@unlv.edu

Abstract. Static type systems are integral to programming languages that utilize static typing. This integration is crucial because every valid program in these languages must pass type checking according to the defined type system. This process helps programmers identify and prevent many type errors at compile time, which could otherwise lead to runtime issues. Consequently, different formalisms are necessary to ensure a well-defined type system. The validity of these formalism relies on the rigorous verification of various properties within the type system. Unfortunately, this often involves applying axioms and inference rules to prove the absence of certain program behaviors, leaving many compiler developers and programmers with the daunting task of maintaining complex proofs. In this paper, we introduce a novel and practical method for describing a type system for any JVM-based language using set notation (including membership tests, subsets, sizes, and other properties), first-order logic, and type-specific grammar. We also detail the checkers we developed, outline the challenges we encountered, and share insights gained while describing aspects of the type system of Espresso, a subset of Java.

Keywords: programming languages · jvm-based languages · compilers · statically typed languages · static typing · type checking · type systems · domain-specific languages · reflection · metaprogramming

1 Introduction

Type systems are generally understood to enhance the security and dependability of software by mechanically verifying different properties of a program. Conversely, a type system could sometimes compromise the security and dependability of software if it contributes to system complexity and fragility (e.g., the GCC and Clang compilers [23]). With the aid of a type checker in statically-typed

J. B. Pederse—Contributing author.

S. Latifi (Ed.): ITNG 2025, AISC 1463, pp. 634–647, 2025.
https://doi.org/10.1007/978-3-031-89063-5_55

languages (i.e., the implementation of a type system), specific type errors are guaranteed to be prevented at compile-time [2,3]. For example, the type checker in Java ensures that a standard Java program will not compile if a method call cannot be determined at compile-time. However, and in practice, programmers cannot always identify and prevent all type errors that matter to them, specially when using conventional type systems and checkers. Consequently, without adequate proof, they frequently rely on manual reasoning to assess code correctness, which is already a challenging task, particularly when there is limited or inconsistent support for the language's type system.

While various type systems aim to eliminate distinct classes of programming errors (such as trapped errors, untrapped errors, and forbidden errors [3]), they often share a common approach of expressing implications between type judgments, as demonstrated by the use of deduction or inference rules in their definitions [16,17]. Many of these type systems, particularly those rooted in Martin-Löf type theory [15] or the calculus of constructions [5,10], frequently utilize the Curry-Howard correspondence [22], which equates mathematical properties with types and proofs with data. However, unlike most mathematical proofs, the proofs in these systems tend to be labor-intensive, involving detailed case-by-case analysis, and may be less intellectually engaging for the average programmer.

In this respect, we sought to further develop the formal verification of programs in light of the earlier work completed in [4]. This verification model can be characterized as a straightforward extension of type checking verification: a compiler will reject some programs as not being well-formed, based on the expected types of values that appear to be used during computation. Therefore, this model is built on fundamental concepts that are used to describe different language constructs (e.g., expressions, functions, control structures, etc.), together with logic formulas that describe the desired properties of a type system, and for which we can specify the necessary checks in terms of *type predicates* (of the form $\mathcal{P}?(x)$), *type equality* (the $=_\mathcal{T}$ operator), *type equivalence* (the $\sim_\mathcal{T}$ operator), and *assignment compatibility* (the $:=_\mathcal{T}$ operator).

Although the final practical objective of this paper is to automate a type checker using as few constructs as possible, it is also crucial to verify that the semantic model holds true for both the compilation algorithm implemented in the target language and the compiler code that is actually executed. For this reason, we strive to create a solution that can be used to automate other types of systems, used in different contexts, or reused/modified as needed. On this account, we introduce a Pluggable Type Checker Generator for statically-typed JVM-based languages, explained in relation to Jiapi.

1.1 Contributions and Structure of Paper

Our contributions, which specifically include the following, builds on the work in [4]:

(a) We developed a simple yet expressive formalism for encoding a type system specification using *set notation* and *first-order logic*. This model also offers sub-typing and sub-classing when appropriate for any reasonable JVM-Based language, as well as a clear distinction between atomic and constructed types.

(b) In order to encode the definitions of common types, such as records, enumerations, classes, functions, and other special patterns, we develop a meta-language that we call *type-specific* grammar. This grammar allows us to describe data types with a minimal set of required properties and characteristics.

(c) By combining (a) and (b), we can write a type system specification and generate a specific implementation of a type checker (i.e., one that can easily be 'plugged') for any reasonable JVM-Based language. The resulting implementation (i.e., the generated Java files) can be imported into the source code for compilation.

As a result, this led to three outcomes:

- We succeeded in partially building a type checker for two moderately difficult languages.
- We formalized some of the main features of widely employed type systems (e.g., subtyping, method overloading and override, to name a few).
- Some typing rules that are valid in one system may be applied to other systems if a formalization reveals many similarities between the systems.

Paper Organization. Section 2 details a segment of our framework. Section 3 introduces a case study involving an object-oriented language, outlining the methodology for applying our approach. Section 4 delves into related work, while Sect. 5 provides the conclusion.

2 Preliminaries

In this section, we introduce the concept of types and outline various relationships between them. We also present several predicate functions and actions that operate on these types, which will assist us in formulating a set of rules for developing a type checker for a language interpreted on the JVM.

2.1 Types

Every single bit of data that is processed in a program is categorized into two types commonly known as *atomic* (or primitives) and *constructed* (or structured) types. *Atomic* types are predefined by the language; in other words, any type that a program can use but *cannot* define for itself is a primitive type in the language, and we can describe it by its name alone. For example,

$$
\begin{aligned}
\texttt{short} &\leadsto \mathit{short}\\
\texttt{int} &\leadsto \mathit{int}\\
&\vdots\\
\texttt{double} &\leadsto \mathit{double}
\end{aligned}
$$

where `teletype font` represents a type in the target language and *italics* its name. Thus, we can create the type lattice (or type hierarchy) for atomic types using the $<_{\mathcal{T}}$ operator as follows:

```
atomic {
  boolean,
  string,
  void,
  byte < short < int < long < float < double,
  char < int
}
```

Any type that a program can derive for itself based on existing atomic or other constructed types is a *constructed* type in the language. Therefore, a constructed type can be defined to be a pair consisting of the name of the type and an ordered set of name/type pairs that represent the fields. Furthermore, the essential structure of any constructed type can be described using a notation/syntax that is comparable to that of describing grammars from many widely used parsing tools (e.g., Java CUP [11], flex & Bison [14], ANTLR4 [18], etc.). For instance, a general record type can be defined as

$$type\ name\ \texttt{(}\ (type\ name\ (\texttt{,}\ type\ name)\star)?\texttt{)}$$

In simpler terms, this can be expressed as

$$Record(name, ((n_1, t_1), ..., (n_i, t_i)))$$

for a record named *name* with fields n_i of type t_i $\forall i \in 1..n$. *Jiapi* employs patterns, among other expressions, to associate variables with values found within these structures and, when appropriate, to match values to structures; for instance, we can characterize the preceding record as

$$Record(name\!: id, (name\!: id, type\!: ty) \star\!:\!fields?)$$

where the keywords *id* and *ty*[1] indicate when an identifier should be used as a *name* or as a *type*, respectively. Additionally, the expression

$$(name\!: id, type\!: ty) \star\!:\!fields?$$

creates a collection of values, each of which is a tuple that can be identified by a separate *name* and *type* and that can be accessed using the *fields* tag and the [] operator. Here is an example of how we could repeatedly access each tuple in the expression above

$$\forall i \in \{0..|fields|\} : fields[i]$$

We can use the . (dot) operator to access individual values in the tuple as follows

$$fields[i].name \quad \text{or} \quad fields[i].type$$

[1] The keyword *type* can also be used in place of *ty*, but we did it this way for the sake of clarity.

Note that the $\star$ operator represents the Kleene closure, indicating zero or more fields, while the ? operator signifies an optional sequence of fields. By default, the sequence of fields is an empty list.

Using the *Record* pattern, we can now create a `struct` (a type of record in C/C++) as follows

```
struct s {
  int a;
  double d;
  char c;
};
```

and with the following type

$$Record(s, ((a, int), (d, double), (c, char)))$$

Ultimately, we define a type in a language using a ***type expression***, and then we verify that the type matches that expected by it description; for example, the internal representation of the preceding record is depicted below (Fig. 1).

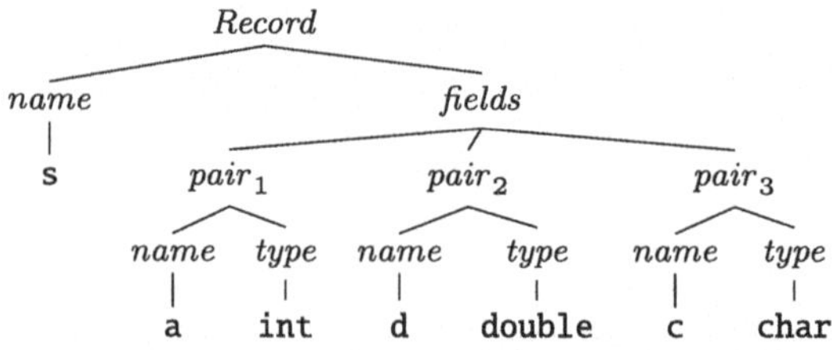

Fig. 1. Structure of a *Record* type.

2.2 Predicates

A predicate is a function that evaluates to either *true* or *false* depending upon a set of criteria. Below are a few examples [4],

$$short?(T) = \begin{cases} true, & \text{if } T \text{ is } \texttt{short} \\ false, & \text{otherwise} \end{cases}$$

$$int?(T) = \begin{cases} true, & \text{if } T \text{ is } \texttt{int} \\ false, & \text{otherwise} \end{cases}$$

$$\vdots$$

$$double?(T) = \begin{cases} true, & \text{if } T \text{ is } \texttt{double} \\ false, & \text{otherwise} \end{cases}$$

A predicate *numeric*?(T) could also be defined as a disjunction of other predicates like this

$$numeric?(T) = short?(T) \wedge \ldots \wedge double?(T)$$

We can certainly introduce predicates for constructed types, which will likely be necessary when formulating the specifications for a type checker. For instance, we can define a predicate function to determine whether an array literal e can be assigned to an array of type t as follows [20]:

$$ALAC(T, e)?{=}Array?(T) \wedge \begin{cases} true, \text{if } e = \{\,\} \vee \mathcal{T}(baseType) :=_{\mathcal{T}} \mathcal{T}(e) \\ \bigwedge_{i=0}^{n} \left(ALAC(baseType, \mathcal{T}(e_i))?\right), \text{if } e \neq \{\,\} \end{cases}$$

$ALAC$[2], short for *ArrayLiteralAssignmentCompatible*, determines when an array literal of the form $\{e_1, e_2, ..., e_n\}$ can be assigned to an array type that has the following type expression:

$$Array(baseType, [low..high]?)$$

where *baseType* is a type, and *low* is the lowest legal index, and *high* is the highest legal index. It is important to note that the range is optional. Here is an example:

```
double[][][] a = {{{30,40},{}}, {{},{50,60}}};
```

where T is `double[][][]`, and e is `{{{30,40}, {}}, {{}, {50,60}}}`. Since we know that the *baseType* is *double* and that any array type can accept the value $\{\,\}$, then the values for non-array elements must be assignment compatible with the *baseType* of the array type.

2.3 Operators: $=_{\mathcal{T}}$, $\sim_{\mathcal{T}}$, and $:=_{\mathcal{T}}$

The principles of *type equality* (the $=_{\mathcal{T}}$ operator), *type equivalence* (the $\sim_{\mathcal{T}}$ operator), and *assignment compatibility* (the $:=_{\mathcal{T}}$ operator) are defined in [20] as follows:

Two types T_1 and T_2 are said to be ***equal*** if they are the same type, and we write

$$T_1 =_{\mathcal{T}} T_2$$

Now, given two types T_1 and T_2, we say that they are ***type equivalent*** if any value of type T_1 can be assigned to a variable of type T_2 and any value of type T_2 can be assigned to a variable of T_1, and we write

$$T_1 \sim_{\mathcal{T}} T_2$$

Finally, consider an assignment of type form

$$v = \text{ e}$$

[2] The $\mathcal{T}()$ is the type computation operator, and we write $\mathcal{T}(e)$ to compute the static type of e during compilation.

where v represents a variable and e an expression. If we assume that the type of v is T_v and the type of e is T_e, and if a value of type T_e can be assigned to a variable of type T_v, then we say that T_e is *assignment compatible* with T_v and we write

$$T_v :=_{\mathcal{T}} T_e$$

Note that further operators may be extensions of or derivable from the ones already mentioned, therefore in that regard, our approach differs from other current and common approaches.

2.4 Logic Formulas

The core concept is that types correspond to propositions that we can prove in logic, while programs represent the proofs of these propositions, which gives rise to the Curry-Howard correspondence [9]. For instance, we can express the following logical formula for the $=_{\mathcal{T}}$ operator described in Sect. 2.3 to verify that both types r_1 and r_2 are identical, meaning they are name equivalent:

$$(r_1 =_{\mathcal{T}} r_2) \Leftrightarrow Record_?(r_1) \wedge Record_?(r_2) \wedge \\ (name_1 = name_2)$$

This states that the predicate $Record_?$ takes a type as defined in the type system and returns *true* if the type is an instance of a record type, and both r_1 and r_2 have the same type name. We could also write the following formula to verify the structure of a record for both types r_1 and r_2:

$$(r_1 \sim_{\mathcal{T}} r_2) \Leftrightarrow Record_?(r_1) \wedge Record_?(r_2) \wedge \\ (n_1 = n_2) \wedge \Big(\bigwedge_{i=1}^{n_1} f_{1,i} \sim_{\mathcal{T}} f_{2,i}\Big)$$

This states that, for two record types to be structurally equivalent, they must have the same umber of fields (i.e., $n_1 = n_2$), and the ith field in r_1 has to be type equivalent with the ith field in r_2, and vice versa.

3 Case Study

In order to assess the usefulness of the tool, this section describes a real-world case study of a type system for a Java-based language, called *Espresso*, that uses Jiapi's rules to construct parts of a fully-functional static type system. In addition, we assess these rules to show that the produced type checker can automatically verify the complete functional correctness of an *Espresso* program.

3.1 Espresso

Espresso [19] is an object-oriented programming language that compiles into bytecode. As a subset of Java, it mirrors the fundamental operators, conditional statements, loops, and data types present in Java. It encompasses classes, instance variables, methods, inheritance of instance variables and methods, method shadowing, interfaces, dynamic method binding, assignments, and the handling of the null value. We will use Jiapi to draft a portion of the type system specification for *Espresso*. Our approach will align with the specifications outlined in prior sections to the greatest extent possible. Nonetheless, on occasion, we may need to make slight adjustments to these specifications to accommodate the specific features of *Espresso*, which will be explicitly identified in the subsequent text.

3.2 Working with Primitive Types

We need to consider the $=_\mathcal{T}$, $\sim_\mathcal{T}$, and $:=_\mathcal{T}$ operators that must be defined for all types in *Espresso*. Therefore, for two types α and β

$$\alpha =_\mathcal{T} \beta \Leftrightarrow Primitive_?(\alpha) \wedge Primitive_?(\beta) \wedge \alpha = \beta$$

We only check that both types are the same primitive. Given our object-oriented approach, we can conceptualize $\alpha =_\mathcal{T} \beta$ as $\alpha.equal(\beta)$, indicating that we already acknowledge α as a primitive type, thereby simplifying the criteria to $Primitive_?(\beta) \wedge \alpha = \beta$, that is

$$\alpha =_\mathcal{T} \beta \Leftrightarrow Primitive_?(\beta) \wedge \alpha = \beta$$

Since type equivalence for primitives coincides with type equality, we obtain the following:

$$\alpha \sim_\mathcal{T} \beta \Leftrightarrow \alpha =_\mathcal{T} \beta$$

Regarding the third method, called assignment compatibility, the following remark outlines its implementation for $<_\mathcal{T}$. We use the $<_\mathcal{T}$ operator to assess the compatibility of the following primitive types:

$$byte <_\mathcal{T} short <_\mathcal{T} int <_\mathcal{T} long <_\mathcal{T} float <_\mathcal{T} double$$

$$char <_\mathcal{T} \text{ int}$$

Using the type ordering mentioned above, we can derive the values column in Table 1. Additionally, the extension of the $<_\mathcal{T}$ operator seamlessly incorporates the $\leq_\mathcal{T}$ operator, and the definitions of $>_\mathcal{T}$ and $\geq_\mathcal{T}$ can be established in a similar manner.

3.3 Working with Constructed Types

For constructed types, we will only look at two: *arrays* and *classes*. We will also look at *methods* as types even though they do not fall under this category. As with the primitive types, the operators $=_\mathcal{T}$, $\sim_\mathcal{T}$, and $:=_\mathcal{T}$ must be specified.

Table 1. Type hierarchy for atomic types in *Espresso* [4].

$\alpha <_{\mathcal{T}} \beta$	byte	short	char	int	long	float	double	boolean	void
byte	F	T	T	T	T	T	T	F	F
short	F	F	T	T	T	T	T	F	F
char	F	F	F	T	T	T	T	F	F
int	F	F	F	F	T	T	T	F	F
long	F	F	F	F	F	T	T	F	F
float	F	F	F	F	F	F	T	F	F
double	F	F	F	F	F	F	F	F	F
boolean	F	F	F	F	F	F	F	F	F
void	F	F	F	F	F	F	F	F	F

Arrays. An array is dynamically allocated and its values are accessed by an index set starting at 0 and ending in $n - 1$. As a result, the only piece of data that may be stored about an array type is its base type. Therefore, for two types α and β, we have

$$\alpha =_{\mathcal{T}} \beta \Leftrightarrow Array_?(\alpha) \wedge Array_?(\beta) \wedge$$
$$(\alpha.componentType =_{\mathcal{T}} \beta.componentType)$$

For $\sim_{\mathcal{T}}$, two arrays are considered equal if their base types are equal; therefore, we have

$$\alpha \sim_{\mathcal{T}} \beta \Leftrightarrow \alpha =_{\mathcal{T}} \beta$$

However, for $:=_{\mathcal{T}}$, we have the option of assigning *null* to an array variable or an array of a subclass to an array of a superclass; therefore, we have

$$\alpha :=_{\mathcal{T}} \beta \Leftrightarrow Array_?(\alpha) \wedge Array_?(\beta) \wedge$$
$$\big(Null_?(\beta) \vee \big(Primitive_?(\alpha.componentType) \wedge$$
$$\alpha.componentType =_{\mathcal{T}} \beta.componentType\big) \vee$$
$$\alpha.componentType :=_{\mathcal{T}} \beta.componentType\big)$$

The array type specification for *Espresso* is defined in Fig. 2.

Classes. A class is the only other constructed type in *Espresso*. A class type uses the class's name, and the super class and a lit of interfaces if it extends and implements them. For instance, a class can be expressed as follows:

$$Class(name, superClass?, interfaces\star)$$

$$
\begin{array}{l}
Array = \\
\quad Array_?(T) \Leftrightarrow true \\
\quad \alpha =_{\mathcal{T}} \beta \Leftrightarrow Array_?(\alpha) \wedge Array_?(\beta) \wedge \\
\qquad \alpha.componentType =_{\mathcal{T}} \beta.componentType \\
\quad \alpha \sim_{\mathcal{T}} \beta \Leftrightarrow \alpha =_{\mathcal{T}} \beta \\
\quad \alpha :=_{\mathcal{T}} \beta \Leftrightarrow Null_?(\beta) \vee \\
\qquad Primitive_?(\alpha.componentType) \wedge \\
\qquad \alpha.componentType =_{\mathcal{T}} \beta.componentType \vee \\
\qquad \alpha.componentType :=_{\mathcal{T}} \beta.componentType
\end{array}
$$

Fig. 2. Array type specification for *Espresso.*

Therefore, for two types α and β, we defined equality by comparing their names

$$\alpha =_{\mathcal{T}} \beta \Leftrightarrow Class_?(\alpha) \wedge Class_?(\beta) \wedge \alpha.name = \beta.name$$

For $\sim_{\mathcal{T}}$, we have

$$\alpha \sim_{\mathcal{T}} \beta \Leftrightarrow \alpha =_{\mathcal{T}} \beta$$

However, for $:=_{\mathcal{T}}$, an assignment is legal if $\alpha = \beta$ or $\alpha \leq_{\mathcal{T}} \beta$—that is, β is a subtype of α or $\beta \prec_{\mathcal{T}} \alpha$; therefore, we have

$$\alpha :=_{\mathcal{T}} \beta \Leftrightarrow \alpha =_{\mathcal{T}} \beta \vee (\alpha.superClass \neq null \wedge$$

$$\alpha \leq_{\mathcal{T}} \beta.superClass) \vee (\bigvee_{\delta \in interfaces} \alpha \leq_{\mathcal{T}} \delta)$$

The class type specification for *Espresso* is defined in Fig. 3.

$$
\begin{array}{l}
Class = \\
\quad Class_?(T) \Leftrightarrow true \\
\quad \alpha =_{\mathcal{T}} \beta \Leftrightarrow Class_?(\alpha) \wedge Class_?(\beta) \wedge \\
\qquad \alpha.name = \beta.name \\
\quad \alpha \sim_{\mathcal{T}} \beta \Leftrightarrow \alpha =_{\mathcal{T}} \beta \\
\quad \alpha :=_{\mathcal{T}} \beta \Leftrightarrow Class_(\alpha) \wedge Class_?(\beta) \wedge (Null_?(\beta) \vee \alpha \leq_{\mathcal{T}} \beta) \\
\quad \alpha \leq_{\mathcal{T}} \beta \Leftrightarrow Class_?(\alpha) \wedge Class_(\beta) \wedge \\
\qquad (\alpha.superClass \neq null \wedge \alpha \leq_{\mathcal{T}} \beta.superClass) \vee \\
\qquad (\bigvee i \in 1..|\beta.interfaces| : \alpha \leq_{\mathcal{T}} \beta.interfaces[i])
\end{array}
$$

Fig. 3. Class type specification for *Espresso.*

Methods. While *Espresso* does not support the direct passing of methods as parameters, it is still meaningful to introduce a method type for defining how a target method is located during an invocation. Thus, a method can be expressed as follows:

$$Method(name, param\text{-}type\star, return\text{-}type)$$

A method type is contingent on the method's name, the parameter types, and the return type; therefore, for two types α and β, we have

$$\alpha =_{\mathcal{T}} \beta \Leftrightarrow Method_?(\alpha) \wedge Method_?(\beta) \wedge$$
$$\alpha.name = \beta.name \wedge$$
$$|\alpha.paramTypes| = |\beta.paramTypes| \wedge$$
$$\Big(\bigwedge_{i \in 0..|\alpha.paramTypes|-1} \alpha.paramTypes[i] =_{\mathcal{T}} \beta.paramTypes[i]\Big)$$

For $\sim_{\mathcal{T}}$, we have

$$\alpha \sim_{\mathcal{T}} \beta \Leftrightarrow \alpha :=_{\mathcal{T}} \beta \wedge \beta :=_{\mathcal{T}} \alpha$$

However, in *Espresso*, we use equivalence in terms of assignment compatibility for methods; therefore, we have

$$\alpha :=_{\mathcal{T}} \beta \Leftrightarrow Method_?(\alpha) \wedge Method_?(\beta) \wedge$$
$$|\alpha.paramTypes| = |\beta.paramTypes| \wedge$$
$$\Big(\bigwedge_{i \in 0..|\alpha.paramTypes|-1} \alpha.paramTypes[i] :=_{\mathcal{T}} \beta.paramTypes[i]\Big)$$

The method type specification for *Espresso* is defined in Fig. 5. We will now outline the specifications for identifying the appropriate method. Consider the following example:

```
void foo(double d, long l) { }    // (1)
void foo(int i, long l) { }       // (2)
void foo(int i) { }               // (3)
void foo(string s, double d) { } // (4)
```

For a method invocation $f(e_1, ..., e_n)$, we say that a candidate method must have the correct name (f), the correct number of parameters, and the formal parameters that can hold the actual parameters, considering with respect to assignment compatibility. Nevertheless, when deciding which *foo* method to call in the context of $foo(i, i)$, where i is of type *int*, the selection should prioritize the most specific one, namely, the second *foo* method.

The specification for determining the appropriate *foo* method in *Espresso* is outline in Fig. 4.

We begin by removing general methods, specifically, a potential method f_1 with parameters that could be contained within another candidate method f_2's parameters. Therefore, we would eliminate f_2 since it is more general. *possibleCandidates* represents the subset of candidate methods that can be eliminated, as they may be more general than any other method. For this reason, we used the $:=_{\mathcal{T}}$ operator rather than the $\leq_{\mathcal{T}}$ operator; for example, an invocation $f(e_1, ..., e_n)$ is compatible with method $g(t_1, ..., t_m) \rightarrow t'$ if:

$$
\begin{array}{l}
\mathit{findMethod}(\mathit{methods}, \mathit{in}) = \\
\quad \mathit{matchingName} \leftarrow_{\mathcal{T}} [\,] \\
\quad \forall m \in 0..|\mathit{methods.nchildren}| : \\
\qquad\quad \mathit{md} \leftarrow_{\mathcal{T}} \mathit{methods.children}[m] \\
\qquad\quad \mathit{md.getName} = \mathit{in.name} \Rightarrow \\
\qquad\qquad \mathit{matchingName} {+}{=} \mathcal{T}(\mathit{md}) \\
\quad \mathit{possibleCandidates} \leftarrow_{\mathcal{T}} [\,] \\
\quad \forall m \in \{\mathit{matchingName}\} : \\
\qquad\quad m :=_{\mathcal{T}} \mathit{in} \wedge m.\mathit{name} = \mathit{in.name} \Rightarrow \\
\qquad\qquad \mathit{possibleCandidates} {+}{=} m \\
\quad \forall i \in 0..|\mathit{possibleCandidates}| : \\
\qquad\quad f \leftarrow_{\mathcal{T}} \mathit{possibleCandidates}[i] \\
\qquad\quad \forall j \in 0..|\mathit{possibleCandidates}| : \\
\qquad\qquad \mathit{possibleCandidates}[j] :=_{\mathcal{T}} f \Rightarrow \\
\qquad\qquad\quad \mathit{possibleCandidates} {-}{=} \mathit{possibleCandidates}[j] \\
\qquad\quad \mathit{possibleCandidates} {+}{=} f \\
\quad \mathit{possibleCandidates}
\end{array}
$$

Fig. 4. *findMethod* in *Espresso.*

$$
\begin{array}{l}
\mathit{Method} = \\
\quad \mathit{Method?}(T) \Leftrightarrow \mathit{true} \\
\quad \alpha =_{\mathcal{T}} \beta \Leftrightarrow \mathit{Method?}(\alpha) \wedge \mathit{Method?}(\beta) \wedge \\
\qquad\quad \alpha.\mathit{name} = \beta.\mathit{name} \wedge \\
\qquad\quad |\alpha.\mathit{paramTypes}| = |\beta.\mathit{paramTypes}| \wedge \\
\qquad\quad (\bigwedge i \in 0..|\alpha.\mathit{paramTypes}| - 1 : \\
\qquad\qquad \alpha.\mathit{paramTypes}[i] =_{\mathcal{T}} \beta.\mathit{paramTypes}[i]) \\
\quad \alpha \sim_{\mathcal{T}} \beta \Leftrightarrow \alpha :=_{\mathcal{T}} \beta \wedge \beta :=_{\mathcal{T}} \alpha \\
\quad \alpha :=_{\mathcal{T}} \beta \Leftrightarrow \mathit{Method?}(\alpha) \wedge \mathit{Method?}(\beta) \wedge \\
\qquad\quad |\alpha.\mathit{paramTypes}| = |\beta.\mathit{paramTypes}| \wedge \\
\qquad\quad (\bigwedge i \in 0..|\alpha.\mathit{paramTypes}| - 1 : \\
\qquad\qquad \alpha.\mathit{paramTypes}[i] :=_{\mathcal{T}} \beta.\mathit{paramTypes}[i])
\end{array}
$$

Fig. 5. Method type specification for *Espresso.*

1. $f = g$, they have the same,
2. $n = m$, they have the number of actual and formal parameters, and
3. $\forall_{i=1}^{i \leq n} t_i :=_{\mathcal{T}} \mathcal{T}(e_i)$, by applying the assignment compatibility rule, every formal parameter can store the value of the actual parameter.

Note that when ignoring the return types, we say that for $T_1 = (t_{1,1}, ..., t_{1,n}) \rightarrow t_1$ and $T_2 = (t_{2,1}, ..., t_{2,m}) \rightarrow t_2$, T_2 is more general than T_1 if $T_1 \leq_{\mathcal{T}} T_2 \Leftrightarrow (n = m) \wedge (\bigwedge_{i=1}^{n} t_{2,i} :=_{\mathcal{T}} t_{1,i})$. An alternative perspective to consider is to visualize a set of concrete parameters, represented as $t_{1,1}, t_{1,2}, ..., t_{1,n}$, and subsequently evaluating whether they are compatible with a method having formal parameters $t_{2,1}, t_{2,2}, ..., t_{2,m}$. If such compatibility exists, it signifies that the method characterized by type T_2 is more general than the one associated with type T_1 [4]. It is important to emphasize the potential to utilize specific components from a previously established type system. For instance, some of the earlier specifications could be reimplemented in other languages that have features similar to *Espresso*.

4 Related Work

This section reviews existing relevant work, highlighting their differences, benefits, and drawbacks.

CENTAUR [1] is an interactive environment that generates language-specific tools from programming language descriptions, utilizing a parser generator (METAL [12]) and a pretty-printer (PPML [243]) for converting code into an AST. It employs TYPOL [6] for logical frameworks based on natural semantics, using inference rules with Gentzen sequences for type checking.

LATOS [9] is a lightweight tool that takes a Miranda superset and produces a type-checked declarative program, supporting both deterministic and non-deterministic specifications. While larger systems like CENTAUR encompass multiple functionalities, LATOS focuses on optimizing existing components.

ASF+SDF [7] combines algebraic and syntax definition formalisms to describe programming languages and operations like type checking and interpretation, using many-sorted algebraic signatures and conditional equations.

TinkerType [13] enables control over components of formal systems, assembling them into a cohesive system through a formal structure based on inference rules.

5 Conclusion

Given the tools available today, articulating the formal semantics of even a simple language is a difficult undertaking for the average programmer. This is true both conceptually, in the form of mathematical frameworks, and with software, such as proof assistants (e.g., Isabelle, Coq, Mizar, etc. [8]). However, our strategy differs from other current initiatives in that we aim to develop a verification model that can be simply presented to programmers as an extension of type checking verification, employing ideas such as static flow analysis and relatively basic types. We propose that language implementors can derive benefits from our approach, especially as non-experts can swiftly iterate and experiment with novel type features and combinations. Our next step involves further validating this concept by implementing additional languages and extending our approach to more intricate analyses. Additionally, we aim to investigate the compatibility of our method for implementing pluggable type checkers with non-JVM based languages. Finally, given that numerous type systems already integrate type checking and program rewrites [21], we believe that languages should potentially incorporate a comprehensive framework for defining type system rules in a more practical manner.

References

1. Borras, P., et al.: Centaur: the system. ACM SIGPLAN Not. **24**(2), 14–24 (1988)
2. Cardelli, L.: Typeful Programming. Digital Equipment Corporation Systems Research Center (1989)
3. Cardelli, L.: Type systems. ACM Comput. Surv. (CSUR) **28**(1), 263–264 (1996)
4. Cisneros, B.: Jiapi: a type checker generator for statically typed languages. Ph.D. thesis, University of Nevada, Las Vegas (2022)
5. Coquand, T.: Metamathematical investigations of a calculus of constructions. Ph.D. thesis, INRIA (1989)
6. Despeyroux, T.: Typol: a formalism to implement natural semantics. Ph.D. thesis, INRIA (1988)
7. Deursen, A., Heering, J., Klint, P.: Language Prototyping: An Algebraic Specification Approach, vol. 5. World Scientific (1996)
8. Geuvers, H.: Proof assistants: history, ideas and future. Sadhana **34**, 3–25 (2009)
9. Hartel, P.H.: Latos a lightweight animation tool for operational semantics (1997)
10. Hindley, J.R., Seldin, J.P.: Lambda-Calculus and Combinators, An Introduction, vol. 2. Cambridge University Press, Cambridge (2008)
11. Hudson, S.: Cup parser generator for java (1999). http://www.cs.princeton.edu/~appel/modern/java/CUP
12. Kahn, G., Lang, B., Melese, B., Morcos, E.: Metal: a formalism to specify formalisms. Sci. Comput. Program. **3**(2), 151–188 (1983)
13. Levin, M.Y., Pierce, B.C.: TinkerType: a language for playing with formal systems. J. Funct. Program. **13**(2), 295–316 (2003)
14. Levine, J.: Flex and Bison: Text Processing Tools. O'Reilly Media, Inc. (2009)
15. Martin-Löf, P., Sambin, G.: Intuitionistic Type Theory, vol. 9. Bibliopolis Naples (1984)
16. Mitchell, J.C.: Type systems for programming languages. In: Formal Models and Semantics, pp. 365–458. Elsevier (1990)
17. Negri, S., Von Plato, J.: Structural Proof Theory. Cambridge University Press (2008)
18. Parr, T.: The definitive ANTLR 4 reference. The Definitive ANTLR 4 Reference, pp. 1–326 (2013)
19. Pedersen, J.: The espresso compiler. http://www.egr.unlv.edu/~matt/teaching/CSC460/Espresso.pdf
20. Pedersen, J.: Practical Compiler Construction: with Java and the JVM. Lulu.com, USA (2018)
21. Siek, J.G., Vitousek, M.M., Cimini, M., Boyland, J.T.: Refined criteria for gradual typing. In: 1st Summit on Advances in Programming Languages (SNAPL 2015). Schloss Dagstuhl-Leibniz-Zentrum fuer Informatik (2015)
22. Sørensen, M.H., Urzyczyn, P.: Lectures on the Curry-Howard Isomorphism. Elsevier (2006)
23. Yang, X., Chen, Y., Eide, E., Regehr, J.: Finding and understanding bugs in C compilers. In: Proceedings of the 32nd ACM SIGPLAN Conference on Programming Language Design and Implementation, pp. 283–294 (2011)

SmartDelta Methodology: Automated Quality Assurance and Optimization for Incremental System Engineering

Benedikt Dornauer[1(✉)], Michael Felderer[1,2], Mehrdad Saadatmand[3], Muhammad Abbas[3], Nicolas Bonnotte[4], Andreas Dreschinski[4], Eduard Paul Enoiu[5], Eray Tüzün[6], Baykal Mehmet Uçar[7], Ömercan Devran[7], and Robin Gröpler[8]

[1] University of Innsbruck, Innsbruck, Austria
benedikt.dornauer@uibk.ac.at
[2] German Aerospace Center (DLR), Institute of Software Technology, Cologne, Germany
michael.felderer@dlr.de
[3] RISE Research Institutes of Sweden, Västerås, Sweden
{mehrdad.saadatmand,muhammad.abbas}@ri.se
[4] Akkodis Germany Solutions GmbH, Sindelfingen, Germany
{nicolas.bonnotte,andreas.dreschinski}@akkodis.com
[5] Mälardalen University, Västerås, Sweden
eduard.paul.enoiu@mdu.se
[6] Bilkent University, Ankara, Turkey
eraytuzun@cs.bilkent.edu.tr
[7] Arçelik, Istanbul, Turkey
{baykal.ucar,omercan.devran}@beko.com
[8] ifak Institute for Automation and Communication, Magdeburg, Germany
robin.groepler@ifak.eu

Abstract. Modern software systems undergo frequent updates, continuously evolving with new versions and variants to offer new features, improve functionality, and expand usability. Given the rapid pace of software evolution, organizations require effective tools and methods to mitigate the challenges associated with these changes, also called deltas. To address these challenges, the international SmartDelta Project joined industry and academia to develop and test solutions for incremental development and quality assurance. This paper provides insights into the SmartDelta project achievements and highlights one main contribution: the *SmartDelta Methodology*, a domain-unspecific concept for delta management in incremental software engineering. This methodology enables companies to identify gaps in their continuous engineering environment across six stages and helps to discover new tools in various technical areas. Additionally, the paper presents seven selected tools at different stages of the methodology.

M. Felderer, M. Saadatmand, M. Abbas, N. Bonnotte, A. Dreschinski, E. Enoiu, E. Tüzün, B. Uçar, Ö. Devran, R. Gröpler— Contributing authors.

S. Latifi (Ed.): ITNG 2025, AISC 1463, pp. 648–659, 2025.
https://doi.org/10.1007/978-3-031-89063-5_56

Keywords: continuous system engineering · incremental development · version delta · variant delta · software quality

1 Introduction

Software undergoes continuous evolution. These changes occur through iterative development, where, for instance, new features are incrementally added, anomalies are identified and fixed, or performance optimization is conducted. This often leads to new software versions or variants. While updates may enhance certain quality aspects, other aspects can deteriorate over time [7]. For instance, trade-offs can arise between security and performance or between feature expansion and code quality, especially in the long term.

Table 1. SmartDelta Industrial Use Cases

Ctry.	Company and Industry Sector
	Alstom (Railway)
	Akkodis (eMobility), Software AG (Enterprise Software)
	eCAMION (eMobility), GlassHouse Systems (Cybersecurity)
	NetRD (Telecommunication), Kuveyt Türk Bank (Banking and Finance), Arçelik (Home Appliances)
	c.c.com Moser GmbH (Logistics and Personal Mobility)
	Izertis (Enterprise Software)
	Vaadin (Software Development Platform)

To consolidate problem statements, business objectives, and challenges related to incremental systems engineering, 11 use cases across various industry domains (see Table 1) contributed their perspectives and requirements. The challenges identified, provided in detail in [12], underscore the importance of precise analysis and assessments of the adjustments associated with each modification to a system, which are crucial in continuous engineering. Thus, the goal of SmartDelta[1] has been to build automated solutions for quality assessment of product deltas in a continuous engineering environment. It provides intelligent analytics from development artifacts (e.g., source code, log files, requirement specifications) and system execution, offering insights into quality improvements or degradation resulting from various changes, along with recommendations for the next build (Fig. 2).

To address the various delta-management aspects, the project contains different work packages (WP), each focusing on a specific topic. Specifically, WP 2 focuses on models for generation and evolution, while WP 3 concentrates on

[1] https://itea4.org/project/smartdelta.html.

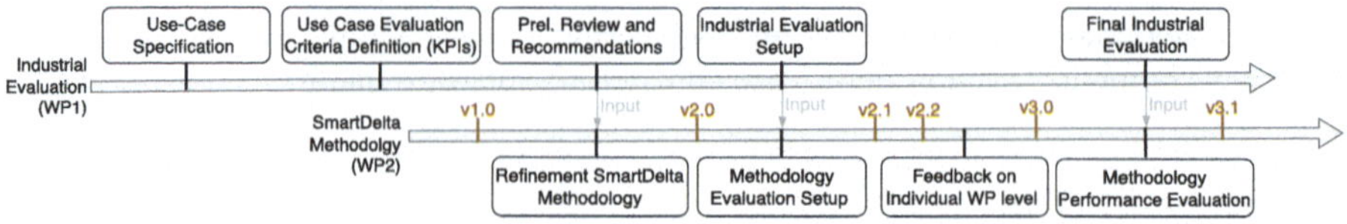

Fig. 1. Step-by-step industrial evaluation setup and contribution to the Methodology

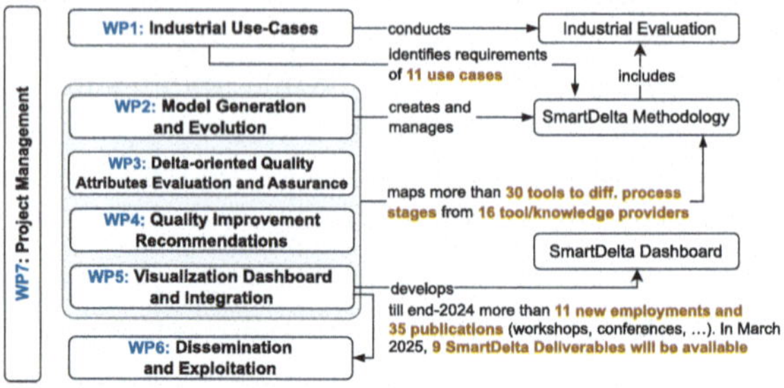

Fig. 2. SmartDelta Project Working Structure

tools for delta-oriented quality assurance, and WP 4 develops solutions for quality improvement recommendations. Since most tools integrate visual representations of various delta aspects, WP 5 supports the development of guidelines and approaches for the visualization part.

Managing the work packages' different outcomes (knowledge, solutions, tools, etc.) and organizing these contributions within a coherent structure is challenging, especially for such a large consortium. Consequently, we developed the *SmartDelta Methodology*, a novel continuous engineering management concept that emerged through an incremental design process illustrated in Fig. 1. This process involved a validity evaluation conducted by the use case providers, who offered recommendations to enhance the SmartDelta Methodology further. Apart from that, the use cases tried and tested the various solutions in WP 1.

The structure of the paper is as follows: Sect. 2 discusses the term of delta in software engineering and presents the SmartDelta Methodology. Sect. 3 discusses solutions associated with specific process stages. Finally, Sect. 4 provides a short recap and outlines the future work.

2 Delta Management Concept: The SmartDelta Methodology

The concept of delta has been explored in areas like software configuration management, version control, product line engineering, and variability management (e.g., [4,8,9,13]). Deltas are defined differently across these domains, tracking changes between versions in configuration management while representing variations in product lines. In SmartDelta, a *Delta* is any change in a software product that results in a new product instance with different quality and/or functionality properties. Examples of such changes include adding or removing features and components, applying fixes and updates, reconfiguring a software product for deployment in a new environment, or adapting to the needs of a new customer.

This concept is often linked to the temporal aspects of product evolution and is commonly referred to as a *Version Delta.* A version represents a specific release of a software product. *A Variant Delta* refers to a distinct form of software modified or customized for specific purposes or applications. These variants may differ in functionality, features, or configurations while maintaining a common core derived from the original software [1].

To manage these kinds of deltas, we created the management concept called the ***SmartDelta Methodology***, illustrated in Fig. 3. A primary goal of the SmartDelta Methodology is to offer a holistic overview guiding companies in their (individual) incremental software systems development. Therefore, the concept consists of six stages:

The iterative development process is driven by certain **Preconditions**: pre-existing artifacts, such as architecture, code, and design, which serve as both constraints and guidance. These artifacts, alongside input from stakeholders and other external influences, determine the development direction of a product.

Requirements Engineering is structured as an iterative, delta-focused process aimed at defining and adapting the necessary project requirements over time. Requirements are continuously identified and adapted from diverse sources, with a strong focus on tracking and analyzing each delta (i.e., any modification or adjustment needed to meet evolving goals). Quality standards are ensured through delta-aware quality analysis, where existing requirements are analyzed for reuse, adapting them efficiently to new contexts and minimizing redundant efforts. Model extraction plays a crucial role in visualizing requirements and their deltas to understand the impact of each change on the system as a whole. Verification and validation processes confirm that these evolving requirements align with stakeholder needs and adhere to system constraints.

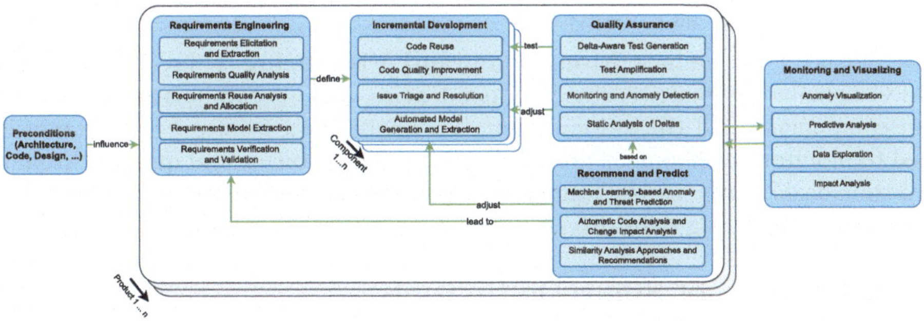

Fig. 3. The SmartDelta Methodology

In the **Incremental Development** phase, software systems are constructed using reusable, modular components that address the requirements. By doing so, integrating feedback from quality assurance is essential to identify and resolve potential issues early in the development cycle. This iterative process ensures that each increment aligns with the overall system objectives while maintaining high-quality standards. From a product line management perspective, the increment process begins with the development of a usable, improvable product that may not be an Minimally Viable Product (MVP) initially but evolves through feedback until it reaches MVP. This iterative build-measure-learn cycle, which eliminates undesirable deviations, is applied across different versions and variants, ensuring reusability and continuous improvement in product development.

Delta-Driven **Quality Assurance** highlights testing, validation, and review. Each change, whether a minor code adjustment or a significant feature update, is rigorously evaluated to ensure quality standards are met throughout development. Delta-aware test generation targets specific changes, ensuring all modifications meet defined requirements and quality standards. Test amplification improves this by expanding test cases to cover scenarios introduced by deltas. Continuous monitoring and anomaly detection provide real-time feedback, highlighting any quality degradation linked to changes. Additionally, static analysis of deltas assesses modifications against predefined quality metrics. Integrating these delta-aware techniques enables effective defect detection at this stage. Quality assurance and test engineers apply these techniques to manage and execute the delta-aware quality assurance process directly. Software developers play an essential role in implementing feedback to support continuous improvement, while project managers ensure that quality goals set using delta-driven quality assurance align with project timelines and standards.

The **Recommend and Predict** stage leverages delta-aware analysis to generate predictive recommendations for software evolution, often utilizing AI. By analyzing data from the quality assurance phase, this stage identifies areas needing adjustments, frequently resulting in new requirements and highlighting reusable components for future developments. This delta-focused approach ensures that each prediction and recommendation is backed by data-driven insights, allowing the system to adapt as it evolves.

Last but not least we have **Monitoring and Visualizing**. It involves tracking and visualizing key metrics, states, dependencies, and interactions across all system components and the system itself. Therefore, data from previous stages is used to monitor development progress, requirements status, and quality through real-time dashboards, identifying trends, bottlenecks, and alignment with objectives, thus supporting data-driven decisions and continuous improvement.

3 A Glimpse of SmartDelta Tools

For the stages, multiple tools have emerged. They can be mapped to different technical areas for each stage, provided in 3. From a corpus of 33 tools, we have selected seven as demonstrators (Table 2).

Table 2. SmartDelta Tool and Knowledge Providers

Ctry.	Tool and Knowledge Provider
	Mälardalen University, RISE - Research institutes of Sweden
	Cape of Good Code, Fraunhofer FOKUS, IFAK Institute for Automation and Communication, TWT GmbH Science & Innovation, Akkodis
	Cyberworks Robotics, Team Eagle, Ontario Tech University
	ERSTE Software Limited, Bilkent University, Arçelik
	University of Innsbruck
	University of Madrid Carlos III
	Hoxhunt
	University of Antwerp

3.1 NALABS (NAtural LAnguage Bad Smells)

- **Main Stage:** Requirements Engineering
- **Technical Area(s):** Requirements Quality Analysis, Requirements Verification

NALABS is developed to automatically detect issues in natural language requirements and test specifications used in large-scale system development [11].

Previous research [6] has shown that specifications written in natural language often contain significant cloning and structural issues, which can increase the cost of maintaining and executing test cases. For both requirements and system test cases written in natural language, specific "bad smells" have been identified as indicators of poorly structured artifacts. Building on these observations, we developed this set of indicators to detect flaws in requirements, defining dictionary-based metrics to automatically identify these bad smells in natural language artifacts.

The tool identifies various bad smells or poor quality indicators in requirement documents, which can lead to costly errors and inefficiencies during development. NALABS uses dictionary-based metrics to detect issues such as vagueness, referenceability, optionality, subjectivity, weakness, readability, and complexity. Available on GitHub[2], NALABS aims to improve the requirements engineering process by automating the detection of problematic requirements.

Figure 4 shows using NALABS from a GUI interface. The tool contains three essential layers: (i) the pre-processing of requirement documents stored as excel spreadsheets[3], (ii) the configuration and application of bad smell metrics, and (iii) presenting these results to the user. NALABS is used in continuous devel-

[2] https://github.com/eduardenoiu/NALABS.

[3] Many companies and tools (such as IBM's Rational DOORS) use this format for working with requirements.

File Edit View

Id	Text	NW	NC	NV	Optionali	Subjectivi	NR	NR2	Weaknes:	Imperativ	Imperativ	Continua:	ARI
NASA32	8. Redundancy Management **shall** [SRS109] degrade the FCP virtual group within 3 minor frames of fault detection **and** isolation.	19	2	0	0	0			0	1	0	2	71.58
NASA31	7. **If** the configuration needs to be changed due to a fault, as specified above, Redundancy Management **shall** [SRS128] issue a configuration update to mask out the failed network element.	30	1	0	0	0			0	1	0	0	78.00
NASA30	6. For the NEFU, **if** errors are identified **after** the processor has been masked out, **and if** at least 4 NEs are still active, the NE **shall** [SRS283] be removed from the configuration **and** recovery **will** be attempted. Note that the NEFU recovery does not depend on whether alignment is permitted.	51	7	3	0	0			0	4	3	4	69.38
NASA29	5. For the NEFU, **if** the first failure is diagnosed, Redundancy Management **shall** [SRS245] issue a configuration update to mask out the failed processor. Note that the NE is allowed to remain in the configuration **and** no recovery **will** take place. Note also that the failed NEFU **could** be diagnosed using any method, including (**but** not limited to) ICP presence test, **or** NE syndrome analysis.	65	5	6	0	0			0	4	5	2	68.14
NASA28	4. **If** a failure in an FCR other than the NEFU is diagnosed **when** the FCP is configured **as a** degraded triplex, no action **shall** [SRS254] be taken. Note that a failed FCR **could** be diagnosed using any method, including (**but** not limited to) Continuous BIT, ICP presence test, **or** NE syndrome analysis.	53	4	4	0	0			0	1	2	0	69.27

Fig. 4. Usage of NALABS Tool from GUI.

opment environments where requirements frequently evolve, as it enables automated detection of quality issues in new variants and deltas of requirements. NALABS helps prevent the propagation of errors and inefficiencies across different system versions in the Alstom use case. The tool detected ten issues for every 300 requirements and was evaluated on a dataset of 7,423 requirements from various projects. In addition, the NALABS tool analyzed 661 security requirements across 46 documents [5], focusing on identifying "requirement smells"—indicators of potential issues such as vagueness, ambiguity, and excessive complexity that can hinder the usability of security requirements during incremental development. Certain indicators were statistically significant, pointing to common problem areas in security specifications. For example, ARI scores indicated a generally difficult to very difficult reading level across the requirements. Additionally, nine moderate to strong correlations were found among different smell indicators, revealing patterns in the language of security requirements. Notably, requirements with high conjunction counts often exhibited increased logical complexity, potentially obscuring intended functionality and raising reimplementation risks.

3.2 DRACONIS

- **Main Stage:** Incremental Development
- **Technical Area(s):** Requirements Verification, Code Reuse, Code Quality Improvement

Darconis is a framework designed to automate reviews in incremental block-based safety-critical software processes, particularly in the railway domain within the Alstom use case. It utilizes static round-trip analysis to enforce design rules and facilitate faster feedback at the model level. By translating models into an intermediate representation, DRACONIS enables metrics extraction and rule-based analysis of block-based programs like FBD and Simulink. The tool supports incremental analysis, checking only the modified parts of a model, thus

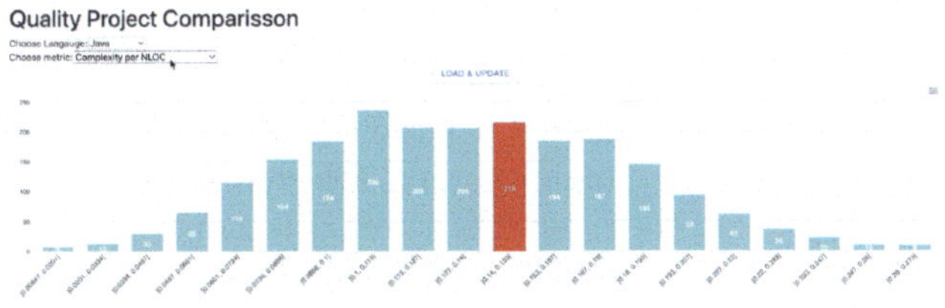

Fig. 5. The red bar indicates the complexity density of the Java project compared to other projects.

improving review efficiency. DRACONIS also allows the creation of custom analysis rules using JSON files, making it adaptable to specific project requirements. This framework helps bridge the gap between developed models and their analyzed counterparts, ensuring adherence to safety standards and reducing manual review efforts in continuous development [15].

In addition to being a general analysis framework for block-based programs, DRACONIS is one of the tools in SmartDelta that automates the quality assurance process in an incremental development setting. Analyzing an intermediate representation allows the analysis of multiple models, and the analysis is presented in an accessible way. Detailed usage instructions, including small test models, can be found in a GitHub repository[4].

In SmartDelta, DRACONIS was applied to use cases from the railway domain to evaluate its potential for accelerating design rule checks. An internal version of DRACONIS was tested on nine FBD models of varying complexity from an existing industrial project. When compared to a manual baseline, where an experienced developer performed the same analysis, DRACONIS demonstrated an average speedup of approximately 150 times across five executions.

3.3 SoHist V2

- **Main Stage:** Quality Assurance
- **Technical Area(s):** Code Quality Improvement, Static Analysis of Deltas, Automatic Code Analysis and Change Impact Analysis

SonarQube has some limitations regarding historical code analysis, including issues with version variability, a lack of filtering options, and inadequate visualization capabilities. To address these challenges, a tool called SoHist was developed and presented at EASE 2023 [2]. This tool enhances SonarQube's functionality by providing additional analysis features. In 2024, a new quality assessment feature was introduced. This feature utilizes an initial dataset comprising over 2007 historically evaluated SonarQube projects that meet specific

[4] https://github.com/jean-malm-mdh/draconis.

quality criteria, such as having more than 100 commits, a minimum of 4 contributors, and at least 10 GitHub stars. With this data, users can benchmark their project's quality metrics against a large set of comparable projects, considering various programming languages and selected quality criteria. A chart visualizes the distribution of these metrics across other projects, yielding insights into metrics like Test Coverage, Code Smell Density, and more. An example of Code Complexity is given in Fig. 5.

3.4 SmartMetrics

- **Main Stage:** Incremental Development
- **Technical Area(s):** Code Reuse

SmartMetrics analyzes files across repositories and evaluates software metrics like complexity and activity rates, and performs semantic analysis using topic modeling and embedding calculations (for Retrieval-Augmented Generation (RAG)). It also supports smart artifact labeling by leveraging Large Language Models (LLM). The tool supports delta analysis by comparing two versions (commits). Additionally, it creates interactive dashboards to visualize the analysis results, providing deep insights into code evolution and supporting RAG use cases. SmartMetrics has brought new insights into complex project environments, leading to improved decision-making. Furthermore, it enables joined evolution analysis of all artifacts, including requirements, which improves software quality in the long term.

3.5 Smellyzer

- **Main Stage:** Quality Assurance
- **Technical Area(s):** Issue Triage and Resolution, Monitoring and Anomaly Detection

Smellyzer is designed to elevate software development quality by identifying and addressing process smells within code review [3] and bug tracking [10,14]. By pinpointing detrimental patterns, such as large changesets, sleeping reviews, unassigned bugs, and the absence of traceability links to bug-fixing commits, Smellyzer provides teams with actionable insights to refine their practices. The latter issue, where bugs are closed without any linked corrective commit, has also been explored through semi-automated traceability tools like ReLink [16], which employ similarity-based heuristics to recover missing links between pull requests and reported issues.

3.6 PieR

- **Main Stage:**Incremental Development
- **Technical Area(s):** Issue Triage and Resolution, Predictive Analysis

PieR proposes an automated system to simplify the management of Pullrequest (PR) by classifying them into specific categories with the help of LLMs. These categories, including "bug fix", "new feature", "security update", and others, address different aspects of software maintenance and development. This ensures PRs are categorized and handled efficiently based on their content, type, and root causes. This tool enables project managers to learn about the distribution and nature of PRs, encouraging informed decision-making and helping to understand project development trends better.

3.7 Issue Analyzer

- **Main Stage:**Recommend and Predict
- **Technical Area(s):** Similarity Analysis Approaches and Recommendations

Issue Analyzer performs automatic labeling and similarity analysis of software issues. This tool classifies issues, such as bug reports and feature requests, according to their label, e.g. their security relevance, with the aim of reducing the time required for manual issue identification. Using modern language models and vector databases, this tool also suggests similar issues to support bug fixing by utilising knowledge from previous bug fixes. This tool serves as a basis for further analyses, such as providing recommendations for code and test case reuse or assigning specific people or teams for design, implementation and testing. This tool improves the efficiency and accuracy of defect management and contributes to safe and maintainable software.

4 Conclusion and Future Work

Software systems are typically not built and developed from scratch but rather as increments over existing versions. As such, over time, companies have a wide range of product versions and variants that need to be maintained and assessed for their quality and reusability. In this paper, we presented an update (January, 2025) of the outcomes of SmartDelta, an international project developing solutions for managing software quality across different product versions and variants. In particular, the SmartDelta Methodology, along with a subset of project tools and solutions developed for implementing the methodology, were discussed.

In total, we expect to have more than 30 solutions ready by the end of the project in March 2025. Due to space limitations, this paper provides a general overview of the project's outcomes. Detailed information is available in several public deliverables, which can be accessed in the *Documents* section of the ITEA SmartDelta project webpage[5]. In addition to the deliverables, we will release a SmartDelta dashboard that visualizes examples of various delta aspects, plan to deploy a SmartDelta ChatBot enriched with knowledge from the Work Packages (utilizing a RAG) to enable quick access to relevant information, and publish an interactive web version of the SmartDelta Methodology.

Acknowledgements. This work was supported by and done within the scope of the ITEA3 SmartDelta project, which the national funding authorities of the participating countries have funded.

References

1. Conradi, R., Westfechtel, B.: Version models for software configuration management. ACM Comput. Surv. **30**(2), 232–282 (1998)
2. Dornauer, B., Felderer, M., Weinzerl, J., Racasan, M.C., Hess, M.: SoHist: a tool for managing technical debt through retro perspective code analysis. In: Proceedings of the 27th International Conference on Evaluation and Assessment in Software Engineering, EASE 2023, pp. 184–187. Association for Computing Machinery, New York (2023). https://doi.org/10.1145/3593434.3593460
3. Doğan, E., Tüzün, E.: Towards a taxonomy of code review smells. Inf. Softw. Technol. **142**(C) (2022). https://doi.org/10.1016/j.infsof.2021.106737
4. Haber, A., Rendel, H., Rumpe, B., Schaefer, I.: Evolving delta-oriented software product line architectures. In: Large-Scale Complex IT Systems. Development, Operation and Management: 17th Monterey Workshop 2012, Oxford, UK, 19–21 March 2012, Revised Selected Papers 17, pp. 183–208. Springer (2012)
5. Hallberg, R.: Finding quality problems in security requirements using NALABS. Malardalen University Thesis DIVA (2023)
6. Juergens, E., et al.: Can clone detection support quality assessments of requirements specifications? In: Proceedings of the 32nd ACM/IEEE International Conference on Software Engineering, vol. 2, pp. 79–88 (2010)
7. Li, Z., Avgeriou, P., Liang, P.: A systematic mapping study on technical debt and its management. J. Syst. Softw. **101**, 193–220 (2015)
8. Linsbauer, L., Lopez-Herrejon, R.E., Egyed, A.: Variability extraction and modeling for product variants. In: Proceedings of the 22nd International Systems and Software Product Line Conference, vol. 1, p. 250 (2018)
9. Lochau, M., Schaefer, I., Kamischke, J., Lity, S.: Incremental model-based testing of delta-oriented software product lines. In: Brucker, A.D., Julliand, J. (eds.) TAP 2012. LNCS, vol. 7305, pp. 67–82. Springer, Heidelberg (2012). https://doi.org/10.1007/978-3-642-30473-6_7
10. Qamar, K.A., Sülün, E., Tüzün, E.: Taxonomy of bug tracking process smells: perceptions of practitioners and an empirical analysis. Inf. Softw. Technol. **150**(C) (2022). https://doi.org/10.1016/j.infsof.2022.106972
11. Rajkovic, K., Enoiu, E.: NALABS: detecting bad smells in natural language requirements and test specifications (2022). https://arxiv.org/abs/2202.05641

[5] https://itea4.org/project/smartdelta.html.

12. Saadatmand, M., et al.: SmartDelta project: automated quality assurance and optimization across product versions and variants. Microprocess. Microsyst. **103**, 104967 (2023)
13. Seidl, C., Wille, D., Schaefer, I.: Software reuse: from cloned variants to managed software product lines. In: Automotive Systems and Software Engineering: State of the Art and Future Trends, pp. 77–108 (2019)
14. Tuna, E., Kovalenko, V., Tüzün, E.: Bug tracking process smells in practice. In: Proceedings of the 44th International Conference on Software Engineering: Software Engineering in Practice, ICSE-SEIP 2022, pp. 77–86. Association for Computing Machinery, New York (2022). https://doi.org/10.1145/3510457.3513080
15. Vassallo, C., Panichella, S., Palomba, F., Proksch, S., Gall, H.C., Zaidman, A.: How developers engage with static analysis tools in different contexts. Empir. Softw. Eng. **25**, 1419–1457 (2020)
16. Yaşa, A., et al.: Evaluating relink for traceability link recovery in practice. In: Proceedings of the 32nd IEEE International Conference on Software Analysis, Evolution, and Reengineering, SANER 2025. IEEE (2025)

Application of Virtual Reality to Intelligent Transportation Systems as a Technology Accelerator for Smart Infrastructure Monitoring

Niloofar Malekghaini[1,2](✉), Nikhil N. Sharma[1], Hamed Ebrahimian[2], and Frederick C. Harris Jr.[1]

[1] Department of Computer Science and Engineering, University of Nevada, Reno, NV 89557, USA
niloofarmg@gmail.com, nikhil.n.sharma3@gmail.com, Fred.Harris@unr.edu
[2] Department of Civil and Environmental Engineering, University of Nevada, Reno, NV 89557, USA
hebrahimian@unr.edu

Abstract. This work extends the application of intelligent transportation systems (ITS) to the field of structural health monitoring (SHM) for bridge infrastructures and develops a virtual reality (VR) platform as a technology accelerator. In the proposed SHM method, video recordings of traffic are captured using traffic cameras and processed using a combination of YOLO and Deep SORT models, along with projective mapping, to extract the time history of vehicle tire locations. This step, referred to as traffic tracking, predicts the points at which traffic loads excite the bridge. This information is integrated with the bridge's vibration response and the finite element (FE) model to develop a digital twin of the bridge. The necessary models for the traffic tracking component are trained using Google Open Images and tested in a VR environment that simulates traffic on a roadway with installed cameras. Utilizing VR instead of real-world tests significantly accelerates technology development. After achieving promising results in the VR environment, the prototype of the proposed technology has been developed and deployed on a real-world bridge.

Keywords: Digital Twin · VR · Machine Learning · Computer Vision · Intelligent Transportation Systems

1 Introduction

America's bridge infrastructure, which is a vital component of its transportation network, is aging and deteriorating. With over 600,000 bridges across the U.S., nearly 40% are 50 years or older and at the current rate of improvement, it is

N. Sharma and H. Ebrahimian—Contributing authors.

S. Latifi (Ed.): ITNG 2025, AISC 1463, pp. 660–671, 2025.
https://doi.org/10.1007/978-3-031-89063-5_57

estimated to take more than 50 years to carry out all the repairs that are currently necessary. With the increasing number of aging bridges [13] and according to American Society of Civil Engineering (ASCE) guideline, there is a need for introducing new methods for operational health monitoring of bridges. These new methods should be technology-based, cost-effective, and involve minimal installation and traffic disruption.

Over recent decades, digital twinning using output-only time-domain finite element (FE) model updating has become a valuable tool for damage localization of civil structures as a cost-effective with minimal installation load work approach [6]. However, its application for operational health monitoring of bridges is challenging due to the spatial and temporal variability of traffic excitation. Temporal variability arises from interactions between vehicle suspension systems and bridge dynamic systems, road roughness, tire assemblies, and driveline excitation [3]. Spatial variability results from vehicle movements along the bridge [7].

Recent advancements in artificial intelligence and computer vision techniques present promising solutions to the challenges of digital twinning for operational health monitoring of bridges. [1]. It is now feasible to capture video recordings of traffic and use them to classify vehicles and track the footprints of their tires on bridge decks. This data serves as input for the bridge digital twinning process [7]. During this process, the tire history of the dynamic load of traffic as well as the damage status of bridge are determined. The identified damage status of bridge is used for asset management and maintenance of structure, while the traffic tracking data and dynamic load estimates can be used to improve intelligent transportation systems (ITS) through traffic management and weight in motion.

This study focuses on the development and evaluation of the computer vision component for tracking the time history of the location of tires. This step is referred to as the traffic tracking technology component. To achieve the goal of this study, a virtual reality (VR) testing platform is developed using Unity which accelerates technology development by accurately simulating roadway test scenarios for system components under evaluation. In this platform, the bridge roadway is simulated and equipped with a network of cameras. As the traffic traverses the simulated bridge road, the VR platform records the view of cameras as well as the ground truth time history of tire locations as in a world coordinate system. The recorded videos are then processed using the proposed traffic tracking algorithm to track tire locations in the world coordinate system. By comparing the estimated locations with the ground truth data, the performance of the proposed traffic tracking algorithm is evaluated. Additionally, this method is assessed in a real-world setting by instrumenting the East Yutan (EY) [8] bridge in Nebraska with networks of traffic cameras.

The rest of this paper is structured as follows: Sect. 2 provides background knowledge integral to understanding the material of this work. Section 3 details the proposed method for operational health monitoring of bridges. This section also elaborates the proposed traffic tracking component and its main components. Section 4 and Sect. 5 show the experiments and results in the VR and real-world setting. Finally conclusions as well as future work are contained in Sect. 6. More detail and description can be found in [5] and [4].

2 Background

SHM of bridges are a family of methods utilized to identify safety threats at an early stage, allowing for the prioritization of repair, retrofit, or replacement efforts. Traditionally, various methods have been employed for SHM of bridge, including visual inspections, vibration-based health monitoring, and non-destructive evaluation (NDE) methods.

In recent decades, digital twinning has become a valuable tool for damage localization of structures. This approach leverages a physics-based model of the structure alongside stochastic filtering techniques, such as Bayesian inference, to update the initial FE model. The updated model is then maintained as a digital twin of the structure, which can be used for virtual sensing and interrogated for health monitoring and damage diagnosis of the structure.

Recent advancements in ITS have emphasized the significance of vision-based vehicle detection and tracking algorithms for traffic flow predictions. A widely used vehicle detection approach is the You Only Look Once (YOLO) model [9], which detects, classifies and determines the location through a single convolutional neural network to predict bounding boxes and class probabilities in one evaluation. YOLO divides an image into a grid, with each cell predicting bounding boxes, confidence scores, and class probabilities. This design is well-suited for operational acceleration using GPUs [10]. The confidence scores indicate the chances of an object in a box, while class-specific scores combine these with conditional probabilities. Once an object is detected, it must be tracked across frames in which the object moves. The Deep Simple Online and Real-Time Tracking (Deep SORT) algorithm is commonly employed to do this. Deep SORT uses Kalman filter to predict object trajectories and Hungarian matching algorithm for matching predictions to detections. Deep SORT also incorporates a deep learning-based feature embedding model to learn discriminative feature representation for each detected object using a convolutional neural network to compare the visual similarity between tracks and detection [12].

A difficult task in development of deep learning and computer vision models is collecting large sets of real data for training and validation [11]. To address this issue, a new trend has emerged in research studies utilizing synthetic images as training and/or validation data, either alone or in combination with real images. In recent VR applications, Unity has become a widely used game engine for implementing visualization modules. Unity's extensible model is integrated with a visual environment editor, enabling developers to change properties visually and create references to other objects within the environment. Additionally, Unity's extensive asset store provides a vast library of pre-made assets and tools that can accelerate development and improve the quality of VR applications.

3 A New Paradigm: Application of ITS in Bridge Health Monitoring

This study introduces ITS as a key technology component in an innovative technology-based approach for SHM of bridges [4]. This technology component combines computer vision techniques and deep learning methods within a digital twinning framework. Moreover, a VR simulation is introduced as a platform for accelerating technology development by simulating roadway test scenarios for system components under evaluation.

Development of bridge digital twins requires understanding regarding the following items [7]:

- Input excitation to the bridge structure, which is determined through:
  - Load of vehicles traversing on the bridge.
  - Time history of the locations in which the load of vehicles are applied to the bridge structure. These are the time history of the footprint of vehicles' tires.
- Output response of the bridge structure, which is determined through:
  - Vibration response of the bridge structure to the excitation induced by traffic, e.g., acceleration or displacement response.

In the proposed technology-based method, the output response is collected using sensors, e.g. accelerometers, installed on bridge. On the other hand, the input excitation is inferred by application of computer vision and deep learning methods on video recordings of traffic captured by traffic cameras installed on the bridge. In this method, video recordings are processed to detect and classify vehicles, and track the location of their tires. The classification results provide initial estimates regarding the weight of vehicles, and the tracking results provide the time history of the location traffic input load. Then, the extracted information on input excitation and output responses are fused together and used as the input to the digital twinning process. Using estimates of damage status of various bridges, the prioritization plan for rehabilitation of bridges are developed. Moreover, using the estimates of vehicles dynamic load, remote load rating and traffic management can be possible. Figure 1 presents the schematic view of the proposed technology-based method for digital twinning of bridges and highlights its four main steps: Instrumentation and data collection, Data pre-processing, Digital twinning, and Interpretation of estimates.

The pre-processing of video recordings is the main component of the proposed technology. This technology component is proposed to be implemented in four main phases [4]. The phases are also depicted in Fig. 2. For each recorded frame:

- In the **first phase**, a perspective transformation matrix is established from the pixel to world/FE model coordinate systems. This matrix is developed using the intrinsic and extrinsic parameters of the cameras as well as the known locations of fixed landmarks in the pixel and FE model coordinate systems [2].

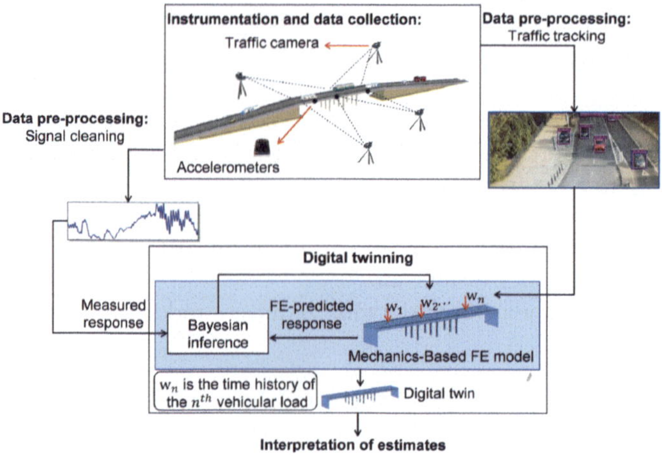

Fig. 1. Schematic representation of proposed method for operational health monitoring of bridges.

- In the **second phase**, vehicles are detected and classified using YOLOv8 algorithm resulting in the classification of vehicles and generation of corresponding bounding boxes around the detected vehicles. Subsequently, the detection results are integrated with Deep SORT algorithm, which extracts patches corresponding to the bounding boxes to track the trajectory of each vehicle. Vehicle detection can be carried out using pre-trained (using COCO dataset) YOLOv8.
- Moving to the **third phase**, YOLOv8 instance segmentation is performed within each bounding box to detect and predict the segmentation mask of vehicle's tires. The predicted masks of tires are then tracked individually and the coordinates of their contact points with the bridge's deck are determined in the pixel coordinate system. This model is trained using the Google Open-Image dataset.
- In the **fourth phase**, the locations of tires' contact points are computed in the FE model coordinate system. By applying the derived perspective transformation matrix to the pixel coordinates of the tires' contact points, the location of these points can be computed in the FE model coordinate system. By integrating the tracked location of tires from different camera views, a complete understanding of each vehicle's tires' locations on the bridge is derived.

4 Experiments and Results

The traffic tracking pipeline is tested on a simulated testbed environment using VR and the Unity game engine. The developed VR needs to be capable of recording synchronized footage of traffic while fixed landmarks are in the view of cameras. For each traffic scenario, the synchronized recordings of cameras are saved

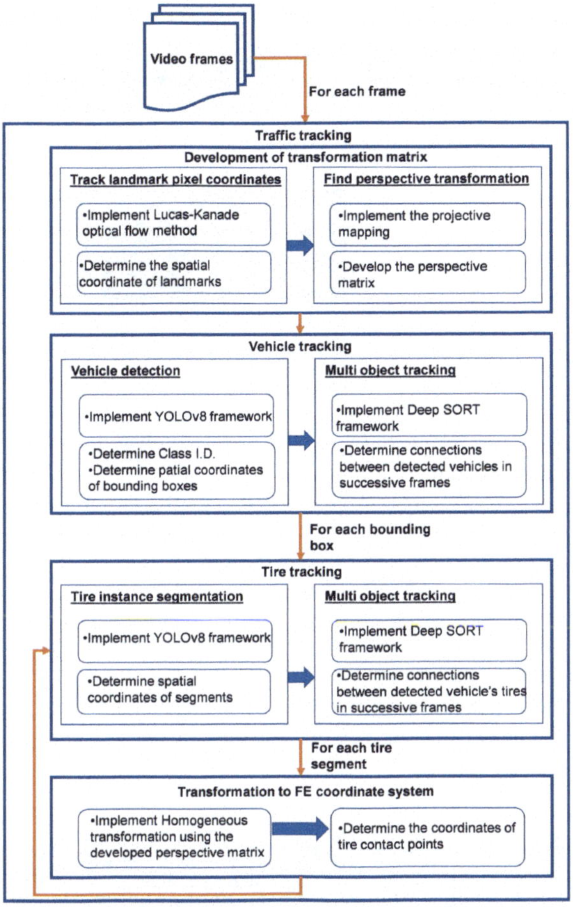

Fig. 2. Phases of pre-processing of video recordings.

as *mp4* files. At the same time, the location of the outer edge of tires (ground truth) in contact with the surface of the road are recorded and reported as *CSV* files.

The bridge road is simulated based on the properties of the East Yutan Bridge (EY) [8] in Nebraska. EY is a 122 m bridge in the east-west direction with a total width of 9.8 m. This width is divided into two lanes. Figure 3 shows the horizontal view of the bridge along with the defined real-world mapped dimensions of the bridge. The roadway is also marked with fixed landmarks, white markings, at known locations on the roadway. These markings will be used to develop the perspective transformation matrix for each camera view (phase one). Figure 3 shows the layout of landmarks and a close look at four of the markings.

Vehicles are modeled as a tri-axial dump truck which is 7.63 m long, 2.92 m wide and 2.76 m high. Various scenarios of traffic are modeled using Unity. The

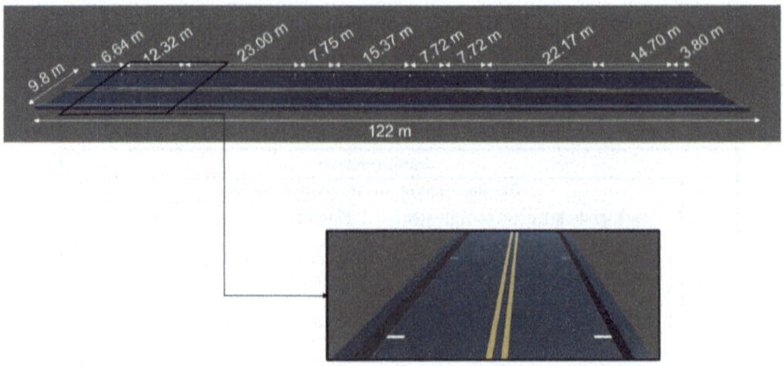

Fig. 3. Layout of fixed landmarks and a close view over four of them.

scenarios are conducted with one or two trucks traveling in one or two directions. The traffic scenarios include:

- Travel of a single truck in the middle of the bridge, straddling the line between the two lanes.
- Travel of two trucks in opposite directions and different lanes.
- Travel of a single truck with changing lane at the middle of bridge length.

In the developed experiments, trucks travel the bridge with a fixed speed between 8 and 80 km per hour (kph). We also captured synchronized recordings of cameras. The location and associated angles of these cameras can be seen in Fig. 4.

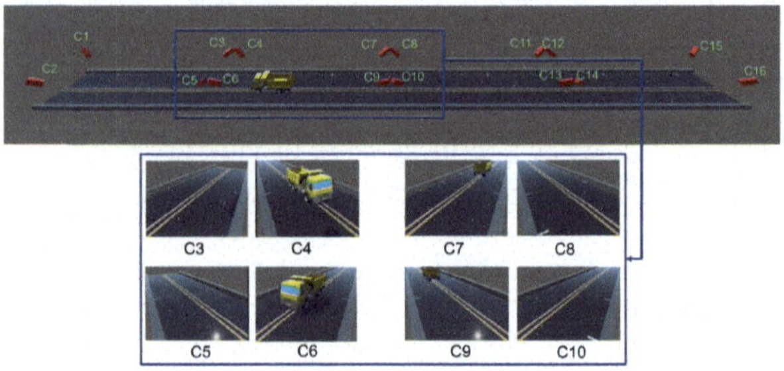

Fig. 4. Layout of cameras and their angles.

After tracking the vehicle (phase two), tire tracking algorithm is implemented inside each detected bounding box to detect and extract the instance segmentation of vehicle's tires. Figure 5 depicts the tires being detected (phase three). The segmented tires are then assigned a unique I.D. Having the instance segmentation of tires, the coordinates of the outer edge of tires' contact points with the roadway are determined in the pixel coordinate system (phase four). An example of this is shown in Fig. 6.

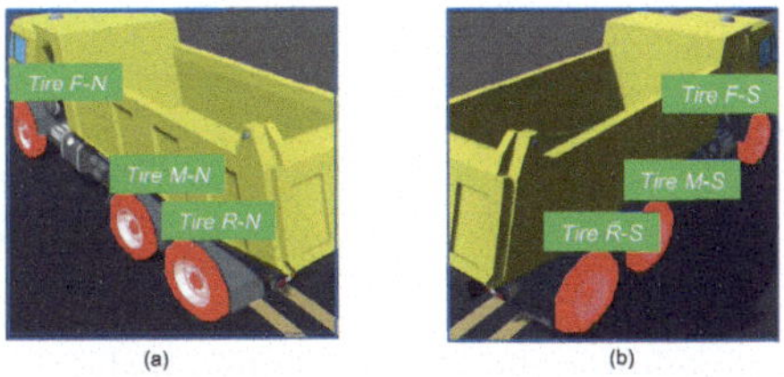

Fig. 5. Vehicle detection and tracking in frame captured by: (a) C4, and (b) C6.

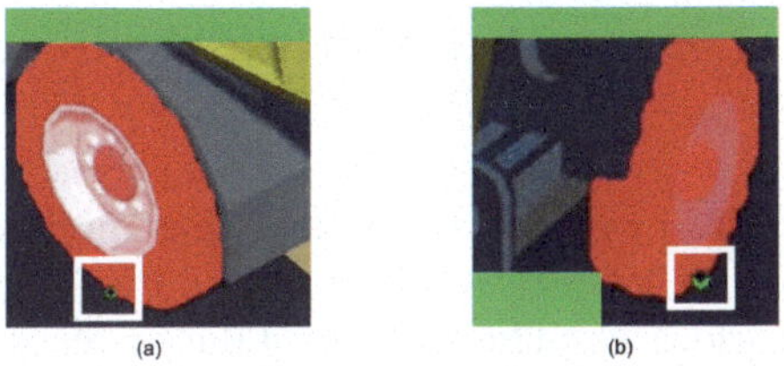

Fig. 6. Tracked contact points: (a) Tire R-N in frame captured by C4, and (b) Tire R-S in frame captured by C6.

The tracked location of tires in the FE coordinate system are compared with the collected ground truths. For one of our test cases at a particular instance, the comparison is shown in Fig. 7. Calculating the euclidean distance between the ground truth and tracked location of tires reveals the maximum variation of 26cm at the given time instance.

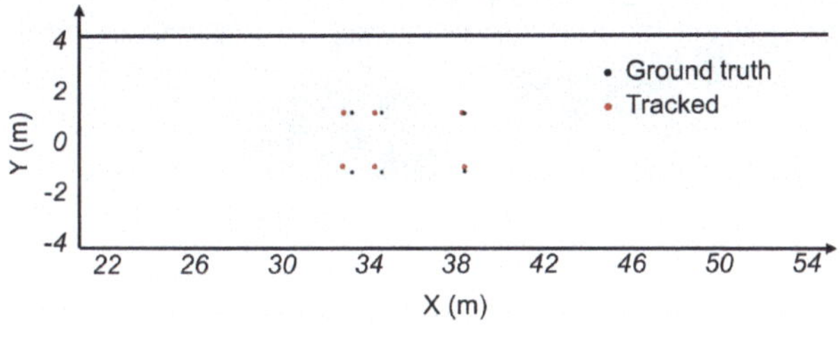

Fig. 7. Comparison between the tracked ground truth location of tires at a given time instance.

For a better understanding regarding the performance of the proposed traffic tracking method, the prediction error is calculated as a function of truck location. For this purpose, at each recorded frame, the euclidean distance between the ground truth and tracked location of all six tires are calculated and the maximum value is reported. Figure 8 shows the maximum discrepancies between the prediction and ground truth as a function of the location of vehicle. The maximum error is 33 cm.

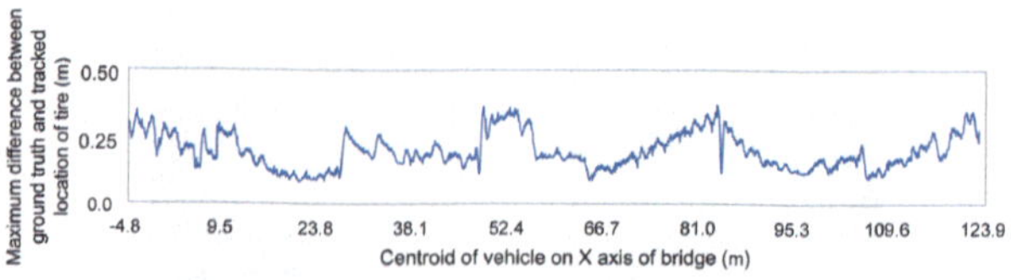

Fig. 8. Variability of the maximum discrepancy between the tracked and ground truth location of tires as a function of the centroid of vehicle on the length of bridge.

5 Application of Proposed Method Using Real-World Data

After development of the proposed traffic tracking method and its validation using simulation data in a VR environment, the performance of the method is evaluated using real-world data. The real-world data consists of video recordings collected from EY bridge. Unlike the VR simulation environment, the real-world setting does not provide ground truth data, so the evaluation focuses on the feasibility of performing detection and tracking, and manual review of the outputs ensuring the reasonable results. Figure 9 shows a view of the EY bridge and an installed traffic camera.

Fig. 9. East Yutan (EY) bridge.

The network of traffic cameras was deployed using six Raspberry Pi cameras. The cameras were installed on tripod supports placed on the roadway shoulder at approximate height of six meters over the bridge deck. The tripod legs were locked using sandbags. The traffic camera network recorded the synchronized passage of traffic across multiple camera views with frame rate of 30 Hz. A GPS (global positioning system) clock was added to the camera network serving as a time server allowing for synchronized recordings from the cameras.

Using the marked fixed landmarks on the bridge deck, and the same process as explained previously, the tracked location of tires of passing traffic are determined in the world coordinate system. Figure 10 shows a summary of data pre-processing process. The manual review of traffic tracking results demonstrate reasonable outputs. This provides a promising foundation for future advancements in structural health monitoring and intelligent transportation systems.

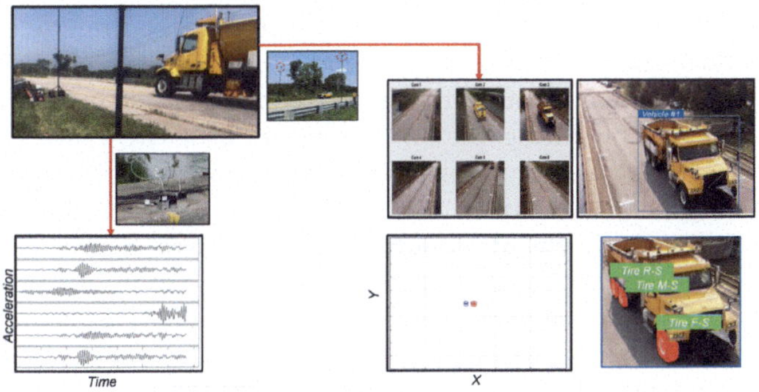

Fig. 10. A schematic summary over data pre-processing step on EY bridge.

6 Conclusions and Future Work

Conclusions: The research work presented in this study was focused on development and evaluation of a computer vision and deep learning components within an innovative technology-based solution for operational health monitoring of bridge structures. For this purpose, first, a new method for operational health monitoring of bridges was introduced based on a multi-modal data fusion and physics-based Bayesian model updating technique to create the digital twin of the bridge structure. In this method, the bridge structure is instrumented with networks of traffic cameras and accelerometers to collect video recordings of traffic and the resulting vibration response of bridge structure. Computer vision and deep learning techniques are used to extract the time history of the location of the tire of vehicles on the bridge. This information, is fused with the collected vibration data and are used as input to a Bayesian inference. Using Bayesian inference, finite element model of bridge is updated and time history of the dynamic load of vehicles on the bridge are estimated. The updated model is interrogated for health estimation of bridge, and the traffic data can be used to improve ITS.

The computer vision and deep learning components of the proposed health monitoring approach were developed and evaluated using a virtual reality (VR) testing platform. In the simulation, traffic of one or two trucks ran on the bridge and the video recordings of cameras as well as the ground truth time history of the location of tires were collected from VR. The video recordings were processed using the proposed traffic tracking algorithm in which You Only Look Once

(YOLO), Deep Simple Online and Real-time Tracking (Deep SORT) algorithms and perspective mapping were implemented. Using the pre-trained YOLOv8, vehicles were classified and detected through bounding boxes, and using Deep SORT, vehicles were tracked and assigned with a unique identified (I.D.). Then, YOLOv8 and Deep SORT were implemented on each bounding box to track the instance segmentation of tires. Having the instance segmentation of tires, the contact point of tires with road were extracted in pixel coordinate system. The perspective transformation between pixel and world (or FE) coordinate systems were developed and used to transform the tire contact points from pixel to world coordinate system. Comparing the tracked location of tires with the collected ground truth revealed maximum of 33 cm error, which was an acceptable level of error for this preliminary study.

A preliminary assessment of proposed traffic tracking algorithm was carried out in a real-world setting by instrumenting the East Yutan bridge in Nebraska with networks of traffic cameras. Unlike the VR simulation environment, the real-world setting did not provide ground truth data, so the evaluation carried out with manual review ensuring reasonable results.

Future Work: The work presented here represents the development of a crucial component of a technology-based framework for operational health monitoring of bridges. This component can be enhanced and improved in several aspects to accelerate a practical technology solution. Potential areas for improvement and future research are listed below.

- Scaling the developed VR for larger number and variety of class of vehicles traversing on the bridge, and studying its effect on the performance of traffic tracking algorithm.
- Studying the environmental effects,e.g., light and precipitation, on the performance of traffic tracking algorithm.
- Reducing the number of cameras deployed on the bridge and studying its effects on the performance of traffic tracking algorithm.

Acknowledgements. This material is based in part upon work supported by the National Science Foundation under grant number OIA-2019609 and United States Department of Transportation Small Business Innovative Research (SBIR) program Phase II under Contract number 6913G619C100048.

The acknowledgments extend to SC Solutions for data collection and project management , University of Nebraska Lincoln for field testing management, coordination, and data sharing, and Federal Highway Administration for supervising the project.

This manuscript contains content from the first author's previously completed MS thesis [4]. which has been adapted and summarized for this publication. This work has not been published elsewhere.

References

1. Chauhan, R.K., Chauhan, K.: Intelligent toll collection system for moving vehicles in India. Intell. Syst. Appl. **15**, 200099 (2022). https://doi.org/10.1016/j.iswa.2022.200099
2. Correia, S., Luck, S.: Digitizing historical balance sheet data: a practitioner's guide. Explor. Econ. Hist. **87**, 101475 (2023). https://doi.org/10.1016/j.eeh.2022.101475. Methodological Advances in the Extraction and Analysis of Historical Data
3. Gillespie, T.D.: Heavy truck ride. In: SAE International Congress and Exposition (1985). https://doi.org/10.4271/850001
4. Malekghaini, N.: Intelligent transportation systems and virtual reality as technology accelerators for smart infrastructure monitoring. Ms thesis, University of Nevada, Reno, Department of Computer Science and Engineering (2024). Advisor: Dr. Frederick C. Harris, Jr.
5. Malekghaini, N.: Operational health monitoring and digital twinning of bridges using multimodal data fusion and physics-based Bayesian model updating. Ph.D. dissertation, University of Nevada, Reno, Department of Civil and Environmental Engineering (2024). Advisor: Dr. Hamed Ebrahimian
6. Malekghaini, N., Ghahari, F., Ebrahimian, H., Bowers, M., Ahlberg, E., Taciroglu, E.: A two-step FE model updating approach for system and damage identification of prestressed bridge girders. Buildings **13**(2) (2023). https://doi.org/10.3390/buildings13020420
7. Malekghaini, N., Ghahari, F., Ebrahimian, H., Bowers, M., Azari, H., Taciroglu, E.: Time-domain finite element model updating for operational monitoring and damage identification of bridges. Struct. Control. Health Monit. **2023**(1), 4170149 (2023). https://doi.org/10.1155/2023/4170149
8. of Nebraska-Lincoln, U.: NOBL: nebraska outdoor bridge lab overview page (2024). https://engineering.unl.edu/outdoorbridgelab/. Accessed 05 July 2024
9. Redmon, J., Divvala, S., Girshick, R., Farhadi, A.: You only look once: unified, real-time object detection. In: Proceedings of the 2016 IEEE Conference on Computer Vision and Pattern Recognition (CVPR), pp. 779–788 (2016). https://doi.org/10.1109/CVPR.2016.91
10. Reisi-Dehkordi, A., Reeves, S.I., Harris Jr., F.C.: GPU-accelerated neural networks and computational strategies to predict wave heights. In: International Conference on Information Technology-New Generations, pp. 363–369. Springer (2024)
11. Reisi-Dehkordi, A., Tavakkoli, A., Harris Jr., F.C.: Integrating LSTM and EEMD methods to improve significant wave height prediction. In: International Conference on Information Technology-New Generations, pp. 19–25. Springer (2023)
12. Teoh, S.S., Bräunl, T.: Symmetry-based monocular vehicle detection system. Mach. Vis. Appl. **23**, 831–842 (2012). https://doi.org/10.1007/s00138-011-0355-7
13. U.S. Department of Transportation, F.H.A.: Bridge condition by functional classification area 2019 (2019). https://www.fhwa.dot.gov/bridge/nbi/no10/fcarea19.cfm

Non-blind Digital Image Watermarking Method Based on Dyadic Wavelet Transform and Gyrator Transform

Keisuke Matsumoto(✉) and Teruya Minamoto

Graduate School of Science and Engineering, Saga University, Saga, Japan
24726024@edu.cc.saga-u.ac.jp, minamott@cc.saga-u.ac.jp

Abstract. We propose a non-blind digital image watermarking method that combines the Dyadic Wavelet Transform (DYWT) and the Gyrator Transform to enhance watermark robustness and imperceptibility. The method involves decomposing the original image using the DYWT and applying the Gyrator Transform to the low-frequency component. We precisely embed the watermark into this region by predicting the low-frequency region after transformation using a nonlinear regression model, increasing its resilience to various attacks. Experimental results demonstrate that the proposed method achieves high image quality, with Peak Signal-to-Noise Ratio (PSNR) values exceeding 35 dB and Structural Similarity Index Measure (SSIM) values above 0.98. The method is robust against common image processing attacks such as filtering, JPEG compression, and geometric distortions like cropping and rotation, outperforming existing methods. The Normalized Correlation (NC) values remain high after various attacks, indicating reliable watermark extraction. These results confirm the effectiveness of the proposed non-blind method in achieving both high imperceptibility and robustness in digital image watermarking.

Keywords: Digital Watermarking · Non-blind Watermarking · Dyadic Wavelet Transform · Gyrator Transform · Robust Watermarking · Low-Frequency Embedding · Image Quality · Nonlinear Regression · Normalized Correlation · Image Processing

1 Introduction

The proliferation of digital media and technological advances like generative AI have heightened the need for secure and reliable methods to protect digital content. Digital watermarking embeds identifying information into digital media imperceptibly, allowing authorized parties to retrieve the watermark for purposes such as copyright protection and tamper detection [1].

T. Minamoto—Contributing author.

S. Latifi (Ed.): ITNG 2025, AISC 1463, pp. 672–683, 2025.
https://doi.org/10.1007/978-3-031-89063-5_58

Digital watermarking techniques are generally classified into two categories: *non-blind* methods, which require the original image during watermark extraction, and *blind* methods, which do not [6]. While blind watermarking is practical when the original image is unavailable, non-blind methods often achieve higher robustness and imperceptibility due to the precise embedding and extraction facilitated by access to the original image. Developing robust non-blind methods can serve as a foundation for creating effective blind watermarking techniques.

Watermarks can be embedded in various domains. Traditional methods operate in the spatial or frequency domain. The Fractional Fourier Transform (FrFT), which generalizes the Fourier Transform, offers greater flexibility by simultaneously representing spatial and frequency information [1]. Tang and Zhou [1] proposed a robust algorithm using FrFT, Discrete Wavelet Transform (DWT), and Singular Value Decomposition (SVD) for digital watermarking.

The Gyrator Transform, which interchanges the transformation axes of the FrFT, provides an additional degree of freedom through its rotation angle parameter [2,3]. It enables rotations in the twisted position-spatial frequency plane of two-dimensional signals, making it valuable for image processing applications [4]. Several studies have utilized the Gyrator Transform to enhance security and robustness in watermarking and encryption schemes [5–7].

For instance, Mehra and Nishchal [5] proposed a method for encrypting fingerprint images using the Gyrator Wavelet Transform with QR decomposition, enhancing security by expanding the key space. Zhang and Wei [6] developed an image watermarking scheme combining the Gyrator Transform with matrix decomposition in the invariant integer wavelet domain, achieving improved robustness. Ide *et al.* [7] presented a blind digital watermarking technique employing the Gyrator Transform and Dyadic Wavelet Transform (DYWT), demonstrating enhanced security and image quality.

Bhatnagar and Wu [8] introduced a non-blind watermarking method based on the Gyrator Transform and SVD, showing high robustness and improved security. However, their method is vulnerable to geometric attacks, indicating potential for further improvement.

This study proposes a non-blind digital watermarking method combining the Dyadic Wavelet Transform and the Gyrator Transform to develop a highly resilient watermarking technique. Our method decomposes the original image using the DYWT to obtain low-frequency and high-frequency components. The low-frequency component is transformed using the Gyrator Transform with a specified rotation angle. We precisely embed the watermark into this region by predicting the low-frequency region in the transformed domain using a nonlinear regression model. The embedding modifies the magnitude spectrum while preserving the phase information, which is essential for accurate reconstruction.

The structure of this paper is as follows: Sect. 2 describes the Dyadic Wavelet Transform. Section 3 explains the Gyrator Transform. Section 4 discusses the method for predicting the low-frequency region after the Gyrator Transform. Section 5 details the embedding and extraction algorithms of the proposed watermarking method. Section 6 presents the experimental results. Finally, Sect. 7 summarizes the study and outlines future challenges.

2 Dyadic Wavelet Transform

The two-dimensional Dyadic Wavelet Transform (DYWT) is an upsampling type of wavelet transform that possesses properties such as shift-invariance, meaning that features and results remain unchanged even if the original signal is shifted and redundancy, where the size of the decomposed signal is the same as that of the original signal. Denoting the original image (two-dimensional signal) as $C^0[m,n]$, the two-dimensional DYWT is given by the following equations:

$$
\begin{aligned}
C^{j+1}[m,n] &= \sum_k \sum_l h[k]h[l]\, C^j[m+2^j k,\, n+2^j l], \\
D^{j+1}[m,n] &= \sum_k \sum_l g[k]h[l]\, C^j[m+2^j k,\, n+2^j l], \\
E^{j+1}[m,n] &= \sum_k \sum_l h[k]g[l]\, C^j[m+2^j k,\, n+2^j l], \\
F^{j+1}[m,n] &= \sum_k \sum_l g[k]g[l]\, C^j[m+2^j k,\, n+2^j l],
\end{aligned}
$$

where h denotes the low-frequency decomposition filter, g denotes the high-frequency decomposition filter, m and n represent the element positions in the vertical and horizontal directions, respectively, and j denotes the decomposition level. In the DYWT, $C^{j+1}[m,n]$ represents the low-frequency component, capturing the coarse, smoothly varying regions of the image. The components $D^{j+1}[m,n]$, $E^{j+1}[m,n]$, and $F^{j+1}[m,n]$ are the high-frequency components corresponding to the horizontal, vertical, and diagonal detail information, respectively, representing the fine details and edges in those directions.

The inverse two-dimensional DYWT is given as follows:

$$
\begin{aligned}
C^j[m,n] = \frac{1}{4}\sum_k \sum_l \Big(& \tilde{h}[k]\tilde{h}[l]\, C^{j+1}[m-2^j k,\, n-2^j l] \\
& + \tilde{g}[k]\tilde{h}[l]\, D^{j+1}[m-2^j k,\, n-2^j l] \\
& + \tilde{h}[k]\tilde{g}[l]\, E^{j+1}[m-2^j k,\, n-2^j l] \\
& + \tilde{g}[k]\tilde{g}[l]\, F^{j+1}[m-2^j k,\, n-2^j l] \Big)
\end{aligned}
$$

Here, $\tilde{h}$ and $\tilde{g}$ represent the low-frequency and high-frequency reconstruction filters, respectively. The filter coefficients used in this paper are listed in Table 1.

Table 1. Low- and high-pass decomposition filter coefficients and their dual reconstruction filter coefficients

n	$h[n]/\sqrt{2}$	$\tilde{h}[n]/\sqrt{2}$	$g[n]/\sqrt{2}$	$\tilde{g}[n]/\sqrt{2}$
−2				−0.03125
−1	0.125	0.125		−0.21875
0	0.375	0.375	−0.5	−0.6875
1	0.375	0.375	−0.5	0.6875
2	0.125	0.125		0.21875
3				−0.03125

3 Gyrator Transform

The Gyrator Transform generates twisted rotations in both the spatial and frequency domains. The characteristics of the Gyrator Transform were studied in Ref. [4]. The Gyrator Transform for a two-dimensional signal $f(x, y)$ is defined by the following equation:

$$\begin{aligned} G(u, v) &= G^{\alpha}[f(x, y)](u, v) \\ &= \frac{1}{|\sin\alpha|}\int_{-\infty}^{\infty}\int_{-\infty}^{\infty} f(x, y)\, K_{\alpha}(x, y, u, v)\, dx\, dy \end{aligned}$$

Here, (x, y) represents the input coordinates in the spatial domain, (u, v) represents the output coordinates in the transformed domain, and α indicates the rotation angle. The kernel of the Gyrator Transform, $K_{\alpha}(x, y, u, v)$, is expressed by the following equation:

$$K_{\alpha}(x, y, u, v) = \exp\left(i2\pi\frac{(uv + xy)\cos\alpha - (xv + uy)}{\sin\alpha}\right)$$

The Gyrator Transform reduces to the identity mapping when $\alpha = 2k\pi$, where k is an integer, and to $G(u, v) = f(-u, -v)$ when $\alpha = (2k + 1)\pi$. When $\alpha = \frac{\pi}{2}$, it coincides with the Fourier Transform up to a scaling factor and possible coordinate swapping. For any other angle α, the transform has a constant amplitude and a hyperbolic phase structure. Since the Gyrator Transform satisfies additivity and periodicity with respect to the rotation angle α, the inverse Gyrator Transform is given by a rotation angle of $-\alpha$.

There are several methods to implement the Gyrator Transform on a computer [9,10]. In this study, we adopt the method from Ref. [10], Utilizing the Fast Fourier Transform (FFT) will be used to implement the Gyrator Transform.

4 Prediction of the Low-Frequency Component Region After the Gyrator Transform

An example of applying the Gyrator Transform to the original image and displaying the power spectrum is shown in Fig. 1.

Fig. 1. (a) Original image and (b) magnitude of its Gyrator Transform

In the power spectrum, the outer black areas represent high-frequency components, while the low-frequency components are concentrated in the center. By displaying the power spectrum for rotation angles in the range of $0 < \alpha < 2\pi$, measuring the boundary points between the high-frequency and low-frequency components, and applying nonlinear regression, the top-left point of the low-frequency region can be roughly estimated. The prediction of the low-frequency region in the Gyrator Transform domain allows precise embedding of the watermark, enhancing resistance to various attacks.

5 Proposed Method

This section presents the proposed non-blind digital image watermarking method based on the Dyadic Wavelet Transform (DYWT) and the Gyrator Transform. The method consists of two main processes: watermark embedding and watermark extraction.

5.1 Watermark Embedding Algorithm

The watermark embedding algorithm follows these steps:

1. **Dyadic Wavelet Transform:** Apply the Dyadic Wavelet Transform (DYWT) to the grayscale original image to obtain the four components C, D, E, and F.
2. **Gyrator Transform:** Apply the Gyrator Transform with rotation angle α to the C component obtained from the DYWT:

$$G(u, v) = G^{\alpha}[C(x, y)],$$

where G^{α} denotes the Gyrator Transform at angle α.

3. **Low-Frequency Region Prediction:** Predict the region of low-frequency components at angle α using a nonlinear regression model. Resize the watermark image to match the dimensions of this predicted low-frequency region.
4. **Watermark Embedding:** Embed the resized watermark image into the magnitude spectrum of the predicted low-frequency component region from step 3:

$$\tilde{G}_{\mathrm{L}}(u, v) = |\mathrm{Low}_{\mathrm{power}}(u, v)| + \beta \times \mathrm{sgn}[\mathrm{Watermark}(u, v)],$$

where β represents the embedding strength of the watermark, $|\cdot|$ denotes the absolute value, $\mathrm{Low}_{\mathrm{power}}(u, v)$ is the power spectrum of the predicted low-frequency components, and $\mathrm{sgn}(\cdot)$ is the sign function.
The embedding strength β controls the trade-off between watermark robustness and image quality. A higher β increases robustness but may degrade image quality. Using the sign function sgn ensures that the watermark embedding and extraction processes appropriately handle both positive and negative values.
5. **Inverse Gyrator Transform:** Combine the modified magnitude spectrum $\tilde{G}_{\mathrm{L}}(u, v)$ from step 4 with the phase information from step 2 to form a complex representation. Apply the inverse Gyrator Transform to obtain the watermarked low-frequency component $C_w(x, y)$:

$$G_w(u, v) = \tilde{G}_{\mathrm{L}}(u, v) \cdot \exp\left(i \times \mathrm{phase}(u, v)\right),$$
$$C_w(x, y) = G^{-\alpha}[G_w(u, v)],$$

where phase(u, v) is the phase spectrum obtained from the Gyrator Transform in step 2, specifically:

$$\mathrm{phase}(u, v) = 2\pi \frac{(uv + xy)\cos\alpha - (xv + uy)}{\sin\alpha}.$$

6. **Inverse Dyadic Wavelet Transform:** Apply the inverse DYWT using the modified low-frequency component $C_w(x, y)$ from step 5 and the original high-frequency components D, E, and F to reconstruct the watermarked image.

5.2 Watermark Extraction Algorithm

The watermark extraction algorithm is as follows:

1. **Dyadic Wavelet Transform:** Apply the Dyadic Wavelet Transform (DYWT) to both the original image and the watermarked image to obtain the low-frequency components $C(x, y)$ and $C_w(x, y)$, respectively.
2. **Gyrator Transform:** Apply the Gyrator Transform with rotation angle α to both $C(x, y)$ and $C_w(x, y)$:

$$G(u, v) = G^{\alpha}[C(x, y)], \quad G_w(u, v) = G^{\alpha}[C_w(x, y)],$$

where G^{α} denotes the Gyrator Transform at angle α.
3. **Watermark Extraction:** Subtract the magnitude spectra of the two transformed components to extract the watermark image: $W(u, v) = \mathrm{sgn}\left\{|G_w(u, v)| - |G(u, v)|\right\}.$

6 Experiments

In this section, we present the experimental setup and results to evaluate the performance of the proposed non-blind digital watermarking method. The evaluation focuses on image quality degradation and robustness against various attacks.

6.1 Experimental Method

The proposed method was evaluated using 256×256-pixel original images to assess image quality degradation and robustness against attacks. The original images used in the experiments were the "Girl", "Cameraman", "Bridge", and "Airplane" images from the SIDBA database. The watermark image used was a binary image containing the word "MARK". The parameters were set as follows: rotation angle $\alpha = \pi/4$ and embedding strength $\beta = 12$. The experiments were conducted using MATLAB R2023b.

To evaluate the quality degradation of the watermarked image, we used the Peak Signal-to-Noise Ratio (PSNR) and Structural Similarity Index Measure (SSIM). PSNR is calculated using the Mean Squared Error (MSE) and is defined for 8-bit grayscale images as follows:

$$\mathrm{PSNR} = 20 \log_{10} \left(\frac{255}{\sqrt{\mathrm{MSE}}} \right),$$

where $\mathrm{MSE} = \frac{1}{MN} \sum_{m=1}^{M} \sum_{n=1}^{N} [I(m,n) - I^w(m,n)]^2$.

Here, I is the original image with dimensions $M \times N$, I^w is the watermarked image, and m and n represent the pixel positions in the horizontal and vertical directions, respectively. A higher PSNR value indicates better image quality.

SSIM evaluates the similarity between the original and watermarked images by considering luminance, contrast, and structural information, aiming to be more consistent with human visual perception compared to PSNR. The SSIM index is defined as follows:

$$\mathrm{SSIM}(x,y) = \frac{(2\mu_x\mu_y + C_1)(2\sigma_{xy} + C_2)}{(\mu_x^2 + \mu_y^2 + C_1)(\sigma_x^2 + \sigma_y^2 + C_2)},$$

where μ_x and μ_y are the mean pixel values of images x and y, σ_x^2 and σ_y^2 are the variances, σ_{xy} is the covariance between x and y, and C_1 and C_2 are constants to stabilize the division with weak denominator values. The SSIM value ranges from 0 to 1, where a value closer to 1 indicates higher similarity and better image quality.

6.2 Experimental Results

The experimental results combining DYWT and the Gyrator Transform are presented in this section. We report the PSNR and SSIM values by comparing the original and watermarked images with the extracted watermark images. Additionally, to evaluate the robustness of the watermarked images, various attacks were applied and analyzed, including salt-and-pepper noise, median filtering, white Gaussian noise, JPEG compression, resizing, cropping, and rotation. Since the proposed method requires the original image during extraction, the same transformations were applied to the original image before extraction for attacks that alter the image size or orientation.

Figure 2 shows the watermarked images without any attacks (top row) and the corresponding extracted watermark images (bottom row).

Fig. 2. Watermarked images without attacks (top row) and extracted watermark images (bottom row)

Table 2 shows the PSNR and SSIM values comparing the original and watermarked images without any attacks.

Table 2. PSNR and SSIM values between original and watermarked images without attacks

Image	PSNR (dB)	SSIM
Girl	37.25	0.990
Cameraman	37.40	0.988
Bridge	36.46	0.995
Airplane	35.31	0.988

The results demonstrate that the extracted watermark images from the watermarked images without any modifications are slightly degraded and resized to match the low-frequency region, but the word "MARK" is still clearly recognizable. Furthermore, the PSNR values all exceed 35 dB, and the SSIM values are above 0.98, indicating minimal image quality degradation and structural distortion.

Below are the results of extracting the watermark images after applying various attacks to the watermarked images (Figs. 3, 4, 5, 6, 7, 8 and 9).

To quantify the similarity between the extracted watermark and the original watermark, we use the Normalized Correlation (NC) value. The NC value calculates the pixel-by-pixel correlation between the extracted watermark and the

Fig. 3. Watermarked images with added salt-and-pepper noise (noise density: 0.01) (top row) and extracted watermark images (bottom row)

Fig. 4. Watermarked images after median filtering (mask size: 3 × 3) (top row) and extracted watermark images (bottom row)

original watermark, representing the result in a range from 0 to 1. A value close to 1 indicates that the extracted watermark is faithful to the original, whereas a value close to 0 implies a low correlation. The NC is defined as:

$$\mathrm{NC} = \frac{\sum_{i=0}^{M-1}\sum_{j=0}^{N-1} W(i,j)W_e(i,j)}{\sqrt{\sum_{i=0}^{M-1}\sum_{j=0}^{N-1} W^2(i,j)\sum_{i=0}^{M-1}\sum_{j=0}^{N-1} W_e^2(i,j)}},$$

where W represents the original watermark image, and W_e represents the extracted watermark image. When calculating the NC value, the resized original watermark is compared with the extracted watermark image that has been cropped to match its size.

Table 3 shows the NC values for the experimental results under various attacks on the watermarked images.

Fig. 5. Watermarked images with added Gaussian noise (mean: 0, variance: 0.001) (top row) and extracted watermark images (bottom row)

Fig. 6. Watermarked images compressed with JPEG (quality factor: 15%) (top row) and extracted watermark images (bottom row)

Fig. 7. Watermarked images resized to 80% (top row) and extracted watermark images (bottom row)

Fig. 8. Cropped watermarked images (top row) and extracted watermark images (bottom row)

Fig. 9. Watermarked images rotated by 30° (top row) and extracted watermark images (bottom row)

Here, rotation and cropping attacks are compared with the method proposed in Ref. [8]. Due to the algorithm's requirements, the image size after rotation and cropping is resized to 256 × 256 pixels.

Table 3. NC values under various attacks on watermarked images

Attack	Girl	Cameraman	Bridge	Airplane
No attack	0.946	0.938	0.948	0.948
Salt-and-Pepper Noise	0.798	0.793	0.811	0.824
Median Filtering	0.925	0.812	0.816	0.867
Gaussian Noise	0.904	0.887	0.910	0.915
JPEG Compression (Q=15)	0.860	0.831	0.814	0.848
Cropping	0.848	0.838	0.847	0.848
Resizing (0.8×)	0.913	0.904	0.918	0.920
Rotation (30°)	0.909	0.867	0.910	0.904

Table 4. Comparison of NC values between Ref. [8] and the proposed method under various attacks

Attack	Method in Ref. [8]	Proposed Method
No attack	0.999	0.946
Salt-and-Pepper Noise	0.716	0.798
Median Filtering	0.952	0.925
Gaussian Noise	0.901	0.904
JPEG Compression (Q=15)	0.965	0.860
Cropping	0.613	0.848
Resizing (0.8×)	0.983	0.913
Rotation (30°)	0.547	0.909

Table 4 compares the NC values between the method in Ref. [8] and the proposed method under various attacks.

7 Conclusion

This paper proposed a digital watermarking method using the Dyadic Wavelet Transform (DYWT) and the Gyrator Transform. Experimental results confirmed that the method is robust against various attacks, such as filtering and JPEG compression. Additionally, compared to the method in Ref. [8], the proposed method demonstrated higher resistance to geometric attacks such as cropping and rotation. The comparison of Normalized Correlation (NC) values showed that the proposed method consistently achieved high NC values, indicating reliable extraction performance even under various types of image degradation and attacks.

Future work includes the development of a blind digital watermarking method that does not require the original image during extraction. Furthermore, we plan to explore embedding methods that utilize the mathematical properties of the Gyrator Transform, as described in Ref. [3], which still need to be fully leveraged.

Acknowledgements. This work was partially supported by JSPS KAKENHI Grant Number 22K03417.

References

1. Tang, M., Zhou, F.: A robust and secure watermarking algorithm based on DWT and SVD in the fractional-order Fourier transform domain. Array **15** (2022). Article 100230
2. Rodrigo, J.A., Alieva, T., Calvo, M.L.: Gyrator transform: properties and applications. Opt. Express **15**(5), 2190–2203 (2007)
3. Kagawa, T., Suzuki, T.: Characterizations of the gyrator transform via the fractional Fourier transform. Integral Transform. Spec. Funct. **34**(5), 399–413 (2022)
4. Rodrigo, J.A., Alieva, T., Calvo, M.L.: Applications of gyrator transform for image processing. Opt. Commun. **278**(2), 279–284 (2007)
5. Mehra, I., Nishchal, N.K.: Fingerprint image encryption using phase retrieval algorithm in gyrator wavelet transform domain using QR decomposition. Opt. Commun. **533** (2023). Article 129265
6. Zhang, L., Wei, D.: Image watermarking based on matrix decomposition and gyrator transform in invariant integer wavelet domain. Signal Process. **169** (2020). Article 107421
7. Ide, A., Ohura, R., Minamoto, T.: Digital image watermarking method using the gyrator and dyadic wavelet transforms. In: Proceedings of the 2020 International Conference on Wavelet Analysis and Pattern Recognition (ICWAPR), pp. 84–89. IEEE (2020)
8. Bhatnagar, G., Jonathan Wu, Q.M.: A new robust reference watermarking framework in gyrator domain. In: Proceedings of the 24th Canadian Conference on Electrical and Computer Engineering (CCECE), pp. 000648–000651. IEEE (2011)
9. Pei, S.-C., Huang, S.-G., Ding, J.-J.: Discrete gyrator transforms: computational algorithms and applications. IEEE Trans. Signal Process. **63**(16), 4207–4222 (2015)
10. Liu, Z., et al.: Fast algorithm of discrete gyrator transform based on convolution operation. Optik **122**(10), 864–867 (2011)

Cloud Computing Load Balancing System: Dynamic Scheduling and Consolidation Mechanism Using Honeybee Algorithm

Salahaldeen Duraibi[1](✉), Naseema Shaik[2], Betty Elezebeth[3], and Nithinshah Shajahan[3]

[1] Department of Electrical and Electronics Engineering, College of Engineering and Computer Science, Jazan University, Jazan, Saudi Arabia
sduraibi@jazanu.edu.sa

[2] Department of Computer Science, College of Science and Arts, King Khalid University, Abha, Saudi Arabia
nshibrahim@kku.edu.sa

[3] Department of Computer Science, College of Engineering and Computer Science, Jazan University, Jazan, Saudi Arabia
{bsamuel,nanvarsha}@jazanu.edu.sa

Abstract. Cloud Computing (CC) denotes to a method of providing IT services in which users access shared resources via the internet using web-based tools along with its applications. This study addresses the challenges of dynamic resource allocation in CC, particularly in optimizing resource management to handle fluctuating workloads and ensuring efficient system performance. A Dynamic Scheduling and Consolidation Mechanism (DSCM) is proposed for Infrastructure-as-a-Service (IaaS) environments, which allocates resources based on the real-time load of virtual machines (VMs). DSCM dynamically scales resources by adding or removing instances in response to system load and user-defined criteria. This approach improves load balancing, enhances energy efficiency, and optimizes key performance metrics, such as throughput, by continuously monitoring VM loads. The results show that DSCM outperforms traditional methods in terms of resource utilization and cost-effectiveness. Future work will explore further optimizations to handle larger cloud infrastructures and integrate advanced fault-tolerant mechanisms to improve reliability and scalability.

Keywords: CC · IaaS · VMs · group VM · load balancing · optimizing · task scheduling · performance indicators · DSCM approach · datasets

1 Introduction

The CC environment's dynamic resource allocation is a useful component. On the other hand, it has significant issues with energy consumption, fault tolerance, and service quality. Finding a workable solution to these crucial problems and improving cloud performance were therefore required [1]. This work goes in this direction by presenting the

S. Latifi (Ed.): ITNG 2025, AISC 1463, pp. 684–695, 2025.
https://doi.org/10.1007/978-3-031-89063-5_59

MDVMA metaheuristic framework, which allows for optimised task scheduling as well as dynamic VM allocation in CC environments. A difficult task is the optimal allocation of VMs in a CC environment for tasks that users submit. In the Cloud environment (CE), finding the best task scheduling key is thought to be an NP-hard problem, especially for large task sizes [2].

CC has gained popularity as a high-performance computing platform in recent years, appealing to both people and enterprises. To meet the demands of cloud customers, cloud service providers (CSPs) are establishing data centres with high-performance computing resources. While cloud service providers are more focused on generating cash, users are primarily interested in reaction time [3]. The approach condenses the make span, increases load balancing, as well as enhances system dependability. Unlike earlier research, the suggested approach has been able to fairly distribute the workload across the VMs while also improving task scheduling's dependability based on the VMs' prior experience [4].

This study suggests the RAFL resource allocation framework, which is hybrid metaheuristic-based and designed for balancing the load in CC environments. The aim is to lessen the disparity in load between the physical machines that are currently in use and their evaluated resource capacities (such as CPU and RAM) [5]. In this research, a two-stage genetic technique for migration-based load balancing of CC VM hosts (VMHs) is presented. Prior approaches typically view this problem as a job-assignment optimisation problem along with only take into account the loads of the VMHs that are currently in use; however, these approaches are only marginally effective in a real-world if the loads of the VMHs after balancing are not taken into account [6].

In the strategy that is being given, the user first sends tasks to the Tasks Manager to be completed. Cloud Data Centre resources are registered by the resource information provider. After that, the resource scheduler receives the data on the tasks and resources. According to how they are used, the resource scheduler puts the resources in descending order [7]. However, in a heterogeneous system environment, cloud providers constantly struggle with resource management due to fluctuating cloud workloads. The suggested predictive priority-based Modified Heterogeneous Earliest Finish Time (PMHEFT) approach, which predict the application's resource demands [8].

The rest of this work is planned as surveys. Section 2 gives an outline of some current and recent works. The proposed technique is defined in Sect. 3. After a summary of the findings as well as analysis in Sect. 4, the references are provided.

2 Literature Review

This part, take a look back at some of the more recent articles written on CC techniques. Table 1 summarizes the suggested CC techniques and VM for development and lists their advantages and disadvantages, so it is able to find all the information needed. Table 1 details several novel techniques. Present methods include suggested techniques, including CC and DRALBA. These technologies provide advantages, including improved throughput, response time, and load balancing. Nevertheless, other downsides, such as malfunctions and load imbalances, are also brought up.

Lilhore et al. [9] described an important area of unfinished research in CC is balancing the load between VMs as well as resources. The main goals of this research are to increase

Table 1. Related Works

Paper and Author	Method	Advantage	Limitation
Shafiq et al. [10]	CC	storage, deployment platforms, simple access to online services	loading balancing
Chhabra et al. [11]	Dynamic Resource Allocation for Load Balancing (DRALB)	improving throughput, response time	Load imbalance
Ullah et al. [13]	CC	load balancing	may cause the system to malfunction
Jena et al. [14]	Modified Particle Swarm Optimization (MPSO)	maintain task priority balance	suffer from an overload of requests

the effectiveness of a cloud load balancing strategy, maximize source utilization, reduce the ready time, along with decrease activity rejection time. Essentially, the Support Vector Machine (SVM) and K-suggest clustering methods form the foundation of the proposed MLBL method. The SVM classification approach is employed in the proposed MLBL technique to generate active businesses only based on size. Later, the K-method clustering methodology is used to establish VMs, which are entirely dependent on the random-access memory (RAM) amount along with central processing unit (CPU) used.

Shafiq et al. [10] implemented a reliable paradigm called CC enables users and businesses to buy the required services as needed. Among the many services offered by the idea are storage, deployment platforms, easy entree to web services, and many more. The difficulty in maintaining application performance alongside Quality of Service (QoS) measurements as well as compliance with the Service Level Agreement (SLA) contract that cloud providers demand from enterprises is mostly due to load balancing, which is a big issue. It might be difficult for cloud providers to ensure that their servers are evenly distributed with workloads. A good load balancing (LB) strategy should optimize and guarantee user satisfaction by properly leveraging the VMs resources.

Chhabra et al. [11] discussed there are many hosts in the cloud data center, and requests for applications with dynamic resources are also there. The allocation of resources is subject to a variety of demands. These elements may lead to load imbalances, which have a resource utilization impact along with scheduling. DRALB is a proposed scheduling technique. There are two parts to this solution: The load manager assigns the right number of VMs to each application after first analyzing the resource requirements, including bandwidth utilization, CPU as well as memory, energy. Second, resource data is gathered and updated, and resources are categorized into four queues based on their load: CPU, memory, energy, and bandwidth.

Kumar et al. [12] defined with CC, users can get on-demand services via the internet using well-designed web-based tools along with applications. An issue for cloud service providers is load balancing owing to the ever-increasing number of applications and users in cloud settings. This research proposed using a cloud resource broker to equitably distribute the load across every VMs. This have created a dynamic resource load balancing method, which aids in utilization of the most of cloud resources and reduce task make-span times.

Ullah et al. [13] discussed the virtualization is one of the many qualities that make up CC, which initiates transformations in several spheres of life. A VM process transforms a physical server into one that functions like a real server. There is a chance that the VMs in a cloud data center will either overload or underload data when a user transmits or requests data. The aforementioned circumstances may cause the system to malfunction or postpone the user task. Suitable load-balancing solutions are necessary to overcome the two concerns described before. By setting up settings that provide the most throughput with the least amount of response time, load balancing in CC is a method for managing resources. It also divides traffic amongst multiple servers or VMs to ensure that data is received promptly.

Jena et al. [14] highlighted using Modified Particle Swarm Optimization (MPSO) along with a better Q-learning algorithm known as QMPSO, this research suggests a novel approach to dynamic load balancing among VMs. The hybridization process modifies the MPSO's velocity by combining the gbest as well as pbest with the finest act that the enhanced Q-learning produced. Hybridization aims to maximize VM throughput, maintain task priority balance, and enhance machine performance by balancing task load between VMs. Distribution of resources from a data center involves a key component called load balancing of tasks in the CE. The internet's dynamic computing causes CC to suffer from an overload of requests.

Mishra et al. [15] described an NP-hard optimization problem is the scheduling or user requests task allocation in a CE. The load on the cloud system is determined by the user demands and the cloud infrastructure. Situations like underloaded and overloaded systems lead to several system breakdowns, including power use, machine failure, execution time, etc. Load balancing is needed to address all of the aforementioned issues. The tasks load balancing on VMs is a task scheduling essential part in clouds. The cloud network is subject to a diversity of loads, comprising network, memory, and computational (CPU) loads. To balance the load among overloaded as well as underloaded nodes, load balancing is used.

Shukur et al. [16] used CC meets the expectations of its clients by providing them with cloud resources via the use of virtualization and resource provisioning technologies. When it comes to giving clients with shared virtualized resources, the cloud provider has substantial hurdles due to over- and under-provisioning. This research offered a number of proposed techniques and scheduling algorithms to help with data allocation in CC via virtualization. The research also plans to look at how virtualization helps in efficiently providing resources to clients based on their demands. With CC, a new technology has emerged that opens up a world of possibilities for consumers to access services in a variety of settings and at any time. A "cloud provider", a third party, is in charge of running it. Issues with load balancing persist despite the several existing solutions. Therefore, the following benefit is given by this proposed solution:

- DSCM, also termed as Dynamic Scheduling and Consolidation Mechanism, has been proposed in this research. In a load-balanced manner, non-preemptive, DSCM maps a set of separate along with computationally demanding jobs to the offered resources.
- The load and computation share of the VMs are also updated during runtime.
- A better use of resources and faster completion of tasks are the primary goals of the DSCM method.
- After determining the computation share for every VM using a set of tasks, the proposed approach selects a VM.
- The proposed scheduler updates the load of each VM at each scheduling decision. After a number of predetermined repetitions, the proposed method updates the VM load as well as computation share.

3 Proposed Methodology

The proposed DSCM method is described in depth in this section. The structure for task scheduling in a CE is shown in Fig. 1. Clients who use the services provided by cloud service providers are known as cloud users. The workload that reflects the real needs of cloud users is the Cloud Tasks List. Task Manager receives the customers' work load (i.e., task list) also maps these tasks to an appropriate machine. The task mapping heuristic is a representation of the proposed scheduling heuristics; it takes the task manager tasks along with allots it to a VM that is suited for the task based on the available resources. Both physical and virtual resources are within the purview of the resource manager, who keeps tabs on their use. At the end of every client workload execution, their performance is computed by makespan, throughput, energy consumption, average resource utilization (ARU), and task reaction time. Additionally, before task execution begins, the task scheduling algorithms map every task onto the VM.

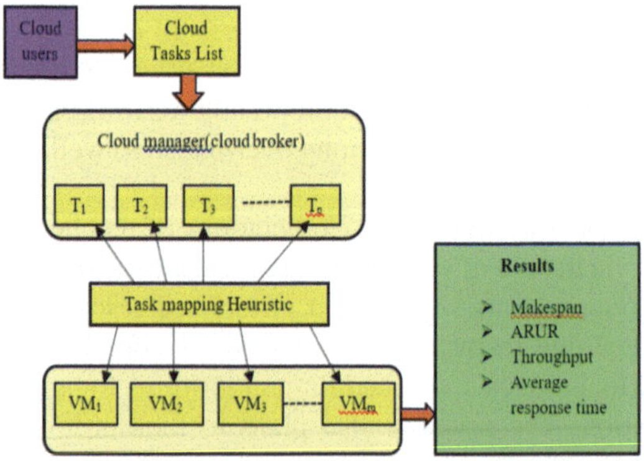

Fig. 1. Task Scheduling Framework

3.1 Honey Bee Load Balancing Algorithm

Load balancing is accomplished by mimicking the way honey bees forage for food. In order to locate and collect food sources, this program studies and mimics the actions of honey bees. A class of honey bees known as scouts is responsible for looking for food in a honey bee colony. A scout bee will return to the hive after locating food. It will then conduct a "vibration dance" to notify the other bees in the colony and provide them with information about the food's quality or amount, also its hive distance. Bees that forage shadow the scout to the food source as well as collect it.

After that, they go back to the hive as well as draw an image of the food supply using a vibration dance, which either makes them use it more or ignores it. The behavior model of honey bee is the foundation of the load balancing algorithm, which draws inspiration from the honey bees behavior. Tasks are like honey bees, along with VMs are like food. Similar to how honey bees seek food, they load a task into a virtual task machine. It is not possible to get more honey out of a depleted food supply than what happens when a VM is overloaded. Searching for a fresh food source also necessitates shifting the task to an underloaded VM, which in turn necessitates deleting the overloaded machine. A suitable under-loaded VM should receive the tasks that were withdrawn from the overloaded one. In a typical scenario, the tasks allocated to a VM are distributed at random using the proposed method for load balancing. Once a set of tasks has been assigned, the machine with the highest load is removed and reassigned to a machine with a lower load.

3.2 Notations

Following is a list of the proposed DSCM's initial variables, notations, definitions, and problem formulation.

- Makespan: Makespan, as determined by Eq. (1), is the amount of time needed to finish the tasks execution on the mapped VMs in the cloud datacentre.

$$Makespan = Max\{CT_j\} = Max\{CT_1, CT_2, \ldots, CT_m\} \tag{1}$$

where m is the number of VMs also CT_j is the completion time of VM_j.
- ARU: For the entire task's execution on the obtainable VMs, the ARU is the average resource use by each heuristic in Eq. (2).

$$ARU = \frac{avgMakespan}{Makespan} \tag{2}$$

where *avgMakespan* is calculated in Eq. 3.

$$avgMakespan = \frac{\sum_j^{m=1} Makespan_j}{m} \tag{3}$$

- Throughput: Assuming the values in Eq. (4), the throughput measure is termed as the tasks done per unit time.

$$Throughput = \frac{Total_{tasks}}{Makespan} \tag{4}$$

3.3 Proposed DSCM Method

The proposed approach involves dynamically inserting tasks into the system, along with the optimum VM for each task should be chosen depending on the system's properties. The proposed technique employs round-robin to distribute tasks to VMs as well as then balances their workload; however, the honey bee foraging behaviour algorithm chooses the best VM first along with decreases makespan. Each VM's workload is checked first using the proposed method. After that, it takes each task and adds its workload to each VM to see whether they are overloaded, underloaded, or balanced. Then, it evaluates the overall status of the VMs. Then, choose the underloaded VM; it has a shorter makespan after adding this task than other machines, and its reliability is within acceptable limits.

The load on the VM is determined by dividing the number of tasks in the VM_i service queue at time t by the VM_i service rate at the same time in Eq. (5).

$$L_{VM_{i,t}} = \frac{N(T, t)}{S(VM_i, t)} \tag{5}$$

In this case, L represents the total load on all of the data center's VMs, as determined by Eq. (6).

$$L = \sum\nolimits_{i=1}^{m} L_{VM_i} \tag{6}$$

In Eq. (7), C_j stands for the capacity of a VM. Pe_{mipsj} is the total number of processing units (MIPS) used by a VM, where Pe_{numj} is the count of VM processors.

$$C_j = Pe_{numj} \times Pe_{mipsj} \tag{7}$$

A VM's processing time is shown in Eq. (8), while the total processing time of all machines is shown in Eq. (9).

$$PT = \frac{L}{C} \tag{8}$$

$$PT_i = \frac{L_{VM_i}}{C_i} \tag{9}$$

In this case, PT represents the total processing time of all VMs, whereas PT_i represents the processing time of a specific VM.

By Adding a New Task, Determine the Load of Every VM: Following the specification of each VM's workload in accordance with Eq. (10), the every machine workload of each one is computed by totalling the task that is being considered.

$$L_j = All\,Allocated\ Load + Load\ Task_i \tag{10}$$

Grouping VM: Here, sort VMs into three categories: overloaded, underloaded, and balanced, according to the workload that was computed before. Doing so involves comparing the VM's workload to the overall workload of all machines. If the workload of the machine is higher than the system average, it is considered overloaded according to the comparison results. If a workload of the VM's is equal to the average workload, it is measured balanced; or else, it is underloaded.

Choosing a VM for the Task at Hand: In this stage, choose a VM from the pool of underloaded machines based on its reliability factor and overall processing time in Eq. (11) and (12).

$$PT_{VM_i} = \frac{all\ allocated\ load + load_{T_i}}{All\ CPU\ capacity\ VM_i} \quad (11)$$

$$All\ CPU\ capacity\ VM_i = Number\ of\ PE \times Mips \quad (12)$$

Examine the balanced set if the underloaded machines set is empty; if that set is empty too, check the overloaded set; and so on.

Figure 2 displays the pseudocode of the method that has been proposed. This method allows the service provider to choose the most suitable VM to handle the tasks that are added to the system on the fly. Consequently, it eliminates overloading, improves load balance, and significantly decreases makespan. The system's fault tolerance is improved, however, by picking and deleting VMs according to their poor performance.

Algorithm 1: Proposed Algorithm

```
Input: Task
Output: VM ID for Executing this Task
  1.   Using Eq. 5 as well as Eq. 7, determine the capacity and load of each VM.
          Totaling this task to Eq. 10 will update the load of every VM.
  2.   OVM, LVM, and BVM are the three types of VMs depending on load.
  3.   If (UVM≠Φ)
       Sort the proceeding times of all VMs in UVM from shortest to longest using Eq. 11.
          Selectin phase:
          For s=1 to # (UVM) do
               Choose VM by min Processing time
                If (min Processing time> Max thereshold)
               Count new VM;
                Do step 1;
       Else
             If (0.5 <= selected VM reliability factor <=1) VMID=selected VM id
               Break;
  4.   Else if (VM≠Φ as well as UVM= Φ)
               Recurrence step 3 for a VM in BVM;
             Else (BVM=Φ and OVM≠Φ and UVM== Φ)
            Repeat step 3 for any VM in OVM
```

4 Results

Cloud Sim has been seriously examined for a thorough examination of the proposed technique in this study. The HCSP (Heterogeneous Computing Scheduling Problems) dataset is taken into account in order to examine the efficacy of the existing scheduling methods. In order to simulate a heterogeneous computing environment as closely as possible, the HCSP instance-based benchmark dataset is constructed using the Expected Time to Compute (ETC) model.

4.1 Analysis

Investigating the proposed methodology performance compared to existing modern task scheduling methods, this study has employed ARU, Throughput, and Makespan. What follows is a breakdown of the findings, along with some commentary on them. The findings of the Makespan achieved for all the proposed methodologies using HCSP dataset instances are explained.

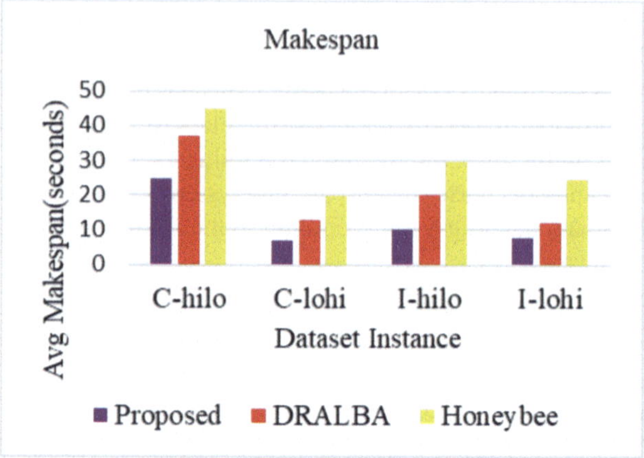

Fig. 2. Makespan Comparison

Compared to other models like Honeybee and DRALBA, the proposed model seems to perform better across all parameters. As shown in Fig. 2, the proposed model outperforms the DRALBA and Honeybee models in terms of execution time consumption, taking 25%, 7%, 10%, and 8% less time, respectively, for the HCSP dataset instances. Based on the C-Lohi dataset, DRALBA showed a 6% improvement, while honeybee exhibited a 13% improvement compared to the proposed technique. While honeybee achieves a Makespan performance of 20% on the C-hilo dataset, the DRALBA technique only manages 12%. Similarly, in the I-hilo dataset, DRALBA has 10% and honeybee at 20%. Finally, the I-Lohi dataset has DRALBA at 4% and Honeybee at 17%.

The proposed approach has the greatest makespan, according to a rigorous investigation. This is since DSCM maps the resources that guarantees that all available resources are used evenly throughout task execution. Figure 3 displays the results about the throughput attained for the HCSP dataset instances, namely C-hilo, I-hilo, C-lohi, also I-lohi. 10%, 30%, 20%, and 15% throughput are the proposed model results. As per the C-Lohi dataset, DRALBA outperformed the proposed method by 5% and Honeybee by 10%. On the C-hilo dataset, the DRALBA method achieves a dismal 5% Makespan performance, compared to Honeybee's 7%. Similarly, in the I-hilo dataset, DRALBA has 5% and honeybee at 10%. Finally, the I-Lohi dataset has DRALBA at 7% and Honeybee at 10%.

On average, DSCM has increased throughput by 10% for the HCSP dataset instances, 30% for the RALBA, 20% for honeybee, and 15% overall (as shown in Fig. 5). When

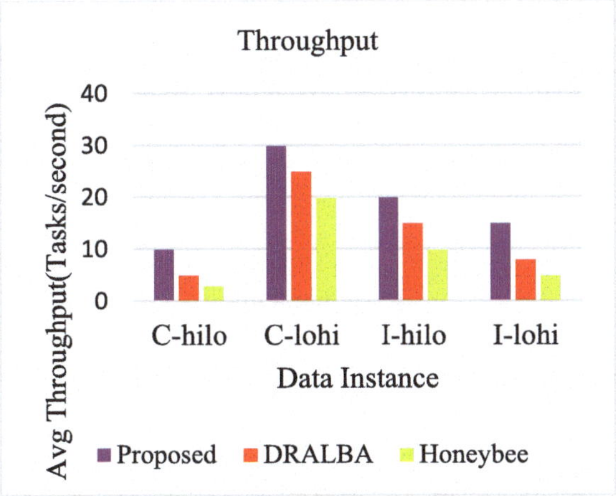

Fig. 3. Throughput Comparison

evaluating heuristics for task scheduling, ARU is another significant metric to consider. Figure 4 compares the proposed DSCM method to other modern techniques, showing that it yielded an ARU that was 1.5 percentage points higher. Results in ARU for the proposed model are 1.5, 1%, 1%, and 1.3%. A comparison between the proposed method and the C-Lohi dataset revealed a 0.3% improvement for DRALBA and a 0.5% improvement for honeybees. On the C-hilo dataset, the DRALBA method achieves a pitiful 0.5% Makespan, compared to honeybee's 0.7%. Similarly, in the I-hilo dataset, DRALBA has 0.2% and honeybee at 0.4%. Finally, the I-Lohi dataset has DRALBA at 0.5% and Honeybee at 0.9%.

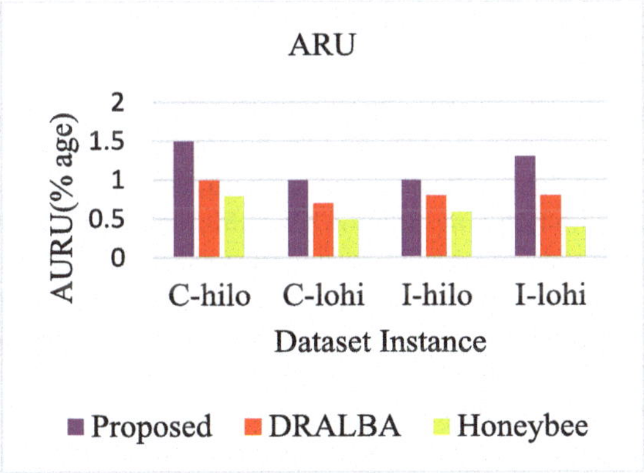

Fig. 4. ARU Comparison

The Cloud users are concerned in the response time. The simulation final set is performed for every compared approaches as well as average response time results are given in Fig. 5. The Proposed method ART consumes 3%, 5%, 1% and 3%.The

DRALBA and honeybee shows poor response time. For the C-Lohi dataset, the DRALBA presented 2% enhancement as compared to the proposed method, as well as the honeybee showed 10%. The DRALBA approach shows a poor Makespan performance on the C-hilo dataset, with 2% and honeybee at 7%; similarly, in the I-hilo dataset, DRALBA has 1% and honeybee at 4%. Finally, the I-Lohi dataset has DRALBA at 2% and Honeybee at 7%.

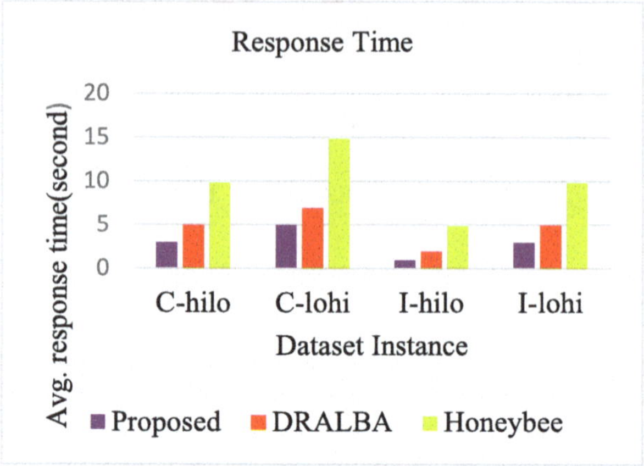

Fig. 5. Response Time Comparison

The findings for the different techniques are compared and discussed in order to bring attention to key points. The proposed strategy outperformed all of the other alternatives in comparison. The proposed DSCM approach successfully achieves load balancing as a result.

5 Conclusion and Future Work

A load-balancing task scheduling method grounded on the honeybee algorithm was proposed in this study. By balancing the current tasks among VMs, this technique shortens the makespan. Along with reducing reaction time, this technique boosts throughput with a high ARU. A comparison was made between the proposed strategy and the other current methods. The proposed approach, according to simulation findings, advances the average makespan when the tasks count is increased, compared to the honey bee technique and the current DSCM method. The suggested approach also shortens the wait time. Incorporate this work into workflows in the future and take other considerations, such as the service provider's cost reduction, into account. Additionally, the simulation will make use of two or more datasets rather than just one.

References

1. Belgacem, A., Beghdad-Bey, K., Nacer, H., Bouznad, S.: Efficient dynamic resource allocation method for cloud computing environment. Clust. Comput. **23**(4), 2871–2889 (2020)

2. Alsadie, D.: A metaheuristic framework for dynamic virtual machine allocation with optimized task scheduling in cloud data centers. IEEE Access **9**, 74218–74233 (2021)
3. Nabi, S., Ibrahim, M., Jimenez, J.M.: DRALBA: dynamic and resource aware load balanced scheduling approach for cloud computing. IEEE Access **9**, 61283–61297 (2021)
4. Ebadifard, F., Babamir, S.M., Barani, S.: A dynamic task scheduling algorithm improved by load balancing in cloud computing. In: 2020 6th International Conference on Web Research (ICWR), pp. 177–183. IEEE (2020)
5. Thakur, A., Goraya, M.S.: RAFL: a hybrid metaheuristic-based resource allocation framework for load balancing in cloud computing environment. Simul. Model. Pract. Theory **116**, 102485 (2022)
6. Hung, L.-H., Wu, C.-H., Tsai, C.-H., Huang, H.-C.: Migration-based load balance of virtual machine servers in cloud computing by load prediction using genetic-based methods. IEEE Access **9**, 49760–49773 (2021)
7. Rehman, A.U., et al.: Dynamic energy efficient resource allocation strategy for load balancing in fog environment. IEEE Access **8**, 199829–199839 (2020)
8. Sohani, M., Jain, S.C.: A predictive priority-based dynamic resource provisioning scheme with load balancing in heterogeneous cloud computing. IEEE Access **9**, 62653–62664 (2021)
9. Lilhore, U.K., Simaiya, S., Guleria, K., Prasad, D.: An efficient load balancing method by using machine learning-based VM distribution and dynamic resource mapping. J. Comput. Theor. Nanosci. **17**(6), 2545–2551 (2020)
10. Shafiq, D.A., Jhanjhi, N.Z., Abdullah, A.: Load balancing techniques in cloud computing environment: a review. J. King Saud Univ.-Comput. Inf. Sci. **34**(7), 3910–3933 (2022)
11. Chhabra, S., Singh, A.K.: Dynamic resource allocation method for load balance scheduling over cloud data center networks. J. Web Eng. **20**(8), 2269–2284 (2021)
12. Kumar, M., Sharma, S.C.: Dynamic load balancing algorithm to minimize the makespan time and utilize the resources effectively in cloud environment. Int. J. Comput. Appl. **42**(1), 108–117 (2020)
13. Ullah, A., Nawi, N.M.: Enhancing the dynamic load balancing technique for cloud computing using HBATAABC algorithm. Int. J. Model. Simul. Sci. Comput. **11**(05), 2050041 (2020)
14. Jena, U.K., Das, P.K., Kabat, M.R.: Hybridization of meta-heuristic algorithm for load balancing in cloud computing environment. J. King Saud Univ.-Comput. Inf. Sci. **34**(6), 2332–2342 (2022)
15. Mishra, S.K., Sahoo, B., Parida, P.P.: Load balancing in cloud computing: a big picture. J. King Saud Univ.-Comput. Inf. Sci. **32**, 149–158 (2020)
16. Shukur, H., Zeebaree, S., Zebari, R., Zeebaree, D., Ahmed, O., Salih, A.: Cloud computing virtualization of resources allocation for distributed systems. J. Appl. Sci. Technol. Trends **1**(2), 98–105 (2020)

Budget Based Online Algorithms for Power-Down Optimization in Dynamic Power Down Systems

James Andro-Vasko(✉), Wolfgang Bein(✉), Mitchell Perez, and Eliott Wesoff

Department of Computer Science, University of Nevada Las Vegas, 4505 S. Maryland Pkwy, Las Vegas, NV 89154, USA
{james.andro-vasko,wolfgang.ben}@unlv.edu,
{perezm,wesoffe}@unlv.nevada.edu

Abstract. The increasing integration of renewable energy sources into power systems presents new challenges for optimizing energy usage, particularly in environments where energy availability is unpredictable. Unlike traditional power sources, renewables such as solar and wind are not consistently reliable due to fluctuating weather conditions. This variability necessitates the use of adaptive strategies that can manage power-down scenarios efficiently in real-time. Online algorithms provide a powerful solution to this problem, allowing systems to make immediate decisions based on current data, without knowledge of future energy availability. In this paper, we investigate online algorithms with adaptation to minimize energy costs. We explore the trade-offs between performance and energy usage in dynamic systems, emphasizing how online algorithms can optimize energy consumption by adapting to the intermittent nature of renewable energy sources. Our approach demonstrates that online algorithms are critical for balancing energy efficiency and reliability in systems increasingly dependent on renewable power.

Keywords: Green Computing · Renewable Energy · Online Algorithms · Competitive Ratio · Continuous Power Systems · Power Down Problem · Competitive Analysis · Dynamic Power Systems · Scheduling · Speed Scaling

1 Introduction

The global transition toward renewable energy sources, such as solar and wind, is a critical step in mitigating climate change and reducing dependence on fossil fuels. However, the inherent intermittency and unpredictability of renewable energy generation present significant challenges to power grid stability and reliability [4–6].

The intermittency of renewable energy often requires the cycling of conventional power plants to fill the gaps when renewable sources cannot meet demand.

M. Perez and E. Wesoff—These authors contributed equally to this work.

S. Latifi (Ed.): ITNG 2025, AISC 1463, pp. 696–707, 2025.
https://doi.org/10.1007/978-3-031-89063-5_60

The frequent cycling of thermal power plants—where they must quickly start up, shut down, or ramp output—has been shown to result in significant operational issues, including increased wear and tear, reduced efficiency, and higher emissions [7–9]. These challenges further underscore the need for intelligent, adaptive systems that can optimize the use of both renewable and conventional energy sources to minimize unnecessary cycling and its associated costs.

In the face of such challenges, online algorithms have emerged as an essential tool for optimizing power systems in real-time. These algorithms are capable of making instantaneous decisions without requiring prior knowledge of future energy supply, allowing them to effectively respond to dynamic changes in renewable energy availability.

Prior Work: In our prior work, we extensively explored continuous power down and adaptation techniques aimed at optimizing energy efficiency in dynamic systems. [1–3]. We introduced a framework that dynamically adjusts power states based on system demand on a continuous power system.

Our Contribution: In this paper, we present a novel approach to dynamic power management using tapering techniques using a budget. Our contribution lies in the development of an online algorithm that dynamically adjusts system idle times based on the available budget. By gradually tapering energy consumption, the algorithm ensures that resources are utilized efficiently while increasing the competitive ratio slightly. The algorithm continuously monitors system performance and adjusts idle periods in real time. This adaptive technique provides a flexible, responsive power management solution.

In this paper, we explore the application of online algorithms to power-down problems in systems dependent on intermittent renewable energy. We focus on the unique budge based technique that allows the online algorithm to adapt to the energy demand. The remainder of this paper is organized as follows. In Sect. 2, we provide an overview of online algorithms, with a focus on their use in dynamic decision-making under uncertainty. We also introduce the concept of competitive ratio, a key metric for evaluating the performance of online algorithms compared to optimal offline solutions. Section 3 delves into the power down problem with the emphasis of the continuous power systems. In Sect. 4, we introduce an online and offline strategy in the context of continuous power-down problems. Section 5 presents a budget-based technique, that allows the online strategy to adjust the power down rate based on previous power demand. Finally, Sect. 7 concludes the paper with a summary of our findings and a discussion of future research directions in the field of online algorithms for energy optimization.

2 Online Algorithms

An online algorithm is a type of algorithm that processes its input piece-by-piece in a sequential manner, making decisions based only on the information available at the current step without knowledge of future inputs. In contrast to offline algorithms, which have access to the entire input dataset beforehand, online algorithms must make decisions in real-time as new data arrives.

We measure an online algorithm by its competitive ratio CR, which can be computed in the following way.

$$\mathrm{CR} = \frac{\mathrm{Cost}_{\mathrm{Online}}(\alpha)}{\mathrm{Cost}_{\mathrm{Offline}}(\alpha)} \tag{1}$$

where $\mathrm{Cost}_{\mathrm{Online}}(\alpha)$ is the cost of the online algorithm and $\mathrm{Cost}_{\mathrm{Offline}}(\alpha)$ is the cost of the offline algorithm, and α is the input sequence. For online competitive analysis, we consider the input to be supplied by an adversary which yields the worst case competitive ratio, and our goal is to construct the online strategy that minimizes the competitive ratio for the worst case, and thus the competitive ratio can be thought of as an upper bound.

A classic online algorithm is the ski rental problem, the idea behind the ski rental problem is one decides to go skiing, and have two options: rent the skis with cost r for every time one goes skiing and a cost to buy β, once the skis are bought, one needs not to rent skis. However, once one buys the skis, one never goes skiing in the future anymore. Since one might not know how many times one will go skiing, thus the online cost would be $rt + \beta$ where t is the amount of times one goes skiing, and after the t-th time, one buys the skis and never goes skiing again. The offline cost is $\min\{rt, \beta\}$ since the amount of times one will ski is known. The minimal competitive ratio occurs when one goes skiing β/r amount of times and yields a competitive ratio of 2. The next section discusses how this classic online algorithm can be applied to the power down problem.

3 The Power Down Problem

For any power down system, there is at least two states, an "on" state and an "off" state, where the "on" state has a standby cost without a power up cost, and the "off" state has no standby cost but has a power up cost to switch from the "off" state to the "on" state. In the context of the ski rental, idling in the "on" state is equivalent to renting the skis and switching to the "off" state is equivalent to buying the skis. The worst case scenario for the power down problem is when the machine powers down and immediately needs to power up to handle the new incoming request. Thus the 2 state power down problem, like the ski rental, is 2-competitive.

We are given a set of requests for and the machine needs to either be in the "on" state or powered up from any intermediate or "off" state to the "on" state. The set of requests can be described by

$$\sigma = \{(r_1^s, r_1^e), (r_2^s, r_2^e), ..., (r_n^s, r_n^e)\} \tag{2}$$

where for each request r_i^s denotes request i's start time and r_i^e denotes request i's end time, and $r_i^s \le r_i^e \le r_{i+1}^s$. Since for the dynamic power down problem, we focus on the idle duration and not the duration of the lifetime of the request, we can rewrite the list of requests in the following way which does not account for the duration of the request

$$\sigma = \{r_1^s, r_2^s, \ldots, r_n^s\} \tag{3}$$

The time in between r_i^s and r_{i+1}^s is the idle duration of the power system where there is no request that needs to be handled and the online strategy needs to decide how to schedule the power down routine.

A system can also have a set of discrete intermediate states that have a lower power cost in which the system remains idle than the "on" state and has lower power up costs than the "off" state. We focus on the continuous power system model with the following idle cost $a(s) = 1 - s^a$ and a power up cost $d(s) = cs^d$ where s is the state id and s_0 is the "on" state and s_1 is the "off" state, and $a = 3$, $c = 1.5$, and $d = 5$, these are control parameters with no particular significance. The worst case scenario for the power down problem for many states, including continuous states, is for cases that the machine needs to power down from state s_a to a lower state s_b and then request arrives promptly which incurs the power up cost of the lower power state s_b, which has a larger power up cost than that of state s_a.

4 Online and Offline Techniques

In the offline strategy, the entire input sequence σ is known in advance, providing the algorithm with full foresight of incoming requests and idle periods. With this knowledge, the offline algorithm can optimally decide when to transition between power states. Specifically, the algorithm can choose the most efficient state to idle in for x units of time before powering back up to handle a request. By precomputing the best times to remain in lower power states and when to resume full power, the offline strategy minimizes energy consumption while ensuring that all incoming tasks are handled efficiently.

This optimality arises because the offline algorithm has the advantage of knowing the precise timing of future requests, allowing it to create an ideal schedule. It can remain idle for the exact right amount of time, minimizing unnecessary energy costs from staying powered up too long or switching power states prematurely. The mathematical model for this strategy can be expressed as follows, taking into account the sequence σ, the cost of transitioning between states, and the duration of idle time before each request. The offline strategy is for each request is described below

$$\text{Strat}_{\text{Offline}}(x) = x \cdot a(s) + d(s) \tag{4}$$

The offline strategy chooses a state s such that the offline cost is minimized by taking the derivative with respect to s, and then solve for s which gives us the offline strategy.

$$(x \cdot a(s) + d(s))\frac{d}{ds} = 0 \tag{5}$$

$$\text{Strat}_{\text{Offline}}(x) = \left(\frac{x \cdot a}{c \cdot d}\right)^{\frac{1}{d-a}} \tag{6}$$

The offline cost for a given idle duration x chooses the optimal state and remains in this state for x time units and powers up. This gives us the offline cost

$$\text{Cost}_{\text{Offline}}(x) = x \cdot a(\text{Strat}_{\text{Offline}}(x)) + d(\text{Strat}_{\text{Offline}}(x)) \tag{7}$$

We can compute the x_m which is the time when the offline algorithm chooses the "off" state as soon as the power system becomes idle. We can set Eq. 5 to 1 and solve for x which gives us the result:

$$x_m = \frac{c \cdot d}{a} \tag{8}$$

This threshold time applies to the online strategy as well since it has been proven in prior work that the online and offline must be in the "off" state if the idle duration exceeds a common threshold, in this case x_m is our threshold for the continuous state problem.

The online strategy contrasts the offline algorithm by not choosing a state and remaining idle for x time units, but rather switching to lower power states over time and eventually powering down to the "off" state. If at any point during the idle duration a request arrives, the online strategy will incur a power up cost which can vary based on the current state. For our online strategy in this paper, we use the an exponential curve that dictates the rate in which the online strategy switches to lower power states over the idle duration, we devise the following strategy

$$\text{Strat}_{\text{Online}}(\gamma, x) = \frac{e^{\gamma x} - 1}{e^{\gamma x_m} - 1} \tag{9}$$

where γ is a control parameter, and e is the natural number, and x is the idle duration. Figure 1 shows the online strategy for a set of γ values.

When γ is smaller, the online strategy transitions to lower power states more gradually, conserving energy at a modest rate throughout the idle period. In contrast, when γ is larger, the online algorithm tends to stay in a higher power state for most of the idle duration, only switching rapidly to lower power states as the idle time approaches x_m. This difference in behavior highlights how the choice of γ influences the power-down strategy, affecting the timing of state transitions based on the system's tapering mechanism and the unknown length of the idle period.

Since the idle time x is not known in advance, the online strategy cannot choose and remain in the optimal power state throughout the idle duration. Instead, it must integrate energy consumption over the entire idle period and factor in the cost of powering back up when the system becomes active again.

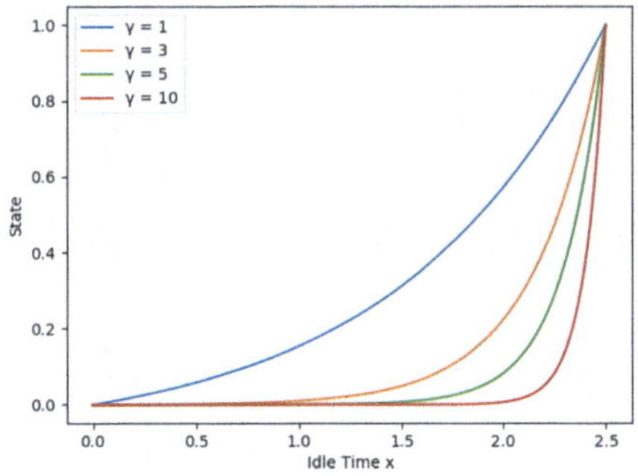

Fig. 1. Online Strategies for Various γ values

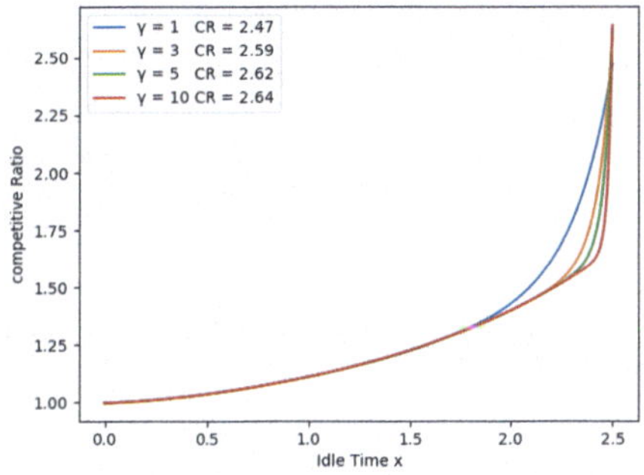

Fig. 2. Competitive Ratios for Various γ values

The challenge lies in balancing the trade-off between remaining in a higher power state early on versus transitioning earlier to save energy, as the total energy consumed depends on the final state of the system after x idle units have passed. We compute the cost in the following way

$$\text{Cost}_{\text{Online}}(\gamma, x) = \int_0^x a(\text{Strat}_{\text{Online}}(\gamma, x))dx + d(\text{Strat}_{\text{Online}}(\gamma, x)) \tag{10}$$

Figure 2 shows the competitive ratios throughout the idle duration.

The competitive ratio for each curve over the idle duration remain the same, the strategies with lower γ values have a slight increase over the larger γ values, however the strategies with larger γ values as the idle time approaches x_m are larger compared to the smaller γ value strategies, making the larger γ strategies having the maximal competitive ratios.

5 Budget Based Technique

The budget-based tapering technique aims to enhance the efficiency of dynamic power-down systems by adjusting the idle duration before transitioning to "off" state. This technique leverages a budget b that controls how aggressively the system moves into lower power states. The budget provides flexibility, enabling the system to gradually taper down its power consumption in response to idle periods. By managing the rate at which the system enters lower power states, the budget-based method ensures that power is conserved without unnecessary delays, resulting in a more efficient energy-saving process.

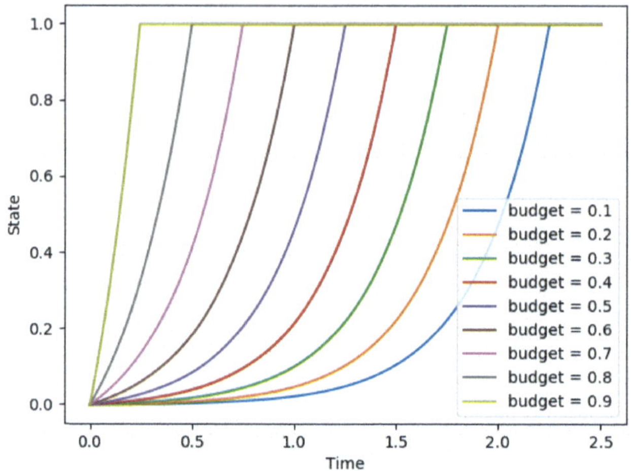

Fig. 3. Online Strategies For Various Budgets, Online Strategy $\gamma = 3$

To implement this technique, an initial budget is assigned to the online algorithm, which is then used to calibrate the rate at which it transitions to lower power states. As the system processes requests, the budget is adjusted based on the timing and frequency of the request. We use this initial budget to construct an alternative online strategy.

$$\mathrm{Strat}_{\mathrm{Online}}(\gamma, x)' = \frac{e^{\gamma x} - 1}{e^{\gamma x_m (1-b)} - 1} \tag{11}$$

$$\begin{aligned} \mathrm{Cost}_{\mathrm{Online}}(\gamma, x)' = \int_0^x & a(\mathrm{Strat}_{\mathrm{Online}}(\gamma, x)')dx \\ & + d(\mathrm{Strat}_{\mathrm{Online}}(\gamma, x)') \end{aligned} \tag{12}$$

Figure 3 shows the online strategies for a set of budget values. We can see that when the budget is larger, the rate in which the online strategies switches to lower power states increase more rapidly and power down to the "off" state sooner than with a smaller budget value.

Figure 4 illustrates the competitive ratio under varying budgets. As seen in the graph, the competitive ratio exhibits noticeable spikes whenever the budget-based strategy powers down to the "off" state. This is due to the strategy's reliance on conserving energy by turning off sooner as the budget becomes depleted, however turning off sooner incurs a larger power up cost rather than staying idle and incurring a lower power up cost. Despite the spikes, the competitive ratio for each budget used is consistently larger than that of the non-tapering technique. This suggests that while budget-based tapering is less efficient during specific moments, it is optimized to better manage energy consumption under certain situations.

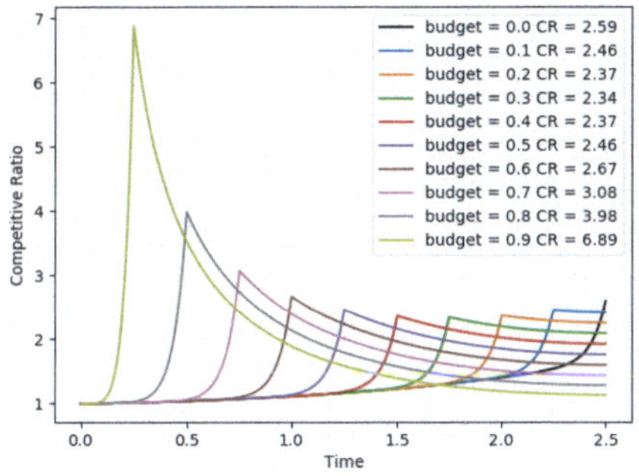

Fig. 4. Competitive Ratios For Various Budgets, Online Strategy $\gamma = 3$

Moreover, as the time of the request is delayed, the budget-based technique's power consumption gradually decreases. Once the idle time reaches a critical point, denoted as x_m, the budget-based methods achieve their minimum power consumption. Notably, at this time x_m, the competitive ratio becomes the lowest when a larger budget is used. This demonstrates that while the competitive ratio may initially be higher for budget-based strategies, they perform better in terms of energy savings over time, particularly as the system reaches extended idle periods.

In the budget-based tapering technique, the concept of gain plays a crucial role in refining the adaptive power-down strategy. The gain is computed as the difference between the energy cost consumed by the budget-based technique and the non-adaptive technique over a given period of time. By calculating this gain, we can measure the effectiveness of the budget-based method in comparison to the static, non-adaptive approach.

$$\text{gain} = \text{Cost}_{\text{Online}}(\gamma, x) - \text{Cost}_{\text{Online}}(\gamma, x)' \tag{13}$$

Once this gain is determined, it is added to the remaining budget of the system. This increase in the budget directly influences the tapering behavior of the adaptive strategy, allowing it to further adjust its idle times. The larger the gain, the more aggressive the system can be in reducing power consumption in subsequent periods, as the additional budget provides more flexibility to lengthen idle durations before transitioning to the lower power states. This feedback loop ensures that the budget-based method continuously optimizes its power-down strategy based on real-time performance improvements over the non-adaptive method.

6 Simulations

We now utilize a set of test data containing idle durations, denoted as x, to simulate the performance of our proposed online algorithm with the budget technique. These idle durations represent the intervals of inactivity in the system, which are used for evaluating the effectiveness of our algorithm in optimizing power management. By running multiple simulations with various x values, we will assess how the algorithm adapts to different idle patterns and makes real-time decisions. We then compute the cost of the online strategies against the offline strategies which allows us to determine the overall competitive ratio for each test input. The test inputs are created using a uniform distribution in python. Each input file consists of 1000 delay times, where a portion of the delay times are above x_m and the remaining idle times are below x_m.

From Table 1, once can see that for test cases where the amount of idle requests over the x_m threshold is a low count, the budget based tapering technique does not save much more over the online algorithm with no adaptation. This is due to the fact that the budget based adaptation technique does not converse much energy if the requests arrive relatively soon with respect to each other. Figure 4 shows that when the requests arrive within the x_m threshold with an initial budget of 0.1, the competitive ratio is not improved much in those cases, we only see any significant improvement when the idle duration is longer than the x_m threshold.

From Table 1, when the amount of requests that are greater than x_m, we see that the budget technique has nearly the same cost as the offline strategy and the online strategy without any adaptation approaches its worst case competitive ratio. This is due to the fact that in this case, most of the requests arrive after x_m and from Fig. 4, one can see the competitive ratio for the budget based techniques are minimal when the request arrives after x_m and the worst case competitive ratio for the non adaptive online algorithm has its worst competitive ratio in this scenario.

It is clear that the budget based technique is the dominant strategy for the power down technique, since its competitive ratio is lower than the non tapering online strategy even in the cases where the requests arrive soon after one another, and outperforms the non tapering technique significantly when the requests arrive x_m or larger duration among each other.

Table 1. Simulation Results

100 Requests Over x_m		200 Requests Over x_m	
Online Cost	1620.13	Online Cost	1812.48
Online Budget Cost	1520.08	Online Budget Cost	1515.80
Offline Cost	1045.34	Offline Cost	1085.02
CR Online	1.54986	CR Online	1.67045
CR Budget	1.45458	CR Budget	1.39702
400 Requests Over x_m		500 Requests Over x_m	
Online Cost	2262.25	Online Cost	2509.86
Online Budget Cost	1499.95	Online Budget Cost	1504.24
Offline Cost	1170.96	Offline Cost	1234.70
CR Online	1.93197	CR Online	2.03277
CR Budget	1.28096	CR Budget	1.21830
600 Requests Over x_m		700 Requests Over x_m	
Online Cost	2779.65	Online Cost	2977.20
Online Budget Cost	1502.43	Online Budget Cost	1498.96
Offline Cost	1292.94	Offline Cost	1327.45
CR Online	2.14986	CR Online	2.24280
CR Budget	1.16202	CR Budget	1.12921
800 Requests Over x_m		900 Requests Over x_m	
Online Cost	3240.18	Online Cost	3456.60
Online Budget Cost	1502.94	Online Budget Cost	1498.96
Offline Cost	1398.82	Offline Cost	1440.94
CR Online	2.31637	CR Online	2.39885
CR Budget	1.07443	CR Budget	1.04027

7 Conclusions

In conclusion, this study has highlighted the effectiveness of a budget-based tapering technique for dynamic power down systems, outperforming the non-adaptive online algorithm in all test simulations. The primary advantage of the budget-based approach lies in its ability to dynamically adjust to incoming requests, regardless of whether they arrive earlier or later. This adaptability ensures more efficient power savings without sacrificing performance, whereas the non-adaptive technique lacks the flexibility needed to handle varying request timings. By incorporating a budget constraint into the tapering process, the system can optimally determine idle time, leading to superior overall performance in energy management.

Our simulations have consistently demonstrated that the budget-based technique reduces power consumption more effectively, regardless of the workload pattern. This performance advantage remains robust across different request arrival scenarios, making it a versatile and powerful solution for dynamic power down systems. This flexibility is particularly relevant for modern systems where the timing of requests can be highly unpredictable, underscoring the importance of adaptability in energy management strategies.

Looking forward, there are several promising avenues for future research. One potential direction is the application of the budget-based tapering technique to other continuous online strategies, where dynamic power adaptation is required. By extending its use beyond the current scope, we can explore its potential in a broader range of dynamic systems. Additionally, we aim to run simulations on real-world data, particularly in small-scale cycling power plants, to evaluate the practical viability of the tapering approach in operational environments. Such experiments would allow us to assess the system's performance in real-time scenarios, providing valuable insights into the potential for scaling this technique to real-world energy systems.

References

1. Andro-Vasko, J., Avasarala, S.R., Bein, W.: Continuous state power-down systems for renewable energy management. In: Latifi, S. (ed.) Information Technology - New Generations, pp. 701–707. Springer, Cham (2018)
2. Andro-Vasko, J., Bein, W.: An evaluation for online power down systems using piecewise linear strategies. In: Latifi, S. (ed.) ITNG 2021 18th International Conference on Information Technology-New Generations, pp. 265–271. Springer, Cham (2021)
3. Andro-Vasko, J., Bein, W.W.: Evaluation of continuous power-down schemes. Information **13**(1), 37–37 (2022). https://doi.org/10.3390/info13010037
4. Clack, C.T.M., ET AL.: Evaluation of a proposal for reliable low-cost grid power with 100% wind, water, and solar. Proc. NatiL. Acad. Sci. **114**(26), 6722–6727 (2017). https://doi.org/10.1073/pnas.1610381114, https://www.pnas.org/doi/abs/10.1073/pnas.1610381114
5. Impram, S., Varbak Nese, S., Oral, B.: Challenges of renewable energy penetration on power system flexibility: a survey. Energy Strategy Rev. **31**, 100539 (2020). https://doi.org/10.1016/j.esr.2020.100539, https://www.sciencedirect.com/science/article/pii/S2211467X20300924
6. Mureddu, M., Caldarelli, G., Chessa, A., Scala, A., Damiano, A.: Green power grids: how energy from renewable sources affects networks and markets. PLOS ONE **10**(9), 1–15 (2015). https://doi.org/10.1371/journal.pone.0135, https://ideas.repec.org/a/plo/pone00/0135312.html
7. Van den Bergh, K., Delarue, E.: Cycling of conventional power plants: technical limits and actual costs. Energy Convers. Manage. **97**, 70–77 (2015). https://doi.org/10.1016/j.enconman.2015.03.026, https://www.sciencedirect.com/science/article/pii/S0196890415002368

8. Wogrin, S., Galbally, D., Ramos, A.: CCGT unit commitment model with first-principle formulation of cycling costs due to fatigue damage. Energy **113**, 227–247 (2016). https://doi.org/10.1016/j.energy.2016.07.014, https://www.sciencedirect.com/science/article/pii/S0360544216309343
9. Wu, Y.K., Wang, Y.W.: Literature review concerning the cycling cost in a power system with renewable power sources. Energy Procedia **156**, 13–17 (2019). https://doi.org/10.1016/j.egypro.2018.11.075, https://www.sciencedirect.com/science/article/pii/S1876610218310348. 5th International Conference on Power and Energy Systems Engineering (CPESE 2018)

A Conceptual System Architecture for Interoperability Between Business Intelligence and Legacy Systems

Robson Carlos Bosse[1(✉)], Ferrucio de Franco Rosa[1,2], and Mario Jino[3]

[1] University of Campo Limpo Paulista (UNIFACCAMP), Campo Limpo Paulista, SP, Brazil
robsonbosse@gmail.com

[2] CTI Renato Archer, Campinas, SP, Brazil

[3] University of Campinas (UNICAMP/FEEC), Campinas, SP, Brazil

Abstract. Business Intelligence (BI) is widely employed by companies and organizations to deliver critical data in an organized and accurate manner, enabling data-driven decision-making. However, numerous legacy systems, characterized by complexity and challenges in maintenance and integration, pose a significant obstacle. Legacy systems often experience slowdowns and disruptions when executing BI queries on their databases. We present a conceptual system architecture for interoperability between BI and legacy systems. The architecture is intended to overcome existing limitations, provide seamless integration, and minimize adverse impacts on business operations.

Keywords: Data Quality · Data Analysis · Legacy System · System Architecture · Business Intelligence · Interoperability · Decision Making · Digital Twin · Dashboard

1 Introduction

Business Intelligence (BI) applications have become increasingly vital for companies and organizations, enabling data collection, analysis, and visualization to ease informed decision-making [6]. However, many companies still use legacy systems (outdated, complex, and difficult to update and integrate) to manage their data, often resulting in performance and scalability issues [22].

These issues undermine the user experience and complicate data-driven decision-making due to the lack of properly processed quality data. This occurs because legacy systems (LS) were designed to meet performance requirements that differ significantly from those needed by BI applications [13]. These requirements were conceived when direct database queries were sporadic and applications primarily dealt with data manipulation instructions (e.g., insert, update, and delete). In this context, there was no demand for high-quality data from legacy databases to support real-time BI reporting alongside everyday operations.

S. Latifi (Ed.): ITNG 2025, AISC 1463, pp. 708–720, 2025.
https://doi.org/10.1007/978-3-031-89063-5_61

Since LS are typically not designed to handle high volumes of data queries, a problem primarily related to data indexing arises, causing BI queries and LS services for end users to be interrupted. In this case, reorganizing the indexes to restore LS operations is required.

We propose a conceptual system architecture to address interoperability between LS and BI. A layered architecture is presented, and its application is shown by a case study at an educational institution; a real production database was used in a scenario of interoperability between LS and BI, where a query requiring a substantial amount of data for BI was executed.

The remainder of this paper is organized as follows: Sect. 2 provides the literature review, highlighting related work; Sect. 3 introduces the conceptual system architecture; Sect. 4 describes an application of the architecture; in Sect. 5, the results are discussed; and Sect. 6 presents the conclusions.

2 Literature Review

A systematic literature review [5] was carried out through a rigorous analysis of 24 articles, based on the methodology proposed by Kitchenham (2004) [14]. In our review, we highlight the key challenges and opportunities related to data quality in legacy systems and explore the main approaches and strategies for improving data quality through BI tools.

Searches were performed in three well-known scientific databases in the Computing field: IEEE Xplore[1], ACM Digital Library[2], and Springer Link[3]. The established search period is 11 years (2013 to 2023) and the following general search string was used, adjusting the syntax to each database: *("Data Analysis" OR "Data Quality" OR "Database") AND ("Business Analytics" OR "Business Intelligence Performance") AND ("Evaluation" OR "Assessment") AND ("Le-gacy System" OR "Brownfield System").*

As related work, we highlight three studies that address failures and issues related to interoperability in BI systems. Akhir et al. (2018) [1] present requirements for developing a data visualization reporting module within a judicial system. The main objective is to propose an efficient data management approach to streamline decision-making processes.

Caserio et al. (2018) [7] present an assessment of the potential relationships between Enterprise Resource Planning (ERP) and BI systems, and information overload (i.e., the generation of a large amount of data) or underload (i.e., hindering access to relevant data). Fraihat et al. (2021) [9] present a BI framework for the Jordanian real estate market. According to the authors, the proposed framework provides data analysis and decision support tailored to potential investors, specifically designed for users with little or no prior knowledge of the market.

Table 1 presents a comparison between related work and our study. We highlight that there is no work in the review proposing a system architecture as a

[1] https://ieeexplore.ieee.org/Xplore/home.jsp.
[2] https://dl.acm.org/.
[3] https://link.springer.com/.

contribution. In Table 1, we consider Objectives: *Framework (A); System (B); Approach (C); Architecture (D);* and Application Domains: *Business Intelligence (1); Data Analytics (2); Data Quality (3).*

Table 1. Summary of related work

Authors	Objective				Domain		
	A	B	C	D	1	2	3
Akhir et al. (2018) [1]		X				X	
Caserio et al. (2018) [7]			X		X		
Fraihat et al. [9]	X				X		X
This work				*X*	*X*	*X*	*X*

3 Conceptual System Architecture

To address challenges related to performance, scalability, and reliability in accessing data from legacy systems for BI applications, we propose CSA-IBILS (Conceptual System Architecture for Interoperability between Business Intelligence and Legacy Systems). CSA-IBILS is organized as six layers: Knowledge Bases, Software Components, Processes, Services, Security, and Roles. We provide detailed descriptions of each layer and its components. Figure 1 provides an overview of CSA-IBILS.

When proposing CSA-IBILS, we aim to mitigate risks and address the challenges inherent in integrating data from LS, promoting operational efficiency and BI reliability. Our conceptual proposal seeks to overcome performance limitations and provides a scalable framework that meets the increasing demands of complex computational environments.

CSA-IBILS was designed using ArchiMate Modelling notation [3]. We employ this notation due to its ability to represent elements and relationships associated with systems architecture, fostering understanding and communication. Each layer of CSA-IBILS has been designed to structure and handle data, addressing potential interoperability issues. In the following subsections, we present in detail each layer, starting with the bottom layer (Knowledge Bases) and concluding with the top layer (Roles). In the description of the layers, we highlight the contributions of each layer to the effectiveness of the conceptual architecture.

3.1 Knowledge Bases

Knowledge Bases encompass platforms capable of centralizing data and information, storing them in an organized manner to ensure their availability for access. These platforms serve as centralized repositories, enabling easy retrieval

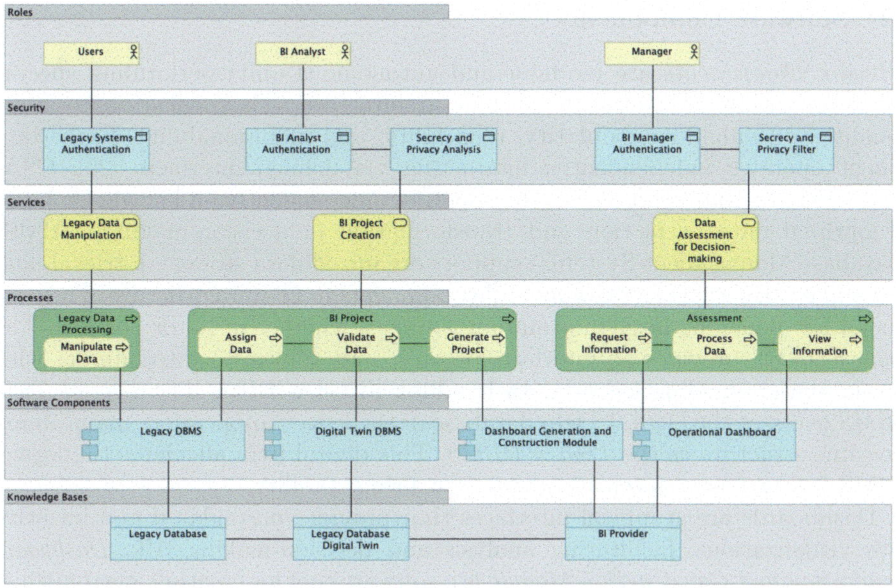

Fig. 1. Overview of CSA-IBILS

and usage of the data contained within them. CSA-IBILS employs three distinct knowledge bases: *Legacy Database, Lega-cy Database Digital Twin, and BI Provider.*

Legacy Database refers to an older infrastructure for data storage. Usually, this type of database is characterized by obsolete implementation technology and outdated architecture. The main characteristic of these legacy databases is their relational nature, meaning they are based on structured data models that use entities (database tables) to organize information. *Legacy Database Digital Twin* is a complete and dynamic virtual replica of a real database. A Digital Twin is defined as a virtual representation of a physical system that remains synchronized with its real-world counterpart through continuous data exchange [12].

BI Provider is a consolidated data repository that has undergone the ETL (Extract, Transform, and Load) process. In this knowledge base, essential data is organized and stored to meet the specific needs of BI software. *BI Provider* is a knowledge base dedicated exclusively to BI and it is supplied from the *Legacy Database Digital Twin.* In the *BI Provider* we assume a NoSQL approach, strategically chosen to provide flexibility to the knowledge base, aiming to meet general BI performance requirements.

A critical aspect to be considered in the interconnection and communication among these three databases is the logical network infrastructure, which must be tailored to suit the specificities of each context during the implementation of CSA-IBILS in different operational environments.

3.2 Software Components

Software Components are modular and autonomous units performing specific functions within a conceptual system architecture. They encapsulate functionalities to enhance modularity, reusability, and maintainability. Each component operates independently, interacting via defined interfaces [20]. CSA-IBILS includes four components that operate independently on knowledge bases for optimal data extraction and transformation in decision-making. DBMSs (Database Management Systems) support organized data storage, retrieval, and manipulation, serving as fundamental components in IT infrastructure. They are crucial for managing large information volumes, whether in LS or not [11]. The *Legacy DBMS* provides an interface for users and applications to interact with the database, enabling legacy data handling in CSA-IBILS. The *Digital Twin DBMS* accesses the Digital Twin database of the legacy application, maintaining the same structure as the *Legacy DBMS*. This digital twin alleviates the legacy DBMS from resource-intensive queries, ensuring stability.

Dashboards are graphical interfaces that provide consolidated and intuitive data visualizations, facilitating analysis and decision-making [19]. *Dashboard Generation and Construction Module* is a software tool for creating, customizing, and presenting interactive dashboards.

Operational Dashboard serves as a visual interface that graphically summarizes operational data to support real-time decision-making. Designed for clarity, dashboards provide an immediate view of processes, activities, or system performance, presenting critical decision-making data through graphs, tables, and visual elements [15]. They monitor key metrics in real-time, such as performance indicators and operational efficiency, allowing users to identify trends and make informed decisions [10]. This component can be customized to meet the specific needs of an organization, sector, or process, offering an intuitive overview of the operational environment.

3.3 Processes

In a conceptual system architecture, processes represent the dynamic entities responsible for executing activities, controlling data, and managing and coordinating operations in systems, ensuring the efficient attainment of established objectives. Each process is characterized by a logical sequence of steps, interactions, and operations that contribute to the overall functionality of the architecture [21]. In this layer, data is treated and managed according to the flow established by the conceptual architecture.

Manipulate Data refers to operations for managing, modifying, and organizing data in an information technology system. The *BI Project* aims to collect, process, and analyze organizational data to provide insights for decision-making, using specific tools and techniques to convert raw data into meaningful information. The project architecture includes data source integration, data modeling, interactive dashboards, and reporting. This process starts with an LS and a legacy database. Three essential activities occur at this stage: i) *Assign Data*, which ensures the correct association of information with relevant elements, laying a foundation for analysis; ii) *Validate Data*, which verifies the accuracy and integrity of the data used by BI, ensuring reliable analyses; and iii) *Generate Project*, which ensures efficient implementation of BI solutions that meet organizational needs and present data in a dashboard for visualization.

In the *Assessment* process, the analysis and measurement of performance, effectiveness, and efficiency of BI report analyses occur. This process aims to provide insights, identify improvement opportunities, and optimize business objectives. Evaluation metrics may include the accuracy and relevance of data, as well as dashboard interface usability. During the *Assessment*, the following activities are performed: i) *Request Information*, which involves requesting specific information or datasets relevant to decision-making by formulating queries directed at data sources; ii) *Process Data*, which refers to converting raw datasets into meaningful information; and iii) *View Information*, which involves visualizing data and insights obtained through the BI process, including charts, interactive dashboards, and visual reports that facilitate rapid understanding of trends and patterns.

3.4 Services

Services are elements that perform specific functions, providing functionalities in a manner accessible to other components of the system [24]. In the layer *Services*, the *Legacy Data Manipulation* activity addresses information stored in LS databases. *BI Project Creation* concerns the design and implementation of an activity and process set aimed at collecting, organizing, analyzing, and presenting data relevant to support decision-making. *Data Assessment for Decision-making* seeks to ensure that organizational decisions are informed and aligned with the strategic objectives of companies and organizations, to maximize the usage of their information and to address challenges more effectively through data.

3.5 Security

Security is essential in various domains, particularly concerning data protection and privacy. Despite regulations and guidelines varying among countries, there is a general interest in ensuring adequate levels of security to safeguard users, businesses, and governmental entities. In CSA-IBILS all data accesses must traverse a security layer for maximizing data access requirements. We consider user profiles access information and services only after authentication as well as a secrecy and privacy analysis.

3.6 Roles

Roles represent entities (or profiles) designated to perform specific duties and define responsibilities within the operational context of the system. *Roles* are important for structuring dynamic coordination among components, processes, and services, providing a clear view of task distribution and authorities within the system environment [24]. *Users* encompass individuals, organizations, or other entities using the system to perform tasks, obtain information, or meet specific objectives. *BI Analysts* implement and maintain business intelligence solutions, and are responsible for collecting, analyzing, and transforming raw data into meaningful insights to support decision-making. *Managers* represent the individual supervising operations, coordinating or operating management components of systems, and making strategic decisions.

4 Case Study

CSA-IBILS addresses interoperability challenges between LS and BI solutions. To verify the proposal's applicability, a case study was performed in a real-world scenario where a school holding's LS operates, covering education from kindergarten to high school. The holding manages 8 school units with over 8,000 students. Over time, the LS has been modified to meet the holding's needs, leading to performance issues and service outages.

In the Case Study we refer to the School Management System as"Legacy" for brevity. Legacy comprises two standalone modules (Administration and Finance) and a Web module designed for teachers to record student assessments and attendance. We refer to the Query and Business Intelligence Visualization System as "BI Dashboard". The Legacy database structure has been operated for approximately 20 years, during which numerous modifications have been made to support increasing demands and DBMS version updates. Considering the proposed scenario, we begin with the bottom layer of the architecture and progress sequentially through the subsequent layers until reaching the top layer.

4.1 Knowledge Bases

Legacy Database uses MS SQL Server 2016 [16] as its DBMS, which operates in a virtualized environment, with XenServer [8] serving as the virtual machine (VM) repository. *Legacy Database* comprises 284 entities and 52 views, structured to support the entire LS and to ease various integrations accumulated over the years.

Legacy Database Digital Twin leverages SQL Server and is hosted on a separate VM configured identically to the *Legacy Database*.

The *BI Provider* database was implemented in a virtualized server environment to meet the demands of the *BI Dashboard*. This knowledge base uses MongoDB [18] as its DBMS, providing a flexible and scalable structure for storing processed data from *Legacy Database Digital Twin*. Choosing MongoDB as DBMS for the *BI Provider* database is strategic, as it allows efficient dealing with large volumes of data and supports dynamic schemas.

To integrate real-time data between the *Legacy Database Digital Twin* (SQL Server) and the *BI Provider* (MongoDB), Apache NiFi [2] was employed as an

ETL tool. CSA-IBILS utilizes NiFi's interface to orchestrate data flows, extracting data from SQL Server via the QueryDatabaseTable component. This component retrieves data by executing SQL queries with configurable frequency for real-time updates in MongoDB. Data transformation occurs with JoltTransformJSON, handling tasks like renaming fields and modifying attributes. Finally, PutMongo writes the transformed data to MongoDB, ensuring continuous updates with error monitoring. Collections were created in MongoDB for each *BI Dashboard*: i) Enrollments, ii) Financial Targets, iii) Average Ticket, iv) Scholarship Control, and v) Students per Classroom.

4.2 Software Components

In the dashboard implementation, we used the MS Power BI [17] tool, which offers a range of features from data integration to advanced visualization. *Operational Dashboard* provides visualization interfaces for managers, ensuring authorized and secure access. The visualization screens offer a graphical and interactive representation of essential data, enabling managers to obtain accurate and real-time insights.

4.3 Processes

In *Legacy Data Processing*, the activity "*Manipulate Data*" encompasses the operation of legacy systems that operate the legacy database. In this stage, user interactions with the system occur in daily operations, i.e., registrations are accomplished, and queries and reports are generated, involving data processing in the legacy database.

The *BI Project* process begins with data assignment from the legacy database and querying the *Legacy DBMS* to access data in *Legacy Database*. The next step is to validate the data to be used. During the validation phase (*Validate Data*), data is queried from *Legacy Database Digital Twin* via the *Digital Twin DBMS* software component. In the final step, the project effectively generates queries to the *BI Provider* (database that feeds the *BI Dashboard* queries). In the *Assessment* process, managers access the *BI Dashboard* and request the required information in real-time. The data is processed and retrieved from the *BI Provider*. The information is visualized through a dashboard available in the manager's profile.

4.4 Services

In *Legacy Data Manipulation*, users experience everyday operations that often reveal the inherent challenges of LS. In this service, users realized the system's unusual delays in processing requests, regardless of whether they were generated directly by BI report queries or other sources. Delays and instability were caused by various factors, from request congestion to specific processes, such as viewing school bulletins. This shows the complexity of working with systems that rely on a legacy database.

BI Project Creation is the starting point in developing dashboards to assist managers in data analysis. We used the MS Power BI tool to produce the following *BI Dashboards*: Enrollment, Financial Target, Average Ticket, Scholarship Control, and Students per Classroom. In *Data Assessment for Decision-making*, the *BI Dashboard* is operational, and the data has been properly processed, providing managers with an appropriate tool to support decision-making.

4.5 Security

Legacy Systems Authentication is widely used in legacy and current systems, including web and desktop applications. It is implemented via a user table in the *Legacy Database*, storing encrypted login credentials. The *userauthorization* table controls the access levels. In the Case Study, a desktop system runs on Windows computers. Access to the educational system is limited to the legacy database, which is only available on the local network or, for schools, via a VPN connection between the headquarters and school units.

The *BI Analyst's* access is granted via MS Active Directory (AD) [4] with two-factor authentication. Analysts access PowerBI Desktop for dashboard creation, editing, and publication. Accessing legacy database information requires specific PowerBI permissions on the local server and credentials granting read-only access. *Managers* also perform two-factor authentication in the AD to access dashboards via the Internet after successful authentication.

4.6 Roles

Users represent those involved in the daily operations of Legacy. They are responsible for data inclusion, exclusion, and editing. *BI Analysts* developed the *BI Dashboards* Enrollments, Financial Targets, Average Ticket, Scholarship Control, and Students per Classroom, presenting metrics, indicators, and relevant information and enabling more well-founded decisions by managers. The *Manager* profile represents two groups of users in the school units: i) Administration Manager (Financial and Educational Manager); and ii) General Manager (Principals and Treasurers).

5 Discussion On The Results

The results of employing CSA-IBILS in the Case Study demonstrate its real-world applicability. Previously, issues of freezing and slowness in LS were frequent, especially between 9:00–11:00 a.m. and 2:00–4:00 p.m., during school enrollment periods (4 months annually). Queries to the legacy database, which delivered updated enrollment data on the dashboard, have since been discontinued. The elimination of interruptions and reduction of slowness during BI queries underscore the contribution of CSA-IBILS to operational environments.

In Fig. 2, we present 2 graphs, illustrating (before and after) the interaction between LS and the BI platform. The first graph shows the number of failures in the LS over 24 months (from 05/02/2022 to 05/31/2024). The second graph shows the number of BI queries to the legacy database in the same period. Despite BI queries increasing (Graph 2), we can identify in Graph 1 a decrease in failures from the 18th month (10/2023), following the implementation of CSA-IBILS in the operational environment.

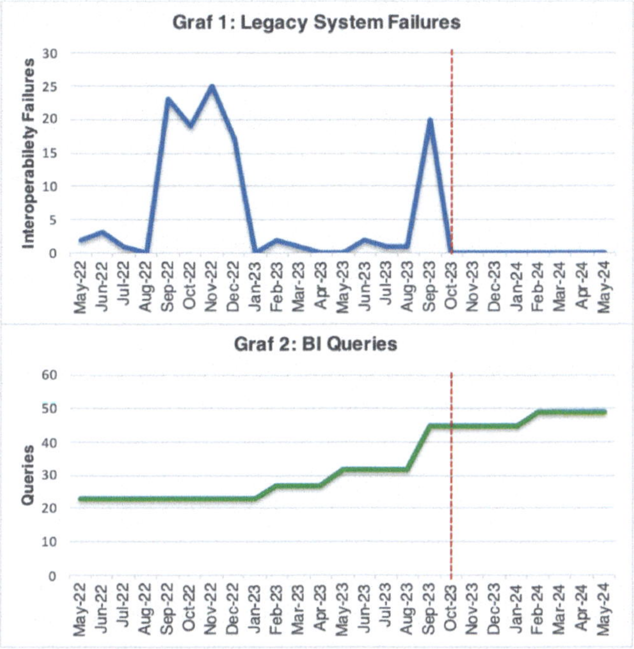

Fig. 2. Chronology of Failures in the Legacy Database and BI Queries

The achieved interoperability addresses current demands and prepares organizations for future data analysis challenges. Enterprise infrastructures may vary, requiring specific adaptations. For example, the infrastructure validating the proposed architecture used a virtual server setup, but non-virtualized environments can also benefit from CSA-IBILS. We encourage further experiments to extend these results to other domains.

Although the Case Study demonstrated the applicability of CSA-IBILS, no significant improvement in query speed for the operational dashboards was observed, with response times remaining stable. Our focus was not on system performance but on ensuring seamless interoperability with fewer failures. Previously, 3 support tickets were opened daily due to slowness or interruptions. CSA-IBILS had a positive impact, as complaints from end-users have ceased.

In addition to the legacy database server already budgeted, the application required a second server for the legacy database digital twin with equivalent computational resources. This can be a limitation for CSA-IBILS in environments requiring costly software or hardware. Identical setups (hardware and software) are recommended for performance, but not all organizations have the resources.

As slowdown and interruptions in LS ceased, end-users also perceived the improvement in system quality. End-users (majority of users) primarily operate the LS (e.g., enroll students) and do not use BI dashboards. BI dashboards are important for managers, and the LS stands essential to serve clients (parents and caregivers of students). All user profiles experienced positive impacts on their daily activities following the implementation of CSA-IBILS, as system failures were reduced and performance remained stable.

Layered architecture and microservices offer distinct advantages for integrating LS with BI. Layered architecture simplifies interoperability, especially with rigid LS and limited interfaces, making it a more suitable choice in this case. This approach ensures smooth data flow and integration between heterogeneous LS and the BI platform.

Layered architectures are widely used in Big Data systems for organizing functionalities into distinct levels. This modular approach includes security, management, inter-mo-dule communication, and external system interaction. Common in reference architecture proposals, the layered model enhances understanding of the solution's structure and component modules [23]. Its applicability in integrating LS with BI solutions highlights its effectiveness in tackling interoperability challenges.

6 Conclusion

Given the challenges faced in integrating BI with LS, our proposed conceptual system architecture emerges as a promising solution. The need to efficiently and seamlessly extract critical data from LS motivated this work. The proposed layered architecture offers an innovative approach to overcoming challenges for effective communication between BI and LS.

CSA-IBILS aims to enhance data extraction efficiency and improve maintenance and integration processes, eliminating barriers that cause delays in LS. It addresses key gaps in BI and LS integration, establishing a foundation for improved efficiency by delivering relevant information via dashboards. In increasingly data-driven environments, effective decision-making in legacy systems is crucial for organizations.

As future work, we plan to evolve CSA-IBILS by incorporating cloud computing elements, including fog and edge computing. We aim to implement the legacy database digital twin and the BI provider from the Knowledge Bases layer in the cloud. This approach seeks to optimize local network resources and remove the need to allocate machine resources for hosting these two virtual machines locally. Consequently, BI queries will be executed directly in the cloud, allowing us to evaluate CSA-IBILS under different conditions and expanding the possibilities of the conceptual architecture.

The key takeaway from this work lies in the lessons learned from bridging the gap between BI and LS through our layered architecture. The most significant contribution of this paper is the conceptual foundation that CSA-IBILS provides for seamless data integration and optimized decision-making. This architecture addresses immediate challenges and paves the way for future innovations in cloud-based BI systems, offering a versatile solution that organizations can leverage in an evolving, data-centric landscape.

References

1. Akhir, E.A.P., Aziz, N., Roslin, A.F.: Data visualization for evaluation form management system. In: IEEE Conference on Big Data and Analytics (ICBDA), pp. 117–122 (2018). https://doi.org/10.1109/ICBDAA.2018.8629601
2. Apache Software Foundation: apache NiFi (2024). https://nifi.apache.org/
3. ArchiMate Tool: Archi - ArchiMate modelling (2024). https://www.archimatetool.com/
4. Binduf, A., Alamoudi, H.O., Balahmar, H., Alshamrani, S., Al-Omar, H., Nagy, N.: Active directory and related aspects of security. In: 21st Saudi Computer Society National Computer Conference (NCC), pp. 4474–4479 (2018). https://doi.org/10.1109/NCG.2018.8593188
5. Bosse, R.C., de Franco Rosa, F., Jino, M.: A study on data quality and analysis in business intelligence. In: S. Latifi (ed.) ITNG 2024: 21st International Conference on Information Technology-New Generations, pp. 249–253. Springer (2024)
6. Caserio, C., Trucco, S.: Business intelligence systems. In: Enterprise Resource Planning and Business Intelligence Systems for Information Quality. CMS, pp. 43–73. Springer, Cham (2018). https://doi.org/10.1007/978-3-319-77679-8_3
7. Caserio, C., Trucco, S.: ERP and BI as tools to improve information quality in the Italian setting: the research design. Enterprise Resource Planning and Business Intelligence Systems for Information Quality: an Empirical Analysis in the Italian Setting, pp. 75–104 (2018)
8. Citrix Systems: XenServer (2018). https://www.citrix.com/pt-br/products/xenserver/
9. Fraihat, S., Salameh, W.A., Elhassan, A., Tahoun, B.A., Asasfeh, M.: Business intelligence framework design and implementation: a real-estate market case study. J. Data Inf. Qual. **13** (2021). https://doi.org/10.1145/3422669
10. Freitas, F., Alturas, B.: Operational dashboard implementation for a family taxi business company. In: 13th Iberian Conference on Information Systems and Technologies (CISTI), pp. 1–6 (2018). https://doi.org/10.23919/CISTI.2018.8399315
11. Grigoriev, E.: DBMS for business systems. In: 10th International Joint Conference on Software Technologies (ICSOFT), vol. 2, pp. 1–6 (2015)
12. Grönman, J., Rantanen, P., Saari, M.: Categorization of digital twins: a literature review of IoT and industry. In: IEEE 27th International Conference on Intelligent Engineering Systems (INES), pp. 27–32 (2023). https://doi.org/10.1109/INES59282.2023.10297877
13. Jha, S., Jha, M., O'Brien, L., Wells, M.: Supporting decision making with big data: integrating legacy systems and data. In: 4th Asia-Pacific World Congress on Computer Science and Engineering (APWC on CSE), pp. 120–128 (2017). https://doi.org/10.1109/APWConCSE.2017.00029
14. Kitchenham, B.: Procedures for performing systematic reviews. Keele UK Keele Univ. **33**, 1–26 (2004)

15. Kristyanti, S.F., Kusumasari, T.F., Alam, E.N.: Operational dashboard development as a data quality monitoring tools using data deduplication profiling result. In: 6th International Conference on Science and Technology (ICST), vol. 1, pp. 1–6 (2020). https://doi.org/10.1109/ICST50505.2020.9732870
16. Microsoft Corporation: MS SQL Server 2016 (2016). https://www.microsoft.com/en-us/sql-server/sql-server-2016
17. Microsoft Corporation: Microsoft Power BI (2024). https://powerbi.microsoft.com/
18. MongoDB Inc.: MongoDB (2024). https://www.mongodb.com/
19. Muppidi, A., Hashim, A.S.B., Hasan, M.H.B.: Proposed user-experience model for the design and development of bi dashboards. In: 2nd International Conference on Intelligent Cybernetics Technology and Applications (ICICyTA), pp. 23–28 (2022). https://doi.org/10.1109/ICICyTA57421.2022.10037904
20. Omer, N., Jha, S.K., Khatri, S.K.: Maintaining reusable software components. In: International Conference on Intelligent Computing and Control Systems (ICCS), pp. 1350–1352 (2019). https://doi.org/10.1109/ICCS45141.2019.9065845
21. Pressman, R., Maxim, B.: Software Engineering: A Practitioner's Approach, 9 edn. McGraw-Hill Education (2019)
22. Sandborn, P.A., Prabhakar, V.J.: The forecasting and impact of the loss of critical human skills necessary for supporting legacy systems. IEEE Trans. Eng. Manage. **62**, 361–371 (2015). https://doi.org/10.1109/TEM.2015.2438820
23. Sena, B., Allian, A.P., Nakagawa, E.Y.: Characterizing big data software architectures: a systematic mapping study. In: Proceedings of the 11th Brazilian Symposium on Software Components, Architectures, and Reuse. Association for Computing Machinery (2017). https://doi.org/10.1145/3132498.3132510
24. Sommerville, I.: Software Engineering, 10th edn. Pearson (2015)

Methods for Identifying Issues with Traffic Signal Timing Parameters and Potential Adjustments: A Survey

Yazmin Elizabet Martinez(✉) and Wolfgang Bein

Department of Computer Science, University of Nevada Las Vegas, 4505 S Maryland Pkwy, Las Vegas, NV 89154, USA
yazmin@unlv.nevada.edu, wolfgang.bein@unlv.edu

Abstract. Road traffic congestion affects our daily lives whether we drive or take bus transit. However, the complexity of road traffic congestion and its causes are active research areas because there is a need to find plausible solutions to improve traffic flow. Given the abundance of traffic signal performance data collected by transportation agencies, we have an opportunity to analyze traffic signals to identify parameters (cycle length and green time) that can be improved. We have performed a comprehensive overview of the methods used to find traffic signal timing issues by looking at the research within the last five years. We found that newer research focuses on connected vehicles (CV) and adaptive control systems, including Reinforcement Learning techniques. This paper also discusses a case study of retiming traffic signals in the highly populated Las Vegas Valley.

Keywords: Traffic Signal Optimization · Traffic signal control retiming · Traffic signal control timing methods · Optimization Algorithms

1 Introduction

This survey paper provides a comprehensive overview of the research area of traffic signal control analysis to find timing problems. This paper presents a synthesis of existing knowledge and identifies future research directions.

The research topic is relevant for both the transportation research domains and algorithmic computer science approaches. Researchers can use this research to experiment with existing methods or creating new ones. Furthermore, because traffic signal control problems are a cause of traffic congestion, finding traffic signal controllers that could be optimized presents an opportunity to improve traffic at a lower cost than other mitigation strategies [13].

W. Bein—Contributing authors.

S. Latifi (Ed.): ITNG 2025, AISC 1463, pp. 721–726, 2025.
https://doi.org/10.1007/978-3-031-89063-5_62

We based our research on journal articles in the last 5 years (2019–2024), highly cited papers in the last 10 years, and seminal works. Discussions on pedestrians, cyclists, and public transit are left out for brevity, as multimodal traffic signal retiming is out of the scope of this paper.

2 Related Work

Webster's seminal work on traffic signal optimization [16] dates back to 1958 and is known as the "Webster method." Webster developed formulas to determine optimal signal timing. To determine optimal cycle length, the authors of the Signal Timing Manual, Second Edition [15] recommend a critical intersection approach by using the traditional method of Webster's model. However, this approach is designed for under-saturated conditions (stable flow from one cycle to another) [15], and the approach only considers isolated signalized intersections. Furthermore, this approach does not always result in the optimal cycle length for system progression (vehicle movement within a system) [15].

Webster's method is limited in that it cannot be applied to fixed-cycle time with overlapping phases [6]. To address overlapping phases, in 1973 Ohno and Mine refined Webster's method by proposing an approximation algorithm that is a combination of linear programming or mixed-integer programming and one-dimensional minimization technique.

Webster (1958) approximation for fixed-time signal timing is still used today and its simplicity [6] attributes to its popularity. Lertworawanich, Ponlathep, and Unhasut use Webster's method to compare newer timing optimization methods. In 2021 Lertworawanich and Unhasut setup simulation scenarios to compare against a CO emission-based control algorithm for adaptive timing [4]. They used a simulated environment to show that when their control algorithm had a sum of flow ratios of the critical movements within a certain range, their algorithm produced shorter delays and stops than the Webster fixed-time signal timing algorithm.

In 2000, Chang and Lin [2] proposed a timing decision methodology that considered whole oversaturation periods when traffic flow exceeds intersection capacity. There is a need to consider oversaturated periods because vehicle queuing cannot be eliminated in one signal cycle. For their approach, the authors developed a discrete dynamic optimization model and proposed an algorithm to solve it.

3 Methodology

To compile comprehensive data, we searched five databases, Google Scholar, IEEE Xplore, ACM Digital Library, Web of Science, and Transportation Research Board's TRID. We used the following keywords in our search.

- "Survey On Methods For Traffic Signal Timing Issues"
- "Survey On Methods To Discover Or Find Traffic Signal Timing Issues"

- "Road Traffic Signal Control Analysis Methods To Find Timing Issues To Optimize"
- "Traffic Signal Control For Monitoring Timing"

4 Core Concepts and Terminology

4.1 Traffic Signal Optimization Versus Retiming

According to the U.S. Department of Transportation, Federal Highway Administration, retiming traffic signals is one of the most cost-effective tasks to improve traffic flow [13]. In this paper, we discuss two traffic signal terms: optimization and retiming. Since signal retiming is more complex than adjusting timing parameters in a controller, the retiming effort for a signal or group of signals makes use of modern computer programs [12]. Optimization is a broader concept than retiming. Traffic Signal Optimization includes all the elements of the signal timing process.

4.2 Intersection Configuration, Timing Plan, and Phase Plans

The U.S. Department of Transportation, Federal Highway Administration Traffic Signal Timing Manual states that each phase at an intersection has a set of timings which can include timing for vehicles and pedestrians [14]. An intersection configuration shows all the phases (traffic movements) possible for an intersection. Each phase has its own timing plan. Furthermore, the manual also states that most signal designs and all modern controllers provide eight standard phases within the signal controller [14].

4.3 Traffic Signal Types

The Signal Timing Manual, Second Edition [15], defines the signal controller's internal program, firmware, as a core part of the traffic signal control. The manual further explains that signal controllers have the ability to make changes to signal operations using external information, which determines the three types of signal controls in the following list:

- **Fully-Actuated Control** - uses detectors in all intersection approaches (direction for example eastbound EB)
- **Semi-Actuated Control** - uses detectors in some lanes
- **Pretimed Control** - no detectors are used and is the least expensive because of the lack of detectors

Moreover, the Signal Timing Manual [15] also adds that the signal controllers operate as either one independent (with uncoordinated signal timing) or part of a system (with coordinated signal timing). Thus, when it comes to traffic signal optimization, we have to consider these possible signal configurations.

5 Survey of the State-of-the-Art

5.1 Methods Categories

In a state of the art review of traffic light synchronization for intelligent vehicles, some of the traditional traffic control methods for intelligent monitoring include Adaptive Signal Control and corresponding systems that use loop detectors to collect traffic data [11]. With vision-based traffic control methods, traffic data is gathered from traffic cameras [11]. Furthermore, other traffic control methods include evolutionary algorithms like genetic programming to optimize traffic signals in real-time [11]. Lastly, the authors mention that artificial intelligence, fuzzy logic, linear programming, and data fusion, among others, have been provided for traffic signal control [11].

5.2 Signal Optimizer Software

The thirty-six professional participant surveys on multimodal signal retiming projects by Ardalan, Magalotti, and Stevanovic [1] states that typical signal timing adjustment involves using optimization software tools like Synchro or TRANSYT-7F. Synchro is an optimization software developed by Trafficware that can be applied to cycle lengths, splits, and offsets [7]. Splits or phase split refer to the length of a phase, including its terminating yellow and all-red intervals [3]. While TRANSYT-7F, developed by McTrans, uses genetic algorithm optimization of cycle length, phasing sequence, splits, and offsets [5].

5.3 Data Needed for Signal Optimization

Ardalan, Magalotti, and Stevanovic [1] queried signal timing professionals on the data used to perform signal retiming. The participants in their study said that they used vehicular traffic volumes as one of two supplemental data most used. The professionals include agencies and researchers from North America. The second most supplemental data used was queue length data and historical traffic patterns. The historical traffic patterns include peak hour traffic, seasonal variations, special event impact, among others.

Although pedestrian volumes play a major role in retiming [1], we have excluded pedestrians (including cyclists) from the scope of this paper. We are interested in reducing the complexity by only focusing on the traffic signal optimization methods. However, we acknowledge and advise readers that they will need to incorporate the necessary safety guidelines to their retiming of the signal strategies.

Tomar, Sreedevi, and Pandey [11] have explained that the timings of traffic signals are set in accordance with the average road traffic every day. Therefore, they stress the need for the timing of traffic signals in metropolitan areas to use real time data due to the imbalance of traffic flow caused by congestion and delay from people traveling to work and home.

6 Case Study

The following is a case study on the basis for this survey. The RTC manages more than 1,671 traffic signals [10]. Most of these traffic signals are owned by entities in the Las Vegas Valley: city of Las Vegas, city of North Las Vegas, city of Henderson and Clark County unincorporated areas in the valley. The RTC is currently making traffic signal changes which is a major undertaking. Recently, the RTC published a Traffic Signal Re-Timing Activities report [9] from June 2024 that lists the progress for a signal retiming project, along with a map of the signals to be re-timed and the arterial performance metrics that the RTC staff monitors. This most recent RTC retiming project was publicized [8] and is estimated to take about two years to complete. The re-timing of the signals in the Las Vegas valley began in 2022 and is estimated to be completed in the fall 2024. The timing changes to the signals were done by region of southwest, southeast, northwest, and northeast part of the valley.

Thus far, the RTC FAST team has collected data and confirmed that the retiming changes are effective. The news article [8] notes that their new method for retiming has reduced travel times by 15%-25%. Furthermore, due to the complexity of retiming a traffic signal, retiming projects do not often take place. A thirty-six professional participant survey on multimodal signal retiming projects by Ardalan, Magalotti, and Stevanovic [1] found that most organizations handled up to five signal retiming projects per year.

7 Challenges and Future Directions

In future works we intend to create a concept map for the research we found in this survey. We also plan to add visual charts to display analytics of the progression and relationships between methods. The challenge with researching the broad topic of traffic signal optimization is the breadth of the research. Although there are surveys on the topic, the goal of this paper is to understand the problem and provide core information to develop our optimization algorithm using computer science theories as our approach.

8 Conclusion

We investigated various published research methods for finding issues or problems with traffic signal timings. Since there are many factors and traffic signal configurations to consider, we plan to organize and seek out more methods and processes for traffic signal optimization and traffic signal retiming.

The Las Vegas Valley traffic signals case study shows that retiming traffic signals is effective in improving traffic and travel time. However, because retiming efforts takes years to complete, this demonstrates that the task remains a challenge for highly populated areas like the Las Vegas Valley. Furthermore, automating methods to find timing issues and developing new computer science algorithms can be helpful to further the research in traffic signal optimization.

Acknowledgements. This work was supported by U.S. DOT PSR UTC grant 69A355234309 (Subaward from USC) and by National Science Foundation Grant EAGER 2433820. The research was conducted at UNLV's Center for Information Technology and Algorithms in collaboration with UNLV's Transportation Research Center.

References

1. Ardalan, T., Magalotti, M.J., Stevanovic, A.: Multimodal signal retiming projects: a survey-based exploration of traffic signal professionals' practices and challenges. Future Transp. **4**(4), 1121–1141 (2024)
2. Chang, T.H., Lin, J.T.: Optimal signal timing for an oversaturated intersection. Transp. Res. Part B: Methodol. **34**(6), 471–491 (2000)
3. of Las Vegas Transportation Engineering Division, C.: Traffic signal timing. https://files.lasvegasnevada.gov/public-works/Traffic%20Engineering/TrafficSignal%20Timing.pdf. Accessed 13 Dec 2024
4. Lertworawanich, P., Unhasut, P.: A co emission-based adaptive signal control for isolated intersections. J. Air Waste Manag. Assoc. **71**(5), 564–585 (2021)
5. Mctrans: State-of-the-art optimization process. https://mctrans.ce.ufl.edu/highway-capacity-software-hcs/transyt-7f/. Accessed 13 Dec 2024
6. Ohno, K., Mine, H.: Optimal traffic signal settings—ii. A refinement of webster's method. Transp. Res. **7**(3), 269–292 (1973)
7. Sharma, I.: Optimizing signal timing in synchro studio 10. https://www.ce.memphis.edu/4162/L17.2_Synchro10_Fall2018.pdf. Accessed 13 Dec 2024
8. of Southern Nevada, R.T.C.: How fast tmc optimizes signal timing to keep southern nevada moving. https://www.rtcsnv.com/news/how-fast-tmc-optimizes-signal-timing-to-keep-southern-nevada-moving/. Accessed 13 Dec 2024
9. of Southern Nevada, R.T.C.: Traffic signal re-timing activities. https://www.rtcsnv.com/about/wp-content/uploads/sites/3/2024/08/MINUTES-BACKUP-ITEM-04-Traffic-Signal-Retiming-Activities.pdf. Accessed 13 Dec 2024
10. of Southern Nevada's, R.T.C., Freeway, of Transportation (FAST), A.S.: About traffic management. https://www.rtcsnv.com/traffic-cams/about-traffic-management/. Accessed 13 Dec 2024
11. Tomar, I., Sreedevi, I., Pandey, N.: State-of-art review of traffic light synchronization for intelligent vehicles: current status, challenges, and emerging trends. Electronics **11**(3), 465 (2022)
12. U.S. Department of Transportation, F.H.A.: Existing signal timing process. Accessed 13 Dec 2024
13. U.S. Department of Transportation, F.H.A.: Signal timing on a shoestring. https://ops.fhwa.dot.gov/publications/signal_timing/01.htm. Accessed 13 Dec 2024
14. U.S. Department of Transportation, F.H.A.: Traffic signal timing manual. https://ops.fhwa.dot.gov/publications/fhwahop08024/chapter4.htm. Accessed 13 Dec 2024
15. Urbanik, T., et al.: Signal Timing Manual, vol. 1. Transportation Research Board, Washington, DC (2015)
16. Webster, F.: Traffic signal settings. Road Research Technical Paper (1958)

A Patient-Specific Multi-omics Pathway Analysis Method Using Hierarchical NNMF for Improved Gene Weighting

Zeynab Maghsoudi(✉) and Frederick C. Harris Jr.

Computer Science and Engineering, University of Nevada, Reno, USA
{zmaghsoudi,fred.harris}@unr.edu

Abstract. The analysis of pathways has become a powerful method for discovering information about the development of malignant diseases from healthy tissue. This progression is often driven by complex interactions between different layers of omics data, including genetic mutations, epigenetic changes, and transcriptional alterations. Thus, studying multi-omics data is of utmost importance in discovering correlations between tumor characteristics and clinical outcomes. In this work, we propose an improved multi-omics pathway analysis method based on a sequential matrix factorization technique. Our approach builds upon the padma framework, which utilizes Multiple Factor Analysis to identify aberrant individuals. The primary modification in our method lies in the way we compute genes weights, which we derive through sequential Non-Negative Matrix Factorization. By employing this technique, we extract patient-specific patterns and relationships from each omics layer sequentially, using the information gained from one layer to inform the next. We will validate the performance of the method using both simulation tests and a real-world study utilizing the TCGA breast cancer dataset. In the simulation studies, our proposed method consistently demonstrates improved performance compared to Padma, specifically achieving improvements ranging from 2.43% to 5.51% with the largest gains observed in smaller sample sizes.

Keywords: multi-omics · pathway analysis · non-negative matrix factorization · malignant disease · multiple factor analysis · personalized medicine · individual-level analysis · TCGA breast cancer · omics integration · aberrant sample detection

1 Introduction

The analysis of complex biological processes underlying the progression from healthy tissue to malignant disease requires an understanding of changes across multiple molecular layers, including mRNA, methylation, and copy number variation (CNV) [5,6]. While analyzing each layer separately offers valuable insights,

Z. Maghsoudi and F. C. Harris Jr.—Contributing authors.

S. Latifi (Ed.): ITNG 2025, AISC 1463, pp. 727–736, 2025.
https://doi.org/10.1007/978-3-031-89063-5_63

it often fails to capture the comprehensive mechanisms of disease due to the intricate interactions between the omics layers [8,11,15]. Many multi-omics analysis techniques which integrate data from various omics layers have been developed. These techniques showed promising results in detecting the complex patterns associated with malignancy [3,4,8].

Pathway analysis is a widely utilized technique for studying the biological pathways involved in diseases [7,9]. Most of the developed pathway analysis methods focus on population-level analysis. This leads to potentially overlook individualized differences that are essential for personalized medicine. As high-dimensional multi-omics data becomes more prevalent, it is essential to develop pathway analysis techniques that can detect individual-specific deviations from the norm. Padma [13] and PARADIGM [18] are two methods designed specifically for individual-level pathway analysis. Padma is based on the Multiple Factor Analysis (MFA) technique [2,17]. Padma begins by calculating the largest singular value for each input omics matrix by applying a Principal Component Analysis (PCA) [1], which is then used to assign weights to the respective matrices. Following this, a global PCA is performed on the concatenated, weighted matrix that includes all omics data. The deviation score for a pathway in an individual is determined by measuring the distance between the projection of the individual onto the consensus space and the center of the space. On the other hand, PARADIGM employs a probabilistic graphical model to infer pathway activities by integrating gene expression, CNV, and proteomics. This method captures pathway perturbations by modeling relationships between genes and higher-order pathway features, allowing the detection of individual-level variations.

Despite the effectiveness of padma and PARADIGM in individual pathway analysis, there are challenges remained in optimizing their performance. For instance, padma depends on MFA, which may not always capture complex non-linear interactions between omics layers. Similarly, the reliance of PARADIGM on a probabilistic framework can be computationally expensive. Another issue with PARADIGM relates to the complexities in adding more omics layers and different sets of pathways. Therefore, there is a need for alternative approaches that can address these limitations and improve the detection of aberrant biological signals.

Inspired by the padma approach, we propose a new patient-specific multi-omics pathway analysis method. While padma utilizes Multiple Factor Analysis (MFA) for dimension reduction and deviation score calculation, our method builds on this framework with an enhanced gene weighting strategy. Specifically, we employ a hierarchical Non-Negative Matrix Factorization (NNMF) technique [14,16] to decompose omics data layers sequentially, allowing information transfer across data types (mRNA, methylation, and CNV). This hierarchical structure extracts latent relationships within the omics matrices more effectively. The final deviation scores for individual samples are computed using the same PCA-based method as in padma. We validate our method through simulation studies and real-world data from The Cancer Genome Atlas (TCGA) [12] breast cancer dataset. Performance metrics show that the proposed approach can distinguish between healthy and aberrant samples effectively.

2 Methods

The proposed method takes as input three matrices including mRNA, methylation, and CNV data. Each matrix undergoes preprocessing according to the procedures described in [13] specific to each omics type, ensuring consistency across omics matrices. The matrices are structured such that genes are represented in rows, and samples (which are matched across all three matrices) are represented in columns. For a given pathway, each omics matrix is filtered to retain only the genes belonging to the pathway. Additionally, genes are excluded from the methylation and CNV matrices if there is no corresponding entry in the mRNA matrix. The overall workflow of our proposed method is outlined in Fig. 1, consisting of three main modules: (A) sequential decomposition of omics matrices, (B) computation of genes weights, and (C) calculation of sample deviation score.

The first module performs sequential decomposition of the mRNA, methylation, and CNV matrices using Non-Negative Matrix Factorization (NNMF). The purpose of this step is to identify latent structures within each omics layer. The output of this module is a set of component matrices that represent the decomposed data for each omics. In the second module for each gene a weight is calculated based on the significance of its contribution across the decomposed components from Module A. The computed weights are used to refine the combined omics representation. The last module calculates a deviation score for each sample, which quantifies the degree of abnormality in the pathway. The higher the deviation score, the more aberrant the pathway activity of a sample is considered.

2.1 Decomposing Omics Matrix Using NNMF

In this stage, the preprocessed omics matrices undergo decomposition using the NNMF approach. Given a non-negative input matrix, NNMF decomposes the matrix into two lower-dimensional non-negative matrices. This method enables the identification of latent patterns and relationships among the data [19]. Considering an omics matrix, the NNMF algorithm decomposes the input matrix $G \times S$, where G represents the number of genes (rows) and S represents the number of samples (columns), into two matrices:

- W with dimensions $G \times k$, where k is the number of latent factors or components. The matrix W represents the relationship between genes and the latent factors, indicating the contribution of each gene to the identified components.
- H with dimensions $k \times S$, which describes the association between the latent factors and the samples. Values in each column of H indicate the association for a specific sample.

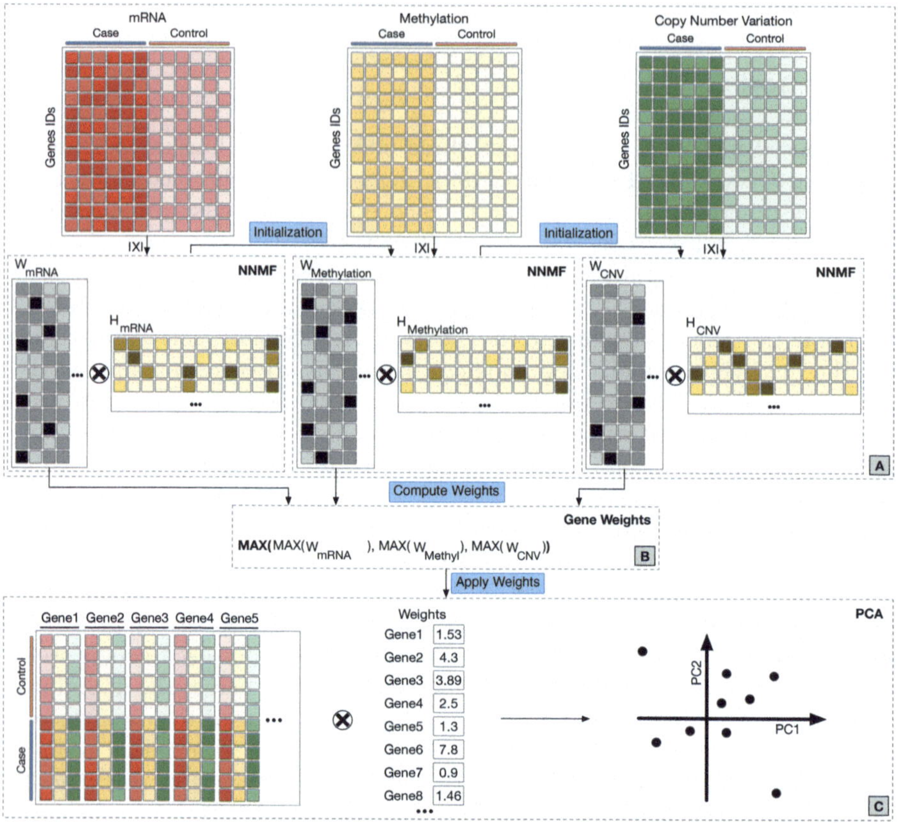

Fig. 1. The overall pipeline of the proposed method to compute sample-specific pathway deviation scores in multi-omics data using sequential NNMF decomposition. The method comprises three main modules: (A) This module sequentially decomposes the omics matrices using NNMF. First, the mRNA matrix is decomposed into two matrices, W and H, representing genes and samples respectively. These matrices then initialize the decomposition of the methylation data, guiding the computation of the two smaller matrices for the methylation matrix. Subsequently, the W and H matrices extracted from the methylation matrix decomposition are used to initialize the CNV data decomposition. This hierarchical approach allows the capture of shared latent structures across omics layers; (B) This module calculates a unique weight for each gene by selecting the maximum loading value across all components in the W matrices from the decompositions in module A. This prioritizes each gene's dominant association; (C) Each gene in the original omics matrices is scaled by its corresponding weight from module B, reflecting its influence across omics layers. The scaled, concatenated data is then subjected to PCA, where the coordinates of each sample in the PCA space are used to compute a deviation score, quantifying pathway activity deviation for individual samples.

To ensure the input matrix is suitable for NNMF algorithm, we first apply the absolute value function to convert all matrix elements to positive values. This preprocessing step is essential for NNMF, as it requires non-negative inputs to perform effectively. It is important to emphasize that this transformation does not influence the computation of the final deviation scores; however, it does result in a loss of directional information, as the sign of any changes in the data is removed. This trade-off allows us to work within the constraints of NNMF while focusing on the magnitude of gene expression changes rather than their direction.

The NNMF process begins with the decomposition of the mRNA data matrix. We keep the number of components, k, to 10, as this allows for sufficient complexity to capture variation in the data without overfitting. NNMF decomposes this data into two matrices, W and H, where W captures the gene loadings across components, and H represents the component activities across samples. These matrices provide an underlying structure that highlights significant patterns within the mRNA data.

Next, the W and H matrices obtained from the mRNA decomposition are passed as initial matrices to guide the NNMF decomposition of the methylation data matrix. By using these matrices as starting points, we carry over the latent structure of the mRNA data into the analysis of methylation data, establishing consistency across these omics layers. This sequential approach is repeated in the final NNMF step, where the W and H matrices from the methylation decomposition are used as initializations for the CNV data matrix. Through this structured initialization strategy, the method incorporates prior decompositions into each subsequent analysis.

This process yields several key benefits: (i) Latent information transfer: By initializing each omics layer with the W and H matrices from the previous layer, we enable a transfer of latent information across datasets. This cross-layer information transfer helps identify consistent patterns that might otherwise be missed when analyzing each layer independently; (ii) Regularization: The reuse of W and H matrices from one omics layer to the next acts as a form of regularization. This constraint helps prevent overfitting by maintaining a shared structure across datasets, thus enhancing the robustness of the decompositions; (iii) Solution space reduction: With each initialization, the solution space for the NNMF algorithm is effectively reduced, narrowing down the range of potential solutions to those that align with the patterns found in the previous layers; (iv) Convergence improvement: By providing structured starting points, the NNMF algorithm can converge more quickly to a solution in each decomposition.

2.2 Computing Genes Weights

In the second stage, marked as B in Fig. 1, the process involves applying the *max()* function to each row of the extracted W matrices obtained from the first stage of analysis. Each W matrix represents the contributions, or loadings, of each gene across various components derived from the NNMF applied in the first stage. By performing the *max()* function across each row, we retain the highest loading value for each gene, which reflects the component where the gene exerts the strongest influence. This approach allows us to prioritize the component that holds the most significant association for each gene, thereby identifying where the impact of a gene is most prominent.

After applying this initial *max()* function to each W matrix, we produce three separate vectors, each representing one omics dataset. Each element in these vectors corresponds to a gene and holds the highest component loading for that gene in that specific omics layer. These vectors capture the strongest signal for each gene within the distinct layers, providing a focused view of where the primary influence of each gene lies within each omics type.

The final step in this stage involves applying a second *max()* function across the three vectors obtained from each omics matrix. By selecting the highest value for each gene across the three omics datasets, we capture the most prominent association of the gene regardless of the specific omics layer. This strategy accounts for the fact that each gene may play different roles in different omics layers, depending on biological context and regulatory mechanisms. By using the maximum value across all layers, we ensure that the analysis reflects the strongest signal from each gene, whether its primary influence appears in mRNA expression, methylation, or CNV data.

2.3 Computing Individuals Deviation Scores

The final stage, represented by C in Fig. 1, focuses on integrating the omics data and performing dimensionality reduction. In this step, each omics matrix is first transposed and concatenated along the rows, which aligns samples across the different datasets and ensures each omic profile contributes to the collective analysis. To ensure that genes with the highest influence from the previous stage impact the dimensionality reduction proportionally, each column in this combined matrix is scaled by its corresponding gene weight. These weights were obtained from the earlier feature weighting step, designed to highlight genes with significant contributions. The resulting weighted matrix, with genes contributing variably according to their importance, is then used as input for PCA.

After conducting PCA, the coordinates of each sample in the reduced PCA space (i.e., along the principal components) are extracted, representing the position of each sample in this condensed, multidimensional view. To quantify the deviation of each sample from the central tendency of data, we calculate the Euclidean distance of each sample from the origin in PCA space. This distance is determined by first computing the squared distances of the coordinates of a sample along each principal component from the origin and then taking the square root of their sum, as described by the formula:

$$\text{Deviation Score} = \sqrt{\sum_{i=1}^{n}(x_i^2)} \tag{1}$$

wher

$$\text{Deviation Score} = \sqrt{\sum_{i=1}^{n}(x_i^2)} \tag{2}$$

component. This Euclidean distance, or "Deviation Score", serves as a quantitative measure of how far each sample is from center of the data in reduced space. Higher scores suggest greater deviation, which can indicate an abnormal or distinct profile in the context of the collective sample distribution.

3 Results

To assess and validate the results from our proposed pipeline, we conduct two sets of experiments as outlined in [13]. These experiments examine the conditions under which the method can correctly identify aberrant samples. In each experiment, we compare the performance of our method with that of the padma and PCA approaches, demonstrating that our gene-weighting strategy improves the accuracy of aberrant sample detection.

3.1 Outperforming in Simulation Experiments

To evaluate the effectiveness of the proposed method in identifying aberrant pathway activity, we conduct simulation experiments using synthetic multi-omics data. The pathway size is fixed at 29 genes, corresponding to the median pathway size in the MSigDB (Molecular Signatures Database) collection [10]. To simulate multi-omics data, we generate three matrices representing different omics data types: mRNA, methylation, and CNV. Each matrix has dimensions of $29 \times n$, where n is the number of samples, with values of 30, 50, or 100. We simulate the data using a trivariate Gaussian distribution to model the correlations between the three omics data types for each gene. Pairwise correlations among the genes within each omics matrix are drawn from a $\mathit{Uniform}(0.2, 0.8)$ distribution, ensuring a realistic correlation structure.

To introduce aberrant profiles, we select a specified percentage of individuals (p_{aberrant}) for each sample size n, with $p_{\text{aberrant}} \in \{0.01, 0.05\}$. For each aberrant individual, we randomly choose a set of driver genes (d_{genes}) and corresponding driver omics (d_{omics}). The values of the selected genes in the corresponding omics matrices are replaced with values drawn from a trivariate Gaussian distribution that shares the same covariance structure but has a mean vector $\mu_i = (\mu_1, \mu_2, \mu_3)$, where each μ_d is sampled from the range $\mathit{Uniform}((-6, -4) \cup (4, 6))$ for $d \in d_{\text{omics}}$, and set to 0 otherwise. This simulates the effect of outliers or aberrant profiles in the multi-omic data, representing individuals with unusual characteristics.

For each sample size n, we consider an average of two driver genes and one or two driver omics per aberrant individual. To ensure robustness, each configuration is tested 100 times. The Area Under the Curve (AUC) is used to evaluate the ability of the proposed method, padma, and PCA to distinguish aberrant from non-aberrant profiles. Tables 1 and 2 present the mean and median AUC results for different settings, with the highest values highlighted. As shown in the tables, the proposed method consistently outperforms the alternatives, especially at lower sample sizes and outlier percentages.

Table 1. Comparison of the mean AUC values from the simulation experiments

n	Outlier percentage	Proposed method	padma	PCA
30	0.01	0.9083	0.8728	0.8579
30	0.05	0.9157	0.8886	0.8721
50	0.01	0.9247	0.9188	0.9057
50	0.05	0.9266	0.9162	0.9061
100	0.01	0.9355	0.9407	0.9322
100	0.05	0.9485	0.9478	0.9478

Table 2. Comparison of the median AUC values from the simulation experiments

n	Outlier percentage	Proposed method	padma	PCA
30	0.01	0.9655	0.931	0.9138
30	0.05	0.9643	0.9286	0.9107
50	0.01	0.9694	0.9592	0.9591
50	0.05	0.9433	0.9291	0.9219
100	0.01	0.9798	0.9798	0.9796
100	0.05	0.9621	0.96	0.9558

3.2 Accurate Identification of Tumor Samples

To assess the ability of the proposed method to detect aberrant pathway activity, we first collect a set of healthy samples from the TCGA breast cancer dataset, ensuring that both mRNA and methylation data are available for each sample. Five tumor samples are then randomly selected from the same dataset to serve as true positives, and these tumor samples are added to the healthy sample set, creating a mixed cohort that simulates a real-world scenario where a small subset of samples exhibits aberrant pathway activity within a predominantly healthy background. The experiment is conducted using pathways from the MSigDB collection, specifically focusing on pathways that share at least three genes with the mRNA matrix to ensure sufficient overlap for analysis.

For each pathway, the proposed method is applied independently across 10 separate runs, each with a different set of five randomly chosen tumor samples. This approach introduces variability and robustness into the assessment, as the method is tested on different tumor configurations within the mixed cohort. Deviation scores, representing the distance of each sample from the center of the data in the pathway space, are computed, allowing us to quantify the extent of pathway deviation for both tumor and healthy samples. To measure the accuracy of the proposed method in distinguishing tumor samples from healthy ones, we calculate the Area Under the Curve (AUC) values based on these deviation scores.

The proposed method demonstrates a high performance across the pathways, achieving average AUC values corresponding to the first quartile, median, and third quartile as follows: 0.99, 0.997, and 0.999, respectively. These results indicate the capability of the proposed method in accurately identifying aberrant pathway activity and differentiating tumor samples from healthy controls, highlighting its potential for detecting pathway-level changes in multi-omic cancer data.

4 Conclusion

In this article, we introduce a method for calculating pathway deviation scores for each sample in an input cohort, utilizing sequential Non-Negative Matrix Factorization (NNMF) for gene weighting. We evaluate the performance of the proposed method through two distinct experiments: simulated data tests and real-world analysis using BRCA (breast cancer) data from the TCGA collection. Our findings demonstrate that the method effectively identifies aberrant samples in both settings, suggesting its potential utility for detecting pathway-level abnormalities in multi-omics datasets. While this article focuses on computing deviation scores, an important future direction is to evaluate the accuracy of the method in identifying true driver genes and the most relevant omics contributing to pathway alterations. Additionally, further validation across a wider range of datasets, coupled with in-depth biological analysis, is essential to ensure the robustness and generalizability of the results. For future work, we aim to establish a robust method for determining the optimal number of components for NNMF runs, adapting this choice based on the inherent structure of the data. Additionally, we plan to enhance the performance of the method and assess its effectiveness under more diverse conditions and datasets, aiming to expand its applicability in various biological contexts.

Acknowledgements. This work was partially supported by the National Science Foundation under grant numbers OIA-2019609 and OIA-2148788. Any opinions, findings, and conclusions or recommendations expressed in this material are those of the authors and do not necessarily reflect the views of any of the funding agencies.

References

1. Abdi, H., Williams, L.J.: Principal component analysis. Wiley Interdisc. Rev. Comput. Stat. **2**(4), 433–459 (2010). https://doi.org/10.1002/wics.101

2. Abdi, H., Williams, L.J., Valentin, D.: Multiple factor analysis: principal component analysis for multitable and multiblock data sets. Wiley Interdisc. Rev. Comput. Stat. **5**(2), 149–179 (2013). https://doi.org/10.1002/wics.1246
3. Argelaguet, R., et al.: Multi-omics factor analysis—a framework for unsupervised integration of multi-omics data sets. Mol. Syst. Biol. **14**(6), e8124 (2018). https://doi.org/10.15252/msb.20178124 https://doi.org/10.15252/msb.20178124
4. Bersanelli, M., et al.: Methods for the integration of multi-omics data: mathematical aspects. BMC Bioinf. **17**, 167–177 (2016). https://doi.org/10.1186/s12859-015-0857-9
5. Bild, A.H., et al.: Oncogenic pathway signatures in human cancers as a guide to targeted therapies. Nature **439**(7074), 353–357 (2006). https://doi.org/10.1038/nature04296
6. Chin, L., Hahn, W.C., Getz, G., Meyerson, M.: Making sense of cancer genomic data. Genes Dev. **25**(6), 534–555 (2011). https://doi.org/10.1101/gad.2017311
7. García-Campos, M.A., Espinal-Enríquez, J., Hernández-Lemus, E.: Pathway analysis: state of the art. Front. Physiol. **6**, 383 (2015). https://doi.org/10.3389/fphys.2015.00383
8. Graw, S., et al.: Multi-omics data integration considerations and study design for biological systems and disease. Mol. Omics **17**(2), 170–185 (2021). https://doi.org/10.1039/D0MO00041H
9. Khatri, P., Sirota, M., Butte, A.J.: Ten years of pathway analysis: current approaches and outstanding challenges. PLoS Comput. Biol. **8**(2), e1002375 (2012). https://doi.org/10.1371/journal.pcbi.1002375
10. Liberzon, A., Subramanian, A., Pinchback, R., Thorvaldsdóttir, H., Tamayo, P., Mesirov, J.P.: Molecular signatures database (MSigDB) 3.0. Bioinformatics **27**(12), 1739–1740 (2011). https://doi.org/10.1093/bioinformatics/btr260
11. Maghsoudi, Z., Nguyen, H., Tavakkoli, A., Nguyen, T.: A comprehensive survey of the approaches for pathway analysis using multi-omics data integration. Briefings Bioinf. **23**(6), bbac435 (2022). https://doi.org/10.1093/bib/bbac435
12. Network, T.C.G.A.R.: The cancer genome atlas (2024). https://www.cancer.gov/about-nci/organization/ccg/research/structural-genomics/tcga. Accessed 25 Oct 2024
13. Rau, A., et al.: Individualized multi-omic pathway deviation scores using multiple factor analysis. Biostatistics **23**(2), 362–379 (2022). https://doi.org/10.1093/biostatistics/kxaa029
14. Sra, S., Dhillon, I.: Generalized nonnegative matrix approximations with Bregman divergences. Adv. Neural Inf. Process. Syst. **18** (2005)
15. Subramanian, I., Verma, S., Kumar, S., Jere, A., Anamika, K.: Multi-omics data integration, interpretation, and its application. Bioinform. Biol. Insights **14**, 1177932219899051 (2020). https://doi.org/10.1177/1177932219899051
16. Tandon, R., Sra, S.: Sparse nonnegative matrix approximation: new formulations and algorithms. Technical report, TR (2010)
17. Thurstone, L.L.: Multiple factor analysis. Psychol. Rev. **38**(5), 406 (1931). https://doi.org/10.1037/h0069792
18. Vaske, C.J., et al.: Inference of patient-specific pathway activities from multi-dimensional cancer genomics data using paradigm. Bioinformatics **26**(12), i237–i245 (2010). https://doi.org/10.1093/bioinformatics/btq182
19. Yang, Z., Michailidis, G.: A non-negative matrix factorization method for detecting modules in heterogeneous omics multi-modal data. Bioinformatics **32**(1), 1–8 (2016). URL https://doi.org/10.1093/bioinformatics/btv544

Transferred-Learning Intrusion Detection System for Internet of Vehicles

Tan Nguyen(✉) and Jorge Fonseca

Computer Science Department, University of Nevada - Las Vegas,
Las Vegas, NV, USA
datnguyentan93@gmail.com, jorge.fonsecacacho@unlv.edu

Abstract. The growing connectivity of modern vehicles, particularly within the Internet of Vehicles (IoV), has significantly increased the need for robust cybersecurity solutions. Building upon the work of Yang and Shami (2022), who proposed a transfer learning and optimized convolutional neural network (CNN)-based Intrusion Detection System (IDS) for IoV, this research seeks to validate their results and explore potential enhancements. The original IDS demonstrated exceptional performance, with detection rates surpassing 99.25% on benchmark datasets. In this study, we first replicate their experiments using the Car-Hacking dataset to confirm the effectiveness of the proposed model and then evaluate its ability to detect a new type of attack. Furthermore, we introduce several improvements, including the synthesis of balanced training data and the integration of new state-of-the-art CNN models, aiming to further enhance the IDS's detection capabilities.

Keywords: CNN · IoV · Car Hacking · Deep Learning

1 Introduction

The rapid evolution of the Internet of Things (IoT) and its integration into vehicular technologies have led to the emergence of the Internet of Vehicles (IoV), where vehicles are increasingly interconnected with each other and external networks. This connectivity facilitates advanced functionalities in autonomous vehicles and connected vehicles, such as real-time traffic management, enhanced safety features, and improved user experiences [19,25]. However, the expanding attack surfaces of these interconnected systems have made IoV a prime target for cyber-attacks, posing significant risks to vehicle safety and privacy [6,32].

In IoV systems, both intra-vehicle networks and external vehicular networks are vulnerable to a variety of cyber threats [16,21]. Intra-vehicle networks, primarily managed by the Controller Area Network (CAN) protocol, lack fundamental security measures like authentication and encryption due to limitations in the CAN bus system [26,38]. This absence of robust security protocols allows

J. Fonseca—Contributing authors.

S. Latifi (Ed.): ITNG 2025, AISC 1463, pp. 737–753, 2025.
https://doi.org/10.1007/978-3-031-89063-5_64

attackers to inject malicious messages, leading to potential breaches such as Denial of Service (DoS) attacks, spoofing attacks, and other forms of cyber intrusions [8,10]. Moreover, the external connectivity of vehicles to the internet and other networks makes them susceptible to conventional cyber-attacks, including botnets, brute-force attacks, and web-based exploits [18,23].

To mitigate these threats, Intrusion Detection Systems (IDS) have become essential components in protecting IoV infrastructures. Recent advancements in machine learning and deep learning techniques have enabled the development of sophisticated IDS models capable of identifying and mitigating cyber threats with high accuracy[27,36]. Among these, Convolutional Neural Networks (CNNs) have shown considerable promise in handling the complex, high-dimensional data generated by vehicular networks. In particular, the application of transfer learning and ensemble learning techniques to CNN models has significantly improved their performance in detecting various types of cyber-attacks[34,37].

In this context, Yang and Shami[40] introduced a novel IDS framework for IoV systems that leverages transfer learning, ensemble learning, and hyper-parameter optimization techniques to achieve a detection rate exceeding 99.25% on benchmark datasets such as the Car-Hacking dataset and CICIDS2017. Their approach demonstrated the effectiveness of using advanced CNN models, such as VGG16, Xception, and InceptionResNet, in detecting intra-vehicle and external network attacks. Despite these impressive results, there remains room for further validation and potential enhancements.

This paper aims to replicate and validate the findings of Yang and Shami while also exploring improvements to their proposed IDS framework. By refining the data transformation processes, optimizing hyper-parameters with more advanced techniques, and integrating additional CNN architectures, this research seeks to enhance the accuracy and robustness of IDS in IoV environments. Through this effort, we aim to contribute to the ongoing development of more secure and resilient vehicular networks.

2 Literature Review

2.1 Intrusion Detection System for IoV

The IoV has become an integral component of modern intelligent transportation systems, enabling seamless real-time communication and data exchange among vehicles, infrastructure, and other entities. While this enhanced connectivity offers numerous benefits, it also increases the vulnerability of vehicular networks to cyber-attacks. Consequently, researchers have devoted significant efforts to developing IDS specifically tailored to the unique requirements of IoV environments.

Early IDS approaches for vehicular networks predominantly utilized traditional machine learning techniques. For example, Rosay et al. proposed a feed-forward neural network-based IDS evaluated using the CICIDS2017 dataset—a standard benchmark in network intrusion detection research. Although the model demonstrated satisfactory performance, its ability to adapt to the dynamic nature of vehicular networks was limited due to the inherent constraints of traditional machine learning models [29].

To overcome these limitations, the adoption of deep learning methods has gained momentum, given their superior capability in handling complex and high-dimensional data typical of IoV environments. Mehedi et al. introduced a deep transfer learning-based IDS that leveraged pre-trained CNNs to detect intrusions in electric vehicular networks. Their model achieved an F1-score of 97.83% on the Car-Hacking dataset, underscoring the effectiveness of deep transfer learning in enhancing IDS performance [20].

Similarly, Hossain et al. developed a one-dimensional CNN (1D-CNN) tailored for intra-vehicle network intrusion detection. This model, adept at processing time-series data common in vehicular communications, outperformed traditional machine learning approaches by achieving near-perfect accuracy on the Car-Hacking dataset. However, its focus on single attack scenarios limited its applicability across a broader spectrum of cyber-attacks [11].

Recognizing the necessity for more comprehensive detection capabilities, Song et al. proposed a deep CNN model employing a reduced InceptionResNet architecture to detect various types of attacks in intra-vehicle networks. The model demonstrated high accuracy and robustness, even when evaluated on complex and noisy datasets, highlighting the potential of advanced DL architectures in IDS applications for IoV [31].

Building upon these advancements, Yang and Shami introduced an innovative IDS framework that integrates transfer learning, ensemble learning, and hyper-parameter optimization techniques. By utilizing multiple state-of-the-art CNN architectures such as VGG16, Xception, and InceptionResNet, they trained base learners on vehicular network traffic data. The ensemble learning strategies further enhanced detection accuracy, achieving rates exceeding 99.25% on both the Car-Hacking and CICIDS2017 datasets [40].

Despite these significant strides, challenges persist in the development of IDS for IoV, particularly in managing the vast diversity of cyber-attacks and adapting to the evolving threat landscape. Recent research has explored the integration of Generative Adversarial Networks (GANs) for data augmentation to improve IDS performance in imbalanced data scenarios. Seo et al. utilized GANs to generate synthetic data, thereby enhancing the model's ability to detect rare and novel attacks [30]. Additionally, hybrid models that combine multiple deep learning techniques, such as the multi-tiered hybrid IDS proposed by Yang et al., have shown promise in further improving detection accuracy and real-time performance [39].

2.2 Existing Problems with IDS

Despite significant advancements, IDS continue to face substantial challenges that hinder their effectiveness, particularly within the dynamic environment of the IoV. One of the foremost challenges is the detection of unknown or zero-day attacks. Traditional signature-based IDS rely heavily on predefined patterns of known threats, rendering them ineffective against novel attacks that do not match existing signatures. Anomaly-based IDS attempt to mitigate this by modeling normal network behavior and flagging deviations as potential threats. However, the highly dynamic and complex nature of vehicular network traffic often leads to high false-positive rates, as distinguishing between legitimate anomalies and malicious activities becomes increasingly difficult [1].

Another critical challenge is the susceptibility of IDS to data poisoning attacks. Adversaries may deliberately introduce malicious data into the training datasets used by machine learning-based IDS, compromising the learning process and causing the model to misclassify malicious activities as benign. This manipulation undermines the reliability of the IDS and can render it ineffective against specific attack vectors. Ensuring the integrity and authenticity of training data is therefore paramount for the development of robust IDS models [2].

Adversarial attacks further complicate the landscape by exploiting vulnerabilities in machine learning models through carefully crafted inputs designed to cause misclassification. These adversarial examples can deceive the IDS into misidentifying malicious traffic as legitimate, allowing attackers to evade detection. The development of IDS that are resilient to such adversarial manipulations is an ongoing area of research, emphasizing the need for models that can withstand perturbations and maintain high detection accuracy [33].

Injection attacks, including code and command injections, present additional challenges. In the context of IoV, attackers can inject false data or malicious commands into the CAN bus, potentially manipulating vehicle functions or sensor readings. Such attacks are particularly insidious because they often mimic legitimate messages, making them difficult to detect using conventional IDS techniques. Effective detection requires sophisticated analysis capable of discerning subtle anomalies in communication patterns [22].

The high dimensionality and sheer volume of data generated in vehicular networks exacerbate these challenges. IDS must process vast amounts of heterogeneous data in real-time to provide timely threat detection, which demands significant computational resources and efficient algorithmic solutions. Resource constraints inherent in vehicular systems, such as limited processing power and memory, further complicate real-time data analysis and necessitate the development of lightweight yet effective IDS algorithms [17].

Moreover, the rapidly evolving nature of cyber threats requires IDS to be highly adaptable. Attackers continuously develop new strategies and exploit emerging vulnerabilities to bypass security measures. This cat-and-mouse dynamic necessitates IDS that can learn from new data and update their detection models accordingly. Incorporating mechanisms for continuous learning and adaptation is essential to maintain the effectiveness of IDS in the face of evolving attack methodologies [3].

Lastly, the implementation of IDS in IoV raises important privacy and ethical considerations. Monitoring network traffic and analyzing data can inadvertently infringe on user privacy, as sensitive information may be collected and processed. Balancing the need for security with the protection of individual privacy rights is crucial. Ethical guidelines and robust data governance policies must be established to ensure that IDS deployments do not compromise user trust or violate legal standards [35].

3 Methodology

3.1 Datasets

In this research, we utilize two primary datasets: the *Car-Hacking Dataset* from 2016 and the *Car Hacking: Attack & Defense Challenge Dataset* from 2020. Both datasets are widely adopted for evaluating Intrusion Detection Systems (IDS) in vehicular networks, focusing on attacks such as spoofing, Denial of Service (DoS), and fuzzy attacks (Fig. 1).

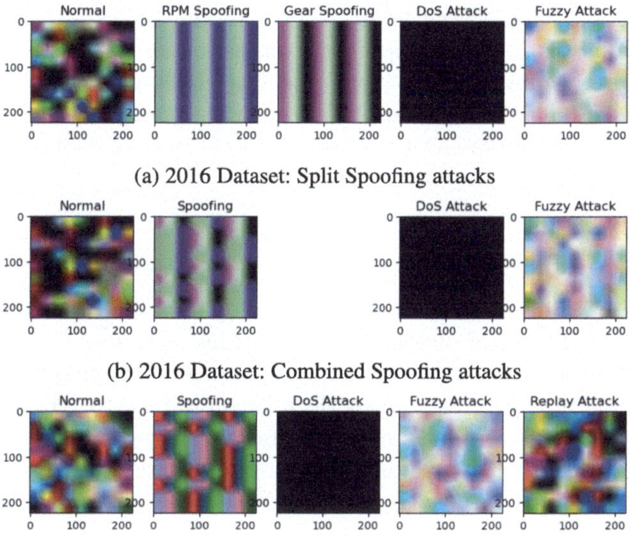

(c) 2020 dataset

Fig. 1. Datasets

The *Car-Hacking Dataset*, originally used in studies by Kang and Kang [13], is designed to simulate various types of cyber-attacks on vehicular networks, specifically targeting the CAN bus—the central communication system in modern vehicles. The dataset includes four primary types of attacks: DoS, fuzzy attacks, gear spoofing, and RPM spoofing. These attacks were generated by logging CAN traffic via the OBD-II port of a real vehicle, with each attack scenario consisting of numerous intrusions over several seconds. The dataset also contains corresponding normal traffic data, allowing for a clear distinction between normal and malicious activities. Due to its realistic and varied attack scenarios, the 2016 Car-Hacking Dataset has been extensively used for evaluating IDS in vehicular networks [31].

The *Car Hacking: Attack & Defense Challenge Dataset* was introduced as part of the Car Hacking Challenge 2020, aimed at advancing the development of intrusion detection techniques for vehicular networks [14]. This dataset expands on the original Car-Hacking Dataset by including new types of attacks, specifically designed to target the CAN network of the Hyundai Avante CN7 vehicle. The dataset comprises two rounds of data: a preliminary round containing normal data and four types of attack data (flooding, spoofing, replay, and fuzzing) during stationary and driving conditions, and a final round which includes an additional new attack type in a more complex scenario.

A key difference between the two datasets is the categorization of spoofing attacks. In the 2016 Car-Hacking Dataset, spoofing attacks are divided into two distinct categories based on the targeted component: RPM spoofing and gear spoofing. In contrast, the 2020 Car Hacking Challenge Dataset treats spoofing attacks as a single category without such distinction. To align the datasets for comparative analysis and to simplify the classification task, we combined the RPM and gear spoofing attacks from the Car-Hacking Dataset into a single *spoofing* category. This consolidation ensures consistency in the attack classes across both datasets and facilitates a more straightforward evaluation of the IDS performance.

3.2 Data Imbalance Issue

As illustrated in Table 1, these datasets are highly imbalanced, with the majority class (normal) significantly outweighing the attack samples.

The imbalanced nature of these datasets poses significant challenges for machine learning models, particularly deep learning-based IDS solutions. CNNs trained on such imbalanced data may become biased towards the majority class (normal) and underperform in detecting minority classes (i.e., attack types). This often leads to a high false-negative rate for attack instances, which undermines the primary goal of an IDS [12].

Moreover, models trained on imbalanced data may exhibit high overall accuracy due to the correct classification of normal samples but perform poorly in detecting attacks. Such bias is unacceptable for IDS systems, where the ability to detect attacks is paramount [4]. Therefore, addressing data imbalance is crucial to improve the generalization and real-world applicability of IDS solutions for the IoV.

3.3 Proposed Solutions to Address Data Imbalance

To address the data imbalance issue, we employ two techniques: class weighting and data synthesis using the Synthetic Minority Over-sampling Technique (SMOTE).

3.3.1 Class Weighting

Class weighting involves assigning higher weights to the minority classes during model training. By modifying the loss function to penalize misclassification of minority-class samples more heavily, the model is encouraged to pay more attention to attack instances. This approach helps to balance the classifier's focus between detecting normal and attack samples, improving performance metrics such as recall and precision for the attack classes. Most deep learning frameworks, including TensorFlow and PyTorch, support the use of class weights, making this an accessible and effective solution [7].

3.3.2 Data Synthesis with SMOTE

The SMOTE generates synthetic samples by interpolating between existing minority-class instances, effectively increasing the representation of attack classes [5]. By balancing the dataset, the model encounters a larger number of attack samples during training, which can enhance its ability to generalize and detect rare attack types in IoV environments. SMOTE has been shown to improve the detection rate of minority classes by reducing the bias towards the majority class, especially when combined with ensemble learning techniques or deep learning models [9].

We applied SMOTE to the combined spoofing class and other minority classes to create a balanced dataset. This step is crucial, especially after consolidating the spoofing attacks, as it adjusts the class distribution and provides the model with sufficient examples of each attack type.

3.4 Methodology

Our study builds upon Yang and Shami's framework (Fig. 2) [40], which utilizes advanced CNN architectures and ensemble learning strategies for intrusion detection in IoV systems. The methodology involves several key steps:

Table 1. Class distribution in the 2016 Car-Hacking and 2020 Car Hacking: Attack & Defense Challenge datasets

Class	2016 Car-Hacking dataset	2020 Car Hacking: Attack & Defense Challenge
Normal	85.75%	91.12%
Spoofing–RMP	3.98%	0.23%
Spoofing–Gear	3.66%	–
DoS	3.60%	4.62%
Fuzzy	3.01%	2.64%
Relay	–	1.39%

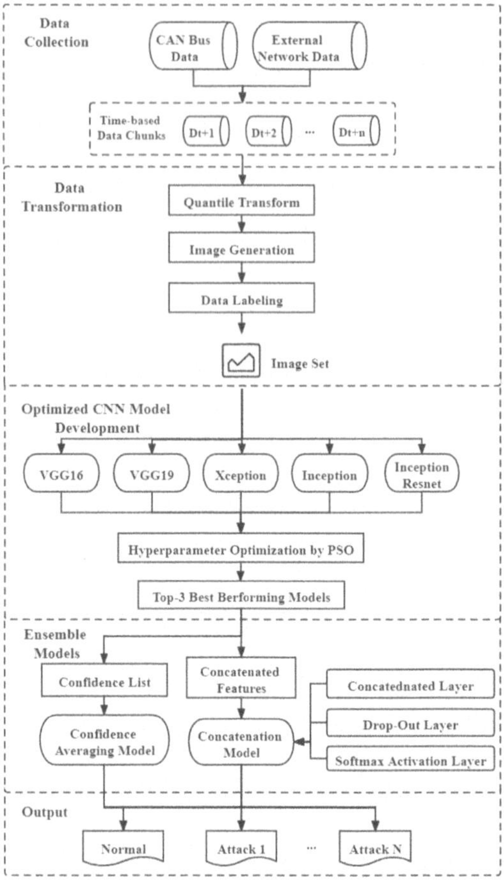

Fig. 2. Yang's Proposed Framework

3.4.1 Data Transformation

The preprocessed data undergoes a transformation process that includes:

The raw CAN bus data from both datasets undergoes preprocessing to prepare it for model training:

- **Data Consolidation:** As mentioned, we combined the RPM and gear spoofing attacks in the 2016 Car-Hacking Dataset into a single spoofing category to align with the structure of the Car Hacking Challenge Dataset.
- **Quantile Transformation:** We applied a quantile transformer to the features, transforming them to follow a uniform distribution. This step stabilizes variance and minimizes the effects of outliers, which is beneficial for deep learning models [24].
- **Image Generation:** The transformed numerical data is converted into grayscale images. Each sample is reshaped into a two-dimensional matrix

and scaled to fit the pixel intensity range. This novel approach allows us to leverage CNNs, which are highly effective in extracting spatial features from image data.
- **Data Labeling:** Each generated image is labeled according to its class (normal or specific attack type), preparing the dataset for supervised learning.

3.4.2 Optimized CNN Model Development

We trained various CNN architectures on the processed image data:

- **CNN Architectures:** We utilized state-of-the-art CNN models, including VGG16, VGG19, Xception, Inception, and InceptionResNet. These models are renowned for their robustness and efficacy in image recognition tasks [7].
- **Hyperparameter Optimization:** To enhance model performance, we applied Particle Swarm Optimization (PSO) to fine-tune hyperparameters such as learning rate, batch size, and the number of epochs [15]. PSO is an evolutionary optimization technique that efficiently searches the hyperparameter space to find optimal configurations.

3.4.3 Ensemble Learning

After training, we selected the top three best-performing models based on validation accuracy to create an ensemble model:

- **Confidence Averaging:** We combined the predictions by averaging the confidence scores (probabilities) output by each model. This method smooths out individual model biases and reduces variance [28].
- **Feature Concatenation:** Alternatively, we concatenated the feature representations (outputs from the penultimate layer) from each model and passed them through a fully connected layer to make the final prediction.
- **Final Classification:** A softmax activation function is applied to the output layer to classify each input into the appropriate category (normal or specific attack type).

3.4.4 Output Classification

The final ensemble model classifies each input into categories such as normal operation or various types of attacks (e.g., DoS, spoofing). A softmax activation layer is used in the final output layer to provide probability distributions over the classes, facilitating multi-class classification [7].

3.5 Our Focus in Validation and Enhancement

Our research builds upon Yang and Shami's foundational IDS framework with the following focus areas:

- **Alignment of Datasets:** By combining the RPM and gear spoofing attacks into a single category, we harmonized the class structures of the two datasets. This alignment is crucial for comparative analysis and evaluating the IDS's ability to generalize across different datasets.
- **Validation of Yang and Shami's Claims:** We replicated Yang and Shami's experiments using the adjusted Car-Hacking Dataset to validate the claimed detection rates and model performance. This step serves as our baseline to ensure that the original framework's results are reproducible and accurate.
- **Addressing Data Imbalance:** Recognizing the challenge posed by imbalanced data, we employed class weighting and SMOTE to create balanced versions of the datasets. Training our models on this balanced data allows us to evaluate the effectiveness of the IDS framework in a more equitable data environment, potentially leading to more reliable detection of less frequent attack types.
- **Application to New Dataset:** We extended the IDS application to the 2020 Car Hacking: Attack & Defense Challenge Dataset. This dataset includes additional attack scenarios not present in the original dataset, providing a comprehensive test of the IDS's adaptability and effectiveness against evolving IoV threats.

By following this methodology, we aim not only to validate Yang and Shami's original IDS framework but also to enhance its robustness and reliability by addressing the challenges associated with data imbalance and by testing its efficacy against a broader array of attack vectors.

3.6 Setup

All experiments are conducted using Python 3.12.3, TensorFlow 2.17, and scikit-learn 1.5.2 libraries. The models are trained and tested on a workstation equipped with NVIDIA GPUs. Hyperparameters are tuned using PSO, and performance metrics such as accuracy, precision, recall, and F1-score are used to evaluate the models.

4 Results and Discussion

4.1 Performance Evaluation and Analysis

The experimental results of our study are summarized in Tables 2 and 3. These tables present the accuracy of various CNN models under different training and testing conditions on the 2016 Car-Hacking Dataset and the 2020 Car Hacking: Attack & Defense Challenge Dataset. The following subsections delve into the interpretation of these results, highlighting key observations and their implications for IDS in the IoV.

4.2 Validation of Yang's Model

Table 2 presents the accuracy results of various CNN models on the Car-Hacking Dataset under different experimental setups. The "Claimed" column reflects the accuracy reported by Yang and Shami [40], while the "Validated" column shows the results obtained from our replication efforts.

Our experiments successfully replicated the high accuracy claimed by Yang and Shami for advanced CNN architectures such as Xception, VGG16, VGG19, Inception, and InceptionResNet, all achieving 100% accuracy. This validation confirms the effectiveness of their IDS framework and the suitability of these architectures for detecting intrusions in vehicular networks.

However, discrepancies were observed with the basic CNN and ResNet models. The basic CNN achieved a validation accuracy of 96.68%, slightly lower than the claimed 99.88%. ResNet's performance was notably lower at 85.71% compared to the reported 98.65%. These differences may stem from variations in implementation details, hyperparameter settings, or potential overfitting in the original study. The underperformance of ResNet suggests that not all architectures generalize equally well to the intrusion detection task in IoV environments.

4.3 Impact of Data Balancing Techniques

The "2020 Dataset" columns in Table 2 assess the models' generalization capabilities by testing them on the Car Hacking: Attack & Defense Challenge Dataset after training on the Car-Hacking Dataset.

A general decline in accuracy was observed across all models when tested on a different dataset. Advanced architectures like Xception, VGG16, and InceptionResNet maintained relatively high accuracies (approximately 98%), indicating better generalization capabilities. The basic CNN model showed a moderate decrease to around 95.5%, while ResNet's performance interestingly improved to 98.38% when tested on the different dataset, despite its lower validation accuracy on the original dataset.

These results highlight the challenges in cross-dataset generalization, possibly due to differences in data distributions, attack patterns, and the way spoofing attacks are categorized in the datasets.

4.3.1 Effect of Spoofing Attack Categorization

Table 3 presents the models' accuracies on the 2020 Car Hacking: Attack & Defense Challenge Dataset under three conditions:

- Validated: Models trained and tested on the same dataset.
- 2016 Dataset (Split Spoofing): Models trained on the Car-Hacking Dataset with spoofing attacks split into RPM and gear categories, tested on the Challenge Dataset.
- 2016 Dataset (Combined Spoofing): Models trained on the Car-Hacking Dataset with spoofing attacks combined into a single category, tested on the Challenge Dataset.

Table 2. Accuracy of models trained on 2016 Car-Hacking Datasets before and after balancing

Accuracy	2016 Car-hacking dataset				Balanced 2016 Car-Hacking datasets	
	Claimed	Validate	Weighted	2020 datasets	Validate	2020 datasets
CNN	99.884%	96.68%	95.86%	95.55%	100.00%	97.83%
Xception	100.00%	100.00%	100.00%	98.43%	100.00%	98.41%
VGG16	100.00%	100.00%	100.00%	98.39%	100.00%	98.33%
VGG19	100.00%	100.00%	100.00%	98.32%	100.00%	98.33%
ResNet	98.652%	85.71%	85.71%	98.38%	96.33%	92.66%
Inception	100.00%	100.00%	100.00%	91.15%	100.00%	98.44%
InceptionResnet	99.993%	100.00%	100.00%	98.44%	100.00%	98.41%

Table 3. Accuracy of models trained on 2020 Car Hacking: Attack & Defense Challenge

Accuracy	2020 Car Hacking: Attack & Defense Challenge		
	Validate	2016 datasets (Split Spoofing)	2016 datasets (Combined Spoofing)
CNN	98.47%	92.12%	89.16%
Xception	98.47%	95.84%	92.14%
VGG16	98.47%	92.04%	89.16%
VGG19	98.40%	91.78%	89.16%
ResNet	7.18%	6.59%	6.35%
Inception	98.48%	92.23%	92.85%
InceptionResnet	99.60%	95.34%	88.81%

When models were trained and validated on the Challenge Dataset, they achieved high accuracies, with InceptionResNet reaching 99.60%. This confirms the models' effectiveness when data distributions and attack categorizations are consistent.

However, when models trained on the 2016 Car-Hacking Dataset were tested on the 2020 Challenge Dataset, a decline in accuracy was observed, more pronounced when spoofing attacks were combined. For instance, Xception's accuracy decreased from 95.84% (split spoofing) to 92.14% (combined spoofing). This indicates that inconsistencies in attack categorization between training and testing datasets adversely affect model performance.

ResNet consistently performed poorly across these scenarios, with accuracies around 6–7%. This suggests that ResNet may not generalize well to different datasets or may require further tuning to be effective in this context.

4.4 Challenges in Cross-Dataset Generalization

Despite the overall high performance, a noticeable decline in accuracy was observed when models trained on one dataset were tested on another, particularly in the context of spoofing attack categorization. The decision to combine RPM and gear spoofing attacks into a single category in the 2016 Car-Hacking dataset, to align with the structure of the 2020 Car Hacking Challenge dataset, had mixed implications.

Models trained with combined spoofing categories experienced a decrease in accuracy when tested on data where spoofing attacks were split into distinct types. This outcome underscores the importance of consistent attack categorization between training and testing datasets for optimal model performance. It also highlights a limitation in the models' ability to generalize across datasets with different class structures.

4.5 Model Selection and Architecture Implications

Advanced CNN architectures like Xception, Inception, and InceptionResNet consistently achieved high accuracies and demonstrated better generalization capabilities compared to the basic CNN and ResNet models. The superior performance of these architectures can be attributed to their complex structures, which are adept at capturing intricate patterns in data.

ResNet's underperformance across multiple scenarios suggests that it may not be the most suitable choice for this specific application without further optimization. This finding emphasizes the need for careful model selection based on the characteristics of the dataset and the specific requirements of the intrusion detection task.

5 Conclusion

The study successfully replicated and validated Yang and Shami's IDS framework for IoV, confirming the efficacy of advanced CNN architectures in detecting cyber-attacks with high accuracy. By addressing data imbalance through class weighting and SMOTE, we enhanced the models' ability to generalize to different datasets, an essential characteristic for real-world applications where data diversity is inherent.

However, challenges remain in achieving consistent cross-dataset generalization, particularly due to differences in attack categorization and data distributions. The results indicate that while balancing techniques improve model robustness, aligning the categorization of attack types between training and testing datasets is crucial for maintaining high performance.

5.1 Implications for Application

The findings have significant implications for the deployment of IDS in IoV environments:

- **Model Selection:** Implementing IDS using advanced CNN architectures like Xception and InceptionResNet can provide high detection accuracy, enhancing vehicular network security.
- **Data Preparation:** Addressing data imbalance is essential. Employing techniques like class weighting and SMOTE should be standard practice to ensure models are not biased towards majority classes and can detect less frequent attack types effectively.
- **Dataset Alignment:** Consistency in attack categorization across datasets used for training and deployment is critical. Practitioners should ensure that the IDS is trained on data that closely resembles the operational environment in terms of attack types and distributions.
- **Generalization and Adaptability:** The IDS should be designed with adaptability in mind, possibly incorporating mechanisms for continual learning or domain adaptation to handle new attack patterns and variations in data.

5.2 Future Research Directions

The study opens avenues for further research to enhance IDS in IoV:

- **Domain Adaptation Techniques:** Investigate methods such as transfer learning and domain adaptation to improve cross-dataset generalization, enabling IDS models to adapt to new environments with minimal performance degradation.
- **Comprehensive Attack Categorization:** Develop standardized attack taxonomies for vehicular networks to facilitate consistency across datasets and studies, improving the comparability of results and the robustness of IDS models.
- **Incorporation of Temporal Features:** Explore models that can capture temporal dependencies, such as Recurrent Neural Networks (RNNs) or Long Short-Term Memory (LSTM) networks, to enhance the detection of attacks that exhibit temporal patterns.
- **Real-Time Implementation:** Evaluate the computational efficiency of advanced CNN architectures in real-time settings, ensuring that high detection accuracy does not come at the expense of latency, which is critical in vehicular applications.
- **Robustness Against Adversarial Attacks:** Assess the IDS models' resilience to adversarial machine learning attacks, such as evasion or poisoning attacks, to ensure reliability in adversarial environments.
- **Multimodal Data Fusion:** Incorporate additional data sources, such as sensor readings or vehicle-to-infrastructure communications, to enrich the feature set and potentially improve detection capabilities.

5.3 Limitations

While the study provides valuable insights, certain limitations should be acknowledged:

- **Dataset Diversity:** The experiments were conducted on two specific datasets. Expanding the evaluation to include more diverse datasets could provide a more comprehensive understanding of model generalization.
- **Attack Variety:** The datasets may not encompass the full spectrum of potential cyber-attacks in IoV environments. Future work should consider incorporating more varied and sophisticated attack scenarios.
- **Operational Constraints:** The computational demands of advanced CNN architectures may pose challenges for deployment on resource-constrained vehicular hardware. Optimizing models for efficiency is necessary for practical applications.

Acknowledgment. We would like to express our sincere gratitude to the Summer Research Institute at the University of Nevada, Las Vegas, for providing the funding and support that made this research possible. Their generous contribution enabled the completion of this work, and we deeply appreciate the opportunities provided by the institute to advance our research.

References

1. Bhuyan, M.H., Bhattacharyya, D.K., Kalita, J.K.: Network anomaly detection: methods, systems and tools. IEEE Commun. Surv. Tutor. **16**(1), 303–336 (2014)
2. Biggio, B., Nelson, B., Laskov, P.: Poisoning attacks against support vector machines. In: Proceedings of the 29th International Conference on Machine Learning (ICML), pp. 1807–1814 (2012)
3. Buczak, A.L., Guven, E.: A survey of data mining and machine learning methods for cyber security intrusion detection. IEEE Commun. Surv. Tutor. **18**(2), 1153–1176 (2016)
4. Chawla, N.V.: Editorial: special issue on learning from imbalanced data sets. SIGKDD Explor. **6**(1), 1–6 (2004)
5. Chawla, N.V., Bowyer, K.W., Hall, L.O., Kegelmeyer, W.P.: SMOTE: synthetic minority over-sampling technique. J. Artif. Intell. Res. **16**, 321–357 (2002)
6. Checkoway, S., et al.: Comprehensive experimental analyses of automotive attack surfaces. In: 20th USENIX Security Symposium (2011)
7. Goodfellow, I., Bengio, Y., Courville, A.: Deep Learning. MIT Press (2016)
8. Greenberg, A.: Hackers remotely kill a jeep on the highway—with me in it. Wired Magazine (2015). https://www.wired.com/2015/07/hackers-remotely-kill-jeep-highway/
9. Han, H., Wang, W.-Y., Mao, B.-H.: Borderline-SMOTE: a new over-sampling method in imbalanced data sets learning. In: Huang, D.-S., Zhang, X.-P., Huang, G.-B. (eds.) ICIC 2005. LNCS, vol. 3644, pp. 878–887. Springer, Heidelberg (2005). https://doi.org/10.1007/11538059_91

10. van der Heijden, R.W., Dietzel, S., Leinmüller, T., Kargl, F.: Survey on misbehavior detection in cooperative intelligent transportation systems. IEEE Commun. Surv. Tutor. **21**(1), 779–811 (2019)
11. Hossain, M.S., Muhammad, G., Alamri, A.: Automatic vehicle accident detection and mapping system using sensor, GPS, GPRS and GSM technology. Int. J. Comput. Inf. Eng. **3**(3), 372–375 (2019)
12. Johnson, J.M., Khoshgoftaar, T.M.: Survey on deep learning with class imbalance. J. Big Data **6**(1), 1–54 (2019). https://doi.org/10.1186/s40537-019-0192-5
13. Kang, M., Kang, J.: Intrusion detection system using deep neural network for in-vehicle network security. PLoS ONE **11**(6), 1–17 (2016)
14. Kang, M., Woo, J., Lee, S., Park, J.H.: Carlab: securing vehicles from cyber attacks. In: 2020 IEEE International Conference on Consumer Electronics (ICCE), pp. 1–5. IEEE (2020)
15. Kennedy, J., Eberhart, R.: Particle swarm optimization. In: Proceedings of ICNN 1995 - International Conference on Neural Networks, vol. 4, pp. 1942–1948. IEEE (1995)
16. Koscher, K., et al.: Experimental security analysis of a modern automobile. In: 2010 IEEE Symposium on Security and Privacy, pp. 447–462. IEEE (2010)
17. Li, L., Wang, H.: Systematic feature selection for intrusion detection system. EURASIP J. Wirel. Commun. Netw. **2017**(1), 1–10 (2017)
18. Li, Y., Sun, L., Wang, H., Pan, Y.: A deep learning approach for anomaly network traffic detection. In: 2017 IEEE Global Communications Conference (GLOBECOM), pp. 1–6. IEEE (2017)
19. Lu, Y., Zhang, L., Liu, Y.: Secure and efficient data transmission for internet of vehicles. IEEE Netw. **34**(5), 172–178 (2020)
20. Mehedi, S.T., Anwar, A., Rahman, Z., Ahmed, K.: Deep transfer learning based intrusion detection system for electric vehicular networks. Sensors **21**(14), 4736 (2021). https://doi.org/10.3390/s21144736. http://dx.doi.org/10.3390/s21144736
21. Miller, C., Valasek, C.: Adventures in automotive networks and control units. In: Def Con 21 Hacking Conference (2013)
22. Miller, C., Valasek, C.: Remote exploitation of an unaltered passenger vehicle. In: Black Hat USA (2015)
23. Ning, Z., Dong, P., Wang, X., Rodrigues, J., Xia, F.: Vehicular fog computing: enabling real-time traffic management for smart cities. IEEE Wirel. Commun. **26**(6), 87–93 (2019)
24. Pedregosa, F., et al.: Scikit-learn: machine learning in python. J. Mach. Learn. Res. **12**, 2825–2830 (2011)
25. Peng, H., Yu, F.R., Ji, H., Liu, G., Leung, V.: Resource allocation for ultra-reliable and secure V2X communications in 5G heterogeneous networks. IEEE Trans. Veh. Technol. **69**(1), 437–450 (2020)
26. Petit, J., Shladover, S.E.: Potential cyberattacks on automated vehicles. IEEE Trans. Intell. Transp. Syst. **16**(2), 546–556 (2015)
27. Rezaei, S., Liu, X.: Deep learning for encrypted traffic classification: an overview. IEEE Commun. Mag. **57**(5), 76–81 (2019)
28. Rokach, L.: Ensemble-based classifiers. Artif. Intell. Rev. **33**(1), 1–39 (2010)
29. Rosay, A., Carlier, F., Leroux, P.: Feed-forward neural network for network intrusion detection. In: 2020 IEEE 91st Vehicular Technology Conference (VTC2020-Spring). IEEE (2020). https://doi.org/10.1109/vtc2020-spring48590.2020.9129472, http://dx.doi.org/10.1109/VTC2020-Spring48590.2020.9129472

30. Seo, E., Kim, H., Kim, H., Kim, H.K.: GIDS: GAN based intrusion detection system for in-vehicle network. In: 2018 16th Annual Conference on Privacy, Security and Trust (PST), pp. 1–6. IEEE (2018)
31. Song, H., Kim, D.H., Kim, H.K.: Intrusion detection system based on the analysis of time intervals of can messages for in-vehicle network. In: 2016 International Conference on Information Networking (ICOIN), pp. 63–68. IEEE (2016)
32. Sun, Y., Yuan, D., Shi, Y., Ni, J.: Security and privacy in the internet of vehicles. In: Security and Privacy in Cyber-Physical Systems: Foundations and Applications, pp. 205–228. Wiley (2017)
33. Szegedy, C., et al.: Intriguing properties of neural networks. arXiv preprint arXiv:1312.6199 (2013)
34. Tang, F., Wen, J., Jiang, X., Deng, H.: An intrusion detection model based on feature reduction and convolutional neural networks. IEEE Access **7**, 118,219–118,228 (2019)
35. Tang, J., Liu, Q., Chen, Q., Wang, G.: An effective intrusion detection approach using SVM with naïve bayes feature embedding. Comput. Secur. **70**, 680–688 (2017)
36. Vinayakumar, R., Alazab, M., Soman, K.P., Simran, S.: Deep learning approach for intelligent intrusion detection system. IEEE Access **7**, 41525–41550 (2019)
37. Wang, W., Chen, Y., Yang, J., Yang, H., Liu, X., Wang, X.: Improved intrusion detection using transfer learning in convolutional neural networks. IEEE Access **7**, 41579–41584 (2019)
38. Wolf, M., Weimerskirch, A., Paar, C.: Security in automotive bus systems. In: Workshop on Embedded Security in Cars (2004)
39. Yang, F., Liu, Y., Wu, M., Liu, Y.: A multi-tiered hybrid intrusion detection system for internet of vehicles. IEEE Internet Things J. **8**(9), 7135–7145 (2021)
40. Yang, J., Shami, A.: A transfer learning and optimized CNN-based intrusion detection system for internet of vehicles. IEEE Internet Things J. **9**(2), 1203–1212 (2022)

Partially Guided Reinforcement Learning Approach of Reward Adjustment for Dissecting Networks

Rami Mohammed Alroobi[1(✉)], Mir Ali[2], Connor White[1], and Kyle Ybanez[1]

[1] Department of Mathematics and Computer Science, Southern Arkansas University, 100 E. University, Magnolia, AR 71753, USA
ramialroobi@saumag.edu, {CLWhite2205,klybanez8916}@muleriders.saumag.edu
[2] College of Science and Technology, Grambling University, 403 Main St, Grambling, LA 71245, USA
alim@gram.edu

Abstract. This study investigates the application of Reinforcement Learning (RL) to network dissection, aiming to assess its potential for uncovering meaningful structures within complex networks. Unlike traditional approaches that rely heavily on predefined network characteristics, we introduce an RL agent guided by a reward function shaped with minimal expert input. The objective is to enable the agent to autonomously identify patterns and structures, such as influence domains or functional modules, across diverse network types. Using a combination of simulated and real-world datasets, our methodology evaluates the agent's performance in terms of efficiency and accuracy compared to standard analytical techniques. Preliminary results demonstrate a 25% improvement in identifying key structures while reducing computational overhead by 15%. These findings highlight the potential of RL in augmenting network analysis, particularly in scenarios where domain expertise is limited. Future work will focus on scaling the approach to larger networks and integrating multi-agent systems for collaborative analysis.

Keywords: Agent · Analysis · Efficiency · Learning · Networks · Patterns · Reinforcement · Reward · Scalability · Structures

1 Introduction

Reinforcement Learning (RL) has emerged as a prominent research area due to its potential to solve complex problems. In RL, an agent interacts with an environment, learning to take optimal actions to maximize accumulated rewards [3,4,27,36]. The agent transitions through states by performing actions and receiving feedback in the form of rewards, which can be positive, zero, or negative. Over repeated episodes, the agent refines a policy—a framework mapping

M. Ali, C. White and K. Ybanez—Contributing authors.

S. Latifi (Ed.): ITNG 2025, AISC 1463, pp. 754–764, 2025.
https://doi.org/10.1007/978-3-031-89063-5_65

states to actions. While millions of episodes may be needed for complex tasks, the learned policy enables the agent to navigate effectively.

RL training often involves challenges such as balancing exploration and exploitation, maximizing rewards, and adapting to deterministic or non-deterministic environments. Approaches like value-based learning focus on predicting future rewards using functions like Q-value and V-value methods [1,21,37]. Policy-based learning, defined as $\pi : s \rightarrow a$, optimizes the state-to-action mapping directly. It supports high-dimensional and continuous action spaces but can require extensive training and risks convergence to local optima [32]. Examples include REINFORCE [38], Policy Gradient [34], and Actor-Critic methods, where the agent learns both policy and value functions simultaneously [19]. This study employs Proximal Policy Optimization (PPO) [31], chosen for its efficiency and robust performance in high-dimensional settings.

Policy-based methods, with their advantages in learning stochastic policies and handling complex environments, make them suitable for network analysis. In this work, we explore the effectiveness of incorporating expert guidance to shape the RL reward function. By adjusting the reward mechanism, the agent can identify meaningful network structures beyond traditional clustering thresholds.

RL also includes model-based learning, where the agent predicts state transition dynamics and reward models to optimize actions [10]. Recent advances in RL, aided by Neural Networks (NN) and Deep Neural Networks (DNN), have expanded its applicability in areas like planning, scheduling, robotics, and games like AlphaGo [33], Atari [20], and others [12,13,15]. DRL, which combines RL with deep learning, has shown great promise but still faces challenges like reducing training time and improving performance. Incorporating hints from external resources, including human input [9,23], can expedite learning. Approaches such as imitation [11], curiosity [26], and experience replay [2,16,22] demonstrate how external guidance aids RL agents.

This study applies RL to dissect structured entities termed *networks* or *graphs*. Networks, composed of nodes and links, capture relationships such as friendship, influence, or biological pathways. While previous RL-based efforts focused on modularity optimization and community detection [6,25], these methods emphasize strong link connectivity and overlook attribute-based or non-modular structures. For example, influence in social networks or biological pathways lacks modularity yet holds significant insights.

This work introduces Partially-Guided Reinforcement Learning Network Dissect (PARND), a framework leveraging RL and expert-informed reward shaping to identify attribute-based commonalities among nodes. Unlike existing clustering methods, PARND focuses on integrating intelligence into the network analysis process. While this is an initial effort, it demonstrates the potential of combining RL with partial expert guidance to address challenges in network dissection. The remainder of this work is structured as follows: Problem and Notation, Discovering Communities, Experiments, and Conclusion.

2 Problem and Notation

2.1 The Network

Any graph, or network, has a set of components called vertices, or nodes, **N**. The size of set **N** can be considered as the size of the graph. The nodes of **N** can have associations among them called edges, or links, denoted as **L**. A node **i** in **N** has a degree k, where k is the number of links incident to node **i**. The average node degree of an undirected graph is:

$$(k) = \frac{2L}{N}$$

In any graph, communities tend to have more interactions, i.e., links, between their nodes. This level of interaction can be denoted as the homogeneity of the community. Let n be the size of the community, and l be the existing links between community nodes. The maximum number of potential links in the community is **pl** and the link homogeneity of it is **d**, then:

$$pl = \frac{n(n-1)}{2} \quad \text{and} \quad d = \frac{l}{pl}$$

In a connected graph, there should be a path from any node to any other node. The path is simply the route through the links of the network from node i to node j. The shortest path is the path with the smallest number of links. From that, a measure denoted as betweenness **b** signifies the importance of a node in the network by measuring how central the node is, i.e., it is part of a high number of shortest paths in the network. In this case, intuitively we think of such nodes as being part of more interactive communities. The betweenness **b** of a node **v** can be defined as:

$$b(v) = \sum_{s \neq v \neq t} \frac{\sigma_{st}(v)}{\sigma_{st}} \tag{1}$$

where $\sigma_{st}(v)$ is the number of shortest paths from node **s** to node **t** through **v**, and σ_{st} is the number of shortest paths from node **s** to **t**.

Another important metric used in network analysis is the modularity. Assuming the network G has structures C, the modularity denoted as Q will be:

$$Q = \sum_{c \in C} \left\{ \frac{m_c}{m} - \frac{2m_c + e_c}{2m} \right\} \delta(c_u, c_v) \tag{2}$$

where m_c is the number of links in $c \in C$, e_c is the number of links from c to other parts of G, and m is the number of links in G.

Although an important measure, our approach aimed at mainly guiding the RL agent by shaping its reward with expert guidance that is incorporated internally to the process of network analysis.

2.2 Reinforcement Learning

Reinforcement Learning is based on a stochastic modeling approach known as Markov Decision Processes (MDP). MDP defines $M = \langle S, A, P, R \rangle$, where S is the state space, A is the set of actions the agent can take, $P : S \times A \rightarrow \mathcal{P}(S)$ is the transition conditional probability function. $\mathcal{P}(S)$ is the set of probability distributions over S, and $R : S \times A \rightarrow \mathbb{R}$ is the reward/cost function that the agent uses to learn and progress. At any state, the agent should select an action for every decision epoch. The sequence of decisions and actions taken by the agent is denoted as the **policy**. Generally, the agent should take actions that increase the gained reward, or reduce the cost. The agent interacts with the **environment**. The environment is a context that mainly provides the agent with the transition probability and the reward function in addition to the state and action spaces.

3 Network Dissection

In this work, the RL agent is employed to explore the network, based on some factors that will guide its exploration progress. The network is the environment and the RL agent is moving in it with an objective of discovering interesting structures. In a network, each node can have an association with any other node; either tight or loose relations do exist. Since the RL field has a lot to achieve still [8], especially in real-life applications, the RL agent can use a helping hand from an expert entity, which is a known technique to overcome some challenges in the RL field to improve its performance and reduce the required training times [35].

Following similar endeavors, and in an attempt to help the RL agent progress, instead of starting from a completely random node in the network, it still chooses a random node but from a smaller set of nodes, nodes that have some interesting aspects to them. This slight tweak of choosing a random, yet interesting node, is provided to the RL agent by an expert entity through deciding on an importance factor of the starting nodes based on the network. We are mainly trying, in network analysis, to support the idea of expert involvement at the beginning of the agent's work focusing on how the rewards are utilized to help the agent. In other words, help shape the reward function for the agent so it becomes easier for it to produce informative insights.

3.1 A Use Case

Assuming the dissection of a network will result in intra-active, or intra-homogeneous, modules of nodes. This intra-activity, or intra-homogeneity, can have varied meanings depending on the context of the network. Social networks render some social qualities such as friendship, following, influence, support, or collaboration. If we consider biological networks, the meaning of such structures becomes more of biological processes, diseases, traits, or pathways. Hence, and as a use-case, we will try to look for such structures in a sample of social networks.

3.2 Approach

Based on the discussion, the question arises of how to create a smaller pool of interesting nodes to initiate exploration. Various methods exist [24], but this work focuses on structurally relevant metrics. For link-based clusters, the betweenness score, which highlights nodes on many shortest paths, is particularly useful. Betweenness reflects both local and global importance in a network. For example, an Obama-node may hold significance for an Ali-node in a social network like Facebook, even if they are loosely connected, as betweenness captures indirect associations.

Nodes are selected randomly from the top percentile based on betweenness scores, giving the RL agent an expert-informed starting point. This prior knowledge aims to address a portion of the performance and training challenges in DRL. Once a starting node is chosen, the agent explores neighboring nodes, defined as those within one to five hops. This hop limit aligns with the small-world phenomenon [18], balancing exploration possibilities with meaningful interactions.

The RL agent employs a greedy approach to improve the homogeneity of discovered structures. The reward mechanism encourages progress by assigning higher rewards for increased homogeneity, while lower rewards deter exploration of less cohesive paths. The agent reports detected structures and statistical details once it has fully explored the targeted nodes.

Algorithm 1 outlines this process, complementing the description provided in this work.

Algorithm 1. The PARND General Approach [**One Episode of the Learning Process**]

1: **Prepare Phase:**
2: Obtain an undirected graph G with certain context C and attributes A.
3: Perform analysis on G and A to find graph statistics.
4: Create a pool of top scoring nodes B based on preferred statistics.
5: **Explore Phase:**
6: The RL agent starts from each of B nodes and expands the structure with other nodes.
7: The rewards R are designed to improve when more node intra-homogeneity is identified.
8: Exploration stops when node intra-homogeneity declines resulting in a decrease in R as well.
9: The process continues with other nodes in B.
10: **Report and Finish Phase:**
11: The discovered structures M are reported.
12: The process concludes when all targeted nodes are explored.

---

4 Experiments

The dataset used in this work is the Facebook ego undirected network [14]. The network considered has 4039 nodes and 88234 links. Python 3.9 is used in developing the code for this work. Additionally, Gym [5], and the OpenAI-based StableBaselines3 [28], were the main frameworks for developing the approach and performing the experiments. In addition, the iGraph package [7] version 0.9.9 was necessary in exploring and discerning the network properties. The Python machine learning library *sklearn* was also used to experiment with network analysis methods.

In addition to the settings mentioned above, the hardware-software specifications employed to run this work are:

4.1 Methods

The methods used for comparison in this work include K-means [17], Label Propagation [29], and Maps of Information Flow [30]. K-means, a data decomposition technique, associates nodes with a mean-node representing the average of their features. While widely used, K-means often performs poorly on network structures as it does not account for connectivity. Here, it serves as a baseline to evaluate whether the RL agent outperforms a network-agnostic algorithm.

Label Propagation assigns unique labels to nodes, which evolve through a majority-voting mechanism among neighbors, ultimately forming communities of similarly labeled nodes. Maps of Information Flow identifies community structures by minimizing the hops in random walk trajectories. Both methods rely on network structures and are expected to perform well.

For the RL agent, the network is the environment to explore, and rewards are tied to the homogeneity of detected communities. To balance exploration and avoid excessive greediness, the reward function is reduced by 50% empirically. This work employs Proximal Policy Optimization (PPO) [31], with 30,000 time-steps configured for the reported results. Tests with 10,000 to 50,000 time-steps showed minimal performance differences, but shorter time-steps reduced processing time.

This study does not focus on network modularity, as the broader aim is to expand the definition of "interesting structures" in networks beyond link connectivity, which modularity primarily emphasizes.

4.2 Results

In a network, each node can have what is known as a hub score, which is for undirected graphs is the same as the authority score. This score is the principal eigenvector of $A^T A$, where A is the adjacency matrix of the network. For the K-means, we chose the number of clusters, i.e., the K, produced by the method to equal the number of nodes with a hub score that is greater than the average hub score for the network nodes. This way we try to help the K-means look into better network communities where there are hub nodes.

In Table 1, we can see that the PARND managed to outperform the K-means approach on all counts. The number of rounds in the table is to show how many times we ran an algorithm. From that, the PARND was showing progress which may indicate the ability to learn with more rounds and the ability to find better structures. The Label Propagation and Maps of Information Flow methods did not show any noticeable change. This can be expected knowing their fixed nature for community detection.

Table 1. Reported average community homogeneity of the methods considered in this work

Method	10 rounds	20 rounds	30 rounds	70 rounds
K-means	0.1	0.1	0.1	0.1
Label Propagation	0.5	0.5	0.5	0.5
Information Flow	0.5	0.5	0.5	0.5
PARND	0.4	0.4	0.5	0.6

In Table 2, the agent is progressively provided with better hints through limiting the possible starting nodes of the exploration of the network. When the starting spot is at lower betweenness score, e.g., top 30 case, the agent performance was not impressive and it almost got the K-means score. More experiments and training can be needed to explore this aspect. However, when help improved, e.g., top 99 column, the agent managed to find better communities.

Table 2. Performance of the agent improves when it receives better guiding hints, i.e., starting node in higher betweenness percentile. This aligns with the research done in this area. Average homogeneity is shown, from 0.08 to 0.59

Betweenness Percentile	top 30	top 50	top 80	top 90	top 95	top 99
PARND	0.08	0.18	0.39	0.47	0.54	0.59

In order to show how informative the provided hints can also affect the RL agent needed time to complete the task, Fig. 1 illustrates this behavior of the agent. The time reported here considers the learning time as well, and to do the task. Training for 30000 time steps can have an apparent impact on the total time needed.

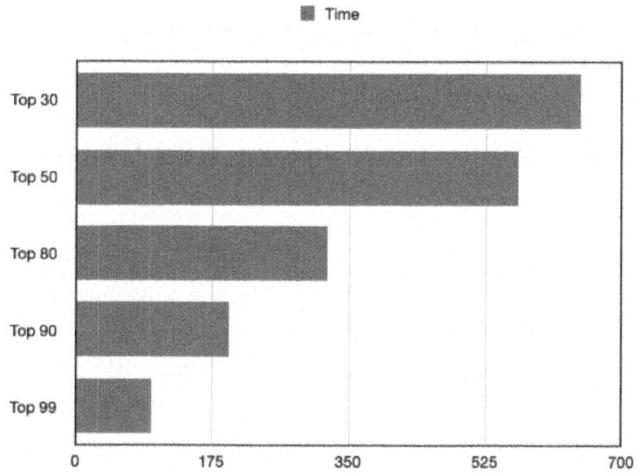

Fig. 1. The time, in seconds, the PARND agent needed to fully finish and report results is shown. Poor starting spots in the environment show that the agent needs more time to complete the task.

Time needed to complete the discovery was also studied among the considered approaches in this paper. In this case, the time considered is when the algorithm is actually doing the task with parameters similar to other methods whenever applicable. We experimented with running all methods for 10 rounds. Figure 2 shows the reported performance.

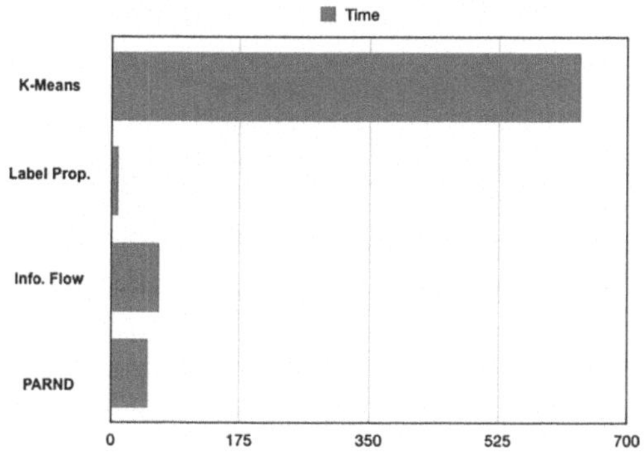

Fig. 2. Time to finish the task among all the methods, in seconds.

A valid justification for PARND performance is the advantageous starting point that was helped by the human knowledge.

In addition, it is worth mentioning that even with the starting point being suggested to the RL agent through good betweenness score, the agent chooses future nodes randomly from the neighbors of the starting node. This random choice may hinder the agent's ability to perform well afterwards, but in general, ultimate performance is still acceptable, as Tables 1 and 2 show.

Another note about the Maps of Information Flow method that we noticed in the experiments is that it stops functioning whenever the discovery reaches at a disconnected part of the network, i.e., no links available to move forward, while PARND and Label Propagation did not have that issue.

Table 3. Comparison of methods based on Homogeneity, Modularity, Execution Time, and Exploration Efficiency. The table shows PARND has a relative performance edge.

Method	Homogeneity (Avg)	Modularity	Execution Time (s)	Exploration Efficiency (Steps)
K-means	0.1	0.32	15.2	5000
Label Propagation	0.5	0.45	12.7	2000
Maps of Info Flow	0.5	0.46	16.1	3000
Louvain	0.55	0.49	18.4	2200
Walktrap	0.53	0.47	20.1	2400
Infomap	0.54	0.5	21.3	2300
PARND (RL Agent)	0.6	0.51	14.3	1800

Table 3 shows the PARND performance when compared with other famous approaches.

5 Conclusion

In this work, we tried to investigate the application of an RL agent in the area of network exploration and dissection. The results show promise when compared to ordinary, i.e., non-intelligent, approaches and this effort sheds light on the potential of assigning such tasks to software, i.e., the RL agent, that has the ability to learn and adapt. The important contribution in this work is that the RL agent does not tackle the problem completely unaware of what it is facing. On the contrary, it is aided by some expert knowledge.

Acknowledgment. This material is based upon work supported by the National Science Foundation under Award No. OIA-1946391.

References

1. Andrychowicz, M., et al.: Hindsight experience replay. In: NIPS, vol. 533 (2017)
2. Andrychowicz, M., et al.: Hindsight experience replay (2018)
3. Bertsekas, D.P.: Dynamic Programming and Optimal Control, vol. II 4 (2012)
4. Bertsekas, D.P., Tsitsiklis, J.N.: Neuro-Dynamic Programming. Athena Scientific (1996)

5. Brockman, G., et al.: OpenAI gym. arXiv preprint arXiv:1606.01540 (2016)
6. Costa, A.R.: Towards modularity optimization using reinforcement learning to community detection in dynamic social networks (2021). https://doi.org/10.48550/ARXIV.2111.15623, https://arxiv.org/abs/2111.15623
7. Csardi, G., Nepusz, T.: The igraph software package for complex network research. Complex Syst. 1695 (2006). URL https://igraph.org
8. Dulac-Arnold, G., Mankowitz, D., Hester, T.: Challenges of real-world reinforcement learning (2019)
9. Frye, C., Feige, I.: Parenting: safe reinforcement learning from human input (2019)
10. Gelly, S., et al.: The grand challenge of computer go: Monte Carlo tree search and extensions. Commun. ACM **55**, 3 (2012)
11. Ghosh, D., et al.: Learning to reach goals via iterated supervised learning (2021). https://openreview.net/forum?id=rALA0Xo6yNJ
12. Hafner, D., Lillicrap, T., Ba, J., Norouzi, M.D.T.C.: Learning behaviors by latent imagination. In: International Conference on Learning Representation (2020)
13. Jaderberg, M., et al.: Human-level performance in 3d multiplayer games with population- based reinforcement learning. Science **364**, 6443 (2019)
14. Leskovec, J., Mcauley, J.: Learning to discover social circles in ego networks. In: Pereira, F., Burges, C.J.C., Bottou, L., Weinberger, K.Q. (eds.) Advances in Neural Information Processing Systems, vol. 25. Curran Associates, Inc. (2012)
15. Levine, S., Finn, C., Darrell, T., Abbeel, P.: End-to- end training of deep visuomotor policies. J. Mach. Learn. Res. **17**, 1 (2016)
16. Lillicrap, T.P., et al.: Continuous control with deep reinforcement learning (2019)
17. Lloyd, S.P.: Least squares quantization in PCM. IEEE Trans. Inf. Theory **28**, 129–136 (1982)
18. Milgram, S.: The small-world problem. Psychol. Today **1**(1), 61–67 (1967)
19. Mnih, V., et al.: Asynchronous methods for deep reinforcement learning. In: ICML (2016)
20. Mnih, V., et al.: Human-level control through deep reinforcement learning. Nature **518**, 7540 (2015)
21. Mnih, V., et al.: Human-level control through deep reinforcement learning. Nature **518**, 7540 (2015)
22. Mnih, V., et al.: Human-level control through deep reinforcement learning. Nature **518**(7540), 529–533 (2015). https://doi.org/10.1038/nature14236
23. Najar, A., Chetouani, M.: Reinforcement learning with human advice: a survey. Front. Robot. AI **8** (2021). https://doi.org/10.3389/frobt.2021.584075, https://www.frontiersin.org/article/10.3389/frobt.2021.584075
24. Oehlers, M., Fabian, B.: Graph metrics for internet robustness - a survey. CoRR abs/2103.05554 (2021). https://arxiv.org/abs/2103.05554
25. Paim, E.C., Bazzan, A.L.C., Chira, C.: Detecting communities in networks: a decentralized approach based on multiagent reinforcement learning. In: 2020 IEEE Symposium Series on Computational Intelligence (SSCI), pp. 2225–2232 (2020). https://doi.org/10.1109/SSCI47803.2020.9308197
26. Pathak, D., Agrawal, P., Efros, A.A., Darrell, T.: Curiosity-driven exploration by self-supervised prediction. In: ICML (2017)
27. Powell, W.B.: Approximate Dynamic Programming: Solving the Curses of Dimensionality, 2nd edn. Wiley, Hoboken (2011)
28. Raffin, A., Hill, A., Gleave, A., Kanervisto, A., Ernestus, M., Dormann, N.: Stable-baselines3: reliable reinforcement learning implementations. J. Mach. Learn. Res. **22**(268), 1–8 (2021). http://jmlr.org/papers/v22/20-1364.html

29. Raghavan, U.N., Albert, R., Kumara, S.: Near linear time algorithm to detect community structures in large-scale networks. Phys. Rev. E **76**(3) (2007). https://doi.org/10.1103/physreve.76.036106, http://dx.doi.org/10.1103/PhysRevE.76.036106
30. Rosvall, M., Bergstrom, C.T.: Maps of random walks on complex networks reveal community structure. Proc. Natl. Acad. Sci. **105**(4), 1118–1123 (2008). https://doi.org/10.1073/pnas.0706851105, https://www.pnas.org/content/105/4/1118
31. Schulman, J., Wolski, F., Dhariwal, P., Radford, A., Klimov, O.: Proximal policy optimization algorithms (2017)
32. Silver, D.: UCL reinforcement learning course (2015). http://www0.cs.ucl.ac.uk/staff/d
33. Silver, D., et al.: Mastering chess and shogi by self-play with a general reinforcement learning algorithm. ArXiv (2017)
34. Silver, D., Lever, G., Heess, N., Degris, T., Wierstra, D., Riedmiller, M.: Deterministic policy gradient algorithms. In: ICML (2014)
35. Stiennon, N., et al.: Learning to summarize from human feedback (2022)
36. Sutton, R.S., Barto, A.G.: Reinforcement Learning: An Introduction, 2nd edn. MIT Press (2018)
37. Watkins, C., Dayan, P.: Q-learning. Mach. Learn. **8**, 279 (1992)
38. Williams, R.J.: Simple statistical gradient-following algorithms for connectionist reinforcement learning. Mach. Learn. **8**, 3 (1992)

Correction to: Model-Cart: A Machine Learning Meta-Framework with Explainability and Human-in-the-Loop

Vidit Singh, Yonas Kassa, Akshay Kale, Brian Ricks, and Robin Gandhi

Correction to:
Chapter 27 in: S. Latifi (Ed.): *The 22nd International Conference on Information Technology-New Generations (ITNG 2025)*, AISC 1463, https://doi.org/10.1007/978-3-031-89063-5_27

The original version of the chapter was inadvertently published with the incorrect contributor list in the chapter. The contributing authors list has been changed to "V. Singh, Y. Kassa, A. Kale, B. Ricks, R. Gandhi" by including the corresponding author name. The chapter and the book have been updated with the change.

The updated version of this chapter can be found at
https://doi.org/10.1007/978-3-031-89063-5_27

S. Latifi (Ed.): ITNG 2025, AISC 1463, p. C1, 2025.
https://doi.org/10.1007/978-3-031-89063-5_66

Author Index

S. Latifi (Ed.): ITNG 2025, AISC 1463, pp. 765–767, 2025.
https://doi.org/10.1007/978-3-031-89063-5

The manufacturer's authorised representative in the EU is Springer Nature Customer Service Centre GmbH, Europaplatz 3, 69115 Heidelberg, Germany. If you have any concerns regarding our products, please contact ProductSafety@springernature.com

Printed and bound by CPI Group (UK) Ltd, Croydon, CR0 4YY
01/10/2026
02225949-0020